Lecture Notes in Computer Science 7729

Commenced Publication in 1973
Founding and Former Series Editors:
Gerhard Goos, Juris Hartmanis, and Jan van Leeuwen

Editorial Board

Jong-Il Park   Junmo Kim (Eds.)

Computer Vision – ACCV 2012 Workshops

ACCV 2012 International Workshops
Daejeon, Korea, November 5-6, 2012
Revised Selected Papers, Part II

 Springer

Volume Editors

Jong-Il Park
Hanyang University
Computer Science and Engineering
222 Wangshimni-ro, Seongdong-gu
Seoul 133-791, Korea
E-mail: jipark@hanyang.ac.kr

Junmo Kim
KAIST
Department of Electrical Engineering
291 Daehak-ro, Yuseong-gu
Daejeon 305-701, Korea
E-mail: junmo@ee.kaist.ac.kr

ISSN 0302-9743   e-ISSN 1611-3349
ISBN 978-3-642-37483-8   e-ISBN 978-3-642-37484-5
DOI 10.1007/978-3-642-37484-5
Springer Heidelberg Dordrecht London New York

Library of Congress Control Number: 2013934227

CR Subject Classification (1998): I.4, I.5, I.2.10, I.2.6, I.3.5, F.2.2, H.5, J.5

LNCS Sublibrary: SL 6 – Image Processing, Computer Vision, Pattern Recognition, and Graphics

Typesetting: Camera-ready by author, data conversion by Scientific Publishing Services, Chennai, India

Printed on acid-free paper

Springer is part of Springer Science+Business Media (www.springer.com)

Preface

The 11th Asian Conference on Computer Vision (ACCV), held in Daejeon, South Korea, during November 5–9, 2012, was accompanied by a series of nine high-quality workshops covering the full range of state-of-the-art research topics in computer vision.

The workshops consisted of six full-day workshops and three half-day workshops. Their topics diversely ranged from traditional issues to novel current trends. On November 5, four workshops took place: the Workshop on Computer Vision with Local Binary Pattern Variants and the Workshop on Computational Photography and Low-Level Vision (both full-day workshops), and the Workshop on Developer-Centered Computer Vision and the Workshop on Background Models Challenge (both half-day workshops). The remaining five workshops were held on November 6: the Workshop on e-Heritage (Electronic Cultural Heritage), the Workshop on Color Depth Fusion in Computer Vision, the Workshop on Face Analysis: The Intersection of Computer Vision and Human Perception, and the Workshop on Detection and Tracking in Challenging Environments (all full-day workshops), and the Workshop on Intelligent Mobile Vision (a half-day workshop).

This year, the workshops received 310 paper submissions, and 78 presentations were selected by the individual workshop committees, yielding an overall acceptance rate of 25%. All contributions to each workshop are published in the two-volume ACCV workshop proceedings. We thank everyone involved in the remarkable programs, committees, reviewers, and authors, for their contributions.

We hope that you enjoy reading these proceedings, which may inspire you to further research.

November 2012

Jong-Il Park
Junmo Kim

International Workshop on Computer Vision with Local Binary Pattern Variants

Local Binary Pattern (LBP) is a simple and efficient texture operator, unifying statistical and structural approaches in texture analysis. It is a powerful gray-scale invariant measure, derived from a general definition of texture in a local neighborhood. Due to its discriminative power and computational simplicity, the LBP operator has become a highly popular approach in various computer vision applications, including facial image analysis, visual inspection, image retrieval, remote sensing, biomedical image analysis, biometrics, motion analysis, environment modelling, and outdoor scene analysis. Especially the use of LBP in biomedical applications and biometric recognition systems has grown rapidly in recent years. LBP has been highly successful in numerous applications around the world and has inspired plenty of new research on related methods. Since the introduction of the basic LBP operator, several variants have been proposed to improve the discriminative power and robustness of the operator. The recent emergence of LBP has also led to significant progress in applying texture methods to various computer vision problems and applications.

This workshop provided a clear summary of the state of the art and discussed the most recent developments on the use of Local Binary Patterns and their variants in different computer vision applications.

The workshop received 45 submissions (16 through direct submission and 29 via dual submission with the ACCV 2012 main conference). Based on the thorough reviews by the program committee, 13 papers were finally selected. Besides the 13 interesting oral presentations, the workshop also included a keynote speech from a pioneer of LBP (Prof. Matti Pietikäinen from the University of Oulu) and a best paper award sponsored by KeyLemon – a leading face recognition software company.

The workshop organizers would like to thank all the participants of this workshop. Many thanks go also to the Program Committee for their efforts during the reviewing process and to the ACCV 2012 workshop chairs and publication chairs who dealt with the organizational aspects of this workshop.

November 2012

Abdenour Hadid
Sebastien Marcel
Jean-Luc Dugelay
Matti Pietikäinen
Mohammad Ghahramani
Stan Z. Li

Program Committee

Janne Heikkilä	University of Oulu, Finland
Norman Poh	University of Surrey, UK
Shengcai Liao	Michigan State University, USA
Hazim Kemal Ekenel	Karlsruhe Institute of Technology, Germany
Jie Chen	University of Oulu, Finland
Caifeng Shan	Philips Research, The Netherlands
Loris Nanni	University of Padua (Padova), Italy
Bill Triggs	Laboratoire Jean Kuntzmann, Grenoble, France
Liming Chen	École Centrale de Lyon, France
Karl Ricanek	University of North Carolina Wilmington, USA
Lior Wolf	Tel Aviv University, Israel
Yunhong Wang	BUAA University, China
André Anjos	Idiap Research Institute, Switzerland
Guoying Zhao	University of Oulu, Finland
Chi-Ho Chan	University of Surrey, UK
Messaoud Bengharabi	Centre de Développement des Technologies Avancées, Algeria
Mohammad Ghahramani	University of Oulu, Finland
Mark Nixon	University of Southampton, UK
John A. Ruiz-Hernandez	University of Oulu, Finland
Zhen Lei	Chinese Academy of Sciences, China
Vili Kellokumpu	University of Oulu, Finland
Fabio Roli	University of Cagliari, Italy
Timo Ahonen	Nokia Research Center, Palo Alto, California
Xilin Chen	Chinese Academy of Sciences, China
Shiguang Shan	Chinese Academy of Sciences, China
Rui Min	Eurecom, France
Yimo Guo	University of Oulu, Finland
Miguel Bordallo	University of Oulu, Finland
Juha Ylioinas	University of Oulu, Finland

Workshop on Computational Photography and Low-Level Vision

Computational Photography is an exciting new field at the intersection of computer vision, computer graphics, and photography. The goal of computational photography is to enhance or extend the capabilities of digital photography to produce new photographs that could not have been taken by a traditional camera. Fundamental low-level computer vision techniques will be particularly useful to this end.

The goal of this workshop is to provide a platform for researchers in computational photography and computer vision to meet and share their ideas on recent trends and research in the two areas.

We received 17 papers and selected 8 papers for publication based on the reviews by the Program Committee. All submissions were reviewed in a double-blind fashion by at least three experts in the area. We thank all the authors who submitted their work. Topics of accepted papers spanned a wide range of areas from camera calibration to vehicle localization. Other topics represented were segmentation and colorimetric correction. We were pleased to have Michael S. Brown (National University of Singapore) as the keynote speaker and also the advisory chair at the workshop. We would also like to express our appreciation to the members of the Program Committee for their remarkable efforts and the quality of the reviews.

November 2012

Jinwei Gu
Yu-Wing Tai
Ping Tan
Sai-Kit Yeung

Program Committee

Workshop on Developer-Centered Computer Vision

The majority of research in computer vision is focused on technology and systems that advance the state of the art. However, there is very little focus on how we can make the state of the art useable by the majority of people. Recently there has been an increased interest in "Vision for HCI" and how we use computer vision to interact with the world. We proposed a parallel theme of "HCI for Vision" for this workshop, looking at how to provide accessible computer vision targeted towards mainstream software developers. We aimed to explore ideas that take existing vision methods and present them in a manner that enables users with varying degrees of vision knowledge to use them.

There has been a relatively recent surge in the number of developer interfaces to computer vision becoming available: OpenCV has become much more popular, Mathworks have released a Matlab Computer Vision Toolbox, visual interfaces such as Vision-on-Tap are available online and specific targets such as tracking (OpenTL) and GPU (Cuda, OpenVIDIA) have working implementations. Additionally, last year, Khronos (the not-for-profit industry consortium that creates and maintains open standards) formed a working group to discuss the creation of a computer vision hardware abstraction layer (tentatively titled CV HAL).

Developing methods to make computer vision accessible poses many interesting questions and will require novel approaches to the problems. DCCV is a half-day workshop aiming to bring together researchers from academia and industry in the fields of vision and HCI to discuss the direction of research into developer-centred computer vision. The workshop included an introductory talk by the organisers followed by presentations of the five accepted papers (24% acceptance), covering mainstream-developer targeted topics as well as more advanced concepts such as algorithm efficiency.

The DCCV organisers would like to thank the ACCV Workshop and Publication Chairs, in particular Junmo Kim and In Kyu Park, for their help and support throughout the workshop organisation process.

November 2012

Gregor Miller
Sidney Fels

Program Committee

Workshop on Background Models Challenge

The detection of moving objects in video sequence is an important task in many video-surveillance systems. As a matter of fact, the output of this very first stage, named *background modeling or background subtraction*, determines the quality of the rest of the pipelines developed for the detection, identification, or tracking of persons, objects, etc. Background modeling is sometimes considered either as a trivial operation, carried out by computing a simple difference between the current frame and a single background image (or with the previous frame, etc.), or a mastered technique that does not need any improvement or development nowadays. In the latter case, very famous methods, such as the Gaussian Mixture Models introduced by Stauffer and Grimson in 1999, are cited, and this is considered sufficient. Unfortunately, these kinds of algorithms are limited in outdoor environments, when used in long-term surveillance applications, because of various unpredictable circumstances such as global variation of luminance, shadows of objects, bad weather, camera tilts, etc.

Since this is a key-point of video-surveillance applications, background subtraction has become a popular topic, and many techniques have been proposed since the 1990s. For the BMC (*Background Models Challenge*), we proposed a new benchmark composed of almost 30 synthetic and real video sequences. Thanks to these data-sets, we were able to propose very complex situations, in various surveillance contexts (human activities or traffic, for example). We also developed a free software (BMC Wizard) to compute relevant criteria for the evaluation of statistical, signal, and structural information from a background subtraction algorithm.

For the BMC, six papers were accepted for publication, and an invited speaker, Thierry Bouwmans, gave a talk on the state of the art of the domain and recent advances in his personal research. We hope that our benchmark will be used as a reference for further research in background modeling. The data-sets and the BMC Wizard will remain available on the BMC website *http://bmc.univ-bpclermont.fr*. Finally, we would like to thank the Program Committee of ACCV for their support in organizing this event.

November 2012

Antoine Vacavant
Laure Tougne
Lionel Robinault
Thierry Chateau

Program Committee

Antoine Vacavant	ISIT, CNRS/Université d'Auvergne, Clermont-Ferrand, France
Laure Tougne	LIRIS, CNRS/Université Lumière Lyon2, Lyon, France
Lionel Robinault	Foxstream company, Vaulx-en-Velin, France
Thierry Chatea	Pascal Institute, CNRS/UBP, Clermont-Ferrand, France
Laurent Lequièvre	Pascal Institute, CNRS/UBP, Clermont-Ferrand, France
Christophe Tournayre	Pascal Institute, CNRS/UBP, Clermont-Ferrand, France
Datta Ramadasan	Pascal Institute, CNRS/UBP, Clermont-Ferrand, France

Workshop on e-Heritage

Digitally archived world heritage sites are broadening their value for preservation and access. Many valuable objects have been decayed by time due to weathering, natural disasters, even man-made disasters such as the Taliban destruction of the great Buddhas in Afghanistan, or the recent destruction by fire of the 600-year-old Great South Gate in Seoul. Cultural heritage also includes music, language, dance, and customs that are fast becoming extinct as the world moves toward a global village. Furthermore, most of the sites still face a problem of accessibility. Digital access projects are necessary to overcome those problems.

Computer vision research and practices have played, and will continue to play, a central role in such cultural heritage preservation efforts. The Workshop on e-Heritage and Digital Art Preservation aimed to bring together computer vision researchers, as well as interdisciplinary researchers, working in areas related to computer vision, in particular computer graphics, image and audio research, image and haptic (touch) research, as well as presentation of visual content over the Web, and education.

In this workshop, eight contributions to the field of e-heritage were presented, covering the areas of automatic character recognition, classification based on shape and image analysis, image enhancement, virtual-reality applications, three-dimensional modeling, and reconstruction. All submissions were double-blind reviewed by at least two experts. We thank all the authors who submitted their work. It was a special honor to have Martial Hebert (Carnegie Mellon University, USA), and Jean Ponce (Ecole Normale Supérieure, France) as the invited speakers at the workshop. We are especially grateful to the members of the Program Committee for their remarkable efforts and the quality of the reviews.

November 2012

Hongbin Zha
Takeshi Oishi
Rei Kawakami
Yunsu Bok
Katsushi Ikeuchi

Program Committee

Olga Bellon	Universidade Federal do Paraná, Brazil
Asanobu Kitamoto	National Institute of Informatics, Japan
Yasuyuki Matsushita	Microsoft Research Asia, China
Shohei Nobuhara	Kyoto University, Japan
Tomokazu Sato	NAIST, Japan
Luciano Silva	Universidade Federal do Paraná, Brazil
Jun Takamatsu	NAIST, Japan
Robby T. Tan	University of Utrecht, Netherlands
Yingqing Xu	Tsinghua University, China
Toshihiko Yamasaki	University of Tokyo, Japan

Workshop on Color Depth Fusion in Computer Vision

The ambition of this workshop was to provide an opportunity to disseminate recent theories, methods, and practical algorithms that explicitly exploit the enormous potential of combining low-resolution depth cameras with high-resolution color cameras for a wide variety of computer vision tasks. The workshop brought together researchers and practitioners from various fields of study: computer vision, robotics, computer graphics, image processing, and sensor architecture.

We received 44 submissions and 12 papers were accepted for single-track oral presentation. We also had an invited demonstration on single-sensor color and depth capturing sensor and applications.

November 2012

Seungkyu Lee
Hyunjung Shim
Ouk Choi
Seung-Won Jung
Radu B. Rusu

Program Committee

Jingu Heo	Samsung Advanced Institute of Technology, South Korea
Howard Leung	City University of Hong Kong, Hong Kong, China
Cech Jan	INRIA-Grenoble, France
Evangelidis Georgios	INRIA-Grenoble, France
Miles Hansard	Queen Mary University of London, UK
Wende Zhang	General Motors, USA
Cha Zhang	Microsoft Research, USA
Xiaoming Liu	Michigan State University, USA
David Liu	Siemens Corporate Research, USA

Workshop on Face Analysis: The Intersection of Computer Vision and Human Perception

The analysis of faces is a very active research area within both the computer vision and the human perception communities, and there is a large array of potential applications and research topics. The two communities have traditionally worked separately, but there are clear benefits to closer collaboration. For instance, humans develop extensive experience in the processing of face identity, age, gender, and non-verbal communication signals such as facial expressions. Thus, tapping into existing knowledge from human facial perception research can enable the targeted design of computer vision facial analysis and synthesis systems with more realistic behavioural facial models and performance. Moreover, computer vision systems can provide many useful tools for research into the human perception of faces such as the generation of photo-realistic and controllable stimuli for perceptual experiments, which enables more subtle manipulation of facial appearance and dynamics than would be possible using natural video capture.

The current state of the art in facial analysis within both disciplines is now quite evolved. To go to the next level, the disciplines need to strengthen their collaboration even further. This workshop provided the forum to enable this step. Its goal was to examine existing work that straddles the border of these communities, and to map out future steps for integrative research.

We received 37 full-paper submissions which underwent a double-blind review. The multi-disciplinary nature of the workshop meant that making decisions regarding papers was more difficult than usual, and so up to 7 reviewers per paper were used (with a minimum of 3 reviewers/paper) to ensure that a balanced assessment was made. A total of 9 papers were selected for the workshop, and are collected in these proceedings.

We were fortunate to have three invited speakers at the workshop, who have all worked extensively at the intersection of computer vision and human perception: Heinrich H. Bülthoff (Max Planck Institute for Biological Cybernetics), Alan Johnston (University College of London) and Darren Cosker (University of Bath). We would like to thank the invited speakers as well as all the members of the Programme Committee for their help in organising and running this event.

November 2012

Paul L. Rosin
David Marshall
Christian Wallraven
Douglas W. Cunningham

Program Committee

Andrew Aubrey	Cardiff University, UK
Heinrich Bülthoff	Max Planck Institute, Germany
Darren Cosker	University of Bath, UK
Nicholas Costen	Manchester Metropolitan University, UK
Hui Fang	Swansea University, UK
Roland Goecke	Australian National University, Australia
Hatice Gunes	Queen Mary University of London, UK
Qiang Ji	Rensselaer Polytechnic Institute, USA
Arvid Kappas	Jacobs University, Germany
Eva Krumhuber	Jacobs University, Germany
Michael Lewis	Cardiff University, UK
Aleix Martinez	Ohio State University, USA
Louis-Philippe Morency	University of Southern California, USA
Marcello Mortillaro	University of Geneva, Switzerland
Alice O'Toole	University of Texas at Dallas, USA
Jason Saragih	CSIRO, Australia
Björn Schuller	Joanneum Research, Austria
Philip Schyns	University of Glasgow, UK
Terence Sim	National University of Singapore, Singapore
Rainer Stiefelhagen	Karlsruhe Institute of Technology, Germany
Barry John Theobald	University of East Anglia, UK
Ian Thornton	Swansea University, UK
Bernie Tiddeman	Aberystwyth University, UK
Massimo Tistarelli	University of Sassari, Italy
Michel Valstar	University of Nottingham, UK
Job Van Der Schalk	Cardiff University, UK
Stefanos Zafeiriou	Imperial College London, UK

Workshop on Detection and Tracking in Challenging Environments (DTCE)

Recent progress in computer vision has opened new possibilities in robust visual tracking and in human and object detection. Although these have a wide range of practical applications, there are still many challenges when applying such algorithms to real-world data. These include: complex crowded environments with many activities, challenging lighting, and frequent occlusions; large variations of pose, motion, and appearance; limited computational resources; and the need for training from large datasets. DTCE 2012 brought together researchers working on these challenging real-world problems to present their recent achievements and provided a place to share their experiences and visions with others.

We received 89 submissions jointly with ACCV and 9 independent submissions; and we selected 15 papers for publication. The review process was double-blind. The independently submitted papers were reviewed by two to three members of the workshop Program Committee, while most of the joint submissions received one independent review from this committee in addition to their three reviews and area chair summary from ACCV. We would like to thank the Program Committee members for their effort in reviewing the papers.

The workshop featured a keynote address by Ming-hsuan Yang of the University of California, Merced, as well as oral presentations of 8 of the accepted papers, and poster presentations of all 15 of the accepted papers.

November 2012

Bohyung Han
Jongwoo Lim
Bill Triggs
Ahmed Elgammal
Jason Corso

Program Committee

Serge Belongie	University of California, San Diego, USA
Terrence Chen	Siemens Corporate Research, USA
Naresh Cuntoor	Kitware, USA
Larry Davis	University of Maryland, USA
Jan Feyereisl	POSTECH, South Korea
Kikuo Fujimura	Honda Research Institute, USA
Abhinav Gupta	Carnegie Mellon University, USA
Iasonas Kokkinos	Ecole Centrale Paris, France
Tony X. Han	University of Missouri, USA
Ser-Nam Lim	GE Global Research, USA
Haibin Ling	Temple University, USA
Sangmin Oh	Kitware, USA
David Ross	Google inc., USA
Yoichi Sato	University of Tokyo, Japan
Zhuowen Tu	MSRA / University of California, Los Angeles, USA
Jianxin Wu	Nanyang Technological Univeristy, Singapore
Ming-hsuan Yang	University of California, Merced, USA
Kuk-Jin Yoon	GIST, Korea
Lei Zhang	Hong Kong Polytechnic University, Hong Kong

International Workshop on Intelligent Mobile Vision (IMV)

With the fast growth of hand-held computing platforms such as smart phones and tablet PCs, computer vision on mobile computing devices has become an important research area. There is tremendous potential for developing computer vision techniques and applications on mobile camera computing devices. In particular, more and more mobile vision applications rely on object and scene recognition or understanding techniques. It is observable that advances in mobile visual-information analyses are closely related to cutting-edge applications in robotics, human-computer interaction, smart sensors, and ubiquitous computing.

The main goal of this workshop was to identify state-of-the-art mobile vision algorithms, systems, and frameworks that are particularly suitable for intelligent visual information processing based on mobile camera computing platforms. In addition to visual information, the integration of vision with additional sensors, such as GPS, accelerometer or gyroscope, or information retrieval transmitted through communication networks, helps to develop an even more intelligent mobile vision application. The associated methodologies and applications are expected to be able to demonstrate the advantages of advanced computer vision techniques based on mobile camera computing devices.

November 2012

Shang-Hong Lai
Chu-Song Chen

Program Committee

Table of Contents – Part II

Workshop on e-Heritage

ACCV Workshop on Color Depth Fusion in Computer Vision

Workshop on Face Analysis: The Intersection of Computer Vision and Human Perception

ACCV Workshop on Detection and Tracking in Challenging Environments (DTCE)

International Workshop on Intelligent Mobile Vision (IMV)

Table of Contents – Part I

Workshop on Computational Photography and Low-Level Vision

Workshop on Developer-Centred Computer Vision

Workshop on Background Models Challenge (BMC)

Historical Document Binarization
Based on Phase Information of Images

Hossein Ziaei Nafchi, Reza Farrahi Moghaddam, and Mohamed Cheriet

Synchromedia Laboratory for Multimedia Communication in Telepresence,
École de technologie supérieure, Montreal (QC), Canada H3C 1K3
hossein_zi@yahoo.com, imriss@ieee.org, mohamed.cheriet@etsmtl.ca

Abstract. In this paper, phase congruency features are used to develop
a binarization method for degraded documents and manuscripts. Also,
Gaussian and median filtering are used in order to improve the final binarized output. Gaussian filter is used for further enhance the output and
median filter is applied to remove noises. To detect bleed-through degradation, a feature map based on regional minima is proposed and used.
The proposed binarization method provides output binary images with
high recall values and competitive precision values. Promising experimental results obtained on the DIBCO'09, H-DIBCO'10 and DIBCO'11
datasets, and this shows the robustness of the proposed binarization
method against a large number of different types of degradation.

1   Introduction

There are many degraded but historically-important old manuscripts and documents distributed across libraries and archives around the world. Some of these
document degradation types are fading of ink, ink bleed-through, show-through
and deterioration of the cellulose structure, among others [1, 2]. Conversion to
binary form is a common and fundamental step in almost all digitization processes. The quality of the binarization step highly affects the performance of the
subsequent document processing steps. Fig. 1 shows some examples of degraded
document images.

In recent years, many approaches have been proposed to binarize the degraded
documents. A problem to all these methods is that their performance is different
against various datasets, as shown in the experimental results section.

This paper proposes a robust method for the binarization of the degraded
documents. The purpose is to maintain a high recall value, while optimizing and
increasing the precision value. In the proposed method, first, phase congruency
features [3] are used to eliminate moderate amount of the background regions
which are easy to identify. Then, the output result is further enhanced using some
postprocessing steps. These steps are developed based on adaptive Gaussian
filter, adaptive median filter and regional minima.

The proposed method is tested on several benchmark datasets, including
DIBCO09 , H-DIBCO10 and DIBCO11 [4–6]. The experimental results show the
robustness of the method across the various degradation types and datasets. A

J.-I. Park and J. Kim (Eds.): ACCV 2012 Workshops, Part II, LNCS 7729, pp. 1–12, 2013.
© Springer-Verlag Berlin Heidelberg 2013

version of the proposed algorithm has been submitted to the H-DIBCO 2012 contest.

The rest of the paper is organized as follows. In section 2, previous work is presented. Phase congruency model is described in section 3. In section 4, feature maps used in this paper are discussed. Binarization model is described in section 5. Section 6 provides comprehensive experimental results for the proposed method. Finally, section 7 draws the conclusions and future prospects.

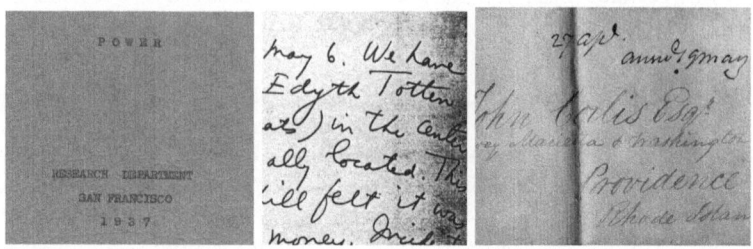

Fig. 1. Three degraded document image samples from DIBCO'11 [6]

2   Related Works

In recent years, binarization of historical documents, as a challenging task, has received a great interest. To encourage the researchers and to advance this field, three competitions (datasets) have been established so far, namely DIBCO'09, H-DIBCO'10 and DIBCO'11, in conjunction with ICDAR 2009, ICFHR 2010 and ICDAR 2011 conferences, respectively [4–6]. In this section, some successful binarization methods, especially those participated in the mentioned competitions, are discussed.

Lu's et al [7] proposed a binarizatition method mainly based on background estimation. First, background of document was estimated via a one-dimensional iterative Gaussian smoothing procedure. Then, for accurate binarization of strokes and sub-strokes, L1-norm gradient image was used. This method ranked first method among 43 algorithms submitted to the DIBCO'09 competition.

Su et al [8] used local maximum and minimum to build a local contrast image. Then, a sliding window was applied across the contrast image to determine local thresholds, where bright pixels indicate foreground and dark pixels refer to background pixels. A version of this method ranked one of the two sharing first-rank winners among 17 algorithms participated in H-DIBCO'10 contest.

Farrahi Moghaddam et al [1] proposed a multi-scale binarization method in which input document was binarized several times using different scales. Then, these output images were combined to form the final output image. Afterward, this method has been extended to the Otsu's method [9] with better results, which is called AdOtsu [10].

Also, combination methods have received great interest, and provided promising results. The goal of these methods is to combine existing methods in order to improve the output assuming different method complement each other. There are two well-known combination schemes. First approach combines several binarization methods with a voting algorithm, like the work proposed in [11]. The second approach combines two or more methods based on some features [12, 13]. Usually combination methods through feature space, divides pixels into three categories: foreground, background and unknown. This partitioning is based on a priori knowledge about the behavior of each method is used.

3 Phase Congruency Model

In this section, phase congruency (PC) model used in this work is presented.

3.1 Phase Congruency Measures

In this section, two feature maps of phase congruency are discussed. These feature maps are maximum moment of phase congruency covariance (MMPCC) and local weighted mean phase angle (LWMPA), respectively, at each pixel of input image. Phase congruency first defined in [14] in terms of a Fourier series expansion of signal. This definition does not provide good localization because this function varies with cosine of phase deviation [3]. Therefore, Kovesi [3] modified and extended the phase congruency to 2D over several orientations and combined the results along different orientations. In phase congruency, points of interest are at those points where the Fourier components are maximum in phase [3, 14].

Let f(x) denote a 1D signal, and M_n^e and M_n^o denote the even-symmetric and odd-symmetric wavelets at a scale n (we will use log-Gabor wavelets). M_n^e and M_n^o will form a quadratic pair. The responses of each quadrature pair of filters forms a response vector:

$$[e_n(x), o_n(x)] = [f(x) * M_n^e, f(x) * M_n^o] .$$ (1)

where * denotes the convolution, and values $e_n(x)$ and $o_n(x)$ are real and imaginary parts of complex valued frequency domain. The local amplitude of the transform at a given wavelet scale is given by:

$$A_n(x) = \sqrt{e_n(x)^2 + o_n(x)^2} .$$ (2)

And, the local phase is given by:

$$\varphi = \tan^{-1}(o_n(x), e_n(x)) .$$ (3)

Let us define:

$$F(x) = \sum_n e_n(x) ,$$ (4)

$$H(x) = \sum_n o_n(x) ,$$ (5)

$$\sum_n A_n(x) = \sum_n \sqrt{e_n(x)^2 + o_n(x)^2} , \qquad (6)$$

where, F(x) and H(x) are computed by summing the even filter convolutions and odd filter convolutions, respectively. Then, the 1D phase congruency is defined as follows:

$$PC(x) = \frac{E(x)}{\sum_n A_n(x) + \epsilon} . \qquad (7)$$

where, $E(x) = \sqrt{F^2(x) + H^2(x)}$ and ϵ is a small positive constant added to the denominator for the cases that Fourier amplitudes are very small. A major problem with PC calculation using equation (7) is its sensitivity to noise. To overcome this problem, Rayleigh distribuation can be used for modeling the distribution the energy [3]. In this paper, noise is considered as:

$$T = \mu_R + k\sigma_R . \qquad (8)$$

where, μ_R is the mean noise response and σ_R is the standard deviation. k is the number of σ_R to be used. In our experiments, k is set to 2. By using this noise model, PC equation is modified as:

$$PC(x) = \frac{\lfloor E(x) - T \rfloor}{\sum_n A_n(x) + \epsilon} , \qquad (9)$$

where, $\lfloor \rfloor$ denotes that the enclosed quantity is equal to itself when its value is positive, and zero otherwise. The value of ϵ is set to 0.0001. Let's define:

$$s(x) = \frac{1}{N} \left(\frac{\sum_n A_n(x)}{\epsilon + A_{max}(x)} \right) , \qquad (10)$$

$$W(x) = \frac{1}{1 + e^{\gamma(c - s(c))}} , \qquad (11)$$

where, s(x) is a fractional measure of spread, W(x) is a phase congruency weighting mean function constructed by applying a sigmoid function to the filter response spread, c is a cut-off value of filter response spread below which phase congruency values become penalized, and γ is a gain factor that controls the sharpness of the cutoff. Also, consider $\Delta\varphi$, which is a more sensitive phase deviation function:

$$\Delta\varphi = \cos(\phi_n(x) - \overline{\phi}(x)) - |\cos(\phi_n(x) - \overline{\phi}(x))| . \qquad (12)$$

Now, we define a new phase congruency as follows:

$$PC_1(x) = \frac{\sum_n W(x) \lfloor A_n(x) \Delta\phi(x) - T \rfloor}{\sum_n A_n(x) + \epsilon} , \qquad (13)$$

where, ϵ is a small positive constant and T is the estimated noise. We use this PC for accurate binarization of strokes and sub-strokes. For generalization to 2D images, 1D response along several orientations is combined. 2D phase congruency is defined as follows by combining various orientations:

$$PC_{2D} = \frac{\sum_o \sum_n W_o(x)\lfloor A_{no}(x)\Delta\phi_{no}(x) - T_o\rfloor}{\sum_o \sum_n A_{no}(x) + \epsilon} \ , \tag{14}$$

where, o denotes the index over orientations.

4 Feature Maps

In this paper, we introduce and use various feature maps (representaions) of the input document image. Some of these representations are based on phase congruency. In general, calculation of these maps is fast. Adaptive Gaussian filter, adaptive median filter and regional minima are also used to obtain three binary representations of the input image. The details of various feature maps are provided in the following subsections.

4.1 Maximum Moment of Phase Congruency Covariance (MMPCC)

As mentioned, maximum moment of phase congruency covariance (MMPCC) map is a measure of the edges strength. The MMPCC map takes values between [0 1], where a larger value means a stronger edge. MMPCC is equal to the 2D PC in equation (14) except that the sum function is replaced with a max function. A method, based on MMPCC, has been proposed in [15] to preprocess historical document images. In preprocessing step, another map, based on MMPCC, is introduced. This binary map is calculated by filling the inner parts of MMPCC map using a proposed gray-level image reconstruction method. Then, Otsu's method is applied on the filled image in order to obtain a binarized map. We found that this step not only can reject some of the badly degraded background pixels, it also keeps the recall value nearly constant as reported in [15].

4.2 Regional Minima

Bleed through degradation is an important and common interfering pattern in the old and historical document images. In this paper, bleed-through is categorized into two classes: i) local bleed-through and ii) global bleed-through. The local bleed-through applies to those degraded pixels where are located under or near the foreground pixels, while the global bleed-through refers to those pixels which are located far from the foreground text. The global bleed-through is one of most challenging degradation because there is no local reference in order to distinguish between the true text and bleed-through. Here, an unsupervised method based on regional minima is proposed to overcome the problem of global bleed-through. A regional minimum M of an image f at elevation t is a connected component of pixels with the value t whose external boundary pixels have a value strictly greater than t [16]:

$$\begin{cases} \forall p \in M, & f(p) =\text{t} \\ \forall q \in \delta^{(1)}(M)\backslash M, & f(q) > \text{t} \end{cases}, \tag{15}$$

where, $\delta^{(1)}(M)$ denote to the elementary dilation of M and $\delta^{(1)}(M)\backslash M$ denotes the difference between two images.

Suppose a document image degraded with global bleed-through. The distribution of regional minima on the foreground is more than bleed-through pixels of background. This pattern is used in an unsupervised approach to remove global bleed-through. Regional minima image is called bwMinima.

4.3   Adaptive Gaussian Filter

An adaptive thresholding method has been proposed by Wellner [17]. In this approach, for each pixel in the image a threshold is selected by calculating local weighted mean along the row, or pairs of rows using a recursive filter. Afterward, Bradley et al [18] modified this approach by using the integral image. They computed local weighted mean along rows and columns.

In this section, a similar approach to that used in [18] is used. However, we used a Gaussian smoothing filter to obtain local weighted mean as the reference value for setting threshold for each pixel. We used a rotationally symmetric Gaussian low-pass filter (Gfilt) of size 60 with $\sigma = 5$, where σ is the standard deviation. Then 2D correlation between Gfilt and input image is computed. In result, a filtered image Filtim is returned. Filtim is the same size as input image and stores local thresholds. A pixel is set to 0 (dark) if the value of that pixel in the input image is less than 95% of corresponding pixel value in Filtim, and it will set to 1 (white) otherwise. We increased the threshold value from 85% [17, 18] to 95% to obtain a near optimal recall value.

4.4   Local Weighted Mean Phase Angle

Second measure of phase congruency is the local weighted mean phase angle (LWMPA) at every point in the image which is calculated using equation (3). The values of this map are between $-\pi/2$ and $+\pi/2$, where a dark line take a value of $-\pi/2$, and a bright line take a value of $+\pi/2$.

4.5   Adaptive Median Filter

It is known that median filter can reject salt and pepper noises in presence of edges. Similar to the method used in section 4.3 for Gaussian filter, local thresholds are computed by applying a 4×4 symmetric median filter for each pixel in the input image. In turns, a filtered image with equal size to the input image is produced medimage, where its pixel values are local thresholds. A pixel is set to 0 (dark) if the value of that pixel in the input image is less than 90% of corresponding pixel value in medimage, and pixel will set to 1 (white) otherwise.

5 Binarization Model

The proposed binarization method is divided into two steps: preprocessing, and main binarization. Fig. 2 shows the flowchart of the proposed binarization model. The following subsections describes of these steps.

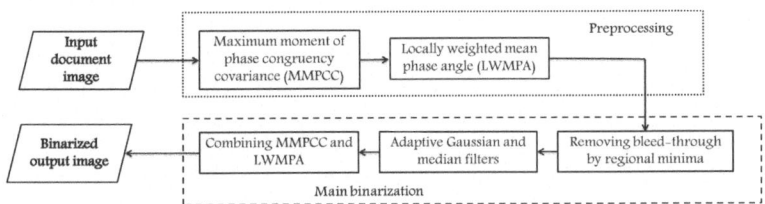

Fig. 2. The overall flowchart of the proposed binarization method

5.1 Preprocessing

Maximum Moment of Phase Congruency Covariance. A number of parameters impacts on the quality of MMPCC output image. Number of 2D log-Gabor filters scales p, the number of orientations of 2D log-Gabor filters r and the number of standard deviations k used to reject noises should be set according to the application. In this paper p, r and k are set with 10, 10 and 2, respectively, based on our experiments. Choosing some smaller values for r and p produces better edge image, but it can result in failure of the filling method. The reason that 10 filter scales produces inaccurate edge image comes from the combination process to form MMPCC, which needs combining more images into one image. Fig. 3 shows a degraded document image and it's over-binarization procedure using MMPCC.

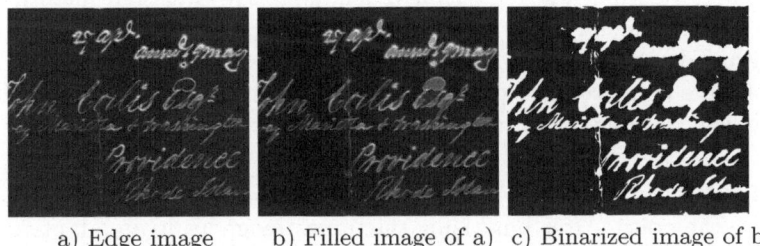

a) Edge image   b) Filled image of a)   c) Binarized image of b)

Fig. 3. A degraded document image and its binarized image obtained by using the phase congruency. This step is able to remove the most of the badly-degraded background.

Local Weighted Mean Phase Angle. Despite promising performance of LWMPA on the strokes and sub-strokes, it's performance for the interior pixels of big objects is low. On the other hand, results of Otsu's method [9] on the input gray-level image for interior pixels of big objects are interesting, while it may lost weak strokes and sub-strokes. Therefore, a combination of LWMPA and Otsu's method produces satisfactory results. The results are combined using logical *or* operator.

5.2   Main Binarization

Removing Global Bleed-through Using Reginal Minima In this section, an unsupervised method based on regional minima is proposed and used to remove global bleed-through. To determine the possibility of bleed-through in images, Otsu's method is used. The steps are listed in Table 1, where the values of $k1$ and k_{bt} control the behavior of the proposed method.

Table 1. Pseudo code for the proposed global bleed-through removing using regional minima

1. Calculate *bwOtsu*. If sum of pixels in *bwOtsu* are k_1 times smaller than sum of pixels in the *bwout* (output binary image of preprocessing in section 5.1), go to step 2, otherwise no global bleed-through is detected by the proposed method.
2. Calculate *bwMinima* using eight-connectivity assumption. Remove regional minima pixels where they are background pixels in the *bwout* image.
3. Divide input image (or bwMinima) into p_1 equal rectangles. In this paper $p_1 = 25$. p_1 should not be a large number.
4. Compute $m_{bw}(i)$ and $m_{Minima}(i)$ for p_1 rectangles, where $m_{bw}(i)$ and $m_{Minima}(i)$ are the average pixels of i-th rectangle in *bwout* and *bwMinima* images, respectively.
5. Let deg=m(bwout)/m(bwMinima) denote the degree of bleed-though in the image, where m denote the mean. The mean values come from step 4.
6. For each rectangle, compute $ratio_i = m_{bw}(i)/m(bwMinima)$. If $ratio_i > k_{bt} \times deg$, add all the pixels lies on the i-*th* rectangle to background.
In this paper $k_{bt} = 1.6$. Now the output of bleed-through correction is ready.

Adaptive Gaussian Filter. First, adaptive histogram equalization is applied on the gray-level input image, histIm. Then, adaptive Gaussian smoothing filter is applied on histIm, bwGauss. The bwGauss map can reject an moderate amount of background pixels as well. However, it introduces a little error on the interior pixels of large connected objects. To avoid this error, bwGauss is first filled by a morphological image filling method in the binary domain.

Adaptive Median Filter. As mentioned in section 4.5, we use the median filter to remove the noise. After binarizing the input image using an adaptive median filter, bwMedian, a simple procedure is used to remove noises. By considering the binarized output image at this step, bwout, a connected component in bwout without any occlusion with bwMedian is rejected from bwout foreground pixels.

5.3   Summary of Binarization Method

In this section, binarization steps used in this paper are summarized into a pseudo code. First, abbreviations used in this paper are listed, and then pseudo code is presented.

MMPCC: Maximum moment of phase congruency covariance which provide an index of the edges strength.

bwMMPCC: Binarized image of MMPCC using Otsu's method.

MMPCCf: Filled image of MMPCC.

bwMMPCCf: Binarized image correspond to the MMPCCf image.

LWMPA: Local weighted mean phase angle which is a measure of phase congruency.

bwLWMPA: Binary form of LWMPA.

bwOtsu: Binarization of gray-level input image using Otsu's method.

histIm: Histogram equalized image of gray-level input image.

bwGauss: Binary image of histIm using adaptive Gaussian filter.

bwGaussF: Filled image of bwGauss.

bwMedian: Binary image obtained by applying median filter.

bwMinima: Regional minima of the input gray-level image using 8 connected structuring element.

bwout: Final binary output image.

Table 2. Pseudo code of the proposed binarization method

Start
0. Calculate MMPCC, MMPCCf, and bwMMPCCf;
1. bwout=bwMMPCCf;
2. bwTemp=OR(bwLWMPA,bwOtsu);
   bwout=AND(bwout,bwTemp);
3. Apply global bleed-through detection and removing listed in Table 1.
4. bwout=AND(bwout,bwGaussF);
5. Label each connected component in bwout (CC)
   for i=1 to (Number of connected component)
   if all pixels in CC(Number) are background pixels in bwMedian
   remove this component from bwout
   endfor
6. bwTemp=AND(not(bwMMPCC),bwLWMPA);
   bwout=AND(bwout,bwTemp);
End

Note: Foreground and background pixels takes the values of "1" and "0", respectively.

6 Experimental Results

The proposed binarization method is evaluated on three different datasets. The datasets used are DIBCO'09 [4], H-DIBCO'10 [5] and DIBCO'11 [6]. These datasets provide a collection of images suffering from various types of degradation. The results of three DIBCO and H-DIBCO competitions show that algorithms show different behavior against different datasets. Therefore, these datasets provide good and also difficult evaluation setup to examine different algorithm. In Fig. 3, outputs of two algorithms are compared with the proposed binarization method. Two methods are top algorithm of DIBCO'09 and top algorithm of DIBCO'11. For objective evaluation, recall, precision, F-measure (F-M), PSNR, distance reciprocal distortion metric (DRD), and misclassification

Fig. 4. A subjective comparison between three binarization methods. From left to right Lu's method (top algorithm of DIBCO'09), top algorithm of DIBCO'11, and the proposed method.

penalty metric (MPM) measures [4–6] are used. Table 3 provides experimental results for the proposed binarization method, while Table 4 provides an objective comparison between the proposed method and some state-of-the-art algorithms, especially top algorithms of the previous contests.

Although the proposed method did not achieved an F-measure higher than the best F-measure in the contests, its main advantage is its robustness across all datasets, which can be seen from its nearly constant F-measure score for all datasets. This shows that the method, and especially phase congruency information, is stable and robust against many types of degradation. The author think that the performance of the proposed method can be improved drastically by modifying of the model and including additional feature maps. The authors also want to emphases that the participants in the contests have not access to the datasets at the time they were developing their methods.

Table 3. Experimental results of the proposed binarization method

Dataset	Recall	Precision	F-measure	PSNR	DRD	MPM
DIBCO'09	92.76	84.80	88.43	17.02	4.63	1.06
H-DIBCO'10	92.62	85.08	88.52	18.07	3.48	0.63
DIBCO'11	90.05	86.40	87.73	17.02	7.34	3.07

Table 4. F-measure comparison between proposed method and some top binarization methods. Best F-Measure of each contest is provided in F-M* row.

Method	DIBCO'09	H-DIBCO'10	DIBCO'11
Proposed method	88.43	88.52	87.73
Lu's method [7]	91.24	86.41	81.67
Su's method [8]	91.06	85.49	85.56
Text block extraction [6, 19]	90.39	86.74	83.58
AdOtsu [10]	91.57	85.30	85.26
Best F-M* (in the contest)	91.24	91.78	88.72

In addition, by applying the proposed preprocessing step as a masking function in conjunction with other binarization methods, outstanding improvement of those binarization methods obtained. To do so, we used the outputs of some algorithms participated in DIBCO'11 (the outputs of all participating algorithms are available[1]. Table 5 show the effect of using the proposed preprocessing step in conjunction with some of top algorithms of DIBCO'11. It can be seen that, for example, the F-measure performance of the first ranked method in the contest is improved by %12 when our preprocessing step is applied. This suggests developing more sophisticated algorithms by combining phase congruency information and other feature maps in order to achieve higher performances. The authors are working on this direction, and will report the results in a report in the future.

Table 5. The effect of using preprocessing step discussed in section 5.1 on the Otsu's method, EBSK [10] and some top methods of DIBCO'11 [6]

Rank in DIBCO'11	F-M	New F-M	Rank	F-M	New F-M
1	80.85	92.16	5	85.21	88.75
2	85.19	90.55	7	79.44	89.12
3	88.72	92.32	Otsu [9]	82.09	88.75
4	83.58	87.71	ESBK [10]	87.98	89.14

7 Conclusion

In this paper, a novel binarization method for degraded document images is proposed based on phase congruency. After a preprocessing step based on phase congruency maps, the main binarization is applied to obtain the final binarized output image. The main binarization step is based on a model which uses some feature maps generated using regional minima, adaptive Gaussian and median filters, among others. The experimental results show the promising performance of the the proposed method. Also, the method has been used as a preprocessing step to improve the performance of other methods, thanks to its high recall values.

---

[1] http://utopia.duth.gr/ĩpratika/DIBCO2011/dibco2011results.htm

References

1. Farrahi Moghaddam, R., Cheriet, M.: A multi-scale framework for adaptive binarization of degraded document images. Pattern Recognition 43(6), 2186–2198 (2010)
2. Special issue on recent advances in applications to visual cultural heritage. IEEE Signal Processing Magazine 12, 234–778 (2008)
3. Kovesi, P.: Image features from phase congruency. Journal of Computer Vision Research 1, 1–26 (1999)
4. Gatos, B., Ntirogiannis, K., Pratikakis, I.: ICDAR 2009 document image binarization contest (DIBCO 2009). In: ICDAR, pp. 1375–1382 (2009)
5. Pratikakis, I., Gatos, B., Ntirogiannis, K.: H-DIBCO 2010 handwritten document image binarization competition. In: ICDAR, pp. 727–732 (2010)
6. Pratikakis, I., Gatos, B., Ntirogiannis, K.: ICDAR 2011 document image binarization contest (DIBCO 2011). In: ICDAR, pp. 1506–1510 (2011)
7. Lu, S., Su, B., Tan, C.: Document image binarization using background estimation and stroke edges. IJDAR 13, 303–314 (2010)
8. Su, B., Lu, S., Tan, C.: Binarization of historical document images using the local maximum and minimum. In: Document Analysis Systems, pp. 159–166 (2010)
9. Otsu, N.: A threshold selection method from gray-level histograms. IEEE Trans. Systems, Man and Cybernetics 9(1), 62–66 (1979)
10. Farrahi Moghaddam, R., Cheriet, M.: AdOtsu: An adaptive and parameterless generalization of Otsu's method for document image binarization. Pattern Recognition 13, 2419–2431 (2012)
11. Gatos, B., Pratikakis, S.J., Perantonis, S.J.: Improved document image binarization by using a combination of multiple binarization techniques and adapted edge information. In: ICPR, pp. 1–4 (2008)
12. Su, B., Lu, S., Tan, C.L.: Combination of document image binarization techniques. In: ICDAR, pp. 22–26 (2011)
13. Su, B., Lu, S., Tan, C.L.: A self-training learning document binarization framework. In: International Conference on Pattern Recognition, pp. 3187–3190 (2010)
14. Morrone, M., Burr, D.: Feature detection in human vision: a phase-dependent energy model. Royal Society of London B 235(1280), 221–245 (1988)
15. Ziaei Nafchi, H., Kanan, H.R.: A phase congruency based document binarization. In: IAPR ICISP, pp. 113–121 (2012)
16. Soille, P.: Morphological Image Analysis, Principles and Applications. Springer (2007)
17. Wellner, P.D.: Adaptive thresholding for the digitaldesk. Tech. Rep. EPC-110 (1993)
18. Bradley, D., Roth, G.: Adaptive thresholding using the integral image. Journal of Graphic Tools 12(2), 13–21 (2007)
19. Messaoud, I.B., Amiri, H., El Abed, H., Margner, V.: New binarization approach based on text block extraction. In: ICDAR, pp. 1205–1209 (2011)

Can Modern Technologies Defeat Nazi Censorship?

Simone Pentzien[1], Ira Rabin[1], Oliver Hahn[1], Jörg Krüger[1],
Florian Kleber[2], Fabian Hollaus[2], Markus Diem[2], and Robert Sablatnig[2]

[1] BAM Federal Institute for Materials Research and Testing
joerg.krueger@bam.de
[2] Computer Vision Lab, Vienna University of Technology
sab@caa.tuwien.ac.at

Abstract. Censorship of parts of written text was and is a common practice in totalitarian regimes. It is used to destroy information not approved by the political power. Recovering the censored text is of interest for historical studies of the text. This paper raises the question, whether a censored postcard from 1942 can be made legible by applying multispectral imaging in combination with laser cleaning. In the fields of art conservation (e.g. color measurements), investigation (e.g. analysis of underdrawings in paintings), and historical document analysis, multispectral imaging techniques have been applied successfully to give visibility to information hidden to the human eye.

The basic principle of laser cleaning is to transfer laser pulse energy to a contamination layer by an absorption process that leads to heating and evaporation of the layer. Partial laser cleaning of postcards is possible; dirt on the surface can be removed and the obscured pictures and writings made visible again. We applied both techniques to the postcard. The text could not be restored since the original ink seems to have suffered severe chemical damage.

1 Introduction

This work presents efforts taken to recover blackened text passages on a postcard sent from the ghetto Piaski in the German-occupied zone of Poland during the WWII and thus subjected to German military censorship. The author of the postcard, Egon Heysemann, was born on November 25, 1925 in Flatow, a small town in West Prussia. Soon after the Nazi rise to power, the life of Jews in this province became unbearable and many moved to Berlin. In Berlin, Egon Heysemann was placed in the Berliner Auerbach orphanage, which opened its doors to the refugees. There he became a close friend of Rolf Rotschild, the addressee. In 1939 Rolf Rotschild succeeded in escaping Germany with a children transport to Sweden, whereas Egon Heysemann was deported to the East on March 28, 1942. On April 1 he arrived in the ghetto Piaski, Poland. On April 9, he sent a postcard to his friend in Sweden, in which he reported on the deportation and asked for help. The censor made unreadable the details of the

J.-I. Park and J. Kim (Eds.): ACCV 2012 Workshops, Part II, LNCS 7729, pp. 13–24, 2013.

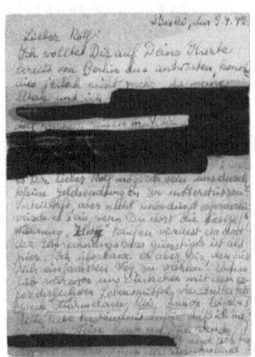

Fig. 1. Postcard, front side   **Fig. 2.** Postcard, reverse side

trip and the situation in the ghetto that served as a transit camp on the way to the death camps. The date and place of Egon Heysemann's death are unknown; this postcard is the last known word received from him. Figure 1 and Figure 2 depict front and reverse side of the postcard. The unreadable, censored parts of the postcard can be seen on the reverse side of the postcard (Figure 2). Barbara Schieb from the German Resistance Memorial Center Berlin and the current owner of the postcard, Walter Frankenstein, entrusted the postcard to the authors for an attempt at text recovery with the help of scientific methods. They kindly permitted us to reproduce the postcard in this paper.

Laser cleaning is a non-contact method to remove unwanted surface layers or coatings. Since the 1970s, laser radiation has been used even to clean works of art. In principle, laser cleaning can be applied selectively, allows in-situ monitoring of the processing, avoids the use of any solvents, and is environmentally friendly. In most cases, pulsed lasers (rather than continuous-wave lasers) are used for cleaning purposes. Different types of lasers (CO_2, Nd:YAG, excimer laser) are employed, covering a wavelength range from about 10 to 0.2 micrometers and pulse-duration regimes from microseconds to nanoseconds. In an ideal case for the application of laser radiation, a light-absorbing contamination layer is located on a non-absorbing object, i.e., the difference between the optical properties of the contamination (to be removed) and original object (to be preserved) should be huge. In the context of this work, a laser working range had to be found that could remove the blackening without any deterioration of the paper matrix material of the postcard, including the original ink of the writing.

MultiSpectral (MS) imaging was originally developed for remote sensing applications and has recently found its way to the field of conservation [1]. In addition to its applicability for art conservation [2], the imaging technique has also proven effective for the analysis of historical and degraded documents [3–6]. Fischer and Kakoulli [1] state that degraded writings that are barely readable under visible light can be made legible by analyzing the text in different spectral ranges. Lettner et al. reassert this statement in [7], where multispectral imaging is successfully applied for the recovery of text in medieval manuscripts.

As in the works cited above, we used multispectral imaging to evaluate whether the censored text regions of the postcard are readable under spectral ranges that are different from visible light. Two unmixing algorithms were applied to increase the contrast of the handwriting.

The paper is structured as follows. The following section explains laser cleaning in general and describes the setup used in detail. Section 3 contains an overview of multispectral imaging and a depiction of the acquisition setup that was used for the imaging of the postcard. The image-processing techniques applied are detailed in Section 4. Section 5 presents the restoration results gained by the laser cleaning procedure and the multispectral image analysis. Finally, concluding remarks are given in the last section of this work.

2   Laser Cleaning and Nondestructive Measurement Techniques

We tried to recover the hidden portion of the text with a consecutive application of a laser cleaning procedure and a multispectral imaging analysis. The theoretical basics of laser cleaning are detailed in the following section, whereas the setup used is described in Section 2.2.

2.1   Laser Cleaning

About 40 years ago, the first attempts to use laser radiation to remove unwanted contamination layers for the preservation of cultural heritage were undertaken [8]. The basic principle of laser cleaning is to transfer laser pulse energy to a contamination layer by an absorption process. This energy deposition process can be highly localized because the laser beam can be focused down to the micrometer scale. The laser energy heats and vaporizes the contamination layer far away from thermal equilibrium. For thick contamination films, a layer-by-layer removal can be employed. The operator has to control the laser's parameter wavelength, pulse duration, pulse repetition rate, and energy density to take the unwanted surface layers off without doing harm to the objects of cultural value.

Today, restorers routinely perform laser cleaning of stone and metal objects. Laser treatment of complex organic materials like paper is still not fully developed for application in conservators' workshops. Cleaning paper is a challenging task because a contamination is to be removed and a fragile organic original material has to be preserved.

First laser cleaning studies of paper objects [9] successfully demonstrated the partial laser cleaning of picture postcards. The results showed that the dirt on the surface, which was particularly resistant to classical dry cleaning (with brushes, eraser blocks, and powders), was removed successfully and the obscured pictures and writings were rendered visible again. To reach such a satisfying laser processing result, laser energy density must stay below the modification threshold of the paper substrate as well as of the coating materials, including paints and inks.

Fig. 3. Laser treatment of the reverse side of the postcard. The bright green laser spot can be seen on the blackened censored part masking the writing.

Various laser cleaning approaches have been tried in the past, especially with respect to the choice of the appropriate laser wavelength to avoid discoloration (yellowing) and mechanical deterioration of the paper substrates. In the nanosecond pulse duration regime, the use of 532 nm wavelength tends to result in the most promising cleaning of the objects and was superior to longer (1064 nm) and shorter wavelengths (355 nm, 266 nm, 248 nm) [10–12]. For the laser treatment of the blackened parts of the postcard, nanosecond laser pulses at 532 nm wavelength were used.

2.2   Laser Cleaning Setup

A prototype laser cleaning system (Bauer + Mück) consisting of a laser cleaning workstation and a remote control system, which is described in detail in [13] has been used. The main component of the cleaning system is an Nd:YAG laser (DINY pQ, IB Laser) emitting laser pulses with a pulse duration of about 8 ns at 532 nm wavelength and a repetition rate of 500 Hz. The laser beam was focused onto the postcard with an F-Theta objective (Sill) with a focusing distance of 480 mm. The laser beam with a diameter ($1/e^2$) of about 0.2 mm in the focus was kept on a flat plane and could be scanned on a pre-defined sample area. This movement of the laser beam makes it possible to vary the number of laser pulses per area. For the experiments to erase the blackening from the postcard, N values between 10 and 60 were used. A laser energy density of about 0.1 J/cm^2 was applied.

Figure 3 shows the spot of the focused laser beam on the upper part of the blackened censored area of the postcard during processing. The grey tube on top of Figure 3 is part of an exhaust system for the removal of the vaporized material.

3   Image Acquisition

In addition to laser cleaning, multispectral imaging is used to make the censored parts of the postcard text readable again. The theoretical principles of

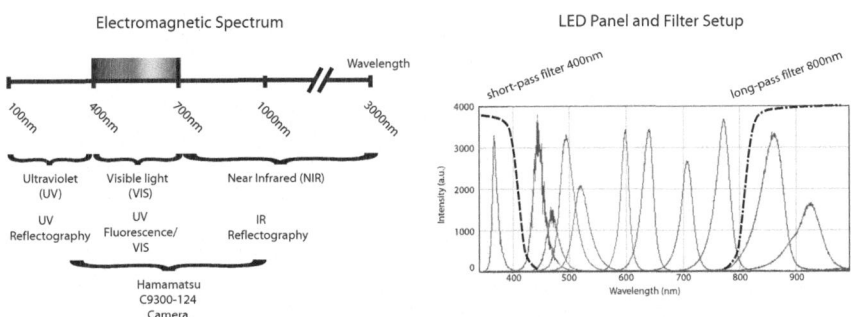

Fig. 4. Filter and LED Panel Spectra

multispectral imaging and related technical terms are introduced in Section 3.1. Section 3.2 depicts the acquisition setup used for the imaging of the postcard.

3.1 Multispectral Imaging

Figure 4 (left) illustrates the electromagnetic spectrum within the range from 100 nm to 3000 nm (UltraViolet (UV), the VISible (VIS) light and Near InfraRed (NIR)). The human eye is sensitive to a wavelength range from approximately 400 (blue) to 700 nm (red). Non-destructive measurements are performed by using image sensors, which have an extended response. Filters (defined by their spectral transmittance) or Light Emitting Diodes (LED) are used to investigate only a predefined spectral range. Since optical filters compromise the spatial resolution of an optical system, narrowband light sources such as LEDs are preferred.

Multispectral imaging systems use non-destructive methods like IR reflectography [14] to give visibility to underdrawings[1] in paintings [2] or to differentiate between different layers of palimpsest texts. IR radiation has the property that it is less scattered than visible light and therefore can penetrate materials that is opaque under VIS illumination [14]. A detailed description of IR reflectography and the mathematical formulation of the reflection are described in [2, 14].

When an object is irradiated by UV light, the light is either reflected from the object (UV reflectography) or absorbed by it to produce fluorescence in the VIS part of the electromagnetic spectrum (UV fluorescence). The reflected and fluorescent light, therefore, appear in different parts of the spectrum. Alternative blocking of each region by a long-wave or short-wave pass filter allows one to measure reflected or emitted light independently from each other. Both types of imaging are especially powerful when one studies objects composed of materials with different optical properties. Historical manuscripts written on parchment

---

[1] Bomford [15] describes underdrawings as *"preliminary drawings on the panel (or canvas) that have been prepared for painting: subsequently they will be covered by the paint layers themselves"*.

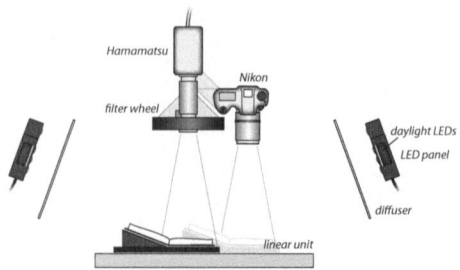

Fig. 5. Schematic setup of the acquisition system

with iron-gall inks present such an example, since these inks do not fluorescence, in contrast to fluorescing parchment. UV fluorescence is used to enhance the contrast and, thus, the readability of palimpsests [2, 3, 16]. Easton et.al. [3] discuss multispectral imaging technique, applied to the Archimedes palimpsest.

3.2 Multispectral Imaging Setup

The acquisition setup used in this work consists of an NIR camera (Hamamatsu C9300-124) and a traditional RGB camera (Nikon D2Xs) mounted next to it. A linear unit allows for a constant shift between the two cameras. The illumination consists of 2 Eureka!Light^{TM} (Equipoise imaging, Archimedes project [17]) LED panels mounted on tripods, which allow the use of 11 different wavelengths, as shown in Figure 4 (right). Additionally, 4 panels with white LEDs (color temperature 5600 Kelvin) are mounted to the left and right of the two Eureka!Light^{TM} panels. To provide a uniformly distributed illumination, diffusers are placed between the lighting and the object. Additionally, a filter wheel with 8 different filters is mounted in front of the Hamamatsu camera. Currently only 2 filters for UV fluorescence (400 nm long-wave pass filter, 40 nm, Full Width Half Maximum) and UV reflectography (400 nm short-wave pass filter, 40 nm, Full Width Half Maximum) are used, whereas the others have been replaced by the LED lighting system.

Figure 5 illustrates the acquisition setup (the system is designed to be transportable with a setup time of approx. 1 hour). The spectra of the Eureka!Light^{TM} LEDs and the transfer function of the filters (illustrated) are shown in Figure 4 (right).

Color images (white LED panels) are captured with the Nikon camera, which provides a resolution of $4288 \times 2848px$. Color images can be taken to provide images for reproductions (facsimile). UV fluorescence/reflectography, and all other wavelengths provided by the Eureka!Light^{TM} are acquired by the NIR (Hamamatsu) camera. A recording area of approx. 150 mm x 110 mm results in approx. 500 dpi. The Hamamatsu C9300-124 camera has a spectral response from UV to NIR (330nm - 1000nm) and a resolution of $4000 \times 2672px$.

Since the various filters on the filter wheel are not perfectly aligned in parallel, a shift between individual images is introduced. The filter influence has been analyzed by [18].

Additionally, if the position/warping of a page changes, the images of different spectral wavelengths are brought out of correct alignment. Thus a registration between consecutive images is necessary. The registration of the images is performed as described in Diem et al. [19].

4 Image Analysis

Spectral unmixing methods are applied to separate the handwriting from the background. The image analysis is applied to the MS images, which have been taken after the laser cleaning.

4.1 Image Enhancement

The MS scan of the postcard can be interpreted as the linear mixture of several patterns, e.g. the handwritten text, the cardboard, the blackening and so forth. We have applied two spectral unmixing approaches in order to separate the handwriting from the remaining patterns. These patterns are also called sources; so-called Blind Source Separation (BSS) methods separate such sources out of mixtures. Several authors - [5], [16], [20]- have proven the applicability of BSS methods for the recovery of degraded handwritten characters. The linear mixture model that we used is mathematically defined by $\mathbf{x}(t) = A\mathbf{s}(t), t = 1, 2, ..., T$, where $\mathbf{s}(t)$ denotes the different sources at pixel t, A is a matrix, containing the mixing coefficients, and $\mathbf{x}(t)$ depicts the observations at pixel t. In our case we have 16 different observations: the postcard was imaged with the Hamamatsu camera at 12 different wavelengths (365, 450, 465, 505, 535, 570, 625, 700, 780, 870, 940 [nm] + white LED) and UV reflectography/fluorescence (400nm LP and SP filter). Additionally, 2 photographs were taken with the Nikon camera (UV fluorescence + color image). Spectral unmixing algorithms aim at estimating the mixing matrix A in order to separate the sources. The problem is solved blindly, since no constraints on the mixing of the signals are made.

The two unmixing techniques, utilized in this work are called Principal Component Analysis (PCA) and Independent Component Analysis (ICA). PCA is a statistical method that transforms the correlated data $\mathbf{x}$ into the uncorrelated data $\mathbf{y}$. The data is projected onto a set of orthogonal axes, whereby the variance of the data is maximized [5]. The transformation is defined by $\mathbf{y} = W\mathbf{x}$ where the transformation matrix W contains k eigenvectors with the largest corresponding eigenvalues of the covariance matrix of $\mathbf{x}(t)$. [16] states that the existence of k dominant eigenvalues indicates that the data consists of k different sources.

In contrast to PCA, the ICA approach [21] finds a linear transformation that maximizes the statistical independence of the sources. Thus, the vector basis – in contrast to the PCA basis – is not necessarily orthogonal. The interested reader is referred to [21] for details on the ICA approach used.

5 Results

The results of the laser cleaning, the MS image acquisition and the BSS techniques applied to the postcard are shown in the following sections.

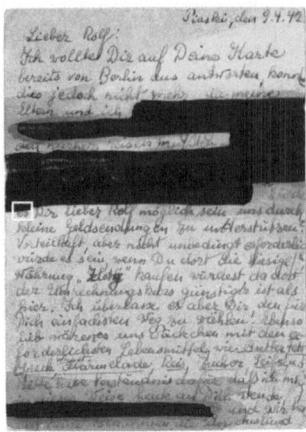

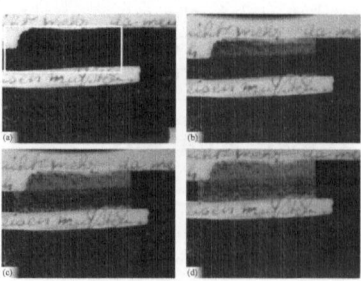

Fig. 7. Reference without laser impact. The white rectangle marks the processing area. Size of the frame 30 mm × 14 mm (a). Laser treated area of the back side of the postcard employing an energy density of 0.1 J/cm^2. N = 20 (b). N = 40 (c). N = 60 (d).

Fig. 6. Reverse side of the postcard with a white rectangle showing the position of the laser treatment

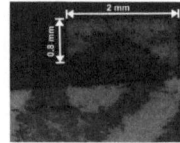

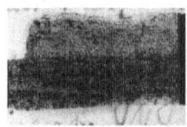

Fig. 8. Detail view of the laser-treated area

Fig. 9. Optical micrograph of the laser cleaned area (0.1 J/cm^2, N = 60)

5.1   Laser Cleaning Results

Figure 6 depicts the reverse side of the postcard with a white rectangle marking the position of the first laser cleaning attempt at the boundary of the blackening. Figure 8 shows the laser cleaning result in detail. An area with dimensions of 2 mm × 0.8 mm was illuminated with an energy density of $0.1 J/cm^2$ and N = 10 laser pulses per spot. It was possible to remove the blackening and to uncover a part of the character "e". The ink, which may have penetrated deeply into the fibers, did not seem to have been affected by the laser treatment.

The first laser cleaning results were encouraging. Thus we processed larger areas of the blackening with the same laser energy density and varying numbers of laser pulses per spot N. Figures 7 (b-d) illustrate the achievements, Figure 7 (a) serves as reference.

Figure 7 (b) shows the laser-treated area of 30 mm × 14 mm after processing with 20 pulses per spot. The brighter appearance of the upper part of the blackening (in comparison with the lower part) after laser treatment indicates that the thickness of the blackening film is inhomogeneous. Obviously, the black layer was thinner at the upper part. After N = 20, the writing was not uncovered.

We therefore repeated the cleaning procedure using the same laser parameters. Figure 7c depicts the result after the second run. A further removal of the black film is obvious and an additional blue line can be seen at the upper part of the area. After a third run (Figure 7 (d)), the whole laser-processed area looks brighter and a second blue line can be observed in the lower part of the area. Additionally, the laser-cleaned area was inspected with an optical microscope. Figure 9 shows the corresponding optical micrograph. Figure 9 clearly indicates that the laser cleaning procedure does not influence the original writing (lower part of Figure 9). The two additional blue lines are visible. The color of these lines and the writing material are different from the original writing. Unfortunately, no characters of the original writing were detected.

5.2   Multispectral Image Acquisition Results

Figure 10 shows a photograph of the postcard, which is illuminated with 570 nm. A detailed view of the region marked by the green rectangle in Figure 10 is given in Figure 11, showing the region imaged at 10 different wavelengths. Like [22], we noticed that the structure of the paper is recognizable in the UV reflectography image. The first four images in Figure 11, which were imaged at lower wavelengths, exhibit more background noise than the remaining images, which were photographed at higher wavelengths. We observed that some fractions of characters which are located at the censored regions - like the vertical stroke in the upper right corner - are most visible at 570 nm or 625 nm. However, only vestiges of the censored text are visible in these images and the censored parts of the writing remain unreadable. Starting at a wavelength of 700 nm, it can be seen that an enlargement of the wavelengths leads to an increase in the reflectivity of the handwriting and hence to a reduced visibility. The text totally disappears if the postcard is illuminated with 940 nm IR light.

5.3   Image Enhancement Results

The effectiveness of BSS techniques has also been demonstrated for the restoration of palimpsest texts [23]. It has been shown that the ICA approach is capable of separating a handwriting pattern from other patterns [23]. Hence, we applied the ICA and the PCA technique to the scan of the postcard to separate the censored text from the blackening. The first eigenvalue of the covariance matrix includes more than 95%. We decided to use the four largest eigenvalues of the PCA for the ICA transformation. The PCA-transformed images are given in Figure 12 and the outputs of the ICA transformation are presented in Figure 13.

It can be seen that the third output of the PCA approach in Figure 12 contains mainly the handwriting, while the handwritten text in the first output of the ICA technique has the greatest background contrast, compared with the remaining images in Figure 13. The ICA output shows fewer portions of the vertical lines (which have been revealed in the laser cleaning procedure) than the PCA result does. If we compare the ICA result with the unprocessed images in Figure 11,

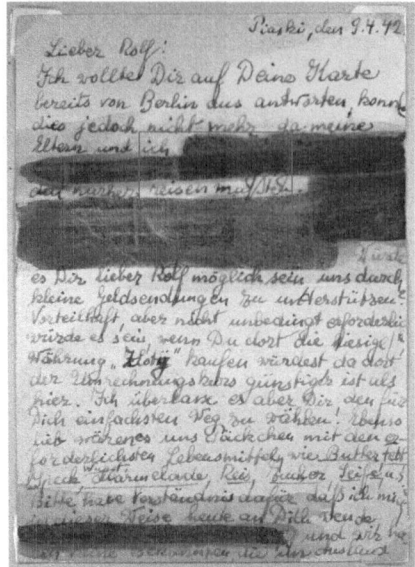

Fig. 10. Postcard imaged at 570 nm

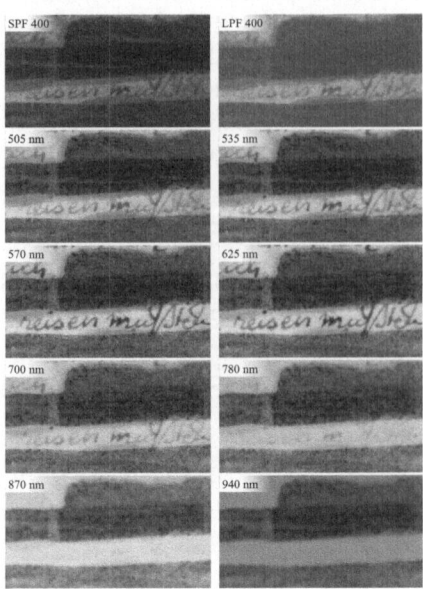

Fig. 11. Postcard portion imaged at 10 different wavelengths

Fig. 12. PCA results

Fig. 13. ICA results

we can see that vestiges of characters are recovered by the spectral unmixing technique. However, the restoration result is modest, since a decipherment of the censored text is still impossible.

6   Discussion

All of the blackened censored areas of the postcard were laser treated. A laser parameter set was chosen to remove the black layer and to preserve the original writing (Figure 8). The thickness of the blackening can be reduced or even completely removed by the laser cleaning procedure (Figure 7). Unfortunately, no original ink characters could be identified after the laser processing (Figure

9). We assume that prior to the blackening of the postcard by the censors, an additional (chemical) treatment of the censored parts was performed resulting in a disintegration of the original ink. Interestingly, additional blue lines were observed under the blackening. These lines consist of other writing materials than the original ink and might have been used as markers by the censors.

Afterwards, multispectral imaging was applied to the laser-processed postcard. The multispectral system makes use of LEDs, which enabled a non-invasive analysis of the postcard at 12 different wavelengths ranging from UV to NIR. The multispectral acquisition did not achieve the desired legibility improvement, since only vestiges of characters are recognizable and these fractions are already best discernible under visible light. We applied two spectral unmixing techniques to the multispectral scan in order to separate the writing from the surrounding patterns and hence to enhance the censored text. By using these approaches, the contrast of some character portions was increased, but the censored text remains unreadable. Future work will be in the direction of FTIR since the remains of the ink might be traceable with this method or MS RTI since traditional laser cleaning and MS image analysis improved the result, however did not make the text readable.

Acknowledgement. We would like to thank Walter Frankenstein and Barbara Schieb (German Resistance Memorial Center, Berlin) for permission to publish the results of scientific investigations of the postcard.

This work was supported by the Austrian Science Foundation (FWF) under grant P23133.

References

1. Fischer, C., Kakoulli, I.: Multispectral and hyperspectral imaging technologies in conservation: current research and potential applications. Reviews in Conservation 7, 3–16 (2006)
2. Hain, M., Bartl, J., Jacko, V.: Multispectral Analysis of Cultural Heritage Artefacts. Measurement Science Review 3 (2003)
3. Easton, R., Knox, K., Christens-Barry, W.: Multispectral Imaging of the Archimedes Palimpsest. In: 32nd Applied Image Pattern Recognition Workshop, AIPR 2003, pp. 111–118. IEEE Computer Society, Washington, DC (2003)
4. Tonazzini, A., Bedini, L., Salerno, E.: Independent component analysis for document restoration. International Journal on Document Analysis and Recognition 7, 17–27 (2004)
5. Rapantzikos, K., Balas, C.: Hyperspectral imaging: potential in non-destructive analysis of palimpsests. In: IEEE International Conference on Image Processing, ICIP 2005, September 11-14, vol. 2, pp. II-618–II-621 (2005)
6. Gippert, J.: The Application of Multispectral Imaging in the Study of Caucasian Palimpsests. Bulletin of the Georgian National Academy of Sciences 175, 168–179 (2007)
7. Lettner, M., Diem, M., Sablatnig, R., Miklas, H.: Registration of Multispectral Manuscript Images as Prerequisite for Computer Aided Script Description. In: 12th Computer Vision Winter Workshop, St. Lambrecht, Austria (2007)

8. Lazzarin, L., Marchesi, L., Asmus, J.: Lasers for Cleaning of Statuary - Initial Results and Potentialities. Journal of Vacuum Science & Technology 10, 1039–1043 (1973)
9. Mäder, M., Holle, H., Schreiner, M., Pentzien, S., Krüger, J., Kautek, W.: Traditional and laser cleaning methods of historic picture post cards. Springer Proceedings in Physics 116, 281–286 (2007)
10. Kolar, J., Strlic, M., Pentzien, S., Kautek, W.: Near-UV, visible and IR pulsed laser light interaction with cellulose. Applied Physics A 71, 87–90 (2000)
11. Kaminska, A., Sawczak, M., Cieplinski, M., Sliwinski, G., Kosmoswski, B.: Colorimetric study of the post-processing effect due to pulsed laser cleaning of paper. Optica Applicata 34, 121–132 (2004)
12. Krüger, J., Pentzien, S., Conradi, A.: Cleaning of artificially soiled paper with 532-nm nanosecond laser radiation. Applied Physics A 92, 179–183 (2008)
13. Kautek, W., Pentzien, S.: Laser cleaning system for automated paper and parchment cleaning. Springer Proceedings in Physics 100, 403–410 (2005)
14. Mairinger, F.: Strahlenuntersuchung an Kunstwerken. E.A. Seemann, Berlin (2003)
15. Bomford, D. (ed.): Art in the making UNDERDRAWINGS in renaissance paintings. National Gallery London (2002)
16. Salerno, E., Tonazzini, A., Bedini, L.: Digital image analysis to enhance underwritten text in the Archimedes palimpsest. International Journal on Document Analysis and Recognition 9, 79–87 (2007)
17. Easton, R.L., Noel, W.: Infinite Possibilities: Ten Years of Study of the Archimedes Palimpsest. Proceedings of the American Philosophical Society 154, 50–76 (2010)
18. Brauers, J., Schulte, N., Aach, T.: Multispectral Filter-Wheel Cameras: Geometric Distortion Model and Compensation Algorithms. IEEE Transactions on Image Processing 17, 2368–2380 (2008)
19. Diem, M., Lettner, M., Sablatnig, R.: Registration of Multi-Spectral Manuscript Images. In: Proceedings of the 8th International Symposium on Virtual Reality, Archaeology and Cultural Heritage, VAST 2007, Brighton, UK, pp. 133–140 (2007)
20. Easton, R., Knox, K.: Digital Restoration of Erased and Damaged Manuscripts. In: Proceedings of the 39th Annual Convention of the Association of Jewish Libraries, Brooklyn, NY (2004)
21. Hyvärinen, A., Oja, E.: Independent component analysis: algorithms and applications. Neural Networks 13, 411–430 (2000)
22. Stuart, B.H.: Analytical Techniques in Materials Conservation. John Wiley & Sons (2007)
23. Hollaus, F., Gau, M., Sablatnig, R.: Multispectral Image Acquisition of Ancient Manuscripts. In: Ioannides, M., Fritsch, D., Leissner, J., Davies, R., Remondino, F., Caffo, R. (eds.) EuroMed 2012. LNCS, vol. 7616, pp. 30–39. Springer, Heidelberg (2012)

Coarse-to-Fine Correspondence Search for Classifying Ancient Coins

Sebastian Zambanini and Martin Kampel

Computer Vision Lab, Vienna University of Technology, Austria

Abstract. In this paper, we build upon the idea of using robust dense correspondence estimation for exemplar-based image classification and adapt it to the problem of ancient coin classification. We thus account for the lack of available training data and demonstrate that the matching costs are a powerful dissimilarity metric to establish coin classification for training set sizes of one or two images per class. This is accomplished by using a flexible dense correspondence search which is highly insensitive to local spatial differences between coins of the same class and different coin rotations between images. Additionally, we introduce a coarse-to-fine classification scheme to decrease runtime which would be otherwise linear to the number of classes in the training set. For evaluation, a new dataset representing 60 coin classes of the Roman Republican period is used. The proposed system achieves a classification rate of 83.3% and a runtime improvement of 93% through the coarse-to-fine classification.

1 Introduction

In ancient times, coins were the usual monetary items and thus everyday objects like they are today [1]. However, nowadays ancient coins are also considered as pieces of art which reflect the individualism of the engravers who manually cut the dies used for minting the coins [2]. Roman coins, for instance, often depict portraits of gods and emperors or historical events, in a similar manner as sculptures or paintings from this era do [2]. Fundamental work of coin experts is the classification of coins according to standard reference books since this provides additional information such as accurate dating, political background or minting place. However, classifying ancient coins is a highly complex task that requires years of experience in the entire field of numismatics [1]. As a substantial part of numismatic coin analysis, coin classification can be supported and facilitated by an automatic image-based system.

The difficulty of ancient coin classification arises from the high number of types (e.g. 550 types are defined for the Roman Republican period [2]) as well as from the high level of intra-class variability. The complexity is additionally increased when the discriminability between the classes is low, as can be seen in Fig. 1 where three coin classes with two samples each are shown. Please note the high global visual similarity between the classes. Nonetheless, one can also see local variations within a class as well as missing parts due to abrasions over the centuries. Apart from abrasions, local spatial variations of features within

J.-I. Park and J. Kim (Eds.): ACCV 2012 Workshops, Part II, LNCS 7729, pp. 25–36, 2013.

Fig. 1. Six coins from the Roman Republican period where (a)-(c) each represents a different class

a class also stem from the fact that different dies from different die engravers were used for minting the coins. For instance, the legends are located differently on the two coins shown in Fig. 1(c). Thus, in contrast to modern coins, image features of ancient coins from the same class can not be aligned and compared by means of a global image transformation.

In this paper we present a method for automatically estimating the visual similarity between two coin images and show how this visual similarity can be used in a coarse-to-fine scheme for the ancient coin classification task. The classification does not rely on machine learning techniques and offline training which eases database extensions as new classes do not involve a re-training of the database. The idea was recently introduced in [3] and we extend this approach towards a rotation-insensitive coarse-to-fine matching, yielding a significant classification speed-up. The main motivation for excluding machine learning techniques from our method is that this way we are less dependent on the availability of a large and representative set of training images. The training of classifiers which provide both a sufficient discriminability for the hundreds of different classes and an adequate representation of the possible variability within a coin class is hindered mainly by the low number of available training samples. Our database of Roman Republican coins from the *Museum of Fine Arts* in *Vienna* is one of the biggest in the world and comprises around 3900 coins. These coins represent 515 different types but for only 237 of them more than three pieces are available. Therefore, instead of trying to cope with the high intra-class variability in a heavy offline classifier training phase we tackle the problem online in the matching stage. We use SIFT flow [4] for this task, but introduce a coarse-to-fine classification scheme which provides a significant decrease of computational time needed for matching against a coin database without loosing the discriminative

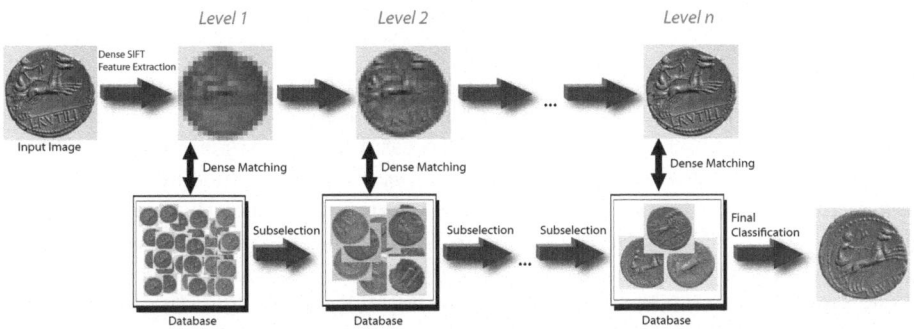

Fig. 2. Schematic illustrating our coarse-to-fine coin classification procedure. Given an input image, a dense set of SIFT features is extracted and matched against the database at the coarsest level. A defined amount of most similar coin images is selected and forwarded to the matching step of the next finer level. This process is continued until the finest level n is reached where the final classification decision is made.

power of SIFT flow for classification. An illustration of our method is shown in Fig. 2.

The work presented in this paper contributes to the research of image-based ancient coin classification, a quite new application field in the area of computer vision. Older methods dedicated to modern-day coins [5–8] have already been shown to be inappropriate for ancient coins [9]. The first approach especially designed for ancient coins was presented by Kampel and Zaharieva [10] with a classification rate of $\sim 90\%$, however by evaluating only three coin types. Their approach is based on matching sparse SIFT keypoints between coin images. Similarity of coin images is then estimated by simply counting the number of matching keypoints while ignoring the geometric configuration of the keypoints. A method based on offline learning was recently proposed by Arandjelović [11]. For feature description also SIFT keypoint detection is used, but geometry is introduced by calculating directional histograms at the keypoint positions. The improvement of adding geometry is shown in the experiments where the proposed method outperforms the bag-of-words approach with a classification rate of 57.2% against 2.4% on a dataset containing 65 Roman Imperial coin classes. This indicates that geometry is an important aspect for ancient coin recognition that we account for by using SIFT flow. This way, we do not depend on the availability of a large set of training images to generalize the intra-class variation: while in [11] between 9 and 160 samples per class were used for training, we show results on a training database of only 1 or 2 samples per class.

The rest of this paper is organized as follows. In Section 2 our SIFT flow based method for coin classification is presented. Results on a dataset representing 60 types of Roman Republican coins are reported in Section 3. The paper is finally concluded in Section 4.

2   Coin Classification Methodology

Our method is based on SIFT flow [4], a method for computing dense pixel-to-pixel correspondences between two images. SIFT flow works by minimizing an energy function which can be exploited to estimate the visual similarity of the images. In a classification scenario, the energy function values between a query image and all class image samples are determined to assign the query image to the class sample with minimum energy. Using this classification scheme, SIFT flow has shown superior results for scene and face classification in scenarios with a low number of available training samples [4]. The motivation to use SIFT flow for coin classification is further rooted in its ability to cope with large image variations in the form they appear within classes of ancient coins. However, we describe some modifications to account for differences of coin location, coin scale and coin rotation. Additionally, we present a coarse-to-fine classification scheme for runtime improvement.

2.1   Insensitivity to Coin Location and Coin Scale by Coin Segmentation

As a dense set of SIFT features with fixed scale has to be computed for SIFT flow computation, we normalize all images to a standard dimension of 150×150 in order to account for scale differences between the coin images. Normalization is achieved by segmenting the coins in the images using a shape-adaptive thresholding approach [12]. The method applies a range and entropy filter to the image which is assumed to provide higher responses at coin regions than on background regions. For the final segmentation mask an optimal threshold value is found by minimizing an objective function describing the circularity of the binary thresholding result.

2.2   Insensitivity to Coin Rotation and Local Spatial Variations by SIFT Flow Image Matching

SIFT flow is based on SIFT features [13] which provide a rotation-invariant description of the local neighborhood by means of gradient orientation distributions. The SIFT features are computed densely over the image, resulting in the so called *SIFT image s*. The pixel-to-pixel correspondences between two SIFT images s_1 and s_2 are represented as a field of flow vectors $\mathbf{w}(\mathbf{p}) = (u(\mathbf{p}), v(\mathbf{p}))$ at grid coordinates $\mathbf{p} = (x, y)$. The optimal correspondences are found by minimizing the following energy function of $\mathbf{w}$:

$$E(\mathbf{w}) = \sum_{\mathbf{p}} |s_1(\mathbf{p}) - s_2(\mathbf{p} + \mathbf{w}(\mathbf{p}))| \tag{1}$$

$$+ \sum_{(\mathbf{p},\mathbf{q}) \in \psi} \min(\alpha |u(\mathbf{p}) - u(\mathbf{q})|, d) + \min(\alpha |v(\mathbf{p}) - v(\mathbf{q})|, d) \tag{2}$$

where ψ contains all four-connected pixel pairs. The energy function is composed of two terms, the *data term* (1) and the *smoothness term* (2), and the influence of the smoothness term is controlled by the parameters α and d. Please note that in our case we do not consider a *small displacement term* in contrast to the original SIFT flow energy function defined in [4]. The reason lies in possible coin rotation differences between the images which demand to allow large pixel displacements in the correspondence search without additional costs. Therefore, without a small displacement term a rotation between image pairs only affects the correspondence search by producing a slightly larger energy in the smoothness term. We quantitatively proof in the experiments in Section 3.2 that the influence of the smoothness term in such cases is negligible and that classification performance is not affected by coin rotation differences. Examples of correspondences found by using the presented energy function can be seen in Fig. 3. Here the query image of Fig. 3(a) is matched against an image of a coin from the same class (Fig. 3(b)), which produces the correspondences visualized in Fig. 3(c) by warping back the image to the query image. We can see that reasonable correspondences have been found despite the variations between the two coins. If we compute SIFT flow for a rotated version of the coin (Fig. 3(d)), the result is almost identical (Fig. 3(e)).

2.3   Runtime Reduction by Coarse-To-Fine Classification

A disadvantage of using SIFT flow for example-based classification is that the runtime is linear to the amount of images in the database. However, SIFT flow itself uses a coarse-to-fine matching scheme for speed-up and better matching results, i.e. correspondences are propagated and re-estimated from coarser to finer levels. We utilize this scheme by selecting only the most similar coin classes at each level for further processing and thus subsequently reduce the amount of possible target coin classes. This way, the computational effort of the whole classification process is reduced as the more costly computations at finer levels have to be conducted only on a subset of coin classes.

More formally, for each sift image s, n pyramid levels $s^{(k)}$ are constructed where $s^{(n)} = s$ and $s^{(k-1)}$ is downsampled from $s^{(k)}$ by a factor of 2. If we denote the set of coin target classes by C and the SIFT flow energy obtained at level k by $E^{(k)}$, classification of a query image s is achieved in the following manner:

1. For all levels k, $k = 1...n$
   (a) Compute SIFT flow energies $E^{(k)}$ between $s^{(k)}$ and all SIFT images of level k of classes C.
   (b) For each class in C, compute the average energy $\bar{E}^{(k)}$ for all its SIFT images in the database.
   (c) Sort all energies $\bar{E}^{(k)}$ and reduce C by selecting only a percentage $\lambda^{(k)}$ of C with lowest energy.
2. Finally, take the class with lowest energy.

(a) Query image   (b) Image of same class   (c) Image (b) warped back
                                                          to query image

(d) Image (b) rotated by 90   (e) Image (d) warped back
degree   to query image

Fig. 3. Results of SIFT flow applied to images of the same class

3   Experiments

For the experiments we use a set of 60 classes of Roman Republican coins [2],
where each class is represented by three coin images of the reverse side. Sample images of all classes are shown in Fig. 4. We report both classification and
runtime performance on our dataset for different subselection values $\lambda^{(k)}$. Additionally, we compare the performance when one or two database images are
available per class. As most of the coins show the same rotation in our database,
the insensitivity against coin rotations is addressed in an individual experiment
where the coin images are artificially rotated. Throughout all experiments, we
used the same empirically determined parameters for SIFT flow matching: dense
SIFT features were computed for a local neighborhood of 12×12 pixels, the number n of pyramid levels was set to 4, and the parameters controlling the influence
of the smoothness term were set to $\alpha = 12$ and $d = 1200$.

3.1   Classification Results

In each classification run, one of the 180 coin images served as query image
and one or two of the remaining images per class served as training images.

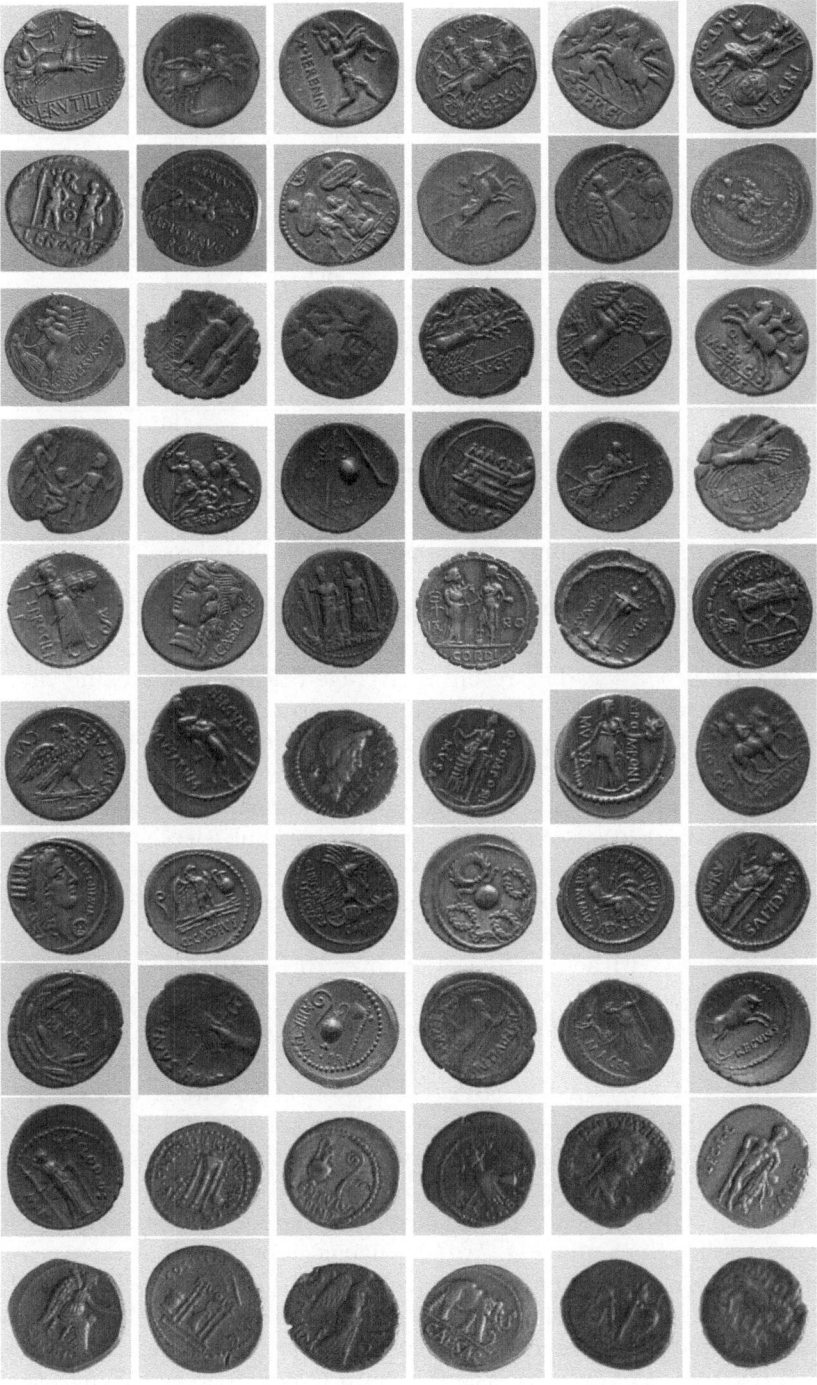

Fig. 4. Sample images of all 60 classes of the evaluation database

Table 1. Classification results

Training set size	$\lambda^{(1)}$	$\lambda^{(2)}$	$\lambda^{(3)}$	Correct classifications	Classification rate	Average classification time
1	100%	100%	100%	257/360	71.4%	235.8s
1	1%	100%	100%	220/360	61.1%	7.1s
1	100%	1%	100%	234/360	65.0%	21.2s
1	100%	100%	1%	257/360	71.4%	70.7s
1	30%	50%	50%	258/360	71.7%	32.5s
1	10%	50%	50%	249/360	69.2%	16.5s
2	100%	100%	100%	150/180	83.3%	471.6s
2	1%	100%	100%	127/180	70.6%	14.1s
2	100%	1%	100%	133/180	73.9%	42.4s
2	100%	100%	1%	141/180	78.3%	141.5
2	30%	50%	50%	150/180	83.3%	65.0s
2	10%	50%	50%	149/180	82.8%	32.9s

This leads to 180 (two training images per class) or 360 classification runs (one training image per class). For runtime evaluation, we measured the average runtime of computing the SIFT flow between two coin images by using the C++ implementation provided by the authors on a standard machine with a quad-core 2.70 GHz processor. The resulting average SIFT flow matching time was 3.93s, where around 3%, 6%, 21% and 70% are needed for the first, second, third and fourth level, respectively. In Table 1 classification results for both training set sizes as well as various values of $\lambda^{(k)}$ are shown. Runtimes are indicated as the time for classifying one coin against our database of 60 classes, without considering feature extraction of the query image. Subselection parameters of $\lambda^{(1)} = \lambda^{(2)} = \lambda^{(3)} = 100\%$ mean that no subselection is performed. Subselection parameters of $\lambda^{(1,2,3)} = 1\%$ mean that only the energies of the first, second or third level, respectively, are used for classification.

One can see that, without subselection, over 70% of the images can be classified correctly with only one training image per class available. Adding a second training image brings a performance improvement of about $7 - 12\%$. Based on the results on this dataset, a reasonable choice for the subselection parameters is $\lambda^{(1)} = 10\%$ and $\lambda^{(2)} = \lambda^{(3)} = 50\%$. The classification rate is very close to the case without subselection (-2.2% for a training set size of 1 and -0.5% for a training set size of 2, respectively), whereas the runtime improvement is around 93%.

In Fig. 5 we show some of our classification results where Fig. 5(a)-(c) depict incorrect classifications and Fig. 5(d)-(f) depict correct classifications. We see that strong abrasions, like in Figure 5(a), as well as the low inter-class variability, like in Figure 5(b), still pose a problem to the system, since the SIFT flow energy becomes less reliable under such conditions. However, also the examples shown in Fig. 5(d)-(f) represent strong abrasions and variations between the images which

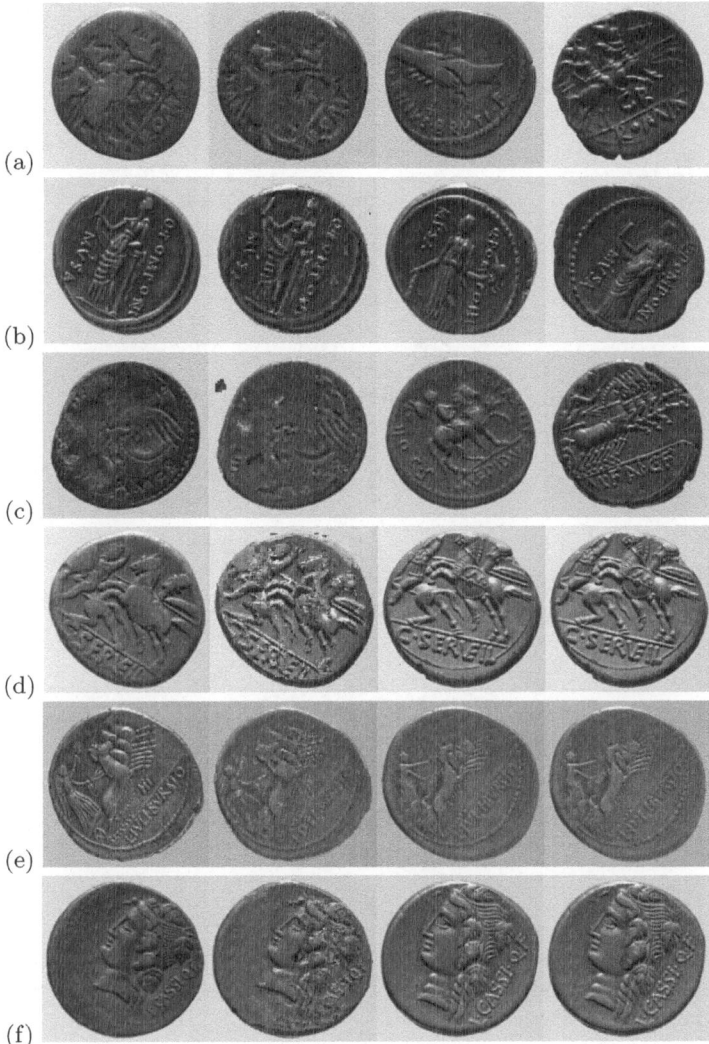

Fig. 5. Six classification results on our dataset. From left to right: query image; most similar image found in the database warped back using the SIFT flow correspondences; original most similar image found; correct most similar image depicting a coin of the same class.

can be dissolved by SIFT flow. Figure 5(c) demonstrates the general limits of image-based ancient coin classification. The query image represents a misprint, which makes it impossible even for human experts to accurately classify the coin if only this coin side is available for examination.

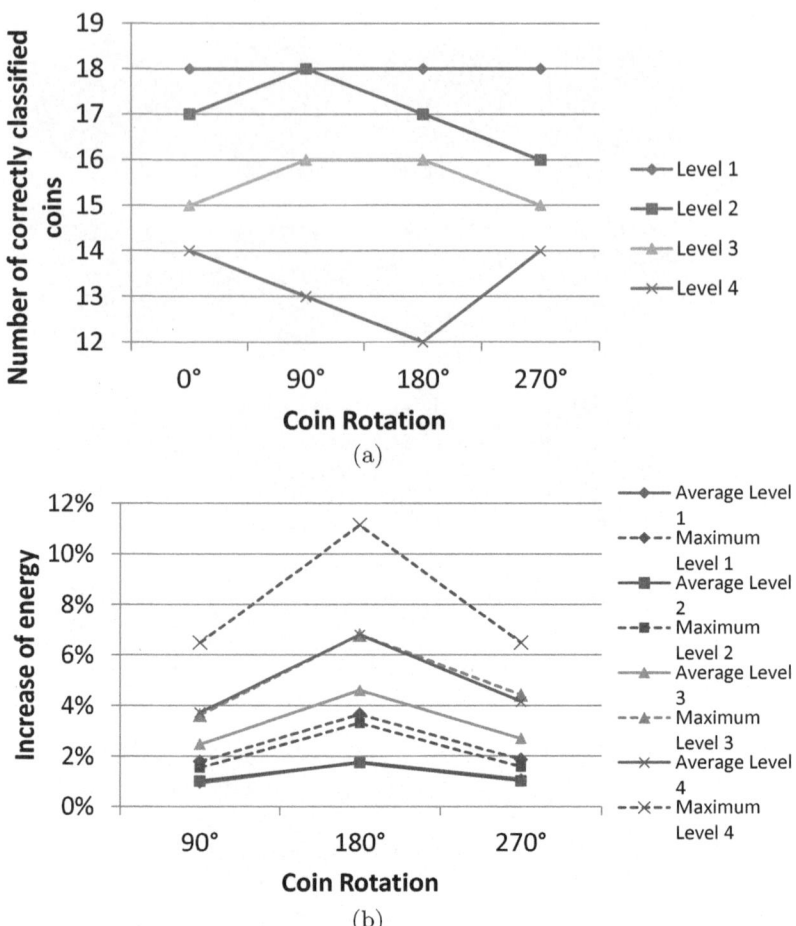

Fig. 6. Results of evaluating sensitivity to coin rotations. (a) Number of correct classifications; (b) Average (solid lines) and maximum increase (dashed lines) of SIFT flow energy between images of the same class.

3.2   Insensitivity to Coin Rotations

In order to assess the sensitivity of our SIFT flow matching to coin rotation differences, we randomly took a query and a training image from 20 coin classes and simulated different coin rotations by rotating the query image in 90 degree steps. Figure 6(a) shows the classification results of all four runs for the four levels of SIFT flow matching. In Fig. 6(b) the average increase of energy due to the additional costs in the smoothness term are plotted. We see that at a coarser level the energy values are more sensitive to coin rotations, thus producing a decrease of classification performance and a higher relative increase of the energy value. Nevertheless, by using a coarse-to-fine classification with subselection parameters

$\lambda^{(1)} = 10\%, \lambda^{(2)} = \lambda^{(3)} = 50\%$, 18 out of 20 classes can be classified correctly for all coin rotation differences. This shows that, although the method is in theory not invariant to coin rotation differences, a high degree of insensitivity is given.

4   Conclusions

In this paper, we have presented a classifier-free system for ancient coin classification. We proposed to use image matching instead of classifier learning for ancient coin classification. The main benefit of such a methodology is that it is less dependent on the number of available training samples as similarities between coins are determined online. This is shown in our experiments where we achieved a classification rate of 71.7% on a dataset with only one training sample available per class.

The major drawback of exemplar-based coin classification is the computational effort since the expensive image matching has to be performed against all coin image samples in the database. We therefore presented a coarse-to-fine scheme that heavily reduces the time needed for classification. In our experiments the average classification time could be reduced from 471.6s to 32.9s, an improvement of about 93%. Additionally, we experimentally proofed the insensitivity of our energy function to coin rotation differences.

In general, our classification results of 83.3% on 60 classes are higher than the ones reported by [11] (57.2% on 65 classes). However, a different dataset was used in the evaluation of [11] and thus no well-founded comparison of classification performance can be presented. As a contribution to other researchers in this field, we make our dataset publicly available[1] which allows for quantitative comparisons of algorithms in the future.

For future research, we plan to further improve our dense correspondence methods for coin similarity estimation. We will focus on a methodology which is less sensitive to appearance variations that arise from different relief heights and lighting conditions. Although due to the efficient optimization scheme SIFT flow is able to handle a large degree of noise in the features, we assume that more adapted features and matching strategies will lead to a significant improvement. We also see potential in using a visual similarity estimation in other forms within the application field of numismatics. Visual similarity estimation can be combined with other methods like symbol or legend recognition for a more extensive classification process. It can also be used for automatic coin hoard grouping where a clustering of coins is performed based on the proposed distance metric.

Acknowledgement. This research has been supported by the Austrian Science Fund (FWF) under the grant TRP140-N23-2010 (ILAC).

---

[1] The dataset is available for download at
http://www.caa.tuwien.ac.at/cvl/people/zamba/

References

1. Grierson, P.: Numismatics. Oxford University Press (1975)
2. Crawford, M.H.: Roman Republican Coinage, 2 vols. Cambridge University Press (1974)
3. Zambanini, S., Kampel, M.: Automatic coin classification by image matching. In: VAST: International Symposium on Virtual Reality, Archaeology and Intelligent Cultural Heritage, pp. 65–72 (2011)
4. Liu, C., Yuen, J., Torralba, A.: Sift flow: Dense correspondence across scenes and its applications. IEEE Transactions on Pattern Analysis and Machine Intelligence 33, 978–994 (2011)
5. Huber, R., Ramoser, H., Mayer, K., Penz, H., Rubik, M.: Classification of coins using an eigenspace approach. Pattern Recognition Letters 26, 61–75 (2005)
6. van der Maaten, L.J., Poon, P.: Coin-o-matic: A fast system for reliable coin classification. In: Proc. of the Muscle CIS Coin Competition, pp. 07–18 (2006)
7. Nölle, M., Penz, H., Rubik, M., Mayer, K.J., Holländer, I., Granec, R.: Dagobert – a new coin recognition and sorting system. In: Proc. of DICTA 2003, pp. 329–338 (2003)
8. Reisert, M., Ronneberger, O., Burkhardt, H.: An efficient gradient based registration technique for coin recognition. In: Proc. of the Muscle CIS Coin Competition, pp. 19–31 (2006)
9. Zaharieva, M., Kampel, M., Zambanini, S.: Image Based Recognition of Ancient Coins. In: Kropatsch, W.G., Kampel, M., Hanbury, A. (eds.) CAIP 2007. LNCS, vol. 4673, pp. 547–554. Springer, Heidelberg (2007)
10. Kampel, M., Zaharieva, M.: Recognizing Ancient Coins Based on Local Features. In: Bebis, G., Boyle, R., Parvin, B., Koracin, D., Remagnino, P., Porikli, F., Peters, J., Klosowski, J., Arns, L., Chun, Y.K., Rhyne, T.-M., Monroe, L. (eds.) ISVC 2008, Part I. LNCS, vol. 5358, pp. 11–22. Springer, Heidelberg (2008)
11. Arandjelovic, O.: Automatic attribution of ancient roman imperial coins. In: Conference on Computer Vision and Pattern Recognition, pp. 1728–1734 (2010)
12. Zambanini, S., Kampel, M.: Robust automatic segmentation of ancient coins. In: 4th International Conference on Computer Vision Theory and Applications (VISAPP 2009), vol. 2, pp. 273–276 (2009)
13. Lowe, D.G.: Distinctive image features from scale-invariant keypoints. International Journal of Computer Vision 60, 91–110 (2004)

Archiving Mural Paintings
Using an Ontology Based Approach

Anupama Mallik, Santanu Chaudhury, Shipra Madan,
T.B. Dinesh, and Uma V. Chandru

Indian Institute of Technology, Delhi, India, International Institute for Art,
Culture and Democracy, Bangalore, India

Abstract. In this paper, we propose an archiving scheme for heritage mural paintings. The mural paintings typically depict stories from folk-lore, mythology and history. These narratives provide content-based correlations between different pieces of art. Our e-heritage scheme for archiving the mural paintings is based on an ontology which captures the background knowledge of these narratives. Media features and patterns derived from the mural content are used to enrich the ontology with multimedia data. We have used the multimedia web ontology language as our ontology representation scheme, as it allows perceptual modelling of domain concepts in terms of their media properties, as well as reasoning with uncertainties. Besides the mural content and its knowledge, the ontology also helps encode other aspects of the mural paintings like their painting style, color, physical location, time-period, etc., which are important parameters of their preservation. We propose a framework to provide cross-modal semantic linkage between semantically annotated content of a repository of Indian mural paintings, and a collection of labelled text documents of their narratives. This framework, based on a multimedia ontology of the domain, helps preserve the cultural heritage encoded in these artefacts.

1   Introduction

Cultural heritage is encoded in a variety of forms - some of which are tangible like monuments, sculpture, paintings, coinage, manuscripts, etc; and others are intangible like folk-lore, mythological stories, performing arts like music and dance. In this paper, we propose an archiving scheme for the preservation of mural paintings. The mural paintings which are found in temples, palaces and other historic monuments, depict stories from folk-lore, mythology and other compositions from history. These stories contain narratives which are associated with these pieces of art. These narratives help provide content-based correlations between artistic renderings in different mural paintings. The preservation of the mural paintings raises issues pertaining to their digitization which are linked to how accurately their images can be captured through high-end cameras, along with issues of scanning and storage. But an important aspect of preservation of heritage is to preserve the background knowledge like the narratives associated

J.-I. Park and J. Kim (Eds.): ACCV 2012 Workshops, Part II, LNCS 7729, pp. 37–48, 2013.

with the content in these artefacts, the artistic style of painting, the time-period to which the murals belong, the history of the temples and the area where they are located. Earlier efforts in e-heritage conservation have focussed more on digitization of the resources and not on their contextual and content-based preservation.

The heritage preservation scheme proposed in this paper uses an ontology to capture the essence and knowledge of the narratives associated with the mural paintings. Such an ontology when enriched with multimedia data, which comes from the digitized images of the mural paintings, helps provide the necessary associations between the high-level semantic concepts of the domain and low-level media content. Besides the mural content and its knowledge, the ontology also helps encode other aspects of the mural paintings like their physical location, the painting tradition followed, the information about the artists and the history of the time-period in which they were painted. We propose an ontology-based framework to provide cross-modal, conceptual linkage between the semantically annotated content of a repository of Indian mural paintings, and a collection of labelled text documents of their narratives. The framework can also be used to develop semantic query, search and browsing interfaces to access other aspects of the mural paintings beside their content, and in this way, helps to provide a semantically integrated holistic view to the user about site or era specific heritages. The scheme proposed in this paper can be extended to other tangible heritage artefacts, such as inscriptions, coinage and manuscripts.

1.1   Related Work

Cultural heritage information of the humankind is distributed in various forms, such as paintings, scriptures, architecture and audiovisual records of performing arts. In recent times, the economics of computing and networking resources have created an opportunity for large-scale digitization of the heritage artifacts for their broader dissemination over the Internet and for preservation [1], [2]. With the increase of collection sizes on the web, correlating different heritage artifacts and accessing the desired ones in a specific use-case context creates a significant cognitive load on the user.

Several research groups [3], [4], [5] have proposed use of ontology in semantic interpretation of multimedia data in collaborative multimedia and digital library projects. While ontology is a useful tool for modeling a conceptual domain, it has not been designed to model multimedia data that is perceptual in nature. In the current approaches, specific computer vision algorithms are used to recognize pre-defined objects or events of interest in the media documents to generate automatic annotations. Domain ontology is used to interpret these conceptual annotations in specific query contexts. Thus, the ontology needs to be customized for the specific annotation scheme followed in the collection. The ontology and the metadata schemes are tightly coupled in these approaches, which necessitates creation of a central metadata scheme for the entire collection and prevents integration of data from heritage collections developed in a decentralized manner.

Most of the e-heritage artefacts are recorded in multimedia format and traditional ontology representation schemes need multimedia records to be annotated for semantic processing. Such annotation is a labor intensive process and a major bottleneck in creating a digital heritage collection. Moreover, the ontology itself needs to be hand-crafted, which is an extremely knowledge intensive effort. In this context, we present an ontology-based framework for semantic annotation of digital heritage artefacts, focussing on Indian mural paintings. One of the key ingredients in our architecture is a cultural heritage ontology that follows a unique representation and a distinct reasoning scheme. The ontology includes descriptions of domain concepts in terms of expected audio-visual features in multimedia manifestations, making it especially suitable for semantic interpretation of multimedia data. We have experimented with a collection of images depicting Indian mural paintings along with a set of text documents containing the narratives in these murals. Starting with a hand-crafted basic ontology for the mural domain, we create a multimedia enriched ontology by using a training set of labelled image segments from mural paintings and labelled text segments from the narratives. Using techniques of semantic matching between the text and image segments, we are able to build co-relations between the two through domain concepts of the ontology. Once an enriched ontology is available, it can be used to provide cross-modal access to mural image segments and text in the story documents. Besides this semantic correlation, the ontology also helps preserve the knowledge about mural paintings by catering to semantic queries pertaining to other aspects of the mural paintings like painting style, location, time-period, etc.

The rest of the paper is organised as follows : Section 2 gives an overview of our ontology based framework for e-heritage preservation. Section 3 explains the advantage of using Multimedia Web Ontology Language(MOWL) to build the multimedia ontology of the murals domain. In section 4, a brief introduction of the mural paintings domain is provided. Section 5 details our implementation of the ontology-based semantic access and cross-Modal retrieval from collections of mural painting images and text narratives. Section 6 concludes the paper with a summary of our findings.

2   Ontology Based Heritage Preservation

In this section, we discuss the proposed ontology-based framework for providing cross-modal, semantic access to repositories of digital Mural paintings and narratives, and the resultant preservation of the e-heritage encoded in these artefacts. Domain experts provide the domain knowledge to create a *Tag Dictionary* of the domain. This *tag dictionary* contains the names of different concepts in the domain and their data instances, and thus can be used as a basis for tagging documents. Using the *tag dictionary*, the domain experts also provide tags and labels for different segments of the heritage collection. We have two heritage repositories - a collection of images of mural paintings, and one of text documents which contain the narratives and stories depicted in these murals.

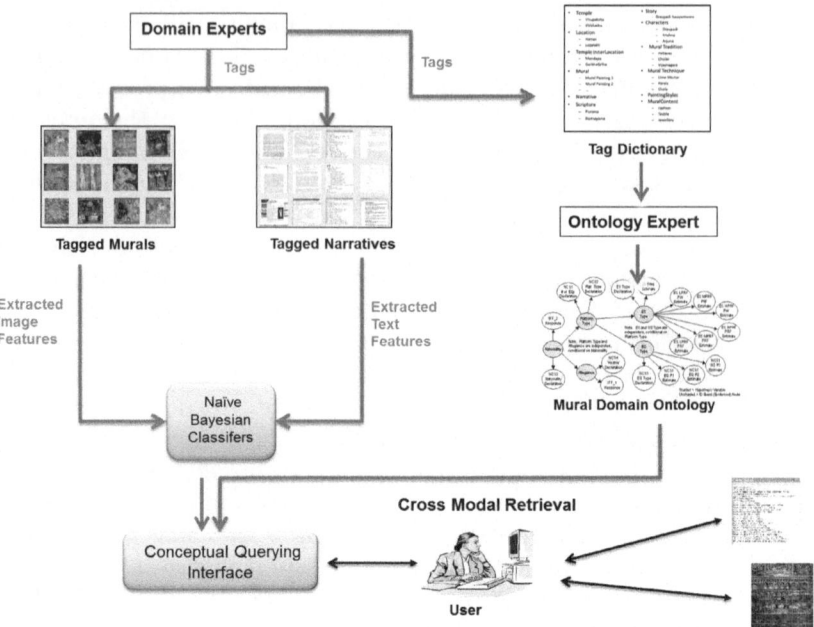

Fig. 1. Ontology-based Framework for Cross-modal, Semantic access to Mural Paintings and Narratives

Once the documents in the two collections have been tagged, media feature extractors are used to extract some image-based and text-based features, in order to provide content-based models of the two sets of documents. An ontology expert analyses the *tag dictionary*, and generates an ontology of the domain. The ontology is enriched with media data from the two collections as image segments and text segments are associated with appropriate semantic concepts in the ontology. This process is semi-automated, as it utilises machine learning of the associations between media content and semantic concepts, to generate automatic labels for media segments, but needs to be curated by a domain expert. Once the annotated repositories are available, and the multimedia ontology has been generated, the framework makes use of Naïve Bayesian classifiers to can generate semantic linkages between the image and text segments, and provide cross-modal access to the repositories. Semantic queries linked to other historic and content-based knowledge contained in the mural images and the narratives, are also effectively answered by querying and searching interfaces based on this framework.

3   Multimedia Ontology Representation

We have used the Multimedia Web Ontology Language (MOWL) [6] for representing the multimedia ontologies used in our experiments. In [7], authors have proposed an ontology-based scheme based on MOWL, for preserving the intangible heritage of Indian classical dance. An ontology encoded in a traditional ontology language, e.g. OWL, uses text to express the domain concepts and the properties. Thus, it is quite straightforward to apply such an ontology for semantic text processing. Semantic processing of multimedia data, however, calls for ontology primitives that enable modeling of domain concepts with their observable media properties. This kind of modeling is called **Perceptual Modeling**, an example of which is shown in figure 2(a). Such modeling needs to encode the inherent uncertainties associated with media properties of concepts too. Traditional ontology languages do not support these capabilities.

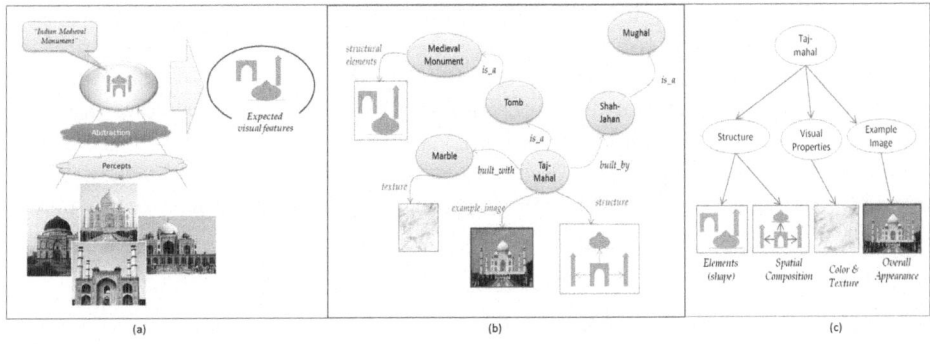

Fig. 2. (a) Perceptual Modeling (b) Multimedia Ontology of Indian Monuments (c) Observation Model of `TajMahal`

In order to support semantic media processing, we use the ontology representation scheme offered by MOWL, that enables encoding of media properties for the concepts in a closed domain. The basic premise of MOWL is a causal model of the world, where real-world concepts (and events) lead to manifestation of media features in multimedia documents. This causal modeling distinguishes MOWL from OWL and other knowledge representation languages. The causal model can be used for abductive reasoning for concept recognition in multimedia data, where the observed media features in a multimedia document can be *causally* explained as manifestations of concepts. MOWL allows encoding of uncertainties which exist in the observation of multimedia data, and in some relations between concepts which are probabilistic. These can be specified as joint probabilities of a concept in relation with several other concepts. This kind of reasoning is useful in concept discovery in documents belonging to multimedia collections. We have used MOWL to encode our domain ontology. MOWL provides the following functionality for a multimedia ontology representation:

- **Concepts and Media Properties**
MOWL distinguishes between two types of entities, namely (a) the *concepts* that represent the real world objects or events and (b) the *media objects* (figure 2(b)) that represent manifestation of concepts in different media forms.
- **MOWL Relations**
Relations between the concepts play an important role in concept recognition. For example, an important cue to the recognition of a medieval monument can be the visual properties of the stone it is built with (as shown in figure 2(b)).
- **Specifying Spatio-temporal Relations**
MOWL defines a subclass of media objects called `ComplexObject` which represents composition of media objects related through spatial or temporal relations.
- **Uncertainty Specification and Reasoning with Bayesian networks**
The knowledge available in a MOWL ontology is used to construct an observation model (OM) for a concept, which is in turn used for concept recognition. For example, figure 2(c)) shows an OM for `Tajmahal`, in terms of its observable media patterns like structural composition of dome and minarets, color and texture of the stone with which it is built. Evidence of one or more of these media patterns found in a multimedia document is sufficient for concept-recognition of `Tajmahal`, with the help of belief propagation in the Bayesian network which is the OM.

4   Mural Paintings Domain

At the time of high Renaissance in Europe, a similar phenomenon of social and cultural resurgence was happening here in the sub-continent, in Vijaynagara empire (current location being the states of Karnataka and Andhra Pradesh in India). Some remnants of the magnificent murals and stories that adorn the ceilings and walls of temples in these states have survived the depredation of time. It is in this context that we seek to preserve the murals and pictorial manuscripts of the Vijaynagara period, a few of which are shown here as digitized images in figure 3. The Indian temples have some of the largest murals in the world, painted across their ceilings and walls. The ceilings are replete with many stories from the epics and puranas, though very few have survived the depredation of time. Each ceiling along the passageway of some of the temples, is like a large panel narrating a story as one moves from the north to the south or east to west. The narrative in each panel flows from one scene to another, amidst many intricate details, to follow the forms into a sequence of expressions by the characters. The imagery of the words from epic poems, the practices of rituals, the oral history, the Sthalapuranas (scriptures linked to a temple defining its architectural elements), remain the source of imagination and an inspiration for many of the temples. Each mural in the temple is further supported by a sub-text and narration and sometimes even a painted manuscript.

Fig. 3. Digitized Images of Mural Paintings from the Hampi temple, India

4.1   Tag Dictionary and Tagging of Murals

Based on domain knowledge derived from domain experts, many of whom are art-historians, an effort has been made to create a *tag dictionary* for the domain of Indian mural paintings. A part of this tag dictionary is shown in figure 4(a). Typically the tag dictionary contains abstract domain concepts like mural, temple, patron, story, narrative, scripture, character, mural technique, painting style, color, etc. The tag dictionary when used for actual tagging of the mural images, contains the data instances of all these concepts - i.e. names of the temples, kings; different kinds of painting style and colors used; names of the epics or scriptures, as well as of the various stories, narratives and characters in these stories. The tag dictionary is used as the basis for providing tags for mural segments in the mural painting repository and for story segments in the narrative text repository. An example of Mural tagging is shown in figure 4(b).

4.2   Ontology of Mural Paintings Domain

A domain-specific ontology models a specific domain or part of the world. In fact, ontologies have proven to be an excellent medium for capturing the knowledge of a domain. Using the *tag dictionary* as a basis, semantic concepts belonging to the domain are determined by an Ontology expert. Relationships between the concepts are defined in the ontology, in such a way that they not only represent the actual, real-life relations but also help model a concept in terms of its perceptual properties. The semantics of these relationships help form the basis for new knowledge derived from the ontology, as required for answering semantic queries. As a very simple example, let the concepts in the mural ontology be : Mural, Temple, Town and King. Some properties defined with these concepts are : Mural is_found_in Temple; Temple is_built_by King; Temple is_located_in Town. Then the ontology helps answer a semantic query of the type : "Found(mural, town) ?", by deriving that if there exists a temple such that (Mural is_found_in Temple) AND (Temple is_located_in Town), then the answer is yes.

Multimedia ontology helps encode media properties of the semantic concepts. In this paper, we have used the MOWL ontology to associate mural image segments to ontology concepts. At the same time text from narratives can also be associated with ontology segments as a media property. This helps build a semantic linkage between the two different modalities, and can be used to provide

(a) Tag Dictionary of the Mural Paintings domain.

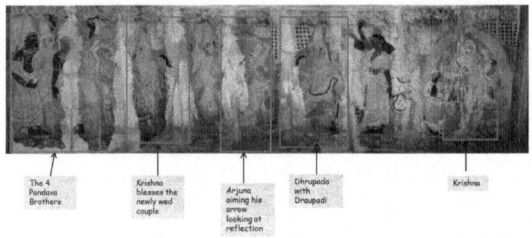

(b) Labels provided for Draupadi Swaya-mawara Panel 1

Fig. 4. Tagging of information in the Mural Paintings Domain

cross-modal access to the heritage collections. The multimedia ontology of the Mural domain is shown in figure 5.

The top layer of the ontology contains concepts like MuralPainting, Temple, MuralTradition, MuralTechnique, PaintingStyle, MuralContent and Narrative. The mural painting typically belongs to a temple, and so to a particular time-period with which the temple is linked. It follows a certain mural tradition and so a typical painting style. It also has a narrative which further has a story. A story contains some characters. The mural also has some other content besides the depiction of the story or narrative. It has some border patterns like floral or geometric patterns, some fashion patterns like hair-styles and jewellery. These media patterns definitely portray the history of the time-period to which the mural belong. The artistic style is also visible in the way characters are painted and placed in the mural. The multimedia ontology can be used to associate these visible patterns with concepts, and thus encode all this background knowledge very effectively. In fact, some of the concepts have mural image segments attached to them through the *hasME* relation which means *has Media Example*.

There are hierarchical relations between subclasses and super-classes as well as between instances and the parent class (shown in black). Other relations which are non-hierarchical (shown in blue) are such that they propagate some media properties from the *domain* of the relation to the *range*. An example of this is the relation *isIn* between the character Shiva and the story ShivaUmaVivaha. The media example image of the story ShivaUmaVivaha, which shows Shiva getting married to Uma, and so has an image of Shiva , is thus also a media example of the concept Shiva. This kind of semantics is not provided by a standard ontology language, and thus makes MOWL the ideal representation to encode such knowledge.

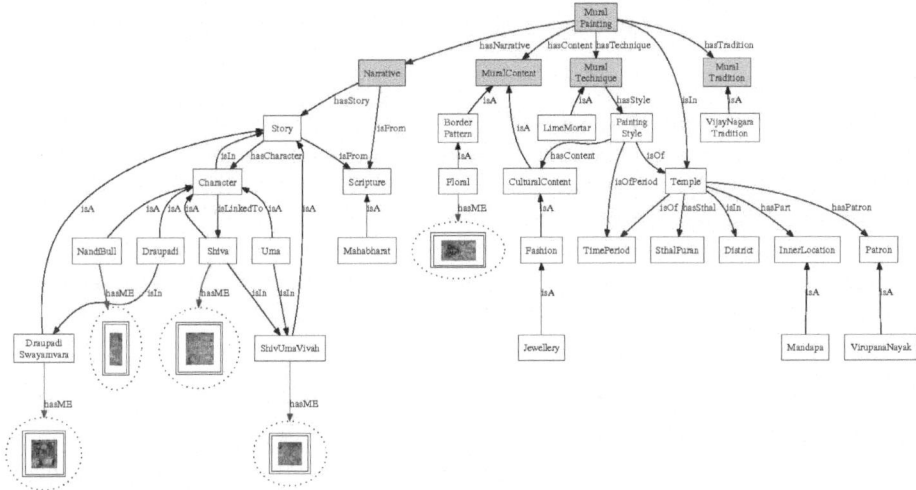

Fig. 5. Graphical view of the Multimedia Ontology of Mural Paintings domain

5   Ontology-Based Cross-Modal Retrieval and Semantic Access

In this section, we show how the multimedia ontology, by its collection-independent modelling of the domain, provides a robust basis for cross-modal access to knowledge of the domain collected in repositories stored in varied media formats. We have a collection of digitized images of mural paintings from the Vijayanagara period. After segmenting, labelling and tagging of the mural images, there were 654 different kinds of image segments. The other repository is of 126 stories from Indian mythology and folk-lore, with 3 different versions of each story on an average. The text segments after labelling and tagging for different characters, episodes, locations, etc. are approximately 1000. The ontology of the mural domain has approximately 173 concepts, 94 of which have media examples and patterns associated with them.

5.1   LDA Based Image and Text Modelling

We have used Latent Dirichlet Allocation (LDA) [8] based topic modelling to model the image and text segments in our repositories.

Text modelling

For the collection of text narratives, we generated segments based on the tagging of the text. Each of these text segments was first pre-processed (removal of stop words, stemming etc.), to get the desired text data set before estimating with GibbsLDA++. This pre-processed data set was then fed to LDA. The parameters used for estimation were $\alpha(50/K$, where K is number of topics), $\beta(0.1)$, number of topics(10), number of iterations(1000), number of most likely words for each topic(1115).

Image modelling

We segmented the mural images into segments based on their labelling by the experts. A bag of SIFT [9] descriptors was extracted from each image segment and the SIFT descriptors extracted from each image were vector quantized with a codebook, producing a vector of visual word counts per image. Besides this *bag of words* representation, we also use a lower-dimensional representation for images, similar to that for text, by fitting an LDA model to visual word histograms and representing images as a distribution over topics.

5.2   Semantic Matching

In the Mural paintings domain, the desired cross-modal retrieval for a text or image query, is to show a story or narrative linked to a mural selected by the user; show image segments from different mural paintings which depict the narrative matching the text query; retrieve mural image segments which depict a particular character from a narrative; and so on. For e.g. user may query for mural segments depicting Shiva, and when those segments are displayed, one of which might be the mural segment depicting ShivaUmaVivaha, the marriage of Shiva and his consort Uma, then she may ask to see the narrative behind the story, and so on.

Let us consider the problem of information retrieval from a database $B = I_1, ...I_B$ of semantically labelled image documents and from a database $N = T_1, ...T_N$ of semantically annotated text segments. The semantic labels in each repository correspond to semantic concepts in the linked ontology. Images and text are represented as vectors in feature spaces R^I and R^T respectively. As illustrated in figure 6, the ontology helps to establish a one-to-one mapping between points in R^I and R^T. The goal of cross-modal retrieval is to return the closest match to a text query $T_q \in R^T$ in the image space R^I, and same for an image query $(I_q \in R^I)$ in the text space R^T.

Semantic matching means to map images and text to representations at a higher level of abstraction, where a natural correspondence can be established. This is obtained by augmenting the database B with a vocabulary $L = 1........L$ of semantic concepts from the domain ontology. Two mappings Π^I and Π^T are then implemented using classifiers of text and images, respectively. Π^T maps a text $T \in R^T$ into a vector Π^T of posterior probabilities $P_{W|T}(w|T)$; where

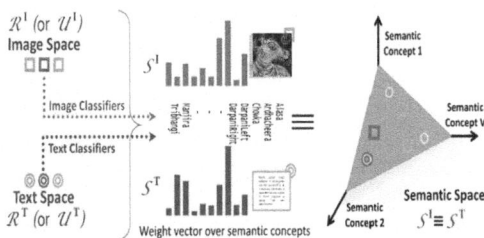

Fig. 6. Semantic matching maps text and images into a semantic space

$w \in 1......L$ with respect to each of the classes in L. The space S^T of these vectors is referred to as semantic space for text, and the probabilities $P_{W|}(w|T)$ as semantic text features. Similarly Π^I maps an image I into a vector Π^I of semantic image features $P_{W|I}(w|I)$; where $w \in 1......L$ in a semantic space for images S^I.

5.3 Cross-Modal Retrieval

Depending on the precise nature of the probability model, Naïve Bayes classifiers can be trained very efficiently in a supervised learning setting. The basic idea of this model is that each category has its own distribution over the codebooks, and that the distributions of each category are observably different. Given a collection of training examples, the classifier learns different distributions for different categories. For both the modalities in our work, posterior probabilities are computed which refer to the semantic concept of the ontology, and thus images and text are mapped to same semantic space. Give any test set from any of the modalities, it is classified to one of the semantic concepts.

KL (Kullback-Leibler) divergence is used to measure the distance between probability distribution of each document with all other documents for both the

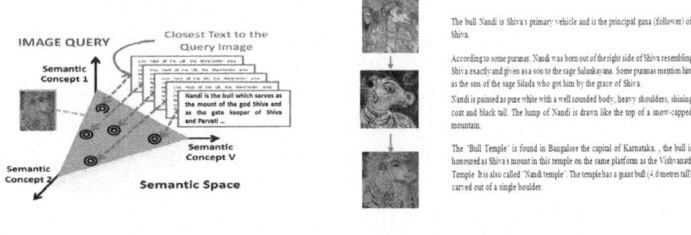

(a) Cross-modal Retrieval using Semantic Matching

(b) Search results using Naïve Bayes classifier and KL distance.

Fig. 7. Cross-Modal Retrieval

modalities. Given a query image I, represented by $\Pi_I \in S^I$, cross modal retrieval will find the text T, represented by $\Pi_T \in S^T$ that minimizes $D_{KL}(I,T) = d(\Pi_I, \Pi_T)$ for some suitable distance measure d between probability distributions. Distance measure considered here is KL (Kullback Leibler) distance. An illustration of cross-modal retrieval using semantic matching and search results using KL divergence after the test set has been classified by Bayesian classifier is given in figures 7(a) and 7(b) respectively.

6   Conclusion

In this paper, we have proposed an ontology-based framework for preservation of heritage encoded in mural paintings and their narratives. The multimedia ontology on which the framework is based, provides the necessary elements for semantic access to digitized heritage artefacts and semantic linkages for cross-modal access to different e-heritage resources stored in various media formats like image, audio, video or text. We have used standard techniques of classification for semantic matching of mural images and their narrative text segments. This preservation scheme can easily be extended to other tangible and intangible heritage artefacts like monuments, manuscripts, coinage, performing arts like dance and folk-theatre.

References

1. Louvre: Louvre museum official website,
   http://www.louvre.fr/llv/commun/home.jsp?bmLocale=en
2. Kalasampada: Kalasampada – digital library: Resources of Indian cultural heritage,
   http://www.ignca.nic.in/dlrich.html
3. Hunter, J.: Enhancing the semantic interoperability of multimedia through a core ontology. IEEE Transactions on Circuits and Systems for Video Technology 13, 49–58 (2003)
4. Hammiche, S.: Semantic retrieval of multimedia data. In: Proceedings of the 2nd ACM International Workshop on Multimedia Databases, pp. 36–44 (2004)
5. Tsinaraki, C., Polydoros, P., Kazasis, F., Christodoulakis, S.: Ontology-based semantic indexing for MPEG-7 and TV-Anytime audiovisual content. Special Issue of Multimedia Tools and Application Journal on Video Segmentation for Semantic Annotation and Transcoding 26, 299–325 (2005)
6. Ghosh, H., Chaudhury, S., Kashyap, K., Maiti, B.: Ontology Specification and Integration for Multimedia Applications. In: Ontologies: A Handbok of Principles, Concepts and Applications in Information Systems. Springer (2007)
7. Mallik, A., Chaudhury, S., Ghosh, H.: Nrityakosha: Preserving the intangible heritage of Indian classical dance. JOCCH 4, 11 (2011)
8. Blei, D.M., Ng, A.Y., Jordan, M.I.: Latent dirichlet allocation. J. Mach. Learn. Res. 3, 993–1022 (2003)
9. Lowe, D.: Distinctive image features from scale-invariant keypoints. International Journal of Computer Vision 20, 91–110 (2003)

Robust Image Deblurring Using Hyper Laplacian Model

Yuquan Xu, Xiyuan Hu, and Silong Peng

Institute of Automation, Chinese Academy of Sciences, Beijing, 100190, P.R. China

Abstract. In recent years, many image deblurring algorithms have been proposed, most of which assume the noise in the deblurring process satisfies the Gaussian distribution. However, it is often unavoidable in practice both in non-blind and blind image deblurring, due to the error on the input kernel and the outliers in the blurry image. Without proper handing these outliers, the recovered image estimated by previous methods will suffer severe artifacts. In this paper, we mainly deal with two kinds of non-Gaussian noise in the image deblurring process, inaccurate kernel and compressed blurry image, and find that handling the noise as Laplacian distribution can get more robust result in these cases. Based on this point, the new non-blind and blind image deblurring algorithms are proposed to restore the clear image. To get more robust deblurred result, we also use 8 direction gradients of the image to estimate the blur kernel. The new minimization problem can be efficiently solved by the Iteratively Reweighted Least Squares(IRLS) and the experimental results on both synthesized and real-world images show the efficiency and robustness of our algorithm.

1 Introduction

Image blur is very common in our daily life and is one of the prime reasons of image degradation in digital photography. In last several years, the single image deblurring has gained considerable attention, a lot of approaches has been propose to solve the ill-posed problem. Given a blurry input image b, most existing deblurring algorithms model the image degradation as a convolution:

$$b = I * k + n \tag{1}$$

b is the input blurry image, I is the latent image we want to recover, k is the blur kernel, n is the additive noise, $*$ denotes the convolution operator. The task of image deblurring is to estimate the clear image I from the given blurry image. Existing image deblurring algorithm can be divided into two classes: when the blur kernel is known, the image deblurring problem is called non-blind image deconvolution and we call blind image deconvolution when the kernel is unknown. Recently image deblurring has made impressive progress, but there are still many challenges in terms of robustness and generality.

When we want to estimate the latent image, the cost function of most previous algorithms have this kind of form:

$$E(I) = ||I * k - b||^\lambda + \phi(I) \tag{2}$$

J.-I. Park and J. Kim (Eds.): ACCV 2012 Workshops, Part II, LNCS 7729, pp. 49–60, 2013.

where λ is always set to 2, which means the n in (1) represents the Gaussian noise. If the blur model is just the trivial convolution, this assumption is reasonable. But in the real case, the image blur model is much more complicate than the convolution model. There may be many kinds of outliers in the real blurry image , such as inaccurate kernel, saturated/clipped pixels, nonlinear camera response curve, non-Gaussian noise, compressed images etc. These outliers will make the blur model not fit the linear convolution model, so treating the n as the Gaussian noise is not suitable here any more. Based on this point, in this paper we analyze two main sources of outliers, artifacts from compression and kernel error, and show that Hyper Laplacian model is more close to the distribution of these non-Gaussian outliers. We then proposed both non-blind and blind image deblurring methods, and use the IRLS to estimate the latent image. We also compare proposed algorithm with other state-of-the-art image deblurring algorithms, including non-blind and blind methods, on both synthetic and real-world images. The results show that our method can handle seriously compressed and blurry images, which other algorithms may lead severe artifacts.

The remainder of this paper is organized as follows. Section 2 discusses the related work, and motivation of this work is detailed in Section 3. In Section 4, we propose our non-blind deconvolution algorithm, and in section 5 we drive the blind image deblurring method. Then we provide some experimental results on both synthetic and real blurred images in Section 6. Section 7 finally presents discussions and the summary of this work.

2   Related Work

In non-blind deconvolution, the blur kernel is assumed to be known. Non-blind image deblurring has been studied for decades in image processing and computer vision fields. Traditional methods including Weiner filtering [1] and Richardson-Lucy method [2][3] are still widely used in the image restoration task nowadays, owing to the quick speed and efficiency. However, these methods are sensitive to image noise and kernel error, so their deconvolution results always suffer from ringing artifacts. To suppress the ringing artifacts, various image priors and regularization have been proposed for image deconvolution. These methods use different forms of ϕ in (6), such as TV regularization [4], natural image statistics [5] [6], sparse image prior [7][8]. There are a few methods dealing with other kind of noise in the blurry image. Bar et al. [9] proposed a data fidelity term based on an L1-norm to handle the impulsive noise. Whyte et al. [10] perform non-blind deblurring of images degraded by camera shake and containing clipped / saturated pixels. Harmeling et al. [11] also proposed a method that deal with the saturated pixels by thresholding input blurry images and mask them out of the deblurring process. Cho et al.[12] analyzed outliers in the non-blind image deblurring and proposed an Expectation-Maximization(EM) method to iteratively refine the outlier classification and the latent image.

Blind deconvolution is a more challenging and ill-posed problem, since the blur kernel is also unknown. Some methods use additional input or hardware to estimate the blur kernel. Yuan et al. [13] use a pair of images, a blurry image and a noisy image, to estimate the kernel. Other work [14] [15] use hybrid camera to aid the image deblurring process. The most ill-posed problem is single-image blind deconvolution, which must both estimate the PSF and latent image form only one input image. Early approaches usually assume the blur kernel obey the simple parametric models such as low-pass filter or uniform linear motion kernel. The first successful single image deblurring method due to the complex motion kernel from shaking camera may be derived by Fergus et al.[5], they used the variation method to solve the ill-posed problem. Followed by their work, [6] [16] [17] [18] [19] [20] presented different approaches to deblur the image. Kim et al. [21] consider nonlinear camera response functions (CRFs) in the context of image deblurring. Tai and Lin [22] propose a technique for jointly denoising and deblurring the images with high level noise.

3   Motivation

Most existing image deblurring method use L2-norm based data fidelity term which assume the n in (1) obey the Gaussian distribution. This assumption will get high quality results when the blurry image contains only a small amount of noise which can be well approximated by a Gaussian distribution. But as we discussed before, in the real case, there are various types of non-Gaussian outliers, which we divide into two categories: kernel error and image error. **Kernel error**: most non-blind image deblurring methods base on the assumption that the input kernel is perfectly accurate. But except

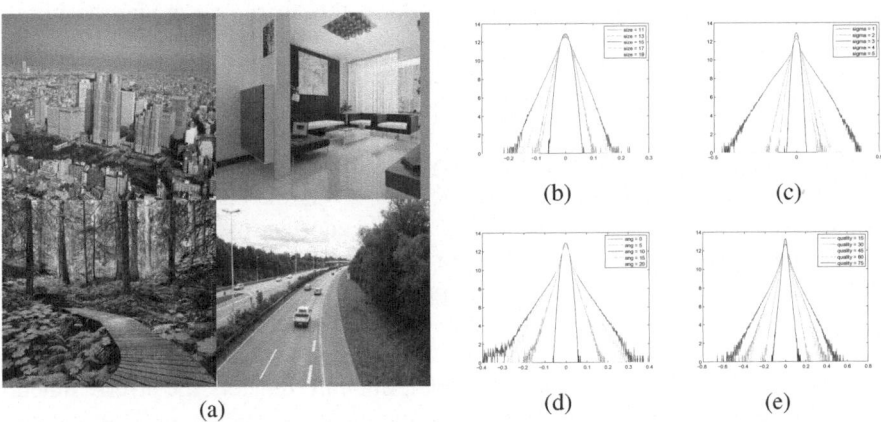

Fig. 1. The curves of the logarithmic density of the error between the images. They are computed using information collected from 20 natural images (a) 4 of 20 natural images, including building, car, trees and human beings. We blurred the natural images using box kernel, Gaussian kernel, and motion kernel with different parameters, and show the curve of error between the blurred images and the reference images in (b) (c) (d). (b) the reference blurry images are blurred by box kernel which size is 15×15 and free of noise, and blur kernels for other images are 11×11, 13×13, 15×15, 17×17, 19×19, and we also add Gaussian noise to the blurry images. (c) the reference kernel is Gaussian kernel with $sigma = 3$ (d) the reference kernel is motion kernel with $angle = 10$, the length of all the motion kernels is set to 20 pixels. (e) We compressed the natural images with different qualities, and show the curve of the compression error.

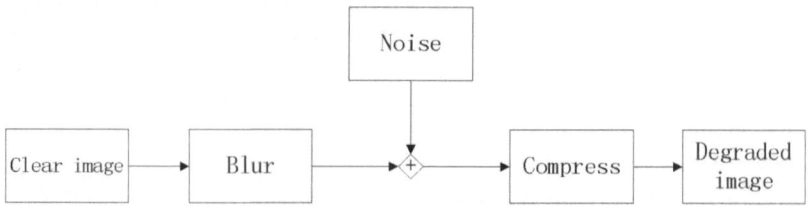

Fig. 2. The blur model of our algorithm

the synthetic examples, the kernel estimated by existing algorithm is almost impossible to be 100 percent accurate, so the input kernel for the non-blind image deblurring more or less has some error. In the blind image deconvolution case, previous methods usually alternative update the blur kernel and the latent image. In this iterative process, the estimated kernel is getting closer and closer to the true value. but in the beginning of the iteration, the error of the estimated kernel may lead the algorithm fail to converge to the ground truth. **Image error**: In the past, most image deblurring algorithm assume the blur model is a trivial convolution as (1). But as mentioned earlier, saturated/clipped pixels, nonlinear camera response curve and artifacts from compression will all lead the blurry image not obey the convolution model any more.

Without taking account of the error in the blur kernel and image, the deblurring methods are difficult to obtain good results. So in this work, we model the blurred image b as:

$$b = g\left(I * (\tilde{k} + \Delta k) + n\right) \tag{3}$$

g represents the post-processing of the blurry image such as clipped pixels, compress, camera response curve. $\tilde{k}$ is the input kernel and Δk is the error between the input kernel and true kernel k, $(k = \tilde{k} + \Delta k)$. We mainly consider two types of errors here, kernel error and compression. In (3), the blur model is no longer the linear convolution model, so the traditional image deblurring approaches may not get high quality results, they will suffer from the ringing artifacts in the restored images. A direct idea is that first we use the inverse function of **g** to process the observed blurry images and then the blur model will change back to the linear model. But unfortunately, the inverse function is difficult to know, especially when **g** denote the image compression. So our starting point is that whether can we use the blur model like (1) as much as possible, which can be effectively solved by existing methods. To illustrate, in Fig 1 we show the curves of the logarithmic density of the error between images satisfying linear convolution model and images with kernel error or compression artifacts. We can find that these two kinds of outliers make the error deviated from the Gaussian model but more close to the Hyper Laplacian model, especially when the kernel error is big or the quality is low. As the result, we combine the image error, kernel error and additive noise into the new 'total noise', which is no longer the Gaussian noise but satisfy the Hyper Laplacian distribution. It is noteworthy that some other work uses the Hyper Laplacian prior as the regularization [7] [8], but we use it as the data fidelity term.

(a)   (b)   (c)

(d)   (e)   (f)

Fig. 3. Image deblurring with the compressed blurred image, inaccurate kernel and Gaussian white noise with std = 5. (a) is the compressed blurred image, the true kernel is the Gaussian kernel with sigma = 2 and sigma = 3 for the input kernel (b) is the result of RL method (c) is the result of [12], (d) is the [8]'s result, (e) is the result of [7],(f) is our result.

4   Non-blind Image Deblurring

In this section, we present our non-blind image deblurring algorithm. We assume the blur model as in Fig 2. Firstly we formulate our objective function. By Bayes' theorem, we can write the posterior probability of the non-blind image deblurring problem:

$$I = \arg\max_I p(I/k, b) \tag{4}$$

$$p(I/k, b) \propto p(b/I, k)p(I) \tag{5}$$

where $p(b/I, k)$ represents the likelihood and $p(I)$ denote the priors on the latent image. Here we choose the sparse prior[7] as the image prior. Since the image noise is modeled as Hyper Laplacian distribution. Our cost function has the form that:

$$I = \min_I E(I) = \min_I ||I * k - b||^\lambda + \gamma_1 ||\nabla I||^{0.8} \tag{6}$$

Where $\nabla I = (\nabla^x I, \nabla^y I)$ and $\nabla^x \; \nabla^y$ represents the gradients of the image in vertical and horizon direction. In the non-blind image deblurring, we set the λ to 0.8 same with the image prior, which is proved to be robust in the experiment. To minimize (6), we

---

Algorithm 1. Non-blind image deblurring

1: **Required**: The blurred image b The input kernel k
2: **Initialization**: $\omega_e, \omega_x, \omega_y$ set to 1
3: **repeat**
4:      Estimate the latent image update (7)
5:      Update $\omega_e, \omega_x, \omega_y$
6: **until** meets the maximum iteration number
7: **return** Restored image

---

use the iteratively reweighted least squares (IRLS) method. The equation (6) can be approximated by:

$$\omega_e |I * k - b|^2 + \gamma_1 (\omega_x |\nabla^x I|^2 + \omega_y |\nabla^y I|^2) \tag{7}$$

Where $\omega_e = |I * k - b|^{(0.8-2)}$, $\omega_x = |\nabla^x I|^{(0.8-2)}$, $\omega_y = |\nabla^y I|^{(0.8-2)}$. When we fixed the $\omega_e, \omega_x, \omega_y$, the (7) is easily to be minimized. So we alternate updating $\omega_e, \omega_x, \omega_y$, and minimizing (7) to find the optimal solution. This iterative scheme is outlined in Algorithm 1

5   Blind Image Deblurring

In this section, we extend our method to solve the blind image deblurring problem. The blind image deblurring methods can be divided into two parts: kernel estimation and restore latent image. We alternately carry out these two steps. Our blind deblurring algorithm utilization the shock filter [23] to predict the edge profiles of the sharp image, which is proved to be a effective method to estimate the complex blur kernel [16] [18]. The cost function for the kernel estimation is:

$$E(k) = ||\nabla \tilde{I} * k - \nabla b||^2 + \gamma_2 ||k||^2 \tag{8}$$

$\tilde{I}$ is the shook filtered image from I. In (8), previous work used two direction gradients to estimate the kernel, when the noise in the blur image is little it can obtain good result. But when the image is compressed to low quality, it does not work well as shown in Fig 4(b), which is the result we use two direction gradients to deblur the severely blurred and compressed image. To make the algorithm more robust we use more direction gradient. In this paper, we use 8 different directions, every 22.5°. Fig 4(c) illustrate the result we use 8 direction gradients, we can see the estimated kernel is more similar with the ground truth. The new energy function is:

$$E(k) = \sum_{i=1}^{8} ||\partial_i \tilde{I} * k - \partial_i b||^2 + \gamma_2 ||k||^2 \tag{9}$$

To restore the image, we use the non-blind algorithm in previous section. Thanks to the robustness of our non-blind algorithm, our blind image deblurring method can solve severely degraded image. In the iteration of the blind deconvolution, the estimated kernel is closer and closer to the real one. In Fig 1, we can find that when the error of

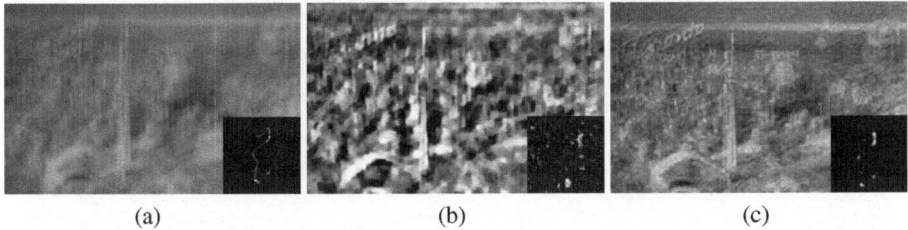

| (a) | (b) | (c) |

Fig. 4. Blind image deblurring with the compressed blurred image using use 8 different direction gradients. (a) is the compressed blurred image and the estimated kernel of our method using the uncompressed image, (b) is the result we use two direction gradients, (c) is the result we use 8 direction gradients.

the kernel is small, the distribution is changed and turned to the Gaussian model. So we change the value of λ in practically we use the sequence of values for parameter λ: 0.5,0.5,0.8,0.8,1,1,1.5,1.5.

6 Experimental Results

To evaluate the effectiveness of the proposed method, we compare our approach with the state-of-the-art deconvolution methods on both synthetic and real-world examples. We implemented our method in Matlab. And the results from other work are generated by their code which all can be downloaded in the internet. We try many different parameters for each algorithm to show the best performance of these algorithms, which make our results more convincing.

6.1 Non-blind

First we test the non-blind deblurring algorithm. The test images are 'Lena' and 'Cameraman' and the image sizes are both 255×255. There are three kind of errors in the blurry image, which are input kernel error, additional Gaussian noise and compressed error. The input wrong kernel and the ground true kernel is showed in Table 1. The standard deviation of the Gaussian noise is set to 5, and the quality of the compressed jpeg image is 20, which is defined in Matlab 'imwrite' command. The peak signal to noise ratio (PSNR) measurement is also used to quantitatively evaluate the quality of

Table 1. The true kernel and the input inaccurate kernel

blur type	true kernel	input kernel
motion	length: 20pixel orientation: 10°	length: 20pixel orientation: 20°
gaussian	$\sigma = 2$	$\sigma = 3$
average	$size = 15 \times 15$	$size = 17 \times 17$

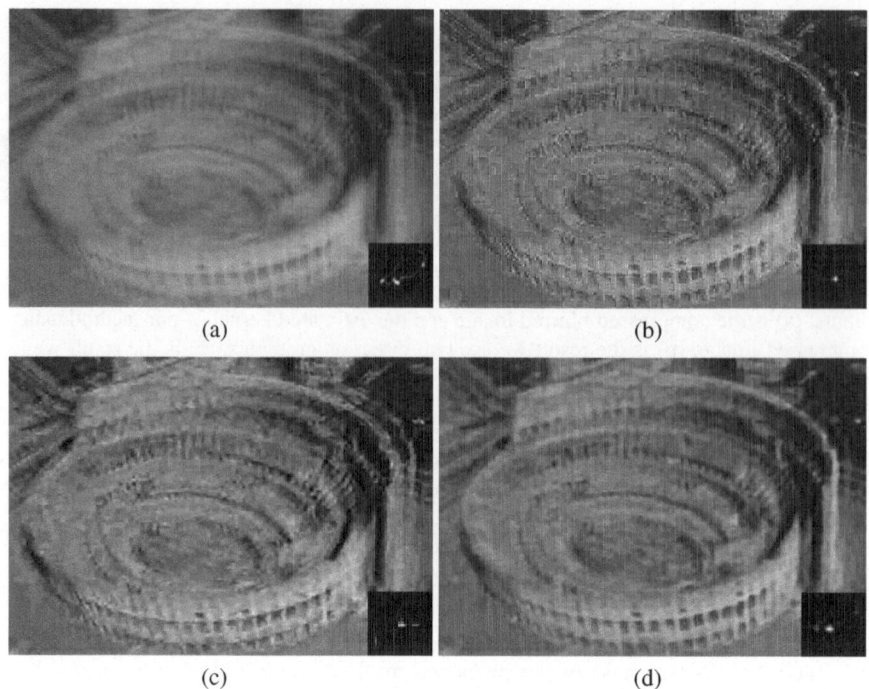

(a)   (b)

(c)   (d)

Fig. 5. Blind image deblurring with the compressed blurred image. (a) is the compressed blurred image and the estimated kernel of our method using the uncompressed image, (b) is the result of Xu and Jia's method[18], (c) is the result of Levin et al's method[19], (d) is our result.

(a)   (b)   (c)   (d)

Fig. 6. Compressed image deblurring results. (a) is the compressed blurred image (b) is the result of Xu and Jia's method[18], (c) is the result of Krishnan et al's method[24], (d) is our result.

the restored results. The PSNR values of the deblurred results for all these methods is showed in Table 2 3 4. Table 2 show the results of which the input kernel is the inaccurate kernel but the blurry images is not compressed. In Table 3 the input kernel is the true kernel and the blurry image is compressed. Table 4 show the results with both inaccurate kernel and compression artifacts. We can find that in most cases our non-blind deblurring approach have the best performance. And the improvement on PSNR value is also consistent with the improvement on visual quality, as in Fig 3. There are least artifacts and most details shown in our result.

Table 2. Wrong Kernel

		Noise std	[12]	[8]	[7]	Proposed
Cameraman	average	0	20.82	20.55	20.77	**21.35**
		5	20.61	20.60	20.56	**21.05**
	gaussian	0	21.88	22.12	22.14	**23.70**
		5	21.69	22.12	22.08	**23.38**
	motion	0	20.94	20.23	20.71	**21.24**
		5	20.70	20.24	20.57	**21.20**
Lena	average	0	22.97	22.44	22.99	**23.11**
		5	22.78	22.46	**22.80**	22.79
	gaussian	0	24.82	24.89	25.13	**27.12**
		5	24.67	24.90	25.00	**26.37**
	average	0	**23.31**	22.31	23.15	23.03
		5	22.99	21.01	22.38	**23.02**

Table 3. Compressed

		Noise std	[12]	[8]	[7]	Proposed
Cameraman	average	0	21.68	21.23	21.59	**21.85**
		5	21.55	21.06	21.46	**21.71**
	gaussian	0	23.58	23.22	23.51	**23.71**
		5	23.42	23.11	23.36	**23.61**
	motion	0	22.42	21.84	22.28	**22.62**
		5	22.29	21.77	22.13	**22.54**
Lena	average	0	23.59	22.94	23.54	**23.70**
		5	**23.43**	22.77	23.40	23.19
	gaussian	0	26.42	25.90	26.42	**26.54**
		5	26.29	25.75	26.28	**26.48**
	average	0	**24.32**	22.94	24.07	24.24
		5	24.09	22.90	23.92	**24.19**

6.2 Blind

In the blind image deblurring, we mainly present the results when the blurry input is compressed. At high compression ratios, the boundaries between the blocks become visible and lead to 'Blocking' artifacts. When we try to deblur the compressed image, this 'Blocking' artifacts will damage the final result in most previous algorithm, but our method can still get reasonable results shown in Fig 5 6 7. Fig 5 show the kernel estimation of [18] and [19], these two blind deblurring algorithms are proved to be the most successful method in spatially invariantly deblurring. But we can see that when

Table 4. Compressed and Wrong kernel

		Noise std	[12]	[8]	[7]	Proposed
Cameraman	average	0	**20.61**	20.33	20.56	20.50
		5	**20.53**	20.27	20.50	20.44
	gaussian	0	21.69	22.04	22.08	**23.38**
		5	21.69	22.04	22.07	**23.24**
	motion	0	20.70	20.19	20.57	**21.56**
		5	20.63	20.11	20.51	**20.96**
Lena	average	0	22.78	22.27	**22.80**	22.79
		5	22.65	22.17	22.69	**22.71**
	gaussian	0	24.67	24.75	25.00	**26.37**
		5	24.66	24.71	24.96	**26.22**
	average	0	22.99	22.21	22.87	**23.22**
		5	**22.81**	21.15	22.67	22.56

(a)　　　　　　(b)　　　　　　(c)　　　　　　(d)

Fig. 7. Blind image deblurring with the compressed blurred image. (a) is the compressed blurred image and the estimated kernel of our method using the uncompressed image, (b) is [6]'s result,(c) is the result of [18], (d) is our result.

the blurred image is compressed, these methods cannot estimate the correct kernel and their results suffer from high level artifacts form the compression. On the other hand, our method can still get reasonable deblurred image without 'Blocking' artifacts and the kernel estimation is similar to the real kernel. In Fig 6 7, we also compare our results with [6] [24], and our algorithm better recovers degraded images.

In summary, our experimental results show that our deconvolution method leads to significantly less artifacts, such as 'Blocking' artifacts than previous approaches.

7   Conclusion and Future Work

In this paper, our contributions are threefold. First, we analyze the noise distribution in the case of inaccurate kernel and compressed image, and we propose Hyper Laplacian distribution other than traditional Gaussian distribution as the noise distribution. Second, we use eight edge filters instead of two gradient operator to make the blind image deblurring more robust when the image is severely compressed and blurred. Last we propose both non-blind and blind image deblurring algorithms and solve the optimization problems efficiently. In future, we would like incorporate our framework to solve the spatially variant problem.

Acknowledgement. This work was supported in part by the National Natural Science Foundation of China under Grant Nos. 60972126, 61101219, 61201375 and the State Key Program of National Natural Science of China under Grant No. 61032007.

References

1. Wiener, N.: Extrapolation, interpolation and smoothing of stationary time series, New York (1949)
2. Richardson, W.: Bayesian-based iterative method of image restoration. Journal of the Optical Society of America 62, 55–59 (1972)
3. Lucy, L.: An iterative technique for the rectification of observed distributions. The Astronomical Journal 79, 745 (1974)
4. Rudin, L., Osher, S., Fatemi, E.: Nonlinear total variation based noise removal algorithms. Physica D: Nonlinear Phenomena 60, 259–268 (1992)
5. Fergus, R., Singh, B., Hertzmann, A., Roweis, S., Freeman, W.: Removing camera shake from a single photograph. ACM Trans. Graph. 25, 787–794 (2006)
6. Shan, Q., Jia, J., Agarwala, A.: High-quality motion deblurring from a single image. ACM Trans. Graph. 73, 73:1–73:10 (2008)
7. Levin, A.: Blind motion deblurring using image statistics. In: Advances in Neural Information Processing Systems, vol. 19, p. 841 (2007)
8. Krishnan, D., Fergus, R.: Fast image deconvolution using hyper-laplacian priors. In: NIPS, vol. 22 (2009)
9. Bar, L., Kiryati, N., Sochen, N.: Image deblurring in the presence of impulsive noise. International Journal of Computer Vision 70, 279–298 (2006)
10. Whyte, O., Sivic, J., Zisserman, A.: Deblurring shaken and partially saturated images. In: Computer Vision Workshops (ICCV Workshops), pp. 745–752. IEEE (2011)
11. Harmeling, S., Sra, S., Hirsch, M., Schölkopf, B.: Multiframe blind deconvolution, super-resolution, and saturation correction via incremental em. In: Proc. ICIP, pp. 3313–3316 (2010)
12. Cho, S., Wang, J., Lee, S.: Handling outliers in non-blind image deconvolution. In: ICCV, pp. 495–502. IEEE (2011)
13. Yuan, L., Sun, J., Quan, L., Shum, H.: Progressive inter-scale and intra-scale non-blind image deconvolution. ACM Trans. Graph. 27 (2008)
14. Ben-Ezra, M., Nayar, S.: Motion-based motion deblurring. IEEE Trans. on PAMI 26, 689–698 (2004)
15. Tai, Y., Du, H., Brown, M., Lin, S.: Image/video deblurring using a hybrid camera. In: CVPR, pp. 1–8 (2008)
16. Cho, S., Lee, S.: Fast motion deblurring. ACM Trans. Graph. 28, 145:1–145:8 (2009)
17. Levin, A., Weiss, Y., Durand, F., Freeman, W.: Understanding and evaluating blind deconvolution algorithms. In: CVPR (2009)
18. Xu, L., Jia, J.: Two-Phase Kernel Estimation for Robust Motion Deblurring. In: Daniilidis, K., Maragos, P., Paragios, N. (eds.) ECCV 2010, Part I. LNCS, vol. 6311, pp. 157–170. Springer, Heidelberg (2010)
19. Levin, A., Weiss, Y., Durand, F., Freeman, W.: Efficient marginal likelihood optimization in blind deconvolution. In: CVPR (2011)

20. Cho, T., Paris, S., Horn, B., Freeman, W.: Blur kernel estimation using the radon transform. In: CVPR (2011)
21. Kim, S., Tai, Y., Kim, S., Brown, M., Matsushita, Y.: Nonlinear camera response functions and image deblurring. In: CVPR (2012)
22. Tai, Y., Lin, S.: Motion-aware noise filtering for deblurring of noisy and blurry images. In: CVPR (2012)
23. Osher, S., Rudin, L.: Feature-oriented image enhancement using shock filters. SIAM Journal on Numerical Analysis 27, 919–940 (1990)
24. Krishnan, D., Tay, T., Fergus, R.: Blind deconvolution using a normalized sparsity measure. In: CVPR (2011)

SVD Based Automatic Detection of Target Regions for Image Inpainting

Milind G. Padalkar[1], Mukesh A. Zaveri[2], and Manjunath V. Joshi[1]

[1] Dhirubhai Ambani Institute of Information and Communication Technology, India
{milind_padalkar,mv_joshi}@daiict.ac.in
[2] Sardar Vallabhbhai National Institute of Technology, Surat, India
mazaveri@coed.svnit.ac.in

Abstract. We are often required to retouch images in order to improve their visual appearance, by removing the visual discontinuities like breaks and damaged regions. Such retouching of images may be achieved by inpainting. Current techniques for image inpainting require the user to manually select the target regions to be inpainted. Very few techniques for automatically detecting the target regions for inpainting are reported in the literature, which are suitable to detect an actual damage or alteration to the given photograph. In this paper, we propose a Singular Value Decomposition (SVD) based novel technique for automatic detection of the damaged regions in the photographed object / scene, for the purpose of digitally restoring them to their entirety using inpainting. Results on an exhaustive set of images suggest that the mask generated using the proposed technique can be suitably used for inpainting purpose to digitally restore the given images.

1 Introduction

Many times we need to improve the visual appearance of a given image by identifying and retouching the visual discontinuities in it. Such a task of digital restoration may be achieved using image inpainting [1–3]. Given an image and a region of interest in it, the task of an inpainting algorithm is to fill up the pixels in that region in a visually plausible manner. Digital restoration of the given images thus consists of two steps viz. (a) selection of the regions to be modified (target regions) and (b) applying a suitable inpainting algorithm on these regions .

The present techniques for inpainting based on propagation of structure [1, 4, 5] and texture [2, 3, 6], require the user to manually select the target regions. Since the manual selection of target regions may vary for different user, the inpainting results being dependent on the target region selection may also vary. By automating the target region detection process, human intervention for inpainting can be avoided. Automatic detection is also useful for reconstruction and repair of digitized 3D models that may be used for creating walk-through applications [7] and on-the-fly inpainting cameras for creating efficient immersive navigation / digital walk-through systems.

J.-I. Park and J. Kim (Eds.): ACCV 2012 Workshops, Part II, LNCS 7729, pp. 61–71, 2013.
© Springer-Verlag Berlin Heidelberg 2013

The literature reports only a few inpainting techniques that also facilitate the automatic detection of target regions [8–11]. Chang et al. [8] proposed a method to detect damage in images due to color ink spray and scratch drawing. Based on the use of several filters and structural information of damages, their method is limited to detection of color ink spray and scratch drawings. Tamaki and Suzuki [9] address the detection of visually less important string-like objects that block user's view of a discernible scene. Their method however is restricted to the detection of only those occluding objects that are long and narrow, and contrasted in intensity with respect to the background. Amano [10] present a correlation based method for detecting defects in images. This method relies on correlation between adjacent patches for detection of defects i.e. small number of regions disobeying an "image description rule", complied by most local regions. The method works well for detecting computer generated superimposed characters having uniform pattern.

All the above mentioned techniques are suitable for detecting actual damaged or alteration caused to a photograph. These techniques do not address the identification of damage in the objects or scenes that are photographed. To the best of our knowledge, the only method that also addresses this issue is proposed by Parmar et al. [11]. Their technique uses matching of edge based features with pre-existing templates to distinguish vandalized and non-vandalized regions in frontal face images of monuments at heritage sites. However, their inpainting results are highly dependent on the selected templates and their method is restricted to frontal face images of monuments. The template creation of both vandalized and non-vandalized regions may not be practically realizable for such images and therefore the detection process may lead to undesired results.

On a similar line, techniques for micro-crack detection in concrete can be found in [12, 13], but one may note that they require special imaging conditions. Recent attempt for crack detection using tensor voting in pavement images can be found in the work by Zou et al. [14]. The performance of their technique is heavily dependent on the accuracy of generation of crack-pixel binary map, that acts as an input to the tensor voting framework.

In this paper we propose a novel technique based on the use of singular value decomposition (SVD) [15], to automatically detect the target regions that are required to be inpainted in the given image. Unlike the techniques reported in the literature that detect an actual external damage or defect due to alteration of a photograph, the proposed method aims to detect damage to the photographed scenes / objects. The damaged areas appear like breaks splitting the objects, developed over a period of time due to environmental effects or due to manual destruction. Such detection followed by inpainting shall enable us to digitally restore the photographed scenes / objects in their integrality. By comparing the similarity of adjacent overlapping patches in the given image, the proposed approach generates a binary map that consists the target region and can be used as an input mask for inpainting.

2   Proposed Approach

Visual discontinuities like cracks / damaged regions in a photographed scene / object attract attention of the human visual system. The damaged areas appear like breaks splitting the objects, developed over a period of time due to natural calamities or due to manual destruction. The visual appearance can be improved by inpainting the given images in which the cracks / damaged regions are the targets to be filled. Such digital restoration shall enable one to view the photographed scene / object in an undamaged form. Figure 1 shows our proposed technique to automatically detect the target regions for inpainting.

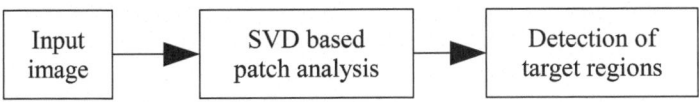

Fig. 1. Proposed approach

A natural image if split into a number of smaller non-overlapping patches, the adjacent patches may exhibit a drastic variations in intensity of corresponding pixels. Thus, even if the adjacent patches contain a single object or scene, measuring similarity between these patches would incorrectly suggest presence of visual discontinuity. On the other hand, if overlapping patches in a sliding window fashion are compared then a high amount of similarity is revealed due to the overlap of most pixels. Here, if a drastic change is encountered between the adjacent overlapping patches then comparison reveals dissimilarity. Patches exhibiting higher amount of dissimilarity are the ones that constitute the visual discontinuity. Thus, the main idea of the proposed method is to compare the overlapping adjacent patches for similarity. The average dissimilarity of the row and column adjacent patches with respect to a patch under consideration helps to reveal the amount of visual discontinuity between the patches. By using a threshold, the damaged areas can then be identified as the ones having higher dissimilarity value.

Given an input image I in the RGB color space, we first transform it into HSV color space and extract the grayscale image I_V that corresponds to the *intensity* image. Now consider a patch Φ_p of size $m \times n$ at pixel $p \in I_V$ with coordinates (x, y). Here, $x = 1, \ldots, M - m$ and $y = 1, \ldots, N - n$, such that $M \times N$ represents the size of the image I_V. The elements of patch Φ_p are rearranged to form a column vector v_p of length $L = mn$ by using lexicographical ordering of pixels.

Now for any two adjacent pixels r and s with respective patches Φ_r and Φ_s, the corresponding vectors v_r and v_s may be compared for similarity. We find the similarity between the vectors v_r and v_s using the geometric interpretation of the SVD model [15] on the matrix having these vectors as its columns. By calculating the similarity between vectors of adjacent patches, we create a similarity matrix S whose elements are then compared with a threshold δ, to detect patches having discontinuities. In the following subsection we discuss patch analysis in the SVD domain.

2.1 Singular Value Decomposition and Patch Analysis

We form a matrix A with the columns as v_r and v_s corresponding to patches Φ_r and Φ_s, and decompose it using SVD such that $A = U\Sigma V^T$. Here U is a $L \times L$ matrix, the columns of which are the eigenvectors of AA^T, V is a 2×2 matrix consisting of eigenvectors of $A^T A$, and Σ is $L \times 2$ matrix of singular values ($\sigma_1 \geq \sigma_2 \geq 0$) at diagonals. We now reduce the size of matrices U to $L \times 2$ and Σ to 2×2, which however does not affect the reconstruction of $A = U\Sigma V^T$.

Now, the rows w_1 and w_2 of matrix $V\Sigma$ reflect the extent to which pixels in the two vectors v_r and v_s have a similar pattern of occurrence [15]. The similarity between columns of A corresponding to patches Φ_r and Φ_s, is therefore given by the cosine of angle between corresponding rows of the matrix $V\Sigma$ as follows.

$$cos(\theta_{rs}) = \frac{w_1 . w_2}{||w_1|| \, ||w_2||} \tag{1}$$

If the vectors v_r and v_s are similar then the angle between them is small, whereas it is large when these two vectors are dissimilar. Therefore, from equation (1) it is clear that $cos(\theta_{rs})$ is nearer to 1 when v_r and v_s are similar, whereas it is nearer to 0 when they are dissimilar. However, here it is observed that since the complete reconstruction of A is possible using the reduced matrices U, Σ and V, the angle obtained by directly considering the vectors v_r and v_s is the same as that obtained between the rows of matrix $V\Sigma$.

Now we consider one more vector v_t as a column of matrix A, where v_t corresponds to a patch Φ_t which is also adjacent to patch Φ_r. If we now decompose A using SVD, we have $A = U\Sigma V^T$ with matrices U, Σ and V of sizes $L \times L$, $L \times 3$ and 3×3, respectively. By discarding the smallest eigenvalue, we reduce the size of matrices U to $L \times 2$, Σ to 2×2 and V to 3×2, which now leads to an approximate reconstruction of matrix A. Such a method is widely used for image compression and noise reduction [16]. If we now consider the angle between rows of the matrix $V\Sigma$ as in equation (1), the true extent of similarity between the corresponding columns of matrix A is still maintained. This is because, we are discarding the eigenvectors corresponding to the smallest eigenvalue. This in turn helps to calculate the true similarity even when the patches are noisy.

One may easily verify this from the following example. Consider the vectors $v_r = [1, 2, 3, 4, 3, 2, 1]^T$ and $v_s = [4, 5, 6, 7, 6, 5, 4]^T$, $v_t = [4, 4, 6, 8, 6, 4, 4]^T$. In SVD domain representation of a matrix having these vectors as its columns, the rows of matrix $V\Sigma$ are obtained to be $w_1 = [6.52, 1.05]$, $w_2 = [14.21, -0.89]$, while $cos(\theta_{rs}) = 0.9756$ and $cos(\theta_{rt}) = 0.9916$. On other hand, if we directly use the vectors v_r, v_s, v_t instead of w_1, w_2, w_3, respectively, we get $cos(\theta_{rs}) = 0.9734$ and $cos(\theta_{rt}) = 0.9807$.

It may be noted that unlike calculating the correlation directly between the actual patches, our method performs the similarity comparison in the SVD domain. By discarding the smallest eigenvalue and the associated eigenvector, the obtained similarity values are robust to noisy patches. We now compare the overlapping patches for similarity by first creating a set E_p corresponding to pixel $p \in I_V$ with coordinates (x, y). The set E_p consists of pixels in the neighbourhood of p having coordinates $(x + 1, y)$ and $(x, y + 1)$. Every patch Φ_q at pixel $q \in E_p$, overlaps with the patch Φ_p, such that Φ_p is a $m \times n$ patch at pixel $p \in I_V$. The size of patch Φ_q is same as that of patch Φ_p. The comparison of patches Φ_q at pixels $q \in E_p$, which overlap with and are row, column adjacent to the patch Φ_p, enables one to simultaneously capture horizontal, vertical and diagonal discontinuities. Thus, pixels having coordinates $(x + 1, y)$ and $(x, y + 1)$ are sufficient to form the set E_p in order to capture visual discontinuities. Including any more pixels in set E_p will induce redundancy leading to processing overhead.

The similarity of all the patches Φ_q corresponding to every pixel $q \in E_p$ with the patch Φ_p is calculated by first arranging the vector v_p and all the vectors v_q corresponding to patches Φ_q as columns of matrix A. After applying SVD on A and reducing the sizes of matrices U, Σ and V as explained earlier, the similarity between columns of A is calculated by using the corresponding rows of matrix $V\Sigma$ as w_1 and w_2 in equation (1). We now create a similarity matrix S such that its element $S(p)$ represents the average similarity value of patch Φ_p with overlapping patches $\Phi_q \forall q \in E_p$ and is calculated as follows.

$$S(p) = \frac{1}{|E_p|} \sum_{q \in E_p} cos(\theta_{pq}), \quad \forall p \equiv (x, y) \in I_V \tag{2}$$

Here $|E_p|$ is the number of pixels in the set E_p.

Once the similarity matrix S is obtained, we use it to binarize I_V for detection of the target regions. If $S(p) < \delta$, we declare the corresponding patches Φ_p and $\Phi_q, \forall q \in E_p$ to be significantly dissimilar. In this manner, all the elements of S are compared with threshold δ to detect dissimilar patches, using which a binary map B is constructed as follows.

$$B(\Phi_p) = \begin{cases} 1, & \text{if } S(p) \leq \delta, \forall p \equiv (x, y) \in I_V, \\ 0, & \text{otherwise,} \quad \text{and} \end{cases}$$
$$B(\Phi_q) = B(\Phi_p), \forall q \in E_p \tag{3}$$

The binary map B generated in this way has the target regions represented by value 1. The elements in S may have different values for different input images. It may be noted that if the overlapping patches in an input image have very high similarity, then the corresponding matrix S may have many elements with values nearer to 1, and therefore a high value of threshold δ could be required for correct detection of patches having discontinuities. Also, variation in threshold value δ, significantly changes the resulting binary map B. Therefore, selection of the threshold δ based on the input image is required for correct detection of target regions, which we describe in the following subsection.

2.2   Selection of Threshold Value δ

In order to select the threshold value δ dynamically for a given image, we consider three quantities derived from the similarity matrix S, viz. the average value $avg(S)$, minimum value $min(S)$ and the maximum value $max(S)$. Since the compared patches are adjacent and also overlap each other, they show high content similarity. Therefore, it is reasonable to assume that the values in S less than the average value $avg(S)$ would definitely correspond to the patches having discontinuities. Thus, the lowest value that δ may take is $avg(S)$.

If the difference between lowest and highest values of S is high, it would mean that the values corresponding to patches with discontinuities are spread over a wider range, while the spread is over a narrow range when the difference is small. If the values in the similarity matrix S vary in a narrow range, then the threshold value δ that detects the patches with discontinuities, would be nearer to $avg(S)$. Thus, we infer that the threshold value is higher than the average value $avg(S)$ and also depends on the minimum $min(S)$ and maximum $max(S)$ values of the similarity matrix S. We set an initial threshold α to be an average of these three terms as given in the following equation.

$$\alpha = \frac{min(S) + max(S) + avg(S)}{3} \tag{4}$$

However, experimentally we found that a correction factor depending on the value α is required for correct detection. Based on our experimentation, we arrive at the following equation that incorporates suitable correction factors to determine the threshold δ.

$$\delta = \begin{cases} \alpha + 0.10, \text{ if } & 0 \leq \alpha < 0.90, \\ \alpha + 0.05, \text{ if } & 0.90 \leq \alpha < 0.95, \\ \alpha + 0.01, \text{ if } & 0.95 \leq \alpha < 0.99, \\ \alpha, \qquad \text{ if } & \alpha \geq 0.99 \end{cases} \tag{5}$$

In this way, the initial threshold α is calculated automatically, based on which an appropriate correction factor is added, to dynamically set the threshold δ depending on the input image.

3   Experimental Results

In this section, we present the results of our proposed technique for automatic region detection, on images downloaded from the Internet [17], as well as on those captured by us. These images contain regions that appear damaged. With our proposed technique, we intend to detect such regions and generate a mask that can be suitably used to inpaint them. Inpainted results using technique proposed in [2] show the suitability of our proposed method to auto-detect target regions for inpainting.

In all our experiments we have considered patches Φ_p of size 3×3. We present the results of our experiments on wall & ceiling images in figure 2, pavement

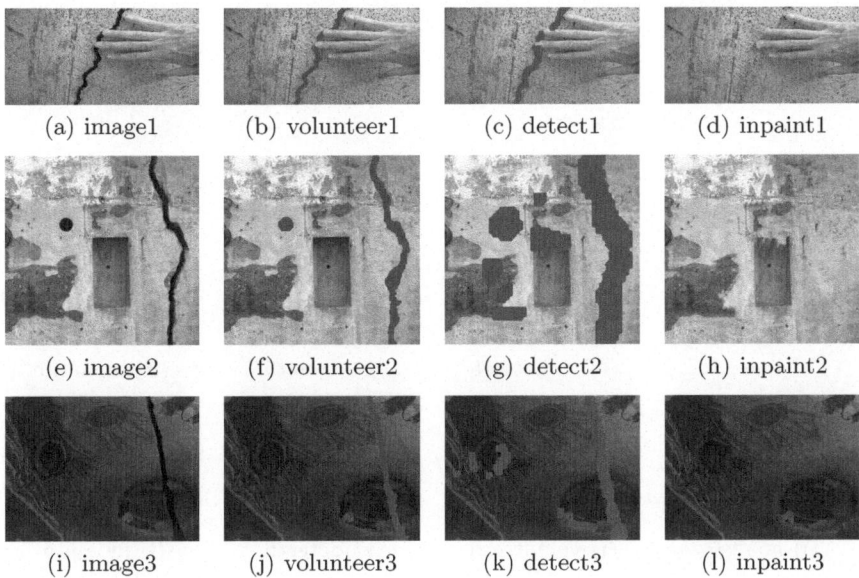

(a) image1	(b) volunteer1	(c) detect1	(d) inpaint1
(e) image2	(f) volunteer2	(g) detect2	(h) inpaint2
(i) image3	(j) volunteer3	(k) detect3	(l) inpaint3

Fig. 2. Wall and ceiling images. (a),(e),(i) input images, (b),(f),(j) target regions selected by volunteers in red color, (c),(g),(k) detected target regions in red color, (d),(h),(l) inpainted results using technique in [2] for regions detected in (c),(g),(k).

images in figure 3. The input images for wall, ceiling and pavement were all downloaded from the Internet [17]. In order to determine the suitability of the resulting masks for the use by inpainting algorithms, we consider the popularly used recall and precision metrics [14] defined as follows.

$$
\begin{aligned}
Recall \quad &= \frac{|Ref \cap Dect|}{|Ref|}, \\
Precision &= \frac{|Ref \cap Dect|}{|Dect|}
\end{aligned}
\tag{6}
$$

Ref are the pixels declared to be in the target regions by volunteers and *Dect* are the pixels detected by the algorithm to be in the target regions. Higher value of *Precision* indicates that a large number of detected pixels indeed belong to the target region, while a higher value of *Recall* indicates that a large number of target pixels have be detected. For a mask to be suitable for use to an inpainting algorithm, it is therefore desired to have the *Recall* value nearer to 1. On the other hand, a low value for *Precision* indicates that more pixels than desired have been detected, which only increase the area to be inpainted and is therefore acceptable. However, if a mask with low *Recall* value is used for inpainting, information from the undetected target regions may propagate inside the detected region, leading to poor inpainting results.

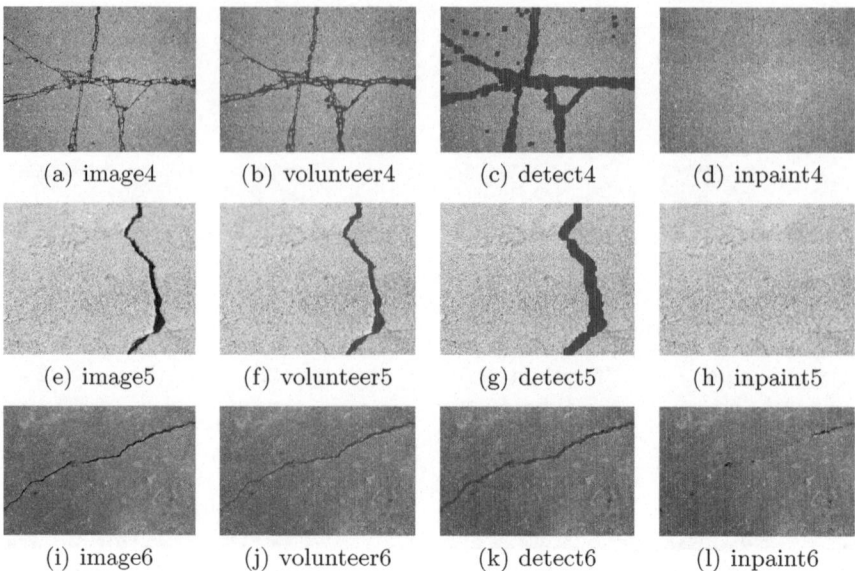

(a) image4   (b) volunteer4   (c) detect4   (d) inpaint4

(e) image5   (f) volunteer5   (g) detect5   (h) inpaint5

(i) image6   (j) volunteer6   (k) detect6   (l) inpaint6

Fig. 3. Pavement images. (a),(e),(i) input images, (b),(f),(j) target regions selected by volunteers in red color, (c),(g),(k) detected target regions in red color, (d),(h),(l) inpainted results using technique in [2] for regions detected in (c),(g),(k).

The performance in terms of *Recall* and *Precision* values for input images in figures 2 and 3 is given in table 1. The results in figures 2 and 3 show that the damaged areas with non-uniform pattern or complex texture have been successfully detected. The detection results obtained using our proposed technique are remarkably similar to the detection performed manually by volunteers and is evident from the performance table 1. We observe that *Recall* value for all the detected target regions in these images is nearer to 1. This clearly indicates that the desired target pixels have been detected. Low *Precision* values indicate that more pixels than desired have been detected. However, from the results we observe that all the desired target regions are covered by the generated mask and its use for inpainting generates visually plausible images.

To the best of our knowledge, no techniques for detection of damaged areas in the photographed scene / object, have been reported in the literature. The nearest technique with which our proposed method can be compared is [10]. The comparative results on images captured by us are shown in figure 4. It may be noted that the results for technique [10] are the best possible, obtained after fine-tuning the parameters. Whereas, the parameter α in our proposed technique is dynamically calculated, depending on the input image.

From the results shown in figure 4 and performance comparison in table 2, it is clear that the desired regions are successfully detected by the proposed method. Although the technique in [10] is good for detection for an alteration to the photograph (like overlay text), our proposed method is comparatively fast

Table 1. Performance of the proposed technique in terms of Recall and Precision

Input	#Target Pixels	Recall	Precision
image1	05414	0.9540	0.6702
image2	02513	0.9988	0.2290
image3	05431	0.9742	0.3708
image4	40741	0.9914	0.4445
image5	05613	0.9984	0.4772
image6	29333	0.8919	0.5781

(a) image7	(b) Amano [10]	(c) detect7	(d) inpaint7
(e) image8	(f) Amano [10]	(g) detect8	(h) inpaint8
(i) image9	(j) Amano [10]	(k) detect9	(l) inpaint9

Fig. 4. Images captured by us. (a),(e),(i) input images, (b),(f),(j) target regions detected using technique in [10], (c),(g),(k) detected target regions in red color, (d),(h),(l) inpainted results using technique in [2] for regions detected in (c),(g),(k).

Table 2. Performance comparison in terms of Recall and Precision

Input	#Target Pixels	Proposed Tech.			Tech. in [10]		
		Recall	Precision	Time (sec)	Recall	Precision	Time (sec)
image7	8217	1.0000	0.5372	4.63	0.1503	0.0093	512
image8	1353	1.0000	0.1749	4.41	0.6438	0.0260	093
image9	3494	0.9531	0.2971	4.51	0.0000	0.0000	109

and more suitable when it comes to detection of damage in the photographed scene / object.

4   Conclusion

In this paper we have presented a technique that can be used to generate an input mask for inpainting algorithms. By comparing overlapping patches in the SVD domain, we form a similarity matrix. The dissimilar patches detected using an image adaptive threshold are used to construct a binary map having the target regions and can therefore be suitably used an input mask for image inpainting. The obtained results on an exhaustive set of images show that the generated masks can be indeed used to inpaint the input images. In future, we aim to extend this detection method to perform simultaneous on-the-fly detection and inpainting, which can be used to build an immersive walk-through system.

Acknowledgement. This work is a part of the project *Indian Digital Heritage (IDH) - Hampi* sponsored by Department of Science and Technology (DST), Govt. of India (Grant No: NRDMS/11/1586/2009/Phase-II). The authors are also grateful to Prof. Toshiyuki Amano, Department of Systems Innovation, Yamagata University, for his valuable inputs and sharing the code of his work in [10].

References

1. Bertalmio, M., Sapiro, G., Caselles, V., Ballester, C.: Image inpainting. In: Proceedings of the 27th Annual Conference on Computer Graphics and Interactive Techniques, SIGGRAPH 2000, pp. 417–424. ACM Press/Addison-Wesley Publishing Co., New York (2000)
2. Criminisi, A., Pérez, P., Toyama, K.: Region filling and object removal by exemplar-based image inpainting. IEEE Transactions on Image Processing 13, 1200–1212 (2004)
3. Pérez, P., Gangnet, M., Blake, A.: Poisson image editing. In: ACM SIGGRAPH 2003 Papers, SIGGRAPH 2003, pp. 313–318. ACM, New York (2003)
4. Grossauer, H.: A Combined PDE and Texture Synthesis Approach to Inpainting. In: Pajdla, T., Matas, J. (eds.) ECCV 2004. LNCS, vol. 3022, pp. 214–224. Springer, Heidelberg (2004)
5. Shibata, T., Iketani, A., Senda, S.: Image Inpainting Based on Probabilistic Structure Estimation. In: Kimmel, R., Klette, R., Sugimoto, A. (eds.) ACCV 2010, Part III. LNCS, vol. 6494, pp. 109–120. Springer, Heidelberg (2011)
6. Padalkar, M.G., Joshi, M.V., Zaveri, M.A., Parmar, C.M.: Exemplar based inpainting using autoregressive parameter estimation. In: Proceedings of the International Conference on Signal, Image and Video Processing, ICSIVP 2012, IIT Patna, India, pp. 154–160 (2012)
7. Shih, T.K., Tang, N.C., Yeh, W.-S., Chen, T.-J., Lee, W.: Video inpainting and implant via diversified temporal continuations. In: Proceedings of the 14th Annual ACM International Conference on Multimedia, MULTIMEDIA 2006, pp. 133–136. ACM, New York (2006)

8. Chang, R.-C., Sie, Y.-L., Chou, S.-M., Shih, T.K.: Photo defect detection for image inpainting. In: Proceedings of the Seventh IEEE International Symposium on Multimedia, ISM 2005, pp. 403–407. IEEE Computer Society, Washington, DC (2005)
9. Tamaki, T., Suzuki, H., Yamamoto, M.: String-like occluding region extraction for background restoration. In: International Conference on Pattern Recognition, vol. 3, pp. 615–618 (2006)
10. Amano, T.: Correlation based image defect detection. In: Proceedings of the 18th International Conference on Pattern Recognition, ICPR 2006, pp. 163–166. IEEE Computer Society, Washington, DC (2006)
11. Parmar, C.M., Joshi, M.V., Raval, M.S., Zaveri, M.A.: Automatic image inpainting for the facial images of monuments. In: Proceedings of Electrical Engineering Centenary Conference 2011, IISc Bangalore, India, pp. 415–420 (2011)
12. Ammouche, A., Riss, J., Breysse, D., Marchand, J.: Image analysis for the automated study of microcracks in concrete. Cement and Concrete Composites 23, 267–278 (2001); Special Theme Issue on Image Analysis
13. Ringot, E., Bascoul, A.: About the analysis of microcracking in concrete. Cement and Concrete Composites 23, 261–266 (2001); Special Theme Issue on Image Analysis
14. Zou, Q., Cao, Y., Li, Q., Mao, Q., Wang, S.: Cracktree: Automatic crack detection from pavement images. Pattern Recognition Letters 33, 227–238 (2012)
15. Deerwester, S., Dumais, S.T., Furnas, G.W., Landauer, T.K., Harshman, R.: Indexing by latent semantic analysis. Journal of the American Society for Information Science 41, 391–407 (1990)
16. Wall, M.E., Rechtsteiner, A., Rocha, L.M.: Singular value decomposition and principal component analysis. In: Singular Value Decomposition and Principal Component Analysis, pp. 91–109. Kluwer, Norwell (2003)
17. Google Images (2012), http://www.images.google.com

Presentation of Japanese Cultural Event Using Virtual Reality

Liang Li[1], Woong Choi[2], Kozaburo Hachimura[1],
Takanobu Nishiura[1], and Keiji Yano[1]

[1] Ritsumeikan University, 1-1-1 Noji-higashi, Kusatsu, Shiga 525-8577, Japan
[2] Gunma National College of Technology, 580 Toribamachi, Maebashi, Gunma
371-8530, Japan

Abstract. With the development of computer graphics and virtual reality technologies, extensive researches have been carried out on digital
archiving of cultural assets. In this paper, we introduce our work on
presenting a traditional Japanese cultural event, namely Yamahoko Parade in Kyoto Gion Festival, using the latest technologies such as 3D
CG modeling, motion capture, high-quality sound recording, vibration
system, immersive virtual environment, and real-time interaction. This
work is one part of the digital museum project, which intends to preserve
and present Kyoto city's culture and tradition.

1 Introduction

Recently, extensive researches have been conducted on digital archiving of cultural properties in the field of e-heritage. For decades, tangible cultural heritage
contents including paintings, historical crafts, archaeological sites, and historical
buildings have been digitally archived. In recent years, there have been attempts
to cover not only individual intangible cultural assets such as dances but also
intangible cultural assets, such as traditional festivals and behaviors of participants in cultural events [1–3].

In this research, we try to virtually represent a traditional Japanese cultural
event called Yamahoko Parade in Kyoto Gion Festival.

Originated from Heian period (794-1185), Gion Festival, which has been registered in the list of "Intangible Heritage of Humanity" by UNESCO, is one of
the most famous festivals in Japan. Every year on July 17, the festival culmi-
nates in a parade of Yamahoko, floats known as "moving museums" because of
their elaborate decorations with centuries-old tapestries, and wooden and metal
ornaments (Fig. 1). The festival is held by the Yasaka Shrine whose parishioners
parade 32 floats to represents each self-governing parish. Approximately 150
thousands spectators from all around the world gather to see the parade every
year.

We generated a content that combines motion and acoustics of the floats,
crews, and spectators, within a virtual platform of "Virtual Kyoto" [4]. In current step, four well-known floats (Fune-hoko, Naginata-hoko, Kanko-hoko, and

J.-I. Park and J. Kim (Eds.): ACCV 2012 Workshops, Part II, LNCS 7729, pp. 72–82, 2013.

Kitakannon-yama) out of thirty-two floats were included in this virtual parade. We also reproduced the motion of four types of crews of Fune-hoko (Hikikata, Ondotori, Kurumakata, Hayashikata) using motion capture technique. We also collected various kinds of data to reproduce the vibration and the whole atmosphere of the event. The Virtual Yamahoko Parade can be experienced with display systems such as immersive virtual environment and head mounted display with real time interaction. This work contributes to the research of digital museum and provides a platform that allows the users to virtually experience the atmosphere of Yamahoko Parade in Kyoto Gion Festival.

Fig. 1. Yamahoko Parade of Kyoto Gion Festival

2   Creation of the Virtual Yamahoko Parade

To generate a CG content of Virtual Yamahoko Parade the following components are required.

- CG models of the Kyoto streets.
- CG models of Yamahoko floats.
- CG models and animations of the parade crews.
- CG models and animations of the spectators.
- Acoustic components of the parade.

We construct the Virtual Yamahoko Parade using a system called Vizard, a product of WorldViz. Vizard enables us to integrate and render CG models, animations, and sounds in a virtual space with real time interaction (Fig. 2).

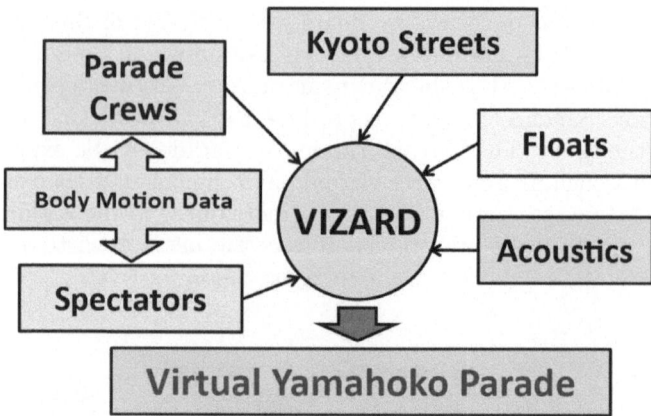

Fig. 2. Composition of the Virtual Yamahoko Parade

2.1   Models of Kyoto Streets

The models of Kyoto streets are taken from the 3D geographical space created for "Virtual Kyoto" [4]. Virtual Kyoto is a platform of Kyoto constructed using the latest geographic information system (GIS) and virtual reality techniques. Virtual Kyoto allows the users to virtually experience Kyoto across time and space at will.

2.2   Models of Yamahoko Floats

A total of thirty-two Yamahoko floats join the parade every year. Currently we selected four floats (Naginata-hoko, Kanko-hoko, Fune-hoko, and Kitakannon-yama) and created their CG models (Fig. 3). The models are created by laser-scanning the structures of the floats or by using the measured drawings of the floats. The textures of the floats are made by capturing photos of the floats during the festival.

2.3   Models and Animations of the Parade Crews

The Virtual Parade includes four kinds of parade crews: Hikikata, who pull ropes to tug the float; Ondotori, who lead the parade with Japanese fans; Kurumakata, who control the float's directions; and Hayashikata, who play instruments on the platform of the float.

We created CG models of the crews using 3ds max and transformed them into Cal3D format to import to Vizard (Fig. 4). The textures of their costumes were obtained from the real costumes. The body action data given to each character model were obtained from actual actions using motion capture technique (Fig. 5).

We arranged the characters of Hikikata, Ondotori, Kurumakata, and Hayashikata on and around the model of Fune-hoko, and created a scene of the parade, in which the float is being towed (Fig. 6).

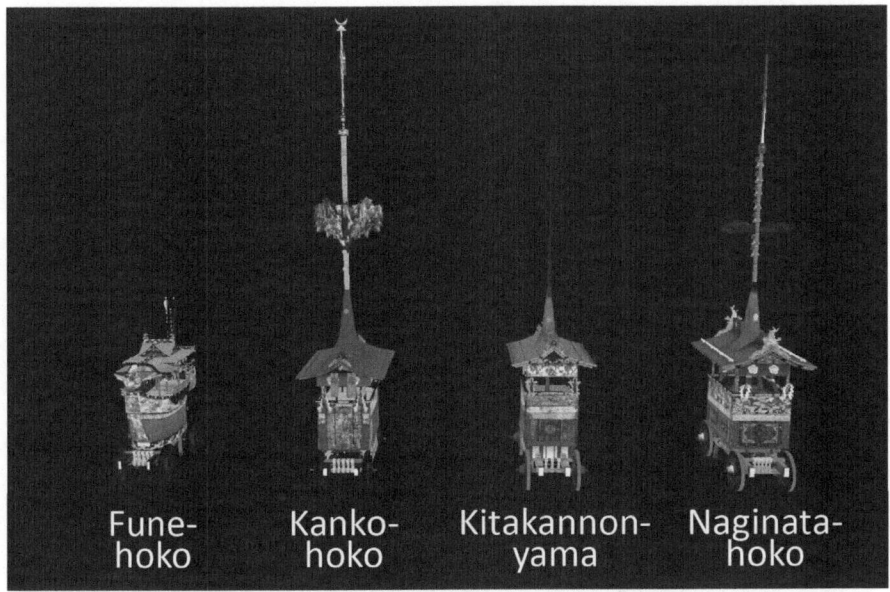

Fig. 3. CG models of the Yamahoko floats

Fig. 4. CG models of the parade crews

2.4 Models and Animations of the Spectators

Every year, over one thousand spectators from all around the world visit Kyoto to watch the parade. In order to recreate the atmosphere the festival, it is very

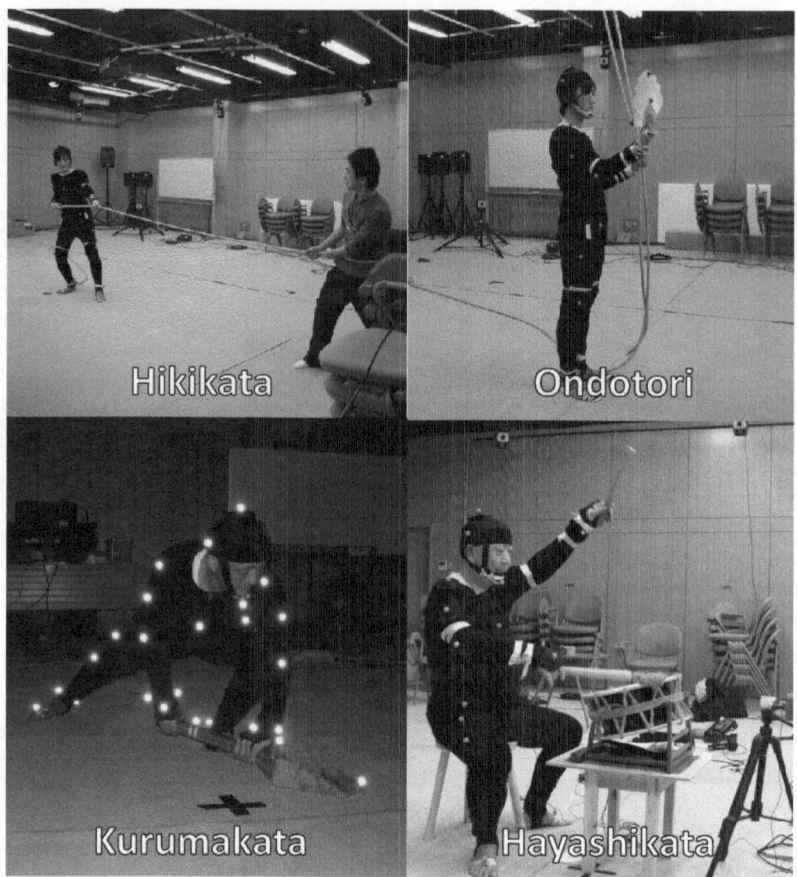

Fig. 5. Motion capture of parade crews

important to reproduce the spectators who gather to join the event. We arranged about 730 characters on both sides of Shijo Street in Virtual Kyoto in an attempt to represent the crowds.

2.5   Acoustics of the Parade

We employed multi-point measurement technique to record and reproduce the music of the parade played with the traditional instruments of drum, flute and bell. Figure 7 shows temporal waveforms of the sounds of each instrument (the maximum amplitude is normalized to 1). On the parade day, we also collected acoustic data by recording audio sources such as creaking sounds of the wheels of the floats, speaking voices of the spectators, an noised made by the crowds.

Construction of a simulation float, shown in Fig. 8, is currently under consideration as an application of surround sound distribution using the omni-directional point-source loudspeakers.

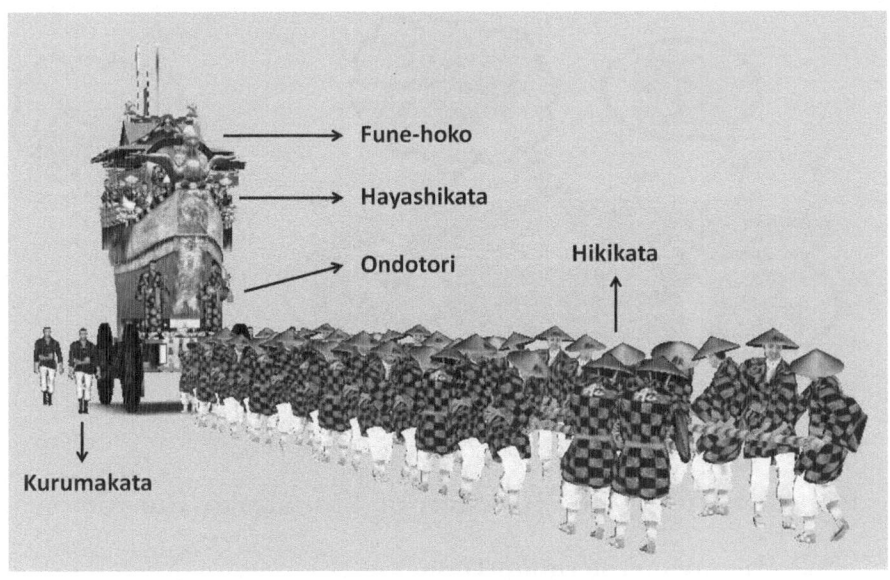

Fig. 6. Fune-hoko and it's parade crews

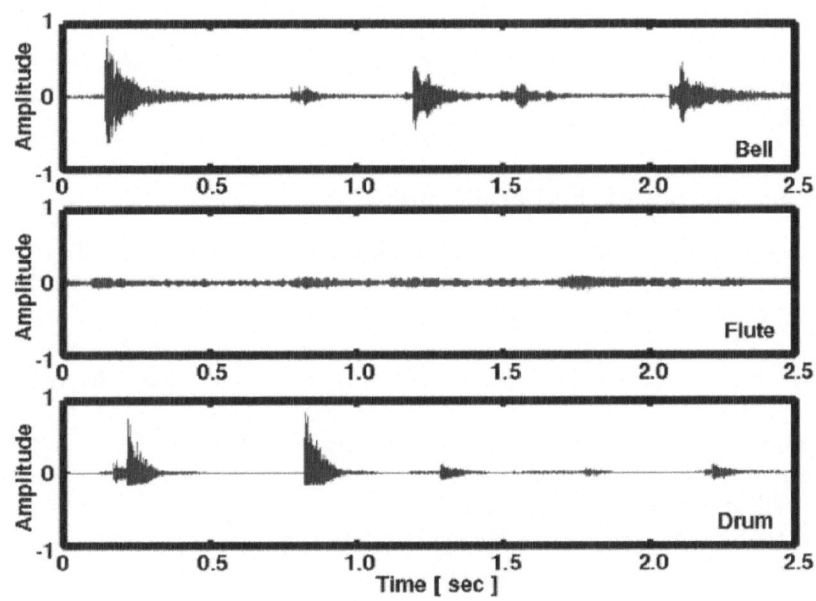

Fig. 7. Waveforms of each instrument

Fig. 8. Replica of float with surround speakers

3   Reproduction the Vibration of Yamahoko Parade

3.1   Collection of Acceleration Data

We collected route data, acceleration data, and sounds of Fune-hoko parade using a GPS logger, acceleration sensors, and a PCM recorder during the rehearsal parade on July 13 and the real parade on July 17, 2011 (Fig. 9). The sampling rates of GPS logger and acceleration sensor are 5Hz and 50Hz respectively. We also captured the scene on the Hayashibutai by a 3D front view camera during the rehearsal parade. We haven't set up the camera in the real parade for the reason of protecting the heritage.

3.2   Visualization of the Parade Route

In this research, the process to visualize the acceleration on Google Earth is as follows.

Firstly, the acceleration data which the fine noise was removed by using band-pass filter is accorded at the initial time of snapped GPS data. The magnitude of the acceleration data at each time was determined by using the 3D version of Pythagoras's theorem.

Secondly, we computed the Root Mean Square (RMS) data of the magnitude data to reduce the effect of acceleration variation from error data. The RMS data of 750 times data corresponding to each GPS data is integrated.

Next, the maximum value is calculated from the integrated RMS data. The integrated RMS data was normalized by the calculated maximum value. The normalized data was changed into the quasi-color and the magnitude of acceleration was colored.

Finally, the information of altitude and color of the bar graphs which are visualized on Google Earth is transmitted into KML data by using MATLAB Google Earth Tool Box.

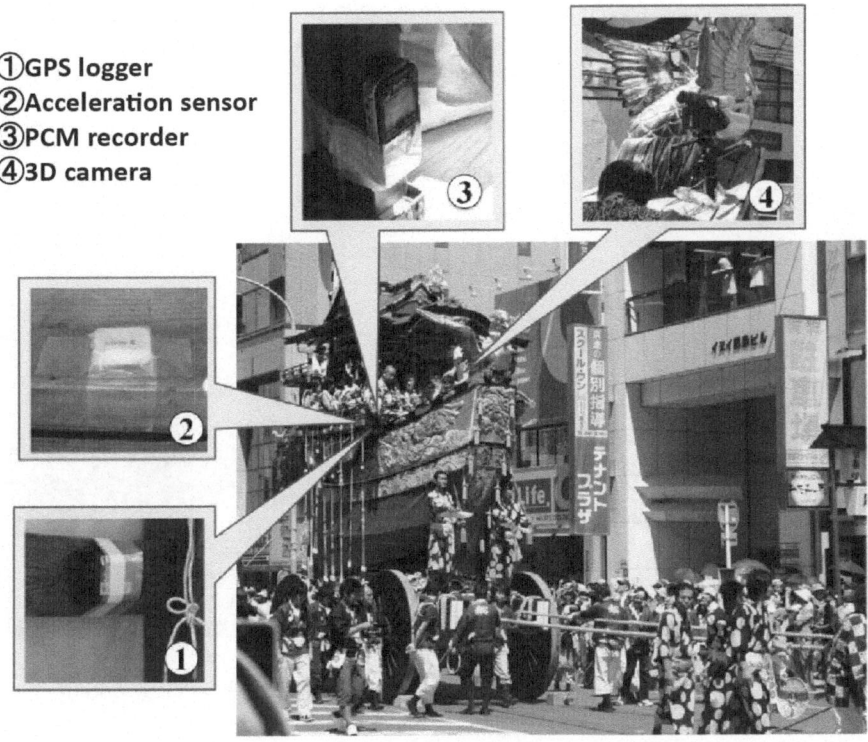

Fig. 9. Data collection

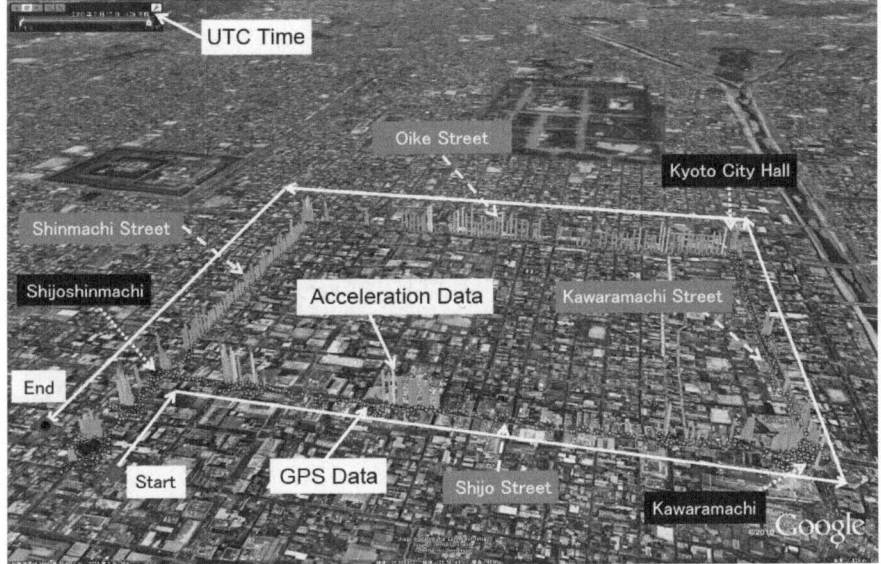

Fig. 10. Visualization of acceleration variation of Yamahoko Parade on Google Earth

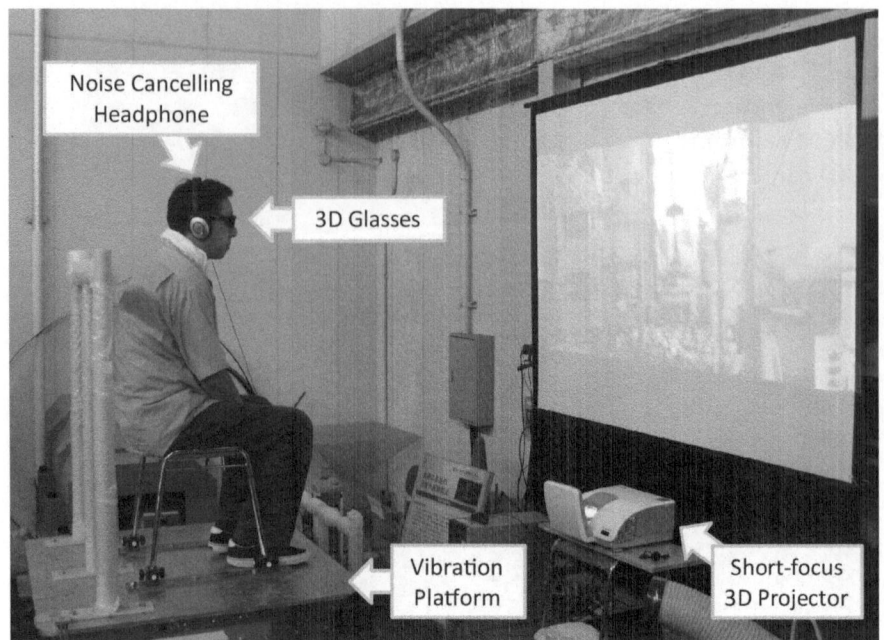

Fig. 11. Experimental Virtual Yamahoko Parade experiencing system

Fig. 12. Immersive virtual environment

As shown in Fig. 10, the acceleration of Fune-hoko at each time was displayed as the bar graph on the GPS data of Google Earth. The high red bar graphs mean the place where the acceleration has a large variation. Conversely, the low

Fig. 13. Virtual environment using HMD

Fig. 14. Screenshots of Virtual Yamahoko Parade

blue bar graphs represent the place where the acceleration has a little variation. Therefore, we noticed that the places where performed Tsujimawashi of Funehoko have the large acceleration variation in Shijoshinmachi, Kawaramachi and City hall of Kyoto compared to that of Shijo street, Kawaramachi street, Oike street and Shinmachi street.

3.3   Reproduction of Vibration

We reproduce the rolling and vibration of Fune-hoko Hayashibutai using a vibration system with 3 degrees of freedom (translation on x, y and z). We transform the acceleration data into displacement data as the input of the vibration system.

To get displacement from acceleration, we calculate the integration in the frequency domain: (1) Remove the mean from the acceleration data. (2) Take the Fourier transform of the acceleration data. (3) Remove the low frequency noise using a high-pass filter. (4) Convert the transformed acceleration data to displacement data by dividing each element by $-\omega^2$, where ω is the frequency. (5) Take the inverse Fourier transform to get the displacement data in time domain. The stopband edge frequency and the passband edge frequency of the high-pass filter were set to 0.2Hz and 0.5Hz respectively.

We integrated vibration, sound, and 3D screen to build an experimental Virtual Yamahoko Parade experiencing system (Fig. 11).

4   Display and Operation of the Virtual Yamahoko Parade

The Virtual Yamahoko Parade can be operated in 3D with high fidelity sounds using a large-scale immersive virtual environment system (Fig. 12). This System is installed at College of Information Science and Engineering at Ritsumeikan University. The Users can interactively control the viewing position and angle in the virtual world at real time with a gamepad. The Virtual Yamahoko Parade can also be displayed with a hand mounted display system (Fig. 13).

Sample screenshots captured under 2D mode are illustrated in Fig. 14.

Acknowledgement. This work was supported in part by the Digital Museum Project in the Ministry of Education, Culture, Sports, Science and Technology, Japan.

References

1. Magnenat-Thalmann, N., Foni, N., Papagiannakis, G., Cadi-Yazli, N.: Real Time Animation and Illumination in Ancient Roman Sites. The International Journal of Virtual Reality 6, 11–24 (2007)
2. Papagiannakis, G., Magnenat-Thalmann, N.: Mobile Augmented Heritage: Enabling Human Life in Ancient Pompeii. The International Journal of Architectural Computing 2, 395–415 (2007)
3. Virtual Vaudeville: http://vvaudeville.drama.uga.edu/
4. Yano, K., Nakaya, T., Isoda, Y.: Virtual Kyoto: Exploring the Past, Present and Future of Kyoto. Nakanishiya, Kyoto (2007)

High-Resolution and Multi-spectral Capturing for Digital Archiving of Large 3D Woven Cultural Artifacts

Wataru Wakita[1], Masaru Tsuchida[2], Shiro Tanaka[1], Takahito Kawanishi[2], Kunio Kashino[2], Junji Yamato[2], and Hiromi T. Tanaka[1]

[1] College of Information Science and Engineering, Ritsumeikan University, Kusatsu, Japan
wakita@cv.ci.ritsumei.ac.jp
[2] NTT Communication Science Laboratories, NTT Corporation, Atsugi, Japan

Abstract. We propose a high-resolution and multi-spectral capturing for digital archiving of large 3D woven cultural artifacts. In the field of digital archive, it is important to measure, model, and represent the shape, color, and texture of the cultural artifact at high-definition, not only physical appearance but haptic impression. The many of the decorative hangings on the Fune-hoko in the "Gion Festival in Kyoto", are very large and stuffed with cotton, so that they have a very noticeable 3D shape. Therefore, a high-resolution and multi-spectral capturing and a large-scale 3D measuring are necessary for the digital archiving of large 3D woven cultural artifact. Then we captured high-resolution images with a two-shot type 6-band image capturing system at low-cost, and modeled woven cultural artifacts in 3D. This paper describes a 3D measurement system with wheel-rail, a capturing system with multi-band camera, and a 3D modeling of large woven cultural artifacts, and show a high-resolution 3D model with multi-band image.

1 Introduction

Recently, research on digital museums received increased attention. In the digital museum project [1], they are working on the digital archiving of the intangible cultural heritage "Gion Festival in Kyoto", focused on the culture of Kyoto, and developed a multi-modal VR exhibition system for 3D woven cultural artifacts [2].

The Yama-hoko float procession of the "Gion Festival in Kyoto" started in the 14th century, which the UNESCO registered on the Representative List of the Intangible Cultural Heritage in 2009. In the Yama-hoko float procession, which has been called "moving museum", thirty-two "Yama" and "Hoko" floats adorned with colorful decorations wind their way through the city. Although 15th-century civil wars destroyed several Yama-hoko, still in use are many of the surviving tapestries and ornaments, imported from Europe and China or made by Japanese craftsmen from the 17th to the 19th centuries. Most of Yama-hoko are kept in private storage warehouse in their town, so it's required to

J.-I. Park and J. Kim (Eds.): ACCV 2012 Workshops, Part II, LNCS 7729, pp. 83–93, 2013.

have action for protect from disasters such as fire or earthquake. Also it is very important to record and analyze the traditional craftsmanship. Therefore, it is necessary to measure and model the decorations used for the floats at high-definition. The decorations used for the floats include embroidery using various materials, like gold thread, cotton, glass, and felt; their texture is highly complex. In particular, the many of the decorative hangings on the Fune-hoko in 32 Yama-hoko are very large and stuffed with cotton, so that they have a very noticeable three-dimensional shape (see Figure 1).

Fig. 1. Fune-hoko and Large 3D Woven Cultural Artifacts

In the field of digital archive, it is important to measure, model, and represent the shape, color, and texture of the cultural artifact at high-definition, not only physical appearance but haptic impression. To archive the shape, color, and texture of the these decorative hangings, a 3D measurement and a high-resolution and multi-spectral capturing are necessary.

Therefore, we developed a 3D scanning system for the large woven cultural artifacts with wheel-rail and laser range scanner, and measured all these. Then we captured high-resolution images with a two-shot type 6-band image capturing system, and modeled woven cultural artifacts in 3D. This paper describes a 3D scanning system, a capturing system with multi-band camera, and a 3D modeling of large woven cultural artifacts, and show a high-resolution 3D model with multi-band image.

2   Related Work

In order to reproduce accurate color, multi-spectral color reproduction technique is used. Using this technique makes it possible to get the object's accurate color and its spectral reflectance which is independent of image capturing device, as well as to estimate illumination condition. Spectral reflectance of object's surface can be estimated by using multi-band data of each pixel respectively. The estimated spectral reflectance enables us to simulate illumination conversion, for example, from daylight to incandescent lamp accurately. Moreover, it allows us to identify the material and colorant.

In previous work, several types of multi-spectral image capturing system are available [3–6]. However, in these systems, hardly any of them have been widely used in business activities and digital archives of cultural heritage, one major reason of which seems that constructing a multi-spectral image capturing system calls for both a large amount of money and special equipments. Recently, affordable commercial multispectral imaging systems such as the 2-CCD camera of JAI have already been available. However, they have a limitation problem of the number of pixels and a depth of field problem. On the other hand, multiple hyperspectral imaging systems are proposed for digital archiving. Cao et al. proposed a prism-mask system for capturing multispectral videos [7]. Their system is capable of capturing frames with high spectral resolution at video rates and allows for different trade-offs between spectral and spatial resolution by adjusting the focal length of the camera. However, their system is low-definition and not considered about large 3D measurement. Kim et al. proposed Visual enhancement of old documents with hyperspectral imaging [8]. They take advantage of hyperspectral images of historical documents to visually enhance the document's content by exploiting additional information provided by the NIR bands. However, their system is not considered about large 3D measurement. Ben-Ezra [9] proposed a digital gigapixel large-format tile-scan camera. Their system can acquire highquality, high-resolution images of static scenes. However, their system is not considered about 3D measurement. To represent digital archived model naturally, traditional 3-band imaging camera is inadequacy. Therefore, various 6-band camera systems are proposed [4, 10] Compared to these, the two-shot type 6-band camera system proposed in this paper uses a cost-efficient commercial digital camera and custom interference filter. Our system can get the image equivalent of large imaging sensor because we use fixed camera and lens.

In previous work about 3D shape, color, and texture, Tonsho et al. proposed a method for extracting gonio-spectral information of three-dimensional objects from multi-band images obtained under several illuminants and reproducing the objects under various kinds of illuminants [11]. However, their system is low-definition and not considered about large 3D measurement. On the other hand, Kim et al. proposed 3D imaging spectroscopy for measuring hyperspectral patterns on solid objects [12]. However, their system is low-definition and not considered about large 3D measurement.

In this work, we contribute a digital archiving of large 3D woven cultural artifact with high-resolution and multi-spectral imaging camera and 3D laser-range scanner.

3   Digital Archiving of Large 3D Woven Cultural Artifacts

In this section, we describe a digital archiving process of the large 3D cultural artifact. Firstly, we measure the shape of the cultural artifact with 3D scanning system, then we analyzed the measurement data, and modeled by the denoising, the interpolating of the lost part, the alignment, and the integration of the overlapped point.

Secondly, we capture the cultural artifact with multi-band imaging camera, then we modeled by alignment. Finally, we model by UV mapping(see Figure 2).

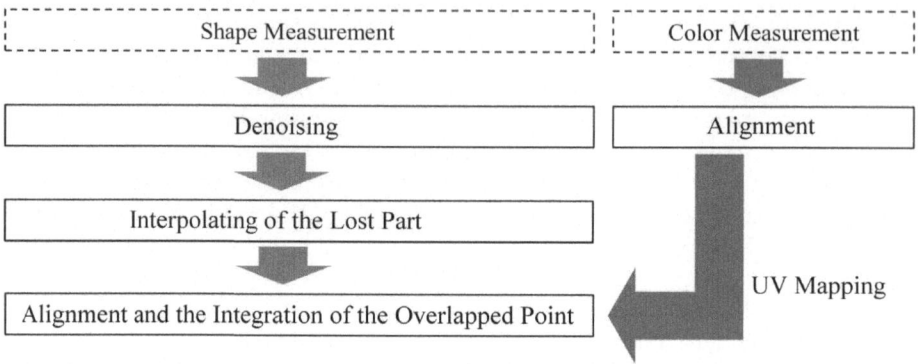

Fig. 2. Digital Archiving Process

3.1   3D Measurement System for Large 3D Woven Cultural Artifacts

In general, for cultural artifacts that disintegrate over time, touching and handling is not permitted. For this reason, when taking measurements of cultural artifacts, methods that involve no contact and no damage are a prerequisite.

The decorations used for the floats include embroidery using various materials, therefore, if one simply uses a laser range scanner to take measurements, data may only be collected for certain portions depending on the strength of the laser, and measurement may not be completed successfully. Also, many of the Fune-hoko hangings are extremely large, approximately 1 m×3 m, and so measurement must be conducted in sections. The final process of combining the data from measurements of individual sections is therefore time-consuming. Choosing the textile cultural artifacts with particularly strong three-dimensionality among the Fune-hoko hangings as our subjects, we have constructed a measurement system that takes into account the process of measurement and ease of combined processing.

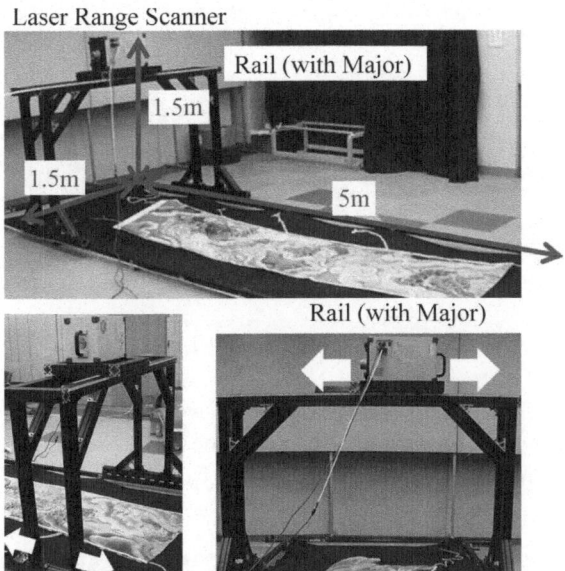

Fig. 3. 3D Measurement System for Large 3D Woven Cultural Artifacts

Figure 3 shows our constructed 3D measurement system for 3D large woven cultural artifacts. This system uses a Konica Minolta VIVID 910 as the laser range scanner.

As described above, the hangings include colorful embroidery in diverse materials like gold thread and cotton, so that measurements cannot be taken successfully unless the laser strength is adjusted to suit the different materials. In order to take thorough measurements, we therefore had to scan multiple times with the laser strength set to several different levels. In VIVID910, it is possible to change from 0 to 255 the laser strength. Figure 3 shows measurements taken with fifteen different levels (30–250) of laser strength. Scanning is repeated to fill its blank space (e.g. white area in Figure 4).

The sections done entirely in gold thread cannot be accurately measured without performing 3–5 scans with the laser strength set low (empirically 7–15), while sections that include dark blue and white thread can only be measured by scanning repeatedly with a wide range (empirically 30–255) of laser strengths. In one scan, the VIVID 910 can measure 640×480 pixels of distance data and color image. The range for setting a measurement input target is 0.6–1.2 m, and it has three types of lens (TELE/MIDDLE/WIDE). We used the TELE lens which was able to be most measured to the high definition as a result of the preliminary experiment, but since this only allows measurement within a narrow range, the required number of measurements increases and integration processing becomes time-consuming. For this reason, in order to simplify the process of integration processing after measurement as much as possible, we laid out a rail so that a wheeled platform could run along it from left to right, and set up the VIVID on

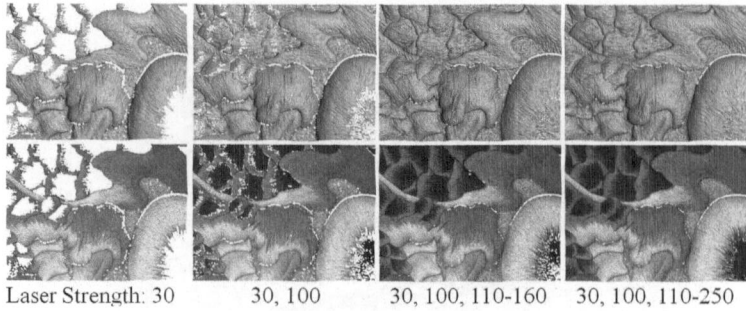

Laser Strength: 30   30, 100   30, 100, 110-160   30, 100, 110-250

Fig. 4. 3D Measurements with Varying Levels of Laser Strength

another wheeled platform on top of the first; we then conducted the measurements without moving the target object, by moving the VIVID at set intervals and recording the necessary shots (see Figure 5).

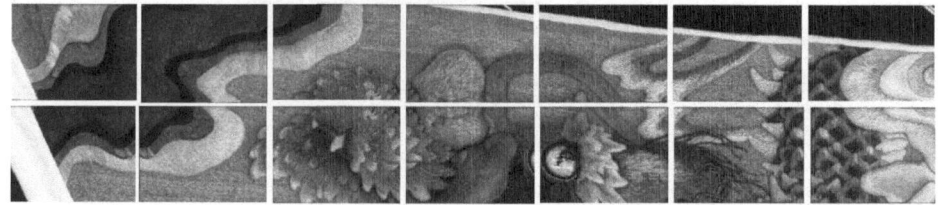

Fig. 5. 3D Measurements with Wheel-rail System

We analyzed the measurement data, and modeled by the denoising, the interpolating of the lost part, the alignment, and the integration of the overlapped point. Figure 6 shows a 3D digital archived woven model. To make this model, we took measurements in 7 vertical sections at 15 cm intervals and 7–18 horizontal sections at 20 cm intervals (a total of 103 sections), for a total of about 600 scans and a total time of about 8 hours. The resolution of this model is about 9 points/mm.

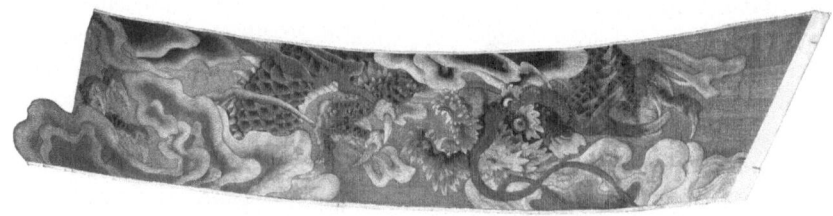

Fig. 6. 3D Model of of Woven Cultural Artifact

3.2   Capturing System with Multi-band Imaging Camera

In order to reproduce accurate color in digital photography, multi-band imaging technology is a practical solution. Then, we constructed ultra high-definition six-band digital camera system and used for archiving some tapestries and ornaments of Fune-hoko. This system can take more than 100 M pixel images and we archived that the resolution is almost 0.02 mm/pixel. Figure 7 shows our two-shot type 6-band image capturing system. This system consists of a consumer 35 mm-format digital camera and a custom interference filter. Attached in front of the camera lens, a customized filter cuts off the left sides, i.e., the short wavelength domain, of the peaks of both the blue and red in original spectral sensitivity. It also cut off the green's right side, i.e., the long wavelength domain. Sliding the filter horizontally by hand, one can capture two images with and without filter alternately. Although the filter can be moved smoothly, there is a few pixels displacement error between two images. This displacement error can be corrected in software [13]. After the correction process, the two 3-band images are combined into a 6-band image. In this system, number of pixels is defined by image sensor. First, an image is taken with the filter in front of the lens. Then, another image is taken without the filter by sliding the filter horizontally by hand. Combining the two RGB color images, a six-band image is synthesized.

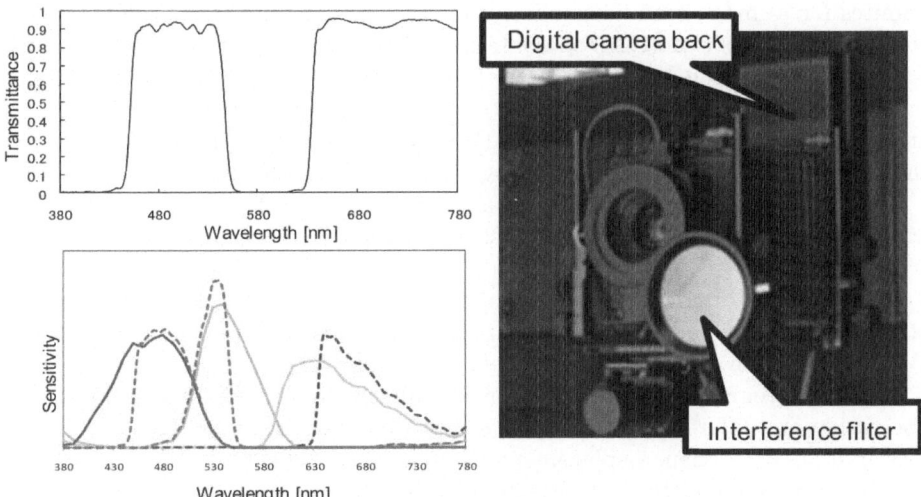

Fig. 7. Two-shot Type 6-band Image Capturing System

The obtained six-band image is converted into the spectral reflectance image based on the Wiener estimation method [14]. And, the spectral reflectance image is converted into RGB image using observing illumination spectrum and display profile data (e.g. tone-curve and primary colors).

When the object (its size is $2\,\mathrm{m}\times2\,\mathrm{m}$) is captured by $100\,\mathrm{M}$ Pixel image sensor, image resolution is $0.2\,\mathrm{mm}$/pixel. To achieve image resolution be $0.01\,\mathrm{mm}$/pixel using this sensor, the object should be divided in to several parts and each image should be capture respectively. The obtained images are synthesized into a large pixel image. In this research, we used two methods to take divided images. The first one which is thought as general way is to move camera or object. Although this method is very simple and not necessary special equipments, it's required to correct effects of lens aberration such as distortion and shading before image synthesis. The second one is to move image sensor instead of moving camera or object. On this method, good quality lens of which image circle size is enough large is required. And also camera which has shift feature of sensor horizontally and verticality is required. A merit of this method is needless to consider effects of lens aberration when image synthesis. Actual way to realize this method is combining a large format camera (e.g. 4×5 camera) and a digital camera back.

In order to estimate spectral reflectance using multi-band image, full spectrum light without emission lines have better to be used. We used the SOLAX^{TM} (SERIC Ltd.) which has spectral power distribution close to natural daylight as illumination light. For avoiding damage of object caused by UV light, UV and IR cut filter is attached in front of lamp. And we used a 4×5 format camera (TOYO-VIEW 45GII), and scanning digital back (DF-S2, Pioneer) is attached as image sensor. This consists of a line CCD sensor of which pixel number is 10,600. Pixel number of scanning image is $185\,\mathrm{M}$ Pixels (10600×17460). It takes about 15 minutes for a image scan (depending to brightness of illumination light), then, almost 30 minutes is necessary to obtain a 6-band image. Note that IR cut filter is changed to custom one which cuts over $800\,\mathrm{nm}$ wavelength in this system. In general digital camera, IR cut filter which cuts over $680\,\mathrm{nm}$ wavelength is attached in front of sensor. However, this degrades S/N ratio of near infrared channel of the 6-band image capturing system used in this research.

Left of figure 7 shows spectral transmittance of custom interference filter attached in front of lens and spectral sensitivities of this 6-band image capturing system.

As for ability of color reproduction accuracy of this camera system, average of color difference between real object and image displayed on LCD monitor is $\Delta E_{ab} = 1.7$ when the target is Macbeth ColorChecker^{TM}.

Although a camera lens for large format camera is used, effects caused by lens aberration are different between image captured around optical axis and image captured off-axis. Then, displacement error corrections between image with filter and image without filter are carried out on each capturing area respectively.

Figure 8 shows a resultant image of a large tapestry which size is $400\,\mathrm{cm}$ width and $80\,\mathrm{cm}$ height. This image was synthesized from three images. The size of the synthesized image is $360\,\mathrm{M}$ pixels ($40,000 \times 9,000$ pixel).

4   3D Modeling Result

Figure 9 shows a comparative result of a multi-band image and a 3D model with multi-band image. Figure 9(c)(d) estimated through the use of a multi-band

Fig. 8. Multi-band Image of Woven Cultural Artifact

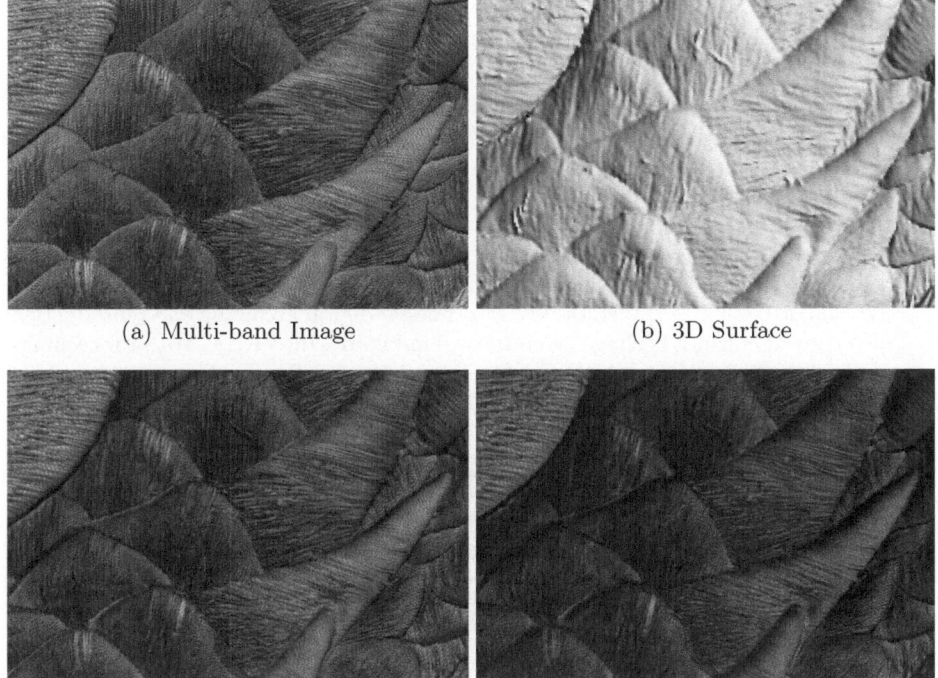

(a) Multi-band Image   (b) 3D Surface

(c) 3D Model with Multi-band Image

Fig. 9. Comparative Result

image, directional light source, and normal vector of 3D surface. As a result, we could estimate the color of the 3D surface of woven cultural artifact.

5   Conclusions and Future Work

This paper described a 3D measurement system with wheel-rail, a capturing system with multi-band camera, and a 3D modeling of large woven cultural artifacts. The many of the decorative hangings on the Fune-hoko in the "Gion Festival in

Kyoto", are very large and stuffed with cotton, so that they have a very noticeable 3D shape. Therefore, a high-resolution and multi-spectral capturing and
a large-scale 3D measuring are necessary for the digital archiving of large 3D
woven cultural artifact. In this work, we developed a 3D scanning system for the
large woven cultural artifacts with wheel-rail and laser range scanner. Then we
captured high-resolution images with a two-shot type 6-band image capturing
system at low-cost, and modeled woven cultural artifacts in 3D. As a result, we
could model a high-resolution 3D model with multi-band image. In future work,
we plan to develop a large-scale and real-time visuo-haptic interaction system for
3D woven cultural artifact exhibition and develop a multi-finger haptic device
and elastic interaction.

Acknowledgement. This research has been conducted partly by the support
of the Digital Museum Project in the Ministry of Education, Sports, Science
and Technology, Japan. We would like to thank the Gion-Matsuri Fune-hoko
Preservation Society, a generous collaborator of this project.

References

1. Tanaka, H.T., Hachimura, K., Yano, K., Tanaka, S., Furukawa, K., Nishiura, T.,
   Tsutida, M., Choi, W., Wakita, W.: Multimodal digital archiving and reproduction
   of the world cultural heritage "gion festival in kyoto". In: VRCAI 2010 Proceedings
   of the 9th ACM SIGGRAPH Conference on Virtual-Reality Continuum and its
   Applications in Industry, pp. 21–28 (2010)
2. Wakita, W., Akahane, K., Isshiki, M., Tanaka, H.T.: A Texture-Based Direct-
   Touch Interaction System for 3D Woven Cultural Property Exhibition. In: Koch,
   R., Huang, F. (eds.) ACCV 2010 Workshops, Part II. LNCS, vol. 6469, pp. 324–333.
   Springer, Heidelberg (2011)
3. Helling, S., Seidel, E., Biehlig, W.: Algorithms for spectral color stimulus reconstruction with a seven-channel multispectral camera. In: Proc. of CGIV (2nd European Conference on Color in Graphics, Imaging and Vision), vol. 2, pp. 254–258
   (2004)
4. Ohsawa, K., Ajito, T., Komiya, Y., Haneishi, H., Yamaguchi, M., Ohyama, N.:
   Six-band hdtv camera system for spectrum-based color reproduction. J. Imag. Sci.
   and Tech. 48, 85–92 (2004)
5. Yamaguchi, M., Murakami, Y., Uchiyama, T., Ohsawa, K.: Natural vision: Visual
   telecommunication based on multispectral technology. In: Proc. of IDW 2000, pp.
   1115–1118 (2000)
6. Tominaga, S., Okajima, R.: Object recognition by multi-spectral imaging with a
   liquid crystal filter. In: Proc. Int. Conf. on Pattern Recognition (2000)
7. Cao, X., Du, H., Tong, X., Dai, Q., Lin, S.: A prism-mask system for multispectral
   video acquisition. IEEE Transactions on Pattern Analysis and Machine Intelligence
   (PAMI) 33, 2423–2435 (2011)
8. Kim, S.J., Deng, F.B., Brown, M.S.: Visual enhancement of old documents with
   hyperspectral imaging. Pattern Recognition 44, 1461–1469 (2011)
9. Ben-Ezra, M.: A digital gigapixel large format tile-scan camera. IEEE Computer
   Graphics and Application 31, 49–61 (2011)

10. Park, J., Lee, M., Grossberg, M.D., Nayar, S.K.: Multispectral imaging using multi-plexed illumination. In: IEEE International Conference on Computer Vision, ICCV (2007)
11. Tonsho, K., Akao, Y., Tsumura, N., Miyake, Y.: Development of goniophotometric imaging system for recording reflectance spectra of 3D objects. In: Proc. SPIE, vol. 4663, pp. 370–378 (2001)
12. Kim, M.H., Rushmeier, H.E., Dorsey, J., Harvey, T.A., Prum, R.O., Kittle, D.S., Brady, D.J.: Imaging spectroscopy formeasuring hyperspectral patterns on solid objects. ACM Trans. Graph. 31, 38:1–38:11 (2012)
13. Takita, K., Aoki, T., Sasaki, Y., Higuchi, T., Kobayashi, K.: High-accuracy image registration based on phase-only correlation. IEICE Trans. Fundamentals E86-A, 1925–1934 (2003)
14. Abe, T., Murakami, Y., Yamaguchi, M., Ohyama, N., Yagi, Y.: Color correction of pathological images based on dye amount quantification. Optical Review 12, 293–300 (2005)

Exploring High-Level Plane Primitives for Indoor 3D Reconstruction with a Hand-held RGB-D Camera

Mingsong Dou[1], Li Guan[2], Jan-Michael Frahm[1], and Henry Fuchs[1]

[1] UNC-Chapel Hill
{doums,jmf,fuchs}@cs.unc.edu
[2] GE Global Research Center
guan@ge.com

Abstract. Given a hand-held RGB-D camera (e.g. Kinect), methods such as Structure from Motion (SfM) and Iterative Closest Point (ICP), perform poorly when reconstructing indoor scenes with few image features or little geometric structure information. In this paper, we propose to extract high level primitives–planes–from an RGB-D camera, in addition to low level image features (e.g. SIFT), to better constrain the problem and help improve indoor 3D reconstruction. Our work has two major contributions: first, for frame to frame matching, we propose a new scheme which takes into account both low-level appearance feature correspondences in RGB image and high-level plane correspondences in depth image. Second, in the global bundle adjustment step, we formulate a novel error measurement that not only takes into account the traditional 3D point re-projection errors, but also the planar surface alignment errors. We demonstrate with real datasets that our method with plane constraints achieves more accurate and more appealing results comparing with other state-of-the-art scene reconstruction algorithms in aforementioned challenging indoor scenarios.

1 Introduction

RGB-D cameras (e.g. Microsoft Kinect), which outputs RGB images and corresponding depth images at video frame rate, are new addition to vision sensors for indoor 3D reconstruction. Recent approaches [1][2][3][4] demonstrated the use of a hand-held RGB-D camera for indoor 3D reconstructions. However, texture-less regions and 3D space with few distinctive geometric structures are challenging places for state-of-the-art 3D reconstruction algorithms. Unfortunately, in indoor environment, such places are very common, for example the white walls and ceiling corners as seen in the middle column of Fig. 1.

In this paper, we propose to extract higher level primitives, i.e. planes, from the RGB-D sensor's depth channel for hand-held RGB-D camera indoor 3D reconstruction. Our major contributions include:

- First, we developed a robust pair-wise matching algorithm across frames via matching of both extracted planes and RGB image visual features. Our evaluation demonstrates that planes better constrain the reconstruction problem in the aforementioned challenging cases in low-texture low geometry information regions.
- Second, planes are compact representations of dense points and have clear associations across frames. This enables us to develop a novel formulation to incorporate

J.-I. Park and J. Kim (Eds.): ACCV 2012 Workshops, Part II, LNCS 7729, pp. 94–108, 2013.

Fig. 1. Results of our system on datasets (a)Rm. FB220, (b)Rm. SN277 and (c)Rm. SN353. Accumulated point clouds (colors are added to distinguish points from different planes), zoomed view, and planes resulted from our algorithm are shown from left to right.

plane correspondences (in addition to visual feature correspondence) to the Bundle Adjustment (BA) [5][6][7], which is usually the final step for 3D reconstruction. The purpose for BA is to globally adjust the recovered camera poses for the best possible reconstruction result.

– In addition, our proposed algorithm results in a piecewise planar representation for planar parts of the scene. Such compact representation of the scene can be deployed for applications such as noise reduction, data compression, etc.

While any primitive representation can be used with our method, a piecewise planar representation is most natural for man-made indoor environments, due to the dominant existence of planar surfaces, such as walls, floors, etc.

We evaluate our approach on several real world indoor datasets. Some have a lot of texture-less regions, which confuse SfM algorithms; some also have areas with significant geometrical ambiguity, such as large pieces of walls, which confuse ICP algorithms. By combining low level appearance features and high level geometric primitives, our algorithm handles these challenges well, and significantly improves the reconstruction results.

1.1 Related Work

Existing indoor 3D reconstruction algorithms fall into two major categories: (1) point cloud based registration such as Iterative Closest Point (ICP) and (2) visual feature based Structure from Motion (SfM).

Newcombe et al. in KinectFusion [2] utilized Iterative Closest Point (ICP) to align one frame's structure with previously captured data. While showing promising accuracy in certain environment settings, the main drawback is that ICP relies only on distinctive geometric information and would confuse when scanning a large piece of wall: the planar point cloud seen in the sensor's field of view cannot be localized in 3D space, because it can arbitrarily move along the wall surface without changing the ICP minimization cost. In order to conquer this drifting problem, salient image features can be extracted from the color channels of the RGB-D data, providing extra constraints to limit the ICP drift in [1]. Additionally, ICP needs a reasonable initialization. [1] uses the result from Structure from Motion to initialize ICP, which tends to fail in challenging scenarios with few image features points detected.

Structure from Motion (SfM) is another popular technique to estimate camera poses. Since only image feature points are used, SfM fails when very few matched features are found for two frames or the extracted features are not well distributed over the scene. As demonstrated later, in the worst case–when the whole room contains very few visually salient features–no frame will be registered together with SfM. Therefore, ICP and SfM algorithms are expected to fail in challenging cases with few image features and little geometric structure information.

However, we notice that additional high level contraints exist in the depth channel. As shown in Fig. 4, although there are no genuine image feature correspondences around the ceiling, the two frames still can be well aligned together if we could extract planes from the depth map and find the correct plane matches in the two frames. By combining high-level depth information and low-level image features, our system handle well previously mentioned challenging situations which cannot be solved alone by ICP or SfM.

In the SfM literature, the final step is usually global error mitigation through Bundle Adjustment. This global adjustment step is crucial for building a model for a large scene such as the whole room. However it is not straightforward to incorporate dense point cloud into the BA framework, since there is no obvious association between points from different frames. Even though ICP could be used to find point associations, it is computational prohibitive considering the large amount of points. It is also unnecessary to find correspondences for all points due to the redundancy in the point cloud. We propose to use plane as compact representation of dense point clouds, and integrate the plane alignment term in the traditional BA formulation.

Plane constraints have been proved useful in other works for 3D reconstruction or 3D mapping. Sinha et al. [8] and Furukawa et al. [9] recovered a piecewise planar representation of the scene during the stereo procedure; Gallup et al. [10] detected the planar surface based on the trained information to improve the results of stereo for urban scene reconstruction. Lee et al. [11] used the constraints from coplanar feature points for visual SLAM. Different from these works, we use planes extracted from relatively accurate depth maps, which serve as an independent piece of information from features in RGB images. Pathak et al. [12][13] also extracted planes from a depth camera and performed plane matching to register frames. However, without the help of salient feature points, planes alone can not determine transformations between frames that contain only simple geometries such as walls. Additionally, they did not exploit the plane alignment constraint for the global error mitigation.

Our work also relates to the research on SfM and Bundle Adjustment ([14][6] [7][15] among others), again, the novelty of our method is that we perform bundle adjustment on both image features and planes.

2   Overview of the System

To build a complete model of a room, we capture a sequence of RGB-D images of the room. The intrinsics of the RGB-D camera is pre-calibrated, thus we only need to estimate the camera poses (extrinsics, namely camera rotations and translations) to reconstruct the room. The main steps of the system are shown in Fig. 2.

For every frame in the sequence, we extract salient image features from RGB color image. SIFT features are used in this paper. We also fit planar surface patches from the depth maps, which is introduced in detail in Section 3. The corresponding 3D coordinate for each image feature point is calculated under its camera coordinate system, given that the depth value at its image location is known. We call these 3D points originating from image features simply as "features" or "points" in order to differentiate them from planes extracted from depth maps. We remove features that locate at discontinuity regions in the depth map, because their depth values are typically inaccurate.

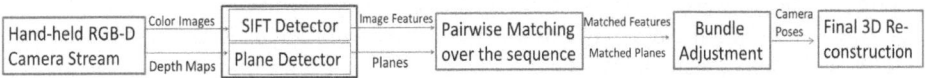

Fig. 2. The flow chart of our 3d reconstruction system

Next, we perform a pairwise matching over the data sequence to find the matched frames. Robust pairwise matching results are achieved by using both planar surfaces and features. For each matched frame pair, the matched features and planes are found, together with their underlying geometric transformation. We will introduce our modified RANSAC matching algorithm that handles both points and planes matches in Section 4. To save computation time, we do not test any two frames for match in the whole data sequence. Instead, similar to [1], the time coherence of the sequence is explored and some key frames are automatically selected during the matching procedure. When a new frame comes, it is only matched against a few neighboring frames before it and all previous key frames.

Finally, we run Bundle Adjustment algorithm on all the matched features and planes from all pairs of the matched frames to obtain their globally optimal camera poses. Our bundle adjustment algorithm is carried out directly in 3D space instead of 2D image space thanks to the 3D location information computed from the depth channel. Different from traditional bundle adjustment, planes provide strong constraints for camera pose estimation in indoor environment. This extended bundle adjustment algorithm is introduced in Section 5. We evaluate our algorithm in Section 6 qualitatively and quantitatively. Our proposed system with plane constraints achieves significant improvement over state-of-the-art reconstruction algorithms in real world indoor datasets.

3   Planar Surface Extraction from Depth Map

Given the camera intrinsics, a depth map can be transformed to a point cloud, and vice versa. Therefore we use these two terms interchangeably. In the literature of plane

| (a)the depth map | (b)the extracted planes in 2D | (c) the plane segments in 3D |

Fig. 3. Planar surface extraction results via voting in plane parameter space

extraction from point cloud, region growing [16] and voting [17] are two popular techniques. Region growing algorithms scatter some initial seeds over the image, and gradually merge neighboring pixels belonging to the same plane. The voting algorithms transform all points into plane parameter space. The peaks in the parameter space correspond to the consensus planes in the original space. We implement a plane voting algorithm. Note that the plane is extracted under the individual camera coordinate system. We represent a plane in 3D space by $\mathbf{n}^T\mathbf{x} - d = 0$, where $\mathbf{n}$ is the plane normal—a 3D unit vector; d is the distance of the plane to the camera optical center. We force the plane normal to point away from the origin, giving us a positive d.

A local plane is fitted for each pixel in the depth map using its neighboring pixels. Then these local planes vote in the plane parameter space to find consensus planes. After finding the peaks in the plane parameter space, we recalculate the plane parameters by fitting planes from all the pixels voting for the same peak. Finally, we assign each pixel to one of the detected planes, or as a non-plane if the distance to all planes is too large. Plane parameters are refined again from their associated pixels. In addition, the convex hull $\{\mathbf{v}_i\}_{i=1}^K$ of a plane segment is found to indicate its boundary. Fig. 3 shows the extracted planes in 2D and 3D.

4   Robust Pair-Wise Matching

Given the detected planes and features for two frames, next we need to find the matched features and matched planes if any, together with the geometric transformation that aligns these two frames. The transformation is represented by a rotation matrix R and a translation vector T, satisfying $X^r = RX^l + T$, where X^l and X^r are 3D points under two camera coordinate systems (left and right).

To reduce the searching space of matched planes and features, the initial feature and plane match set are found and then refined. The initial feature match set is computed by checking the similarity of SIFT descriptors. It is much more difficulty to find the initial plane match set, since the appearance of the same plane in two frames might be dramatically different when each frame contains different parts of the same plane. Fortunately, the relative angles between planes are constant even when viewed from different positions, which is a useful clue to find the plane match set, although some ambiguities still exist. Hence, instead of finding one initial plane match set, multiple plane matching

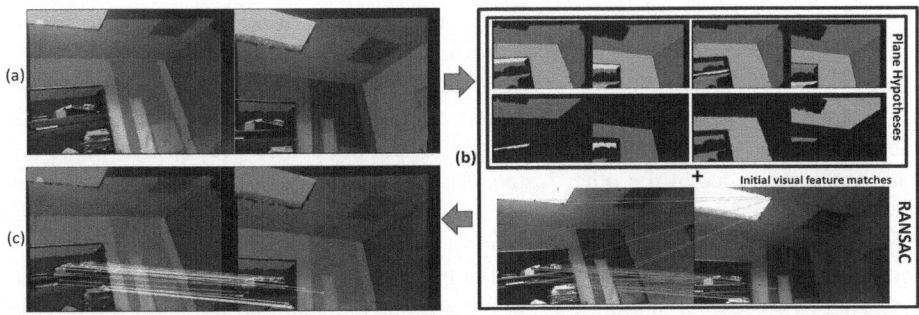

Fig. 4. Robust Pairwise Matching of Both Planes and Features. (a) two input frames and detected planes. (b) four plane matching hypotheses are shown on top (same color indicates a plane match), and the initial feature matches are shown on bottom. When searching for plane hypotheses, only plane appearance and angles between planes are used, resulting in some plane mismatches. RANSAC is used to find the underlying geometrical transformation between two frames and the matching features and matching planes as well, as shown in (c).

hypotheses are found due to the aforementioned ambiguities on plane correspondence. Each plane matching hypothesis consists of a set of tentatively matching planes.

RANSAC [18][19] is a powerful technique to refine the match set and find the underlying transformation from an initial match set. Since we have two different kinds of match sets, the conventional RANSAC is extended to take both the feature match set and plane match set as input and output the refined matching pairs. Since there are multiple plane matching hypotheses and RANSAC takes only one plane match set as input each time, we run RANSAC multiple times on every plane matching hypothesis together with initial feature match set. Within multiple RANSAC-refined match sets, we pick the one that collects most supports from the rest pairs in the match set. The extended RANSAC algorithm is discussed in detail in Sec. 4.2. An example of the pairwise matching is shown in Fig. 4.

4.1 Plane Matching Hypothesis

Relative plane angle and plane appearance similarity are used as plane matching criterion. Even though the viewing location and orientation of the moving camera are different when capturing any two frames, the relative angles between the observed planes of the static scene are constant. Moreover, the appearance of a plane in the RGB channel in one frame should be somewhat similar to its corresponding plane in the other frame. For a plane segment in one frame, a joint histogram of hue-saturation $\mathbf{h}^{HS}$ is calculated from RGB channel to represent its color information; an intensity histogram $\mathbf{h}^{I}$ is used to represent its texture information. The appearance similarity of the two planes is defined as the overlapping area of their histograms, i.e., $\sum_i \min(\mathbf{h}_1^{HS}(i), \mathbf{h}_2^{HS}(i)) + \sum_i \min(\mathbf{h}_1^{I}(i), \mathbf{h}_2^{I}(i))$. Based on these two clues—relative angles and appearance—we designed an algorithm to find a set of plane matching hypotheses.

Finding two plane subsets with same relative plane angles is equivalent to finding a rotation matrix that rotates the plane normals in one set to the plane normals in the other set. Theoretically, a brute force search on the rotation matrix space is possible. That is, we apply all possible rotation matrices on the planes of one frame, and find a subset of planes in this frame that have corresponding planes with similar plane normals and appearance in the other frame. We call a subset of planes in one frame and the matching planes in the other frame as a **plane matching hypothesis**. Generally, more than one reasonable plane matching hypothesis would be found. For instance, as shown in Fig. 4(a), three planes are mutually perpendicular, yielding totally six plane matching hypotheses assuming all three planes have the similar appearance. To eliminate some hypotheses, we constrain the rotation angle within a threshold, given the practical assumption that two nearby frames should not rotate too much.

However, the above brute force algorithm is intractable, since it searches a 3D space for the rotation matrix. Similar to [13], we take advantage of the fact that a rotation matrix can be decomposed into a rotation axis and a rotation angle about the axis. We repetitively pick one pair of planes $\langle \mathcal{P}_i, \mathcal{P}_j \rangle$ that have similar appearance from two frames, and find a rotation matrix R that rotates normal vector $\mathbf{n}_i$ to $\mathbf{n}_j$, i.e., $R\mathbf{n}_i = \mathbf{n}_j$. After applying R on all the planes in the first frame, $\mathcal{P}_i$ and $\mathcal{P}_j$'s normals are aligned. Most likely by now only $\mathcal{P}_i$ and $\mathcal{P}_j$ are aligned, but not any other planes in the two views. Next, we rotate the planes of the first frame about axis $R\mathbf{n}_i$ to see if some other planes have matching planes in the second frame. In this way, we search along only one dimension—the rotation angle about axis $R\mathbf{n}_i$.

4.2   Run RANSAC on one Plane Matching Hypothesis and the Feature Match Set

The initial feature match set might contain spurious pairs, as might a plane matching hypothesis. We extend standard RANSAC method to find the underlying geometric transformation between two frames. When a feature match or plane match fits a candidate transformation, it supports this transformation (details are given later). Our RANSAC algorithm returns the transformation that collects most supports from the putative feature matching set and plane matching hypothesis.

Randomly Sample Matched Pairs. To perform RANSAC, we need to randomly sample a minimum number of matched pairs from initial match set to determine a rigid transformation candidate. We have two kinds of matched pairs–matched 3D features and matched planes. As elaborated in [12], three planes with linearly independent normal vectors uniquely determine a transformation; similarly three non-collinear features uniquely determine a transformation. Additionally, two nonparallel planes and a feature also uniquely determine a transformation, as do two features and a plane with a different normal direction than the vector connecting the two feature points. Our randomly sampled match subsets are evenly distributed over the above four cases—three planes, three features, two planes with one feature, and two features with one plane.

Calculating the Transformation from Pairs of Matches. Inside each loop of RANSAC, the transformation needs to be calculated from the randomly sampled matches (the seeds). We consider a general case, where there are n pairs of matched planes $\mathcal{S} = \{\langle \mathcal{P}_i^l, \mathcal{P}_j^r \rangle\}$,

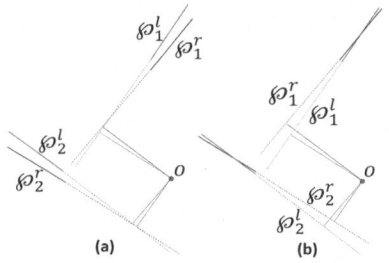

Fig. 5. The closeness on planes parameters shown in (a) does not equal to the closeness of plane segments shown in (b). In (b), even though each pair of matched planes have much larger difference on the origin-to-plane distance d, they are better aligned comparing with (a). The solid line segments denote the plane segments, and o is the origin of the world coordinate.

and m pairs of matched features $\mathcal{T} = \{\langle \mathbf{f}_i^l, \mathbf{f}_j^r \rangle\}$. A transformation $\langle R, T \rangle$ is estimated such that when applying it to the planes and features from one frame, the overall distance between matched items should be minimized. That is,

$$\min \sum_{\langle \mathcal{P}_i^l, \mathcal{P}_j^r \rangle \in \mathcal{S}} D_{pln}^2(Q(R, T, \mathcal{P}_i^l), \mathcal{P}_j^r) + \sum_{\langle \mathbf{f}_i^l, \mathbf{f}_j^r \rangle \in \mathcal{T}} D_{pt}^2(R\mathbf{f}_i^l + T, \mathbf{f}_j^r). \tag{1}$$

where $Q(\cdot)$ applies transform $\langle R, T \rangle$ to $\mathcal{P}_i^l$, while $D_{pln}(\cdot)$ and $D_{pt}(\cdot)$ are distance functions for planes and features respectively. The distance between features is simply defined as Euclidean distance. However, the Euclidean distance of plane parameters $\langle \mathbf{n}, d \rangle$ is not a good measurement of the plane distance as shown in Fig. 5, mainly because the planes in our case actually mean "plane segments" which have boundaries.

Distance between Plane Segments. Instead of measuring Euclidean distance between plane parameters, we measure the distances from the boundary points (convex hull) of one plane segment to its matched plane. Ideally, we could use the sum of the point-to-plane distances from all points in one plane segment to the other plane, since these points are the direct observations of a plane. However it is computationally more expensive given the number of points in a plane segment. In addition, the convex-hull-to-plane distance is the upper-bound of the sum of point-to-plane distances[1].

Let's assume we are measuring the distance between two plane segments $\{\mathcal{P}_i^l, \mathcal{P}_j^r\}$ under the transformation $\langle R, T \rangle$. Their plane parameters are $\langle \mathbf{n}_i^l, d_i^l \rangle$ and $\langle \mathbf{n}_j^r, d_j^r \rangle$ in their individual camera coordinate systems, and the vertices on the convex hulls of two plane segments are $\{\mathbf{v}_{i,k}^l\}_{k=1}^{K_1}$ and $\{\mathbf{v}_{j,k}^r\}_{k=1}^{K_2}$. The upper-index l and r indicates

---

[1] Each observed point p assigned to a plane could be represented by a linear combination of the convex hull vertices $\{v_i\}$, i.e., $p = \sum_i c_i v_i + \epsilon$; where $\sum_i c_i = 1$ and ϵ accounts for the fact that the point p might not lie exactly on the fitted plane. Since the point-to-plane distance is a convex function, the weighted sum of convex-hull-vertex-to-plane distances is statistically the upper-bound of the distance from an observed point to the world plane. Taking all the points associated with one plane into consideration, the sum of squared point-to-plane distances (upper-bound) is $c \sum_i w_i * \mathcal{D}^2(v_i)$, where c is the number of the points; $\mathcal{D}(\cdot)$ measures point-to-plane distance; and w_i is the overall weight on a vertex satisfying $\sum w_i = 1$.

which frame the data belong to, and we will ignore them whenever it does not cause confusion. As shown in [12], after applying the transformation $\langle R, T \rangle$ to plane $\mathcal{P}_i^l$, its plane parameter will be $\langle R\mathbf{n}_i, \ d_i + \mathbf{n}_i^T R^T T \rangle$, and its convex hull vertices will be $\{R\mathbf{v}_{i,k} + T\}_{k=1}^{K_1}$. Then the distance between plane segments in Eq. 1 is defined as,

$$
\begin{aligned}
D_{pln}^2 = &\sum_{i=1}^{K_1} w_{i,k} \| \mathbf{n}_j^T R\mathbf{v}_{i,k} + \mathbf{n}_j^T T - d_j \|_2^2 + \\
&\sum_{i=1}^{K_2} w_{j,k} \| \mathbf{n}_i^T R^T \mathbf{v}_{j,k} - \mathbf{n}_i^T R^T T - d_i \|_2^2.
\end{aligned}
\tag{2}
$$

The first item in the right side of the equation is the perpendicular distance between the transformed vertices on the convex hull of the first plane and the second plane, while the second term is the distance between the vertices of the second plane and the transformed first plane. Each point-to-plane distance is given a weight $w_{i,k}$, satisfying $\sum_k w_{i,k} = 1$, because vertices on the convex hull are not a uniform sample of the plane boundary. Larger weights are given to vertices further away from their neighboring vertices along the convex hull.

Find the Supporting Pairs and Choose the Best Transformation. For each transformation candidate calculated from the randomly selected matched pairs, we count how many other matching pairs fit this transformation. It is straightforward to use Euclidean distance when testing a pair of features and use the distance defined in Eq. 2 to measure the distance between planes after transformation. We also check if two plane segments overlap after transforming one plane's segment to the other plane's image space. Only when a pair of planes are close enough and have overlap under one transformation, we say they fit the transformation.

5 Extended Bundle Adjustment of Feature Points and Planes

After performing the above robust pairwise matching algorithm on the frames of the depth map and RGB image sequence, we have a large number of feature/plane match sets. Given all these pairwise match sets, matching features/planes over the whole sequence are linked together. A set of linked features $\{\mathbf{f}_k^i\}_{i \in \mathcal{C}_k}$ is called a **feature track**, corresponding to the same 3D point $\mathbf{p}_k$ in a world coordinate system. The notation $\mathbf{f}_k^i$ represents a 3D feature point in the k-th visual feature track from the i-th frame, and $\mathcal{C}_k$ is the set of frame indices where all the indexed frames have feature corresponding to the 3D world point $\mathbf{p}_k$. Similarly from the plane matching sets, a number of **plane tracks** can be found. Each plane track is composed of a set of linked planes $\{\mathcal{P}_j^i\}_{i \in \mathcal{D}_j}$ from various frames corresponding to the same world plane $\mathcal{Q}_j$. The notation $\mathcal{P}_j^i$ represents a plane in the j-th plane track and extracted from i-th frame, while $\mathcal{D}_j$ is the set of frame indices where all the indexed frames have the extracted planes corresponding to the world plane $\mathcal{Q}_j$.

Problem Statement. After the pairwise matching on a sequence, we have M plane tracks $\{\{\mathcal{P}_j^i\}_{i \in \mathcal{D}_j}\}_{j=1}^M$, and K feature tracks $\{\{\mathbf{f}_k^i\}_{i \in \mathcal{C}_k}\}_{k=1}^K$. The planes and features inside the tracks are represented under their own camera spaces. The unknowns in our problem are the camera poses $\{R_i, T_i\}_{i=1}^N$ for N frames in the sequence, the plane parameters $\{\mathbf{n}_j, d_j\}_{j=1}^M$ for M planes $\{\mathcal{Q}_j\}$ in the world, and K point locations $\{\mathbf{p}_k\}_{k=1}^K$ in the world. The camera pose of the i-th frame is represented by rotation matrix R_i and a 3D translation vector T_i, which transform a point X_{wld} in world space to the i-th camera space coordinate X_i via $X_i = R_i X_{wld} + T_i$.

Cost Function. Based on the extracted feature and plane tracks, we adjust simultaneously the camera poses and the parameters of the world planes and world points to make sure that the desired world planes and world points are as close as possible to the planes and points detected in each frame. The cost function to be minimized is:

$$\frac{c}{N_{pln}} \sum_{\{i,j|i \in \mathcal{D}_j\}} c_j^i D_{pln}^2 \left(Q(R_i, T_i, \mathcal{Q}_j), \mathcal{P}_j^i\right) + \\ \frac{1-c}{N_{pt}} \sum_{\{i,k|i \in \mathcal{C}_k\}} D_{pt}^2 \left(Q(R_i, T_i, \mathbf{p}_k), \mathbf{f}_k^i\right), \tag{3}$$

where $D_{pln}(\cdot)$ measures the distance between a detected plane and the world plane, while $D_{pt}(\cdot)$ measures the distance between an observed feature and a world point. $Q(\cdot)$ transforms a point or a plane in the world space to a certain camera space given the camera pose. Constant c weights the effects of plane tracks against feature tracks; N_{pln} and N_{pt} are the number of planes in all plane tracks and the number of points in all feature tracks respectively. c_j^i is the weight on the plane in the plane track and equals to the number of associated pixels in a plane divided by the average number of pixels among planes in all plane tracks.

Again we use the Euclidean distance to measure D_{pt}, i.e., $D_{pt}^2\left(Q(R_i, T_i, \mathbf{p}_k), \mathbf{f}_k^i\right) = \|R_i\mathbf{p}_k + T_i - \mathbf{f}_k^i\|_2^2$. For distance between a detected plane and a world plane, as in Eq. 2, we measure the convex-hull-to-plane distance, i.e., $D_{pln}^2\left(Q(R_i, T_i, \mathcal{Q}_j), \mathcal{P}_j^i\right) = \sum_h w_{j,h}^i \|\mathbf{n}_j^T R_i^T \mathbf{v}_{j,h}^i - \mathbf{n}_j^T R_i^T T_i - d_j\|_2^2$, where $\{\mathbf{v}_{j,h}^i\}$ are the vertices on the convex hull of the plane $\mathcal{P}_j^i$, and $\langle \mathbf{n}_j, d_j \rangle$ are the plane parameters of $\mathcal{Q}_j$. Putting everything together, we have the cost function,

$$\frac{c}{N_{pln}} \sum_{\{i,j,h|i \in \mathcal{D}_j\}} c_j^i w_{j,h}^i \|\mathbf{n}_j^T R_i^T \mathbf{v}_{j,h}^i - \mathbf{n}_j^T R_i^T T_i - d_j\|_2^2 + \\ \frac{1-c}{N_{pt}} \sum_{\{i,k|i \in \mathcal{C}_k\}} \|R_i\mathbf{p}_k + T_i - \mathbf{f}_k^i\|_2^2. \tag{4}$$

Note that the camera pose for the very first camera is fixed at the origin with an identity rotation matrix in the above function.

Statistically speaking, the noise on observed points comes from variant independent factors, for example, errors from the 2D SIFT detector, depth map, camera calibration, thus we assume it is normally distributed based on Central Limit Theory. Hence, L2 norm is used to measure the error of an observed feature in the above cost function. As to the measurement of the error on a plane segment, we use the convex-hull-to-plane distance as a compromise since it is computationally prohibitive to use the sum of the point-to-plane distances given the number of points. As stated earlier, this convex-hull-to-plane distance is statistically the upper-bound of the error of the points on the plane segment. Since this error (or noise) also comes from various independent factors, we assume it is normally distributed and L2 norm is used.

Initialization and Optimization. A general sparse Levenberg-Marquardt solver [20] is used to minimize the cost function defined in Eq. 4. Since there are only three degrees of freedom in a rotation matrix and two degrees of freedom in a normal vector, the Rodriguez representation and spherical coordinates are used respectively when performing the optimization. Although there are three sets of unknowns, only the camera poses need to be initialized. The initial parameters for the world planes and world points are estimated from their corresponding tracks with the given camera poses. The camera poses are initialized one by one with the pairwise matching results.

Plane Track Refinement. Since we do not match every possible frame pair, some large planes tend to have several disjoint plane tracks instead of one complete track. These plane tracks need to be merged together. We compute the plane distance between estimated planes in the world space returned by bundle adjustment according to Eq. 2; if some of them are closely located, we merge them into one plane and also merge their plane tracks accordingly. Moreover, we delete a detected plane from its plane track if its distance to the corresponding world plane is beyond a threshold d. If more than half the planes of one plane track do not fit its estimated world plane, we delete the entire plane track. After refining the plane tracks, we rerun the bundle adjustment algorithm, and repeat the above procedure several times until the plane tracks no longer change.

6   Experiments

We evaluate our algorithm on four datasets of indoor office settings captured in real-world environment: SN353, SN277, FB220 and LAB. All the datasets except "LAB" have considerably fewer visual features. Some statistics on the datasets when running our algorithm are shown on Table 1. In each data set, 200 to 500 frames of RGB-D data are captured with significant overlaps (please check the supplemental video for what the datasets look like). Among four datasets, "FB220" is the most challenging one, since the number of detected feature is dramatically fewer than others. On average, there are only around 10 matched points per matched frame pair (the number "46" shown under

Table 1. Statistics on four datasets. Under column "#frames", the number of frames in the sequence is recorded before the slash, while the number of frames registered with others is shown after the slash. The third and fourth column give the average number of all detected features and matched features respectively. Column "#plane tracks" gives the number of the detected plane tracks provided to BA and also the number of planes tracks after refinement in BA. Column "#planes in tracks" gives the total number of planes in all tracks before and after BA.

dataset	#frames	#features per frame	#matched features	#feature tracks	#plane tracks	#planes in tracks
SN353	228/191	328	143	6066	83/52	1179/916
SN277	381/350	497	129	10695	155/72	1896/1520
FB220	360/280	137	46	3357	138/64	1516/1232
Lab	180/180	694	267	10636	85/46	1236/858

Table 2. The quantitative measurements of our algorithm. Inside each cell, the average error and the standard deviation are provided.

dataset	point proj error(cm)	plane proj error(cm)	zero angles(°)	right angles(°)
SN353	1.46± 1.37	1.74± 2.03	-0.60± 2.95	89.97± 2.14
SN277	2.01± 1.83	1.68± 1.58	-0.22± 0.80	89.94± 1.70
FB220	2.38± 1.63	2.36± 2.15	-1.17± 2.41	89.96± 2.42
Lab	1.91± 1.77	1.91± 1.66	-0.59± 4.27	89.61± 2.11

the forth column for "FB220" is the average number of the features in one frame that have matches in all other frames), which can hardly lead to robust registration (camera pose transformation between two frames) with traditional SfM methods. As shown in the second column of Table 1, some frames are not registered with others and thus abandoned. This is because: (1)a frame is only matched with key frames and some adjacent frames; (2)the white balance and gain of the RGB channel on Kinect is in "auto" mode and cannot be disabled with current drivers, so the appearance of objects (mainly walls and ceilings) across frames changes dramatically, especially when pointing the camera towards a light (which explains why some ceiling parts in the following results disappear), leading to missing matches between frames.

The plane/feature trade-off coefficient c in Eq. 4 is set to 0.1 empirically for all the experiments presented here. The registered point clouds for different rooms using our algorithm are shown in Fig. 1, along with the planes delivered by BA. The point clouds shown here come directly from the depth camera and are not further processed to reduce the noise, while a volumetric method such as [2] could be used to fuse all the depth map together given the camera poses.

Comparison to Structure from Motion (SfM) Algorithms. To compare our method with state-of-the-art SfM algorithms, we ran Bundler [5][6] on all datasets. Since camera poses from SfM are determined up to a scale, we need to the find this scale compared to the one used by the depth camera. SfM outputs a sparse point cloud which can be projected to image space to extract the depth values from the depth cameras. Hence by comparing the depth values from the depth camera and those from Bundler, we have the relative scale.

Not surprisingly, Bundler fails on "FB220", giving camera poses for only three frames out of 360 frames. Bundler does give results for "SN353" and "SN277", but there are dramatic misalignments between frames as shown in Fig. 6(a), while Bundler gives visually almost perfect results for dataset "Lab" with rich features. Clearly Bundler does not work well on dataset with relatively few features, and fails when features are very sparse. Another observation is that the frames in "Lab" have significant depth variance than frames in other three scenes. For example, in dataset "FB220" many frames only capture the side of wall, which is the degenerated case for Bundler and thus results in inaccurate camera poses.

Comparison to ICP Method. We compared our system with RGBD-ICP algorithm [1]. Since RGBD-ICP also uses visual features to constrain the planes from drifting along the plane surface direction arbitrarily, it achieved better result than Bundler. But our

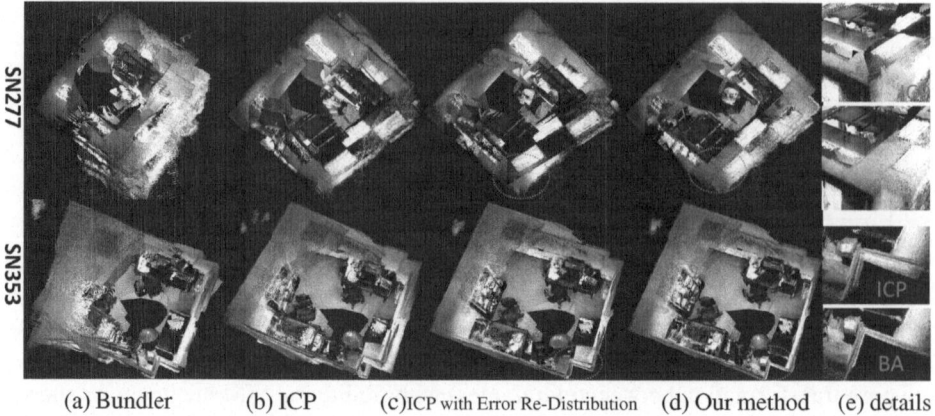

(a) Bundler   (b) ICP   (c)ICP with Error Re-Distribution   (d) Our method   (e) details

Fig. 6. Comparison of our algorithm with other methods. The top-down view of the room is shown. (e) gives detailed comparison of ICP with error distribution and our BA algorithm at circled parts in (c) and (d).

algorithm comfortably outperformed RGBD-ICP algorithm as shown in Fig. 6(b), since the error accumulation problem is not handled in RGBD-ICP, while we address that with our new bundle adjustment formulation. We also compared our system with KinectFusion[2] which uses only the depth information to align one frame with previously accumulated data. And as expected, KinectFusion did not result in a reasonable output on our dataset either, due to its lack of constraints to handle the drifting in some frames with simple geometry structure.Since our reconstruction input frames are temporally down-sampled from the original 30 fps hand-held Kinect streams, the adjacent frames normally have noticeable amount of camera pose differences. Hence, for RGBD-ICP and KinectFusion to work, a decent initial camera pose is required. The transformation returned by our robust pairwise matching algorithm introduced in Sec. 4 is used for camera pose initialization.

Comparison to ICP with Global Error Mitigation. As shown in [1], the accumulated error can be re-distributed globally. To fairly compare our algorithm to the ICP algorithm with error distribution, we use the same frame matching strategy, that is, to match one frame against all previous key frames and its adjacent frames. After collecting all the pairwise transformations $\{\langle R_{j \leftarrow i}, T_{j \leftarrow i} \rangle | (i, j) \in \mathcal{E}\}$, as in [21], we distribute the error over the graph by minimizing $\sum_{(i,j) \in \mathcal{E}} \|R_{j \leftarrow i} - R_j R_i^T\|_2^2 + \|T_{j \leftarrow i} - T_j + R_{j \leftarrow i} T_i\|_2^2$. As shown in Fig. 6(c), after distributing the accumulated error, ICP gives decent results. However, some details are not preserved as well as using our method. For example, the ceiling in "SN277" is distorted as shown in Fig. 6(c). Note that both ICP with and without error re-distribution have to use our robust pairwise matching algorithm with planes and features for initial registration, otherwise both ICP algorithms would fail frequently in texture-less regions.

Quantitative Measurement of Errors. To evaluate our method quantitatively, we measure the projection error of feature points and planes in all datasets. The estimated 3D world points and world planes are projected to the camera coordinate system at each frame to compare them with extracted planes and features. Additionally, the relative angles between some planes in the room, such as walls, ceilings and floors, are known (zero angle or right angle), and these angles serve as the ground truth for the measured angles between the world planes delivered by our system. All the measurements are listed in Table 2. Although the error on the dataset "FB220" is slightly bigger than others, the projection error is considerably small and the measured angles are fairly close to the ground truth.

Running Times. Our single thread program takes 2.5 to 3 seconds to extract planes from one frame, another 1.5 seconds to extract SIFT features on a desktop PC with 3.0 CPU Hz. Depending on how many planes and features are in the frames, it takes up to a few seconds to finish one pairwise matching. In our four datasets, the pairwise matching on a whole sequence takes two to five hours. We do not perform incremental bundle adjustment, and instead we perform BA on all frames directly. Generally it takes less than fifty iterations to converge with the default parameters of the chosen Levenberg-Marquardt solver [20]. We run bundle adjustment and the plane track refinement repeatedly. The whole BA procedure takes 5 to 20 minutes on a dataset. We expect significant computation speed acceleration with optimized code or on parallel processing unit such as GPU. This remains one of the future directions of this research work.

7 Conclusions

In this paper, we present a complete pipeline of indoor 3D reconstruction with a handheld RGB-D camera. Given indoor settings, we explore the commonly available high level plane constraint to achieve better reconstruction quality. By combining both low level feature correspondences and high level plane primitive, we significantly improves the reconstruction result in challenging cases with low-texture or low-geometry information. More specifically, we demonstrate to use the plane primitive in robust pairwise matching even when few salient feature points are detected or they are not well distributed. This compact representation of dense points for planar parts of the scene also help us incorporate these constraints into the BA framework to globally mitigate the error. Real world data sets show that our method significantly improves the reconstruction quality over state-of-the-art scene reconstruction methods and the measured error is very small comparing to the ground truth.

Acknowledgement. This work was supported in part by CISCO and by the BeingThere Centre, a collaboration of UNC Chapel Hill, ETH Zurich, NTU Singapore, and the Media Development Authority of Singapore.

References

[1] Henry, P., Krainin, M., Ren, X., Herbrt, E., Fox, D.: Rgb-d mapping: Using depth cameras for dense 3d modeling of indoor environments. ISER (2010)

[2] Newcombe, R.A., Izadi, S., Hilliges, O., Molyneaux, D., Kim, D., Davison, A.J., Kohli, P., Shotton, J., Hodges, S., Fitzgibbon, A.: Kinectfusion: Real-time dense surface mapping and tracking. In: IEEE ISMAR (2011)

[3] Neumann, D., Lugauer, F., Bauer, S., Wasza, J., Hornegger, J.: Real-time rgb-d mapping on the gpu using the random ball cover data structure. In: IEEE ICCV/CDC4CV (2011)

[4] Lieberknecht, S., Huber, A., Ilic, S., Benhimane, S.: Rgb-d camera-based parallel tracking and meshing. In: ISMAR (2011)

[5] Snavely, N., Seitz, S.M., Szeliski, R.: Modeling the world from internet photo collections. International Journal of Computer Vision (2007)

[6] Snavely, N., Seitz, S.M., Szeliski, R.: Photo tourism: Exploring image collections in 3d. ACM Transactions on Graphics (2006)

[7] Crandall, D., Owens, A., Snavely, N., Huttenlocher, D.P.: Discrete-continuous optimization for large-scale structure from motion. In: CVPR (2011)

[8] Sinha, S.N., Steedly, D., Szeliski, R.: Piecewise planar stereo for image-based rendering. In: ICCV (2009)

[9] Furukawa, Y., Curless, B., Seitz, S.M., Szeliski, R.: Manhattan-world stereo. In: CVPR (2009)

[10] Gallup, D., Frahm, J.-M.: Piecewise planar and non-planar stereo for urban scene reconstruction. In: CVPR (2010)

[11] Lee, G.H., Fraundorfer, F., Pollefeys, M.: Mav visual slam with plane constraint. In: IEEE Int. Conf. on Robotics and Automation (2011)

[12] Pathak, K., Birk, A., Vaskevicius, N., Poppinga, J.: Fast registration based on noisy planes with unknown correspondesces for 3-d mapping. IEEE Transactions on Robotics (2010)

[13] Pathak, K., Vaskevicius, N., Poppinga, J., Pfingsthorn, M., Schwertfeger, S., Birk, A.: Fast 3d mapping by matching planes extracted from range sensor point-clouds. IROS (2009)

[14] Pollefeys, M., Van Gool, L., Vergauwen, M., Verbiest, F., Cornelis, K., Tops, J., Koch, R.: Visual modeling with a hand-held camera. IJCV (2004)

[15] Steffen, R., Frahm, J.-M., Förstner, W.: Relative Bundle Adjustment Based on Trifocal Constraints. In: Kutulakos, K.N. (ed.) ECCV 2010 Workshops, Part II. LNCS, vol. 6554, pp. 282–295. Springer, Heidelberg (2012)

[16] Poppinga, J., Vaskevicius, N., Birk, A., Pathak, K.: Fast plane detection and polygonalization in noisy 3d range images. IROS (2008)

[17] Borrmann, D., Elseberg, J., Lingemann, K., Nuhter, A.: The 3d hough transform for plane detection in point clouds: A review and a new accumulator design. 3D Research 02, 1330–1334 (2011)

[18] Raguram, R., Frahm, J.-M., Pollefeys, M.: A Comparative Analysis of RANSAC Techniques Leading to Adaptive Real-Time Random Sample Consensus. In: Forsyth, D., Torr, P., Zisserman, A. (eds.) ECCV 2008, Part II. LNCS, vol. 5303, pp. 500–513. Springer, Heidelberg (2008)

[19] Fischler, M.A., Bolles, R.C.: Random sample consensus: A paradigm for model fitting with applications to image analysis and automated cartography. Comm. of the ACM 24 (1981)

[20] Lourakis, M.I.A.: Sparse Non-linear Least Squares Optimization for Geometric Vision. In: Daniilidis, K., Maragos, P., Paragios, N. (eds.) ECCV 2010, Part II. LNCS, vol. 6312, pp. 43–56. Springer, Heidelberg (2010)

[21] Grisetti, G., Grzonka, S., Stachniss, C., Pfaff, P., Burgard, W.: Efficient estimation of accurate maximum likelihood maps in 3d. IROS (2007)

A Fusion Framework of Stereo Vision and Kinect for High-Quality Dense Depth Maps

Yucheng Wang and Yunde Jia

Beijing Lab of Intelligent Information Technology, School of Computer Science,
Beijing Institute of Technology, Beijing 100081, P.R. China
{wangyucheng,jiayunde}@bit.edu.cn

Abstract. We present a fusion framework of stereo vision and Kinect for high-quality dense depth maps. The fusion problem is formulated as maximum a posteriori estimation of the Markov random field using the Bayes rule. We design a global energy function with a novel data term, which provides a reasonable, straight-forward and scalable way to fuse stereo vision and the depth data from Kinect. Particularly, visibility and pixelwise noises of the depth data from Kinect are taken into account in our fusion approach. Experimental results demonstrate effectiveness and accuracy of the proposed framework.

1 Introduction

Depth acquisition is one of the most fundamental and challenging problems in computer vision. It receives much attention in human computer interaction, 3D scene reconstruction, augmented reality and view synthesis [1,8,13]. The growing desire for depth sensing in these fields has promoted a widespread use of depth sensors.

Many kinds of depth sensors, such as stereo-based RGBD imager [5,9], Time-of-Flight (ToF) sensor, and Kinect, have the ability of capturing grayscale (color) images and corresponding per-pixel depth values of dynamic scenes simultaneously at a video rate. However, each depth acquisition method has its own limitation in some aspect where some other approaches may be effective. Passive stereo vision fails in textureless regions, repeated patterns, and occlusion, because there exist multiple local minima in disparity optimization where active methods excel. In contrast, active approaches like ToF sensor and Kinect, can deal with those problems by emitting signals [6], but they are low-resolution, noisy and become unreliable interfered by the ambient light and scene reflectance where passive stereo methods work well. ToF sensors have biases on object albedo which cause difficulties in depth sensing. Kinect provides a mechanism that refuses to estimate depth value of pixels with low reliability(as shown in Fig. 1(b)). Laser scanner is of high resolution and accuracy, but usually expensive, not mobile or too slow.

Hence, combining depth information from different kinds of methods undoubtedly makes the results more robust and with higher quality. Since the color images utilised in stereo vision are of higher resolution and better texture than

J.-I. Park and J. Kim (Eds.): ACCV 2012 Workshops, Part II, LNCS 7729, pp. 109–120, 2013.

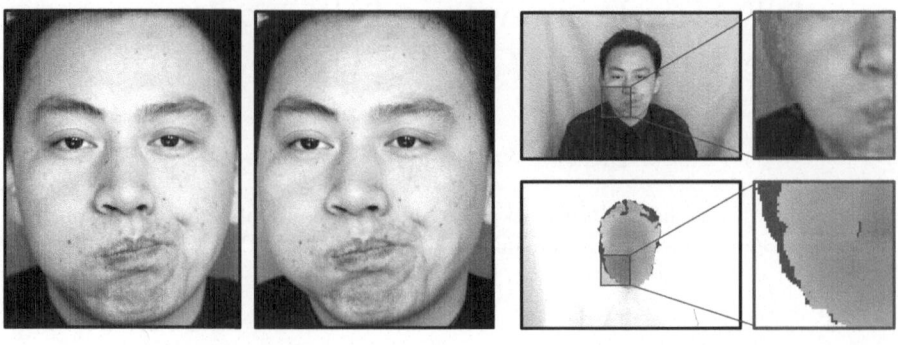

(a) The stereo pair   (b) Kinect: color image and depth map

Fig. 1. (a) the stereo pair with 1100×1400 pixels; (b) the Kinect can generate a color image along with a corresponding depth map of 640×480 pixels. The depth values of the pixels in red region are not available in the close-up image. This figure is best viewed in color.

that of the depth sensors like Kinect, it is reasonable to fuse the depth data from depth sensors with color cameras to produce corresponding high-resolution depth maps of images from color cameras. Kopf [12] *et al.* presented a bilateral up-sampling method which employs some pixels with known disparity to recover a corresponding disparity map of a high-resolution guidance image via bilateral filter technique. Wang [16] *et al.* utilized sparse ground control points with depth values obtained in different ways in stereo vision framework for high-quality depth acquisition.

In this paper, we propose a framework to fuse stereo vision and depth sensor for 3D sensing of a scene. The fusion problem is formulated as Maximum A Posteriori (MAP) estimation of the Markov Random Field (MRF) which integrates the geometrical information from stereo vision and Kinect naturally. Compared with existing works, the per-pixel depth estimates from sensors are considered independent from each other and modeled as real depth values adding pixelwise noises which follow Gaussian distributions around the true values in our method.

Thus, a novel data term and a conventional smoothness term are defined in a global energy function. We employ the graph cuts named α-expansion [4] to minimize the energy function. With visibility constraint, our fusion method can produce robust results when the depth values of some 3D points in the scene are not visible in the field of view(FoV) of Kinect. Moreover, the proposed framework is scalable for incorporating multiple depth sensors.

The main contributions of this paper include: (1) a novel data term, which provides a reasonable, straight-forward and scalable way to combine stereo vision, the depth data from depth sensors, and the pixelwise noises of the depth data; (2) a fusion framework of Kinect and stereo vision with visibility constraint.

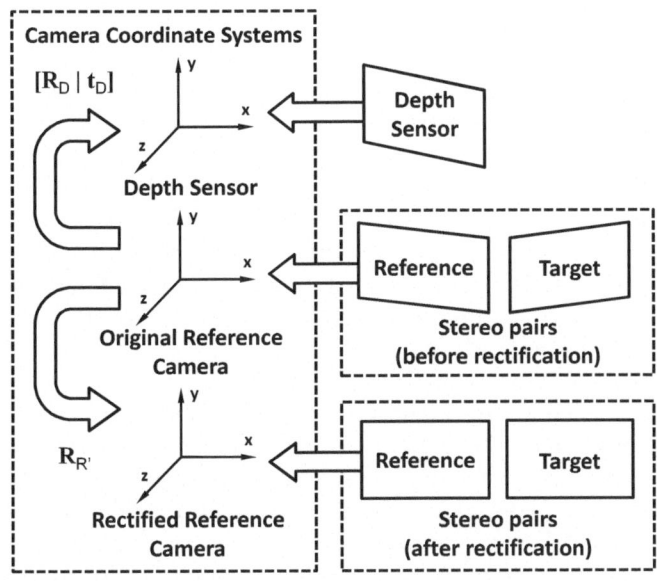

Fig. 2. Geometrical model

2   Camera Calibration

We denote the depth sensor Kinect and the reference camera as D and R , respectively. There are three camera coordinate systems involved in the model: camera coordinate system of the depth sensor, camera coordinate systems of the reference camera before and after epiploar rectification. The coordinate of a 3D point $\mathbf{M}$ in these camera coordinate systems are denoted as $\mathbf{M}_D$, $\mathbf{M}_R$ and $\mathbf{M}_{R'}$, respectively.

The intrinsic parameters of the depth sensor and the reference camera after the epipolar rectification are denoted as $\mathbf{K}_D$ and $\mathbf{K}_{R'}$, meanwhile the extrinsic parameter to transfer between original camera coordinate systems of the reference camera and the depth sensor is denoted as $[\mathbf{R}_D|\mathbf{t}_D]$ and the affine transformation of the reference image for rectification is denoted as $\mathbf{R}_R$. These parameters can be estimated only using the grayscale (color) images generated by the depth sensor and reference camera with the standard calibration techniques [3,7]. Particularly, radial distortion and tangential distortion of these sensors has been estimated and removed.

After calibration, we can obtain the correspondence of different camera and image coordinate systems (as shown in Fig.2). Let $I_{R'}$ be set of pixels in reference image, given a pixel $p \in I_{R'}$ with its image homogeneous coordinate $\widetilde{\mathbf{m}}_{R'} = [x_{R'}, y_{R'}, 1]^T$ and a disparity value d_p, we can calculate its corresponding coordinate $\mathbf{M}_{R'} = [X_{R'}, Y_{R'}, Z_{R'}]^T$, $\mathbf{M}_D = [X_D, Y_D, Z_D]^T$, and $\widetilde{\mathbf{m}}_D = [x_D, y_D, 1]^T$, and vice versa.

3   Problem Formulation

The corresponding disparity map of image is a grid of disparity pixels, where each disparity pixel implicitly encodes the depth to a 3D scene point along the measurement ray through the image pixel. The depth map has a similar definition but it explicitly records the depth values. Suppose we have a stereo image pair $I = \{I_{\mathsf{R}'}, I_{\mathsf{T}'}\}$, where $I_{\mathsf{R}'}$ and $I_{\mathsf{T}'}$ are the rectified reference and target images, and a depth map $\widehat{Z}_{\mathsf{D}}$ estimated by the depth sensor. Since the resolution of the color image from a CCD camera is much higher than that from a depth sensor, our goal of fusion is to compute the dense disparity map $D_{\mathsf{R}'}$ of the reference images for a high-resolution result.

The spatial smoothness constraint is most widely used prior in the stereo literature, which encourages neighboring pixels to have similar disparities based on the assumption that the scene is locally smooth. We employ the MRF in the fusion framework because the Markov property is very suitable for modeling hypothetically piecewise-smooth disparity map. It provides a natural way to integrate the geometrical information from other depth sensors with stereo vision in the disparity optimization. Among the energy minimization approaches, some new algorithms such as loopy belief propagation and graph cuts are presented and proved to be powerful [15] in the last decade. Thus we formulate the fusion problem as MAP estimation of the MRF and solve it with graph cuts [4].

Using the Bayes rule, the posterior probability over $D_{\mathsf{R}'}$ given I and $\widehat{Z}_{\mathsf{D}}$ can be computed by $P(D_{\mathsf{R}'}|I, \widehat{Z}_{\mathsf{D}}) \propto P(I|D_{\mathsf{R}'})P(\widehat{Z}_{\mathsf{D}}|D_{\mathsf{R}'})P(D_{\mathsf{R}'})$. The optimal disparities $D_{\mathsf{R}'}$ are obtained by finding the MAP estimate which is equivalent to a global energy function of the form:

$$
\begin{aligned}
E(D_{\mathsf{R}'}) &= -\left(\log(P(I|D_{\mathsf{R}'})) + \log(P(\widehat{Z}_{\mathsf{D}}|D_{\mathsf{R}'}))\right) - \log(P(D_{\mathsf{R}'})) \\
&= E_d + \lambda_s \cdot E_s
\end{aligned}
\tag{1}
$$

with $\lambda_s \geq 0$. When λ_s equals 0, the computation turns to be finding the local optimal without smoothness prior. E_d is the data term from both stereo and depth sensor which is different from the data term given in literature [15], and E_s is smoothness terms. λ_s controls the rate of date term and smoothness term in the energy function.

3.1   Data Term

In our MAP-MRF fusion framework, the data term E_d is given by

$$
E_d = (1 - \alpha_{dd}) \cdot E_{ds} + \alpha_{dd} \cdot E_{dd}
\tag{2}
$$

with $1 \geq \alpha_{dd} \geq 0$. E_{ds} and E_{dd} are the components of data term from stereo vision and Kinect respectively. α_{dd} controls the rate of geometrical information from stereo and depth sensor in the data term, is unfixed and determined using the method in literature [17].

Stereo Vision Component. E_{ds} comes from stereo vision and measures how well the disparity map $D_{\mathsf{R}'}$ agrees with the input rectified stereo image pair I. We define the stereo matching data term E_{ds} as the sum of per-pixel color difference. Given a pixel $p \in I_{\mathsf{R}'}$ and a disparity d_p. Our data term from stereo matching is defined as

$$E_{ds} = \sum_{p \in I_{R'}} C_{dd}(p, d_p) \tag{3}$$

where $C_{dd}(\cdot, \cdot)$ is the Birchfield-Tomasi stereo matching cost [2] which is insensitive to image sampling.

Kinect Component. The definition of E_{dd}, which encodes the prior of geometric information from the Kinect, is the vital part of our fusion model. Given a pixel $p \in I_{\mathsf{R}'}$ with its image homogeneous coordinate $\tilde{\mathbf{m}}_{\mathsf{R}'} = [x_{\mathsf{R}'}, y_{\mathsf{R}'}, 1]^T$ and a disparity value d_p, we can calculate $\mathbf{M}_{\mathsf{R}'}$, $\mathbf{M}_{\mathsf{D}} = [X_{\mathsf{D}}, Y_{\mathsf{D}}, Z_{\mathsf{D}}]^T$, $\tilde{\mathbf{m}}_{\mathsf{D}}$, and the corresponding depth value on $\widehat{Z}_{\mathsf{D}}$. $\widehat{Z}_{\mathsf{D}}(p, d_p)$ is considered as the observation of the ground truth $Z_{\mathsf{D}}(p, d_p)$. We regard depth values $Z_{\mathsf{D}}(\cdot, \cdot)$ and $\widehat{Z}_{\mathsf{D}}(\cdot, \cdot)$ as functions of variables p and d_p. We assume that observation $\widehat{Z}_{\mathsf{D}}(p, d_p)$ from the depth sensor is given by $Z_{\mathsf{D}}(p, d_p)$ with additive Gaussian noises so that

$$\widehat{Z}_{\mathsf{D}}(p, d_p) \sim \mathcal{N}(Z_{\mathsf{D}}(p, d_p), \sigma^2) \tag{4}$$

where σ^2 is the variance of Gaussian noise, and the variance is subjected to many factors, e.g. depth, object reflectance, temperature, etc. In our fusion framework, pixelwise variances of the noises at different depth values have been taken into consideration. In the literature [11], the noises of the depth data from Kinect has been quantified, and $\sigma = k \cdot Z_{\mathsf{D}}^2$, i.e. errors increase quadratically with depth values increase. Thus, we have

$$P(\widehat{Z}_{\mathsf{D}}(p, d_p) | Z_{\mathsf{D}}(p, d_p)) = \frac{1}{\sqrt{2\pi} k \cdot Z_{\mathsf{D}}^2} e^{-\frac{(\hat{z}_{\mathsf{D}} - z_{\mathsf{D}})^2}{2(k \cdot Z_{\mathsf{D}}^2)^2}} \tag{5}$$

with $k = 1.5 \times 10^{-5}$ (1/cm).

In the fusion process, the depth value of a 3D point from Kinect may be unavailable. There are two main reasons causing $\widehat{Z}_{\mathsf{D}}(p, d_p)$ to be unavailable: (1) the corresponding $\mathbf{M}_{\mathsf{D}}$ is not visible in the FoV of the depth sensor; (2) Kinect provides a mechanism which refuses to give the depth estimate of corresponding pixels with low reliability (as shown in Fig.1(b)). A pixel $p \in I_{\mathsf{R}'}$ has its disparity $d_p \in [d_{min}, d_{max}]$, where d_{min} and d_{max} are minimum and maximum values of disparity level. We denote unavailable $\widehat{Z}_{\mathsf{D}}(p, d_p)$ as $\widehat{Z}_{\mathsf{D}}(p, d_p) = \emptyset$. Let $A_{\mathsf{R}'}$ be the set of pixels in the reference image whose corresponding 3D points of different disparity values are all available on Kinect, and we name $A_{\mathsf{R}'}$ the Available Set. Thus, we have

$$A_{\mathsf{R}'} = \{p | \forall d_p, \widehat{Z}_{\mathsf{D}}(p, d_p) \neq \emptyset\} \tag{6}$$

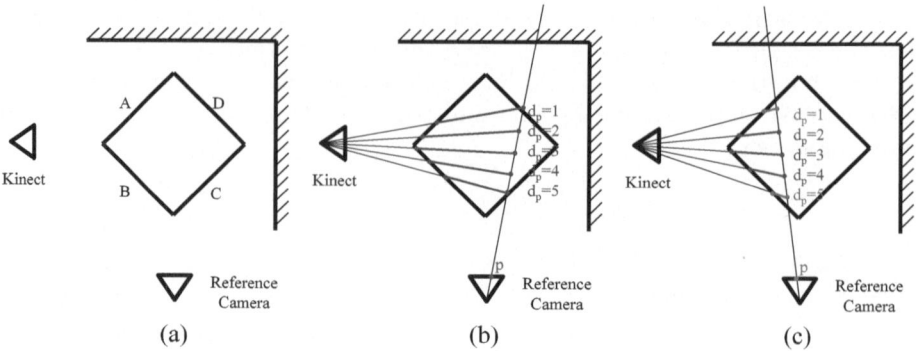

Fig. 3. The visibility model. (a) The object has four surfaces: A, B, C and D. Surface C and D are invisible to Kinect, while surface A and D are invisible to reference camera. (b) There are five disparity levels. All the corresponding 3D points of the pixel p with disparity values $d_p \in [d_{min}, d_{max}]$ are not visible to Kinect. (c) The corresponding 3D points of the pixel p with some disparity values($d_p = 1, 5$) are visible to Kinect. The point with $d_p = 1$ is occluded by the point with $d_p = 5(\tilde{d}_p = 5)$.

Furthermore, we have the Visible Set

$$V_{R'} = \{p | p \in A_{R'} \cap \exists d_p, P(\hat{Z}_D(p, d_p) | Z_D(p, d_p)) > T_z\} \tag{7}$$

When there exists d_p of the pixel p making $P(\hat{Z}_D(p, d_p) | Z_D(p, d_p))$ larger than a threshold T_z, it means that the corresponding points of the pixel with this disparity values d_p are visible in FoV of Kinect. In other words, if all the $P(\hat{Z}_D(p, d_p) | Z_D(p, d_p))$ are smaller than the threshold T_z, all the corresponding points of the pixel with disparity values $d_p \in [d_{min}, d_{max}]$ are all invisible(as shown in Fig.3(b)). In this situation, Kinect cannot provide any useful depth information to this pixel in the reference image for fusion.

For a pixel $p \in V_{R'}$ has multiple d_p satisfying $P(\hat{Z}_D(p, d_p) | Z_D(p, d_p)) > T_z$, we select 3D point with the largest disparity value as the visible point, and other points are occluded and invisible(as shown in Fig.3(c)). Therefore the visible disparity value is

$$\tilde{d}_p = max\{d_p | p \in V_{R'} \cap P(\hat{Z}_D(p, d_p) | Z_D(p, d_p)) > T_z\} \tag{8}$$

Thus, for a pixel $p \in I_{R'}$ with disparity d_p, its cost function is defined as

$$C_{dd}(p, d_p) = \begin{cases} |d_p - \tilde{d}_p|, & \text{if } p \in V_{R'} \\ 0, & \text{otherwise} \end{cases} \tag{9}$$

And the component from Kinect in the data term is

$$E_{dd} = \sum_{p \in I_{R'}} C_{dd}(p, d_p) = \sum_{p \in V_{R'}} |d_p - \tilde{d}_p| \tag{10}$$

3.2 Smoothness Term

The smoothness term comes from a prior that the disparity map is piecewise-smooth. Let $N_{\mathsf{R}'}$ denote the set of the neighboring pixel pairs in $I_{\mathsf{R}'}$, we employ a traditional smoothness energy term

$$E_s = \sum_{\{p,q\} \in N_{\mathsf{R}'}} \min(\omega_{pq} \cdot |d_p - d_q|, T_s) \tag{11}$$

where ω_{pq} measures the similarity of color of pixels p and q, and is computed using the method given in literature [16]. With respect to discontinuity preserving, the threshold T_s is also included in our smoothness term.

4   Experimental Results

We verify the effectiveness of the proposed fusion framework by recovering the face geometry (as shown in Fig. 1). Fig. 1(a) gives the rectified reference and target images captured from CCD cameras, which provide geometrical information for stereo matching. Fig. 1(b) shows the color image and associated depth map from Kinect. We take them as the input of the fusion framework.

Fig. 4 shows the results estimated by the fusion framework using different energy functions. From left to right, parameter α_{dd} is fixed at 0.0, 1.0, or adaptively chosen using the method given in literature [17]. When $\alpha_{dd} = 0.0$, it is a traditional global stereo matching method. Similarly, when $\alpha_{dd} = 1.0$, a high-resolution Kinect-like result is generated. By setting α_{dd} ranging from 0.0 to 1.0, we can obtain a robust result fusing stereo vision and Kinect. The difference between top and bottom row is whether the smoothness prior is considered where parameter $\lambda_s = 0$ or 20. Without smoothness prior, our method turns out to be a winner-takes-all strategy to find local best estimates of disparity. When the smoothness prior is taken into consideration, we can see that the results of bottom row are smoother and more reasonable than that of top row. However, Fig. 4(d) using stereo vision still fails in the textureless regions and repeated patterns. In Fig. 4(b) the depth values of a hole on the left side of the nose is unavailable, while in the Fig. 4(e) it is filled with the help of the smoothness prior. Our fusion approach(as in Fig. 4(f)) produces a more natural and reasonable result with higher quality than using stereo vision or Kinect only.

Moreover, we compare our method with the bilateral up-sampling method [12] and the sparse ground control point method [16](as shown in Fig.5). The pixels with depth value on $\widehat{Z}_{\mathsf{D}}$ are back-projected to 3D space, and then projected on $I_{\mathsf{R}'}$. Thus, we can calculate the disparity of some pixels on $I_{\mathsf{R}'}$ and take these pixels which are called sparse ground control point in literature [16] as the input of the approaches.

in the blue box of the disparity maps using Laplace Filter to check the structural details of the results generated by different approaches. As can be seen, the result of bilateral up-sampling is too over-smoothing; the depth

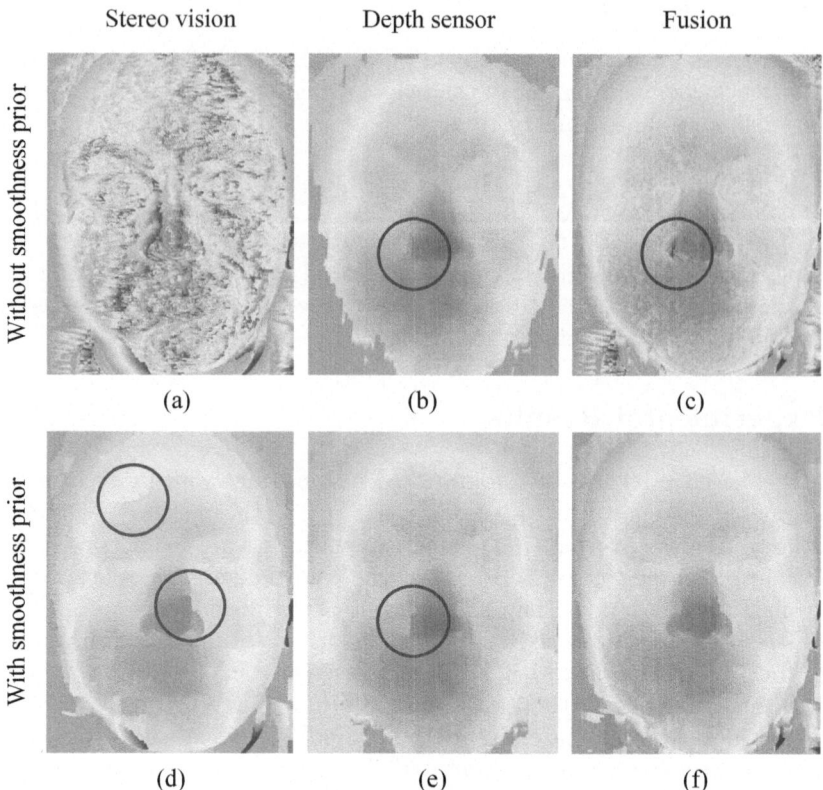

Fig. 4. Disparity maps estimated by our fusion framework using different energy functions. From left to right, the results are generated with different parameters, which employs the components of stereo vision, Kinect, or fusion of them in the data term, respectively. The results in the top row are without smoothness prior, while the results in the bottom row are with smoothness prior. (a) Stereo vision without smoothness prior; (b) Kinect without smoothness prior; (c) fusion without smoothness prior; (d) stereo vision with smoothness prior; (e) Kinect with smoothness prior; (f) fusion with smoothness prior. Red pixels have larger disparity values, while green pixels have smaller values.

value computed by the method sparse GCPs unnaturally flips on the region of the nose; only our framework preserves reasonable fine structural details of human face without over-smoothing them.

For quantitative evaluation of the method on the ground truth data, we employ the "Fountain-P11" multi-view stereo sequence publicized by Strecha *et al.* [14] as our test data(as shown in Fig.6). The range data is obtained from a laser scanner. In addition to 3D point clouds, there are also 11 high resolution color images(numbered 0-10) shot at multiple viewpoints with their

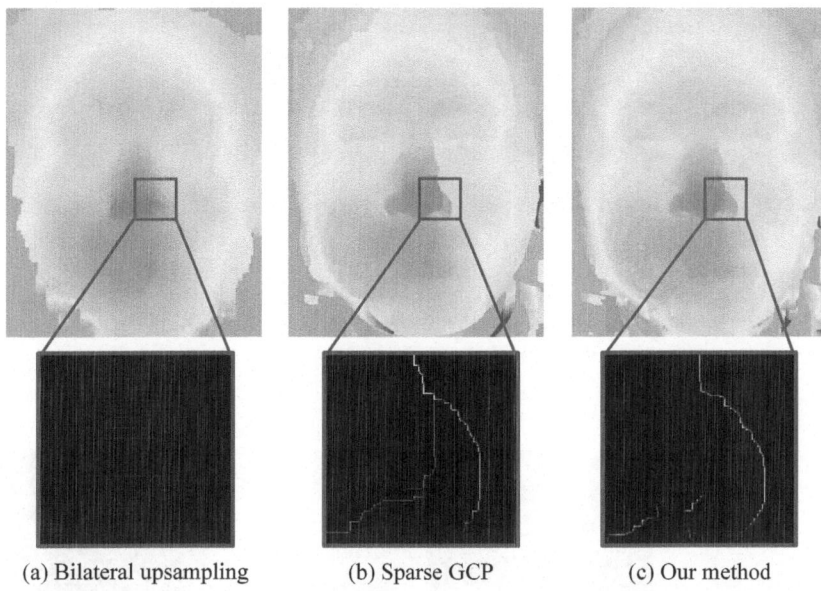

(a) Bilateral upsampling   (b) Sparse GCP   (c) Our method

Fig. 5. Images in the top row are the disparity maps using different methods for comparison. In the bottom row, we check whether the edges on disparity maps are associate with reference image via the Laplace filter in the blue boxes. (a) the result of Bilateral Up-sampling method; (b) the result of Sparse GCPs method; (c) the result of our method.

corresponding camera parameters provided in the dataset. We select the center image 5 as the reference image, and use its left neighboring image 6 as target image. Due to memory and speed consideration, the color images are down-sampled to 1536×1024, which is half of their original size after rectification. In order to simulate the color image and depth map from Kinect, we down-sample the color image from the camera 4 to 384×256, which is a rather low-resolution image size. Furthermore, the corresponding depth map is calculated and added some Gaussian noise whose parameters is given using the Eq. (5).

We focus on recovering the structure of a fountain at the center of the images, and the depth space is quantized into 150 levels. For each method, we calculate the depth maps and error maps which record the absolute difference between the estimated depth maps and the ground truth. As shown in Fig.7, our method preserves both better structure details and sharper discontinuities at object boundaries than other approaches. We consider a pixel to be a bad pixel when the absolute difference between its estimated depth value and ground truth is larger than a threshold T_b, and the ratio of bad pixels in a depth map is denoted as r_b. Fig.8 depicts the ratio r_b at different thresholds T_b using different approaches, our method has the lowest ratios of bad pixels at different thresholds

(a)   (b)

Fig. 6. The test data. (a) the stereo pair; (b) the synthetic color image and depth map from Kinect, the close-up shows that the depth map is rather noisy which makes it sticky to preserve the details at boundary of objects.

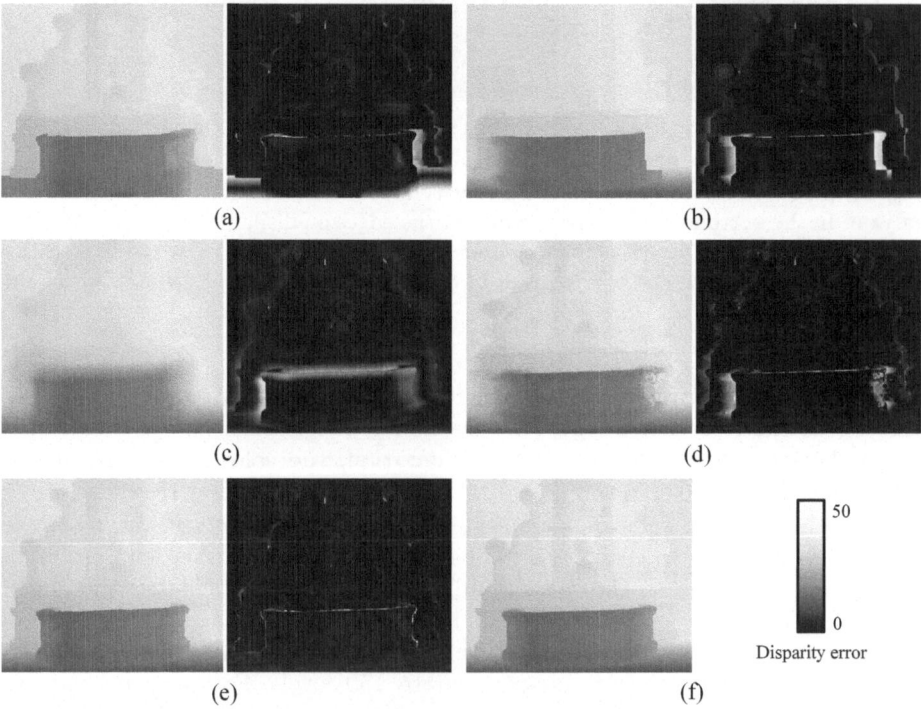

(a)   (b)

(c)   (d)

(e)   (f)

Disparity error

Fig. 7. (a)-(e) the depth maps and error maps generated by stereo vision, Kinect, bilateral up-sampling, sparse GCPs, and our method respectively; (f)the ground truth. For depth maps, red pixels have larger disparity values, while green pixels have smaller values.

and reaches an extreme low ratio which is less than one percent when $T_b = 9$. Note that for all the methods we have tuned appropriate parameters to enable a fair comparison.

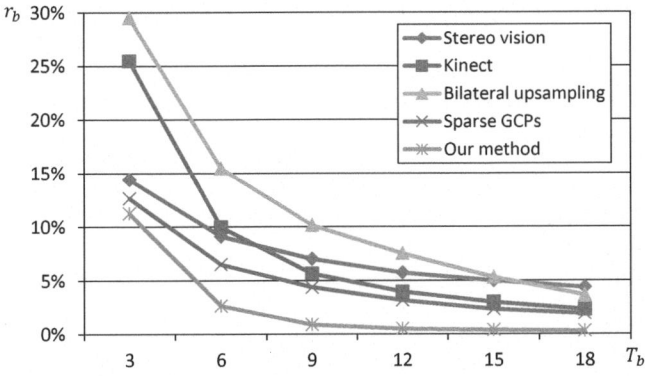

Fig. 8. The ratio r_b of bad pixels at different values of threshold T_b

5   Conclusion

In this paper, a fusion framework which merges cues from stereo vision and depth sensors for high-quality dense depth maps has been presented. Unlike previous works, visibility and noises of the depth data from depth sensors has been considered in our fusion framework. Results show that the proposed framework is more effective than using either stereo vision or depth sensor only. Future work includes implementing our fusion approach on hardware (such as GPU and FPGA) for a higher speed of computation[10].

Acknowledgement . This work was supported in part by the Natural Science Foundation of China(NSFC) under Grant No. 60905006 and NSFC-Guangdong Joint Fund under Grant No. U1035004.

References

1. Beeler, T., Bickel, B., Beardsley, P., Sumner, B., Gross, M.: High-quality single-shot capture of facial geometry. ACM Transactions on Graphics 29, 40 (2010)
2. Birchfield, S., Tomasi, C.: A pixel dissimilarity measure that is insensitive to image sampling. PAMI 20, 401–406 (1998)
3. Bouguet, J.Y.: Camera calibration toolbox for matlab (2010),
   http://www.vision.caltech.edu/bouguetj/calib_doc/
4. Boykov, Y., Veksler, O., Zabih, R.: Fast approximate energy minimization via graph cuts. PAMI 23, 1222–1239 (2001)
5. Chen, L., Jia, Y., Li, M.: An fpga-based rgbd imager. Machine Vision and Applications 23, 513–525 (2012), doi:10.1007/s00138-011-0334-z
6. Freedman, B., Shpunt, A., Machline, M., YoelArieli, A.: Depth mapping using projected patterns. U.S. Patent 2010/0118123 A1 (2010)
7. Hartley, R., Zisserman, A.: Multiple View Geometry in computer vision. Cambridge University Press (2000)

8. Izadi, S., Kim, D., Hilliges, O., Molyneaux, D., Newcombe, R., Kohli, P., Shotton, J., Hodges, S., Freeman, D., Davison, A., Fitzgibbon, A.: KinectFusion: real-time 3d reconstruction and interaction using a moving depth camera. In: Proceedings of the 24th Annual ACM Symposium on User Interface Software and Technology, pp. 559–568 (2011)
9. Jia, Y., Zhang, X., Li, M., An, L.: A miniature stereo vision machine (MSVM-III) for dense disparity mapping. In: ICPR (2004)
10. Kalarot, R., Morris, J.: Comparison of FPGA and GPU implementations of real-time stereo vision. In: CVPR Workshop (2010)
11. Khoshelham, K.: Accuracy analysis of kinect depth data. In: ISPRS Workshop Laser Scanning (2011)
12. Kopf, J., Cohen, M., Lischinski, D., Uyttendaele, M.: Joint bilateral upsampling. ACM Transactions on Graphics 26 (2007)
13. Oikonomidis, I., Kyriazis, N., Argyros, A.: Efficient model-based 3D tracking of hand articulations using kinect. In: BMVC (2011)
14. Strecha, C., Von Hansen, W., Van Gool, L., Fua, P., Thoennessen, U.: On benchmarking camera calibration and multi-view stereo for high resolution imagery. In: CVPR, pp. 1–8 (2008)
15. Szeliski, R., Zabih, R., Scharstein, D., Veksler, O., Kolmogorov, V., Agarwala, A., Tappen, M., Rother, C.: A comparative study of energy minimization methods for markov random fields with smoothness-based priors. PAMI 30, 1068–1080 (2008)
16. Wang, L., Yang, R.: Global stereo matching leveraged by sparse ground control points. In: CVPR (2011)
17. Zhu, J., Wang, L., Yang, R., Davis, J., Pan, Z.: Reliability fusion of time-of-flight depth and stereo geometry for high quality depth maps. PAMI 33, 1400–1414 (2011)

Robust Fall Detection by Combining 3D Data and Fuzzy Logic

Rainer Planinc and Martin Kampel

Vienna University of Technology, Computer Vision Lab
Favoritenstr. 9-11/183-2, A-1040 Vienna, Austria
{rainer.planinc,martin.kampel}@tuwien.ac.at

Abstract. Falls are a major risk for the elderly and where immediate help is needed. The elderly, especially when suffering from dementia, are not able to react to emergency situations properly, thus falls need to be detected automatically. An overview of different classes of fall detection approaches is presented and a vision-based approach is introduced. We propose the use of a Kinect to obtain 3D data in combination with fuzzy logic for robust fall detection and show that our approach outperforms current state-of-the-art algorithms. Our approach is evaluated on 72 video sequences, containing 40 falls and 32 activities of daily living.

1 Introduction

Wild et al. [1] show that the mortality of fallers is higher compared to other elderly. Moreover, if elderly are not able to get up on their own again they may lie on the floor for hours, until help is provided [1]. Noury et al. [2] have shown that getting help quickly after a fall reduces the risk of death by over 80% and the risk of hospilization by 26%. Furthermore, elderly suffering from dementia are not able to react to emergency situations properly [3]. Hence, the aim of assistive systems is not only to assist, but also to reduce the cognitive load on the user [4]. This motivates the introduction of a fall detection system, which is able to detect falls and raise alarms automatically. Moreover, these systems boost the confidence of elderly in living independently [5]. The contribution of this paper is to present an overview of current state-of-the-art fall detection approaches and to introduce a robust vision-based fall detection system using 3D data obtained by the Kinect in combination with fuzzy logic.

Fall detection systems can be divided into three major approaches [5]: wearable devices, ambient devices and camera-based (or vision-based) approaches. Figure 1 shows an overview of the three major approaches including divisions for each of these approaches into smaller and thus more specific approaches.

Wearable devices broadly used to assist elderly are panic buttons, which need to be worn (e.g. on the wrist) by the elderly and pressed if an emergency situation occurs and help is needed [6]. These devices have the main drawback that elderly need to push the button actively - if they are not able to push the button (e.g. due to the lost of consciousness), help can not be provided. Hence, wearable sensors

J.-I. Park and J. Kim (Eds.): ACCV 2012 Workshops, Part II, LNCS 7729, pp. 121–132, 2013.

detecting falls automatically have been developed (e.g. [7–11]). These wearable sensors detect the body orientation, the impact of falling (using accelerometers) or the amount of activity/movement. Särelä et al. [11] combine a panic button (i.e. button on the wrist) together with a movement sensor to detect emergency situations automatically if the user is not able to push the button any more. Noury et al. [10] combine the measurement of the impact together with the measurement of the body orientation and the vibrations on the body surface to build a fall detection device which they called "actimeter". The main advantage of wearable devices are costs, as such systems are cheap - the main disadvantage is that sensors need to be worn, which is very intrusive [5].

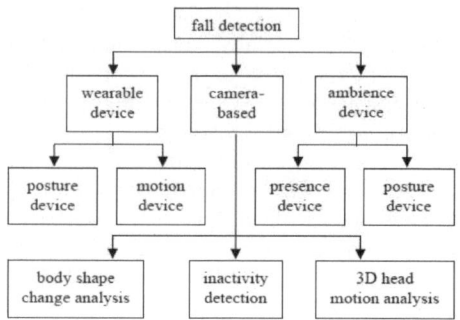

Fig. 1. Classification of fall detection approaches taken from [5]

Ambient devices are multiple sensors which are installed within the flat [5], turning the flat into a smart home [12] being able to support elderly living alone at home [13]. Approaches and sensors used in this field are very broad, including measuring the vibration of the floor to detect falls [14, 15], detecting falls by using pressure mats [6, 16] or motion sensors [16]. Ambient sensors are not intrusive, as they can be hidden within a smart home, but have the drawback of a high false alarm rate [5].

Vision-based systems are able to overcome limitations of other sensor types [17], but raise privacy issues. Hence, in contrast to Xinguo [5], we propose to not record any video data in order to respect concerns about privacy. Vision-based systems can be distinguished between systems using 2D images and systems using 3D data (e.g. obtained by multiple cameras [18] or 3D sensors [19]). To overcome limitations of multiple cameras (e.g. calibration is needed) and 3D sensors (e.g. availability and costs) we propose to use the Kinect as a vision-based 3D sensor for fall detection.

In contrast to the focus on the fall event mentioned by Xinguo [5], we neither focus on the fall event nor do we constrict a fall to time constraints (i.e. the fall process lasts from x to y seconds). Hence we propose to automatically raise an alarm if a person is detected to be on the ground and is not able to get up any more, as this is the situation where help is needed - independently of the reason for being on the floor (falling or lying down on purpose). Hence, if a person lies down on the floor on purpose and is not able to get up again, an alarm will be raised since help is needed anyway. Furthermore our approach combines and benefits from all sub categories of vision-based approaches defined by Xinguo [5]: body shape change analysis is done by analysing the major orientation of the person, whereas a person lying on the ground is seen as inactivity analysis, since the person is not moving or getting up. Furthermore, 3D motion analysis is done by tracking the person's skeleton position in a 3D environment over time.

The rest of this document is structured as follows: Section 2 provides an overview of state-of-the-art approaches in the field of vision-based fall detection. The methodology of our fall detection approach is introduced in Section 3, an empirical evaluation is presented in Section 4. Finally, a conclusion is drawn in Section 5.

2 State-of-the-Art

2.1 Body Shape Change Analysis

The shape of a person implies the orientation and thus is used to distinguish whether a person is in an upright position or not. The use of the bounding box aspect ratio (width to height ratio) to detect falls is proposed by Anderson et al. [20]. If people are in an upright position, the bounding box aspect ratio is bigger than one (i.e. *height* > *width*). In case of a fall, the ratio changes to a value smaller than one (i.e. *height* < *width*). Another approach presented by Rougier et al. uses information of an approximated ellipse instead of a bounding box [21]. Falls are detected by analyzing the orientation of the ellipse as well as the ratio of the major axis of the ellipse. Figure 2 illustrates these two approaches and depicts the shape of a person during a normal activity and during a fall. Furthermore, the corresponding bounding boxes and ellipses to analyze the bounding box aspect ratio and the orientation of the ellipse are illustrated. The use of a bounding box and an approximate ellipse for fall detection is feasible, but depends on the quality of the background segmentation. Assuming that the background segmentation yields in robust results, the fall detection also yields in robust results. A fall into the direction of the camera only using 2D images cannot be recognized by both approaches, as the change of orientation of the person cannot be detected.

Approaches not using 3D sensors reconstruct 3D information for humans from silhouettes gained by different camera views [22]. The human is represented by the use of voxels allowing to identify different states (upright, on–the–ground and in–between), depending on the shape of the person. The quality of this approach also depends on the quality of background segmentation, but it has the main drawback of needing a calibrated camera setup.

Zambanini et al. [18] propose a method to detect falls by using multiple cameras, and they distinguish between an uncalibrated camera-setup and a calibrated camera-setup. When using an uncalibrated camera-setup, scene analysis is performed on each camera individually. Afterwards, the individual results are combined to get an overall decision. In contrast, if information from multiple cameras using a calibrated camera-setup is combined to reconstruct the person in 3D space, the combination takes place at an early stage. Feature extraction is done on the 3D reconstruction of the person and a decision whether a fall occurred or not is made afterwards. Compared to other works (e.g. [23]), their system is not vulnerable to low-quality images (e.g. high noise and low resolution) as only basic information (i.e. silhouettes) are extracted from the image anyway. Using a calibrated camera-setup results in a higher accuracy than using an uncalibrated camera-setup, but it is practically not possible to calibrate the cameras if they are installed in an elderly person's flat or house.

Time-of-Flight cameras [24] are generating depth maps and can be used for fall detection [25]. Jansen et al. [19] mention the higher accuracy in contrast to stereo vision and propose a system for pose recognition discriminating the poses standing, sitting or lying by thresholding the height of the centroid. They state that their approach works in nursing homes reliably, but not in real homes due to false alarms.

2.2   Inactivity Detection

Unusual inactivity can be determined by tracking people from an overhead position [26, 27]. Therefore, zones with low activity (and little motion) are identified automatically and marked as an inactivity zones (e.g. sofa). Unusual inactivities are detected by analyzing the motion. If the amount of motion is below a threshold and occurred outside of the learned inactivity zones, this event is defined to be an unusual inactivity (e.g. person is lying on the floor). Inactivity detection is only able to detect falls indirectly by the lack of motion. Therefore it is important to ensure that the system is able to handle new situations (e.g. a chair is moved to a new position, thus moving the inactivity zone) properly.

A combination of applying a statistical model of inactivity zones and shape-based fall detection is introduced by Zweng et al. [28]. A so called accumulated hitmap models areas with low and high activities. In combination with their shape-based fall detection, the robustness of their approach is enhanced.

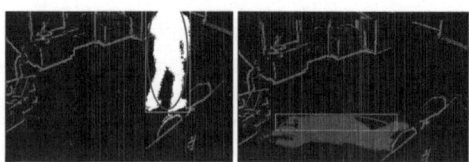

Fig. 2. Analysis of the bounding box aspect ratio and the orientation of the ellipse to detect falls

2.3   3D Motion Analysis

3D head motion analysis by using stereo vision sensors to detect falls is used by Belbachir et al. [29]. These biologically-inspired sensors feature a massively parallel pre-processing and reduce the amount of data in comparison to stereo vision cameras as they are not frame-based, but event-based. Hence, the motion of people can be determined and the position of the person can be extracted. A fall is detected by tracking the position and velocity of the head, as they assume the position of the head changes rapidly during a fall. Another approach by Rougier et al. [30] uses 3D information obtained by one single camera to track the head of the person and to obtain its trajectory. Not only the head position but also the motion speed is taken as an indicator for falls as the motion speed is assumed to be higher during a fall than during activities of daily living.

The approaches of Zambanini et al. [18], Belbachir et al. [29] and Rougier et al. [30] consider motion speed to detect falls, as they assume that the velocity is higher during a fall than during activities of daily living. From our point of view this assumption should not be made, as falls can also occur slowly and thus are not detected using these approaches.

In contrast to the definition introduced by Xinguo [5], we do not restrict 3D motion analysis to the head of a person, as other body parts (e.g. centroid) are analyzed as well. An approach using Time-of-Flight cameras detects moving regions within the 3D points cloud in a first step [25]. The person (foreground) is segmented from the background and - in contrast to other works analyzing the head position - the distance of the person's centroid to the ground floor is analyzed. This results in an efficiency of 80% and a reliability of 97.3% when using a centroid-ground floor distance of 0.4 meters as threshold [25]. Furthermore they propose to extract the skeleton from the depth data to analyse the orientation of the person's spine.

Since the introduction of the Kinect sensor in 2010, a new 3D sensor is available. Smisek et al. [31] analyzed the depth resolution and accuracy of the Kinect. Evaluation shows that regarding multi-view reconstruction the Kinect overperforms a Time-of-Flight sensor (SwissRanger SR-4000) and the quality is almost equal to a reconstruction using a 3.5 Mega Pixel SLR Stereo approach. Using the Kinect sensor for fall detection is proposed by Rougier et al. [32], but they focus on low level vision tasks like foreground / background segmentation and detecting the ground plane. Their proposed fall detection algorithm analyzes the distance between the centroid of the body and the ground floor as well as motion speed. Mastorakis et al. [33] use the Kinect and the 3D bounding box to detect falls. Falls are detected by analyzing the velocity of the person (i.e. falls occur if the velocity is higher than a threshold) as well as by assuming that a fall is followed by an inactivity period (i.e. no motion after a fall). Motion speed is not a suitable feature for fall detection as the motion speed is not necessarily high during a fall. Furthermore motion can occur after a fall since elderly might be able to crawl on the floor.

3 Methodology

Zweng et al. [28] show that the accuracy of their fall detection approach is higher when using a 3D reconstruction of the person, but having the main drawback of needing a calibrated camera setup. Therefore we propose to use a 3D reconstruction of a person, but using the Kinect instead of multiple cameras. Due to the use of infrared light, the Kinect also works during the night, when falls of elderly occur (e.g. when going to the bathroom in the dark). Furthermore, changing lighting conditions (e.g. switching the lights on and off) does not affect the results of the Kinect. Hence, the results (e.g. foreground/background segmentation, tracking) are more robust when using the Kinect than using standard IP cameras.

Our fall detection approach combines body shape analysis together with inactivity detection and 3D motion analysis. A fall is detected by analyzing the body orientation and the height of the spine. If a person is detected to be on the floor and is not able to get up on her/his own within a specified time, an alarm is triggered (inactivity detection).

The workflow of our approach is shown in Figure 3. Starting with a depth image obtained by the Kinect, skeleton information is extracted and the ground plane is estimated by OpenNI [34]. The skeleton information provided by OpenNI is optimized for being in an upright position (since it was developed for the use with the Xbox), but also works in different positions (e.g. lying on the floor). Based on the coordinate data of the skeleton, features to determine the pose of the person (i.e. orientation of the body and distance to the ground) are extracted. A final decision about the pose of the person is made by applying fuzzy logic. This simplistic approach is chosen to reduce the computational load of our algorithm and therefore there is no need for special system requirements when running our approach. The evaluation shows that our approach already yields in reasonable results.

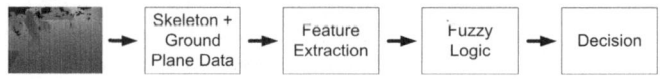

Fig. 3. Workflow

3.1 Feature Extraction

The depth image of a person includes the skeleton information of the shoulder (center), spine and hip (center). Since the coordinates are not in 2D but in 3D space and the ground plane equation is known, the pose of the person can be illustrated relative to the ground floor, depicted in Figure 4. Therefore the major axis of the person is estimated by approximating the coordinates of the shoulder, spine and hip by a line. This line is approximated by calculating the mean slope between these three skeleton coordinates. Afterwards the following features are calculated:

- *Similarity between the body orientation and the ground plane:* the pose is estimated by calculating the similarity of the person's orientation and the ground floor. If the orientation of the person is parallel to the ground floor, the person is defined to be "lying" (either on the floor or on the bed). If the major orientation is orthogonal to the ground floor, the person is in the position "upright". Although this is only an approximate approach, experimental results show that this is already sufficient to detect falls with a high accuracy.
- *Spine distance to the ground floor:* the distance between the spine and the ground floor is calculated, allowing to determine whether the person is lying on the floor or e.g. on the bed. The integration of this feature is essential, since otherwise it is not possible to determine if a person is lying on the bed or on the floor, which results in false alarms.

The use of the similarity of the body orientation and the ground floor is illustrated in Figure 4a and Figure 4b: in contrast to a person being in an upright position (shown in Figure 4a), a person lying on the floor is shown in Figure 4b. Therefore the similarity of the body orientation can be used as a feature to distinguish between these poses. The need for analyzing the spine distance is illustrated in Figure 4c and Figure 4d: Figure 4c shows a person sitting on a chair, whereas a person is sitting on the floor is illustrated in Figure 4d.

Fig. 4. Person (a) being in an upright position, (b) lying on the floor, (c) sitting on a chair and (d) sitting on the floor

3.2   Pose Estimation Based on Fuzzy Logic

Similar to Anderson et al. [22] and Zweng et al. [28], pose estimation is based on confidence values for the poses "upright", "in between" and "lying on the floor". In contrast to Zweng et al., our pose estimation and fall detection is only based on features introduced in Section 3.1 and motion speed is not taken into consideration. This is done in order not to constrain the fall event, but to be able to detect a variety of falls - even those, which occur slowly. To be able to differentiate between our three defined poses, trapezoidal functions [35] for the poses are created by finding thresholds empirically. Figure 5a depicts the trapezoidal function of the posture confidence depending on the body orientation. The posture confidence with respect to the ground plane distance is shown in Figure 5b.

Posture confidences for the orientation and the height of the body are calculated independently in the first step. To get an overall decision whether a fall occurred, the confidence values are combined by calculating the arithmetic average. This combination results in three confidence values for the poses "upright", "in between" and "lying on the floor". The final decision whether a fall occurred is made by thresholding the confidence values. Eliminating outliers is achieved by analyzing the average of pose confidences over time (e.g. 50 frames).

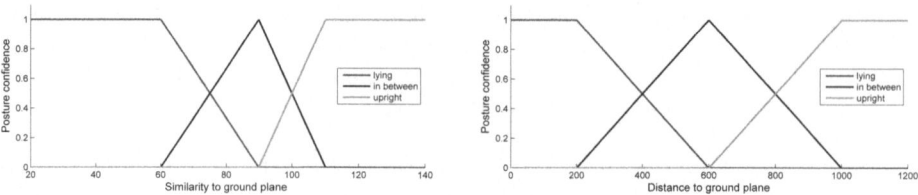

Fig. 5. Definition of fuzzy boundaries for the (a) orientation similarity and (b) spine-ground distance

4   Evaluation

Falls are simulated in a way that is similar to the definition of falls by Noury et al. [2], but using an extended version of scenarios, depicted in Table 1. The additionally added scenarios are "sitting down on a chair and falling while getting up", "lying down to a bed and falling out of the bed" and "falling into camera direction". These scenarios are added to enhance the quality of evaluation. Furthermore two scenarios are taken out of the original definition of Noury et al. since we do not agree with the uniqueness of the outcome. The modification results in 18 different sequences, containing ten falls and eight no-falls. These scenarios are simulated by two subjects, simulating each scenario twice. This results in an overall set of 72 videos, containing 40 falls and 32 no-falls.

Experiments have shown that the pose "in between" is not necessarily needed for evaluation, as analyzing only two poses is sufficient. We present our results using a precision-recall curve alternatively to the ROC curve [36] since it is not possible to specify the number of negative samples as the overall number of negative samples is not known in advance (a negative sample in our dataset is specified as a "no fall" event). Since our algorithm is frame-based, "no fall" events occur in each sequence (even if it is a sequence containing a fall event), since most of the frames are "no falls" and only a few frames show the fall respectively a person lying on the floor. Hence we cannot define all frames not containing a fall as "negative samples", but we cannot define only 32 negative samples either. The precision-recall curve in Figure 6 is generated by varying the thresholds $t_{upright}$ and t_{lying}.

Table 1. Definition of scenarios, similar to Noury et al. [2]

Category	Description	Outcome
Backward fall	Ending sitting	Positive
	Ending lying	Positive
	Ending in lateral position	Positive
	With recovery	Negative
Forward fall	With forward arm protection	Positive
	Ending lying flat	Positive
	With rotation, ending in lateral position (left or right)	Positive
	With recovery	Negative
Lateral fall (to the left or right)	Ending lying	Positive
	With recovery	Negative
Neutral	To sit down on a chair, then to stand up	Negative
	To lie down on the bed, then to stand up	Negative
	Walking	Negative
	To bend down, pick something up, then to rise up	Negative
	To cough or sneeze	Negative
Additional sequences	To sit down on a chair, then fall while getting up	Positive
	To lie down on the bed, then to fall out of the bed	Positive
	Fall into camera direction	Positive

Our approach results in an accuracy of 98.6% on 72 videos, resulting in one FP in the whole dataset. This FP occurs due to a tracking error after a fall, since the person is not tracked correctly while getting up again. Hence, a second fall is detected within the same sequence but as this fall does not occur in the time interval specified in the ground truth annotation, it is marked as a FP.

These results show that our approach outperforms other state-of-the-art approaches (e.g. [25],[28]). Although the evaluation is not based on the same dataset (due to the lack of dataset/code availability), the evaluation setting is similar (laboratory setting is similar to Zweng et al.[28]). Furthermore, similar to

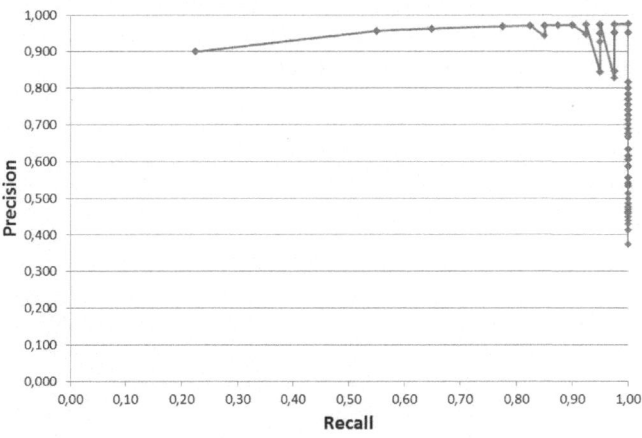

Fig. 6. Precision-recall curve of our fuzzy fall detection approach

Zweng et al.[28] the fall scenarios defined by Noury et al.[2] are used. Our approach results in only one FP on the whole dataset whereas the ROC curve from Zweng et al.[28] indicates a higher number of FP. Furthermore, the use of the Kinect offers practical advantages: it is robust to changing lighting conditions, also works also during the night and the installation in real homes is simplified by using only one sensor without the need for a complex calibration.

5   Conclusion

This article discussed state-of-the-art approaches based on different classes of fall detection. We introduced the combination of 3D tracking data obtained by the Kinect together with fuzzy logic for fall detection. Evaluation showed that this approach results in a high accuracy, being able to detect falls robustly. Our proposed approach was evaluated on a dataset of 72 video sequences and outperformed other state-of-the-art approaches. The fall dataset was recorded in cooperation with medical scientists and care taker organizations and is publicly available[1].

Acknowledgement. This work is supported by the European Union under grant AAL 2010-3-020. The authors want to thank the whole *fearless* project team, since techniques and ideas of many researchers involved are used.

References

1. Wild, D., Nayak, U.S., Isaacs, B.: How dangerous are falls in old people at home? British Medical Journal (Clinical Research Ed.) 282, 266–268 (1981)
2. Noury, N., Rumeau, P., Bourke, A.K., OLaighin, G., Lundy, J.E.: A proposal for the classification and evaluation of fall detectors. Biomedical Engineering and Research IRBM 29, 340–349 (2008)
3. Leikas, J., Salo, J., Poramo, R.: Security Alarm System Supports Independent Living of Demented Persons. Gerontechnology: A Sustainable Investment in the Future. Technology and Informatics 48, 402–405 (1998)
4. Lubinski, R.: Dementia and Communication. B.C. Decker, Inc. (1991)
5. Yu, X.: Approaches and principles of fall detection for elderly and patient. In: 10th International Conference on e-health Networking, Applications and Services (HealthCom 2008), pp. 42–47 (2008)
6. Miskelly, F.G.: Assistive technology in elderly care. Age and Ageing 30, 455–458 (2001)
7. Boissy, P., Choquette, S., Hamel, M., Noury, N.: User-based motion sensing and fuzzy logic for automated fall detection in older adults. Telemedicine Journal and e-Health: the Official Journal of the American Telemedicine Association 13, 683–693 (2007)
8. Doukas, C., Maglogiannis, I., Tragas, P., Liapis, D., Yovanof, G.: Patient Fall Detection using Support Vector Machines. In: Boukis, C., Pnevmatikakis, A., Polymenakos, L. (eds.) Artificial Intelligence and Innovations 2007: From Theory to Applications. IFIP, vol. 247, pp. 147–156. Springer, Boston (2007)

---

[1] http://fall.fearless-project.eu

9. Lin, C., Hsu, H., Lay, Y., Chiu, C., Chao, C.: Wearable device for real-time monitoring of human falls. Measurement 40, 831–840 (2007)
10. Noury, N., Barralon, P., Virone, G., Boissy, P., Hamel, M., Rumeau, P.: A smart sensor based on rules and its evaluation in daily routines. In: Proceedings of the 25th Annual International Conference of the IEEE Engineering in Medicine and Biology Society, vol. 4, pp. 3286–3289 (2003)
11. Sarela, A., Korhonen, I., Lotjonen, J., Sola, M., Myllymaki, M.: Ist vivago reg; - an intelligent social and remote wellness monitoring system for the elderly. In: Proceedings of the 4th International IEEE EMBS Special Topic Conference on Information Technology Applications in Biomedicine, pp. 362–365 (2003)
12. Scanaill, C., Carew, S., Barralon, P., Noury, N., Lyons, D., Lyons, G.: A Review of Approaches to Mobility Telemonitoring of the Elderly in Their Living Environment. Annals of Biomedical Engineering 34, 547–563 (2006)
13. Chan, M., Campo, E., Estève, D., Fourniols, J.Y.: Smart homes - current features and future perspectives. Maturitas 64, 90–97 (2009)
14. Alwan, M., Rajendran, P.J., Kell, S., Mack, D., Dalal, S., Wolfe, M., Felder, R.: A Smart and Passive Floor-Vibration Based Fall Detector for Elderly. In: IEEE International Conference on Information & Communication Technologies: from Theory to Applications, ICTTA, vol. 1, pp. 1003–1007 (2006)
15. Litvak, D., Zigel, Y., Gannot, I.: Fall detection of elderly through floor vibrations and sound. In: 30th Annual International Conference of the IEEE Engineering in Medicine and Biology Society, EMBS 2008, vol. 2008, pp. 4632–4635 (2008)
16. Zhang, Z., Kapoor, U., Narayanan, M., Lovell, N.H., Redmond, S.J.: Design of an Unobtrusive Wireless Sensor Network for Nighttime Falls Detection. In: Annual International Conference of the IEEE in Engineering in Medicine and Biology Society, EMBC, pp. 5275–5278 (2011)
17. Mihailidis, A., Carmichael, B., Boger, J.: The Use of Computer Vision in an Intelligent Environment to Support Aging-in-Place, Safety, and Independence in the Home. Gerontechnology 2, 173–189 (2002)
18. Zambanini, S., Machajdik, J., Kampel, M.: Early versus Late Fusion in a Multiple Camera Network for Fall Detection. In: 34th Annual Workshop of the Austrian Association for Pattern Recognition (ÖAGM 2010), Zwettl, Austria, vol. 819862, pp. 15–22 (2010)
19. Jansen, B., Temmermans, F., Deklerck, R.: 3D human pose recognition for home monitoring of elderly. In: Conference of the IEEE on Engineering in Medicine and Biology Society, Lyon, pp. 4049–4051 (2007)
20. Anderson, D., Keller, J., Skubic, M., Chen, X., He, Z.: Recognizing falls from silhouettes. In: 28th Annual International Conference of the IEEE Engineering in Medicine and Biology Society, EMBS 2006, New York, pp. 6388–6391 (2006)
21. Rougier, C., Meunier, J., St-Arnaud, A., Rousseau, J.: Fall detection from human shape and motion history using video surveillance. In: 21st International Conference on Advanced Information Networking and Applications Workshops, AINAW 2007, Niagara Falls, vol. 2, pp. 875–880 (2007)
22. Anderson, D., Luke, R.H., Keller, J.M., Skubic, M., Rantz, M., Aud, M.: Linguistic Summarization of Video for Fall Detection Using Voxel Person and Fuzzy Logic. Computer Vision and Image Understanding 113, 80–89 (2009)
23. Aghajan, H., Wu, C., Kleihorst, R.: Distributed Vision Networks for Human Pose Analysis. In: Mandic, D., Golz, M., Kuh, A., Obradovic, D., Tanaka, T. (eds.) Signal Processing Techniques for Knowledge Extraction and Information Fusion, pp. 181–200. Springer, US (2008)

24. Oggier, T., Lehmann, M., Kaufmann, R., Schweizer, M., Richter, M., Metzler, P., Lang, G., Lustenberger, F., Blanc, N.: An all-solid-state optical range camera for 3D real-time imaging with sub-centimeter depth resolution (SwissRanger). In: Proceedings of SPIE, vol. 5249, pp. 534–545. SPIE (2004)
25. Diraco, G., Leone, A., Siciliano, P.: An active vision system for fall detection and posture recognition in elderly healthcare. In: Design, Automation Test in Europe Conference Exhibition (DATE), Dresden, pp. 1536–1541 (2010)
26. McKenna, S.J., Charif, H.N.: Summarising contextual activity and detecting unusual inactivity in a supportive home environment. Pattern Analysis and Applications 7, 386–401 (2005)
27. Nait-Charif, H., McKenna, S.: Activity summarisation and fall detection in a supportive home environment. In: Proceedings of the 17th International Conference on Pattern Recognition, ICPR, vol. 4, pp. 323–326. IEEE (2004)
28. Zweng, A., Zambanini, S., Kampel, M.: Introducing a Statistical Behavior Model into Camera-Based Fall Detection. In: Bebis, G., Boyle, R., Parvin, B., Koracin, D., Chung, R., Hammoud, R., Hussain, M., Kar-Han, T., Crawfis, R., Thalmann, D., Kao, D., Avila, L. (eds.) ISVC 2010, Part I. LNCS, vol. 6453, pp. 163–172. Springer, Heidelberg (2010)
29. Belbachir, A.N., Lunden, T., Hanák, P., Markus, F., Böttcher, M., Mannersola, T.: Biologically-inspired stereo vision for elderly safety at home. e & i Elektrotechnik und Informationstechnik 127, 216–222 (2010)
30. Rougier, C., Meunier, J., St-Arnaud, A., Rousseau, J.: Monocular 3d head tracking to detect falls of elderly people. In: 28th Annual International Conference of the IEEE on Engineering in Medicine and Biology Society, EMBS 2006, New York, pp. 6384–6387 (2006)
31. Smisek, J., Jancosek, M., Pajdla, T.: 3D with Kinect. In: IEEE International Conference on Computer Vision Workshops, ICCV Workshops, pp. 1154–1160. IEEE Computer Society Press, Los Alamitos (2011)
32. Rougier, C., Auvinet, E., Rousseau, J., Mignotte, M., Meunier, J.: Fall Detection from Depth Map Video Sequences. In: Abdulrazak, B., Giroux, S., Bouchard, B., Pigot, H., Mokhtari, M. (eds.) ICOST 2011. LNCS, vol. 6719, pp. 121–128. Springer, Heidelberg (2011)
33. Mastorakis, G., Makris, D.: Fall detection system using Kinects infrared sensor. Journal of Real-Time Image Processing (2012)
34. Shotton, J., Fitzgibbon, A., Cook, M., Sharp, T., Finocchio, M., Moore, R., Kipman, A., Blake, A.: Real-time human pose recognition in parts from single depth images. In: IEEE Conference on Computer Vision and Pattern Recognition, CVPR, pp. 1297–1304 (2011)
35. Zadeh, L.: Fuzzy sets. Information and Control 8, 338–353 (1965)
36. Davis, J., Goadrich, M.: The relationship between Precision-Recall and ROC curves. In: Proceedings of the 23rd International Conference on Machine Learning, ICML 2006, pp. 233–240. ACM Press, New York (2006)

KinectAvatar: Fully Automatic Body Capture Using a Single Kinect

Yan Cui[1], Will Chang[*], Tobias Nöll[1], Didier Stricker[1]

[1] Augmented Vision, DFKI
william.y.chang@gmail.com

Abstract. We present a novel scanning system for capturing a full 3D human body model using just a single depth camera and no auxiliary equipment. We claim that data captured from a single Kinect is sufficient to produce a good quality full 3D human model. In this setting, the challenges we face are the sensor's low resolution with random noise and the subject's non-rigid movement when capturing the data. To overcome these challenges, we develop an improved super-resolution algorithm that takes color constraints into account. We then align the super-resolved scans using a combination of automatic rigid and non-rigid registration. As the system is of low price and obtains impressive results in several minutes, full 3D human body scanning technology can now become more accessible to everyday users at home.

1 Introduction

Three-dimensional geometric models of real world objects, especially human models, are essential for many applications such as design, virtual prototyping, quality assurance, games, and special effects. However, highly trained artists are mainly responsible for performing this modeling task, often using specialized modeling software. While 3D scanning technology has been available as an alternative for some time, it is still not used very widely for capturing models. This is because scanning devices are still expensive and often require expert knowledge for operation. And many scanning systems are limited for only rigid objects.

In this work, we present a novel approach to full 3D human body scanning which employs a single Microsoft Kinect [1] sensor. The Kinect has a variety of advantages over existing 3D scanning devices. It can capture depth and image data at video rates without a significant dependence on specific lighting and texture conditions. Also the Kinect is compact, low-price, and as easy to use as a normal video camera.

The challenge in using a single Kinect for scanning is that it provides relatively low-resolution data (320×240) with a high noise level. With these data characteristics, it is difficult to produce high-quality 3D models. Devices such as the Kinect were designed for use in object detection and natural user interfaces, not for high-quality 3D scanning. In addition, since we observe our subject only from a single, fixed location, movement of the subject is unavoidable during scanning. Therefore, we must estimate and compensate for this motion in order to combine scans from different viewpoints and produce a complete model.

---

[*] Author has no affiliation at this time.

J.-I. Park and J. Kim (Eds.): ACCV 2012 Workshops, Part II, LNCS 7729, pp. 133–147, 2013.

We claim that data captured from a single Kinect is sufficient to produce a good quality full 3D human model, without any additional equipment or shape priors (e.g. a turntable, additional Kinects, or a database of human body models). A key benefit of our algorithm is that it only relies on the input data and does not require an explicit shape model of the scanned subject. This allows us to reproduce personalized detail geometry such as faces and clothing. Our pipeline leverages prior work in depth super-resolution and articulated non-rigid registration. The contributions are

- a pipeline for automatically scanning full human body using a single Kinect,
- an improved depth super-resolution algorithm, taking color constraints into account,
- and a unified formulation of rigid and non-rigid registration under a probabilistic model, improving upon prior work [2] [3] for registration quality in high noise scenarios and for solving the loop closure problem.

In the following section, we discuss the relationship of this paper to previous work. Afterwards, we give technical details of our scanning system and verify our main claim by demonstrating results on several real-world datasets.

2   Related Work

Recent developments in low cost real-time depth cameras have opened up a new field for 3D content acquisition. In particular, several publications about 3D shape scanning with the Kinect have appeared. Henry et al. [4] build dense 3D maps of indoor environments, and Newcombe et al. [5] present a real-time 3D scanning system for arbitrary indoor scenes. These projects focus mainly on scanning static scenes of indoor environments.

Specifically concerning the problem of human body scanning, Cui et al. [6] try to scan a human body with one Kinect. However, they do not use the color information, and fundamental proplem is that they can not handle non-rigid movement, so the reconstructed results in the arm and leg parts are not of high quality. The work by Weiss et al. [7] estimates the body shape by fitting the parameters of a SCAPE model [8] to depth data and image silhouettes from a single Kinect. In contrast, our work does not require a prior shape model and relies mainly on registration. Thus, our method reproduces personalized details such as faces or dresses from the scans. Most recently, Tong et al. [9] present a system to scan a body using three Kinects and a turntable. They also utilize a global non-rigid registration algorithm which uses a rough template constructed from the first depth frame. While we share some similarities, our system setup is much simpler (using only a single Kinect), and our global registration works directly with the data without requiring a rough template.

As the raw Kinect depth data is of low resolution with high noise levels, a smoothing algorithm should be applied as a pre-processing step. Newcombe et al. [5] apply a bilateral filter [10] to the raw Kinect depth map to obtain a discontinuity preserved depth map with reduced noise. Schuon et al. [11] develop a super-resolution algorithm (LidarBoost) to improve the depth resolution and data quality of a ToF range scan, and Cui et al. [2] further develop this method. In this paper, we compare these existing methods and provide a new super-resolution algorithm for the Kinect depth and color data. This improves the resolution, reduces the noise, and preserves shape detail.

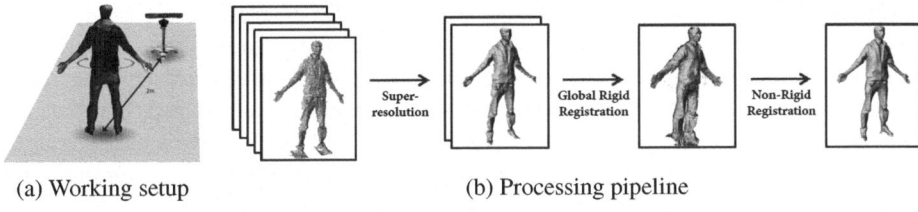

(a) Working setup   (b) Processing pipeline

Fig. 1. Outline of our working setup and processing pipeline

The global rigid registration problem is an important topic in object scanning. Iterative Closest Points (ICP) and its variants [12] can solve the local rigid alignment problem. Global rigid alignment techniques [13][14][15] can be used to register the scans against each other and solve the well-known loop closure problem. However, the input data used in these algorithms are acquired using structured light or laser scanning, which has high resolution and little noise. Cui et al. [2] present a probabilistic scan alignment approach that explicitly takes the sensor's noise characteristics into account. However, this method can only solve a local alignment task. We improve upon their approach and develop a probabilistic global alignment algorithm.

Even after the global rigid alignment, the scans do not align well because of the human bodies' non-rigid deformation during scanning (especially in the arms and legs). Aligning these scans is a challenging task that often requires high quality scan data and small changes of the pose in each scan [3][16][17][18][19][20][21]. While most of these approaches mainly deal with high-resolution scan data, we propose a robust non-rigid global alignment that can work well even with the Kinect's resolution and noise levels. Since the human body is largely articulated, we build on the global articulated registration work by Chang and Zwicker [3] and significantly improve the algorithm using a probabilistic scan alignment approach robust to the noise of the scans.

3   Overview

First, we give an overview of our scanning system, easily built as shown in Fig. 1a. The user stands before a Kinect in a range of about 2 meters, such that the full body falls in the Kinect's range of view. Then, the user simply turns around 360 degrees for about 20 to 30 seconds while maintaining an approximate "T" pose.

While scanning, we capture $s = 10$ frames as one view "chunk" with depth maps $\mathcal{D}_\ell = (D_1, \ldots, D_s)$ and color maps $\mathcal{C}_\ell = (C_1, \ldots, C_s)$. We then stop 0.5 seconds, capture another chunk, stop, capture, and so on until the human body has turned completely. The capture process yields about 10 raw data frames for each view, with the body turning approximately 30 degrees between the views. This ensures that there are enough overlapping areas for registration while providing full 360 degree coverage. The depth maps and color maps are calibrated and aligned (calibration off-line). In total, we get K chunks, where $K = 30 \sim 40$. Since the Kinect captures at about 30 FPS, the displacement is relatively small within a single chunk.

The resulting range data is then processed by our reconstruction algorithm, comprised of the following steps (Fig. 1b). First, a super-resolution step takes each chunk as input, and produces a single super-resolved depth scan with both greater depth accuracy and resolution (Sec. 4). Second, global rigid and non-rigid alignment steps combine the super-resolved scans into a final model (Sec. 5). Here, by "global" we mean that all of the frames in the sequence are aligned together. After the alignment is complete, we reconstruct a single mesh using Poisson mesh reconstruction [22] as a post processing step. Finally, textures are created with the Kinect's raw color data (Sec. 6). Our results show that this system can capture impressive 3D human shapes with personalized geometric details. In subsequent sections, we describe each step of the system in more detail.

4   Super-Resolution

We apply our super-resolution algorithm to each chunk of depth images, yielding K new super-resolved depth maps H_ℓ with much higher X/Y resolution and less noise. Our algorithm is largely based on Cui et al [23].

First, we align all depth and color maps of a chunk to its middle frame using optical flow. This is sufficiently accurate since the maximum viewpoint displacements throughout the entire chunk are typically one to three pixels. This corresponds to a turning speed of aobut 30 seconds per revolution by the user. Second, we extract a high-resolution denoised depth map H_ℓ from the aligned low resolution depth and color maps by optimizing the following objective function:

$$\min_{H_\ell} \mathcal{E}_{\text{data}}(D_1, \ldots, D_s, C_1, \ldots, C_s, H_\ell) + \gamma \mathcal{E}_{\text{reg}}(H_\ell) , \tag{1}$$

$$\mathcal{E}_{\text{data}} = \sum_{k=1}^{s} \| W_k \circ (H_\ell - f(D_k)) \|^2$$
$$W_k = \frac{1}{C_k - \frac{1}{s}\sum_{k=1}^{s} C_k} . \tag{2}$$

Here, we upsample the depth data by a factor $\beta = 2$ in both X and Y dimensions. After the optimization, the depth map H_ℓ is converted into a 3D point cloud $Y_\ell = \{y_j \mid j = 1 \ldots \beta^2 N_X N_Y\}$ using the Kinect's intrinsic parameters after optimization (N_X, N_Y are the resolution of the raw depth data).

To explain the optimization further, $\mathcal{E}_{\text{data}}$ measures the agreement of H_ℓ with the low resolution depth maps. Here, f is a function, which upsamples the low resolution $\mathcal{D}_\ell$ to higher resolution of $\beta^2 N_X N_Y$ and align to the center depth frame. We performed experiments to determine the best resampling strategy. It turned out that a nearest neighbor sampling from the low resolution images is preferable over any type of interpolated sampling. Interpolation implicitly introduces unwanted blurring that leads to a less accurate reconstruction of high-frequency shape details in the superresolved result. W_k is a per-pixel weight that measures the quality of the optical flow alignment, ∘ denotes element-wise multiplication, and $\|\|^2$ denotes sum of the square normal for each pixel. The $\frac{1}{\cdot}$ notation for W_k means that we invert the value per pixel. Since the human is

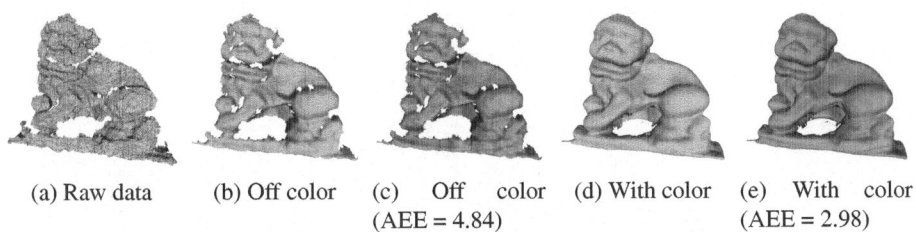

(a) Raw data   (b) Off color   (c)   Off   color   (d) With color   (e)   With   color
                                   (AEE = 4.84)   (AEE = 2.98)

Fig. 2. Super-resolved mesh and error distribution results, produced without the color constraint (b,c) and with the color constraint (d,e). AEE denotes average Euclidean error, computed by averaging the error values over all points in the mesh.

moving in 3D space but optical flow just considers the movement in 2D image space, some points are not aligned properly in each chunk. The weight term W_k quantifies this by producing a larger value if the color in each raw frame is similar to the average color; otherwise, the point is not correctly aligned and the weight is smaller. Finally, the $\mathcal{E}_{\text{reg}}$ term is a feature-preserving smoothing regularizer tailored to the Kinect depth data, with γ as the weighting coefficient. We use the same definition as [23], with their four different versions of regularizers that are inspired by image processing counterparts [24]: linear, square nonlinear, isotropic nonlinear and anisotropic nonlinear. Compared to previous work [23], the new contribution is the weight term W_k which allows the optimization to the color into account as an additional constraint.

Our implementation uses the Euler-Lagrange equation to transform the optimization problem into a linear system of equations, which we then solve using Gauss-Seidel [23]. The runtime is shown in Tab. 1 with a C++ implementation.

Fig. 2 compares the difference between super-resolution without the color constraint (Fig. 2b) and with the color constraint (Fig. 2d). Incorporating the color constraint gives a smoother surface while preserving important shape detail. It also produces a result that compares more favorably to ground truth Fig. 5c. Here, the errors are shown in the color-coded error plots (Fig. 2c and Fig. 2e, error range in Fig. 5e). The average Euclidean error (AEE) with the color constraint is smaller as a result.

We also demonstrate the effect of the color constraint for a human scan. Fig. 3b shows results using the LidarBoost [11] filter (square regularization), and Fig. 3c shows results using the anisotropic nonlinear filter [23]. In general, anisotropic performs best because it employs a diffusion tensor instead of a simple square regularization. Notice that for both filters, the result that uses the color constraint gives a smoother appearance while preserving important shape detail. We thus use the anisotropic nonlinear operator with the color constraint to super-resolve the point clouds.

5   Global Registration

We consider the human body as articulated, with rigid structures connected by joints. Aligning such point clouds is challenging, especially considering that the Kinect depth data has much noise even after super-resolution. We solve this special global registration

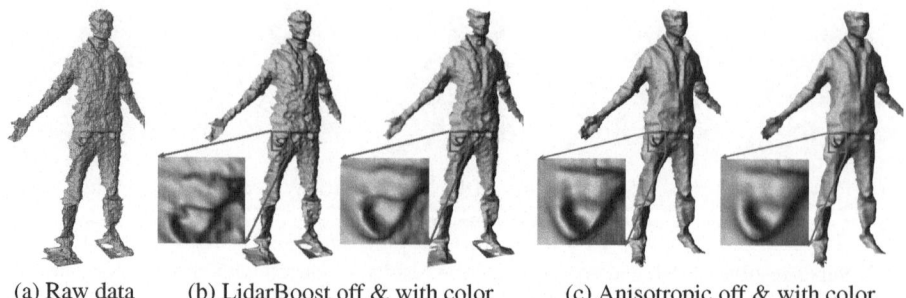

(a) Raw data   (b) LidarBoost off & with color   (c) Anisotropic off & with color

Fig. 3. Super-resolved meshes using the LidarBoost and Anisotropic filters. The images compare results with and without the color constraint.

problem inspired by two approaches. First, we incorporate the maximum-likelihood formulation described by Myronenko et al. [25] and Cui et al. [2] which explicitly takes into account the sensor's noise characteristics. Second, we employ the articulated model described in Chang and Zwicker [3] to describe the non-rigid motion. Here, the surface motion is expressed in terms of an articulated model. We extend their ideas to our setting and develop an approach for global rigid and non-rigid registration for noisy Kinect data.

5.1 Problem Setup

The input to registration are 3D point clouds $Y_f = \{y_{f,n} \mid n = 1, \ldots, N_f\}$ from each super-resolved frame $f \in H_\ell$. Here, N_f is number of points in frame f.

We need to solve for the motion of each frame so that all frames align to each other. To parameterize the motion, we define a set of rigid transformations $M_f^0, M_f^1, M_f^2, \ldots$ for each frame f. This set will describe the motion of the frame from its original location. Since there are multiple transformations per frame, we will need to define which transformation associates with each point. We define this association as a label $i(n)$: the index of the transformation assigned to $y_{f,n}$. Therefore, $M_f^{i(n)}$ will be the transformation for the point $y_{f,n}$. This parameterization of the motion is essentially an articulated model, since the points divide into rigid parts according to their label [3].

When we set up the problem in this way, the alignment task then reduces to solving for the transformations and labels that give the best alignment possible. We model a likelihood function describing the quality of the alignment, and we maximize this likelihood for all frames simultaneously to produce the result.

Conformal Geometric Algebra. We use an exponential map based on conformal geometric algebra to express the transformations. The Euclidean transformations of a point X into X' in the conformal space caused by a motor M and a reverse motor $\tilde{M}$ is approximated as:

$$X' = MX\tilde{M} = E + e_\infty(x - l \cdot x - m), \tag{3}$$

where E is the identity matrix, e_∞ is the point at infinity, and $l, m \in \mathbb{R}^3$ corresponds to the rotation and traslation in the conformal geometric algebra, respectively. This formulation has the advantage that we only have 6 degrees of freedom (DoF). A formulation with a $3 \times 3 \in \mathbb{R}$ matrix and a $3 \times 1 \in \mathbb{R}$ translation vector would have 12 variables instead. A further advantage is that we can gain linear equations with respect to minimizing an energy function using the motor. For notational convenience, we use the expression $\mathcal{T}_f^{i(n)}(X)$ to denote the transformation of point X using $M_f^{i(n)}$ $(\mathcal{T}_f^{i(n)}(X) = M_f^{i(n)} X \tilde{M}_f^{i(n)})$. For more background information for Eqn. 3 and its optimization, please refer to the references [26][27][28].

Deformation Model. To aid the optimization, we find neighbor relations between the rigid parts and constrain such neighbors to be joined together at a common location. These locations we call "joints" and infer them during the optimization. Specifically, we use ball joint relations (3 DoF) between rigid parts, expressed by a single point denoted $y_{f,ab} \in \mathbb{R}^3$ relating two rigid transformations M_f^a and M_f^b. The joints constrain neighboring transformations to agree on this common ball joint location. This ensures that the rigid parts stay connected and do not drift away. We estimate the joint locations automatically using the same technique as in previous work [3].

Difference to Previous Work. Our problem setup is similar to that of Chang and Zwicker [3]. However, there are notable differences in our work. First, we do not subsample the frames but instead define a label for all points in all frames. Therefore, we use the mesh connectivity in each frame to define neighborhood relationships between the points.

Second, we do not define a reference frame or employ a dynamic sample graph. The transformations operate directly on the scanned points so that every scan aligns to all others. In addition, since we associate the labels directly to the scan points, we obtain a separate, independent segmentation per frame (based on the motion occuring in the frame). This means that, unlike prior work, all frames are moving independently. Also, we have a separate set of joint locations per frame, whereas the joint locations were defined on the reference frame previously. Changing the problem setup in this manner allows us to optimize the alignment of all frames in a single optimization step.

Probabilistic Model. The key ingredient for making the registration robust to noise and outliers is the probabilistic modeling of the point clouds. We consider each Y_f to be generated from a Gaussian Mixture Model (GMM), with density as follows [2][25]:

$$p(x) = \sum_{n=1}^{N_f} \frac{1}{N_f} p(x \mid n) \text{ with } p(x \mid n) \propto N(y_{f,n}, \sigma^2 I) . \tag{4}$$

Simply speaking, $p(x)$ gives the probability that $x \in \mathbb{R}^3$ is generated by Y_f. As the equation shows, we model the GMM using the original point set Y_f and center each multi-variate Gaussian $N(y_{f,n}, \sigma^2 I)$ at the scanned points $y_{f,n}$. All Gaussians share the same isotropic covariance matrix $\sigma^2 I$, with I as a 3×3 identity matrix and σ^2 as the variance in all directions.

5.2   Energy Function

We define a measure of how well the point sets align using the following function:

$$\arg\min_{\mathcal{M},\mathcal{L}} E_{\text{data}}(\mathcal{M},\mathcal{L}) + \lambda E_{\text{reg}}(\mathcal{M},\mathcal{L}), \tag{5}$$

where $\mathcal{M}$ is the entire transformation set and $\mathcal{L}$ are the labels for all the points of the K frames. E_{data} measures the alignment distance between points in each frame, E_{reg} constrains the labels for a smooth segmentation and also neighboring transformations to agree on a common joint location. λ is weighting coefficient.

Data Term E_{data}. To achieve a closed model, all frames should be aligned globally with minimal distance. For each pair of frames f and g, the alignment task is performed by minimizing the negative log-likelihood based on the probabilistic model:

$$E_{\text{data}}(\mathcal{M},\mathcal{L}) = -\sum_{f,g}\sum_{n=1}^{N_f}\log\sum_{m=1}^{N_g}\exp\left(\frac{\left\|\mathcal{T}_f^{i(n)}(y_{f,n}) - \mathcal{T}_g^{j(m)}(y_{g,m})\right\|^2}{-2\sigma_{f,g}^2}\right), \tag{6}$$

where $i(n)$ is the index of the transformation assigned to the point $y_{f,n}$, and $j(m)$ the same for the point $y_{g,m}$ but based on the segmentation of frame g. The variance $\sigma_{f,g}^2$ of mixture components is estimated separately for each point using

$$\sigma_{f,g}^2 = \frac{1}{N_f N_{\text{near}}}\sum_{n=1}^{N_f}\sum_{m\in\text{Near}(y_{f,n})}^{N_{\text{near}}}\left\|\mathcal{T}_f^{i(n)}(y_{f,n}) - \mathcal{T}_g^{j(m)}(y_{g,m})\right\|^2, \tag{7}$$

where $m \in \text{Near}(y_{f,n})$ denotes the indices of the nearest N_{near} points in frame g for point $y_{f,n}$. In our experiments, we use a value of $N_{\text{near}} = 20$.

Regularization term E_{reg}. There are two parts for the regularization term E_{reg} [3]. The first part is a smoothness term for the labels, which constrains neighboring points n, m to have a similar label to ensure a smooth segmentation result. If the label is not the same ($i(n) \neq i(m)$), we apply a penalty $I(\cdot) = 1$ which is added to E_{reg}.

The second part is the joint constraint which ensures that the rigid parts do not drift away from each other. Each ball joint specifies that its point $y_{f,ab}$ should move to the same location when applying M_f^a and M_f^b. With α providing a relative weighting of the two constraints, the resulting energy function is

$$E_{\text{reg}}(\mathcal{M},\mathcal{L}) = \sum_f\left(\underbrace{\sum_{(n,m)\in f} I(i(n)\neq i(m))}_{\text{Label Constraint}} + \alpha\underbrace{\sum_{\text{Joints }(a,b)}\left\|\mathcal{T}_f^a(y_{f,ab}) - \mathcal{T}_f^b(y_{f,ab})\right\|^2}_{\text{Joint Constraint}}\right). \tag{8}$$

5.3   Expectation-Maximization

Therefore, for each of the K frames, we minimize the above energy to get a the set of transformations per frame. We use an iterative Expectation Maximization (EM) like procedure to find a maximum likelihood solution of Eqn. 5.

During the E-step, the best alignment parameters from the previous iteration are used to compute an estimate of the posterior $P_{f,g}^{\text{old}}(m \mid y_{f,n})$ of mixture components using Bayes theorem [2]. This posterior is a matrix of dimension $N_g \times N_f$, where each matrix entry p_{mn} gives a conditional probability for a pair of points $(y_{f,n}, y_{g,m})$ from frame f and g. Computing this matrix is intensive and would spend about 10 hours for two frames. It turns out that most entries are zero, and a relatively small number of pairs are actually close enough to yield a non-zero probability. Therefore, we only consider the N_{near} closest points in frame g for each point in frame f. As mentioned earlier, we use $N_{\text{near}} = 20$ which we empirically determined to be sufficient for our experiments. In addition, we compute the posterior matrix values only for four frames before and four frames after Y_f. These eight neighbor frames were enough for the registration, since the subject turns continuously in our experiments. Using these approximations, we can simultaneously optimize the alignment of all frames in a reasonable computation time.

During the M-step, the new alignment parameter values are found by minimizing the negative log-likelihood function, or more specifically, its upper bound Q which evaluates to:

$$Q\left(\mathcal{M}, \mathcal{L}\right) = \sum_{f,g} \left(\sum_{n=1}^{N_f} \sum_{m=1}^{N_g} P_{f,g}^{\text{old}}\left(m \mid y_{f,n}\right) \frac{\left\| \mathcal{T}_f^{i(n)}(y_{f,n}) - \mathcal{T}_g^{j(m)}(y_{g,m}) \right\|^2}{2\sigma_{f,g}^2} \right). \tag{9}$$

Here, the variances $\sigma_{f,g}^2$ in Eqn. 7 are continuously recomputed which is similar to an annealing procedure, in which the support of the Gaussians is reduced as the point sets get closer. Since the deformation parameters change after each M-step, the variances are recomputed after the M-step update.

The EM procedure converges to a local minimum of the negative log-likelihood function. To improve convergence, we optimize in two phases. In the first phase, we perform rigid registration by setting all labels $i(n)$, $n = 1, \ldots, N_f$ to be the same within each frame Y_f. Thus, each frame is exactly one rigid part. We iterate the E and M steps until the transformations converge, yielding a rigid registration of all frames.

After completion of the first phase, we move to the second phase. This time, we run the EM procedure once again, but relaxing the restriction on the labels. Thus, while the E-step remains the same, we solve for both labels and transformations in the M-step. Iterating the E and M steps in this fashion completes the non-rigid registration.

5.4   Non-Rigid M-Step

Since the rigid registration is essentially the same as previous work, we refer to the references for further information [2][3][28]. Instead, we describe the non-rigid M-step optimization in a little more detail.

For the non-rigid registration with joint constraints, we need to minimize the whole term given in Eqn. 9, including both the labels and the transformations. Using the rigid transformation results from the first phase as the initial input of the non-rigid energy, we perform the M-step in two sub-steps iteratively until convergence.

Sub-Step 1. Fix labels $\mathcal{L}$ and solve for transformations $\mathcal{M}$. For E_{data}, the labels $i(n)$ and $j(m)$ are fixed, and the variables are the transformations M. For the regularization

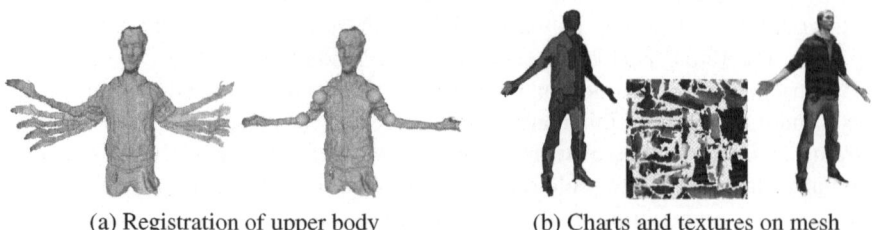

(a) Registration of upper body   (b) Charts and textures on mesh

Fig. 4. (a) Challenging task for the non-rigid registration. Here, the joints are shown as blue balls. (b) Texture mapping.

E_{reg}, only the joint constraint remains, since the labels are fixed. We use the same optimization method as the rigid registration, except that the joint constraints are added as additional terms, and we solve for more transformations simultaneously.

Sub-Step 2. Fix transformations $\mathcal{M}$ and solve for labels $\mathcal{L}$. The labels $i(n)$ are the variables that we are solving for, and this affects the location of the points because it changes which transformation is being applied. Therefore, the goal is to re-segment the points to yield a better registration. In E_{reg}, the joint constraint can be ingnored, and only the label constraint is left to ensure that the number of segmented parts in each frame is not too high. We solve the resulting discrete optimization problem using the α-expansion algorithm [29].

6   Texture Mapping

After the registration is complete, we reconstruct a closed mesh and apply textures to reproduce the person's appearance. In order to assemble a texture, we first compute a 2D parameterization for the reconstructed mesh. Since 3D surfaces cannot be mapped to 2D without any form of distortion, we segment the mesh in regions homeomorphic to discs that can be unfolded to the 2D domain without exceeding a certain threshold of distortion. These regions are called *charts* (Fig. 4b). We use the method of Sander et al. [30] that automatically segments the mesh into charts. Then we compute the 2D parameterization of each chart independently. To obtain a global non-overlapping parameterization of the whole mesh we pack the 2D parameterizations of each chart into a common *texture atlas* [31] (Fig. 4b).

Based on the non-rigid global alignment, we build a textured depth map for each view. This map provides the combined depth and texture information (after non-rigid registration) for each pixel of the depth camera. For each pixel p in the texture map we compute a corresponding 3D point P on the reconstructed surface using the parameterization. We project P into each textured depth map and use the difference in depth to decide whether P is visible in the respective view. The final color of p is assembled as the weighted average of the texture information from each view where P is visible. We use the scalar product of the surface normal and the viewing direction as weight. Fig. 4b depicts our texture mapping result.

7   Results

First, we demonstrate the accuracy of our rigid global registration with a static Lion model (Fig. 5). A local ICP algorithm optimizes the transformations frame by frame, so the registration drifts and loops are not closed properly (shown in Fig. 5a from a top view). However, our result (Fig. 5b) computes the transformation based on a global energy function and properly solves the loop closure problem. We also quantitatively measure the reconstruction quality with a laser-scanned ground truth model. Fig. 5d shows the final reconstruction result and the accuracy of the reconstruction compares favorably to the ground truth Fig. 5c, as seen in the color-coded error plots (Fig. 5d, 5e). While the ICP registration yields an error above 1cm for 50% of the points, our result yields an error below 3mm for 90% of the points.

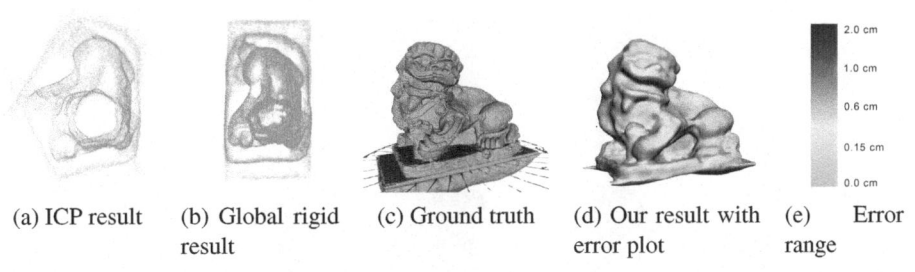

(a) ICP result   (b) Global rigid result   (c) Ground truth   (d) Our result with error plot   (e)   Error range

Fig. 5. Global rigid registration for a Lion model (height 18cm). The rightmost two figures show a comparison with ground truth.

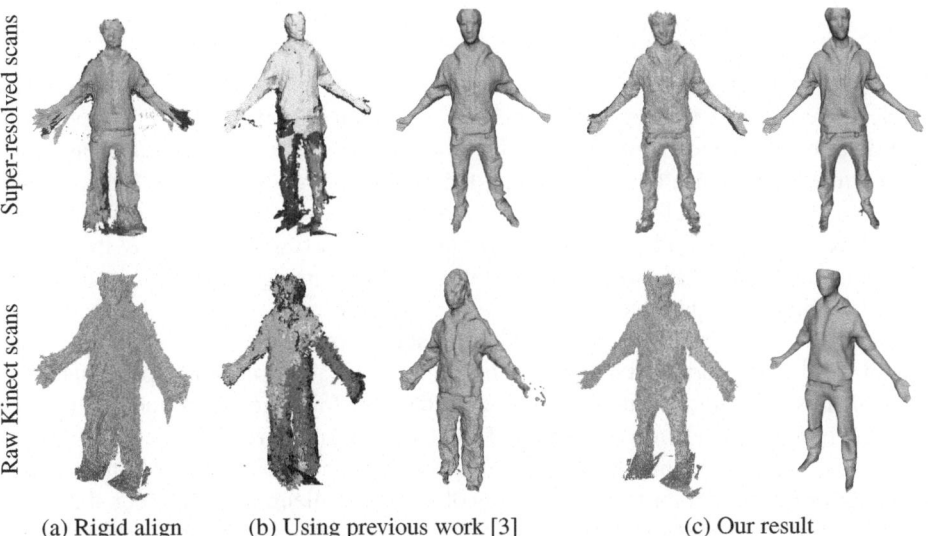

(a) Rigid align   (b) Using previous work [3]   (c) Our result

Fig. 6. Comparing global non-rigid registration for super-resolution data (top row) and raw Kinect data (bottom row). We show the registration and the reconstructed mesh for both cases. The total energy (Eqn. 5) for (b)-top: 66.82, (b)-bottom: 128.93, (c)-top: 4.18, (c)-bottom: 23.74.

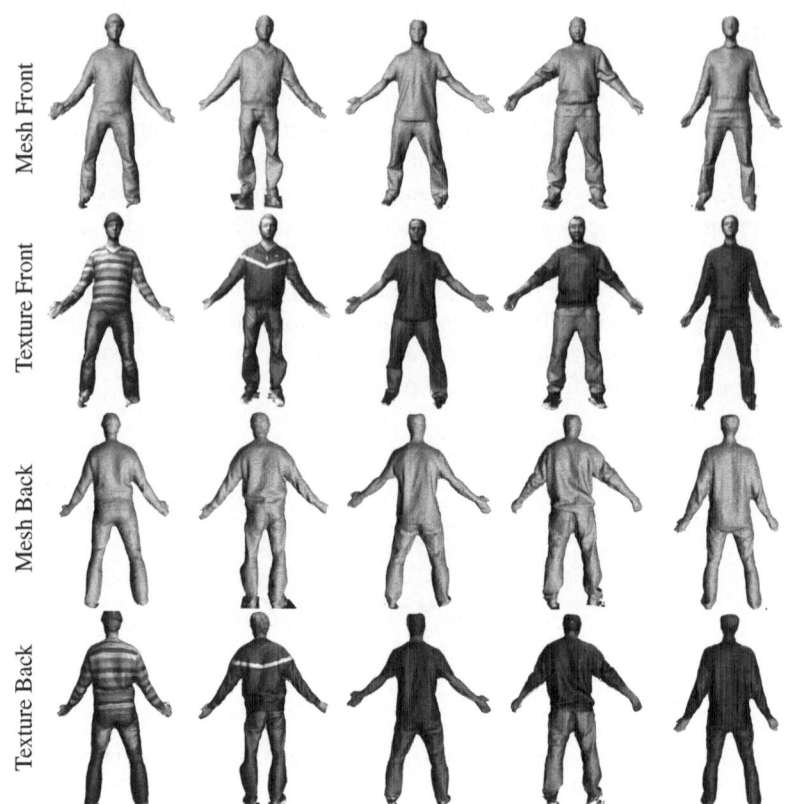

Fig. 7. Results of full 3D body scanning with a single Kinect

Next, we compare the non-rigid alignment algorithm to previous work. We perform two types of experiments: one testing the alignment of super-resolved scans, and one testing in extreme noise conditions using the raw Kinect scans (Fig. 6). By raw data, we mean the raw output from the Kinect depth sensor without the super-resolution step. In both cases, our method gives more accurate alignment results, especially in the arm and hand regions that have much movement. Therefore, our algorithm is more robust to noise in a high-noise data scenario.

The most important reason our method is successful is the probabilistic distance model. This works very well in the presence of noise. In addition, unlike previous work [3] which uses a sliding window to solve for the transformations, our method always takes all frames into account and can perform loop closure more effectively.

In most of our experiments, the real non-rigid displacement for each frame is not large. However, we demonstrate the results of a more challenging task in Fig. 4a. Here, the Kinect has captured five frames of a upper body human model with waving arms. The displacement between each frame is about 20 pixels, and the width of the arm is only about 10 pixels. Even in this case, the non-rigid algorithm can still find the correct registration and joint positions.

Table 1. Runtime for each processing part and biometric error measurements. Timings and measurements are averaged among eight people.

Frames	S. R.	Rigid	Non-Rigid	Poisson	All
36	28 sec	110 sec	620 sec	68 sec	826 sec (13.8 min)
Neck-Hip	**Shoulder W.**	**Arm L.**	**Leg L.**	**Waist Girth**	**Hip Girth**
2.1cm	1.0cm	2.3cm	3.1cm	3.2cm	2.6cm

Finally, Fig. 7 shows our scanning results on five users. Our result reproduces the whole human structure well (especially the arms and legs), and can reconstruct detailed geometry such as the face structure and the wrinkles of the clothes. To evaluate the accuracy of the reconstruction, we compare the biometric measurements of the reconstructed human models with data from the actual people in Tab. 1. The values are the average absolute distance among eight people.

We also show average runtime statistics in Tab. 1. The whole processing time for each model is about 14 minutes on average, using an Intel(R) Xeon 2.67GHz CPU with 12GB of memory. Note that 90% of the time in our method is used for computing closest points. Previous work on human body reconstruction [7] can only capture nearly naked human bodies and spends nearly one hour of computation time, and prior work on articulated registration [3] computes the registration frame by frame in K minimization steps, taking nearly two hours to compute.

8   Conclusion and Future Work

In this paper, we demonstrate that a full 3D human body model can be scanned with a single Kinect, which at first glance seems completely inappropriate for the task. Currently, the system requires the user to maintain a relatively awkward "T" pose while turning. Too much motion in the arms and legs can throw the registration off. And there are three main cases caused possilbe failure: 1) Not enough overlapping area for corresponding views. 2) The non-rigid movement segmentations are similar shape. 3) The segmentations change a lot in different views. To make this process more comfortable, we plan to investigate more sophisticated noise and deformation models to be able to handle larger user movements. In addition, we would like improve our algorithm to run in real-time. This would provide real-time feedback to the user about inaccurate or unfilled areas during the scanning process.

References

1. Microsoft, http://www.xbox.com/en-US/kinect
2. Cui, Y., Schuon, S., Chan, D., Thrun, S., Theobalt, C.: 3D shape scanning with a time-of-flight camera. In: IEEE Proc. CVPR (2010)
3. Chang, W., Zwicker, M.: Global registration of dynamic range scans for articulated model reconstruction. ACM Trans. Graph. 30 (2011)

4. Henry, P., Krainin, M., Herbst, E., Ren, X., Fox, D.: RGB-D mapping: Using depth cameras for dense 3D modeling of indoor environments. In: RGB-D: Advanced Reasoning with Depth Cameras Workshop in Conjunction with RSS (2010)

5. Newcombe, R.A., Kim, D., Kohli, P.: Kinectfusion: Real-time dense surface mapping and tracking. In: ISMAR (2011)

6. Cui, Y., Stricker, D.: 3D shape scanning with a kinect. ACM SIGGRAPH Posters (2011)

7. Weiss, A., Hirshberg, D., Black, M.J.: Home 3D body scans from noisy image and range data. In: Proc. IEEE Intl, Conf. Computer Vision, CVPR (2011)

8. Anguelov, D., Srinivasan, P., Koller, D., Thrun, S., Rodgers, J., Davis, J.: SCAPE: shape completion and animation of people. ACM Trans. Graph. (SIGGRAPH) 24, 408–416 (2005)

9. Tong, J., Zhou, J., Liu, L., Pan, Z., Yan, H.: Scanning 3D full human bodies using kinects. IEEE Transactions on Visualization and Computer Graphics (Proc. IEEE Virtual Reality) (2012) (to appear)

10. Tomasi, C., Manduchi, R.: Bilateral filtering for gray and color images. In: Proceedings of the Sixth International Conference on Computer Vision (1998)

11. Schuon, S., Theobalt, C., Davis, J., Thrun, S.: Lidarboost: Depth superresolution for ToF 3D shape scanning. In: Proc. CVPR (2009)

12. Besl, P.J., McKay, N.D.: A method for registration of 3-D shapes. IEEE PAMI 14, 239–256 (1992)

13. Benjemaa, R., Schmitt, F.: A Solution for the Registration of Multiple 3D Point Sets Using Unit Quaternions. In: Burkhardt, H., Neumann, B. (eds.) ECCV 1998. LNCS, vol. 1407, pp. 34–50. Springer, Heidelberg (1998)

14. Bergevin, R., Soucy, M., Gagnon, H., Laurendeau, D.: Towards a general multi-view registration technique. IEEE PAMI 18, 540–547 (1996)

15. Weise, T., Wismer, T., Leibe, B., Van Gool, L.: Online loop closure for real-time interactive 3D scanning. Computer Vision and Image Understanding 115 (2011)

16. Mitra, N.J., Flory, S., Ovsjanikov, M., Gelfand, N., Guibas, L., Pottmann, H.: Dynamic geometry registration. In: Symposium on Geometry Processing, pp. 173–182 (2007)

17. Huang, Q., Adams, B., Wicke, M., Guibas, L.: Non-rigid registration under isometric deformations. Computer Graphics Forum 27, 1449–1457 (2008)

18. Wand, M., Adams, B., Ovsjanikov, M., Berner, A., Bokeloh, M., Jenke, P., Guibas, L., Seidel, H.-P., Schilling, A.: Efficient reconstruction of non-rigid shape and motion from real-time 3D scanner data. ACM Trans. Graph. (2009)

19. Popa, T., South-Dickinson, I., Bradley, D., Sheffer, A., Heidrich, W.: Globally consistent space-time reconstruction. Computer Graphics Forum (Proceedings of SGP) 29 (2010)

20. Li, H., Luo, L., Vlasic, D., Peers, P., Popović, J., Pauly, M., Rusinkiewicz, S.: Temporally coherent completion of dynamic shapes. ACM Trans. Graph. 31 (2012)

21. Tevs, A., Berner, A., Wand, M., Ihrke, I., Bokeloh, M., Kerber, J., Seidel, H.-P.: Animation cartography - intrinsic reconstruction of shape and motion. ACM Trans. Graph. (to appear)

22. Kazhdan, M., Bolitho, M., Hoppe, H.: Poisson surface reconstruction. In: Proceedings of the Fourth Eurographics Symposium on Geometry Processing, SGP 2006, pp. 61–70 (2006)

23. Cui, Y., Schuon, S., Thrun, S., Stricke, D., Theobalt, C.: Algorithms for 3D shape scanning with a depth camera. IEEE T-PAMI (2012)

24. Weickert, J., Hagen, H.E.: Visualization and Processing of Tensor Fields. Springer, Berlin (2006)

25. Myronenko, A., Song, X., Carreira-Perpinan, M.: Non-rigid point set registration: Coherent Point Drift. In: NIPS, vol. 19, p. 1009 (2007)

26. Murray, R.M., Li, Z., Sastry, S.S.: A Mathematical Introduction to Robotic Manipulation. CRC (1994)

27. Rosenhahn, B.: Pose estimation revisited. PhD thesis, Universität Kiel (2003)

28. Cui, Y., Hildenbrand, D.: Pose estimation based on geometric algebra. In: GraVisMa (2009)
29. Boykov, Y., Kolmogorov, V.: An experimental comparison of min-cut/max-flow algorithms for energy minimization in vision. IEEE PAMI 26, 1124–1137 (2004)
30. Sander, P.V., Wood, Z.J., Gortler, S.J., Snyder, J., Hoppe, H.: Multi-chart geometry images. In: Proceedings of the Eurographics/ACM SIGGRAPH Symposium on Geometry Processing, SGP, pp. 146–155 (2003)
31. Nöll, T., Stricker, D.: Efficient packing of arbitrary shaped charts for automatic texture atlas generation. Computer Graphics Forum 30, 1309–1317 (2011)

Essential Body-Joint and Atomic Action Detection for Human Activity Recognition Using Longest Common Subsequence Algorithm

Sou-Young Jin and Ho-Jin Choi

Dept. of Computer Science, KAIST
291 Daehak-ro, Yuseong-gu, Daejeon 305-701, Korea (South)

Abstract. We present an effective algorithm to detect essential body-joints and their corresponding atomic actions from a series of human activity data for efficient human activity recognition/classification. Our human activity data is captured by a RGB-D camera, i.e. Kinect, where human skeletons are detected and provided by the Kinect SDK. Unique in our approach is the novel encoding that can effectively convert skeleton data into a symbolic sequence representation which allows us to detect the essential atomic actions of different human activities through longest common subsequence extraction. Our experimental results show that, through atomic action detection, we can recognize human activity that consists of complicated actions. In addition, since our approach is "simple", our human activity recognition algorithm can be performed in real-time.

1 Introduction

Human activities are complicated. It often consists of many different atomic actions such as hand stretching, leg lifting, and/or head moving. Human activity recognition, on the other hand, tries to recognize high level semantic meaning of human actions such as walking, running, jumping. Majority of previous approaches [1–14] recognize human activity by considering different part of human actions as a whole to build a human activity classifier/model. Essentially, such approaches often fail when a human performs complicated actions that are largely deviated from the "normal" activities in their training data.

In this paper, in contrary to previous approaches that use a whole body actions to recognize human activities, we propose an alternative approach that recognizes human activity by detecting essential atomic actions of different human activity classes in training data. Our approach is motivated by the observation that only certain part of atomic actions of a human is sufficient to recognize a human activity. For example, to recognize a "walking" activity, we only require to recognize the body movement and the atomic actions of legs where actions of upper body of a human is irrelevant. This observation also applies to many common human activities such as hand waving, jumping and sitting.

To demonstrate our idea on human activity recognition, we develop an algorithm that works on the human skeleton data. The human skeleton data can

J.-I. Park and J. Kim (Eds.): ACCV 2012 Workshops, Part II, LNCS 7729, pp. 148–159, 2013.
© Springer-Verlag Berlin Heidelberg 2013

be captured by a RGB-D camera and/or by any other motion capturing system that can gather human skeleton data. For our convenience, we use the Kinect [15] and its SDK to get human skeleton data [16]. We collect many skeleton data of different activity classes, i.e. boxing, jumping, hand waving, sitting/standing. Each training example is labeled manually with the class label. Our goal is to detect the atomic actions within each class such that we can recognize activities by just detecting similar atomic actions in testing video. We assume essential atomic actions of each class are the actions that are repeated by themselves frequently (with variations) in training data.

Our algorithm starts by converting human skeleton data into symbolic sequence representation. The symbolic sequence representation can tolerate inter-class variations of atomic actions, which allows us to recognize human activity in a robust manner. In addition, since we assume essential atomic actions are repeated actions in training data, we convert our problem into a problem that finds the longest common subsequences in training data that are repeated by themselves within and among different training examples within the same activity class. In testing phase, the learnt longest common subsequences will be compared with the symbolic sequence representation of testing data to classify if a testing example belongs to an activity class defined in training phase. Since our recognition algorithm only involves a very simple operation for common subsequent string detection, our algorithm can recognize human activity in real-time. Our experimental results also show that our approach out-performs previous approaches that use the whole body actions for activity recognition.

The remainder of this paper is organized as follows. Section 2 presents the related works. Section 3 describes our algorithm in details. Section 4 demonstrates the effectiveness of our approach through experiments. Section 5 concludes our paper.

2   Related Works

There is a great deal of work targeting activity recognition. We discuss examples most relevant to our approach that recognize human activity using human pose information and refer the reader to the following recent articles by Poppe [17] and Aggarwal and Ryoo [18] for more thorough surveys.

With an adventure of Microsoft Kinect [15], state-of-the-art pose estimation algorithm [16] can predict human pose in terms of 3D body-joint positions from a single depth image. Such information is very useful in human activity recognition. In Sung et al. [19], they use the pose information from Kinect to extract features such as body pose features, hand position and motion information for activity recognition. Their body pose features compose of 10 torso related body-joint orientations, such as relative position between foot/hand and torso, relative position between head and hip, and relative positions between head and torso. For each consecutive frame in a video, they calculate the changes of these body-joint angles and train an activity recognizer using hierarchical maximum-entropy Markov models for each labeled action.

Similar to Sung et al. [19], Tran et al. [20] also use Kinect data for activity recognition. They introduced a polar space which measures the distance and angle orientations of different body part from the center of a body. A 2D histogram is created to capture the frequency of body parts being observed at each different quantized locations. A classifier is trained using the 2D histogram as features to recognize different human activity. While both Sung et al.[19] and Tran et al. [20] demonstrated some successes in human activity recognition using Kinect data, these two approaches consider human activity as a whole body actions. When there is an "abnormal" activity, such as a person is waving hand during walking, these approaches often produce less than satisfactory results.

Meanwhile, there are also approaches that try to recognize human activity by considering the importance of different body parts. Ryoo and Aggarwal [21] proposes a description-based approach for activity recognition. Their basic idea is to use a context-free grammar (CFG) to represent and encode the hierarchical structure of human activity. In each frame, the system extracts poses and gestures for each body part: head, upper-body, lower-body, and hand position. If all of required poses and gestures are recognized and the time intervals for them satisfy the relationships described in the representation, the system deduces the action is occurred. Unlike other previous works, Ryoo and Aggarwal treat body parts separately so that poses and gestures are recognized in each body part. However, in their training process, it requires manual encoding on poses or gestures for each body part as well as the time intervals for each atomic action using their CFG syntax.

Chakraborty et al. [22] propose the ensemble of body-part detectors using hidden Markov model. Similar to our motivation, they observe that not all body-parts contribute equally to all action classes. Thus, their approach is especially robust in distinguishing between similar actions since it only considers the certain body-parts that has major contribution to the actions. Nevertheless, their approach still requires manual labeling to characterize different body-parts for different human activity class.

Comparing our work with previous works in [21] and [22], our approach only requires manual label of different activity class. It is automatic in detecting essential body-joints and atomic actions that are necessary for activity recognition. Moreover, our symbolic sequence representation is robust in body-joint angle orientation and it is efficient in activity recognition after training.

3 Algorithm

In this section, we present our algorithm for activity recognition from human skeleton data. We will first present our representation on how to convert skeleton data into our symbolic sequence representation. Next, we will present our main algorithm on how to learn essential body-joint and atomic action from training data using the Longest Common Subsequence (LCS) algorithm [23, 24]. Finally, we describe how we can detect human activity via sequence matching.

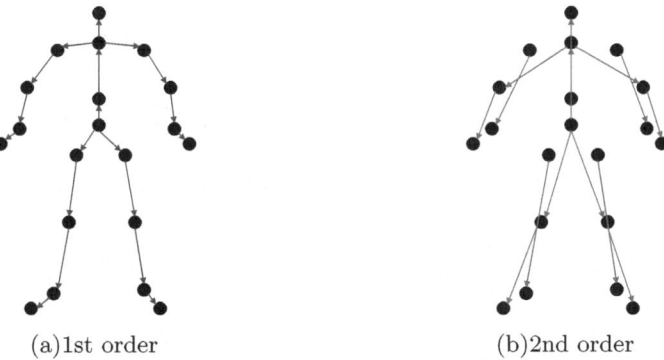

(a)1st order   (b)2nd order

Fig. 1. The skeleton data from Kinect SDK contains 19 different body-joint vectors. Each body-joint vector captures the vector orientation from one joint to another joint. On top of the first order neighborhood provided by the SDK, we defined a second order neighborhood that connects adjacent vectors. The body-joint angle of second order vector can be easily computed from the body-joint angles of the first order vector. In total, we have 19 first order vectors and 14 second order vectors which capture human pose information at time t.

3.1   From Skeleton to Symbolic Representation

The skeleton data from Kinect SDK has 19 different body-joint vectors and the center location of human body as illustrated in Figure 1(a). These body-joint vectors encode the vector orientations of different body parts from one body-joint to another body-joint in 3D world-coordinate. Since human actions usually contain different amount of variations, using the original body-joint angles for training can be sensitive. Here, we propose to convert these body-joint angles into a symbolic representation by quantizing the angles from continuous domain into a discrete symbol as illustrated in Figure 2. For each body-joint angle in 3D world coordinate, we project the angle onto xy-plane and yz-plane respectively. The projected angles are then quantized into eight discrete symbols. Hence, we quantize body-joint angles in 3D world coordinate into 64 discrete symbols which is robust enough to encode human action while tolerates small variations across different examples of the same activity. Figure 3 shows an example on how to encode the body-joints angles of left shoulder, left elbow, left wrist, and left hand, into symbols for the left hand waving action at time t and time $t + 1$.

Besides body-joint angles, we also need to consider the change of human body location as it also provides very useful information in activity recognition. Similar to body-joint angle, we measure the location difference of "hip center" body-joint between the current and the next frame and then convert the movement vector into our symbolic representation. Figure 4 illustrates our process. We quantize the location difference similar to body-joint angle. After that, we further quantize the magnitude of movement vector to "no movement", "small movement" and "large movement" by two thresholds. The thresholds were set empirically. In our experiments, we found that setting the first threshold as half of average

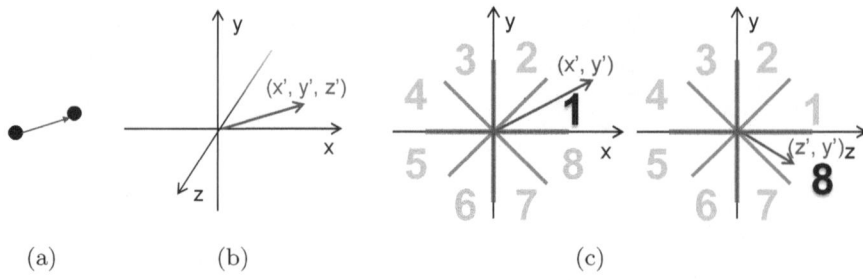

Fig. 2. We convert a continuous angle in 3D space into discrete symbol according the projected angles in xy-plane and yz-plane

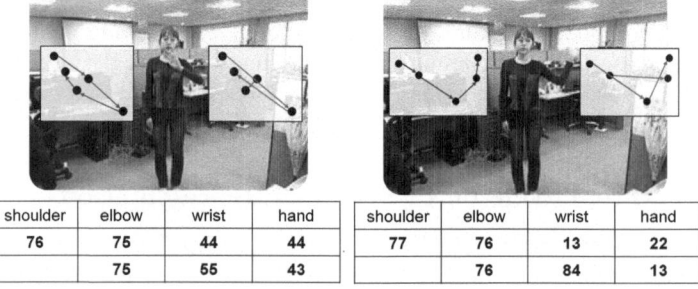

	shoulder	elbow	wrist	hand	shoulder	elbow	wrist	hand
1st order adjacent vector	76	75	44	44	77	76	13	22
2nd order adjacent vector		75	55	43		76	84	13

Fig. 3. Body-joint symbols for one item of data containing a "waving" action

movement in training data, and the second threshold as three times larger than the first threshold produce good recognition accuracy.

After converting skeleton data into our symbolic representation, we obtain 34 sequences where 19 sequences corresponding to the first order body-joint angles, 14 sequences corresponding to the second order body joint angles, and 1 sequences corresponding to human body movement. Since we are interested in "active" actions, we remove symbols that are identical in consecutive frame as illustrated in Figure 5. The compact representation of motion sequences allows us to focus on only the moving part of human activity for training and testing.

3.2 Essential Body-Joints and Atomic Action Detection

From the symbolic representation of motion sequences, we want to detect the essential body-joints and the atomic action that can be used for activity recognition. We define the essential body-joints to be the body-joint vectors that are necessary and sufficient for recognizing certain activity. For example, to recognize a "walking" activity, we only require the body-joint vectors of hips and legs and the human body movement vector. The body-joint vectors of head and

previous
current
y
x
z
large movement
small movement
no movement
(a)
(b)
(c)

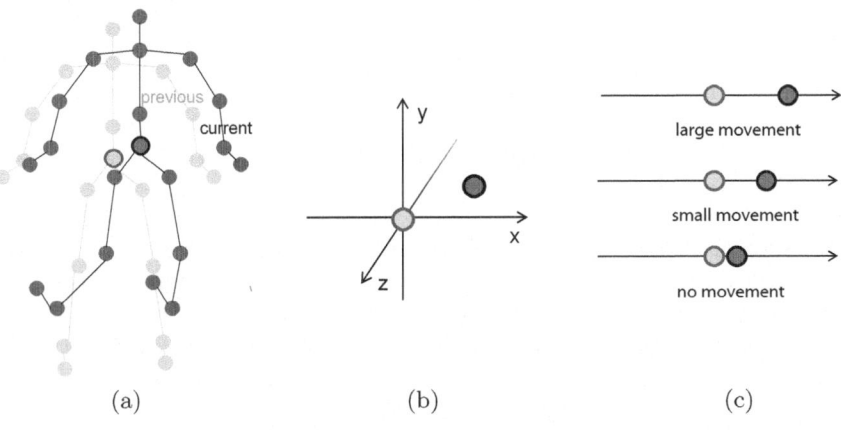

Fig. 4. (a) We compute the movement vector by measuring the location difference of "hip center" body-joint between consecutive frames. (b) The movement vector is converted into symbolic representation according to the movement vector direction. (c) We additionally add another symbol to encode magnitude of the movement vector.

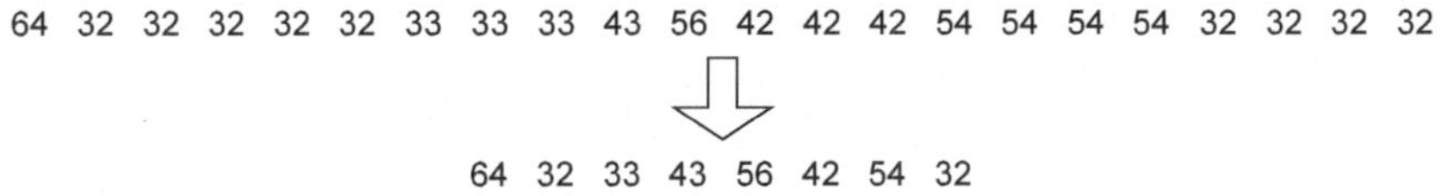

64 32 32 32 32 32 33 33 33 43 56 42 42 42 54 54 54 54 32 32 32 32
64 32 33 43 56 42 54 32

Fig. 5. We remove identical symbols in consecutive frames to adopt a compact representation of motion sequences

hands are less important or even irrelevant. This observation also applies to many other human activity that only involves certain parts of body actions but not the whole body actions. We define the atomic action to be the pattern of actions of essential body-joints that can be used to recognize certain activity. Using the "walking" activity as example, this activity involves legs lifting and legs stretching to follow certain sequences. By detecting atomic actions of essential body joints, one can evaluate and recognize if a person has performed certain activity or not.

We adopt the Longest Common Subsequence (LCS) algorithm [23, 24] to serve our purpose. Among the labeled training data, we compare the current longest common subsequence of each body-joint to the symbolic motion sequence of new data of the same body joint:

$$atomic(a,i) = \begin{cases} LCS(atomic(a,i-1), i\text{-}th\ example) & \text{if } i > 0 \\ i\text{-}th\ example & \text{if } i = 0 \end{cases} . \qquad (1)$$

where a is index of body-joint, i is index of training examples, and $atomic(a, i)$ is the atomic action of body-joint a learnt from the first i-th training examples. This process continues until we process all training examples. After performing the LCS algorithm, some body-joints may have no common sub-sequences. We discard these body-joints as we believe these are irrelevant body-joint. Accordingly, we consider a body-joint is essential only if the learnt atomic actions have three or more symbols in its sequences.

During the recognition phase, we build a finite state machine for each atomic action of essential body-joint in each training category. When a new testing sequence comes, the skeleton data can be converted into our symbolic representation in real-time and then verified by the finite state machine. The benefit of finite state machine is that it can effectively filter out consecutive identity symbols in testing caused by slow motion of activity. Hence, our algorithm is robust to the speed variations of human activity.

4   Experimental Results

We evaluate our algorithm in this section using real-world examples. We will first describe our training data. Next, we will evaluate if our algorithm can successfully detect essential body-joints by comparing with ground truth essential body-joints. Finally, we test our learnt essential body-joints and atomic actions to recognize human activity with challenging examples.

4.1   Training Data

We build our own training data by using Kinect to capture real-world examples. Our training data consists of four different activity categories: "boxing", "jumping", "hand waving" and "sitting/standing". For each activity category, we collect two different type of examples: Single Action (SA) examples and Composite Action (CA) examples. In the single action training data, only the essential body-joints are moving while the other body-joints are static. In composition action training data, not only the essential body-joints are moving but the other body-joints are also free for moving. Figure 6 shows examples of single action and composite action training data. We assign label manually to each training example. For composite action training examples that consist of multiple action categories, multiple labels has been given. Within each category, we collect 16 training examples with different persons performing the same activity.

4.2   Evaluation on the Detected Essential Body-Joints

Our first experiment evaluates the effectiveness of our method on essential body-joints detection. We detect the essential body-joints for each category using SA and CA training data separately. Figure 7 shows our detected essential body-joints for different activity category. The skeletons on the left hand side of each category are the results learnt from SA training data, and the skeletons on the

(a) Single Action (SA) example of a "hand waving" action.

(b) Composite Action (CA) example that has both "hand waving" and "walking" actions.

Fig. 6. Examples of our training data

right hand side of each category are the results learnt from SA training data. The detected essential body-joints were highlighted.

With the SA training data, not only the essential body-joints were detected. Some inessential body-joints were also detected. For example, the body-joints related to two arms should not be detected as essential for a "jumping" action. However, the system identifies some body-joints around arms since the subjects habitually move their arms to perform a "jumping" action. When the system uses the CA training data, the result of essential body-joint detection is better in most action categories. However, although both left and right arms are essential for a "boxing" action, only two body-joints in the right arm - 'wrist right' and 'hand right' - are detected as essential. Though not all essential body-joints are selected for a "boxing" action, the system is able to recognize the action with only these two body-joints. Table 1 shows the learnt atomic action sequences for the "boxing" action and the "hand waving" action. Both wrist and hand were detected as essential body-joints, yet our learnt atomic action sequences are different. Our algorithm can successfully detect different body-joints and atomic actions for different activity category and the "correct" essential body-joints in each activity were included.

4.3 Activity Recognition Accuracy

We evaluate and compare our activity recognition accuracy with approach from Tran et al. [20]. In [20], Tran et al. describe video data by a polar histogram which each histogram bin represents the frequency of certain body-parts appeared in the

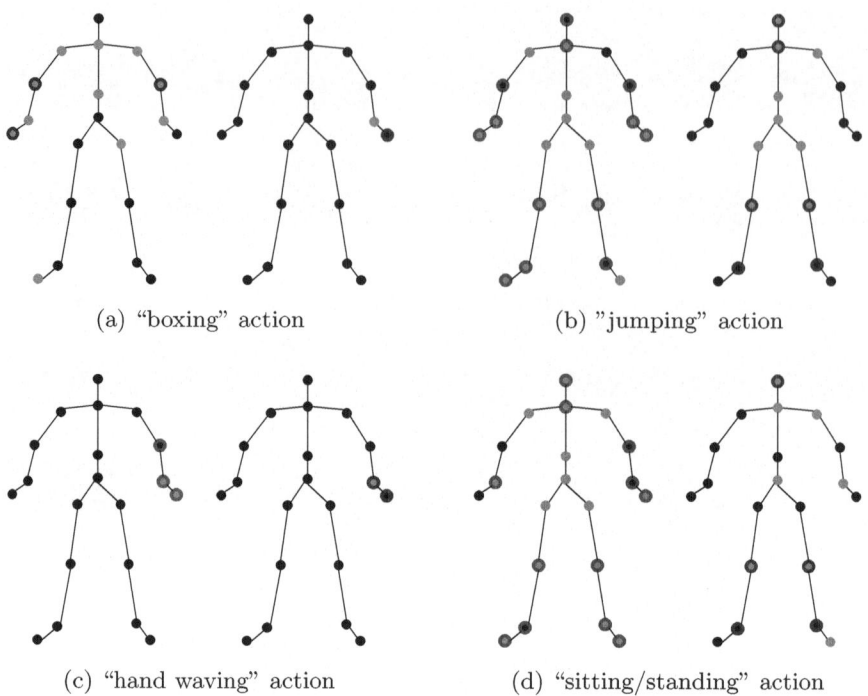

(a) "boxing" action   (b) "jumping" action

(c) "hand waving" action   (d) "sitting/standing" action

Fig. 7. Results of essential body-joint detection. Every left skeleton indicates the detected body-joints when the SA dataset is used while every right skeleton is associated with those when the CA dataset is used. A body-joint is filled with a color, if the body-joint's 1st order adjacent vector is selected. In addition, a body-joint's edge is colored if the 2nd order adjacent vector is selected.

Table 1. Comparison of the common sequences between a "boxing" action and a "waving" action. Although the detected essential body-joints are similar, atomic action sequences for each selected body-joint differ from each action.

Body joint		When the system detects essential body joints using CA training data	
Name	Adjacent order	"boxing" action	"hand waving" action
wrist right	1st	44 55 44	33 23
hand right	2nd	44 55 44	23 33
wrist right	2nd		14 34 24

histogram bin area. We call this approach a Whole Body-Joint (WBJ) approach since they use all available body-part actions to build their polar histogram for activity recognition. While our approach only uses the detected essential body-joints for activity recognition.

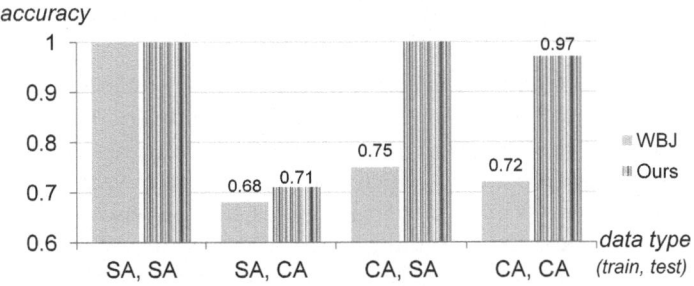

Fig. 8. Activity recognition accuracy. Our approach out-perform WBJ approach when the training/testing data contains composite action.

Similar to previous experiments, we again separate the SA and CA data to train and test the activity recognition accuracy of our approach and the WBJ approach. Note that we adopt the leave-one-out cross validation. Figure 8 compares the average accuracies. In the axis of data type, the first letter is related to the type of training data while the second is associated with that of test data. As we can observe from Figure 8, when training or test data contains composite actions, our approach performs much better than the WBJ approach.

Meanwhile, we have also observed that our approach does not work well when training data is SA but the testing data type is CA. This is because some of inessential body-joints were detected and were used to recognize activity. This result also confirms that essential body-joint detection is necessary for activity recognition. Consequently, our approach performs better when using CA data for training.

Fig. 9. We show seven frames from the CA training data where the ground truth labels are both "hand waving" and "sitting/standing". WBJ approach fails this example, while our approach can still recognize both labels.

Figure 9 shows a challenging testing example. This example contains both "hand waving" and "sitting/standing" activity simultaneously. WBJ approach fails to recognize the labels of this example since their approach uses a whole body actions to recognize activity which is not applicable to this challenging composite action. On the other hand, our approach only uses essential body joints to recognize activity. Both "hand waving" and "sitting/standing" labels were successfully detected using our proposed algorithm.

5   Conclusion

We have presented a simple yet effective algorithm to automatically detect essential body-joints and atomic action sequences for human activity recognition. To handle within class activity variations, we convert continuous skeleton data into discrete symbolic representation which can tolerate small variations of activity performed by different people. After that, we detect essential body-joints and atomic action sequence from training data by extracting the longest common subsequence among the activity examples within the same activity category. In recognition phase, we build a finite state machine for each learnt atomic action sequences of detected essential body-joints to recognize human activity in real-time.

While majority of previous approaches focus on whole body actions of human activity, our approach focuses on the atomic actions of essential body-joints. The experimental results show that our observation that only essential part of body actions is necessary and sufficient for activity recognition is valid. With essential body-joint detection, our approach can outperform recent WBJ activity recognition algorithm on challenging examples which contain composite actions with multiple labels. Our current approach assumes the training data does not contains any noise or outliers, e.g. wrong labels. Such outliers can potentially damage our atomic action detection algorithm. In the future, we shall study how to incorporate probability model into our algorithm to increase the robustness for detecting atomic action sequences. We shall also study how to incorporate some advanced feature selection techniques into our algorithm for essential body-joint detection.

Acknowledgement. This work was supported by the National Research Foundation (NRF) grant (No. 2012-0001001) of Ministry of Education, Science and Technology (MEST) of Korea. This work was also supported by the KUSTAR-KAIST Institute, Korea, under the R&D program supervised by the KAIST.

References

1. Wang, L., David Suster, D.: Recognizing human activities from silhouettes: Motion subspace and factorial discriminative graphical model. In: IEEE Conference on Computer Vision and Pattern Recognition, pp. 1–8 (2007)
2. Weinland, D., Boyer, E.: Action recognition using exemplar-based embedding. In: IEEE Conference on Computer Vision and Pattern Recognition, pp. 1–7 (2008)
3. Wang, Y., Huang, K., Tan, T.: Human activity recognition based on r transform. In: Workshop of IEEE Conference on Computer Vision and Pattern Recognition for Visual Surveillance, pp. 1–8 (2007)
4. Souvenir, R., Babbs, J.: Learning the viewpoint manifold for action recognition. In: IEEE Conference on Computer Vision and Pattern Recognition, pp. 1–7 (2008)
5. Wang, L., Suter, D.: Informative shape representations for human action recognition. In: International Conference on Pattern Recognition, pp. 1266–1269 (2006)

6. Huang, F., Xu, G.: Viewpoint Insensitive Action Recognition Using Envelop Shape. In: Yagi, Y., Kang, S.B., Kweon, I.S., Zha, H. (eds.) ACCV 2007, Part II. LNCS, vol. 4844, pp. 477–486. Springer, Heidelberg (2007)

7. Cherla, S., Kulkarni, K., Kale, A., Ramasubramanian, V.: Towards fast, view-invariant human action recognition. In: Workshop of IEEE Conference on Computer Vision and Pattern Recognition for Human Communicative Behaviour Analysis, pp. 1–8 (2008)

8. Souvenir, R., Babbs, J.: Learning the viewpoint manifold for action recognition. In: IEEE Conference on Computer Vision and Pattern Recognition, pp. 1–7 (2008)

9. Weinland, D., Ronfard, R., Boyer, E.: Free viewpoint action recognition using motion history volumes. Computer Vision and Image Understanding 104, 249–257 (2006)

10. Huang, W., Wu, Q.M.J.: Human action recognition based on self organizing map. In: IEEE International Conference on Acoustics Speech and Signal Processing, pp. 2130–2133 (2010)

11. Ahmad, M., Lee, S.W.: Variable silhouette energy image representations for recognizing human actions. Image and Vision Computing 28, 814–824 (2010)

12. Abdelkader, M.F., Abd-Almageed, W., Srivastava, A.: Silhouette-based gesture and action recognition via modeling trajectories on riemannian shape manifolds. Computer Vision and Image Understanding 115, 439–455 (2011)

13. Jia, K., Yeung, D.Y.: Human action recognition using local spatio-temporal discriminant embedding. In: IEEE Conference on Computer Vision and Pattern Recognition (2008)

14. Gorelick, L., Shechtman, E., Irani, M., Basri, R.: Actions as spatio-temporal shapes. IEEE Transactions on Pattern Analysis and Machine Intelligence 29, 2247–2253 (2007)

15. Corporation Microsoft.: Kinect for xbox 360 (2010)

16. Shotton, J., Fitzgibbon, A., Cook, M., Blake, A.: Real-time human pose recognition in parts from single depth images. In: IEEE Conference on Computer Vision and Pattern Recognition (2011)

17. Poppe, R.: A survey on vision-based human action recognition. Image and Vision Computing 28, 976–990 (2010)

18. Aggarwal, J., Ryoo, M.: Human activity analysis: A review. ACM Computing Surveys 43, 1–43 (2011)

19. Sung, J., Ponce, C., Selman, B., Saxena, A.: Unstructured human activity detection from rgbd images. In: IEEE International Conference on Robotics and Automation, pp. 842–849 (2012)

20. Tran, K., Kakadiaris, I., Shah, S.K.: Part-based motion descriptor images for human action recognition. Pattern Recognition 45, 2562–2572 (2012)

21. Ryoo, M., Aggarwal, J.: Semantic representation and recognition of continued and recursive human activities. International Journal of Computer Vision 82, 1–24 (2009)

22. Chakraborty, B., Bagdanov, A.D., Gonzalez, J., Roca, X.: Human action recognition using an ensemble of body-part detectors. Expert System (2011)

23. Hirschberg, D.S.: Algorithms for the longest common subsequence problem. Journal of the ACM 24, 664–675 (1977)

24. Bergroth, L., Hakonen, H., Raita, T.: A survey of longest common subsequence algorithms. In: 7th International Symposium on String Processing and Information Retrieval, pp. 39–48 (2000)

Exploiting Depth and Intensity Information for Head Pose Estimation with Random Forests and Tensor Models

Sertan Kaymak and Ioannis Patras

Queen Mary, University of London, UK
{s.kaymak,i.patras}@eecs.qmul.ac.uk

Abstract. Real-time accurate head pose estimation is required for several applications. Methods based on 2D images might not provide accurate and robust head pose measurements due to large head pose variations and illumination changes. Robust and accurate head pose estimation can be achieved by integrating intensity and depth information. In this paper we introduce a head pose estimation system that employs random forests and tensor regression algorithms. The former allow the modeling of large head pose variations using large sets of training data, while the latter allow the estimation of more accurate head pose parameters. The combination of the above mentioned methods results in more robust and accurate predictions for large head pose variations. We also study the fusion of different sources of information (intensity and depth images) to determine how their combination affects the performance of a head pose estimation system. The efficiency of the proposed framework is tested on the Biwi Kinect Head Pose dataset, where it is shown that the proposed methodology outperforms typical random forests.

1   Introduction

Human-centered user interfaces can greatly gain from accurate 3D head pose estimation. The need for accurate, resolution independent methods that operate in real-time is therefore crucial. The majority of work on head pose estimation was on databases that contain few head pose variations. Limited work has been conducted on data containing large head movements. However, the accuracy of head pose estimation algorithms that employ 2D images might be affected by several factors, such as illumination changes and large head pose variations. In order to deal with it, a system could employ a 3D sensor and acquire in that way a depth image, that is less affected by illumination changes and large head pose variations.

In this paper we aim at estimating in a more robust and accurate way the head pose parameters. To achieve that we exploit both intensity and 3D depth information. Our system is comprised of two parts. In the first part, we attempt head pose estimation from intensity images and depth data using random regression forests as in [1]. More precisely we extract fixed size patches from the

J.-I. Park and J. Kim (Eds.): ACCV 2012 Workshops, Part II, LNCS 7729, pp. 160–170, 2013.
© Springer-Verlag Berlin Heidelberg 2013

face region and learn a regression model using these patches. In the second part, we employ tensor regression as in [4]. To this end we find a function that maps tensorial representations to output parameters.

In Figure 1 we can see an outline of the proposed head pose estimation system. First, the captured depth data are mapped onto the intensity image for head pose estimation. Afterwards the face region is detected. In this study, face region is assumed to be known. After face region is determined on both intensity and depth data, we generate regression models based on random forest and tensor models. The model generation is performed in two stages. In the first stage, random regression forest is trained similarly to [1], [2] and leaf nodes are constructed. In standard random forests, the leaf nodes consist of head location and orientation parameters. In this study, the leaf nodes contain extracted patches in addition to head location and orientation parameters. In the second stage, a tensor model for regression at each leaf node is generated using patches and corresponding head orientation parameters. A tensor model is represented with a weight vector and a bias and stored at each leaf node. The higher rank Support Tensor Regression model is used to map head regions to head orientation parameters. In the testing part, patches are densely extracted from face regions of intensity image and depth data and passed to the random forest. The tensor model provides estimates of head pose parameters at each leaf node.

The remainder of this paper is as follows. Related work is discussed in Section 2. The proposed technique which is based on the combination of random forests and tensor regression models is presented in Section 3. In Section 4, the proposed methodology is evaluated using a Kinect database. Finally, in the final section, conclusions are drawn.

2 Related Work

Head pose estimation approaches can be divided in three main categories: those that employ 2D images, those that employ 3D depth data and those that combine 2D images and 3D depth data.

Head pose estimation approaches that employ 2D images can be further divided into two groups: 2D appearance based methods and 2D feature based methods. The 2D appearance based techniques analyze the entire head region. Osadchy et al. [9] proposed a Convolutional Neural Network based large head pose estimation system which allowed mapping face images to head pose parameters and achieved a near real-time performance (5 fps). The modeling of facial regions was also used for tracking using statistical techniques such as Active Appearance Models (AAMs) [12], multi-view AAMs [13], and 3D Morphable Models [14],[15] and Constrained Local Models [16]. 2D feature based methods are based on facial feature detection for the head orientation calculation. Vatahska et al. [10] proposed detecting facial features and estimating head orientation using the detected locations of the points in three stages. First, the head pose was classified as frontal, left and right profile using a face detector. Afterwards, the facial features were detected by training AdaBoost classifiers with Haar-like

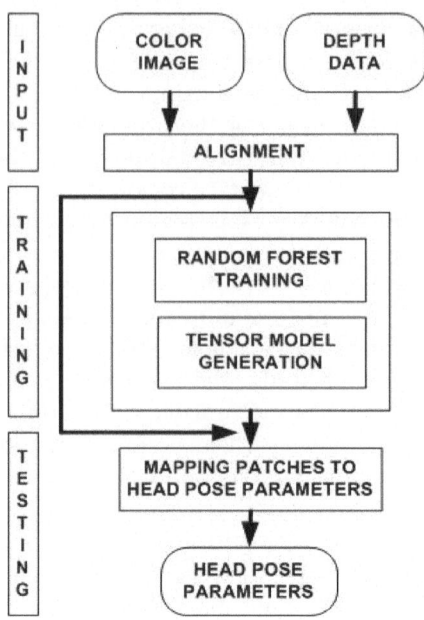

Fig. 1. Head Pose Estimation system

features. Finally, the locations of detected features were mapped to head orientation parameters using neural networks. Whitehill et al. [11] proposed a method in which the orientation of a head was calculated using locations of the node tip and both eyes.

The second category of head pose estimation systems uses 3D depth data. The system proposed by Breitenstein et al. [5] allowed head pose estimation from depth data in real time. The real time performance has been achieved using a GPU. Large 3D face depth data in different head poses was stored into the GPU memory and a unknown depth data was recognized by comparing with the stored data. Real-time head pose estimation techniques also employed Random Regression Forests, like in Fanelli et. al. [1], [2]. In [1] the authors generated large 3D synthetic faces and trained Random Forests for continuous head pose estimation. They also extended this technique and achieved joint classification and regression for head pose estimation in [2]. This system allowed extracting patches from the upper body region of a person from depth data and only patches which belonged to the head region were used to estimate the head pose in real time. The estimation was performed using low resolution data captured by Microsoft Kinect Camera.

The third category involves the combination of 2D and 3D data. Seemann et al. [7] presented a head pose estimation system based on neural networks. Grayscale and depth data were used as inputs to the neural networks to calculate the head pose. Morency et al. [8] calculated a prior model of the face using intensity and depth images. This model was used to calculate the absolute difference in pose for each new image.

Random forests [3] are a powerful method which efficiently addresses classification and regression problems. They are capable of using large training data while they are easy to implement. It was shown in [3] that growing trees in the forest by randomly selecting a subset of training images from the training set provides a strong regressor. The random selection of features in each node is also of great importance.

Tensor based learning approaches are known as more powerful than vector-based approaches [4]. The use of vector representations results in several problems. First the structure information is lost. Second, high dimensional data might cause over fitting problems to appear. In [4], tensor learning for regression was proposed. Two mapping functions were learned using Canonical (CANDE-COMP)/Parallel factors (PARAFAC) decomposition, [6]. The square loss and e-sensitive loss functions were studied together with Frobenius norm. The reported results showed that tensors resulted in improved accuracy in terms of angular error for head pose estimation.

3   Random Forest with Tensor Learning

Random regression forests, [1],[2], consist of several trees and allow continuous mapping from input patches to output parameters. During a random forest construction each tree is constructed separately and independently. For each tree, fixed size patches are extracted from a subset of training data. Starting from the root node, patches are divided into two groups at each non leaf node according to a binary test that is determined during training time. The splitting process provides several subsets called leaf nodes. The leaf nodes are generated using location and orientation vectors. These models are based on the multivariate Gaussian distributions which are calculated and stored at each leaf node.

In this study, the higher rank Support Tensor Regression model (hrSRT), [4], that was proposed instead of the multivariate Gaussian distribution for regression is used. The hrSTR models were generated and stored at each leaf nodes of the trees.

3.1   Feature Channels

Data fusion is proposed in order to increase robustness and accuracy. More precisely, features are generated from both intensity image and dense depth values. Captured color images are converted to gray scale images for head pose estimation.

Intensity. Ten different feature channels are generated. The first channel contained raw gray scale values, while the remaining nine channels contained HoG like features. Gray values are used as a feature channel as extra computation to obtain them is not required. HoG like features are used in the fusion process as they provide good estimates for head pose parameters [17].

Depth. Raw depth values are used as one of the feature channels in the fusion process as their values provide accurate estimates with random regression forests [2].

3.2   Random Forest Training

A number of aligned gray scale images and depth data with head location and orientation parameters are used for the construction of a forest. An example of aligned depth data and grayscale image can be seen in Figure 2. For each tree construction, a subset of aligned intensity and depth data is selected and several fixed size patches are extracted from both intensity image and depth data. HoG features are also calculated from intensity images and fixed size patches are also extracted from images which contain HoG features. In this study, the face region of a person is assumed to be known and the patches are extracted only from this region.

A tree in the forest can be represented by $T=\{T_t\}$ while a subset of patches can be represented by $P_i = (I_i, \theta_i)$. $I_i = (I_i^1, ..., I_i^C)$ represents patches which are extracted from intensity images, intensity feature images and depth data. C denotes the number of channels. $\theta = \{\theta_x, \theta_y \theta_z, \theta_{ya}, \theta_{pi}, \theta_{ro}\}$ represents a vector which describes the distance between the path center and head center together with the Euler angles.

The parameters of a tree in each leaf node are defined by generating several tests and selecting the best parameter set according to a binary test defined by $t_{F_1, F_2, \tau}$

$$| F_1 |^{-1} \sum_{q \epsilon F_1} I^a(q) - | F_2 |^{-1} \sum_{q \epsilon F_2} I^a(q) > \tau \tag{1}$$

where I^a is a feature channel and F_1 and F_2 are rectangular regions determined randomly within the regions of depth patch. τ represents the threshold.

After determining the best test that maximizes the optimization function, patches are split based on test parameters at each leaf node. The splitting process stops and a leaf node is defined when a minimum number of patches existing in the node or the predefined depth level is reached. Leaf nodes are defined by parameters of multivariate Gaussian distribution.

The optimization function which is maximized at each non leaf node is the information gain defined by

$$U_R(P \setminus t^k) = H(P) - (w_L H(P_L) + w_R H(P_R)). \tag{2}$$

where H(P) is differential entropy of the set P and $w_{i=L,R}$ is the ratio of patches sent to each child node.

The equation becomes

$$U_R(P \setminus t^k) = log(\Sigma^v + \Sigma^a) - \sum_{i=\{L,R\}} w_i log(\Sigma_i^v + \Sigma_i^a) \tag{3}$$

where Σ^v and Σ^a are the covariance matrices of the offset vectors and rotation angles.

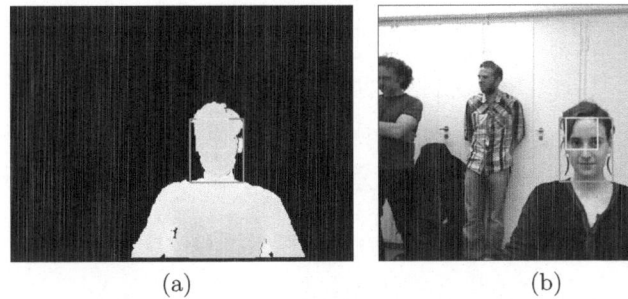

(a)   (b)

Fig. 2. Aligned depth data and gray scale images. (a) Captured depth data. The green bounding box shows head/face region in which fixed size depth patches are extracted. The yellow bounding box shows the extracted fixed size depth patch. (b) Corresponding gray scale image. The green bounding box shows head/face region in which fixed size gray scale patches are extracted. The yellow bounding box shows the extracted fixed size gray scale patch.

3.3   Tensor Learning Model for Regression

Higher rank Support Tensor Regression models are generated at each leaf node. These models provide mapping between sets of patches and head orientation parameters. A tensor model is defined as

$$y = f(\mathcal{X}; \mathcal{W}, b) = \langle \mathcal{X}, \mathcal{W} \rangle + b, \qquad (4)$$

where $\mathcal{X} \in \mathbb{R}^{I_1}$ represents the feature channel patch tensor of 1-mode, and $\mathcal{W}$ is the weight tensor of 1-mode data tensor $\mathcal{X}$. The scalar b is the bias. The weight tensor is defined as the sum of R rank-one tensors and given as

$$\mathcal{W} = \sum_{r=1}^{R} \boldsymbol{u}_r^{(1)} \circ \boldsymbol{u}_r^{(2)} \circ \cdots \circ \boldsymbol{u}_r^{(M)} \triangleq [\![\boldsymbol{U}^{(1)}, \boldsymbol{U}^{(2)}, \cdots, \boldsymbol{U}^{(M)}]\!], \qquad (5)$$

Model parameters ($\Theta = \{ \boldsymbol{U}^{(1)}, \boldsymbol{U}^{(2)}, \cdots, (\boldsymbol{U})^M, b \}$) at each leaf node are learned by minimizing the regularized empirical risk function. This function is minimized by using a set of labeled 1-mode feature channel patch tensors $\{\mathcal{X}_i, y_i\}_{i=1}^N$ and the associated Euler angles, y_i. The risk function is given by

$$L(\Theta) = \frac{1}{2} \sum_{i=1}^{N} l(y_i, f(\mathcal{X}_i; \Theta)) + \frac{\lambda}{2} \psi(\Theta) \qquad (6)$$

where $l(\cdot)$ is the ϵ-insensitive loss function and $\psi(\cdot)$ is the Frobenius norm regularization term. The Frobenius norm requires a priori selection of the rank R of the tensor weight which is calculated by performing cross validation on the data at each leaf node.

The objective function to be minimized is given by

$$\min_{U^{(j)},b,\xi,\hat{\xi}} \quad \frac{1}{2}\mathrm{Tr}(\widetilde{U}^{(j)}\widetilde{U}^{(j)\mathrm{T}}) + C\sum_{i=1}^{N}(\xi_i + \hat{\xi}_i), \tag{7a}$$

$$s.t. -y_i + \mathrm{Tr}(\widetilde{U}^{(j)}\widetilde{X}_{i(j)}^{\mathrm{T}}) + b \geq \epsilon + \hat{\xi}_i,$$

$$y_i - \mathrm{Tr}(\widetilde{U}^{(j)}\widetilde{X}_{i(j)}^{\mathrm{T}}) - b \leq \epsilon + \xi_i,$$

$$\epsilon \geq 0, \ \xi_i \geq 0, \ \hat{\xi}_i \geq 0, \ \forall i = 1, 2, \cdots, N. \tag{7b}$$

The parameters, $\widetilde{U}^{(j)}$ and $\widetilde{X}_{i(j)}$, can be vectorized since

$$\mathrm{Tr}(\widetilde{U}^{(j)}\widetilde{U}^{(j)\mathrm{T}}) = \|\mathrm{vec}(\widetilde{U}^{(j)})\|^2,$$

$$\mathrm{Tr}(\widetilde{U}^{(j)}\widetilde{X}_{i(j)}^{\mathrm{T}}) = [\mathrm{vec}(\widetilde{U}^{(j)})]^{\mathrm{T}}[\mathrm{vec}(\widetilde{X}_{i(j)})], \tag{8}$$

and the problem can be solved using a typical SVMs/SVR optimizer. After calculating $\widetilde{U}^{(j)}$, $U^{(j)}$ is calculated as below:

$$U^{(j)} = \widetilde{U}^{(j)}B^{-\frac{1}{2}}. \tag{9}$$

3.4 Head Pose Estimation Using Random Regression Forest with Tensor Learning

Head pose estimation using random regression forests with tensor regression models is conducted at two stages. First, fixed size patches are densely extracted from the feature channels according to the channel information stored at non leaf nodes. Patches from feature channels are directed using binary tests calculated during training time to a set of leaf nodes. All patches and corresponding tensor models are considered from a forest. A number of Euler angles are calculated using patches that reach leaf nodes. The mean shift algorithm with spherical kernel is applied to the grouped Euler angles in order to determine maximum Euler vector and remove outliers. L2 norm of the Euler angles is calculated in order to group Euler angles and remove outliers. The remaining Euler angles are averaged to obtain estimated head pose.

4   Experiments

The performance evaluation of the proposed head pose estimation system was studied by conducting experiments on the publicly available Biwi Kinect Head Pose database [1],[2]. This dataset was created using Kinect sensor. It consists of depth data and RGB images of upper body region of 20 different people

(14 men and 6 women) that turn their heads in different directions. 24 sequences were generated while some people were recorded twice. All images are annotated with head center locations and rotation angles. The rotation angles range is approximately between ± 75 ° for yaw, ± 60 ° for pitch and ± 50 ° for roll. The approximate rotation and translation errors of this dataset is reported about 1 mm and 1 degree.

The dataset was partitioned into 18 sequences of 18 subjects as training set and 2 sequences of 2 subjects as test set. Two methods were trained using training set. A random forest was constructed by generating 7 trees. Each tree was generated using 3000 sample images.

Random Forest Training. During the tree generation, the parameters were set according to previous experimental observations. More precisely, the values of the parameters used to train the random forest were set as follows: the patch and sub-patch sizes were equal to 60x60 and 10x10 pixels, respectively, the maximum tree depth was 15, the minimum number of patches required for a split was 20 and the number of tests generated at each leaf node was 10000. The stride was set equal to 4 and the maximum variance to 1500. For the random forest and random forest with tensor experiments, the patch size was set equal to 80x80 and the number of patches extracted from the face/head region was 20.

Tensor Model Training. The tensor models training is performed for each leaf node after the random forest is trained. Their performance depends on the rank (R) and the regularization parameters (C) used. The best parameter values were selected by employing cross validation.

4.1   Random Forest with Tensor Models

In this section, we report experiments in order to compare the performance of random forests and random forests combined with tensor regression. The features employed for the experiments were the patches acquired from the depth images. The achieved results can be seen in Figure 3.

More precisely, in Figure 3 (a), the acquired accuracy is plotted against the percentage of retained leaves. As we can see, the proposed framework that combines random forests with tensor regression outperforms the typical random forests. In Figure 3(b), we present the Mean Angle error (MAE) against the percentage of leaves. Random forests, when combined with tensor regression achieve lower MAE than that of typical random forests. We should note here that the head pose estimation was considered successful when the estimated head angle error was less than 10 degrees. In Figure 3(c), we report the accuracy of head orientation for different values of thresholds. Once again the proposed framework that combines Random Forest with tensor regression outperforms the typical random forests.

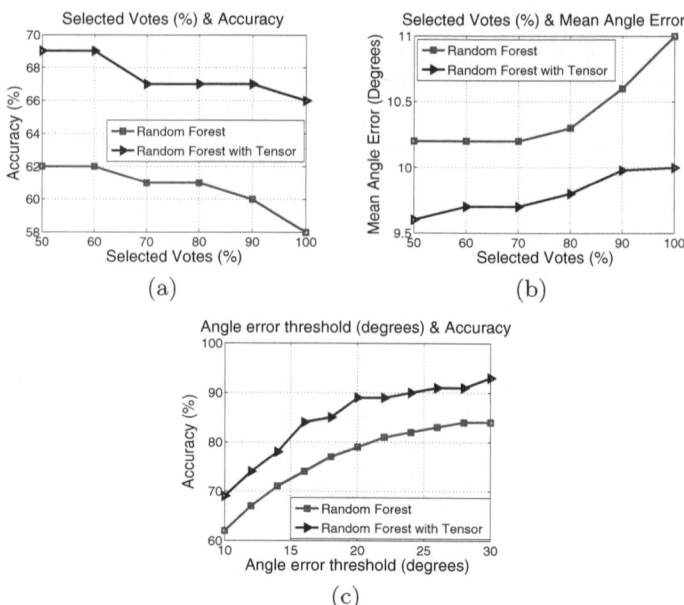

Fig. 3. The performance of typical random forests and random forests combined with tensor regression. (a) Accuracy against the percentage of selected votes. (b) Average angular error against the percentage of retained votes. (c) Accuracy of head orientation estimation for different angle thresholds. The accuracy values were calculated when 50 % votes were selected.

4.2   Data Fusion Using Random Forest

In this section, we study the performance of each type of information (intensity and depth data). More precisely, for intensity data we created ten feature channels, with the first one containing the raw gray values and the remaining 9 channels containing HoG descriptors. For depth data we considered one channel that contained raw depth values.

The acquired results are reported in Figure 4. As can be seen, both higher accuracy and lower MAE were achieved when different sources of information and more than 80 % of the retained votes were used. Table 1 presents mean and standard errors calculated using 2-fold cross validation. In the first row we report mean and standard errors using only depth values while in the second row we report mean and standard errors using depth values, gray scale values and HOG features. A forest is generated using 3000 depth data for the construction of each of seven trees for each fold. As can be seen in Table 1, the fusion of different sources of information provided lower mean angular error.

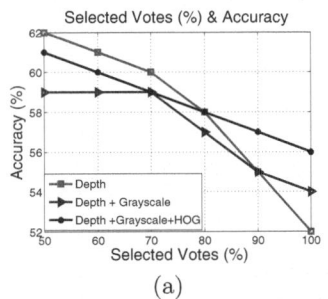
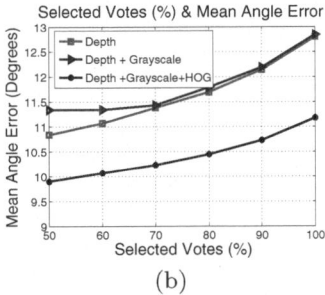

Fig. 4. The performance for different types of information. Depth data, depth data combined with gray values and depth data, combined with gray values and HoG descriptors. (a) Accuracy against the percentage of selected votes. (b) Corresponding average angular error against the percentage of retained votes.

Table 1. Errors of head orientation in terms of mean and standard deviation. Errors are computed using a 2-fold cross validation.

	MAE	Pitch Error	Yaw Error	Roll error
Depth	12.7	4.8 ± 5.3	8.1 ± 9.1	6.5 ± 7.4
Depth+Grayscale+HOG	12.3	5.0 ± 5.3	7.4 ± 7.9	6.6 ± 7.4

4.3 Conclusion

In this paper we present a novel framework for head pose estimation that combines random forests and tensor regression schemes. More precisely, we create random trees and extend them so that they include a tensor regressor at each leaf. In that way we combine the advantages of both methods, thus being able to process large sets of training data by generating strong predictions at each leaf node. We also study the effect of fusion of multiple sources of information has on the performance of a head pose estimation system. The efficacy of our method was demonstrated on the publicly available Biwi Kinect Head Pose Database. The experiments showed that the proposed framework that combines random forests and tensor regression outperforms typical random forests.

Acknowledgement. This work is supported by the EPSRC grant Recognition and Localization of Human Actions in Image Sequences (EP/G033935/1).

References

1. Fanelli, G., Gall, J., Van Gool, L.: Real Time Head Pose Estimation with Random Regression Forests. In: Computer Vision and Pattern Recognition, CVPR, pp. 617–624 (2011)

2. Fanelli, G., Weise, T., Gall, J., Van Gool, L.: Real Time Head Pose Estimation from Consumer Depth Cameras. In: Mester, R., Felsberg, M. (eds.) DAGM 2011. LNCS, vol. 6835, pp. 101–110. Springer, Heidelberg (2011)
3. Breiman, L.: Random Forests. Machine Learning 45, 5–32 (2001)
4. Guo, W., Kotsia, I., Patras, I.: Tensor Learning for Regression. IEEE Transactions on Image Processing 21, 816–827 (2012)
5. Breitenstein, M.D., Kuettel, D., Weise, T., Van Gool, L., Pfister, H.: Real-time face pose estimation from single range images. In: IEEE Conference on Computer Vision and Pattern Recognition, CVPR 2008, pp. 1–8 (2008)
6. Kolda, T.G., Bader, B.W.: Tensor Decompositions and Applications. SIAM Review 51, 455–500 (2009)
7. Seemann, E., Nickel, K., Stiefelhagen, R.: Head pose estimation using stereo vision for human-robot interaction. In: Proceedings of the Sixth IEEE International Conference on Automatic Face and Gesture Recognition, pp. 626–631 (2004)
8. Morency, L.P., Sundberg, P., Darrell, T.: Pose estimation using 3D view-based eigenspaces. In: IEEE International Workshop on Analysis and Modeling of Faces and Gestures, AMFG 2003, pp. 45–52 (2003)
9. Osadchy, M., Cun, Y.L., Miller, M.L.: Synergistic Face Detection and Pose Estimation with Energy-Based Models. J. Mach. Learn. Res. 8, 1197–1215 (2007)
10. Vatahska, T., Bennewitz, M., Behnke, S.: Feature-based head pose estimation from images. In: 2007 7th IEEE-RAS International Conference on Humanoid Robots, pp. 330–335. IEEE (2007)
11. Whitehill, J., Movellan, J.R.: A discriminative approach to frame-by-frame head pose tracking. In: 8th IEEE International Conference on Automatic Face Gesture Recognition, FG 2008, pp. 1–7 (2008)
12. Cootes, T.F., Edwards, G.J., Taylor, C.J.: Active appearance models. IEEE Transactions on Pattern Analysis and Machine Intelligence 23, 681–685 (2001)
13. Ramnath, K., Koterba, S., Xiao, J., Hu, C., Matthews, I., Baker, S., Cohn, J., Kanade, T.: Multi-view AAM fitting and construction. International Journal of Computer Vision 76, 183–204 (2008)
14. Blanz, V., Vetter, T.: A morphable model for the synthesis of 3D faces. In: Proceedings of the 26th Annual Conference on Computer Graphics and Interactive Techniques, pp. 187–194. ACM Press/Addison-Wesley Publishing Co. (1999)
15. Storer, M., Urschler, M., Bischof, H.: 3d-mam: 3d morphable appearance model for efficient fine head pose estimation from still images. In: 2009 IEEE 12th International Conference on Computer Vision Workshops, ICCV Workshops, pp. 192–199. IEEE (2009)
16. Cristinacce, D., Cootes, T.: Feature detection and tracking with constrained local models, pp. 929–938 (2006)
17. Murphy-Chutorian, E., Trivedi, M.M.: Head Pose Estimation and Augmented Reality Tracking: An Integrated System and Evaluation for Monitoring Driver Awareness. IEEE Transactions on Intelligent Transportation Systems 11, 300–311 (2010)

Dynamic Hand Shape Manifold Embedding and Tracking from Depth Maps

Chan-Su Lee, Sung Yong Chun, and Shin Won Park

214-1 Dae-dong, Gyeongsan-si, Gyeongsangbook-do, 712-749, Korea(ROK)
Department of Electronic Engineering, Yeungnam University
{chansu,whiteyongi,psw0085}@ynu.ac.kr

Abstract. Hand shapes vary for different views or hand rotations. In addition, the high degree of freedom of hand configurations makes it difficult to track hand shape variations. This paper presents a new manifold embedding method that models hand shape variations in different hand configurations and in different views due to hand rotation. Instead of traditional silhouette images, the hand shapes are modeled using depth map images, which provides rich shape information invariant to illumination changes. These depth map images vary for different viewing directions, similar to shape silhouettes. Sample data along view circles are collected for all the hand configuration variations. A new manifold embedding method using a 4D torus for modeling low dimensional hand configuration and hand rotation is proposed to model the product of three circular manifolds. After learning nonlinear mapping from the proposed embedding space to depth map images, we can achieve the tracking of arbitrary shape variations with hand rotation using particle filter on the embedding manifold. The experiment results from both synthetic and real data show accurate estimations of hand rotation through the estimation of the view parameters and hand configuration from key hand poses and hand configuration phases.

1 Introduction

There has been a great deal of research on natural human computer interaction in virtual environments, augmented reality, smart devices, and games. Input devices using human body motion are available for games and smart devices like smart TV. Their functionality, however, is limited and far from natural manipulation and signing gestures. The hand is one of the most effective interaction tools due to its dexterous functionality in both communication and manipulation [1]. One of the key components in the understanding of object manipulation and hand signing is in determining articulated hand configurations without the need to wear complicated devices, such as gloves. The estimation of hand motion from 2D image and 3D depth has become a popular research area in computer vision and pattern recognition studies [2–4].

Two of the main approaches in hand configuration estimation are the model-based and learning-based methods. In the model-based approach, complex 3D

J.-I. Park and J. Kim (Eds.): ACCV 2012 Workshops, Part II, LNCS 7729, pp. 171–182, 2013.

models or simple 2D cardboard models are used to generate possible hand pose samples. Hand pose can be determined by searching for the best matched sample generated from a given model. An articulated hand model requires more than 20 DOFs in the model-based approach even though actual fingers have interdependency and constraints depending on the applications. Searching these high dimensional spaces is very challenging and difficult to achieve real-time tracking. Recently methods to reduce the dimensionality of the configuration space or focus on specific task like grasping are used to eliminate difficulties of dimensionality of hand motion configuration space. The learning-based approach maps the shape or appearance of a hand directly into the hand configuration, which requires to learn multi-modal and many-to-one mapping due to self-occlusion and similarity in different views. Specialized map [5] and other regression models are applied to learn the mapping.

Recently, alternative manifold-based approaches have been applied to articulated human motion tracking that utilize the intrinsic constraints of body configuration through low dimensional manifold embedding [6, 7]. Gaussian Latent Variable Models (GPLVM) are applied to learn the mapping between the intrinsic low dimensional body configuration space and the observed feature space in [8]. For hand pose estimation, manifold learning based approaches were applied for hand posture estimation [9], where an Isometric Self-Organizing Map(ISOSOM) is used for an effective 3D hand pose estimation using a low dimensional nonlinear manifold. The hand rotation occurring during a hand configuration change between two key poses is modeled using a cylindrical manifold embedding [10].

However, the modeling of articulated hand configurations in a low dimensional manifold space is still a challenging problem due to the high degree of variation in hand configurations and change in the shape or appearance occurring in different views caused by a hand pose change or hand rotation. In addition, hand configuration changes can occur smoothly from one hand pose to another. Distinctive hand poses can be represented by key hand poses. The question then becomes: how can we model variations from one key hand pose to another taking into account the changes in the view by hand rotation?

This paper presents a new manifold embedding method used to model various key hand configurations and their variations over time (dynamics), and the view changes caused by hand rotation by extending torus manifold embedding to a higher dimensional manifold. Previous torus manifold embedding methods have been applied to model cyclic human motion like walking [11]. The motion manifold is homeomorphic to a circle or part of a circle. The model cannot applied to articulated multiple hand key pose and its variations over time into another key pose. Cylindrical manifold embedding is therefore used to model the configuration change between two key poses [10]. However, it is still a difficult problem to extend the model to multiple key hand poses. The use of 4D torus manifold embedding, which is the product of three one dimensional circle manifold, is proposed to model the key pose variations and simultaneous configuration changes between key poses, and the circular hand rotation in each configuration.

In addition, this paper presents a new shape modeling method using a 3D depth map instead of shape silhouettes to extract hand shapes in variant illumination while still preserving the shape information details. It is difficult to robustly extract hand shapes that match the observed shape with the learned generative models. The 3D depth map can easily distinguish the front of a hand from other objects and backgrounds. A new normalization method from the extracted depth map to compare real images at different depths is also discussed. The distinctions between hands and arms are a challenging problem since they are connected and have similar distances from the camera. To distinguish between arms and hands, we utilized skin color information assuming that the subjects wear long-sleeve shirts, which cover the arms. The experiment results using synthesized and real data show accurate simultaneous estimations of hand configurations and hand rotation.

2   Modeling Arbitrary Hand Pose Variations

All the possible hand pose variations would need to be captured and used to model very accurate hand configuration spaces. However, it is not possible to capture all of the possible hand pose configurations due to the high degrees of freedom of hand configurations, and it may not be necessary to do so.

2.1   Problem Formulation: The Articulated Hand Pose Transition

We can model arbitrary hand configuration variations from one hand configuration to another as a transition between two distinctive key hand configurations. Therefore, when we collect all of the possible key poses (configurations) and their pairwise transitions, we can model the arbitrary hand configuration variations between key poses. Two typical hand configurations are an open hand and a closed hand. If we choose the closed hand as the base hand configuration, then all extreme of the key hand pose can be represented by an open hand with different degree of finger bending (phase). When we determine all of the transition phases between the closed hand and the other extreme key hand poses, we can then model arbitrary in-between pose variations by the combination of two pose transitions between the closed hand and the other key poses.

2.2   Data Collection

We used graphic tools to collect synthesized hand configuration depth map data from multiple simultaneous views. Since we need to model the shape variations not only by hand configuration but also by hand rotation, we need to collect shape variation data along view circle for each given hand configuration. When the closed hand is given as a base hand shape, the extreme key hand configuration can be modeled by the combination of unbend fingers. When we assume each finger is able to independently choose two extreme choices, fully bent and unbent, we end up with 31 cases ($2^5 - 1$). However, each finger motion is not independent. We selected 12 sample key poses in this experiments. For each key pose, we generated 18 samples for the configuration change from a closed hand to the key pose and back to the closed hand. For each given configuration, we sampled view variations from rotation by 15 degree intervals; a total 24 view sample shapes were collected for each configuration. Therefore, a total 5,184 samples were collected in this experiment. Figure 1 (a) shows the 12 selected key poses, (b) the configuration variations in a fixed view and key pose, and (c) the view variations in a fixed configuration. Since this synthetic data is used for learning hand configurations and view shape model from depth map, the model is user-independent and can be applied for any subject in different view and configuration using the extracted depth map with normalization.

(a)

(b)

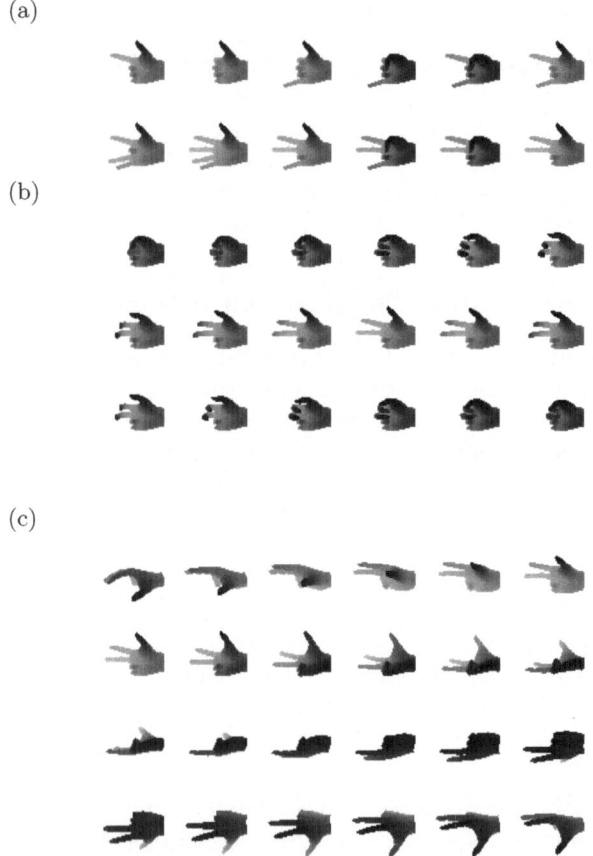

(c)

Fig. 1. The sample shape sequences of the depth map: (a) key hand poses (with fixed view and configuration), (b) hand configuration (with fixed key pose and configuration), and (c) hand rotation (with fixed key pose and configuration)

2.3   The Data-Driven Manifold Embedding

From the collected dataset, we applied linear and nonlinear dimensionality reduction techniques in order to determine the low dimensional representation of the hand shape manifold space. Figure 2 shows the 3-dimensional manifold embedding based on (a) PCA and; (b) the Isomap from the collected depth map sample sequences. These low dimensional embedding has a partially circular manifold structure due to view and configuration variations. However, it is hard to model or parameterize the embedding space based on view circle or hand configuration parameters directly from the embedding spaces. More compact manifold representation with still representing view and configuration separately is required to track hand configuration and view shape variations and estimate configuration and view efficiently.

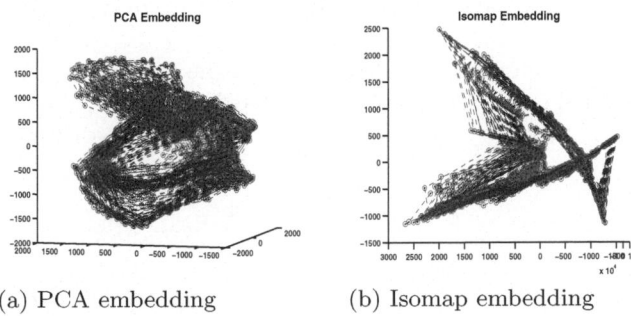

(a) PCA embedding   (b) Isomap embedding

Fig. 2. The data-driven manifold embedding results from the collected depth map sample sequences

3   The Extended Torus Manifold Embedding Method

When a hand configuration sequence changes from the base hand pose (closed hand) to a specific key hand pose; and returns back to the base hand pose, the sequence can be modeled as a circular one dimensional manifold; the sequence is homeomorphic to circle. Additionally, the shape variations along the view circle can be modeled as another independent circular manifold. Therefore, the transition from the base pose to one specific key pose with view circles can be modeled with a torus manifold [11] when the configuration returns from the target key pose to the base pose. Here, the key pose is not fixed but uses arbitrary hand configurations; in this experiments, 12 sample key poses were used. The problem addressed was how to extend the model to cover various key poses to model arbitrary hand configurations and their transitions.

3.1   The 4D Torus Manifold Embedding Method

We used a key hand configuration manifold, independent of view and hand transitions, which represents various key pose variations;modeled by a circular manifold. Since it is independent of view or shape deformation, the key hand configuration can be embedded along a circular manifold based on the pairwise distance amongst the key hand poses. The hand configuration can be represented by the joint angles of each finger joint. Given the key hand configuration angles, distance between two key hand configuration can be measured using the average pairwise distance of the key hand configuration joint angles as

$$d(\boldsymbol{q}_i, \boldsymbol{q}_j) = \frac{1}{\theta_N} \sqrt{\sum_{k=1}^{\theta_N} (\theta_k^{\boldsymbol{q}_i} - \theta_k^{\boldsymbol{q}_j})^2}, \tag{1}$$

where $\theta_k^{\boldsymbol{q}_i}$ is the kth angle component of the $\boldsymbol{q}_i$ key pose; and θ_N is the total number of angle element used to represent key hand configuration. Equation 1 determines the pairwise distance measure between two i'th and j'th key pose $\boldsymbol{q}_i$, and $\boldsymbol{q}_j$ as the average Euclidian distance of each angle element θ_k, where $k = 1, \cdots, \theta_N$.

Given a hand configuration distance metric, the optimal key hand pose configuration embedded on circular manifold can be defined as the shortest-closed loop needed to travel through all of the key hand poses by the distance metric [12] as

$$E = arg \min_Q \sum_{i=1}^{N_k} d(\boldsymbol{q}_i, \boldsymbol{q}_{i+1}), \tag{2}$$

where $Q = q_i \in [1, N_k] | i = 1, \cdots, N_{k+1}$ and $\boldsymbol{q}_i \neq \boldsymbol{q}_j$ for $i \neq j, \boldsymbol{q}_1 = \boldsymbol{q}_{N_k+1}$.

We thereby obtain three independent circle manifolds (S1), which are independent each other. Similar to torus manifold embedding [11], which is the product of two circular manifolds ($S1 \times S1$) and embedded in 3 dimensional Euclidian space with two independent parameters, these three circular manifold product($S1 \times S1 \times S1$) can be represented by 4 dimensional Euclidian space with three independent parameters as follows;

$$w = (r_3 + (r_2 + r_1 cos(\theta))cos(\phi))cos(\psi), \tag{3}$$
$$x = (r_3 + (r_2 + r_1 cos(\theta))cos(\phi))sin(\psi), \tag{4}$$
$$y = (r_2 + r_1 cos(\theta))sin(\psi), \tag{5}$$
$$z = r_1 sin(\theta), \tag{6}$$

where $r_1, r_2, and r_3$ are the radius of each circular manifold and θ, ϕ, ψ are parameters for the key hand pose, hand configuration phase between the base and key hand pose, and hand rotation (view angle) respectively. Therefore, the collected data for each key pose with the hand configuration variations between the base hand pose and key hand pose with circular view change in each configuration can be mapped one-to-one to this extended torus manifold. Figure 3 shows a projection of the 4D torus manifold into three 3-dimensional spaces (w, x, y), (w, x, z), and (w, y, z) for the sample dataset. Each subset of the product of the three circular manifolds contains characteristics of the product of two circular manifolds.

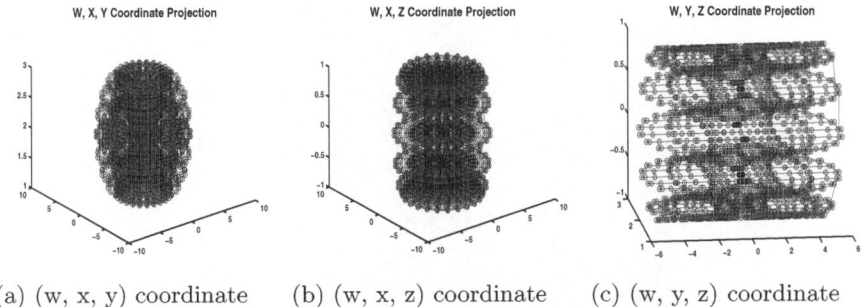

(a) (w, x, y) coordinate   (b) (w, x, z) coordinate   (c) (w, y, z) coordinate

Fig. 3. 3D projections of the embedded 4D torus manifold

3.2 The Nonlinear Generative Model with 4D Manifold Embedding and Reconstruction

Nonlinear Generative Models: Nonlinear mapping can be learned between the embedding space and original shape images of the hand with hand configuration and view variations. In order to normalize the depth map independent of hand size or distance from the camera, we determine the mean distance from the depth map and normalize it by dividing it by the distance value of the estimated standard deviation of the given depth map frames. The nonlinear mapping is obtained by embedding points $x_i = (w_i, x_i, y_i, z_i)$ to the corresponding hand shape $y_i \in R^n$, where n is the column stacked observation distance map image dimension. We used 73×66 gray image representations for the depth map and $n = 4818$ dimensions. The nonlinear mapping can be learned using the generalized radial basis function (GRBF) similar to in [11]. Alternatively, GPLVM [13] can be applied using initialization with 4D torus parameterization and an additional modification of the embedding space. Figure 4 shows the interpolation of the new sequences from the trajectories from the 4D torus embedded manifold.

Tracking by Particle Filter: Using the proposed 4D torus manifold embedding method and its generative model, the hand shape with hand configuration changes and hand rotation can be estimated based on the 4D torus embedding points. The estimation of key hand pose (θ), temporal configuration (ϕ), and hand rotation (ψ) parameters. Particles on the manifold can be used to achieve tracking on the manifold. Hand state $\boldsymbol{X}_t = (\theta_t, \phi_t, \psi_t)$ is estimated from a given observation $\boldsymbol{Y}_t$ using particle filter as:

$$P(\boldsymbol{X}_t|\boldsymbol{Y}^t) \propto P(\boldsymbol{Y}_t|\boldsymbol{X}_t) \int_{\boldsymbol{X}_{t-1}} P(\boldsymbol{X}_t|\boldsymbol{X}_{t-1})P(\boldsymbol{X}_{t-1}|\boldsymbol{Y}^{t-1})d\boldsymbol{X}_{t-1},$$

where likelihood $P(\boldsymbol{Y}_t|\boldsymbol{X}_t)$ is measured based on the Euclidian distance of the observation sample normalized distance map and generated sample normalized distance, transition $P(\boldsymbol{X}_t|\boldsymbol{Y}^t)$ assuming a random walk on the manifold space. The mode or the MAP(maximum a posterior) of each parameter is estimated from the particle filtering.

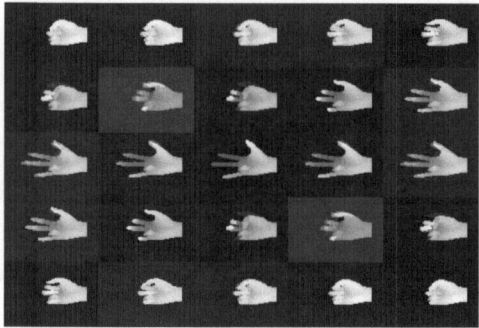

(a) Hand configuration interpolation with fixed view and key pose

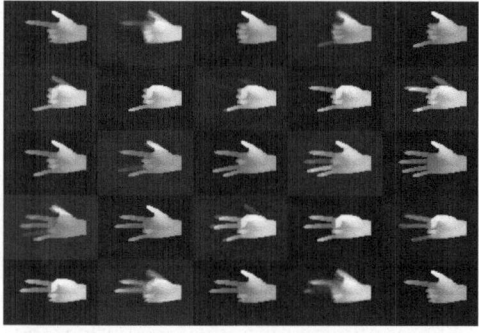

(b) Key pose interpolation with fixed view and configuration

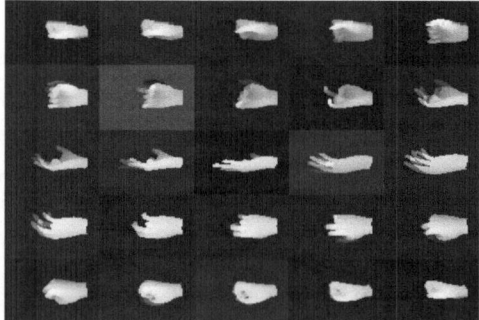

(c) Simultaneous key pose, view, and configuration interpolation

Fig. 4. The synthesis of new sequences from the trajectory of the 4D torus embedding

Reconstruction: The hand rotation or view can be computed directly from the estimated hand rotation parameter, ψ_t by simple scaling if it is necessary. In the case of body configuration, we need to estimate one body configuration from the estimated key hand pose (θ_t) and the configuration phase (ϕ_t). Even though we can obtain discrete key poses and hand configurations, the reconstruction need to use a continuous interpolation in the intermediate parameters. We obtain the interpolation function based on Radial Basis function for a given key hand pose parameter and phase of the configuration parameter to the joint angles of hand configuration by:

$$f(\theta_t, \phi_t) = \Omega(q_1, \cdots, q_{N_k}) \tag{7}$$

This allows the synthesis of the hand configuration from any estimated key hand pose parameter and phase of the configuration. Through the combination of the view and configuration parameters, the estimated hand pose can be generated using animation toolkits or graphics programs. $Poser^{\circledR}$, a character animation tool, was used in our experiments to synthesize new hand shape from the estimated view and configuration parameter.

4   Experimental Results

We performed experiments on two data sets: synthetic data set, a real data set from Kinect. In the synthesized data set, we generated a sequence of hand pose variations with hand rotation. The generated data sequence was preprocessed by a similar normalization procedure used for the training sequence. For the Kinect data set, both the depth information and color information were used to exclusively extract the hand components.

4.1   The Synthetic Data

For the synthesized data, we generated 44 frames with simultaneous view and hand configuration variations. The generated data sequence was rearranged to have the same center of gravity for each foreground shape, resized into 66×73 images, and column stacked into 4818 dimension vector. 20 particles for key pose(θ_t), 15 particles for phase of hand configuration (ϕ_t) and 15 particles for hand rotation (ψ_t), for a total of 4500 $(20 \times 15 \times 15)$ particles were used for the estimation of the 4D torus embedding from the observed shapes. Figure 5 (a) shows the sample input sequence for test, Figure 5 (b) shows the reconstruction result based on the estimated MAP parameters, Figure 5 (c) shows the estimated configuration parameters, Figure 5 (d) shows the estimated view parameters, Figure 5 (e) shows the estimated key pose parameters with ground true parameters and Figure 5 (f) shows the 3D reconstruction result using estimated view, configuration, and key pose parameters using Poser. Estimate parameters show accurate result for hand rotation and hand configuration phase. In case of key pose, several inaccurate estimation of key pose is occurred.

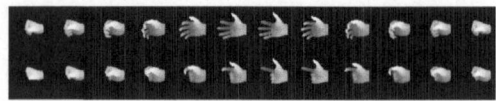

(a) Test sequence samples

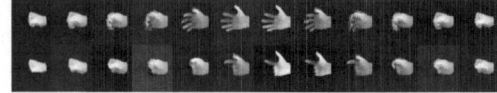

(b) Reconstructed samples

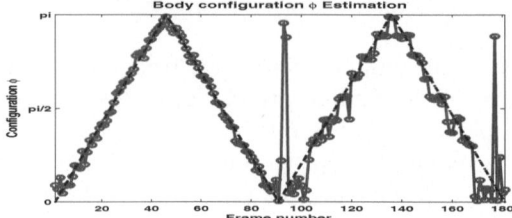

(c) Estimated configuration parameters: blue-ground truth, red-estimated

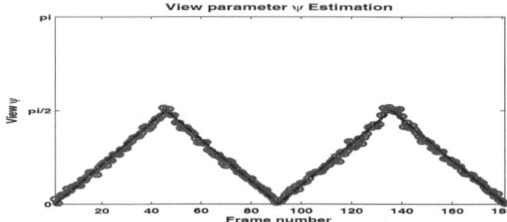

(d) Estimated view parameters: blue-ground truth, red-estimated

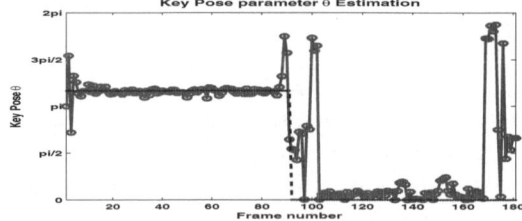

(e) Estimated key pose parameters: ground truth(blue), estimated(red)

(f) 3D reconstruction based on estimated key pose, view and configuration parameters

Fig. 5. Test and reconstructed sequences with estimated parameters

(a) Original image samples

(b) Original depth samples

(c) Reconstructed samples

Fig. 6. The test sequence and reconstructed sequence from the real data

4.2   The Real Data from Kinect

We used both the depth information and skin color information to obtain robust hand shape extraction. In order to extract the hand shape from the real data, the skin color or foreground image after background subtraction can be used. Still, it is not easy to extract accurate hand spaces from real data. Through this combination of depth and skin color, we can extract accurate hand shape from the test sequence.

At first, we used a hand detector, which was developed based on the cascade Adaboost [14] for several static hand shape separately. This detector is used to initialize ROI (Region of Interest), which was used to define boundary for shape depth segmentation and skin color estimation. In the case of the depth segmentation, we specified the depth (z axis) boundary to extract only the hand area (arm area could be included due to similar distance with hand) and the skin color based on back projection [15] are extracted to find hand area. By logical operation and morphological operation for the two type of extracted binary hand shapes, we can extract accurate hand shape for recognition. The extracted hand shape images were normalized similar to the synthetic ones. Figure 6 shows the sample sequence used for the hand tracking, estimated and reconstructed shapes from the estimated sequences.

5   Conclusions

In this paper, we presented a new framework to estimate hand configuration and hand rotation for various hand shapes using higher order torus manifold embedding.

The proposed method can be applied for accurate estimation of hand configuration and rotation during object manipulation in a virtual environment. It can also be used for sign language recognition especially for hand spelling based on estimated hand configuration in various view directions.

Acknowledgements. This research was financially supported by the Ministry of Education, Science Technology (MEST) and National Research Foundation of Korea(NRF) through the Human Resource Training Project for Regional Innovation and by Basic Science Research Program through the National Research Foundation of Korea(NRF) funded by the Ministry of Education, Science and Technology (2012R1A1A105003830).

References

1. Erol, A., Bebis, G., Nocolescu, M., Boyle, R.D., Twombly, X.: Vision-based hand pose estimation: A review. CVIU, 52–73 (2007)
2. Hamer, H., Schindler, K., Koller-Meier, E., Van Gool, L.: Tracking a hand manipulating an object. In: Proc. of ICCV, pp. 1475–1482 (2009)
3. Hamer, H., Gall, J., Urtasun, R., Van Gool, L.: Data-driven animation of hand-object interactions. In: Proc. of FGR, pp. 360–367 (2001)
4. Oikonomidis, I., Kyriazis, N., Argyros, A.: Full dof tracking of a hand interacting with an object by modeling occlusions and physical constraints. In: Proc. of ICCV, vol. 6, pp. 2088–2095 (2011)
5. Rosales, R., Athitsos, V., Sclaroff, S.: 3d hand pose reconstruction using specialized mappings. In: Proc. of ICCV, pp. 378–387 (2001)
6. Urtasun, R., Fleet, D.J., Fua, P.: 3d people tracking with gaussian process dynamical models. In: Proc. of CVPR, pp. 238–245 (2006)
7. Lee, C.S., Elgammal, A.: Coupled visual and kinematic manifold models for tracking. IJCV 87, 118–139 (2010)
8. Tian, T.P., Li, R., Sclaroff, S.: Articulated pose estimation in a learned smooth space of feasible solutions, pp. 50–57 (2005)
9. Guan, H., Feris, R.S., Turk, M.: The isometric self-organizing map for 3d hand pose estimation. In: Proc. of FGR, pp. 263–268 (2006)
10. Lee, C.S., Park, S.W.: Tracking hand rotation and grasping from an ir camera using cylindrical manifold embedding. In: Proc. of ICPR, pp. 2612–2615 (2010)
11. Elgammal, A.M., Lee, C.S.: Tracking people on a torus. IEEE Trans. Pattern Anal. Mach. Intell. 31, 520–538 (2009)
12. Fan, G., Zhang, X.: Video-based human motion estimation by part-whole gait manifold learning. In: Wang, L., Zhao, G., Cheng, L., Pietikainen, M. (eds.) Machine Learning for Vision-Based Motion Analysis, pp. 215–261. Springer (2011)
13. Lawrence, N.D.: Gaussian process models for visualisation of high dimensional data. In: Proc. of NIPS, pp. 329–336 (2004)
14. Viola, P., Jones, M.: Rapid object detection using a boosted cascade of simple features. In: Proc. of CVPR, pp. 511–518 (2001)
15. Imagawa, K., Lu, S., Igi, S.: Color-based hands tracking system for sign language recognition. In: Proc. of FGR, p. 462

View-Invariant Object Detection
by Matching 3D Contours

Tianyang Ma, Meng Yi, and Longin Jan Latecki

Dept. of Computer and Information Sciences,Temple University, Philadelphia, USA
ma.tianyang@gmail.com, {Mengyi,latecki}@temple.edu

Abstract. We propose an approach for view-invariant object detection directly in 3D with following properties: (i) The detection is based on matching of 3D contours to 3D object models. (ii) The matching is constrained with qualitative spatial relations such as above/below, left/right, and front/back. (iii) In order to ensure that any matching solution satisfies these constraints, we formulate the matching problem as finding maximum weight subgraphs with hard constraints, and utilize a novel inference framework to solve this problem. Given a single view of an RGB-D camera, we obtain 3D contours by "back projecting" 2D contours extracted in the depth map. As our experimental results demonstrate, the proposed approach significantly outperforms the state-of-the-art 2D approaches, in particular, latent SVM object detector, as well as recently proposed approaches for object detection in RGB-D data.

1   Introduction

Since the beginning of computer vision, the researchers have realized that 3D information makes object detection and recognition simpler and more robust than using 2D image information only. In particular, contours of 3D objects have been utilized in object recognition many decades ago, e.g., [1,2], since they offer a view invariant representation of 3D objects. Moreover, in contrast to 3D surfaces, 3D contours offer a simpler 1D like representation of complex shapes in 3D like chairs or other man-made objects. However, extraction of 3D contours from single 2D images or stereo image pairs turned out to be a challenging problem. Only due to recent progress of RGB-D sensors, robust extraction of 3D contours became possible. However, we still face the problem of matching of 3D contours. The main challenges are intra class object variance, e.g., everyday objects like chairs come in different sizes and shapes, and occlusion.

Contour is an important cue for human to recognize objects, and has been widely used in 2D single-view object detection in [3,4,5]. While contour has certain advantages, such as its low computation cost and its invariance to color and texture changes, it varies significantly under different viewpoints. This challenges most of current state-of-the-art shape-based detection approaches on a multi-view object detection task. As early computer vision approaches, we address this challenge by directly working with contours of 3D objects instead of their 2D projections. In our approach, we still utilize the fact that contours

J.-I. Park and J. Kim (Eds.): ACCV 2012 Workshops, Part II, LNCS 7729, pp. 183–196, 2013.
© Springer-Verlag Berlin Heidelberg 2013

of 3D objects project to 2D contours. It allows us for efficient recovery of 3D contours from 2D contours extracted from depth maps. This is possible thanks to Kinect, which is the most popular RGB-D camera. Since depth information can be obtained from a single view of a given scene, it is possible to recover 3D point cloud representing object surfaces. Depth map certainly provides more information that a single RGB image, and has proved to boost the performance of object recognition methods [6].

Object detection in 3D point clouds is an active research topic in the robotic community. There objects are recognized by directly matching 3D point clouds or by fitting surfaces to 3D point clouds. While surfaces are appropriate models for certain object classes, e.g., a ball, it is very hard if impossible to model object classes like chairs with surfaces alone. Contours appear to be a very suitable representation for RGB-D images. We observe that contours of 3D objects project to contours in 2D images. This in particular means that we can obtain 3D contours by lifting back contours from 2D images to 3D.

The processing flow of the proposed approach is illustrated in Fig. 1. After obtaining an RGB and depth images of a single view of a scene with Kinect, we first run Canny edge detector on the depth map. By linking the edge pixels, we obtain 2D edge fragments shown overlaid on the depth map in Fig. 1(b) with different colors. Since for each pixel in the depth map we can recover the 3D point that projects to it (with exception of out of range readings), we can "back project" each edge fragment to a set of 3D points, which we call 3D contour fragment. In Fig. 1(c) we see the 3D points recovered form the depth map in (b); for clarity of visualization the floor points are not shown. In Fig. 1(d) we show the 3D contour fragments in different colors. Each 3D contour fragment is represented with a set of 3D line segments fitted to the 3D points "back projected" from the corresponding 2D edge fragment. While one can recognize there the 3D contours of the two chairs and the stand, there are also many other contours present. They represent edges of walls and the background clutter.

After this preprocessing phase, we are ready for the proposed object detection. The 3D contours that belong to two detected chairs are shown in red and green in Fig. 1(e). All other 3D contours are shown in cyan. The detection is obtained by matching the model chair shown in Fig. 1(f) to all 3D contours in (e). In our system we used only one extremely simplistic model chair, as shown in (f), in order to demonstrate the power of matching 3D contours. The main challenges addressed by the proposed approach are intra class variability of 3D contours and occlusion. Occlusion and self-occlusion results in missing parts of 3D contours, which makes their matching challenging. To address these challenges we utilize the fact that geometric relations between 3D contours have more expressive power, and consequently, are less ambiguous compared to 2D.

We propose to solve the object detection by 3D matching problem by finding maximal weight subgraphs (MWSs) that satisfy mutex constraints. An example result is shown in Fig. 2. There for each of the three detected chairs, we mark with the same color their 3D segments and the corresponding model segments. We observe that the three chairs vary in shape and size, and all are substantially

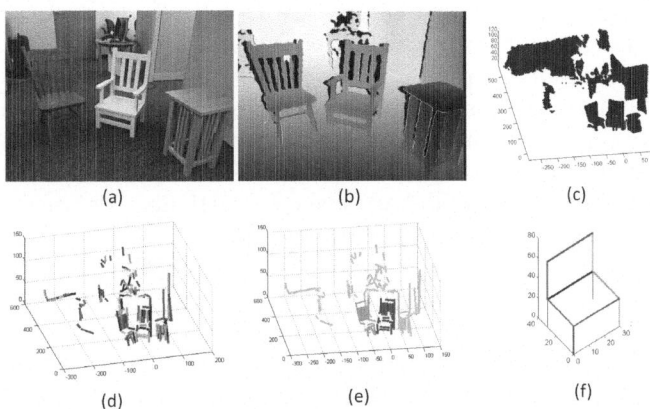

Fig. 1. An RGB image in (a) and the corresponding depth map in (b). The 3D points recovered from (a) are shown in (c). We recover 3D contour fragments, shown in different colors in (d) from edge fragments in (b). The line segments of two detected chairs in (d) are shown in green and red in (e). They are detected by matching segments of a single model shown in (f) to the segments in (d).

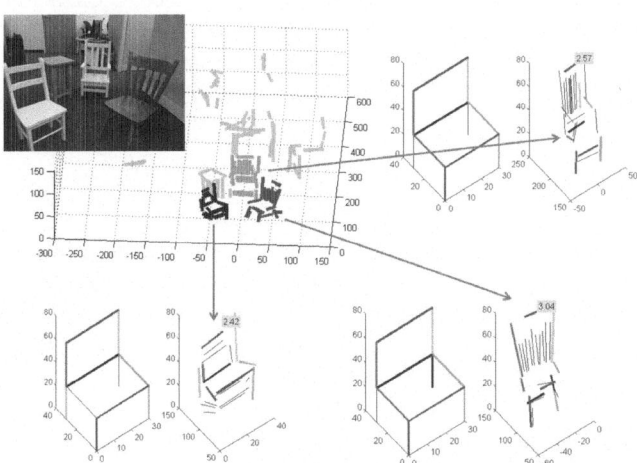

Fig. 2. A recovered 3D scene from a single RGB-D image. Contours of 3D objects are represented with 3D line segments. Object detection is performed by finding MWSs in the correspondence graph composed of pairs (model segment, 3D scene segment). We mark with the same colors the corresponding segments for three detected chairs shown in red, green, and blue in the 3D scene.

different form our single model chair. Moreover, due to self-occlusion, and since some edge fragments are not detected in the 2D depth images, all three chairs have some missing parts. The proposed matching approach is able to robustly deal with these challenges. This is possible due to our inference framework for finding MWSs that allows us to enforce hard, mutual exclusion (mutex) constrains. The mutex constraint, which express qualitative spatial relations such as above/below as well as prohibit grouping 3D contours that are too far from each other, eliminate the majority of impossible matching configurations. This allows us to obtain correct detections with weak shape similarity relations, which in turn allow us to tolerate a significant shape and size variance of 3D contours representing objects in the same shape class. In particular, we use only one chair exemplar in our experiments on chair detection.

We compute the MWSs on the correspondence graph composed of all pairs (model segment, 3D scene segment). As shown in Fig. 1(f), our exemplar chair is composed of 11 line segments. If we have 200 segments in a given 3D scene, for example, then the correspondence graph has 2200 nodes. In order to detect MWSs in this graph, we initialize with one correspondence, and compute a MWS that contains this correspondence, i.e., we have 2200 initializations. Then we sort the MWSs according to their weights. The three detected chairs in Fig. 2 represent MWSs with three highest weights. As can be seen the subgraphs have 8 to 10 nodes. Thus, our inference framework is capable of finding very small MWSs in graphs with a few thousand nodes.

In Sec. 2, we review related works. In Sec. 3, we introduce our shape representation and matching, also how to formulate the object localization problem as finding maximal weight subgraph with mutex constraints. In Sec. 4, a formal definition of maximal weight subgraph with mutex constraints will be given and an algorithm we used to solve it is described. Experiment results are shown in Sec. 5.

2   Related Work

There are some recent works utilizing 3D contour information to perform object detections in range images. Stiene et al. [7] proposed a detection method in range images based on silhouettes. Drost et al. [8] use a local hough-like voting scheme that uses pairs of points as features to detect rigid 3D objects in 3D point clouds. Hinterstoisser et al. [9] proposed a multimodal template matching approach based on RGB-D data that is able to detect objects in highly cluttered scenes.

In a very early work, Ponce et al. [10] established a 3D object recognition framework, where objects are collections of small (planar) patches, their invariants, and a description of their 3D spatial relationship. Ferrari et al. [11] proposed a method to compute feature tracks densely connecting multiple model views of a single object. In [12], Implicit Shape Model [13] and [11] are combined, and activation links for transferring votes across views are used to address the object detection from arbitrary viewpoints. Savarese and Fei-Fei [14] propose a compact model of an object by linking together diagnostic parts of the objects from different viewpoints. Instead of recovering a full 3D geometry, parts are mutually connected by homographic transformation in this approach. More recently, a probabilistic approach to learning affine constraints between object parts is introduced in [15]. In [16], discriminative part-based 2D detectors and generative 3D representation of the object class geometry which can be learned

from a few synthetic 3D models are combined. Yan et al. [17] collect patches
from viewpoint-annotated 2D training images and map them onto an existing
3D CAD model. In [18], a 3D implicit shape model is obtained via sparsely annotated 2D feature positions. Payet and Todorovic [19] proposed a shape-based
3D object recognition method, in which a few view-dependent shape templates
are jointly used for detecting object occurrences and estimating their 3D poses.

A recent work by Janoch et al. [20] explores different options on how to utilize
the depth information from RGB-D cameras to improve the detection accuracy of
objects seen from different viewpoints. They call Deformable Part Model (DPM)
[21] applied to depth images Depth HOG, and conclude that Depth HOG is never
better than HOG on the original 2D image. The best performing system on their
dataset is a linear combination of DPM running on the original image with the
size distribution of a given object class, which is modeled with a single Gaussian.
We call this system DPM-SIZE.

View-invariant object detection can also be addressed by directly using single
2D images, i.e., no 3D contour or surface reconstruction is attempted prior to
the detection. Recent approaches of this type include [12,16,22]. While 2D single-
view object detection methods can be used to addressed the task by combining
the outputs of classifiers trained for different object views, such approaches are
argued to be only effective when there are sufficient single-view detectors to
cover all possible viewpoints [12]. However, this strategy requires a lot of training samples, and many independent detectors may lead to a substantial increase
in the number of false-positives. In order to obtain a better multi-view object detector, many methods made an effort to learn a generative model by combining
2D appearance and geometric viewpoint information [15,23,16]. While promising results are obtained by such methods, they suffer from ambiguous 2D local
features and lack of direct modeling of 3D viewpoint geometry.

In general graph matching frameworks [24], while local features' similarity
(unary potential) and geometric relations between them (binary potential) are
usually considered, very coarse qualitative geometric constraints such as above/
below, or left/right do not draw much attention. We demonstrate in our work
that using mutex constraints to enforce these qualitative geometric constraints
makes our method more robust to the noise, and therefore, able to generate
higher quality solutions.

3   Object Detection by Matching 3D Contours

In order to obtain contours of 3D objects form a given RGB-D image, we first
find edge fragments in the depth map. They are obtained by linking edge pixels
obtained by the Canny edge detector to 2D curves. Then we lift each 2D edge
fragment back to a 3D curve. Let C be a single edge fragment. We first dilate it
with a dilation radius of 2 pixels. Then we find the set of 3D points Z that project
to pixels in dilated C. Finally we iteratively fit 3D line segments to points in Z. We
run RANSAC to fit a line and identify the inlier points and outlier points. Then we
repeat this process for the outlier points until the number of outlier points is lower

than a threshold. Hence we represent each 3D curve Z as a set of 3D line segments, and consequently, we represent 3D contours obtained from a given RGB-D image as set of line segments in 3D. An example is shown in Fig. 1(d).

Object detection in the proposed approach is formulated as finding configurations of line segments recovered from a given RGB-D image that are similar to the line segment configuration of the exemplar modeling a given shape class. Thus, we need to identify a subset of 3D line segments that best matches the exemplar. This computation is formulated here as finding maximum weight subgraphs (MWS) in a weighted correspondence graph. We begin with definitions of pairwise similarities of line segments.

3.1   Similarity of 3D Vectors

We use a set of straight line segments $\mathcal{S} = \{B_1E_1, \cdots, B_nE_n\}$ to approximate object contours in 3D, where B_i is the beginning point and E_i is the endpoint of segment B_iE_i. An example is shown in Fig 1 (b). Since the line segments are oriented, they are vectors in 3D, and from now on we treat them as vectors. For the model contour each line segment is represented with just one vector. In contrast, each contour line segment in 3D image is represented by two vectors that differ by their orientation.

Although we know the exact size of objects in 3D, the size of objects in the shape shape class may still vary significantly. To obtain a size-invariant vector representation, we characterize each B_iE_i by its angle with a reference vector r defined as

$$\angle(B_iE_i, r) = \arccos(\frac{B_iE_i \cdot r}{||B_iE_i||\ ||r||}) \in [0, \pi] \tag{1}$$

We take vector $r = [0, 0, 1]$ representing the z-axis as the reference vector. Since 3D objects are supported by the floor, which is represented as xy-plane, the representation in (1) is invariant to the rotation around the z-axis. This means it is invariant to object location on the floor, under the assumption that the object is standing on the floor. To simplify the notation, we omit the direction r below when possible, and use $\angle B_iE_i$ to represent the angle of vector B_iE_i with z-axis.

Given the above angle-based segment representation, we treat two vectors as similar if they have similar angles with the z-axis. We compute this similarity value as

$$\psi(B_iE_i, B_jE_j) = \exp(-\frac{(\angle B_iE_i - \angle B_jE_j)^2}{\sigma^2}) \tag{2}$$

where σ represents the tolerance of angle differences (it is set to $\frac{\pi}{3}$ in all our experiments).

3.2   Similarity of Vector Configurations

Let $\mathcal{E} = \{B_1^eE_1^e, \cdots, B_m^eE_m^e\}$ be 3D vectors that represent an exemplar (model) of a given shape class, and let $\mathcal{S} = \{B_1^sE_1^s, \cdots, B_n^sE_n^s\}$ be 3D vectors representing the vectors of the recovered 3D scene.

We construct a weighted association graph $G = (V, A)$ with $V = \mathcal{E} \times \mathcal{S}$. Hence each node represents a correspondence $u = (B_i^e E_i^e, B_j^s E_j^s)$ between a model vector i and an image vector j. Consequently, there are $N = m \times n$ nodes in the graph.

We define now the entries of the adjacency matrix A. If $u = v = (B_i^e E_i^e, B_j^s E_j^s)$, then $A(u, u) = \psi(B_i^e E_i^e, B_j^s E_j^s)$, which simply the similarity of the angle with z-axis of both vectors. Given a pair of different correspondences $u \neq v$, where $u = (B_i^e E_i^e, B_j^s E_j^s)$ and $v = (B_k^e E_k^e, B_l^s E_l^s)$, the weight $A(u, v)$ between nodes u and v represents the consistency of the their assignments. We measure it by computing the similarity of the spatial configuration of exemplar vectors $B_i^e E_i^e, B_k^e E_k^e$ to the configuration of the 3D scene vectors $B_j^s E_j^s, B_l^s E_l^s$. For this we consider new vectors that join their start points. For example, in Fig. 3 vectors $B_i^e E_i^e, B_k^e E_k^e$ are the cyan lines in the model, and the new vector $B_i^e B_k^e$ is marked with the black dashed line while the new vector $E_i^e E_k^e$ is marked with the red dashed line. The same colors are used for the corresponding vectors in the 3D scene. The similarity of this configuration is determined by the similarity of the angles between the corresponding dashed vectors:

$$A(u, v) = \psi(B_i^e B_k^e, B_j^s B_l^s) \cdot \psi(E_i^e E_k^e, E_j^s E_l^s). \tag{3}$$

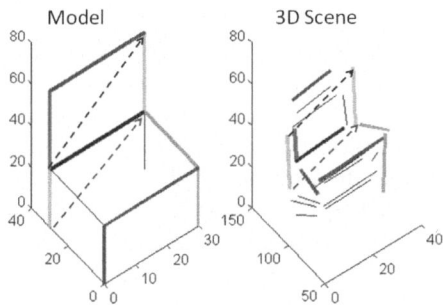

Fig. 3. Similarity of the two configurations of cyan lines is defined as similarity of the angles between two black dashed vectors and between two red dashed vectors

3.3   Mutex Constraints between Contour Vectors

Compared to other graph matching frameworks, the key and unique property of our formulation is usage of qualitative spatial constraints, such as above/below or left/right or front/back. For example, if for a given pair $u = (B_i^e E_i^e, B_j^s E_j^s)$ and $v = (B_k^e E_k^e, B_l^s E_l^s)$, the model vector $B_k^e E_k^e$ is above vector $B_i^e E_i^e$, then we require the same for the corresponding vectors in the 3D scene, i.e., $B_k^s E_k^s$ should be above $B_l^s E_l^s$. By enforcing the qualitative geometric relations in the correspondence computation, we can significantly improve the solution quality. In particular, the matching becomes robust to significant variance in shape and size of objects form a given class.

We define a symmetric mutex relation $M \subseteq V \times V$ between vertices of the graph defined in Section 3.2. It is represented with a binary matrix $M \in \{0, 1\}^{N \times N}$. If $M(u, v) = 1$ then the two vertices u, v cannot belong to the same maximum clique. In other words, mutex represents incompatible vertices that cannot be selected together. Since a vertex cannot exclude itself, we set $M(u, u) = 0$ for all vertices $u \in V$.

Given a pair of two vertices representing the correspondences $u = (B_i^e E_i^e, B_j^s E_j^s)$ and $v = (B_k^e E_k^e, B_l^s E_l^s)$, where $u \neq v$, $M(u, v)$ represents the compatibility of the the spatial relations between vectors $B_i^e E_i^e$ and $B_k^e E_k^e$ in the model, and $B_j^s E_j^s$ and $B_l^s E_l^s$ in the 3D scene. For example, if $B_i^e E_i^e$ is above $B_k^e E_k^e$ in the model and $B_j^s E_j^s$ is below $B_l^s E_l^s$ in the scene, then $M(u, v) = 1$. One the other hand, if $B_j^s E_j^s$ is also above $B_l^s E_l^s$, then $M(u, v) = 0$. Similarly, $M(u, v) = 1$ if front/back or left/right spatial relations are violated.

In order to define M without checking different cases, we project the 4 points $B_i^e, E_i^e, B_k^e, E_k^e$ to vectors $B_i^e E_i^e$ and $B_k^e E_k^e$ in the model and the 4 points B_j^s, E_j^s, B_l^s, E_l^s to vectors $B_j^s E_j^s$ and $B_l^s E_l^s$ in the scene. Then we check whether the two 1D orders on the projection lines are compatible. If yes, we set $M(u, v) = 0$, and if not, we set $M(u, v) = 1$. We skip the technical details, since they only require elementary 3D geometry and the limited space.

4   Maximum Weight Subgraphs with Mutex Constraints

Given the weighted correspondence graph G, we formulate the problem of localizing objects in images as finding constrained maximum weight subgraphs. Each node in our graph is a matching between an image vector and a model vector. Therefore, a configuration of nodes, i.e., subgraph, corresponds to a set of selected scene vectors matched to the model shape. Hence each subgraph represents a configuration of 3D scene vectors and the corresponding configuration of model vectors. The unary and binary potentials in Section 3.2 are defined so that the more similar are both configurations the larger is the weight of their subgraph, which is just the sum of unary and binary potentials. Therefore, maximum weight subgraphs identify the instances of the model exemplar present in a given 3D scene recovered from a single RGB-D image.

Formally, the input is a weighted graph $G = (V, A)$, where $V = \{v_1, \ldots, v_N\}$ is the set of nodes representing the matches between model segments and image segments, N is the number of nodes, and A is a symmetric $N \times N$ affinity matrix, defined in Section 3.2, with all nonnegative entries, i.e., $A_{ij} \geq 0$ for all $i, j = 1, \ldots, N$. The selected matches are identified with and indicator vector $\mathbf{x} = (x_1, \ldots, x_N) \in \{0, 1\}^N$, where a given match v_i is selected if and only if $x_i = 1$.

We are also given a symmetric mutex relation $M \subseteq V \times V$ between vertices of the graph defined in Section 3.3. The mutex relation M imposes constraints on the indicator vector $\mathbf{x} \in \{0, 1\}^N$: if $M(i, j) = 1$, then $x_i + x_j \leq 1$. This formulation is equivalent to the requirement $\mathbf{x}^{\mathbf{T}} M \mathbf{x} = 0$.

We find the contours belong to the target object by solving the following maximization problem

$$\text{maximize } f(\mathbf{x}) = \mathbf{x}^T A \mathbf{x} \text{ s.t. } \mathbf{x} \in \{0,1\}^n \text{ and } \mathbf{x}^\mathbf{T} M \mathbf{x} = 0. \tag{4}$$

The goal of (4) is to select a subset of vertices of graph G such that f is maximized and the mutex constraints are satisfied. Since f is the sum of pairwise affinities of the elements of the selected subset, the larger is the subset, the larger is the value of f. However, the size of the subset is limited by mutex constraints. The problem (4) is a combinatorial optimization problem and is NP-hard [25].

By setting $W = A - \gamma M$, with a large positive γ we reformulate problem (4) into the following form:

$$\text{maximize } \mathbf{x}^\mathbf{T} W \mathbf{x} = \mathbf{x}^\mathbf{T} A \mathbf{x} - \gamma \mathbf{x}^\mathbf{T} M \mathbf{x} \text{ s.t. } \mathbf{x} \in \{0,1\}^n. \tag{5}$$

Finally, we relax (5) to

$$\text{maximize } \mathbf{x}^\mathbf{T} W \mathbf{x} = \mathbf{x}^\mathbf{T} A \mathbf{x} - \gamma \mathbf{x}^\mathbf{T} M \mathbf{x} \text{ s.t. } \mathbf{x} \in [0,1]^n. \tag{6}$$

We utilize the algorithm described in [26] to solve problem (6). Its key property is that if $\gamma > \max_i \sum_j A_{ij}$ and if the solution $\mathbf{x}^*$ is discrete, then $\mathbf{x}^*$ is guaranteed to satisfy all mutex constraints, i.e., $(\mathbf{x}^*)^\mathbf{T} M \mathbf{x}^* = 0$.

Although this algorithm solves the relaxed problem (6) the obtained solutions were discrete in all of our experiments. Hence the solutions satisfy all mutex constraints.

Since the algorithm in [26] converges to a local optimum, multiple initializations are required to increase the change of getting a globally optimal solution. In our implementation, we initialize from every node in the graph. More precisely, for every $u \in V$ we set $(\mathbf{x}_{(0)})_u = 1$ and $(\mathbf{x}_{(0)})_i = 0$ for all $i \neq u$, where $\mathbf{x}_{(0)}$ denotes the initial vector $\mathbf{x}$. Starting from the $\mathbf{x}_{(0)}$, we obtain a maximal subgraph indicated by a binary vector $\mathbf{x}^*$. $\mathbf{x}^*$ is a local maximizer of $\mathbf{x}^T A \mathbf{x}$ while satisfying $\mathbf{x}^{*T} M \mathbf{x}^* = 0$.

Therefore, we obtain N maximal subgraphs in total. Since there may be duplicated subgraphs among these N maximal subgraphs, we perform a non-maximum suppression over these subgraphs according to their $\mathbf{x}^T A \mathbf{x}$ values. Finally, we take the remaining subgraphs as object detections.

We are not only able to find out which contours in image belong to a detected object, but more importantly, we are also able to establish a correspondence of these contours. This is very important in some applications, such as robot manipulation.

To obtain the location of the object, i.e., its bounding box, in RGB image. We project the 3D segments back to RGB images, and compute the bounding box.

5   Experiments

Chair is an icon object class that has gained much attention form the beginning of AI. Although humans have no problem in identifying chairs, until today no artificial system is able to cope with chair detection. Chair detection is a challenging problem for most computer vision, detection algorithms [27], considering that the chair shape in 2D images varies significantly due to different viewpoints and due to resulting perspective distortion. Moreover, chairs come in different shapes and sizes. Therefore, we focus our performance evaluation on chair detection. We selected a stand as the second object class, since it is visually very similar to the chair in that it usually has 4 legs supporting a flat rectangular surface on top. The main difference is that the stand does not have any back support and its legs are longer, e.g., see the left image in Fig. 4.

We collected a dataset containing 109 RGB-D images captured with the Kinect sensor. It contains a total of 213 chairs shown from many different view points and 40 stands. Our dataset also contains other objects that may be confused with chairs and stands like tables and trash cans as can be seen in Fig. 4. Moreover, may objects are occluded and are shown in many different views.

 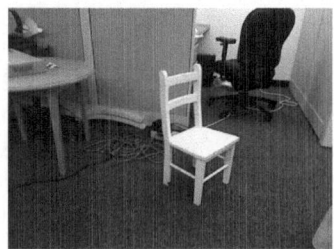

Fig. 4. Example images in our chair-stand dataset

In order to demonstrate that our dataset is very challenging and in order to compare to state-of-the-art object detectors, we compare the performance of our approach to DPM by Felzenszwalb et al. [21] and to DPM-SIZE recently proposed in Janoch et al. in [20]. DPM-SIZE augments DPM with depth information. It utilizes the expected object sizes in 3D scenes to boost DPM performance. We also compare to the popular contour based detection method PAS by Ferrari et al. [3]. For a quantitative evaluation, we use recall-precision curves and average precision (AP) computed as described in [28].

The detection results of chairs are summarized in Fig. 5. The proposed approach achieves a significantly better AP value compared to DPM and to DPM-SIZE. Our AP is nearly 30% higher than the second best performing method DPM-SIZE [20]. Moreover, the fact that DPM-SIZE, DPM, and PAS have all very low recall clearly demonstrates that these methods cannot cope with significant view changes and perspective distortions. This comes at no surprise for DPM and PAS, since both methods are based on 2D image analysis. In contrast, the direct matching of 3D contours in 3D allows us to overcome the challenges of view changes and of perspective distortion. We stress that our approach does not require any training, as opposed to the other three approaches, and we only have one extremely simplistic chair model. Moreover, our chair model is not extracted from the test dataset.

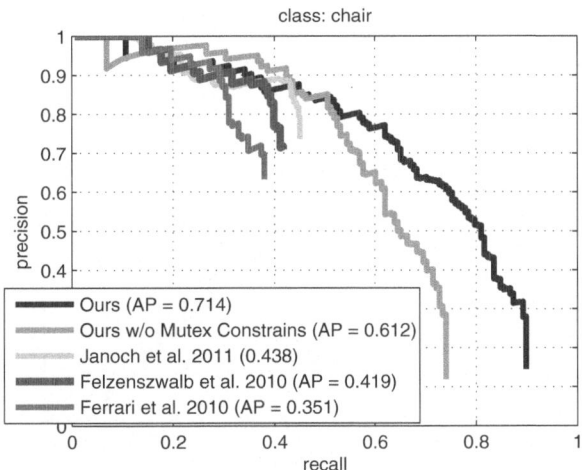

Fig. 5. Recall-Precision and AP comparison for the class chair

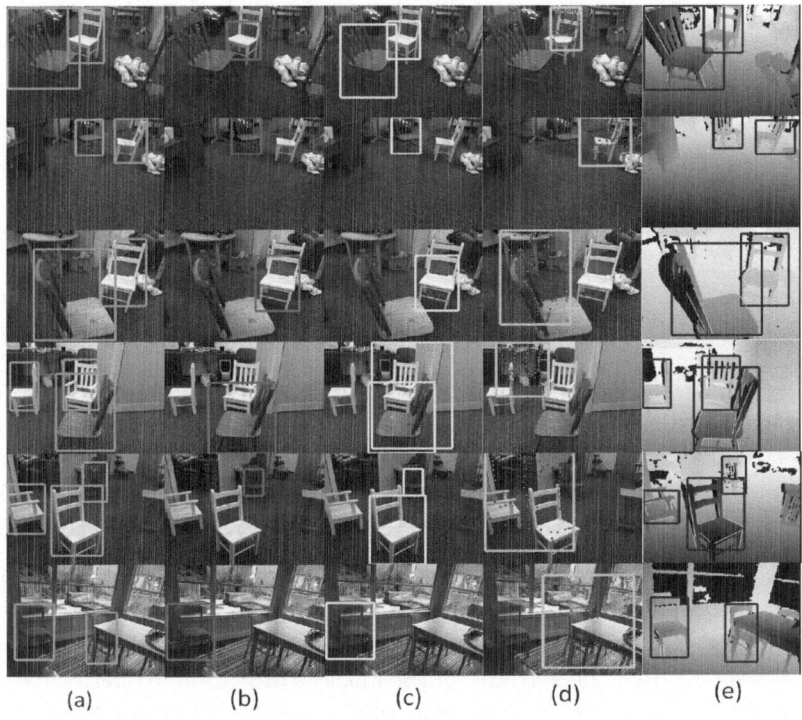

Fig. 6. Some chair detection results. (a) ground truth, (b) DPM [21], (c) DPM-SIZE [20]. (d) PAS [3] with transformed model shown with dots, and (e) The proposed method with results shown on depth map to stress that they are obtained in 3D.

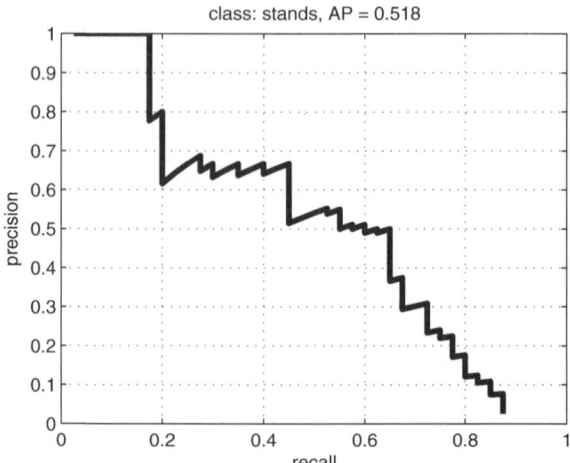

Fig. 7. Recall-Precision and AP of our detector with mutex constraints on class stand

The significance of the qualitative mutex constraints is demonstrated by the fact that the performance of our method drops by 10% when these constraints are not used. This in turn illustrates the importance of the utilized inference framework.

In Fig. 6, we show some detection results. As seen in Fig. 6(b), DPM [21], DPM-SIZE [20], PAS [3] missed many chairs. Adding 3D information about expected object sizes in the 3D scenes (DPM-SIZE [20]) is able to improve the performance of DPM, but still some chairs are missed. The main reason is that the initial detection is still performed in the 2D images (using sliding window processing of DPM).

We use the already trained version of DPM, which is publicly available on the authors' webpage. DPM [21] attempts to solve the object detection problem by using a multiple components object model, and each component is aimed to capture the object appearance under certain view-point. The 2D chair appearance model of DPM is trained using images from [28] with thousands of chairs. We also tried to train DPM detector on half of our dataset and test on the other half as opposed to using the trained detector from images in [28]. This process yields a much worse AP of 0.01. However, the DPM detector is able to get 0.96 AP on training images. This again demonstrates how challenging is significant view point variance, and perspective distortion to state-of-the-art 2D object detectors. The expected size of the chair for DPM-SIZE was learned as described in [20]. We trained it on a random half of our dataset and test on the other half. This process was repeated 10 times. We also used the software of the authors of PAS [3] to perform experiments on our chair dataset. A shape is learned automatically using this software, following the same procedure as for size training of DPM-SIZE.

Since there does not exist any trained version of DPM for the class stand and our dataset exhibits too large view variance for training DPM, we only report the result of our detector with mutex constraints on the class stand in Fig. 7.

6   Discussion and Future Work

We only used one simplistic chair model, which differs in both size and shape from the various chairs captured in our dataset. This allows us to demonstrate the robustness of the proposed 3D matching framework. Our matching framework is also robust to occlusion, and of course, it is not influenced by view point changes. Similarly we only used one simplistic stand model.

However, more 3D contour models are needed to capture the intra class variability. In particular, some chairs may only have one leg like the office chair shown in the right image in Fig. 4. Such models can be easily learned by clustering training objects using the proposed similarity measure.

One of the biggest challenges of our 3D contour-based object detection are objects without clear 3D contours like humans or sofas. For such objects it is still possible to extract occluding contours from the RGB-D data, and those contours exhibit significantly lower variation than contours extracted form 2D RGB images. Also the contour detection problem in RGB-D images is significantly simpler. However, the 3D occluding contours exhibit larger variation than intrinsic 3D contours of objects like chair or stand. Our future work will focus on matching the occluding 3D contours.

Acknowledgement. This work has been supported by AFOSR FA9550-09-1-0207 and by NSF grants BCS-0924164, IIS-0812118, and OIA-1027897.

References

1. Barrow, H., Tenenbaum, J.: Interpreting line drawings as three-dimensional surfaces. Artificial Intelligence 17, 75–116 (1981)
2. Lowe, D.G.: Three-dimensional object recognition from single two-dimensional images. Artificial Intelligence 31(3), 355–395 (1987)
3. Ferrari, V., Jurie, F., Schmid, C.: From images to shape models for object detection. International Journal of Computer Vision 87, 284–303 (2010)
4. Shotton, J., Blake, A., Cipolla, R.: Multiscale categorical object recognition using contour fragments. IEEE Trans. Pattern Anal. Mach. Intell. 30, 1270–1281 (2008)
5. Opelt, A., Pinz, A., Zisserman, A.: Learning an alphabet of shape and appearance for multi-class object detection. International Journal of Computer Vision 80, 16–44 (2008)
6. Bo, L., Lai, K., Ren, X., Fox, D.: Object recognition with hierarchical kernel descriptors. In: CVPR, pp. 1729–1736 (2011)
7. Stiene, S., Lingemann, K., Nuchter, A., Hertzberg, J.: Contour-based object detection in range image. In: Third International Symposium on 3D Data Processing, Visualization and Transmission (2006)
8. Drost, B., Ulrich, M., Navab, N., Ilic, S.: Model globally, match locally: Efficient and robust 3d object recognition. In: IEEE Computer Society Conference on Computer Vision and Pattern Recognition, pp. 998–1005 (2010)
9. Hinterstoisser, S., Holzer, S., Cagniart, C., Ilic, S., Konolige, K., Navab, N., Lepetit, V.: Multimodal templates for real-time detection of texture-less objects in heavily cluttered scenes. In: IEEE International Conference on Computer Vision, pp. 858–865 (2011)

10. Ponce, J., Lazebnik, S., Rothganger, F., Schmid, C.: Toward true 3d object recognition. In: Congres de Reconnaissance des Formes et Intelligence Artificielle (2004)
11. Ferrari, V., Tuytelaars, T., Van Gool, L.J.: Integrating multiple model views for object recognition. In: CVPR, pp. 105–112 (2004)
12. Thomas, A., Ferrari, V., Leibe, B., Tuytelaars, T., Schiele, B., Van Gool, L.J.: Towards multi-view object class detection. In: CVPR, pp. 1589–1596 (2006)
13. Leibe, B., Leonardis, A., Schiele, B.: An Implicit Shape Model for Combined Object Categorization and Segmentation. In: Ponce, J., Hebert, M., Schmid, C., Zisserman, A. (eds.) Toward Category-Level Object Recognition. LNCS, vol. 4170, pp. 508–524. Springer, Heidelberg (2006)
14. Savarese, S., Li, F.F.: 3d generic object categorization, localization and pose estimation. In: ICCV, pp. 1–8 (2007)
15. Sun, M., Su, H., Savarese, S., Li, F.F.: A multi-view probabilistic model for 3d object classes. In: CVPR, pp. 1247–1254 (2009)
16. Liebelt, J., Schmid, C.: Multi-view object class detection with a 3d geometric model. In: CVPR, pp. 1688–1695 (2010)
17. Yan, P., Khan, S.M., Shah, M.: 3d model based object class detection in an arbitrary view. In: ICCV, pp. 1–6 (2007)
18. Arie-Nachimson, M., Basri, R.: Constructing implicit 3d shape models for pose estimation. In: ICCV, pp. 1341–1348 (2009)
19. Payet, N., Todorovic, S.: From contours to 3d object detection and pose estimation. In: ICCV, pp. 983–990 (2011)
20. Janoch, A., Karayev, S., Jia, Y., Barron, J.T., Fritz, M., Saenko, K., Darrell, T.: A category-level 3-d object dataset: Putting the kinect to work. In: ICCV Workshops, pp. 1168–1174 (2011)
21. Felzenszwalb, P.F., Girshick, R.B., McAllester, D., Ramanan, D.: Object detection with discriminatively trained part-based models. IEEE Transactions on Pattern Analysis and Machine Intelligence 32, 1627–1645 (2010)
22. Savarese, S., Tuytelaars, T., Van Gool, L.J.: Special issue on 3d representation for object and scene recognition. Computer Vision and Image Understanding 113, 1181–1182 (2009)
23. Liebelt, J., Schmid, C., Schertler, K.: Viewpoint-independent object class detection using 3d feature maps. In: CVPR (2008)
24. Berg, A.C., Berg, T.L., Malik, J.: Shape matching and object recognition using low distortion correspondences. In: CVPR, pp. 26–33 (2005)
25. Asahiro, Y., Hassin, R., Iwama, K.: Complexity of finding dense subgraphs. Discrete Applied Mathematics (2002)
26. Ma, T., Latecki, L.J.: Maximum weight cliques with mutex constraints for video object segmentation. In: CVPR (2012)
27. Grabner, H., Gall, J., Van Gool, L.J.: What makes a chair a chair? In: CVPR, pp. 1529–1536 (2011)
28. Everingham, M., Van Gool, L.J., Williams, C.K.I., Winn, J., Zisserman, A.: The pascal visual object classes (voc) challenge. International Journal of Computer Vision 88, 303–338 (2010)

Human Detection with Occlusion Handling by Over-Segmentation and Clustering on Foreground Regions

Li Wang, Kap Luk Chan, and Gang Wang

School of Electrical and Electronic Engineering
Nanyang Technological University

Abstract. Two-dimensional image based human detection methods have been widely used in surveillance system. However, detecting human in the presence of occlusion is still a challenge for such image based systems. In this paper, a human detection method aiming to handle occlusions by using the depth data obtained from 3D imaging methods, such as those easily acquired from the Microsoft Kinect depth sensor, is proposed. In the context of surveillance setting, background subtraction on the depth data can be used to extract foreground regions which may correspond to humans. The proposed method analyzes the 3D data of the foreground regions using a "split-merge" approach. Over-segmentation and clustering are preformed on foreground regions followed by the height validation. Experimental results demonstrate that the proposed method outperforms two state-of-art human detection methods.

1 Introduction

Human detection plays an important role in a wide range of computer vision applications, such as video surveillance and human-computer interaction. Although many methods on human detection have been reported, the problem of occlusion is still a challenge. Recently, off-the-shelf affordable depth sensors, such as the Microsoft Kinect depth sensor, become available. The depth data is in fact a less ambiguous representation of 3D objects due to depth coherence, making it a better option for 3D object detection and human is one of such objects. This paper demonstrates the use of depth data for human detection in the presence of occlusions by using Kinect depth data.

The motivation of the method is that 3D data could be more discernable when coming to separate occluded persons than 2D image based features due to the fact that occluded persons occupy different depth values. The straightforward idea is to cluster pixels in a foreground region into segments according to depth values of pixels. However, the number of pixels in a foreground region could be larger than 10^4. Many clustering algorithms cannot be easily applied to this kind of large-scale data due to the possible computational problem and the large ambiguity on determining the number of clusters. Therefore, we propose to detect occluded persons using the "split-merge" approach by which the foreground

J.-I. Park and J. Kim (Eds.): ACCV 2012 Workshops, Part II, LNCS 7729, pp. 197–208, 2013.

region is firstly over-segmented based on depth data and then the over-segmented subregions like the "superpixels" in [1] are clustered into segments based on 3D data, which in this paper refers to the spatial coordinates (x, y) in the image plane and the depth value d. Although the idea is intuitive, clustering subregions is nontrivial due to the fact that depth values on one person could vary because of the typical surveillance camera setup, i.e. looking down to the scene at an angle instead leveled. This makes depth-plane slicing not feasible as the depth ranges of two persons in occlusion could overlap. To conquer this difficulty, we propose a boundary based similarity metric for clustering subregions adequately into segments. To validate that a segment corresponding to a human, we use a height constraint on object segments in view of incomplete observation of a human and hence the difficulty of constructing partial human models to address infinitely possible occlusions. This constraint has proven to work effectively in our surveillance setting.

2   Related Work

In the past, some human detection methods [2] [3] often employ features representing human and learn the discriminative classifiers to separate between human and non-human. However, they often fail to detect occluded human due to using non-occluded human in training data. Subsequently, many methods [4] [5] [6] are proposed to handle occlusions. Most recently, Felzenszwalb et al. [7] propose to discriminatively train deformable part based models to detect occluded human. However, when the occlusion is severe, Felzenszwalb's method cannot work well as shown in the first and the third rows of Fig. 3.

Before the prevalence of Kinect depth cameras, human detection methods [8] [9] [10] had already made use of depth information. After Kinect depth cameras becoming popular, many methods [11] [12] [13] [14] are proposed to detect human by using Kinect depth data. These methods can be divided into two categories, the discriminative methods [11] [12] and the generative ones [13] [14]. In the generative methods, Xia et al. [14] recently propose to use a 2-D head contour model and a 3-D head surface model to detect seed regions and then detect human by using a region growing algorithm.

Our proposed method requires 3D data and performs in an generative manner. Hence, Xia's method [14] is most relevant to and will be compared with the proposed one. To further evaluate the effect of using depth data rather than color images, Felzenszwalb's method [7] will be compared with the proposed one.

3   The Proposed Method for Human Detection

The proposed method is illustrated in Fig. 1. First, the background subtraction is applied on the depth data to obtain foreground regions. Then, the gradient based watershed transform [15] is used in each foreground to obtain the over-segmented subregions. Subsequently, a boundary based similarity metric is proposed to construct a graph including all subregions. Spectral clustering [16] [17] is then

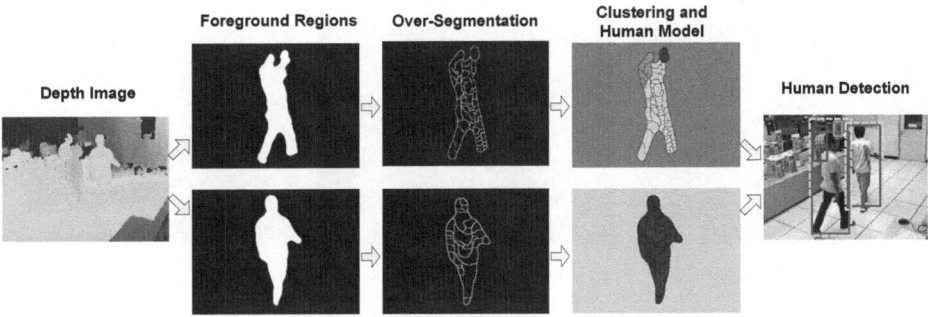

Fig. 1. The illustration of the proposed method for human detection. In the output, the bounding boxes with yellow dashed lines refer to the foreground regions; the bounding boxes with different colors and solid lines refer to the detected persons.

employed to cluster subregions into segments based on object depth coherence using a boundary based similarity metric. Lastly, a constraint on height values of persons in the scene is used as a human validation model to determine a segment as a person.

3.1   The Boundary Based Similarity Metric

Given the subregions in a foreground region as shown in Fig. 2(a), a fully-connected graph $\mathcal{G} = (V, E)$ can be constructed for clustering subregions into segments. Each node $i \in V$ denotes a subregion in the foreground region and each edge (i, j) denotes the pairwise similarity between two subregions. Due to the fact that the depth values on one person could vary in a certain range and the depth values on different persons could overlap with each other, it is nontrivial to define a similarity metric which is capable of clustering subregions correctly into segments each of which is likely to be a person. In this subsection, a novel boundary based similarity metric will be described.

To compute the similarities between subregions, we use the 3D coordinates at each pixel which includes the spatial coordinates (x, y) in the image plane and the depth value d. Moreover, boundaries are used to represent subregions due to the intuition that the boundary is able to depict the connectivity between two subregions. Additionally, it can reduce the computational cost because of using pixels on boundaries of subregions rather than all pixels in subregions. Therefore, similarities between subregions are computed based on their boundaries.

Given two boundaries B_i and B_j, a straightforward way of computing the similarity between them is to compute the average similarity of all pairs of points on these two boundaries. However, if two adjacent boundaries are on one person, the depth values of all pixels on them could be significantly different, which could yield to a low average similarity between these two boundaries. Obviously, the similarity metric does not correctly represent the relationship between these two adjacent boundaries on one person. Moreover, if two adjacent boundaries are on two persons, the depth values of all pixels on them could overlap with each other,

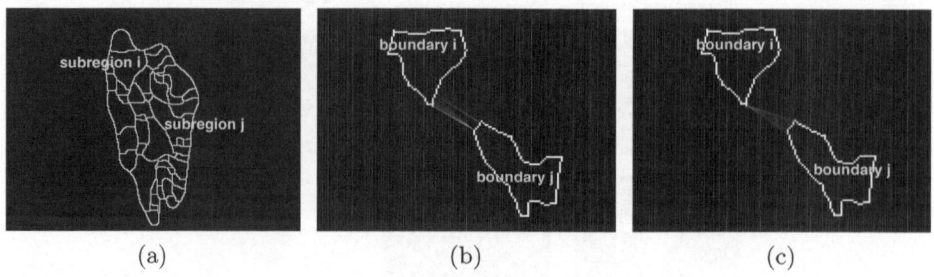

Fig. 2. The illustration of the boundary based similarity metric. (a) Over-segmented subregions in a foreground region; (b) The subset of K ($K = 10$) pairs of points from the ith to the jth boundary; (c) The subset of K ($K = 10$) pairs of points from the jth to the ith boundary.

which could lead to a high average similarity between these two boundaries. The similarity metric makes a mistake again.

To obtain the correct similarity metric, a subset of K pairs of points on two boundaries is used to compute the similarity. The subsets from B_i to B_j and from B_j to B_i are illustrated in Fig. 2(b) and Fig. 2(c) respectively. For example, the subset $(p_k^i, p_k^j)_{k=1,\ldots,K}$ from B_i to B_j is defined as follows

$$(p_k^i, p_k^j)_{k=1,\ldots,K}^{p_k^i \in B_i, p_k^j \in B_j} = \min_{p_i \in B_i}^{K} \min_{p_j \in B_j}^{1} d_{xy}(p_i, p_j), \tag{1}$$

where K indicates the number of point pairs in the subset and $d_{xy}(p_i, p_j)$ denotes the Euclidean distance between points with only considering spatial coordinates (x, y) in the image plane. Moreover, $\min_{p_j \in B_j}^{1}$ means to select the least pairwise distance from each point on B_i to all points on B_j. Then, $\min_{p_i \in B_i}^{K}$ means to select K least pairwise distances among points on B_i. Similarly, the subset $(\tilde{p}_k^i, \tilde{p}_k^j)_{k=1,\ldots,K}$ from B_j to B_i is defined as follows

$$(\tilde{p}_k^i, \tilde{p}_k^j)_{k=1,\ldots,K}^{\tilde{p}_k^i \in B_i, \tilde{p}_k^j \in B_j} = \min_{p_j \in B_j}^{K} \min_{p_i \in B_i}^{1} d_{xy}(p_i, p_j). \tag{2}$$

For two adjacent subregions, the point pairs in two subsets are very close in the image plane. If two subregions are on one person, the similarity by using the subsets will be high due to the fact that depth values on one person usually change smoothly in local regions. Although there exists the case that body parts on one person could be self-occluding and hence the depth values could change rapidly in local regions, the consequent low similarity will not affect the global connectivity among subregions on the same person. It means that the proposed boundary based similarity metric does not require that similarities between subregions on one person must always be high. It emphasizes more about the global connectivity among subregions on one person. If two subregions are on two persons, the similarity by using the subsets will be low because that depth values in the neighboring region between two persons change significantly. Therefore, the subsets are able to adequately represent relationships between subregions.

Since there are two subsets between two subregions, the average similarity between B_i and B_j is defined as follows

$$S(B_i, B_j) = 0.5 * (S_{B_i \to B_j} + S_{B_j \to B_i}),$$ (3)

where $S_{B_i \to B_j}$ and $S_{B_j \to B_i}$ indicate the similarities based on two subsets from B_i to B_j and from B_j to B_i respectively. Furthermore, $S_{B_i \to B_j}$ and $S_{B_j \to B_i}$ can be defined respectively as follows

$$S_{B_i \to B_j} = \frac{1}{K} \sum_{k=1}^{K} \exp\left(-(p_k^i - p_k^j)^T \text{Diag}(\sigma)(p_k^i - p_k^j)\right),$$ (4)

$$S_{B_j \to B_i} = \frac{1}{K} \sum_{k=1}^{K} \exp\left(-(\tilde{p}_k^i - \tilde{p}_k^j)^T \text{Diag}(\sigma)(\tilde{p}_k^i - \tilde{p}_k^j)\right),$$ (5)

where p_k^i, p_k^j, $\tilde{p}_k^i$ and $\tilde{p}_k^j$ refer to the 3D coordinates $[x, y, d]^T$ of points on boundary B_i and B_j. (x, y) denotes the coordinate in the image plane and d denotes the depth value. K is the number of point pairs in the subset and is set to 10. The digonal matrix $\text{Diag}(\sigma)$ is defined as follows

$$\text{Diag}(\sigma) = \begin{bmatrix} \frac{1}{2\sigma_1^2} & 0 & 0 \\ 0 & \frac{1}{2\sigma_1^2} & 0 \\ 0 & 0 & \frac{1}{2\sigma_2^2} \end{bmatrix},$$ (6)

where σ_1 and σ_2 control the neighborhood widths of coordinates (x, y) in the image plane and depth values d respectively. σ_1 and σ_2 are set to 10 and 100 respectively.

Eq. 4 and Eq. 5 compute similarities between points by using not only depth values d but also coordinates (x, y) in the image plane. It is because adding coordinates (x, y) into the similarity computation will decrease similarities between subregions not adjacent to each other, which can separate subregions with similar depth values but on different persons. Meanwhile it will not decrease similarities between adjacent subregions on one person because points in subsets between adjacent boundaries are close in the image plane. The proposed similarity metric has proven to work effectively for clustering subregions adequately into different segments which may correspond to humans.

3.2 Clustering Subregions into Segments

With the boundary based similarity metric, a graph can be constructed on subregions and spectral clustering [16] is then employed to cluster them into segments. The affinity matrix A is defined as follows

$$A_{ij} = A_{ji} = S(B_i, B_j).$$ (7)

Algorithm 1. Clustering subregions into segments

Input: N Subregions in a foreground region, the number k of segments.
1. Compute the affinity matrix A as Eq. 7.
2. Compute the Laplacian matrix L as Eq. 8.
3. Select the k eigenvectors $u_1, \ldots, u_k$ of L corresponding to the k smallest eigenvalues of L.
4. Let $U \in R^{N \times k}$ be the matrix including the eigenvectors $u_1, \ldots, u_k$ as its columns.
5. Cluster row vectors $u_i, i = 1, \ldots, N$ of the matrix U into k clusters $C_1, \ldots, C_k$ by using the k-means algorithm.
6. Assign the ith subregion to the segment $S_j, j = 1, \ldots, k$ if and only if the ith row of the matrix U belongs to the cluster C_j.
Output: Segments $S_i, i = 1, \ldots, k$.

The Laplacian matrix L is defined like in [17] as follows

$$L = D^{-\frac{1}{2}}(D - A)D^{-\frac{1}{2}}, \tag{8}$$

where $D = \text{Diag}(d_1, \ldots, d_N)$ with N is the number of subregions and $d_i = \sum_{j=1}^{N} A_{ij}$. The spectral clustering algorithm used for clustering subregions is shown in Algorithm 1.

The spectral graph theory [18] believes that if there are k clusters, the first k smallest eigenvalues of the Laplacian matrix are close to 0. Therefore, a threshold value T_{eigval} on eigenvalues is used to define which eigenvalues are close to 0. T_{eigval} is set to 0.01. Then, the number of eigenvalues close to 0 is regarded as the number k of segments in a foreground region.

3.3 The Constraint on Height Values of Persons in the Scene

A constraint on height values of persons in the scene is proposed to eliminate the false positive detections caused by the noisy background or the over-segmented part bodies. This can be regarded as a human model in the proposed method.

The unit of the height values is pixel. The height values of persons in the scene occupy a certain range which can be modeled as a Gaussian distribution. The probability that a detection bounding box with the height value h indicates a person is defined as follows

$$p(h; \mu_h, \sigma_h) = \frac{1}{\sigma_h \sqrt{2\pi}} \exp\left\{-\frac{1}{2}\left(\frac{h - \mu_h}{\sigma_h}\right)^2\right\}, \tag{9}$$

where μ_h ($\mu_h = 143$) and σ_h ($\sigma_h = 32$) denote respectively the mean and the standard deviation of height values of persons in the training data. The training data includes 38 frame images in which 5 persons go through the scene. If the probability at a height value is smaller than the threshold T_{prob}, the corresponding detection bounding box will be eliminated. T_{prob} is set to 0.0001.

4   Experimental Results

In this section, the proposed dataset used to evaluate the performance is firstly described. Then, the accuracy evaluations including qualitative and quantitative results are given to show that the proposed method outperforms two state-of-art human detection methods [7] and [14]. At last, the efficiency performances on the proposed method and two state-of-art human detection methods are presented.

4.1   Dataset

A new dataset is built for evaluating the capability of handling occlusions in human detection. The extend of occlusions in the dataset can be observed in Fig. 3. The dataset consists of color images and the corresponding depth data, both of which are taken by a Kinect camera in a lab environment. The Kinect camera is mounted on a tripod and elevated to the roof so that the Kinect camera can look down to the view of the scene. It is trying to imitate the surveillance camera setting. Due to the fact that students and staff in the lab go to have dinner together, they could be occluded seen by the Kinect camera which captures data for 8 days. From the original data, 297 frame images with occlusions are manually selected to create the dataset challenging for human detection. 1066 persons in the dataset are manually labeled in bounding boxes. The ground truth can be used to evaluate performances of methods. The following experimental results are reported based on the proposed dataset.

4.2   Accuracy Evaluation

The qualitative results from Felzenszwalb's method [7] and the proposed one are shown in Fig. 3. The first and the third rows refer to the results from Felzenszwalb's method which only uses the color images. Their results show that some occluded persons cannot be detected. The second and last rows show the corresponding results from the proposed method which handles the occlusions well for human detection.

Besides the qualitative results, the quantitative results are also given. The accuracy measurement used in the detection competition of PASCAL VOC 2012 is employed to quantitatively evaluate the performance of the proposed method. The measurements used in this paper include precision, recall and the average precision. Details can be found in [19].

The first objective of the quantitative results is to show that the boundary based similarity metric and the constraint on height values of persons significantly affect the performance of the proposed method. The first baseline of the proposed method is to simply use centers of subregions to compute the similarity without the constraint on height values of persons in the scene. The second baseline is to use all pairs of points on boundaries of subregions to compute the similarity without the height constraint. The third baseline is to use the K pairs of points on boundaries of subregions to compute the similarity without the height constraint. The proposed method uses the proposed boundary based

Fig. 3. The illustration of the qualitative results. The first and the third rows refer to the results from Felzenszwalb's method [7] which only uses the color image data. The second and the last rows refer to the corresponding results from the proposed method which uses the 3D data.

similarity metric with the height constraint. In Fig. 4, it can be observed that the precision and recall of the first baseline (the purple curve) is significantly improved to the second one (the yellow curve) due to the use of all pairs of points on boundaries of subregions. It substantiates that boundaries are representative of subregions. Moreover, the precision and recall of the third baseline (the cyan curve) are higher than the second one because K pairs of points between boundaries are more adequate for encoding the relationships between subregions than all pairs of points. Furthermore, with the constraint on height values of persons, the precision and recall of the third baseline (the cyan curve) can be improved to the proposed method (the red curve).

The quantitative results also demonstrate that the recall of Felzenszwalb's method [7] is lower than the proposed one although the precision of Felzenszwalb's method is high. This is because that Felzenszwalb's method fails to

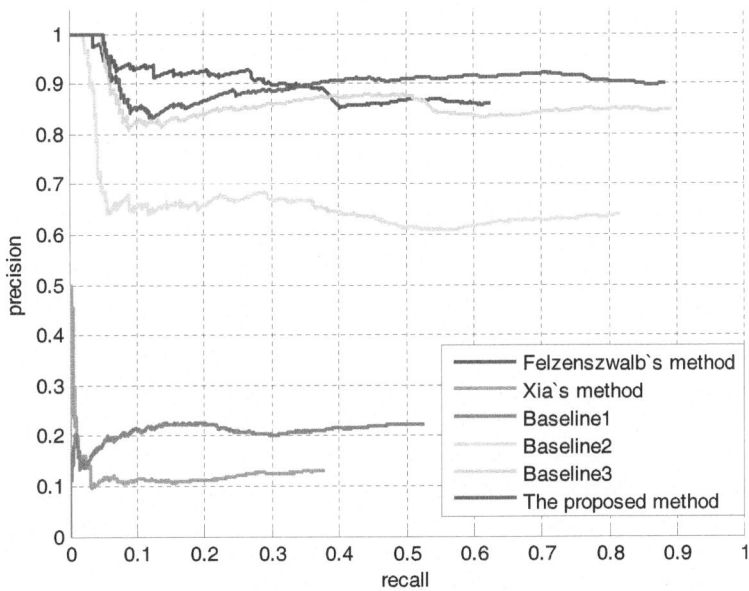

Fig. 4. The plot of precision versus recall on different methods

detect some occluded persons while the proposed one can handle these occlusions well. Moreover, Xia's method [14] obtains low precision and recall as shown in Fig. 4. The reasons are threefold. The first is that the 3D hemisphere model does not work well on our data such that false positives cannot be eliminated. The second is that two or more seed regions on one person can be detected, which yields to multiple detections for one person. These two reasons cause the low precision. The last one is that the parameter controlling the termination of the region growing algorithm is so sensitive that Xia's method cannot detect accurate regions of human. This reason causes the low recall.

To better evaluate the performance of the proposed method, Table 1 compares the average precisions on different methods. The measurement of the average precision considers both precisions and recalls. It means that to achieve a high score for the average precision, precisions are required to be high even when recalls are also high. From Table 1, it can be observed that the proposed method outperforms not only three baselines but also two state-of-art human detection methods on the average precision.

4.3 Efficiency Evaluation

Table 2 gives the run-time performances of different methods on the proposed dataset which contains 297 frame images. The term of "Total run-time" in the table refers to the time required to process all frame images in the proposed dataset. The term of "Average run-time" is computed as follows

$$\text{Average run-time} = \frac{\text{Total run-time}}{297}. \tag{10}$$

Table 1. The average precisions on different methods

Methods	Precision	Recall	Average Precision
Xia's method [14]	12.94%	37.80%	5.28%
Felzenszwalb's method [7]	86.05%	62.48%	56.71%
Baseline1 of the proposed method	22.29%	52.63%	11.84%
Baseline2 of the proposed method	64.02%	81.61%	55.05%
Baseline3 of the proposed method	84.99%	89.21%	78.16%
The proposed method	90.06%	88.37%	81.69%

It is necessary to mention that Xia's method [14] and the proposed method are performed on a PC with the specification of Quad-Core 3.30 GHz CPU and 8 GB RAM while Felzenswalb's method [7] is performed on a PC with the specification of Quad-Core 2.83 GHz CPU and 4 GB RAM. The reason for using two PCs is that Xia's method and the proposed method run on the windows system while Felzenswalb's method runs on the linux system. It is fair to compare the efficiency performance between Xia's method and the proposed one because two methods run on the same PC. It can be found from Table 2 that the proposed method performs faster than Xia's method because the proposed method does not require the region growing which is computational expensive. Since Felzenswalb's method and the proposed method run on different PCs, it is not fair to compare the run-time performances between these two methods. However, from the statistical results in Table 2, it can be concluded that the performance of the proposed method is comparable to the one of Felzenswalb's method.

Table 2. The run-time performances on different methods

Methods	Total run-time (sec)	Average run-time (sec)
Xia's method [14]	6367	21.4
Felzenszwalb's method [7]	3230	10.9
The proposed method	1630	5.5

5   Conclusions

In this paper, a human detection method with occlusion handling is proposed. The method firstly over-segments foreground regions into subregions and then clusters subregions into segments which may correspond to humans. This "split-and-merge" process can detect occluded persons by using the proposed boundary based similarity metric based on 3D coordinates of pixels. Moreover, a challenging dataset with human occlusions is built by using a Kinect camera for evaluating the proposed method. The qualitative and quantitative results demonstrate that the proposed method outperforms two state-of-art human detection methods on the proposed dataset.

In the future, human tracking could be developed based on the detection results derived from the current method. The objective is to facilitate the use of Kinect cameras in the video surveillance.

Acknowledgement. The authors would like to acknowledge the Ph.D. scholarship from School of Electrical and Electronic Engineering in Nanyang Technological University. The authors are also grateful to the useful suggestions from Dr. Wang Junyan, Miss Zhang Xiaoyan and Mr. Wang Bing on this work.

References

1. Ren, X., Malik, J.: Learning a classification model for segmentation. In: IEEE International Conference on Computer Vision, pp. 10–17 (2003)
2. Viola, P.A., Jones, M.J., Snow, D.: Detecting pedestrians using patterns of motion and appearance. In: IEEE International Conference on Computer Vision, pp. 734–741 (2003)
3. Dalal, N., Triggs, B.: Histograms of oriented gradients for human detection. In: IEEE Computer Society Conference on Computer Vision and Pattern Recognition, pp. 886–893 (2005)
4. Wu, B., Nevatia, R.: Detection of multiple, partially occluded humans in a single image by bayesian combination of edgelet part detectors. In: IEEE International Conference on Computer Vision, pp. 90–97 (2005)
5. Lin, Z., Davis, L.S., Doermann, D.S., DeMenthon, D.: Hierarchical part-template matching for human detection and segmentation. In: IEEE International Conference on Computer Vision, pp. 1–8 (2007)
6. Wang, X., Han, T.X., Yan, S.: An hog-lbp human detector with partial occlusion handling. In: IEEE International Conference on Computer Vision, pp. 32–39 (2009)
7. Felzenszwalb, P.F., Girshick, R.B., McAllester, D.A., Ramanan, D.: Object detection with discriminatively trained part-based models. IEEE Transactions on Pattern Analysis and Machine Intelligence 32, 1627–1645 (2010)
8. Xu, F., Fujimura, K.: Human detection using depth and gray images. In: IEEE Conference on Advanced Video and Signal Based Surveillance, pp. 115–121 (2003)
9. Gavrila, D.M., Munder, S.: Multi-cue pedestrian detection and tracking from a moving vehicle. International Journal of Computer Vision 73, 41–59 (2007)
10. Ikemura, S., Fujiyoshi, H.: Real-Time Human Detection Using Relational Depth Similarity Features. In: Kimmel, R., Klette, R., Sugimoto, A. (eds.) ACCV 2010, Part IV. LNCS, vol. 6495, pp. 25–38. Springer, Heidelberg (2011)
11. Spinello, L., Arras, K.O.: People detection in rgb-d data. In: IEEE/RSJ International Conference on Intelligent Robots and Systems, pp. 3838–3843 (2011)
12. Choi, W., Pantofaru, C., Savarese, S.: Detecting and tracking people using an rgb-d camera via multiple detector fusion. In: IEEE International Conference on Computer Vision Workshops, pp. 1076–1083 (2011)
13. Gill, T., Keller, J.M., Anderson, D.T., Luke III, R.H.: A system for change detection and human recognition in voxel space using the microsoft kinect sensor. In: IEEE Applied Imagery Pattern Recognition Workshop, pp. 1–8 (2011)
14. Xia, L., Chen, C.C., Aggarwal, J.K.: Human detection using depth information by kinect. In: IEEE Computer Society Conference on Computer Vision and Pattern Recognition Workshops, pp. 15–22 (2011)

15. Gonzalez, R.C., Woods, R.E.: Digital image processing. Prentice-Hall, Inc., Upper Saddle River (2006)
16. Shi, J., Malik, J.: Normalized cuts and image segmentation. IEEE Transactions on Pattern Analysis and Machine Intelligence 22, 888–905 (2000)
17. Ng, A.Y., Jordan, M.I., Weiss, Y.: On spectral clustering: Analysis and an algorithm. In: Advances in Neural Information Processing Systems, pp. 849–856 (2001)
18. Chung, F.R.K.: Spectral graph theory. In: CBMS Regional Conference Series in Mathematics, vol. 92, pp. 1–212. American Mathematical Society (1997)
19. Everingham, M., Van Gool, L.J., Williams, C.K.I., Winn, J.M., Zisserman, A.: The pascal visual object classes (voc) challenge. International Journal of Computer Vision 88, 303–338 (2010)

Spin Image Revisited: Fast Candidate Selection Using Outlier Forest Search

Young-Woon Cha, Hwasup Lim, Seong-Oh Lee,
Hyoung-Gon Kim, and Sang Chul Ahn

Imaging Media Research Center
Korea Institute of Science and Technology, Seoul, Korea
{ywcha,hslim,solee,hgk,asc}@imrc.kist.re.kr

Abstract. Spin-images have been widely used for surface registration and object detection from range images in that they are scale, rotation, and pose invariant. The computational complexity, however, is linear to the number of spin images in the model data set because valid candidates are chosen according to the similarity distribution between the input spin image and whole spin images in the data set. In this paper we present a fast method for valid candidate selection as well as approximate estimate of the similarity distribution using outlier search in the partitioned vocabulary trees. The sampled spin images in each tree are used for approximate density estimation and best matched candidates are then collected in the trees according to the statistics of the density. In contrast to the previous approaches that attempt to build compact representations of the spin images, the proposed method reduces the search space using the hierarchical clusters of the spin images such that the computational complexity is drastically reduced from $O(K \cdot N)$ to $O(K \cdot \log N)$. K and N are the size of the spin-image features and the model data sets respectively. As demonstrated in the experimental results with a consumer depth camera, the proposed method is tens of times faster than the conventional method while the registration accuracy is preserved.

1 Introduction

Object detection is one of the most important tasks in computer vision research and there has been remarkable progress during the last decade. However, several challenging issues still remain in terms of generalization to unseen examples. One of such issues is viewpoint change, for which researchers have tried to develop new 2D image descriptors or classifiers robust to scale, rotation, and deformations.

When it comes to detecting objects on 3D space using 3D models, i.e. mesh model or point clouds, the viewpoint variation does not change the form of object representation. Nowadays, commercially cheap range sensors such as Microsoft Kinect and Asus Xtion are available and broadly used for computer vision and robotics areas. One of the most fascinating applications is real-time 3D reconstruction and modelling introduced in KinectFusion [1], which allows us to promptly acquire 3D models and 3D scenes of real world. The advent of

J.-I. Park and J. Kim (Eds.): ACCV 2012 Workshops, Part II, LNCS 7729, pp. 209–222, 2013.

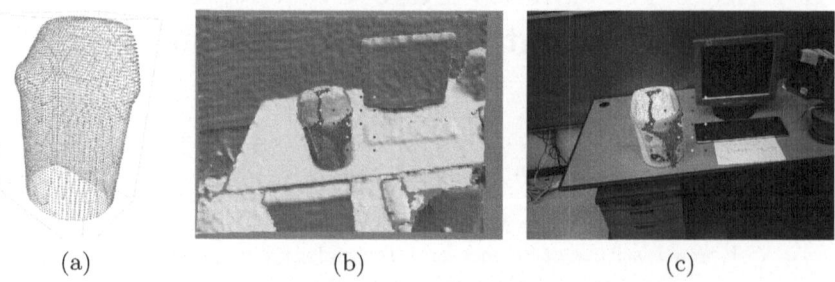

(a)   (b)   (c)

Fig. 1. Object detection through viewpoint estimation. 1a is a point cloud model. 1b and 1c are the detection result overlaid with the normal map of the depth image and the color image. Detailed explanation can be seen in Sec. 4.

such range sensors is encouraging to develop object detection methods using 3D models to resolve the viewpoint variation.

3D Object detection can be reformulated as viewpoint estimation, namely to find the pose of the objects from the given 3D point clouds acquired from range data, for example, as shown in Fig. 1. The point clouds and the input depth images were generated using KinectFusion. Here their vertex coordinates are invariant to scale in the input range data.

There were a few attempts to incorporate depth data with color images for viewpoint estimation. Hinterstoisser et al. [2] proposed the multi-modal method based on the intensity and range orientation templates. This method performs in real time due to efficient template representation for fast linear memory comparison but requires a set of 2D images captured at whole viewpoints including entire rotations and scales beforehand for viewer-oriented representation. In this case, it is more desirable to construct templates based on the object-oriented representation using 3D models.

Johnson and Hebert [3] suggested the 3D surface matching method using spin images. A spin image represents a 3D surface at a point into the 2D histogram invariant to rotation. In their work, object detection is carried through surface registration as follows. First, the correspondence pairs between the points in the input image and the ones in the object are found by comparing their spin images. Then, the possible transformations are estimated by grouping the correspondences using geometric consistency. The object is assumed to be detected when sufficiently large neighbor points around the correspondence group are aligned using the transformation.

The simple ways to find correspondences are to search the top-n similar points from the model points or to threshold similarities using a constant value, but these rules involve incorrect correspondences or miss reliable ones so that it leads to a decrease in accuracy. This problem is illustrated in Fig. 2. For this reason, the method in [3] assumed that more plausible corresponding points have extremely larger similarity values, so-called outlier values, than those of other model points. This assumption necessitates the use of outlier detection methodology.

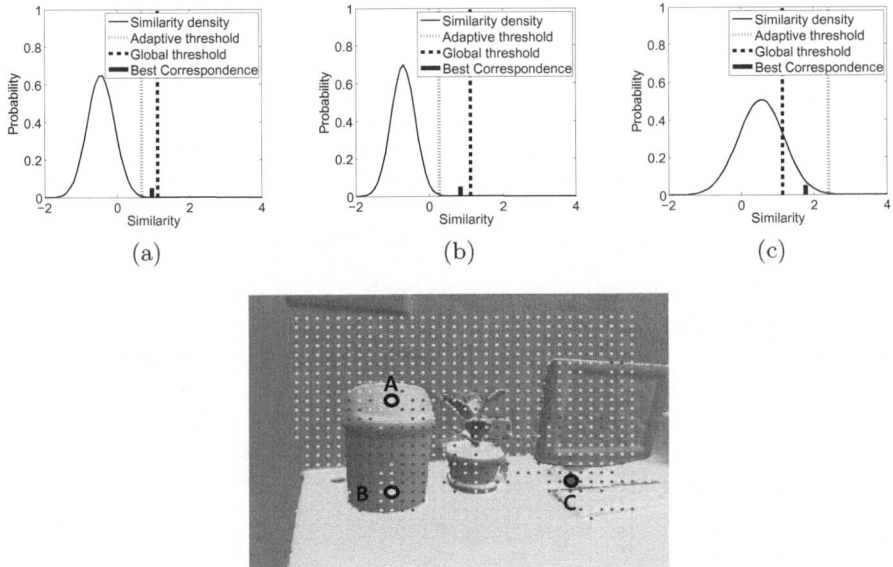

Fig. 2. Correspondence detection using a similarity threshold. The red and yellow points were uniformly sampled from the scene (bottom). Red points in the scene indicate that there are some correspondences from the trash can model, while yellow points are not. The points A and B were misclassified as false-negatives. C is an example of false-positives. As shown in the first row, similarity measurements do not directly mean the better correspondence.

The exhaustive search algorithm from large feature sets to detect outlier points requires huge computational costs. In practice, a 3D model consists of ten thousands of points and each descriptor of such points contains hundreds of dimensional vectors and comparing all the descriptor pairs makes infeasible to real-time applications. Its complexity is $O(K \cdot N)$. The K and N are the size of the spinimage features and the model data sets (viewpoints) respectively. However, whole descriptor sets should be considered every time for accurate pose estimation.

To reduce this difficulty, Johnson and Hebert [4] proposed the principal component analysis for spin image compression but it entails degradation of the accuracy. Dinh and Kropac [5] presented the multi-resolution spin image, which lowers the complexity by $O((K/c) \cdot N)$ where c is a constant. Unfortunately, these methods still depend on the exhaustive search in the large data sets.

This paper revisits the spin image for 3D surface matching and proposes an efficient search algorithm for finding correspondences from extensive data sets with no exhaustive search. We employ the k-nearest neighbor (KNN) search [6] and the outlier detection method [7] to reduce the searching cost from $O(K \cdot N)$ to $O(K \cdot \log N)$. The experiments show that the proposed method effectively estimates correspondences using sampled inliers and significantly reduces the running time.

This paper organizes as follows. Section 2 discusses related works and preliminaries. Section 3 explains the suggested approach called *Outlier Forest Search* and its properties. Section 4 demonstrates the performance of the proposed method on viewpoint estimation of everyday objects.

2   Related Work

2.1   Object Detection Using Spin Image Descriptors

A given 3D model represented by point clouds, which consists of vertices and their normal vectors, can be re-casted into point descriptors at each vertex. The point descriptor indicates each object-oriented viewpoint of the model. The spin image representation [3] is the compact descriptor to describe 3D surfaces, $R^3 \rightarrow R^2$. The spin image at each point has the base with the direction of the vertex normal and its tangent plane as shown in Fig. 3. The spin image coordinates are formally defined as,

$$S_p(q) \rightarrow (\alpha, \beta) = (\sqrt{\| p - q \|^2 - (n \cdot (p - q))^2}, n \cdot (p - q)) \tag{1}$$

Neighborhood points are binned according to axial circulation. The similarity measurements, which are the outcomes of comparison between spin images of the sample point in the input image and entire spin images in the object, are summarized into a similarity histogram. Outlier points with relatively strong correspondence can be identified by the statistical test from the similarity density. Then the object's pose is estimated by grouping correspondences and performing surface registration.

Another works for improving the 3D object detection using spin images were suggested in order to enrich description powers. The covariance descriptors [8] exploited more geometric features and showed normal features are more important than others. Spin images were also used for the content based 3D objects retrieval and indexing techniques were used for enhancing retrieval efficiency [9].

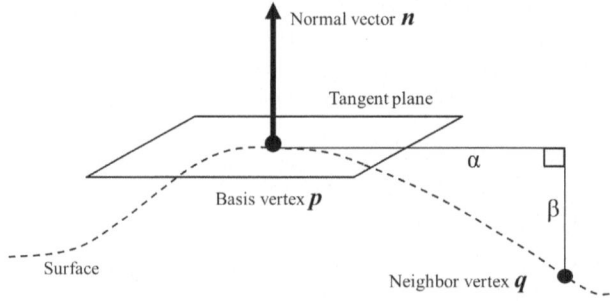

Fig. 3. The concept of the spin image histogram. The neighbor points are binned according to their (α, β) coordinates.

2.2   Data Indexing and Outlier Detection

Content-based image retrieval (CBIR) approaches try to index image sets in advance so that they can search the most similar image with the query quickly and confidently. Sorting based indexing methods such as KD-tree [10] and R-tree [11] are the broadly exploited data structures for indexing images. These methods, however, rely on aligning data sets recursively according to one of the dimensions of data with the largest variance. Thus for high dimensional data such as SIFT descriptors [12] whose dimensions are usually 128 or 256, clustering based indexing is more suitable since it considers whole dimensions of the visual data at each step. For this purpose a hierarchical k-means clustering is introduced for building the vocabulary tree [6]. Thanks to its index structure based on a tree, it costs only $O(\log N)$ for searching top-n ranked data using the approximate k-nearest neighbor search (KNN).

All top-n ranked data do not always have distinctively high similarities compared with the rest in the tree as shown in Fig. 2. Instead detecting outliers, defined as an anomaly compared with other data, are required for selecting reliable correspondences in the presence of variations. Ramaswamy et al. [7] suggested an efficient method to find outliers by partitioning data sets.

3   Correspondence Detection Using Outlier Forest Search

This paper handles the correspondence detection problem between the point in the input image and the ones in the object by comparing their spin images. Each sample in the input image should be compared with all spin image descriptors captured at each vertex of the object in order to attain the exact pose estimation, which is an unacceptable operation for real-time applications. As mentioned earlier finding the top-n best correspondences or thresholding according to the

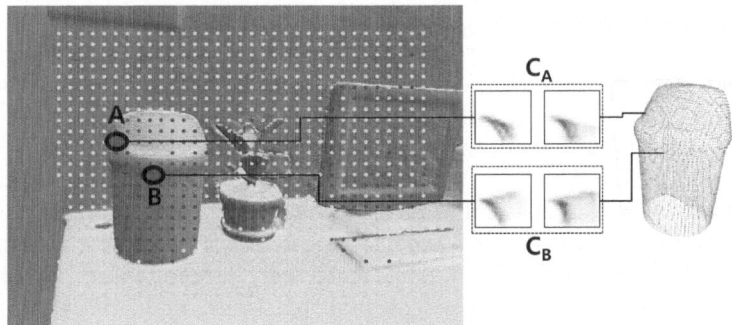

Fig. 4. Correspondence detection result. The red and yellow points were uniformly sampled from the scene (left). Red points in the scene indicate that there are some correspondences from the model (right), while yellow points are not. The points A and B in the scene were matched from the trash can model by comparing their spin image descriptors (center), respectively.

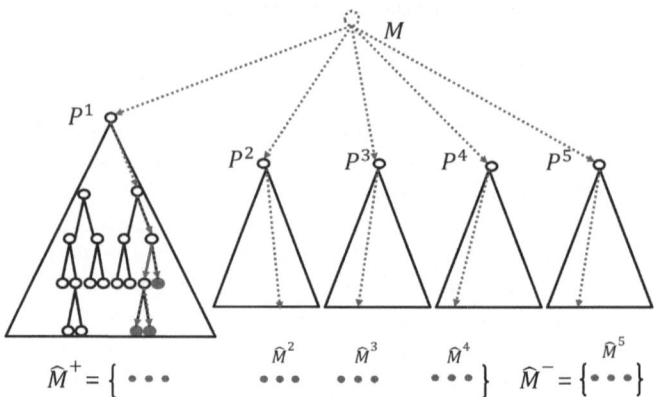

Fig. 5. Outlier forest search. P^j represents the separated data partitions from the object descriptor set M. During multi-path KNN sampling denoted by dotted red arrows, top-n ranked $\hat{m}_z^j \in \hat{M}^j$ are sampled from each P^j.

constant similarity value are not desirable ways because they cause unnecessary correspondence pairs or rejects correct ones.

This irrelevance is due to the fact that degrees of the noise or shapes of the distribution in descriptors influences change in similarity measurements. Therefore, it is necessary to develop the method to not only find correspondences efficiently from large data sets but also can adaptively threshold similarities. To achieve the goal, this paper combines the methods of indexing visual descriptors using the hierarchical k-means clustering [6] and the outlier detection using sampled inliers. Searching on a tree or a forest brings up logarithmic computations for searching, and sampled inliers gathered from separate trees (data partitions) allow to reasonably approximate the inlier distribution.

Fig. 4 shows a correspondence detection process. In the training phase, the spin image descriptors M of the model are generated using the given model point clouds and the set is indexed into multiple trees (a forest) using hierarchical k-means clustering. During the detection phase, a given scene is uniformly sampled and scene spin image descriptors S are calculated for detection.

For each scene point descriptor $s_i \in S$, the top-n most similar $\hat{m}_z \in \hat{M}$ are identified using KNN. The hat, "ˆ" is denoted to represent samples acquired by KNN and thus $\hat{M} \subset M$. Instead of using a constant threshold, we declare pairs of (s_i, m_z) as correspondences by the statistical test for adaptive thresholding, assuming that the reliable (s_i, m_z) has extremely large similarity than other pairs. Using indexed M structured as multiple trees, multi-path sampling is carried out to detect inliers and candidate outliers simultaneously as illustrated in Fig. 5. The sample similarity density $\hat{f}$ is estimated by assuming a single modal Gaussian distribution using inliers M^+. Finally, the candidate outliers M^- are declared as outliers O if they are placed far from $\hat{f}$.

Since the proposed approach uses point clouds of the object and the input scene reconstructed by KinectFusion [1], the spin image descriptors $\tilde{h}_i$ are sensitive to noise and mesh resolution. To be independent to them, the descriptors $\acute{h}_i$ should be normalized as

$$\acute{h}_i = \frac{\tilde{h}_i}{\sum \tilde{h}_i} \tag{2}$$

We denote $\acute{h}_i$ as h_i hereafter for the simplicity. Eq. 2 ensures h_i to be a likelihood map.

3.1 Forest Training

The given M is separated into partitions $P^j \subset M$ using k-means clustering. Then, hierarchical k-means clustering [6] is performed for each P^j. This forest indexing is illustrated in Fig. 5. In experiment, even a few partitions gave equivalent detection results with the exhaustive search in large models. For each P^j, 15 branching factor was used for the hierarchical k-means clustering, recommending 15 to 30 branching factors. The superscript stands for partition index, and the subscript indicates descriptor index in the corresponding set.

3.2 Outlier Candidates Detection

When comparing two descriptors between s_i and m_z, L^2-like distance measures are better than L^1 distance measures for robustness [4,13]. As used in the background modeling [14], the inverse of the χ^2-square distance is employed as the similarity measure of $L(s_i, m_z)$ or simply $L(m_z)$,

$$L(s_i, m_z) \equiv -2 \cdot \log C(s_i, m_z) \tag{3}$$

$$C(h_i, h_j) \equiv \frac{1}{2} \sum_{k=1}^{K} \frac{[h_i(k) - h_j(k)]^2}{h_i(k) + h_j(k)} \tag{4}$$

Eq. 3 transforms χ^2-square distance in Eq. 4 into similarity values. χ^2-square distance is equivalent to the likelihood ratio test by normalizing expected distance as in the denominator in Eq. 4. The negative log scaling in Eq. 3 is important because it emphasizes samples with extremely small distances to have exponentially large similarity values while smoothing the others. This fact leads the reasonable assumption that similarity measurements are distributed under a single modal Gaussian.

As illustrated in Fig. 5, multi-path top-n KNN sampling finds $|\hat{M}^j| = n$ for each P^j in order of similarities. $\hat{m}_z^j \in \hat{M}^j$ denotes the z-th similar descriptor with s_i in P^j. Identifying the inlier sample set $\hat{M}^+$ or the outlier candidate sample set $\hat{M}^-$ is easily calculated by comparing $\hat{m}_1^j \in \hat{M}^j$ as,

$$\hat{M}^- \equiv \{\hat{m}_z \in M^r | \arg\max_r L(\hat{m}_1^r) \in \hat{M}\} \tag{5}$$

$$\hat{M}^+ \equiv \{\hat{m}_z \in M^j | j \neq r\} \tag{6}$$

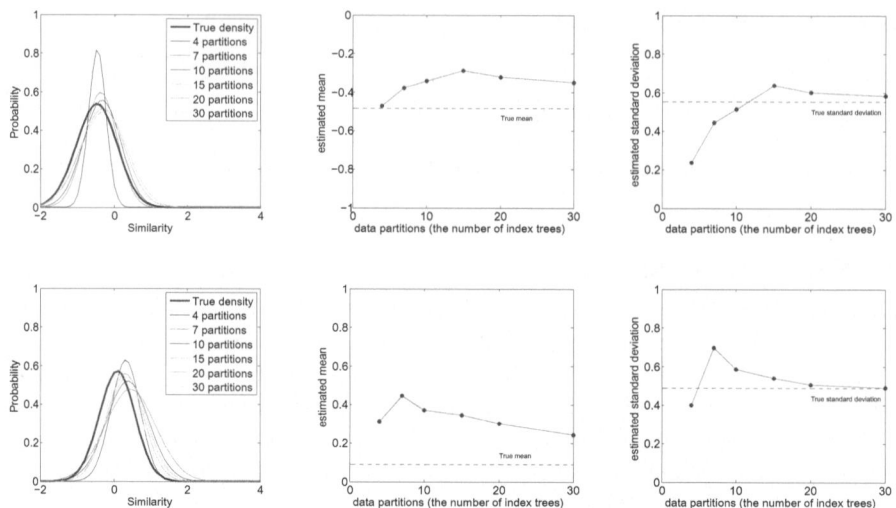

Fig. 6. Inlier similarity density estimation in altering the partition number k. Each row shows estimated similarity densities at the regions A and B in Fig. 4. The first column compares the true Gaussian density and estimated ones. The second and third columns compare the means and the standard deviations respectively.

3.3   Inlier Density Estimation and Outlier Detection

In the view of background modelling, inlier distribution function $\hat{f}$ is estimated from $\hat{\boldsymbol{M}}^{+}$ (background) under the assumption of a Gaussian distribution and can be used to decide if $\hat{m}_z^r \in \hat{\boldsymbol{M}}^{-}$ (foreground) is an outlier. Since $\hat{m}_z^j \in \hat{\boldsymbol{M}}^{+}$ is a representative from the different sized $\boldsymbol{P}^j$, we estimate the weighted sample mean $\hat{\mu}$ and the unbiased weighted sample standard deviation $\hat{\sigma}$ according to the partition weight t^j using the cardinality $|\boldsymbol{P}^j|$ as,

$$t^j = \frac{|\boldsymbol{P}^j|}{|\boldsymbol{M}^+|} \quad \text{s.t.} \quad \sum_{j=1}^{k-1} t^j = 1 \tag{7}$$

$$w_z^j = \frac{t^j}{n} \quad \text{s.t.} \quad \sum_{j=1}^{k-1}\sum_{z=1}^{n} w_z^j = 1. \tag{8}$$

$$\hat{\mu} = \sum_{j=1}^{k-1}\sum_{z=1}^{n} w_z^j \cdot L(\hat{m}_z^j) \tag{9}$$

$$\hat{\sigma} = \sqrt{\frac{1}{1 - \sum_j (t^j)^2} \sum_{j=1}^{k-1}\sum_{z=1}^{n} w_z^j \cdot (L(\hat{m}_z^j) - \hat{\mu})^2} \tag{10}$$

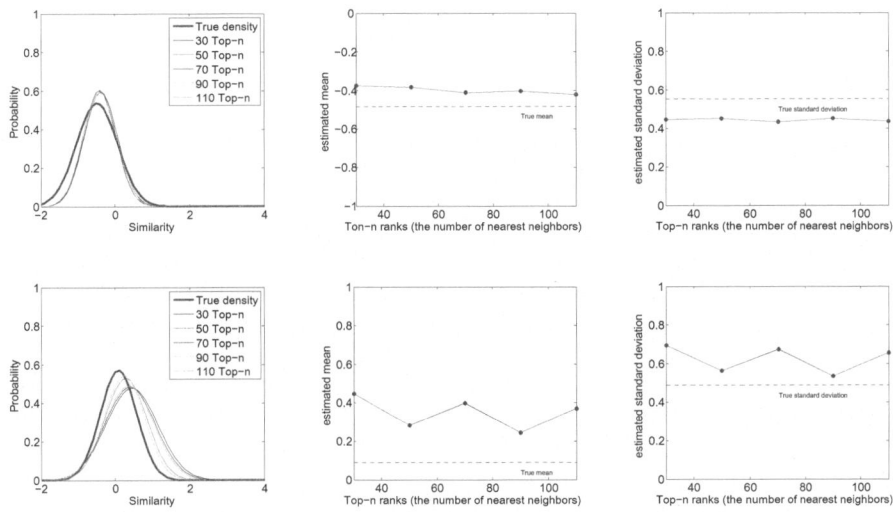

Fig. 7. Inlier similarity density estimation in changing the top-n nearest neighbor number. Each row shows estimated similarity densities at regions A and B in Fig. 4. The first column compares the true Gaussian density and estimated ones. The second and third columns compare the means and the standard deviations respectively.

Let n be the parameter for the top-n KNN sampling and k is the number of trees. Note that the first term of Eq. 10 makes the $\hat{\sigma}$ the unbiased estimator. Each partition $\boldsymbol{P}^j$ is regarded as a single sample of the conventional unbiased estimator because the KNN sampling is performed for each $\boldsymbol{P}^j$.

The estimated outliers $\boldsymbol{O}$ are detected by calculating how much each $\hat{m}_z^r \in \boldsymbol{M}^-$ is not able to statistically belong to $\hat{f}$. Thus an adaptive threshold $\hat{T}$ is calculated as,

$$\hat{T} = \hat{\mu} + \alpha \cdot \hat{\sigma} \tag{11}$$

$$\boldsymbol{O} \equiv \{\hat{m}_z^r | L(\hat{m}_z^r) > \hat{T}\} \tag{12}$$

The outlier sample set $\boldsymbol{O}$ is declared as the correspondences of s_i. In practice, we recommend $\alpha = 6$ for using $\hat{T}$ as the upper bound of T obtained from the population density f.

3.4 Properties of Parameters

This section investigates how the k and n affect the performance in running time and accuracy, relating to the exhaustive search. Suppose that the top-n KNN sampling almost always samples the true top-n samples. Under this assumption, the samples $\hat{m}_z^j \in \hat{\boldsymbol{M}}$ always hold Eq. 13, which means that $L(\hat{m}_z^j)$ is the upper

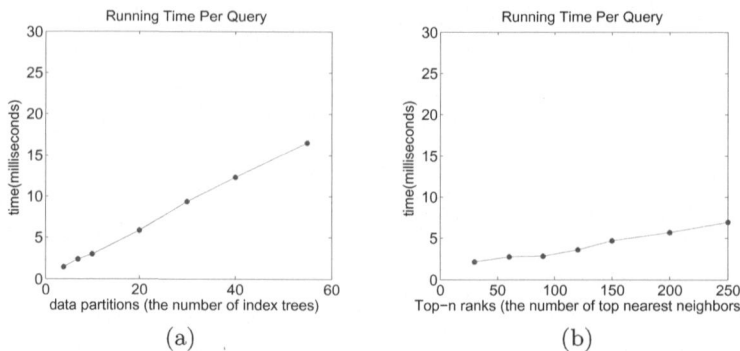

Fig. 8. Effects of parameter selection to the running time. 8a and 8b show how the variations of partition numbers k and top-n nearest neighbor numbers affect the running time respectively.

bound of $L(m_z^j)$. Therefore, $\hat{\boldsymbol{M}}$ are representative samples. Fig. 6, 7 shows that the estimated means always have lager values than their true means.

$$L(\hat{m}_z^j) > L(m_l^j) \in \boldsymbol{P}^j \tag{13}$$

Another property is that the $\hat{f}$ approaches to f as k or n increases to N in Eq. 8 and N is $|M|$. As k grows, t^j decreases by the definition of k-means clustering. As k or n increases, w_z^j decreases as well. When w_z^j is reduced to $1/N$, the sample $\hat{\mu}$ and $\hat{\sigma}$ converge to the population μ and σ. Therefore the following equation holds.

$$\lim_{k \to N \text{ or } n \to N} w_z^j = \frac{1}{N} \tag{14}$$

It is worth mentioning that the degree of convergence by k is much faster than that of n. The more k increases, the more $\hat{M}$ are uniformly sampled by the multi-path sampling. However, the increment of n appends nearest neighbor samples first due to the KNN sampling strategy. Once k is large enough, small n is sufficient for preserving the estimation accuracy.

Fig. 6 shows that $\hat{f}$ approaches to f as k increases. Note that very small k causes inadequately small $\hat{\sigma}$ so that $\hat{T}$ can be smaller than T. In experiment, $k > 7$ is enough to be the upper bound of the exhaustive search. Alternatively large α in Eq. 11 can be chosen for small k. In Fig. 7 with the increase in n, $\hat{f}$ does not sufficiently reach to f due to the KNN sampling.

Fig. 8 shows that the k is more critical for the running time because it is related to the branching overhead of the multi-path search. The smaller k instead of smaller n is preferred since large k significantly increases the running time.

4   Experiments

In this section, the proposed method is demonstrated by comparing the running time and the retrieval performance with the exhaustive search. The proposed method is applied to the viewpoint estimation of objects using spin images [3] and results are illustrated in Fig. 1 and 9. The models such as the trash can, the chair, the potted plant, and the monitor comprise 7892, 16873, 9236, and 16,095 vertices, respectively. 640×480 depth signals were captured from Asus Xtion. The models and input scenes were reconstructed using KinectFusion [1] with same parameters, so their mesh resolutions are always same. 15×15-sized spin image descriptors were exploited with a 90 degree support angle. The proposed method was implemented using C^{++} and tested with Intel i7-3610QM 2.3GHz computer with 16GB memories.

The blue points in the second column in Fig. 1, 9 represents uniformly picked samples on the input image. The yellow points indicate correct correspondences detected by outlier search from each model descriptor set, and initial transformations were calculated using them [15]. The blue regions denote grouped

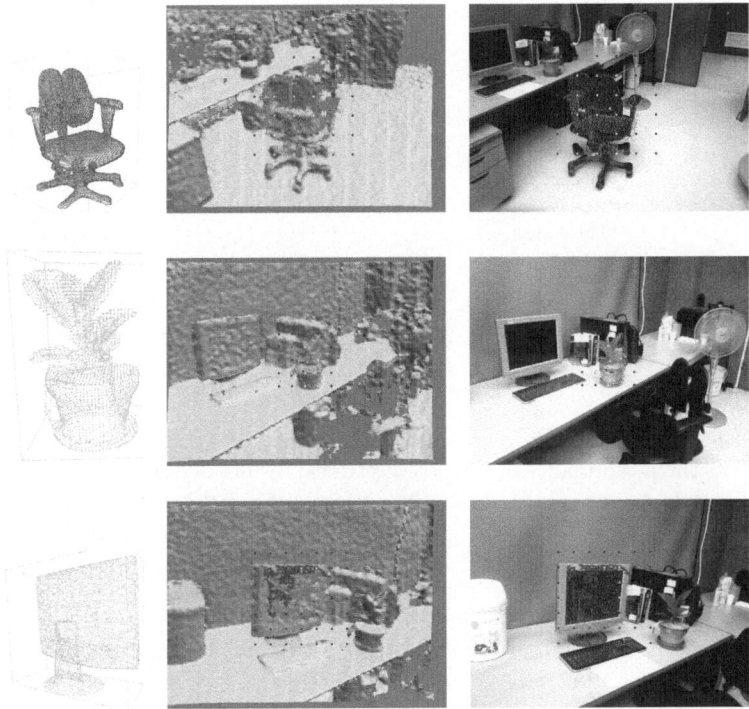

Fig. 9. Object recognition and its viewpoint estimation result. The first columns represent each point cloud model. The second and third columns indicate the detection results overlaid with the normal map of the depth image and the color image.

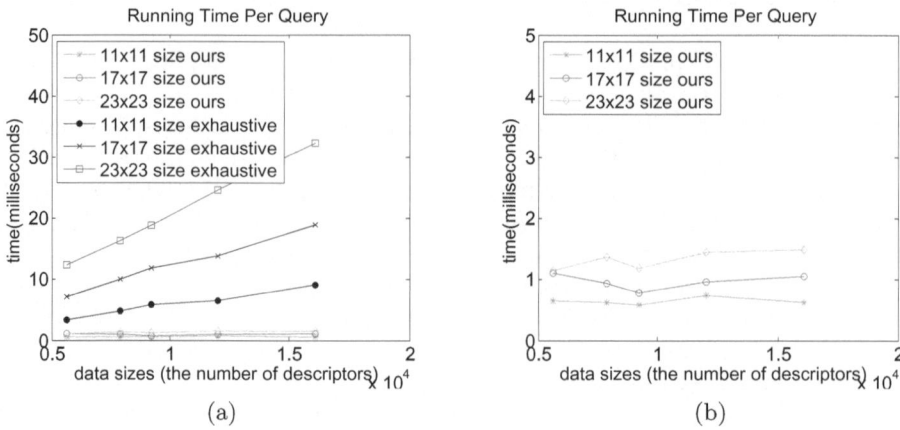

Fig. 10. Running time comparison between the exhaustive search [3] and ours with different-sized spin image descriptors

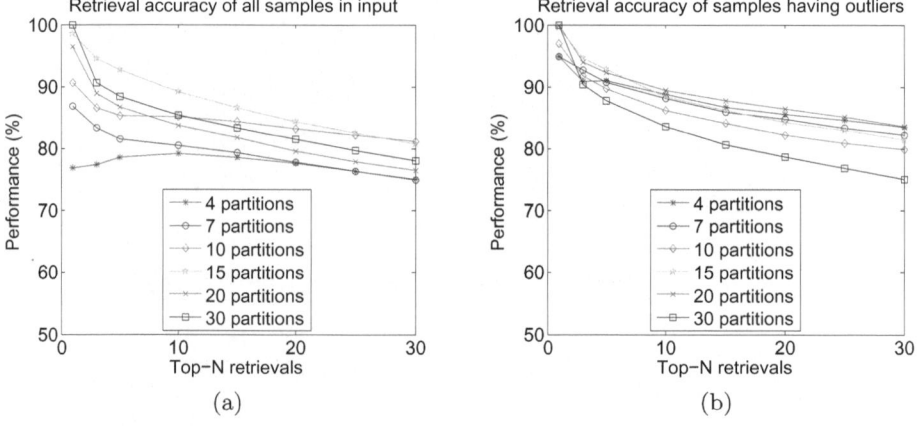

Fig. 11. Retrieval performance comparison between the exhaustive search and ours. (11a) depicts the performance of the top-n retrieval results of ours compared with those of the exhaustive search. (11b) shows the performance of the outlier retrieval.

points with the transformations based on the geometric consistency [3]. The green points describe the estimated poses.

The suggested search algorithm drastically reduces the running times as shown in Fig. 10, where k and n were set to 5 and 30 respectively. Around 800 query samples were uniformly sampled from the input image. The running time per query means the elapsed time for finding correspondences per each sample in the input image from the model data sets. Note that the proposed method is tens of times faster than the exhaustive search. Even ours with 23×23-sized descriptor is much faster than the exhaustive search with the much smaller-sized descriptors in

Fig. 10a. In Fig. 10b it is shown that the outlier forest search requires approximately constant times regardless of the number of descriptors.

Fig. 11 shows the retrieval comparison of the retrieved samples using the proposed method and the exhaustive search per each query. The horizontal axis indicates the percentage of matched top-n responses between two algorithms. Fig. 11a is generated by all queries in the input image, yellow and red points in Fig. 4. Fig. 11b is the performance about the queries that contains outliers, in other words, solely red points in Fig. 4. The hierarchical k-means indexing [6] gives higher performances when similar responses exist in data sets. Consequently reliable correspondences are still retrieved through KNN search on indexed trees with no exhaustive search.

5 Conclusion

This paper suggests the efficient correspondence search algorithm for the object detection using 3D surface matching with spin images. The method, called *Outlier Forest Search*, approximates the exhaustive search by combining hierarchical k-means clustering and sampling for outlier detection. The experiments show that the method significantly reduces the computational cost in large part, and high dimensional features and preserves the retrieval performance at the same time.

One of the disadvantages of this method is difficulty in finding optimal k. In experiment, $k > 7$ is a reasonable choice and fairly robust. Although the retrieval performance of the approximation is decreased if the query does not have similar response in the data set, reliable correspondences are still not lost.

According to its properties, Outlier Forest Search is the upper bound, more strict, of the exhaustive search and converges to it as k increases. Since the indexing techniques can be used for other visual features, the proposed method can be applied to other descriptors.

References

1. Newcombe, R., Izadi, S., Hilliges, O., Molyneaux, D., Kim, D., Davison, A., Kohli, P., Shotton, J., Hodges, S., Fitzgibbon, A.: Kinectfusion: Real-time dense surface mapping and tracking. In: 2011 10th IEEE International Symposium on Mixed and Augmented Reality (ISMAR), pp. 127–136. IEEE (2011)
2. Hinterstoisser, S., Cagniart, C., Ilic, S., Sturm, P., Navab, N., Fua, P., Lepetit, V.: Gradient response maps for real-time detection of texture-less objects. IEEE Transactions on Pattern Analysis and Machine Intelligence (2011)
3. Johnson, A., Hebert, M.: Recognizing objects by matching oriented points. In: Proceedings of the 1997 IEEE Computer Society Conference on Computer Vision and Pattern Recognition, pp. 684–689. IEEE (1997)
4. Johnson, A., Hebert, M.: Using spin images for efficient object recognition in cluttered 3d scenes. IEEE Transactions on Pattern Analysis and Machine Intelligence 21, 433–449 (1999)

5. Dinh, H., Kropac, S.: Multi-resolution spin-images. In: 2006 IEEE Computer Society Conference on Computer Vision and Pattern Recognition, vol. 1, pp. 863–870. IEEE (2006)
6. Nister, D., Stewenius, H.: Scalable recognition with a vocabulary tree. In: 2006 IEEE Computer Society Conference on Computer Vision and Pattern Recognition, vol. 2, pp. 2161–2168. IEEE (2006)
7. Ramaswamy, S., Rastogi, R., Shim, K.: Efficient algorithms for mining outliers from large data sets. ACM SIGMOD Record 29, 427–438 (2000)
8. Fehr, D., Cherian, A., Sivalingam, R., Nickolay, S., Morellas, V., Papanikolopoulos, N.: Compact covariance descriptors in 3d point clouds for object recognition. In: 2012 IEEE International Conference on Robotics and Automation (ICRA), pp. 1793–1798. IEEE (2012)
9. Assfalg, J., Bertini, M., Bimbo, A., Pala, P.: Content-based retrieval of 3-d objects using spin image signatures. IEEE Transactions on Multimedia 9, 589–599 (2007)
10. Bentley, J.: Multidimensional binary search trees used for associative searching. Communications of the ACM 18, 509–517 (1975)
11. Guttman, A.: R-trees: a dynamic index structure for spatial searching, vol. 14. ACM (1984)
12. Lowe, D.: Distinctive image features from scale-invariant keypoints. International Journal of Computer Vision 60, 91–110 (2004)
13. Jou, F., Fan, K., Chang, Y.: Efficient matching of large-size histograms. Pattern Recognition Letters 25, 277–286 (2004)
14. Sheikh, Y., Shah, M.: Bayesian modeling of dynamic scenes for object detection. IEEE Transactions on Pattern Analysis and Machine Intelligence 27, 1778–1792 (2005)
15. Gelfand, N., Mitra, N., Guibas, L., Pottmann, H.: Robust global registration. In: Proceedings of the Third Eurographics Symposium on Geometry Processing, p. 197. Eurographics Association (2005)

MRF Guided Anisotropic Depth Diffusion for Kinect Range Image Enhancement

Karthik Mahesh Varadarajan and Markus Vincze

Vienna, Austria

Abstract. Projected texture based 3D sensing modalities are being increasingly used for a variety of 3D computer vision applications. However, these sensing modalities, exemplified by the Microsoft Kinect Sensor, suffer from severe drawbacks that hamper the quality of the range estimate output from the sensor. It is well known that the quality of reconstruction of the 3D projected texture for range estimation is a function of the material properties of objects in the image. Objects colored black, yellow or deep red often do not reflect the texture in a manner suitable for the detector to estimate the range values. Furthermore, shiny or highly reflective objects can also scatter the projected texture patterns. Objects with skewed surface orientation, occlusions, object self-shadows and intra-object mutual shadows, transparency and other factors also create problems with projected texture reconstruction. In order to alleviate these concerns, depth interpolation techniques have been used in the past. These techniques, however, create loss of depth structures crucial for segmentation and detection processes. In order to alleviate these concerns, we present a novel MRF based color- depth fusion algorithm which uses information from the RGB sensor of the Kinect and couples it with the depth content to produce fine structure, high fidelity depth maps. This algorithm can be implemented in hardware on the Kinect device, thereby improving the depth resolution, fidelity of the sensor while eliminating range errors and shadows.

Keywords: Kinect Sensor, Range Enhancement, Depth Diffusion.

1   Introduction

3D data acquisition for real-time applications has been progressing towards sensing modalities such as RGB-D sensors, in particular the Microsoft Kinect Sensor. The advantages of using such a high-frame rate projected texture active sensing approach are numerous in comparison with traditional approaches such as Stereo, Laser and Lidar. While providing superior reconstruction in most types of scenarios including those with areas of homogeneous texture that lack stable features, without actively changing the visible spectrum of the scene, projected light texture based RGB-D cameras produce several artifacts in the 3D reconstruction. These arise from the factors such as material properties of the scene being captured - the absorption spectrum of the material matching the wavelength of the projected light, transparency/ very high or very low reflectance of the surface, besides factors such as surface orientation skew

J.-I. Park and J. Kim (Eds.): ACCV 2012 Workshops, Part II, LNCS 7729, pp. 223–235, 2013.
© Springer-Verlag Berlin Heidelberg 2013

with respect to the sensing angle, occlusions, shadows between the projector and the sensor and sparsity of the projected texture that results in a loss of fine structures. In order to alleviate these concerns, it is necessary to estimate the depth pixels in regions where the sensor response is zero as well as to refine the depth pixel values at regions of fine structure. To this end, we present a novel Markov Random Fields (MRF) based Anisotropic Depth Diffusion algorithm that predicts regions of reliable depth inter/extrapolation as well as regions of fine texture followed by real-time depth message passing using the information from the information from the RGB component of the RGB-D data for depth refinement. This algorithm can also be implemented on the sensing device thereby producing enhanced RGB-D data with the same hardware. Reconstructed results shown demonstrate the efficiency of the approach in terms of reconstruction accuracy as well as visible surface conformance.

2   Depth Estimate Errors in RGB-D Projected Texture 3D Sensors

The following list presents the most common reasons behind poor depth estimation in RGB-D sensors.

1. Shadows: Shadows are regions that are occluded by objects and hence are not seen by the texture projector. However, since the optical detector is positioned at a small distance to the projector, these areas are visible to the detector, but the range at these locations cannot be estimated. An example of a shadow region is shown around the handles of the bag in the Bags image.

2. Steep Orientations: Regions which have a steep orientation (near normal) to the camera plane have projected texture that cannot be resolved by the sensor. This is due to the finite resolution of optical detector. Range estimation in such areas is thus heavily hampered. An example of such a problem is demonstrated in the case of the sides of the bag in Bags3 image as well as sides of the blue box in BoxesGroup image.

3. Black Regions: Black surfaces do not reflect sufficient light and hence the projected texture is lost on these surfaces. As a result, reconstruction of the range data is often erroneous or unavailable. Examples of such scenarios are shown in Combo, Crockery, PhoneScale images.

4. Metallic/Shiny Regions: Shiny or highly reflective regions also cause problems in the detection of the projected texture. Examples are shown in the Canister, Cassette, Cylinder and Handles images.

5. Plastic/Porcelain Surfaces: Smooth surfaces also cause diffusion of the projected texture. As a result the reconstruction from low confidence in the range estimation process. Examples are shown in the figures BlueBox, and BowlPlates.

6. Transparent/ Translucent Regions: Objects made of transparent or translucent materials such as glass or plastic also do not reflect the projected texture in the manner necessary for reconstruction. As a result these objects also produce erroneous range estimates. Examples are shown in Combo image.

7. Mutual Shadows or Rainshadow Regions: Regions between objects can be invisible from the projector viewpoint, though these can be visible from the optical detector. As a result a shadow region is created. Conventionally depth diffusion algorithms fail under such scenarios as the smooth the region using data from the objects, while being agnostic the intermediate object/ negative space. This is exemplified in the BoxofObjects example on the top right corner.

8. Yellow Regions: Yellow objects of a certain fluorescence also do not create the necessary reprojection and as a result cause erroneous range estimates. This is demonstrated in the Basket image.

9. Bright Light Sources: Sources of light also create diffusion of the project texture and hence cause problems in reconstruction

10. Fine Structures: Fine structures such as cup handles are often lost due to the poor resolution of the projector. These structures are crucial for reliable estimation of object boundaries and also for object detection/ categorization. This is exemplified in figures ComboKnife and Greeks.

11. Red Regions: Red objects also produce wrong estimates of range values. These are shown in the Basket and HandleStructs examples.

12. Other Holes: Noise in the system can also create holes that are to be removed

3   Algorithm

In order to alleviate these concerns, we present a noise removal and depth propagation algorithm for surface reconstruction. Given the recent surge in the development of RGB-D sensors, a host of methods are available for detecting and tracking RGB-D features across multiple frames as well combining these frames to yield dense 3D point clouds. However, there is a dearth of techniques for RGB-D refinement. It is crucial to employ a depth estimate propagation or diffusion algorithm to generate best approximation surface curvature in these regions for visualization. In this paper, we extend the Depth Diffusion using Iterative Back Substitution scheme to Kinect like RGB-D sensor data for real time surface reconstruction.

Inference and Message Passing techniques are integral to surface generation algorithms. Dynamic Programming, Belief Propagation, Relaxation, Diffusion, Graph Cut etc. have been used in the past for these purposes, though not in the context of the Kinect sensor. These techniques are further unsuitable for diffusion of extremely sparse depth data (such as - homogenous surfaces). Propagation of gross errors in the initial data can significantly affect end reconstruction. Traditional scene agnostic filtering schemes are slow and ineffective on surfaces with sparse data points. Besides, most of these algorithms are based on assumptions or dependence on coplanarity or curvature metrics of data points

Traditional diffusion approaches such as [3] are unsuitable since they do not take into account the boundaries of objects and hence result in loss of structural content. Markov Random Fields (MRFs) [1, 2] or Variational methods based approaches are well suited for the purpose. It is crucial to deploy a depth estimate propagation or diffusion algorithm to generate best approximation surface curvature in these regions. In this paper we develop a technique based on both [1,2] and [3] that is well adapted to the situation at hand. The Real-Time Depth Diffusion algorithm using Iterative Back Substitution [3] is used for surface regeneration with Kinect like RGB-D sensor data in this paper.

3.1   MRF Based Structural Region Selection

Diffusion is normally carried out in all regions with missing data in the case of [3]. However, such an approach is unsuitable since it creates smoothing of structural boundaries. To avoid this, in our paper, we present a scheme which takes into account the neighborhood connectivity of regions with missing range data. Firstly, a low complexity super-pixel segmentation of the color image is carried out. This is followed by detection of neighborhood superpixels for each region with missing range values. A Gaussian Mixture Model (GMM) based likelihood model is created for each neighborhood region and a graph-cut is employed to determine the neighborhood region which is closest in terms of the GMM model with the one containing the missing range values. The equations corresponding to this MRF formulation are presented below.

The energy condition to minimized (for given classification of foreground or background) is given as

$$E: \{0, 1\}^n \rightarrow R$$

$$E(x) = \sum_{i \in P} \theta_i(x_i, z_i) + \sum_{i,j \in N} \theta_{i,j}(x_i, x_j, z_i, z_j)$$

where x is the unknown classification (either foreground or background with respect to the region properties of the current neighbor under analysis), and z the given image. Both unary and pairwise potentials are used in this process. Also, the neighborhood system N can be 4- or 8-connected. In our algorithm, we choose N as 4 for the case of simplicity. While the unary term is used to describe the likelihood based on the RGB color values of individual pixels in the neighborhood region, the pairwise terms demonstrate texture relationships. The unary term is computed via Gaussian Mixture models, which in turn is estimated using the Expectation Maximization (EM) Algorithm. The multivariate normal distribution for the mixture component k is given by

$$N_k(y|\mu_k, \sigma_k) = \frac{1}{\sqrt{2\pi\sigma_k{}^2}} e^{\frac{(y-\mu_k)^2}{2\sigma_k{}^2}}$$

$$N_k(x|\mu_k, \Sigma_k) = \frac{1}{(2\pi)^{\frac{D}{2}}|\Sigma_k|^{\frac{1}{2}}} e^{(-\frac{1}{2}(x-\mu_k)^T\Sigma_k{}^{-1}(x-\mu_k))}$$

Where $|\Sigma_k|$ is the determinant of Σ_k and D is the dimensionality of x. The value of x at i is determined as the weighted sum of the multi-variant Gaussian distributions.

$$V(x_i) = \sum_{k=1}^{K} W_k N_k(x_i|\mu_k, \Sigma_k)$$

Where $\sum_{k=1}^{K} W_k = 1$. Using a latent variable y to represent the mixture component,

$$V(x_i) = \sum_{k=1}^{K} p(y_k)p(x_i/y_k)$$

Where $p(y) = \prod_{k=1}^{K} W_k^{y_k}$ and $p(x|y) = \prod_{k=1}^{K} N_k(x|\mu_k, \Sigma_k)^{y_k}$

The E-Step of the EM algorithm computes the responsibility that a mixture component takes for explaining an observation x_i

$$\tau(y_k) = p(y_k = 1|x_i) = \frac{p(y_k = 1)p(x_i|y_k = 1)}{\sum_{m=1}^{K} p(y_m = 1)p(x_i|y_m = 1)}$$
$$= \frac{W_k N_k(x_i|\mu_k, \Sigma_k)}{\sum_{m=1}^{K} W_m N_m(x_i|\mu_m, \Sigma_m)}$$

While the M-Step re-estimates the parameters of the mixture models as

$$\mu_k{}^{new} = \frac{\sum_{i=1}^{N} \tau(y_{ik}) x_i}{\sum_{i=1}^{N} \tau(y_{ik})}$$

$$\Sigma_k{}^{new} = \frac{\sum_{i=1}^{N} \tau(y_{ik})(x_i - \mu_k{}^{new})(x_i - \mu_k{}^{new})^T}{\sum_{i=1}^{N} \tau(y_{ik})}$$

$$W_k^{new} = \frac{1}{N} \sum_{i=1}^{N} \tau(y_{ik})$$

The unary potentials are generated as the log likelihood cost of each of the pixels in the test region with respect to the background and foreground Gaussian mixture models. FG and BG models are indeed chosen from the labels of the super-pixels by a binary selection to permit for better classification of pixels in the empty regions through GraphCut minimization.

$$Cost_{FG} = -2 \log p(x|y = FG)$$
$$Cost_{BG} = -2 \log p(x|y = BG)$$

while the pairwise term is computed using

$$\theta_{i,j}(x_i, x_j, z_i, z_j) = |x_i - x_j|(\lambda_1 - \lambda_2 e^{-\beta \|z_i - z_j\|_2}$$

$$\beta = \frac{2}{\sum_{i,j} |z_i - z_j|}$$

The pairwise term helps reduce linking of neighbors that are dissimilar. In the above equations λ_1 represents the standard Ising prior, and λ_2 the contrast sensitive term. The Energy function to be minimized (given by $E(x)$), is solved by using a Graph Cut Algorithm based on Max-Flow Min-Cut Approach [4].

Based on the neighborhood connectivity/ similarity relationships, a diffusion mask is created, which creates hard constraints across boundaries between unrelated neighborhoods while maintaining low boundary conditions for related neighborhood regions. Using the anisotropic diffusion approach presented in [3], it is possible to obtain a real-time employing the neighborhood region boundary conditions.

The real-time depth diffusion is simulated using the heat flow equation. The Partial Differential Equation representing the flow of heat in a 2 dimensional isotropic medium is given by (according to [3])

$$\frac{\partial u(r,t)}{\partial t} = c\left(\frac{\partial^2 u(r,t)}{\partial x^2} + \frac{\partial^2 u(r,t)}{\partial y^2}\right)$$

where, $u(r, t)$ represents the heat measured in the two dimensional space r(x,y) at time t. If c is not a constant or varies in the space of the depth map dimensions, the equation becomes anisotropic. This equation can be also used to represent depth diffusion, where $u(r, 0)$ represents the original depth values and $u(r, t_{ss})$ final depth values obtained after diffusion (at steady state). This equation is equivalent to

$$\frac{\partial u(r,t)}{\partial t} = c\nabla^2 u(r,t)$$

where, ∇^2 is the Laplacian operator. This yields the formulation for anisotropic diffusion. Boundary conditions on are imposed on the segments using a nearest neighbor approach. The input 2D image for the above equation takes the form of the masked depth image.

$$\frac{\partial u(r,t)}{\partial t} = \nabla.[c(\|\nabla I(r,0)\|)\nabla u(r,t)]$$

where, $u(r, 0) = D(r)$ is the original depth map and $c(\|\nabla I(r,0)\|)$ is the anisotropic diffusion coefficient computed from the neighborhood relationships of the regions.

Using the tuple (i, j) for the row and column indices of the image, we have

$$\frac{\partial u(r_{(i,j)}, t)}{\partial t} = u(r_{(i,j)}, t+1) - u(r_{(i,j)}, t)$$
$$= \varphi\big[\, c(\|\nabla I(r_{(i-1,j)}, 0)\|) . \nabla u(r_{(i-1,j)}, t)$$
$$+ \; c(\|\nabla I(r_{(i,j-1)}, 0)\|) . \nabla u(r_{(i,j-1)}, t)$$
$$+ \; c(\|\nabla I(r_{(i+1,j)}, 0)\|) . \nabla u(r_{(i+1,j)}, t)$$
$$+ \; c(\|\nabla I(r_{(i,j+1)}, 0)\|) . \nabla u(r_{(i,j+1)}, t)\big]$$

where, the constant $\varphi \le 0.25$ controls the overall rate of diffusion. In the steady state,

$$\left(\frac{1}{\varphi}\right) . u(r_{(i,j)}, t_{ss})$$
$$= -\big[\, c(\|\nabla I(r_{(i-1,j)}, 0)\|) . u(r_{(i-1,j)}, t_{ss})$$
$$+ \; c(\|\nabla I(r_{(i,j-1)}, 0)\|) . u(r_{(i,j-1)}, t_{ss})$$
$$+ \; c(\|\nabla I(r_{(i+1,j)}, 0)\|) . u(r_{(i+1,j)}, t_{ss})$$
$$+ \; c(\|\nabla I(r_{(i,j+1)}, 0)\|) . u(r_{(i,j+1)}, t_{ss})\big]$$
$$= 0$$

Representing $(1/\varphi)$ as λ and linearizing the tuple indices, (5) can be reduced to a matrix system. A sample matrix for a 3x3 depth image, is shown in (6)

$$
\begin{bmatrix}
\lambda & c_{12} & c_{21} & & & & & & \\
c_{11} & \lambda & c_{13} & c_{22} & & & & & \\
 & c_{12} & \lambda & & c_{23} & & & & \\
c_{11} & & & \lambda & c_{22} & c_{31} & & & \\
 & c_{12} & & c_{21} & \lambda & c_{23} & c_{32} & & \\
 & & c_{13} & & c_{22} & \lambda & & c_{33} & \\
 & & & c_{21} & & & \lambda & c_{32} & \\
 & & & & c_{22} & & c_{31} & \lambda & c_{33} \\
 & & & & & c_{23} & & c_{32} & \lambda
\end{bmatrix}
\begin{bmatrix}
u_{11} \\ u_{12} \\ u_{13} \\ u_{21} \\ u_{22} \\ u_{23} \\ u_{31} \\ u_{32} \\ u_{33}
\end{bmatrix}
=
\begin{bmatrix}
0 \\ 0 \\ 0 \\ 0 \\ 0 \\ 0 \\ 0 \\ 0 \\ 0
\end{bmatrix}
\tag{6}
$$

$$Ax = B \tag{7}$$

This linear equation system forms a block-tridiagonal matrix system with fringes. In the above equation (matrix A), the blocks are denoted by red squares, the tri-diagonals by the blue and violet indices along with the main diagonal, the upper fringe in green and lower fringe in orange. The system is solved using the Iterative Back Substitution (IBS) algorithm, specialized for tridiagonal matrices with fringes, with Multi-grid optimization for run-time enhancement.

4   Results

Results are presented for each of the error scenarios described in Section 2. In order to compare the results of range image enhancement, the algorithm in [3] has been used as a benchmark. The various stages of the pipeline are shown in Fig 1. Visual quality differences with the approach in [3] are shown for several test scenarios in Figure 2. It can be seen that the structural definition of the objects is well preserved in the case of

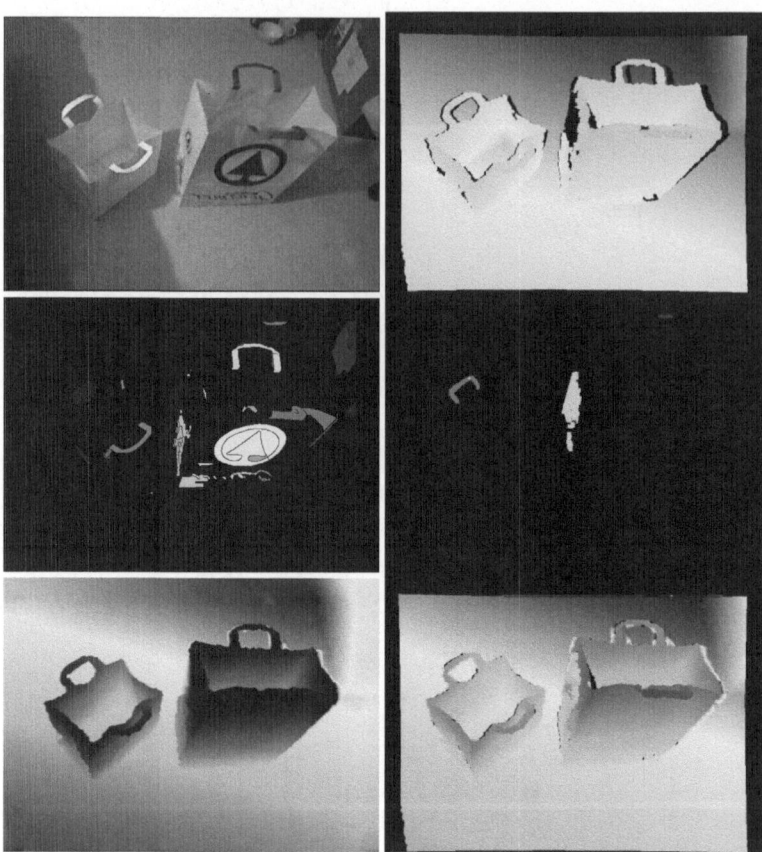

Fig. 1. Stages of MRF based Depth Diffusion (Left to Right and Top to Bottom): (a) Input color image, (b) depth image, (c) detected depth neighbors using super-pixel segmentation, (d) MRF based detection of "related" neighbors – Note correct boundary estimation with respect to neighbors (esp. the large missing region on the side on the bag is linked to the bag structure and separated from the background) (e) Diffusion using the algorithm in [3]. (f) Results using MRF Diffusion – Note clear structure preservation on the side of the bag as opposed to (e).

our approach while such information is lost due to excessive smoothing with the algorithm in [3]. Furthermore, the segmentation based variant of the algorithm in [3] is not suited for the application at hand since the depth refinement in our algorithm is intended to serve as a pre-processing stage that can be integrated into the Kinect sensor hardware for increased depth resolution and fidelity.

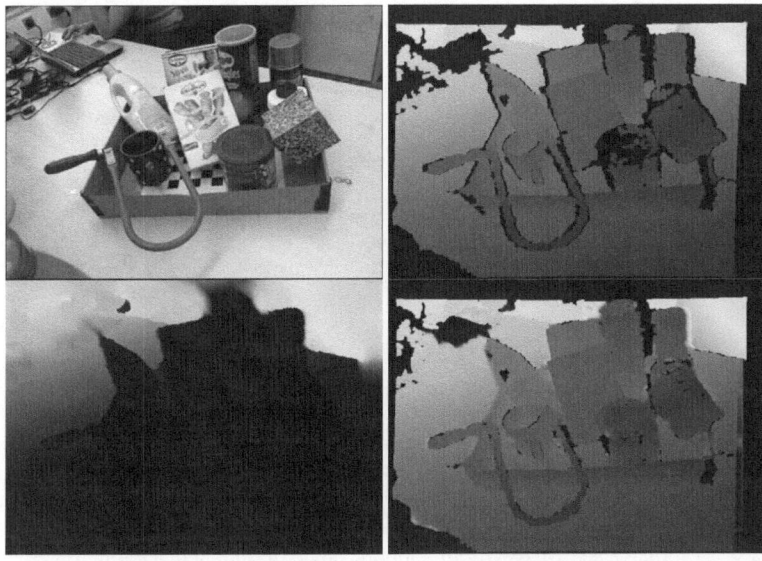

Image set: BoxofObjects: Note preservation of boundaries between the two cylindrical objects (mutual-shadow region) with MRF Diffusion

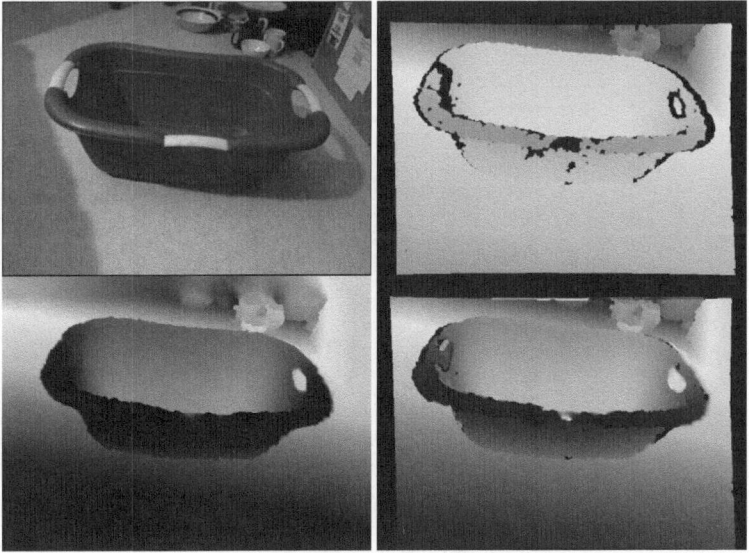

Image set: Basket: Yellow/red regions with holes

Fig. 2. (a) Input color image, (b) raw depth map, (c) Depth Diffusion [3] and (d) MRF based Depth Diffusion

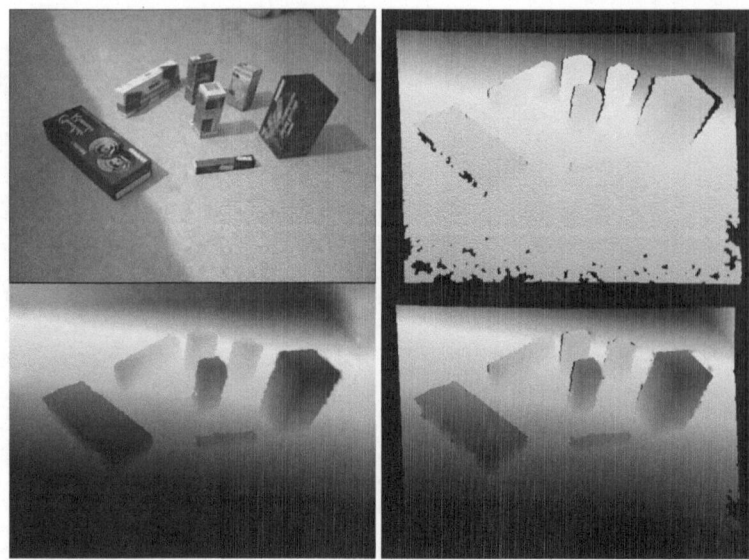

Image set: BoxesGroup: Oblique angle of optical detection

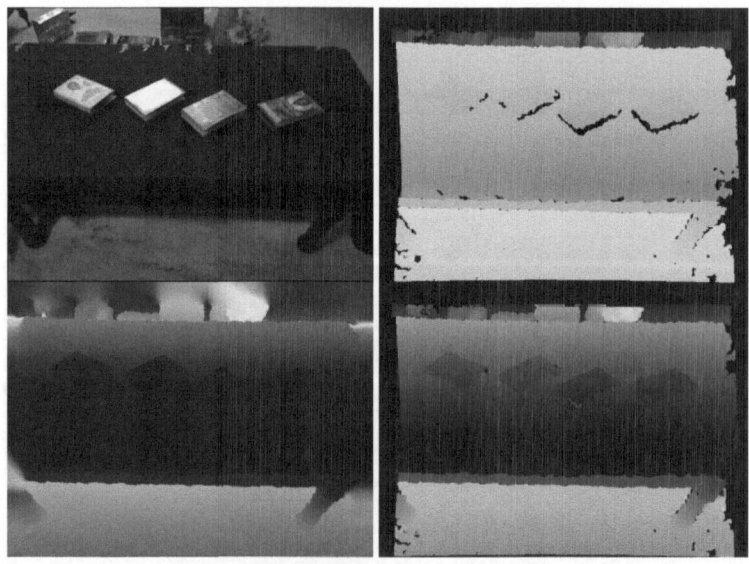

Image set: Cassette: Shiny/plastic surfaces, dark surfaces

Fig. 2. (*continued*)

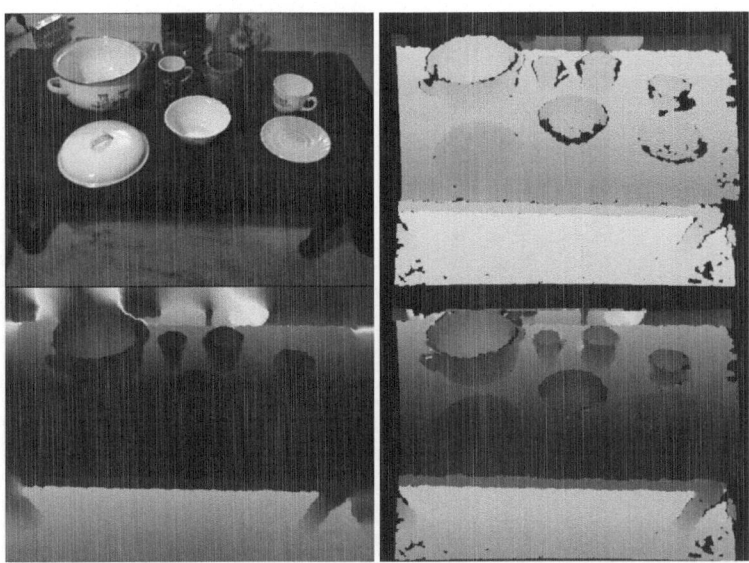

Image set: Crockery: Plastic objects

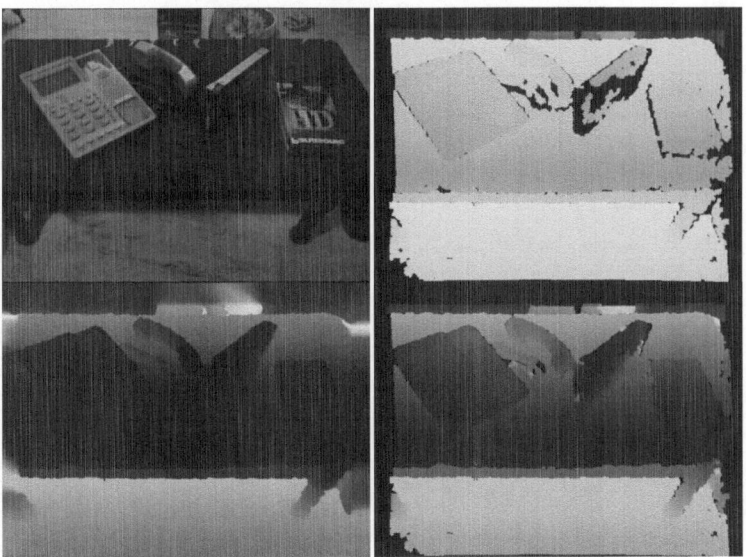

Image set: Phone-Scale: Dark objects, oblique angles

Fig. 2. (*continued*)

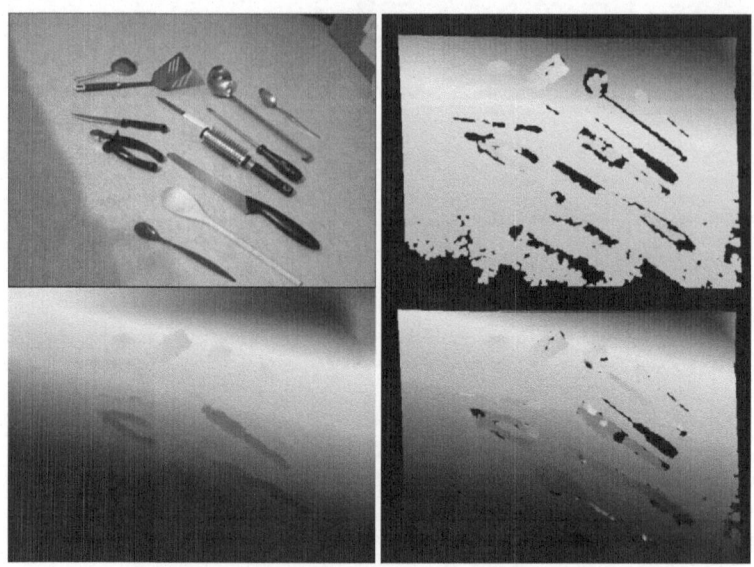

Image set: Handles: Fine structures, dark regions

Image set: BlueBox: plastic objects, oblique views

Fig. 2. (*continued*)

5 Conclusion

In this paper, we have presented a novel algorithm for the refinement of depth data from an RGB-D sensor such as the Microsoft Kinect. Projected texture sensors suffer from several limitations in the reconstruction process owing to material properties, projection angle and sensor resolution constraints as a result of which it is common to encounter missing data from RGB-D sensors. In this paper, we present a novel RGB-D MRF Depth Diffusion algorithm that can be integrated with the Kinect hardware in order to obtain high fidelity range maps from the current sensor hardware. The algorithm has been tested under varied conditions and results compared with the state of the art diffusion algorithms. In the current version, regions with no filled super pixel neighbors tend to produce cavities in the diffused output. Fine grained resolution of these regions forms future work.

References

1. Diebel, J., Thrun, S.: An application of Markov random fields to range sensing. In: Neural Information Processing Systems, NIPS 2005 (2005)
2. Dolson, J., et al.: Upsampling range data in dynamic environments. In: IEEE Computer Vision and Pattern Recognition, CVPR 2010 (2010)
3. Varadarajan, K.M., Vincze, M.: Real time depth diffusion for 3D Surface Reconstruction. In: International Conference on Image Processing, ICIP 2010 (2010)
4. Boykov, Y., Kolmogorov, V.: An Experimental Comparison of Min-Cut/Max-Flow Algorithms for Energy Minimization in Vision. IEEE Transactions on Pattern Analysis and Machine Intelligence 26(9), 1124–1137 (2004)

A Priori-Driven PCA

Carlos Thomaz[1], Gilson Giraldi[2], Joaquim Costa[3], and Duncan Gillies[4]

[1] Department of Electrical Engineering, FEI, Sao Paulo, Brazil
[2] National Laboratory for Scientific Computing, Rio de Janeiro, Brazil
[3] Department of Applied Mathematics, University of Porto, Portugal
[4] Department of Computing, Imperial College London, UK

Abstract. Principal Component Analysis (PCA) is a multivariate statistical dimensionality reduction method that has been applied successfully in many pattern recognition problems. In the research area of analysis of faces particularly, PCA has been used not only as a pre-processing step to produce accurate analytical model for automated face recognition systems, but also as a conceptual framework for human face coding. Despite the well-known attractive properties of PCA, the traditional approach does not incorporate high level semantics from human reasoning which may steer its subspace computation. In this paper, we propose a method that allows PCA to incorporate such semantics explicitly. It allows an automatic selective treatment of the variables that compose the patterns of interest, performing data feature extraction and dimensionality reduction whenever some high level information in the form of labeled data are available. The method relies on spatial weights calculated, in this work, by separating hyperplanes. Several experiments using 2D frontal face images and different data sets have been carried out to illustrate the usefulness of the method for dimensionality reduction, interpretation, classification and reconstruction of face images.

1 Introduction

Principal Component Analysis (PCA) [1, 2] is the best known multivariate statistical linear method for dimensionality reduction and has been applied successfully in many pattern recognition problems to reduce the computational costs, mitigate the curse of dimensionality and improve the classification performance.

In the research area of analysis of faces particularly, PCA has been used not only as a pre-processing step to produce accurate and computational efficient model for automated face recognition systems [3, 4], but also as a conceptual framework for human face reasoning and coding [5–8]. However, despite the well-known attractive properties of PCA in both computer vision and human perception communities, incorporating prior information in its process remains a challenge. In face recognition, without prior information important content-based features represented by principal components with small eigenvalues may be discarded reducing the accuracy of the automated representation. Analogously, PCA with no prior knowledge is a non-supervised algorithm unable to

J.-I. Park and J. Kim (Eds.): ACCV 2012 Workshops, Part II, LNCS 7729, pp. 236–247, 2013.
© Springer-Verlag Berlin Heidelberg 2013

convey the different visual cues that create, for instance, a separate human perception coding for facial identity and expression [9].

The development of techniques that bring together dimensionality reduction and prior knowledge can be performed in the framework of supervised learning approaches. However, when only a small number of labeled samples are available, multivariate supervised dimensionality reduction methods tend to perform poorly due to overfitting [10]. Such problem has been addressed recently by a number of works related to semi-supervised dimensionality reduction methods [11–15]. A common issue to all semi-supervised learning techniques is how to optimize the regularization parameters necessary to blend supervised and non-supervised information often represented by local and global scatter matrices.

In this paper, we address this issue through separating hyperplanes. We propose a spatially weighted form of PCA that incorporates domain knowledge and generates an embedding space that preserves the optimality properties of dimensionality reduction and interpretability of the standard PCA. Unlike other similar projection approaches that either formulate their solutions in conjunction with parametric [16–18] or non-parametric [19] models or are restricted to the number of groups of patterns available [20], in our method a separating hyperplane, based on a discriminant criterion, is computed and discriminant weights are determined and used to generate the spatially weighted PCA subspace. In this sense, weights incorporate separately the prior knowledge extracted from the labeled data and can be systematically computed through any hyperplane direction. The approach is a simple way of allowing an automatic selective treatment of the variables that compose the patterns of interest, performing data feature extraction and dimensionality reduction whenever some high level semantics in the form of spatial weights are available.

2   A Priori-Driven PCA

Let an $N \times n$ training set matrix X be composed of N input samples (or patterns of interest, such as face images) with n variables (or attributes, such as pixels), that is, $X = (\mathbf{x}_1, \mathbf{x}_2, \ldots, \mathbf{x}_N)^T$. This means that each column of matrix X represents the values of a particular variable observed all over the N samples. Let this data matrix X have covariance matrix

$$S = \frac{1}{(N-1)} \sum_{i=1}^{N} (\mathbf{x}_i - \bar{\mathbf{x}})(\mathbf{x}_i - \bar{\mathbf{x}})^T, \tag{1}$$

where $\mathbf{x}_i = [x_{i1}, x_{i2}, \ldots, x_{in}]^T$ and $\bar{\mathbf{x}}$ is the grand mean vector of X.

It is a proven result that the set of m $(m \leq n)$ eigenvectors of S, which corresponds to the m largest eigenvalues, minimizes the mean square reconstruction error over all choices of m orthonormal basis vectors [21].

To note explicitly the spatial association between the jth and k^{th} variables, we can rewrite the sample covariance matrix S described in equation (1) in order to indicate the position of each variable in the N samples. When n variables are observed on each sample, the sample variation can be described by the following sample variance-covariance equation [22]:

$$S = \{s_{jk}\} = \left\{ \frac{1}{(N-1)} \sum_{i=1}^{N} (x_{ij} - \bar{x}_j)(x_{ik} - \bar{x}_k) \right\}, \tag{2}$$

for $j = 1, 2, \ldots, n$ and $k = 1, 2, \ldots, n$. The covariance s_{jk} between the jth and k^{th} variables reduces to the sample variance when $j = k$, $s_{jk} = s_{kj}$ for all j and k, and the covariance matrix S contains n variances and $\frac{1}{2}n(n-1)$ potentially different covariances [22].

It is clear from equation (2) that the variable deviations from the mean have the same importance in the standard sample covariance matrix S formulation. In other words, all the n variables are equally weighted. However, there are situations where this should not be the case, particularly in pattern recognition problems where some parts of the samples might be more informative than others.

2.1   Weighted Sample Covariance

The well-known Pearson's sample correlation coefficient between the jth and k^{th} variables is defined as [22]:

$$r_{jk} = \frac{s_{jk}}{\sqrt{s_{jj}}\sqrt{s_{kk}}} \tag{3}$$

$$= \frac{\sum_{i=1}^{N}(x_{ij} - \bar{x}_j)(x_{ik} - \bar{x}_k)}{\sqrt{\sum_{i=1}^{N}(x_{ij} - \bar{x}_j)^2}\sqrt{\sum_{i=1}^{N}(x_{ik} - \bar{x}_k)^2}},$$

for $j = 1, 2, \ldots, n$ and $k = 1, 2, \ldots, n$. It is important to note that $r_{jk} = r_{kj}$ for all j and k.

From equation (3), it is clear that the sample correlation coefficient is a normalized version of the sample covariance, where the product of the square roots of the sample variances, known as the sample standard deviations, provides the spatial normalization of the sum of the variable deviations from the mean [22]. In other words, r_{jk} is a measure of the linear association between two variables that does not allow variables with larger variance or scale to dominate the corresponding deviations from the mean and, consequently, the subspace calculation of PCA.

In our method, we want to give higher importance to the variables which characterise a class of interest. However, the variables that vary most are not necessarily the ones that allow best interpretation of the sample groups. Therefore, we need to define a measure of association between variables, based on the Pearson's sample correlation coefficient, which uses the notion of spatial weights and is more or less dominant depending on the values of each spatial weight.

Extending equation (3), we can define a weighted sample covariance r_{jk}^* between the jth and k^{th} variables by [23]

$$r_{jk}^* = \frac{(\sqrt{w_j}\sqrt{w_k})s_{jk}}{\sqrt{s_{jj}}\sqrt{s_{kk}}} \qquad (4)$$

$$= \frac{\sum_{i=1}^N \sqrt{w_j}(x_{ij} - \bar{x}_j)\sqrt{w_k}(x_{ik} - \bar{x}_k)}{\sqrt{\sum_{i=1}^N (x_{ij} - \bar{x}_j)^2}\sqrt{\sum_{i=1}^N (x_{ik} - \bar{x}_k)^2}},$$

for $j = 1, 2, \ldots, n$ and $k = 1, 2, \ldots, n$. The spatial weighting vector

$$\mathbf{w} = [w_1, w_2, \ldots, w_n]^T \qquad (5)$$

is such that $w_j \geq 0$ and $\sum_{j=1}^n w_j = 1$, where each w_j measures the spatial power of the jth variable. Thus, when n variables are observed on N samples, the weighted sample covariance matrix R^* can be described by

$$R^* = \{r_{jk}^*\} = \left\{ \frac{\sum_{i=1}^N \sqrt{w_j}(x_{ij} - \bar{x}_j)\sqrt{w_k}(x_{ik} - \bar{x}_k)}{\sqrt{\sum_{i=1}^N (x_{ij} - \bar{x}_j)^2}\sqrt{\sum_{i=1}^N (x_{ik} - \bar{x}_k)^2}} \right\}, \qquad (6)$$

for $j = 1, 2, \ldots, n$ and $k = 1, 2, \ldots, n$. The weighted sample covariance r_{jk}^* between the jth and k^{th} variables is equal to w_j when $j = k$, $r_{jk}^* = r_{kj}^*$ for all j and k, and so the matrix R^* is a $n x n$ symmetric matrix.

Let R^* have respectively P^* and Λ^* eigenvector and eigenvalue matrices, that is,

$$P^{*T}R^*P^* = \Lambda^*. \qquad (7)$$

The set of m ($m \leq n$) eigenvectors of R^*, that is, $P^* = [\mathbf{p}_1^*, \mathbf{p}_2^*, \ldots, \mathbf{p}_m^*]$, which corresponds to the m largest eigenvalues, defines a new orthonormal coordinate system for the training set matrix X and is called here as the *spatially weighted principal components*.

In the last years, weighted PCA techniques have been proposed [24, 23, 25, 26] to obtain a consistent subspace representation of the original data in the presence of noise, outliers and missing data. However, a key remaining issue for the weighted PCA methods in general is how to automatically compute the optimal weights to combine low level features, such as colour, shape and texture inherent to problems like face image analysis, with high level semantics, such as labeled information from human reasoning. In other words, the remaining question is: how can we define spatial weights w_j to incorporate prior knowledge? Our approach is to define a systematic method to compute the weights from labeled data.

2.2   The Spatial Weights

We propose the idea of using the discriminant weights given by statistical separating hyperplanes as the spatial weights of the weighted sample correlation matrix defined in equation (6). The models need some labeled data of N pairs

$$(\mathbf{x}_1, y_1), (\mathbf{x}_2, y_2), \ldots, (\mathbf{x}_N, y_N), \tag{8}$$

where the $\mathbf{x}_i \in \Re^n$ denote the ith training observations and y_i are scalars that correspond to the classification labels. For simplicity and without loss of generality, we concentrate on two-class problems, that is, $y_i \in \{-1, 1\}$.

One way to define the parametric spatial weights is provided by Linear Discriminant Analysis (LDA) [27, 21]. LDA depends on all of the data, even points far away from the separating hyperplane and its main objective is to find a projection vector $\mathbf{w}_{lda}$ that maximizes the Fisher's criterion [21]:

$$\mathbf{w}_{lda} = \arg\max_{\mathbf{w}} \frac{|\mathbf{w}^T S_b \mathbf{w}|}{|\mathbf{w}^T S_w \mathbf{w}|}. \tag{9}$$

The S_b and S_w matrices are the between-class and within-class scatter matrices. The vector $\mathbf{w}_{lda}$ defines the normal vector of the hyperplane that best separates the two classes.

Alternatively, to allow the investigation of spatial discriminant weights determined by non-parametric separating hyperplanes, we can use the Support Vector Machine method [28] based on the risk-minimization approach. The primary purpose of SVM is to maximize the width of the margin between two distinct sample classes [28]. Given a training set as described in the formulation (8), the SVM method seeks to find the hyperplane defined by

$$f(\mathbf{x}) = (\mathbf{x} \cdot \mathbf{w}) + b = 0, \tag{10}$$

which separates positive and negative observations with the maximum margin. It can be shown that the solution vector $\mathbf{w}_{svm}$ is defined in terms of a linear combination of the training observations, that is,

$$\mathbf{w}_{svm} = \sum_{i=1}^{N} \alpha_i y_i \mathbf{x}_i, \tag{11}$$

where α_i are non-negative coefficients obtained by solving a quadratic optimization problem with linear inequality constraints. Those training observations $\mathbf{x}_i$ with non-zero α_i lie on the boundary of the margin and are called support vectors [28].

2.3   The Step-by-Step Algorithm

The main steps for calculating the spatially weighted principal components $P^* = [\mathbf{p}_1^*, \mathbf{p}_2^*, \ldots, \mathbf{p}_m^*]$ of an $N \times n$ training set matrix X composed of N input samples with n variables can then be described as follows:

1. Calculate the spatial weighting vector $\mathbf{w} = [w_1, w_2, \ldots, w_n]^T$ using some labeled data and a separating hyperplane method, as described in the previous sub-section;

2. Normalize $\mathbf{w}$ such that $w_j \geq 0$ and $\sum_{j=1}^{n} w_j = 1$, that is, replace w_j with $\frac{|w_j|}{\sum_{j=1}^{n} |w_j|}$;

3. Standardize all the n variables of the data matrix X such that the new variables have $\bar{x}_j = 0$ and $s_j^2 = s_{jj} = 1$, for $j = 1, 2, \ldots, n$. In other words, calculate the grand mean vector

$$\bar{\mathbf{x}} = \frac{1}{N} \sum_{i=1}^{N} \mathbf{x}_i = (\bar{x}_1, \bar{x}_2, \ldots, \bar{x}_n)$$

and the vector of variances $(s_1^2, s_2^2, \ldots, s_n^2)$, where

$$s_j^2 = \frac{1}{(N-1)} \sum_{i=1}^{N} (x_{ij} - \bar{x}_j)^2,$$

and replace x_{ij} with z_{ij} given by

$$z_{ij} = \frac{x_{ij} - \bar{x}_j}{\sqrt{s_j^2}}$$

for $i = 1, 2, \ldots, N$ and $j = 1, 2, \ldots, n$;

4. Spatially weigh up all the standardized z_{ij} variables using the normalized weighting vector $\mathbf{w}$ calculated in step 2, that is

$$z_{ij}^* = z_{ij} \sqrt{w_j};$$

5. The spatially weighted principal components P^* are then the eigenvectors corresponding to the m largest eigenvalues of $(Z^*)^T Z^*$, where

$$Z^* = \{\mathbf{z}_1^*, \mathbf{z}_2^*, \ldots, \mathbf{z}_N^*\}^T.$$

3 Experimental Results

We have divided our experimental results into two parts. Firstly, we have investigated the usefulness of the priori-driven principal components in recognizing samples compared to the standard PCA and the corresponding separating hyperplanes. Then, in the second part, we have analyzed the effectiveness of the new principal components in reconstructing samples compared to the standard PCA.

The following two-group separation tasks have been performed using frontal face images: (a) Gender experiments (female versus male samples); (b) Facial expression experiments (non-smiling versus smiling). The goal of the gender experiment is to evaluate the method proposed on a discriminant task where the differences between the groups are evident. The facial expression experiment poses an alternative analysis where there are subtle differences between the groups.

In all experiments, the total number of training examples N is limited and significantly less than the dimension of the feature space, that is, $N \ll n$. To address this problem for the Fisher's criterion, we have calculated the leading eigenvector $\mathbf{w}_{lda}$ by using two different approaches. The first approach, based on the Zhu and Martinez method [29], replaces S_w with the $n \times n$ identity matrix and $\mathbf{w}_{lda}$ becomes simply the leading eigenvector of S_b. The other, based on the Maximum uncertainty Linear Discriminant Analysis (MLDA) proposed by Thomaz et al. [30], considers the issue of regularizing the S_w estimate with a multiple of the identity matrix.

3.1   Recognition Rate

We have used two publicly available data sets to evaluate the classification performance of the spatially weighted principal components: FEI [31] and FERET [32]. The FEI data set is composed of 200 subjects (100 men and 100 women). Each subject has two frontal images (one with a neutral or non-smiling expression and the other with a smiling facial expression). In total 400 images were used to perform the gender and expression experiments. In the FERET database, we have used 200 subjects (107 men and 93 women). Each subject has two frontal images (one with a neutral or non-smiling expression and the other with a smiling facial expression), also providing a total of 400 images to perform the gender and expression experiments. We adopted the 10-fold cross validation method to evaluate the classification performance of all the methods. Throughout all the classification experiments, we have assumed that the prior probabilities and misclassification costs are equal for both groups. On the PCA subspace, the mean of each class has been calculated from the corresponding training images and the Mahalanobis distance from each class mean has been used to assign a test observation to either groups. In all the standard and weighted PCA experiments, we have considered different numbers of principal components to calculate the recognition rates of the corresponding methods implemented. Additionally, as benchmark measures, we have also calculated the classification performance of the separating hyperplanes on the corresponding original spaces.

Figure 1 shows the recognition performance of the 10-fold cross validation of the gender experiments using the FEI and FERET databases and different numbers k of principal components selected by the corresponding largest eigenvalues. The horizontal dashed lines denote the separating hyperplanes' classification accuracies using all the original features available without any dimensionality reduction. It can be seen that even in such experiments where the differences between the sample groups are not subtle, the use of prior information given by labeled samples improves the discriminant power of the principal components, allowing similar or higher average recognition rates with the same number of components. For instance, in the gender experiments using the FEI database, all the spatially weighted PCA methods consistently outperform the standard

PCA when the number of principal components retained has been higher than 10, that is, when $k \geq 10$. In the gender experiments using the FERET database, which is composed of frontal face images not as well aligned as in the FEI database, the superiority of the spatially weighted PCA is less evident, but still it is possible to see a better classification performance than the standard PCA when using few principal components, that is, when $5 \leq k < 40$. Additionally, it is possible to see on the boxplots of Figure 1 that the top recognition rates of the spatially weighted principal components are comparable to or higher than the separating hyperplanes, but less sensitive to the parametric (Zhu&Martinez and MLDA) or non-parametric (SVM) discriminant information used, particularly for the FEI experiments.

The importance of allowing a priori-driven treatment of individual pixels and, consequently, minimizing the potential problem of discarding information related to subtle group differences on the first components of the standard PCA can be seen in Figure 2. In both FEI and FERET face databases, the average recognition rates of the spatially weighted principal components are much higher than the standard ones when the original dimensionality of the data is considerably

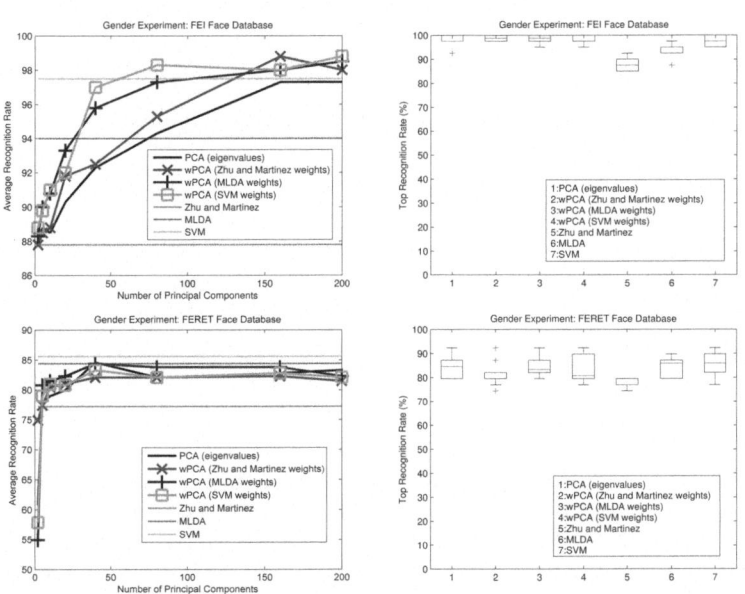

Fig. 1. Gender recognition performance of spatially weighted PCA (wPCA) compared to standard PCA using the FEI and FERET databases with 10-fold cross validation. On the left there are the average recognition rate curves using different numbers of principal components. As reference values, the horizontal dashed lines denote the corresponding separating hyperplane classification accuracies using all the original features without any dimensionality reduction. On the right there are boxplots of the top recognition rates achieved on a specific number of principal components for each method considered.

reduced. For example, when using only $k = 5$ spatially weighted principal components in the FEI face database, it is possible to achieve an average recognition rate of approximately 92% compared to 55% of the standard PCA. A significant improvement in classification performance is illustrated as well in the expression experiments using the FERET face database, where the spatially weighted and the standard principal components have achieved respectively approximately 71% and 57% with $k = 40$ components, for instance. In the expression experiments, more remarkably, the top recognition rates of the spatially weighted PCA are also comparable to or higher than the separating hyperplanes, but much less sensitive to the choice of using parametric or non-parametric spatial discriminant weights and consequently less prone to overfitting.

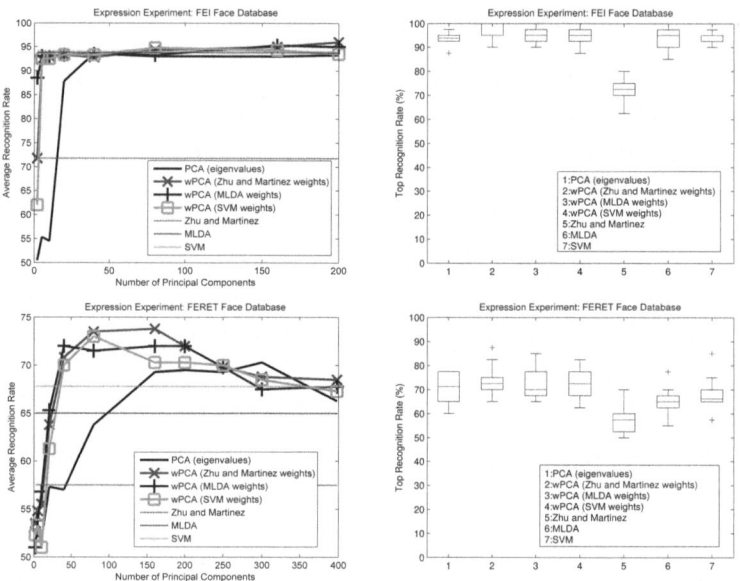

Fig. 2. Expression recognition performance of spatially weighted PCA (wPCA) compared to standard PCA using the FEI (top) and FERET (bottom) databases with 10-fold cross validation. On the left there are the average recognition rate curves using different numbers of principal components. As reference values, the horizontal dashed lines denote the corresponding separating hyperplane classification accuracies using all the original features without any dimensionality reduction. On the right there are the boxplots of the top recognition rates achieved on a specific number of principal components for each method considered.

3.2   Reconstruction

The reconstruction task cannot be performed by separating hyperplanes because both parametric and non-parametric classifiers retain only the information necessary to discriminate the classes, which is not enough to represent them back in the original feature space. In terms of the spatially weighted principal components, however, we can carry out an overall reconstruction process similar to unweighted PCA, but more efficient for making predictions especially on the major axes of projection because of the spatial weights control over the individual pixels within the face images.

Figure 3 shows two examples of the correlations between an image and its reconstruction. Data is given for the whole image and three smaller parts (eyes, nose and mouth) exclusively. The subjects are a male smiling (left on Figure 3) and a female with neutral expression (right on Figure 3) taken from the FEI database. The images are projected into the eigenspace and then reconstructed using 5, 10, 20, 40, 80, 160, 320 and all principal components. Three different methods were used for each experiment which are, from top to bottom: standard PCA, spatially weighted PCA using gender and spatially weighted PCA using expression. The spatial weights were calculated using the MLDA method and the other two separating hyperplanes considered in the previous subsection gave similar correlation results.

It is possible to see that the eigensubspace composed of the weighted principal components tends to reconstruct first the most informative parts of the face images for predicting differences relating to the choice of spatial weights. For example, on the left part of Figure 3 the image to be reconstructed is of a smiling face. Hence it is the region around the mouth that carries the most important discriminant information. Using spatially weighted PCA, with the expression discriminant weights we need only 5 weighted principal components to reconstruct the mouth with high correlation (> 0.7). Standard PCA needs at least 20 components to reconstruct the mouth correctly. The feature space weighted by gender information does not

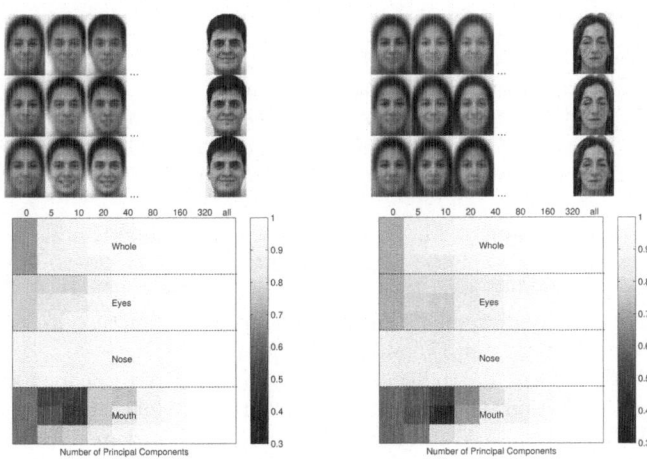

Fig. 3. Correlations between parts of smiling male (left) and non-smiling female (right) images and their reconstructions using different numbers of principal components and the following feature spaces (from top to bottom): standard PCA, PCA weighted by gender and PCA weighted by expression. The images show some of the partial reconstructions.

focus on large changes in the mouth, but contains information on other parts of the faces that better describe the main differences between male and female. It overtakes the standard PCA in reconstruction accuracy with around 40 components. A similar behavior can be observed on the right part of Figure 3, but now exemplifying a non-smiling sample reconstruction.

4 Conclusion

We have proposed a priori-driven PCA method using a modification of the Pearson's correlation formula that incorporates domain knowledge and generates an embedding space that preserves the properties of dimensionality reduction and interpretability of the standard PCA, without jeopardizing its inherent straightforward and simple calculation. This approach might be particularly useful for visual analytics and human perception experiments because it not only provides a more flexible form of data compression unlimited by the number of separating groups or classes, but also extracts relevant features in low dimension spaces providing better understanding and interpretation of the data for any specific *a priori* information of interest.

Acknowledgement. The authors would like to thank the support provided by LNCC, FAPESP (grants 2005/02899-4 and 2009/53556-0), CNPq (grants 473219/04-2 and 472386/07-7) and CAPES (grant 094/2007).

References

1. Pearson, K.: On lines and planes of closest fit to systems of points in space. Philosophical Magazine 2, 559–572 (1901)
2. Jolliffe, I.: Principal component analysis. Springer Series in Statistics (2002)
3. Sirovich, L., Kirby, M.: Low-dimensional procedure for the characterization of human faces. Journal of Optical Society of America 4(3), 519–524 (1987)
4. Turk, M., Pentland, A.: Eigenfaces for recognition. Journal of Cognitive Neuroscience 3, 71–86 (1991)
5. Hancock, P.J.B., Burton, A.M., Bruce, V.: Face processing: Human perception and principal components analysis. Memory and Cognition 24, 26–40 (1996)
6. O'Toole, A.J., Deffenbacher, K.A., Valentin, D., Mckee, K., Huff, D., Abdi, H.: The perception of face gender: The role of stimulus structure in recognition and classification. Memory and Cognition 26, 146–160 (1997)
7. Burton, A.M., Bruce, V., Hancock, P.J.B.: From pixels to people: a model of familiar face recognition. Cognitive Science 23, 1–31 (1999)
8. Calder, A.J., Burton, A.M., Miller, P., Young, A.W., Akamatsu, S.: A principal component analysis of facial expressions. Vision Research 41, 1179–1208 (2001)
9. Calder, A.J., Young, A.W.: Understanding the recognition of facial identity and facial expression. Nature Reviews: Neuroscience 6, 641–651 (2005)
10. Guyon, I., Elisseeff, A.: An introduction to variable and feature selection. Journal of Machine Learning Research, 1157–1182 (2003)
11. Zhang, D., Zhou, Z.H., Chen, S.: Semi-supervised dimensionality reduction. In: Proc. of the 2007 SIAM Intern. Conf. on Data Mining (2007)

12. Sugiyama, M., Idé, T., Nakajima, S., Sese, J.: Semi-Supervised Local Fisher Discriminant Analysis for Dimensionality Reduction. In: Washio, T., Suzuki, E., Ting, K.M., Inokuchi, A. (eds.) PAKDD 2008. LNCS (LNAI), vol. 5012, pp. 333–344. Springer, Heidelberg (2008)
13. Cai, D., He, X., Han, J.: Semi-supervised discriminant analysis. In: Proceedings of the International Conference on Computer Vision, ICCV 2007 (2007)
14. Lim, G., Park, C.H.: Semi-supervised dimension reduction using graph-based discriminant analysis. In: Inter. Conf. on Comp. and Inf. Tech., vol. 1, pp. 9–13 (2009)
15. Sun, D., Zhang, D.: A new discriminant principal component analysis method with partial supervision. Neural Processing Letters 30, 103–112 (2009)
16. Tipping, M.E., Bishop, C.M.: Mixtures of probabilistic principal component analysers. Neural Computation 11, 443–482 (1999)
17. Yu, S., Yu, K., Tresp, V., Kriegel, H.P., Wu, M.: Supervised probabilistic principal component analysis. In: Proc. of the ACM SIGKDD Intern. Conf. (2006)
18. Das, K., Nenadic, Z.: An efficient discriminant-based solution for small sample size problem. Pattern Recognition 42, 857–866 (2009)
19. Maszczyk, T., Duch, W.: Support Vector Machines for Visualization and Dimensionality Reduction. In: Kůrková, V., Neruda, R., Koutník, J. (eds.) ICANN 2008, Part I. LNCS, vol. 5163, pp. 346–356. Springer, Heidelberg (2008)
20. Wang, F., Wang, X., Li, T.: Beyond the graphs: Semi-parametric semi-supervised discriminant analysis. In: IEEE Conference on Computer Vision and Pattern Recognition, CVPR 2009, pp. 2113–2120 (2009)
21. Fukunaga, K.: Introduction to Statistical Pattern Recognition. Academic Press, New York (1990)
22. Johnson, R., Wichern, D.: Applied Multivariate Statistical Analysis. Prentice Hall, New Jersey (1998)
23. Costa, J.F.P., Silva, I., Silva, M.E.: Time dependent principal component analysis of time series data. In: IASC (2007)
24. Forbes, K., Fiume, E.: An efficient search algorithm for motion data using weighted pca. In: Proc. of the 2005 ACM SIGGRAPH/Eurographics Symp. on Comp. Anim., pp. 67–76 (2005)
25. Skocaj, D., Leonardis, A., Bischof, H.: Weighted and robust learning of subspace representations. Pattern Recogn. 40, 1556–1569 (2007)
26. Zhang, Y., Qiu, Z., Sun, D.: Palmprint identification using weighted pca feature. In: Proc. of the Inter. Conf. on Signal Processing, pp. 2112–2115 (2008)
27. Devijver, P., Kittler, J.: Pattern Classification: A Statistical Approach. Prentice-Hall (1982)
28. Vapnik, V.N.: Statistical Learning Theory. John Wiley & Sons, Inc. (1998)
29. Zhu, M., Martinez, A.M.: Selecting principal components in a two-stage lda algorithm. In: CVPR 2006, pp. 132–137 (2006)
30. Thomaz, C.E., Kitani, E.C., Gillies, D.F.: A maximum uncertainty lda-based approach for limited sample size problems - with application to face recognition. Journal of the Brazilian Computer Society 12, 7–18 (2006)
31. Thomaz, C.E., Giraldi, G.A.: A new ranking method for principal components analysis and its application to face image analysis. Image and Vision Computing 28, 902–913 (2010)
32. Philips, P.J., Wechsler, H., Huang, J., Rauss, P.: The feret database and evaluation procedure for face recognition algorithms. Image and Vision Computing 16(5), 295–306 (1998)

The Face Speaks: Contextual and Temporal Sensitivity to Backchannel Responses

Andrew J. Aubrey[1], Douglas W. Cunningham[2], David Marshall[1],
Paul L. Rosin[1], AhYoung Shin[3], and Christian Wallraven[3]

[1] School of Computer Science and Informatics, Cardiff University, Cardiff, UK
[2] Brandenburg Technical University Cottbus, Germany
[3] Korea University, Korea

Abstract. It is often assumed that one person in a conversation is active
(the speaker) and the rest passive (the listeners). Conversational analysis
has shown, however, that listeners take an active part in the conversation, providing feedback signals that can control conversational flow. The
face plays a vital role in these *backchannel responses*. A deeper understanding of facial backchannel signals is crucial for many applications in
social signal processing, including automatic modeling and analysis of
conversations, or in the development of life-like, effective conversational
agents. Here, we present results from two experiments testing the sensitivity to the context and the timing of backchannel responses. We utilised
sequences from a newly recorded database of 5-minute, two-person conversations. Experiment 1 tested how well participants would be able to
match backchannel sequences to their corresponding speaker sequence.
On average, participants performed well above chance. Experiment 2
tested how sensitive participants would be to temporal misalignments
of the backchannel sequence. Interestingly, participants were able to estimate the correct temporal alignment for the sequence pairs. Taken
together, our results show that human conversational skills are highly
tuned both towards context and temporal alignment, showing the need
for accurate modeling of conversations in social signal processing.

1 Introduction

The face and head are a crucial aspect of human communication as they contain a wealth of non-verbal cues and are key indicators of emotional state. This
has led to a large body of work on facial expression (including head motion)
perception and production. Although conversational analysis is traditionally a
cognitive science endeavor, there is a growing interest in the automatic recognition and synthesis of conversational behavior, particularly for the creation of
virtual conversation agents. Recent reviews, [1], [2], provided a detailed overview
of this new research field of social signal processing.

In terms of facial expression research, the majority of work is based on the
so-called universal expressions (happiness, sadness, anger, disgust, fear and surprise) defined by Ekman [3]. With the exception of happiness, however, these

J.-I. Park and J. Kim (Eds.): ACCV 2012 Workshops, Part II, LNCS 7729, pp. 248–259, 2013.
© Springer-Verlag Berlin Heidelberg 2013

expressions do not occur with high frequency in *everyday* conversations. In recent years there has been an effort to examine other expressions that occur in conversations with a higher frequency (such as thinking, agreeing, being confused, being bored, *etc.*): [4–8]. Conversational expressions are not limited to movements of facial muscles, they also include global head motion and orientation (*e.g.* to indicate agreement or disagreement) [9] and gaze [10] (*e.g.* to indicate the addressee of a question). Which regions of the face are necessary and sufficient for expression recognition was investigated in [7]. They showed that the motion of different face regions contribute a varying amount to recognition performance. A clear advantage of dynamic over static stimuli for conversational expressions was demonstrated in experiments conducted in [5]. Modeling of conversational expressions therefore needs to take into account the temporal aspects of facial movements.

The term *backchannel* was coined by Yngve [11] and is used to describe the exchange of signals from the listener(s) to the speaker. These signals, which can control conversational flow, are short visual (*e.g.* nod) or vocal (*e.g.* "uh-huh") signals that the listener uses to indicate understanding, disgust, a desire to speak, or interest in the conversation, for example. While backchannel signals can be considered a subset of all feedback signals [12], this work is only concerned with visual backchannel signals (specifically, those of the face, head, and shoulders).

Until recently, studies on backchannel signals have primarily used static facial expression stimuli, such as the work by Baron-Cohen *et al.* [4]. Recently, Wehrle *et al.* [8] compared dynamic and static expressions. Even though the dynamic data were synthesised, results showed that static stimuli were more easily confused than dynamic. Bavelas *et al.* [10] found that periods of mutual gaze increased the likelihood of a backchannel occurring. In [13], the effect of quantity, timing and type of backchannel was investigated. Participants were asked to rate whether the reaction of an artificial listener to a real speaker was human-like. Several interesting results were obtained. Too many or too few backchannels per minute reduced the quality of the listener, furthermore, a lower and upper limit of 6 and 12 per minute respectively was suggested. Nods were often more appropriate than vocal signals and the timing of the backchannel influences how human-like the listener was perceived.

The goal of the present work is to further our understanding of the perceptual sensitivity to backchannel responses in conversations. More specifically, we will present two experiments that aim to test the contextual and temporal sensitivity for processing of facial feedback signals. These experiments are intended to provide important contributions towards full spatio-temporal modeling of the non-verbal facial (and head-related) information channel in conversations, complementing previous research on conversational facial expressions (*e.g.*, [7]). These kinds of models will be indispensable for the creation of virtual agents or artificial listeners that can display human-like listener behaviors [6, 14].

The remainder of the paper is organized as follows: Section 2 describes the database from which we derived the backchannel responses, Sections 3 and 4 discuss the two experiments, and Section 5 provides a brief conclusion.

2   Database

The database contains natural conversations obtained by recording both speaker and listener in a non-scripted conversation.

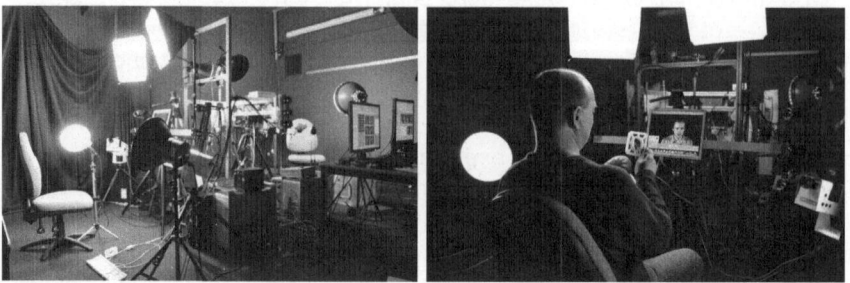

Fig. 1. Left) Setup of recording equipment, Right) View of person opposite during use

2.1   Recording Equipment

To capture the conversations in as natural a setting as possible, two audio-video recording systems were set up as shown in Fig. 1(a). The equipment used to capture *each side* of the conversation contained the following: a 3dMD dynamic scanner captured 3D video, a Basler A312fc firewire CCD camera captured 2D color video, and a microphone placed in front of the participant out of view of the camera captured sound (at 44.1KHz). A view from one side of the setup is shown in Fig. 1(b). In this paper only the 2D recordings are used; the 3D system setup and subsequent processing of that data is the subject of future work.

To ensure all audio and video could be reliably synchronized, each speaker had a hand-held buzzer and LED (light emitting diode) device, used to mark the beginning of each recording session. A single button controlled both devices and simultaneously activated the buzzer and LED. No equipment was altered between the recording sessions, except for the height of the chair to ensure the speaker's head was clearly visible by the cameras.

2.2   Recording Methods

The full dataset consists of 30 conversations, each lasting five minutes and containing two people. There were 16 speakers in total, 12 male and 4 female between the ages of 25 and 56. Prior to the recording session each speaker was asked to fill out a questionnaire. The questions simply required a response on a five point scale from strongly dislike (1) to neutral (3) to strongly like (5) and was aimed at finding out how strongly the speakers felt about possible conversation topics. The questionnaire was used to suggest topics to each pair of speakers for which they had similar or dissimilar ratings, and could if they desired be used as a basis for their conversation. Examples of the topics covered in the questionnaire are the like or dislike of different genres of music (rap, opera, jazz, rock etc), literature (poetry, sci-fi, romance biographies etc), movies, art, sports (rugby, football, ice hockey, golf etc), technology (smartphones, tablets), games, television and current affairs. However, the speakers were not restricted to the topics suggested. All participants were fluent in the English language.

2.3 Backchannel Sequences

Eleven short video sequences containing a mixture of speakers and listeners were chosen. Each sequence consisted of a "main channel" (a speaker) and a "backchannel" (a listener's concurrent non-verbal response). The sequences were chosen so that all main channel clips contained a short, easily understandable segment of a conversation. In order to allow us to systematically vary the synchronization between main and backchannel, all backchannel clips were constrained so that they contained one main visible response (possibly followed by several other smaller ones) and no speech for a period of several seconds. This constraint reduced the total number of possible sequences considerably. Sequences 1, 2, 3, 4, 8, 9, and 10 were about movies. Sequences 5 and 6 were about literature. Sequence 7 was about games. Figure 2 (Section 3.3) shows who was involved in each sequence.

3   Experiment 1: Sensitivity to Context

To investigate how well participants can pick the "correct" backchannel response given a main channel spoken segment, five possible main-channel/backchannel pairings were shown to twenty-one participants for each of the eleven sequences. In addition to trying to identify the correct matching, participants also to evaluate each main-channel/backchannel pairing along several dimensions.

3.1 Methods

Stimuli. For each of the eleven sequences, we picked four plausible alternate clips using the same listener. Thus, Sequence 1 contained a seven second long snippet of the conversation between S2 (as speaker) and S5 (as listener). The four alternate backchannels also had S5 as a listener. The alternate sequences also contained only one main visible response and no speech for a period of several seconds. The visible response was also to have roughly the same length as the main backchannel, although this was not strictly enforced. This further reduced the possible number of usable backchannel sequences. In some cases the alternate sequences had similar behaviour to the original backchannel (e.g., alternate sequences of agreement or of laughter). In other cases, the alternate sequence was very different, but still a very plausible response. The backchannel of the four alternate sequences were manually synchronized with the main channel. The participants only heard the audio from the main channel.

Procedure. The twenty-one participants were seated one at a time in front of a computer screen and asked to wear headphones. The experimental chamber was lit with normal daylight. Once the participant was seated and any initial questions were answered, written instructions for the experiment were then presented. Once participants indicated that they understood, the experiment began (controlled by Psychtoolbox3).

The eleven sequences were shown to the participants in random order, with each participant receiving a different random order. Evaluation of each sequence consisted of three phases, all of which must be completed before another of the eleven sequences could be evaluated. In the first phase, icons representing each of the five main channel-backchannel pairings were shown. Clicking on an icon started a full screen presentation (with the videos of the speaker and the listener being shown side by side) of that particular pairing after which participants were returned to the icons. Participants could watch the pairings as often as they wanted, in any order they wanted. Once all five pairings had been seen at least once each, participants could continue to the second phase.

In the second phase, participants were asked to decide which of the five pairings was the original main-channel/backchannel pairing. In the third phase, participants were shown each of the pairings again, one at a time. After watching the pairing, they were asked to rate it on four different Likert-type scales. The first three scales used the following terms for the five levels "(1) fully inappropriate", "(2) somewhat inappropriate", "(3) neutral", "(4) somewhat appropriate", "(5) fully appropriate". The first scale asked "How appropriate was the timing of the response?". The second scale asked "How appropriate was the intensity of the response". The third asked "How appropriate was the contents of the response?". The fourth scale asked "How humorous (in terms of the conversational expressions rather than comic dialogue) was the WHOLE conversation?" and used the levels "(1) fully non-humorous", "(2) somewhat non-humorous", "(3) neutral", "(4) somewhat humorous", and "(5) fully humorous".

After the experiment participants were thanked for their participant, paid, debriefed, and any questions were answered.

3.2   Results and Discussion

Overall, people were able to find the correct backchannel and the ratings showed some surprising similarity to the pattern of recognition choices.

3.3   Recognition Performance

With an average performance of 41%, recognition accuracy was significantly above chance; $t(20) = 6.29, p < 0.0001$.

Everyday experience would suggest that some people are more sensitive to the natural flow of a conversation than others. This is reflected in the accuracy results. Most participants were able to correctly identify the original pairing around 40% of the time. Indeed, all but four participants were well above the 20% chance level. Yet, some participants were much more accurate (*e.g.*, participant

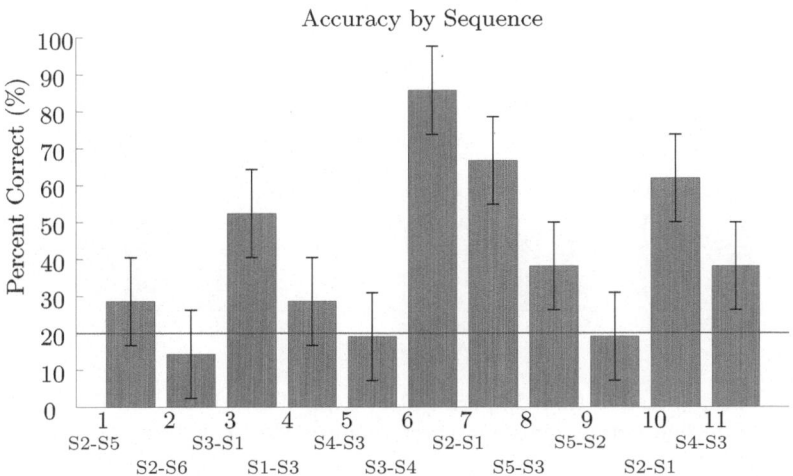

Fig. 2. Accuracy results from Experiment 1, by sequence. The letters indicate the speaker-listener pair. The horizontal line represents chance performance. The error bars represent the 95% confidence interval.

19 at 73%) while others were much worse (*e.g.*, participants 5 and 9, both at 9%).

In Figure 2, the responses are plotted by sequence. There is considerably more variation between the sequences than between the participants. Some sequences (such as Sequence 6) were recognized by almost all participants, while others (Sequence 9) were rarely recognized. Indeed, for at least 5 of the 11 sequences, performance was at or near chance levels! Thus, it seems that the overall recognition rates is being driven by a few exceptional sequences.

The variation between sequences might be due any of a number of reasons, including the degree of talent of the speaker or listener, the topic, or even the poor quality of the alternatives. Eleven sequences is not, however, a large enough sample to conclusively determine why some sequences were better than others. It is clear, however, that some the recordings of speakers as well as of some listeners were associated with higher high recognition performance. (see Figure 3). This is consistent with everyday experience: some people are better conversationalists than others.

As can be seen in Figure 4 (the leftmost bar in each cluster is always the correct response), some alternatives were more plausible than others. On the other hand, a low quality of the alternatives cannot explain most of the accuracy performance. In most sequences one of the incorrect responses was often chosen, suggesting that these false backchannels did indeed share some characteristics with the proper backchannel. In fact, in 6 of the 11 sequences one false backchannel was chosen more often than the correct alternative.

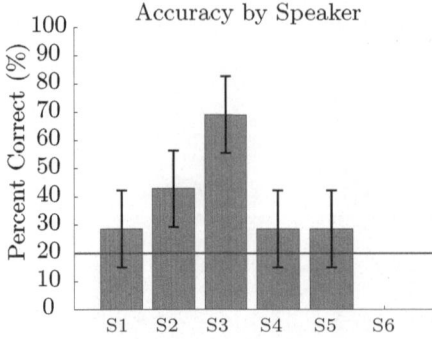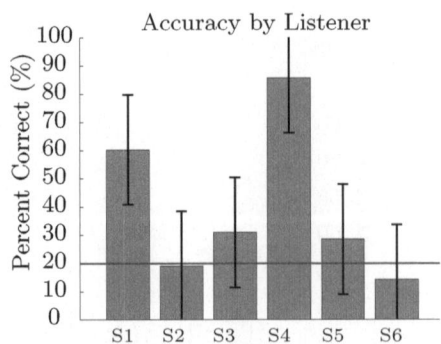

Fig. 3. Accuracy results from Experiment 1, by speaker (Left) and listener (Right). The horizontal line represents chance performance. The error bars represent the 95% confidence interval.

3.4   Rating Performance

Figures 5 – 7 show the ratings for appropriateness of *timing, intensity*, and *content* (respectively) for all 55 backchannel clips (5 backchannels for 11 sequences). Overall, the pattern of results on the three rating scales is very similar – not only to each other, but also to the pattern of choices in the recognition task. For example, in Sequences 6 and 7, the original sequence was chosen very often in the recognition task and received high ratings on all three scales, while the remaining responses were chosen less often and received lower ratings. Likewise in Sequences 1 and 3, the 1st and 3rd alternatives were chosen very often and received proportionally higher ratings. Perhaps the biggest anomaly is Sequence 10, where alternatives 1 and 5 were rated equally high on all three scales, but alternative 5 was rarely chosen in the recognition task.

Sequences 2 and 8 prove to be the exceptions to the rule of rating similarity. In Sequence 2, alternative 4 was chosen most often while the other four alternatives were rarely chosen, which is reflected somewhat in the *intensity* ratings and very much so in the *content* ratings. All five alternatives, however, were rated equally appropriate in terms of *timing*. In Sequence 8, alternatives 1 and 3 were chosen rather often and received high *timing* and *content* ratings. In contrast, the five alternatives received similar *intensity* ratings.

The *humour* ratings diverge from the results of the other tasks quite a bit. Although there is a similarity between the *humour* ratings and the choice frequencies in the recognition task for Sequences 3, 7 and 8, the rest are either slightly anomalous or simply not diagnostic.

To further examine the relationship between performance on the tasks, we correlated all four rating dimensions with recognition performance to investigate any potential linear relationships in the data. Correlations were $r_{timing} = 0.73, r_{intensity} = 0.71, r_{content} = 0.77$, and $r_{humour} = 0.54$. Apart from the humour dimension, every rating dimension therefore carries some information

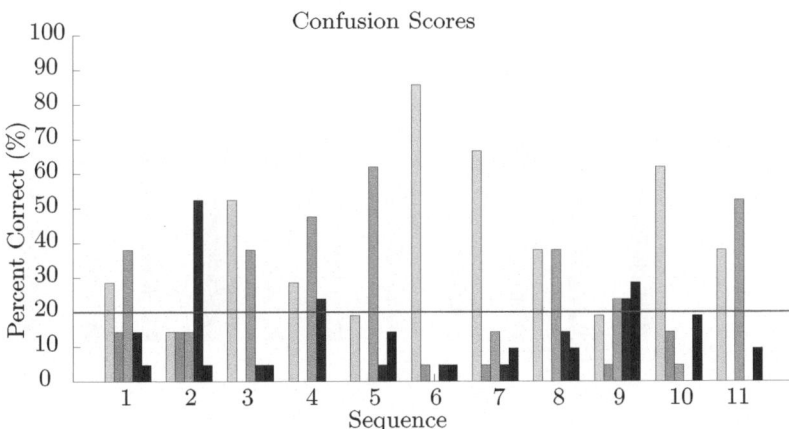

Fig. 4. Confusions in Experiment 1. The horizontal line represents chance performance.

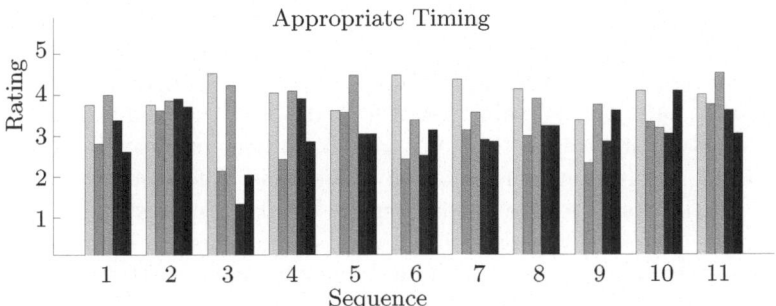

Fig. 5. Timing rating results from Experiment 1, by Sequence

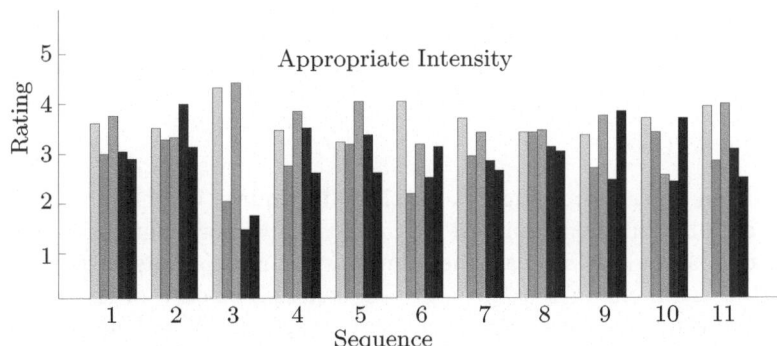

Fig. 6. Intensity rating results from Experiment 1, by Sequence

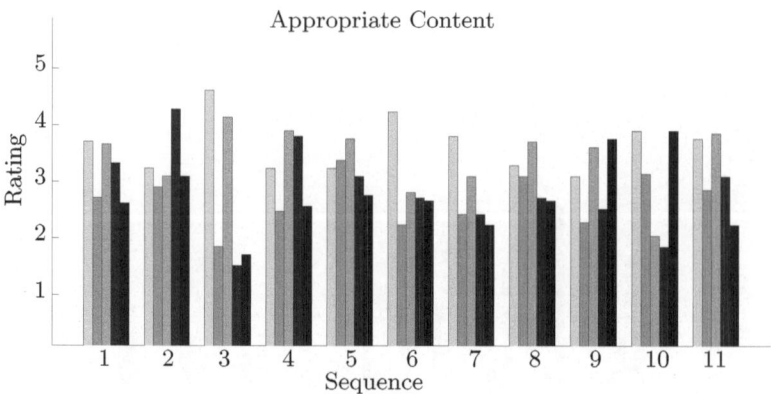

Fig. 7. Content rating results from Experiment 1, by Sequence

about the recognition performance.We then conducted a linear regression with all four ratings as predictors on the recognition performance to see the contribution of each rating in a joint model. The resulting equation from the regression was: performance = 55 + timing * 8.65 + intensity * (-10.19) + content * 21.23 + humour *4.45. The r^2 value for this model is $r^2 = 0.63$ indicating a good prediction performance. In this joint model, content carries the highest weight in predicting the recognition performance outcome. The absolute weight of both timing and intensity in the prediction are similar. Interestingly, intensity receives a negative weight, suggesting a reverse relationship. Finally, as could be seen already from the correlation data, humor has the lowest weight in the joint model.

4   Experiment 2: Sensitivity to Synchronization

Experiment 1 focused on investigating high-level, contextual sensitivity to backchannel responses. In Experiment 2 we focus on sensitivity to timing.

4.1   Methods

Stimuli. In order to investigate sensitivity to timing, we chose to measure psychometric functions in a standard psychophysical experiment (using the method of constant stimuli). The baseline stimuli for this experiment consisted of the original 11 main-channel/backchannel sequences. For each of the 11 sequences, we created 6 more backchannel sequences – each the same length – by shifting the selection window backwards and forwards in time. We chose temporal offsets of -45 to +45 frames in 15 frame intervals (-1.5 to +1.5 seconds in 0.5 second intervals). The 77 sequences were repeated three times each in completely randomized order, yielding a total of 231 trials. This repetition is standard procedure in experimental design, see [15] for further details.

Procedure. The experiment used the same hardware setup as in Experiment 1. A different set of 20 participants were recruited for this experiment. Once the participants was seated comfortably, the experiment began with instructions shown on-screen.

Each of the 231 main-channel/backchannel pairs was shown in random order to the participants. Participants were instructed to carefully watch the videos and to decide whether the backchannel response was too early or too late. The experiment lasted about 45 minutes. After the experiment, participants were rewarded for their participation, debriefed and any remaining questions were answered.

4.2 Results and Discussion

Psychometric functions were fitted to each participant's data using psignifit version 2.5.6, a software package which implements the maximum-likelihood method described in [16]. The fitted functions were used to derive two important psychophysical parameters: the point of subjective synchronicity (which is the time offset at which both main-channel and backchannel perceptually appear synchronized), and the just-noticeable-difference (which is the time difference between two sequence pairs that will be noticed as different). Note that the use of the fitted curve to derive the PSS and JND means that the JND can lie outside of the observed stimulus range. The data of four of the 20 participants proved to be anomalous (e.g, the thresholds diverged from the mean by more than one standard deviation).

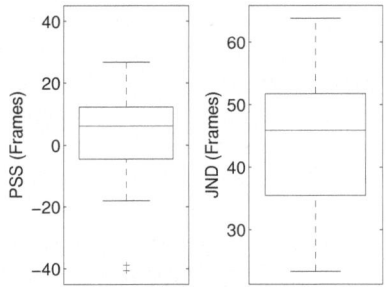

Fig. 8. Boxplots showing the distributions of (left) the point of subjective synchronicity and (right) the just-noticeable-difference

The distribution of the point of subjective synchronicity (PSS) is shown in Figure 4.2). Its median is consistently around +6.11 frames . That is, participants felt that the backchannel properly matched the main-channel when the back channel lagged by 6 frames (about 200 milliseconds). Given the difficulty of the task, it seems that people are remarkably sensitive to the correct timing between a speaker and a listener.

One potential reason for the observed lag might lie in the high cognitive load imposed by the task. Usually when watching a conversation, we do not explicitly pay attention to the timing. By bringing this element of a conversation to conscious attention, participants may need additional cognitive resources (and thus additional time) to process the videos in a manner that is less automatic than usual. It is also possible that the delay is related to saccades as participants were required to redirect their gaze and attention from the speaker to the listener, when the listener begins to become more active (or is expected to become more active). It is well known that an intentional shift of gaze focus takes at least 200 ms [17].

The distribution of the just-noticeable-difference (JND) is shown in Figure 4.2. Its median is 45.9 frames, corresponding to 1.5s. This means that in order to reliably detect time offset differences between two sequences (in either direction), they would need to be shifted by 45 frames. The lowest JND among all participants was 0.7 seconds, the highest 2 second. Hence, whereas participants on average can detect the veridical time offset, their sensitivity to *changes* in synchronization is on the order of 1.5 seconds. Although this difference may seem large at first glance, one has to bear in mind that the JND did does not test the *detection* of backchannel responses, but rather measures how well *changes to the synchronization of two speakers* could be judged - something which imposes much more complex demands on conversation processing.

5   Conclusion

In this paper, we presented results of two experiments on contextual and temporal sensitivity to backchannel responses. Using a newly recorded database of natural conversations, we extracted several speaker-listener interactions containing non-verbal, visual backchannel responses (of the face, head, and shoulders) of a listener. In Experiment 1 we found that participants were well able to identify the correct backchannel response among a list of alternative responses - the success of this match, however, depended crucially on both speaker and listener. The additional dimensions that were analyzed correlated with recognition performance and we were able to predict recognition performance reasonably well using a joint linear model. A more detailed model using additional, important dimensions for conversational analysis, however, still needs to be investigated. In Experiment 2, we examined sensitivity to time offsets in the backchannel response. We found that participants' points of subjective synchronicity were on average almost veridical. Their sensitivity (as measured by the just-noticeable-difference) was around 1.5s, which is fairly good considering the complexity of the speaker-listener interaction. The two experiments here represent the start of our investigations into full spatio-temporal models of how facial expressions and facial gestures are used in conversational contexts. In future work, we will be constructing full active-appearance models of the speakers and listeners in the database. These will be used to create video sequences modified to freeze certain facial parts (*e.g.*, [7]), to warp the timing of the backchannel responses, *etc.* With these modified sequences, we can conduct more detailed experiments on the sensitivity to physical changes for facial expressions in conversational contexts.

References

1. Vinciarelli, A., Pantic, M., Heylen, D., Pelachaud, C., Poggi, I., D'Errico, F., Schroeder, M.: Bridging the Gap between Social Animal and Unsocial Machine: A Survey of Social Signal Processing. IEEE Transactions on Affective Computing 3, 69–87 (2012)
2. Gatica-Perez, D.: Automatic nonverbal analysis of social interaction in small groups: A review. Image and Vision Computing 27, 1775–1787 (2009)
3. Ekman, P.: Universal and cultural differences in facial expressions of emotion, pp. 207–283 (1972)
4. Baron-Cohen, S., Wheelwright, S., Jolliffe, T.: Is there a "language of the eyes"? evidence from normal adults, and adults with autism or asperger syndrome. Visual Cognition 4, 311–331 (1997)
5. Cunningham, D.W., Wallraven, C.: Dynamic information for the recognition of conversational expressions. Journal of Vision 9 (2009)
6. Pelachaud, C., Poggi, I.: Subtleties of facial expressions in embodied agents. The Journal of Visualization and Computer Animation 13, 301–312 (2002)
7. Nusseck, M., Cunningham, D.W., Wallraven, C., Bülthoff, H.H.: The contribution of different facial regions to the recognition of conversational expressions. Journal of Vision 8 (2008)
8. Wehrle, T., Kaiser, S., Schmidt, S., Scherer, K.R.: Studying the dynamics of emotional expression using synthesized facial muscle movements. Journal of Personality and Social Psychology 78, 105–119 (2000)
9. Heylen, D.: Challenges ahead: head movements and other social acts during conversations. In: Joint Symposium on Virtual Social Agents, pp. 45–52 (2005)
10. Bavelas, J.B., Coates, L., Johnson, T.: Listener responses as a collaborative process: The role of gaze. Journal of Communication 52, 566–580 (2002)
11. Yngve, V.: On getting a word in edgewise. In: Papers from the Sixth Regional Meeting of the Chicago Linguistic Society, pp. 567–578 (1970)
12. Schröder, M., Heylen, D., Poggi, I.: Perception of non-verbal emotional listener feedback. In: Proceedings of Speech Prosody, Dresden, Germany (2006)
13. Poppe, R., Truong, K.P., Heylen, D.: Backchannels: Quantity, Type and Timing Matters. In: Vilhjálmsson, H.H., Kopp, S., Marsella, S., Thórisson, K.R. (eds.) IVA 2011. LNCS, vol. 6895, pp. 228–239. Springer, Heidelberg (2011)
14. Poppe, R., Truong, K.P., Reidsma, D., Heylen, D.: Backchannel Strategies for Artificial Listeners. In: Allbeck, J., Badler, N., Bickmore, T., Pelachaud, C., Safonova, A. (eds.) IVA 2010. LNCS (LNAI), vol. 6356, pp. 146–158. Springer, Heidelberg (2010)
15. Cunningham, D.W., Wallraven, C.: Experimental Design: From User Studies to Psychophysics. A K Peters/CRC Press (2011)
16. Wichmann, F.A., Hill, N.J.: The psychometric function: I. Fitting, sampling and goodness of fit. Perception and Psychophysics 63, 1293–1313 (2001)
17. Trottier, L., Pratt, J.: Visual processing of targets can reduce saccadic latencies. Vision Research 45, 1349–1354 (2005)

Virtual View Generation Using Clustering Based Local View Transition Model

Xi Li[1], Tomokazu Takahashi[2], Daisuke Deguchi[3],
Ichiro Ide[1], and Hiroshi Murase[1]

[1] Graduate School of Information Science,
Nagoya University, Japan
[2] Department of Economics and Information,
Gifu Shotoku Gakuen University, Japan
[3] Information and Communications Headquarters,
Nagoya University, Japan

Abstract. This paper presents an approach for realistic virtual view generation using appearance clustering based local view transition model, with its target application on cross-pose face recognition. Previously, the traditional global pattern based view transition model (VTM) method was extended to its local version called LVTM, which learns the linear transformation of pixel values between frontal and non-frontal image pairs using partial image in a small region for each location, rather than transforming the entire image pattern. In this paper, we show that the accuracy of the appearance transition model and the recognition rate can be further improved by better exploiting the inherent linear relationship between frontal-nonfrontal face image patch pairs. For each specific location, instead of learning a common transformation as in the LVTM, the corresponding local patches are first clustered based on appearance similarity distance metric and then the transition models are learned separately for each cluster. In the testing stage, each local patch for the input non-frontal probe image is transformed using the learned local view transition model corresponding to the most visually similar cluster. The experimental results on a real-world face dataset demonstrated the superiority of the proposed method in terms of recognition rate.

1 Introduction

Due to its wide range of potential real-life applications such as identity authentication, intelligent surveillance, human-computer interface and so on, face recognition has been one of the most active research topics in the biometric field within the computer vision and the pattern recognition communities [1]. Unlike other biometric techniques such as fingerprint recognition, palm print recognition or iris recognition, face recognition is inherently a passive and non-intrusive technique that has the advantage of not requiring cooperative subjects. That is to say, a practical face recognition system is supposed to have the ability to recognize the face of an uncooperative subject in an arbitrary situation and uncontrolled environment setting, even without the target subject noticing.

J.-I. Park and J. Kim (Eds.): ACCV 2012 Workshops, Part II, LNCS 7729, pp. 260–271, 2013.
© Springer-Verlag Berlin Heidelberg 2013

This advantage of environment setting generality also poses great challenges to the problem of face recognition because as the viewing condition changes, the captured face appearances might vary too drastically to be easily identified. Within the past several decades, many methods have been proposed for face recognition. However, most of those traditional methods can successfully recognize faces only when face images are captured under constrained condition and controlled environment, for example recognize frontal faces with normal expressions and typical indoor illuminations, which are usually unrealistic in many real-life application scenarios. Usually the performance of these traditional methods will degrade greatly when face images are captured in unconstrained conditions caused by factors such as varying viewpoints, illumination changes, occlusions, aging, expressions and poses.

This work studies the problem of face recognition across poses, where each subject has a frontal gallery face image stored in the database and the probe image is not necessarily frontal. It is of great interest in many real-world face recognition application scenarios such as surveillance systems, where the subjects are either indifferent or uncooperative, so the captured face images are usually low-resolution and non-frontal. Pose variation has been identified as one of the prominent difficult problems in the research of face recognition [1]. The major difficulty of the cross-pose face recognition is that the intra-person appearance differences caused by rotation are often larger than the inter-person differences. That is to say, the distance between appearance vectors of two faces of different persons under similar viewpoints is much smaller than that of the same person under different viewpoints. This phenomenon makes the traditional face recognition methods such as eigen-face [2] or fisher-face [3] infeasible. Obviously one straightforward method for cross-pose face recognition is to actively compensate pose variations by providing gallery views in each rotation angles to recognize rotated non-frontal probe views. This can be achieved by first collecting and preparing multiple real-view templates beforehand for every known individual in each specific pose condition. Although the number of required real gallery images can be reduced by proper quantization on the rotation angles due to the fact that general face recognition algorithms are able to tolerate small pose variations to some extent, the tedious process of collecting multiple face images in different poses for real-view based matching is still unfavorable and even impractical in some cases. For example, in the application of airport security surveillance systems, there is only one frontal passport photo per person that could be collected and stored in the database.

Previously, both 3D model based methods [4][5] and 2D appearance based methods [6][7][8][9][10] have been proposed for pose invariant face recognition. 3D Morphable Model [4] is a typical 3D model based method for pose invariant face recognition. The 3D morphable model is built using the principal component analysis of the 3D facial shapes and textures obtained from laser scanner devices where inter-person pixel correspondences are established using the optical flow on the 3D surfaces. The 3D Morphable Model can then realize recognition either by transforming non-frontal face images to frontal view or by directly performing

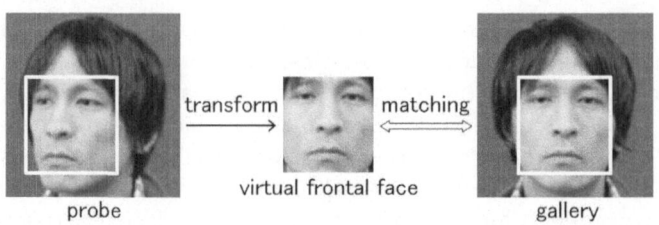

probe

virtual frontal face

gallery

Fig. 1. Cross-pose face recognition based on virtual view generation

the recognition by using the coefficients of the morphable model. But usually it is difficult to detect dense facial feature points that are accurate enough for the model fitting from low-resolution surveillance camera images.

Among the 2D appearance based methods, one of the successful approaches is to first generate a virtual frontal view by applying pose transformation on any given non-frontal face view. The View-Transition Model (VTM) [6] is a noteworthy work for pose transformation that can construct human appearance models for different poses which have proper texture information from a limited number of input images. The VTM method transforms views of an object between different poses by linear transformation of pixel values in images. For each pair of poses, a transformation matrix is calculated from image pairs of the poses of a large number of training data. The VTM was further extended to Local VTM (LVTM) in a patch-wise way [7] and it was shown that a more satisfactory face recognition result can be achieved using the virtual frontal face view generated by the local patch based LVTM than the original global patch based LVTM. This paper further extends the LVTM and presents a framework for face recognition across poses based on virtual frontal view generation using the LVTM with local patches clustering, which is denoted as c-LVTM hereafter. The proposed c-LVTM can describe the inherent transforming relationship between pixel values of patch pairs in a more precise way, thus more realistic virtual frontal face images can be generated and a higher recognition rate can be obtained. The experimental results on a real-world face dataset demonstrated the superiority of the proposed method.

The rest of this paper is organized as follows: in section 2, the underlying principle of the original VTM for pose transformation and the LVTM based face recognition methods are introduced briefly. Section 3 describes the proposed clustering based local VTM method (c-LVTM) in detail. Section 4 introduces the experimental result and section 5 is the summary.

2   Cross-Pose Face Recognition by Virtual Frontal View Generation

Instead of directly classifying the probe non-frontal face image, VTM or LVTM based cross-pose face recognition methods firstly synthesize a virtual frontal face view before a general face recognition procedure is applied, as shown in Fig. 1.

Both the VTM and the LVTM methods use a general training image dataset consisting of faces of a large number of individuals viewed from both frontal and various profile angles. The linear transformations learned from the training dataset are applied to the probe non-frontal face images, either in a global way as in the VTM or in a local patch based way as in the LVTM, to generate the counterpart virtual frontal face image that is then fed into a general traditional face recognition engine.

More specifically, given a training multi-pose face image dataset $\Theta\{\mathbf{Q}_\phi^1, ..., \mathbf{Q}_\phi^N$, $\mathbf{Q}_{\theta_1}^1, ..., \mathbf{Q}_{\theta_L}^1, ..., \mathbf{Q}_{\theta_1}^N, ..., \mathbf{Q}_{\theta_L}^N\}$, where N is the number of training subjects, $\mathbf{Q}_\phi^n$, $(n = 1, ..., N)$ represents the frontal face image for the n-th subject as a vector which is a column vector that has pixel values of the image as its elements and $\mathbf{Q}_{\theta_l}^n, (l = 1, ..., L, n = 1, ..., N)$ represents the non-frontal face image for the n-th subject with the pose rotation angle θ_l. For an input probe non-frontal face image $\mathbf{P}_{\theta_l}$, the purpose is to generate its virtual frontal image $\mathbf{P}_\phi$ using the linear transformation learned from the training dataset. The VTM can be applied for virtual frontal face generation by one or any number of input images. However, in the interest of simplicity, we describe the frontal face generation algorithm for one non-frontal face input image only and assume that the training dataset consists of frontal-nonfrontal face image pairs with a single rotation degree θ. The VTM calculates the linear transformation $\mathbf{T}$ beforehand using the training dataset by solving the following equation [6]:

$$\left[\mathbf{Q}_\phi^1 \cdots \mathbf{Q}_\phi^N\right] = \mathbf{T}\left[\mathbf{Q}_\theta^1 \cdots \mathbf{Q}_\theta^N\right] \tag{1}$$

Then the VTM generates $\mathbf{P}_\phi$, which denotes the virtual frontal face image for the probe image, from the input non-frontal probe face image $\mathbf{P}_\theta$ as follows:

$$\mathbf{P}_\phi = \mathbf{T}\mathbf{P}_\theta \tag{2}$$

Faces of two persons might have similar parts although these faces are not in total similar. Thus transforming the input face image using the information of the entire face image of other individuals might degrade the characteristics of the input individual's face. In order to solve this problem, the VTM was further extended in a local patch based way called Local View Transition Model (LVTM) [7], which achieves face pose transformation by synthesizing a face image from partial face image patches. That is to say, instead of transforming directly the entire global face image, the LVTM transforms face patches that are partial images of a face image for each location in the face image.

Let $\mathbf{q}_{\phi(x,y)}$ and $\mathbf{q}_{\theta(x,y)}$ represent face patches with patch center location at (x, y) of corresponding frontal and non-frontal global face image planes $\mathbf{Q}_\phi$ and $\mathbf{Q}_\theta$ respectively. The LVTM learns the location specific linear transforms $\mathbf{T}_{(x,y)}$ in a similar way with the VTM as follows:

$$\left[\mathbf{q}_{\phi(x,y)}^1 \cdots \mathbf{q}_{\phi(x,y)}^N\right] = \mathbf{T}_{(x,y)}\left[\mathbf{q}_{\theta(x,y)}^1 \cdots \mathbf{q}_{\theta(x,y)}^N\right] \tag{3}$$

Local Patches

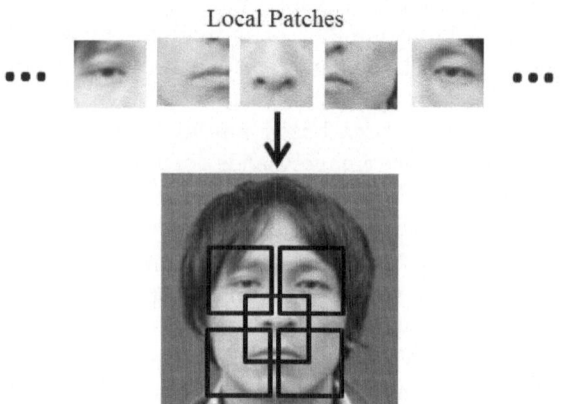

Fig. 2. Face image synthesis by local patches aggregation

It should be noted that the LVTM transforms each local area of an image while the VTM transforms the entire area of an image. Then the virtual frontal appearances for each local patches can be generated as follows:

$$\mathbf{P}_{\phi(x,y)} = \mathbf{T}_{(x,y)} \, \mathbf{p}_{\theta(x,y)} \tag{4}$$

After this, the LVTM synthesizes an output frontal face image $\mathbf{P}_\phi$ from all the transformed local patches $\mathbf{p}_{\phi(x,y)}$. The pixel values of regions where face patches are overlapped are calculated by averaging the pixel values of the overlapped patches as illustrated in Fig. 2. Experimental results showed that the LVTM can achieve a higher recognition rate than that of using VTM for pose transformation [7].

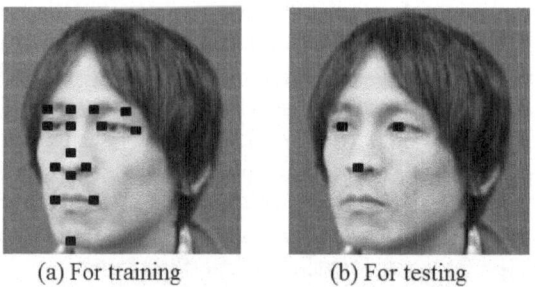

(a) For training   (b) For testing

Fig. 3. Affine alignment using landmarks. Different strategies are used for training and testing stages. (a) In the training stage, in order to learn the linear transformations more accurately, the face images are finely affine aligned using multiple (15) landmarks labeled manually. (b) While in the testing stage, the input probe face image is only roughly affine aligned using three landmarks (left eye, right eye, and nose tip), which can be easily detected by any off-the-shelf facial point detectors.

3 Virtual View Generation Using Clustering Based LVTM (c-LVTM)

The key point of VTM-like methods is the underlying linear relationship in the frontal and non-frontal face image pairs. Next we will show that the accuracy of the appearance transition model and the recognition rate can be further improved by better exploiting the inherent linear relationship between frontal-nonfrontal face image patch pairs. This is achieved based on the observation that variations in appearance caused by pose are closely related to the corresponding 3D structure, and intuitively, frontal-nonfrontal patch pairs from more similar local 3D face structures should have a stronger linear relationship. Thus for each specific location, instead of learning a common transformation as in the LVTM, in the proposed c-LVTM, the corresponding local patches are first clustered based on the appearance similarity distance metric and then the transition models are learned separately for each cluster. We assume that those patches with similar 3D shapes and thus similar 2D appearances should have a more precise linear mapping relationship. For the purpose of describing the relationship of frontal-nonfrontal pairs more precisely, it is better to learn the transformations specific for each cluster separately, rather than learning just a single common linear mapping using all the patch pairs for a specific location. As Fig. 3(a) shows, in order to learn the linear transformations in a precise way, the training face image pairs are finely affine aligned using multiple landmarks. However in the testing stage, as Fig. 3(b) shows, the input probe face image is only roughly affine aligned using three landmarks (left eye, right eye, and nose tip), which can be easily detected by any off-the-shelf facial point detectors.

More specifically, we first cluster the local patches $\mathbf{q}_{\theta(x,y)}$ for each location (x,y) into K clusters based on the appearance similarity using the Normalized Cross-Correlation score[11], where cluster k has c_k samples as $\{\mathbf{q}^1_{\theta(x,y)}, ..., \mathbf{q}^{c_k}_{\theta(x,y)}\}$. Then for each cluster, the corresponding linear transformation $\mathbf{T}^k_{(x,y)}$, which is both location specific and local 3D structure specific, is learned as follows,

$$\left[\mathbf{q}^1_{\phi(x,y)} \cdots \mathbf{q}^{c_k}_{\phi(x,y)} \right] \tag{5}$$
$$= \mathbf{T}^k_{(x,y)} \left[\mathbf{q}^1_{\theta(x,y)} \cdots \mathbf{q}^{c_k}_{\theta(x,y)} \right], \quad (k=1,...,K)$$

In the testing stage, the probe non-frontal face image is first roughly affine aligned, for example using only three landmark points at left eye, right eye and mouth, which can be easily obtained using any standard facial feature point detector. Then for each local patch of the input non-frontal face image $\mathbf{p}_{\theta(x,y)}$, the most visually similar cluster in the training set is searched in the neighborhood regions $([x-\epsilon, x+\epsilon], [y-\epsilon, y+\epsilon])$ space of a specific location (x,y). If we denote the most visually similar patch found resides in the k_{opt}-th cluster of location $(x_{\mathrm{opt}}, y_{\mathrm{opt}})$, then

$$\mathbf{P}_{\phi(x,y)} = \mathbf{T}^{k_{\mathrm{opt}}}_{(x_{\mathrm{opt}}, y_{\mathrm{opt}})} \mathbf{P}_{\theta(x,y)} \tag{6}$$

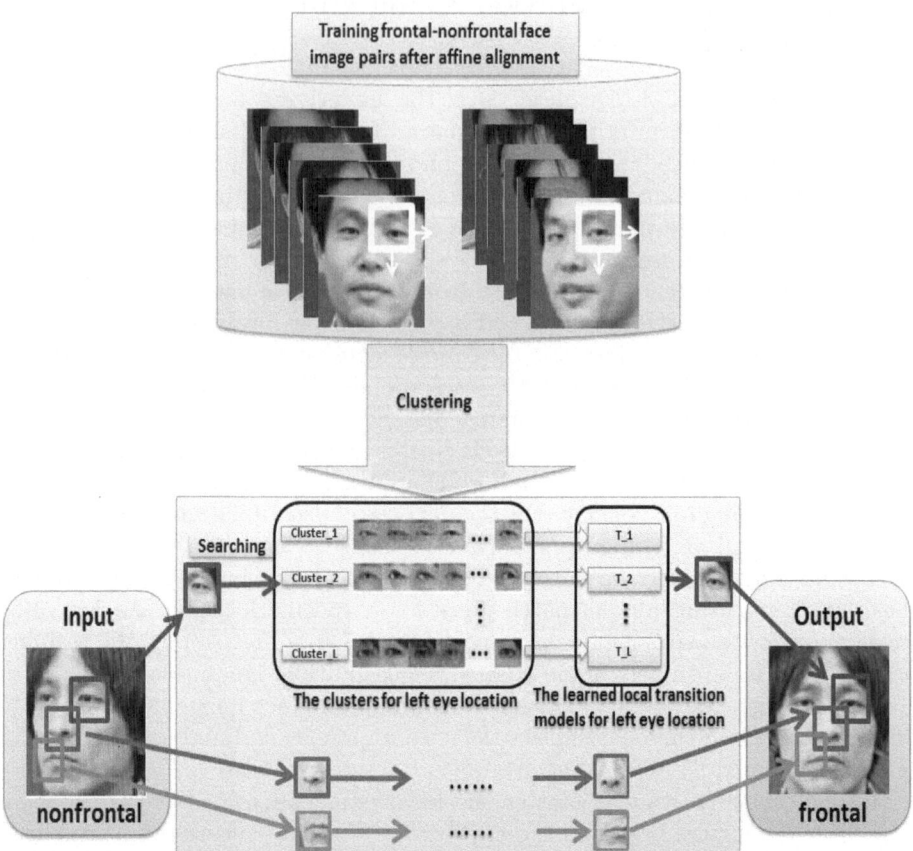

Fig. 4. The illustration of main steps of the proposed c-LVTM method. The steps of the appearance clustering based local transition models computation and the optimum transition model searching are depicted by taking the local patches located on the left eye as an example. First, the local patches location on the left eye are clustered into clusters of **cluster_1, cluster_2,..., cluster_L** based on appearance similarity. Then for each cluster, the local transition models **T_1, T_2,..., T_L** are computed using the corresponding local patches. Then for left eye local patch of the input non-frontal face image, the most visually similar clusters in the training set is searched in the neighborhood regions and local transition model corresponding to the most visually similar patch found is used to perform the transformation. The final transformed global frontal face image is the aggregation of all transformed local patches where the pixel values of the overlapped patches are averaged.

The final transformed global frontal face image is aggregated from $\mathbf{p}_{\phi(x,y)}$ in a similar way as in the LVTM. The main idea of the appearance clustering based local transition models computation and the optimum transition model searching is illustrated in detail in Fig. 4 and the flowchart of the proposed c-LVTM is described in Fig. 5.

The differences between the VTM, the LVTM and the proposed c-LVTM are illustrated in Fig. 6. The VTM learns a global linear mapping on the holistic face image plane. The LVTM learns location specific linear mapping for each local patch. The proposed c-LVTM learns linear mappings that are both location specific and local 3D structure specific.

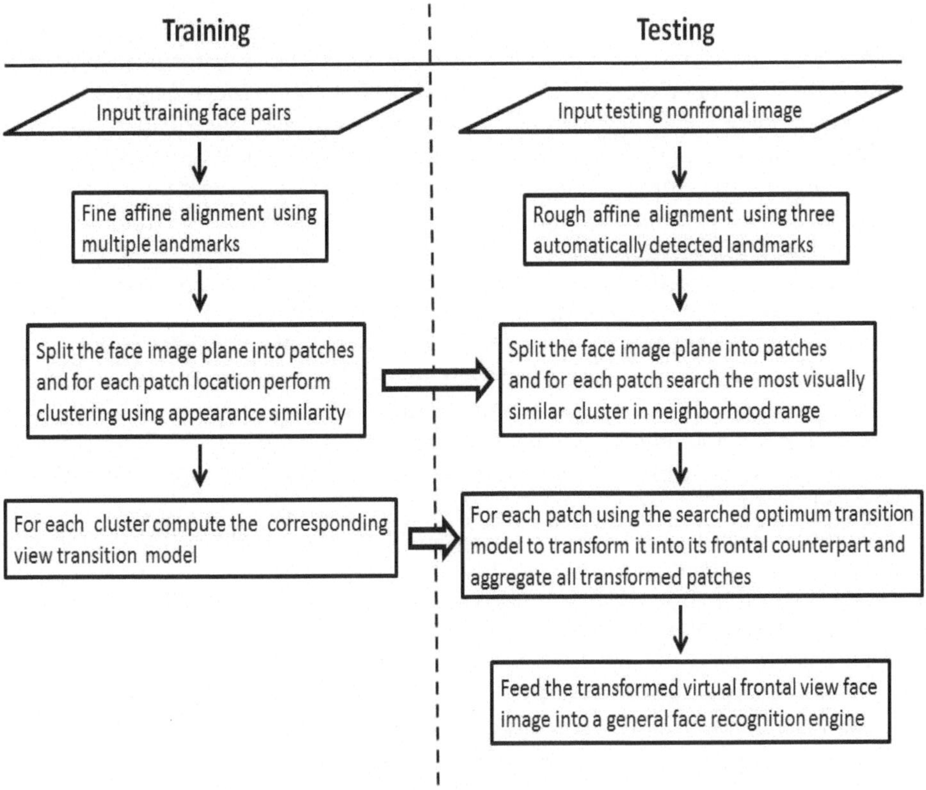

Training   **Testing**

Input training face pairs   Input testing nonfronal image

Fine affine alignment using multiple landmarks   Rough affine alignment using three automatically detected landmarks

Split the face image plane into patches and for each patch location perform clustering using appearance similarity   Split the face image plane into patches and for each patch search the most visually similar cluster in neighborhood range

For each cluster compute the corresponding view transition model   For each patch using the searched optimum transition model to transform it into its frontal counterpart and aggregate all transformed patches

Feed the transformed virtual frontal view face image into a general face recognition engine

Fig. 5. The flowchart of the proposed c-LVTM method

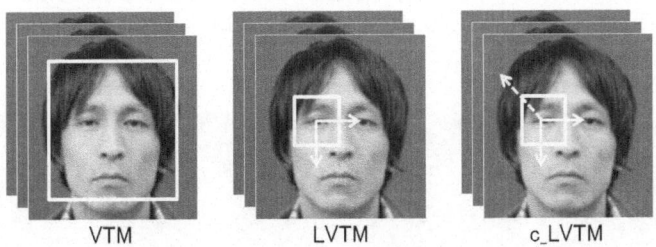

VTM   LVTM   c_LVTM

Fig. 6. The difference in how the image patterns are selected for the transition model computation between VTM, LVTM and the proposed c-LVTM method

4   Experiment

We used a subset of the face image dataset provided by SOFTPIA JAPAN to demonstrate the effectiveness of the proposed method. The subset consists of 250 individuals' images. They were taken with horizontal angles varying from -30 degrees to 30 degrees at 10 degrees interval as shown in Fig. 7. We compared the performance of using input images directly, the VTM, the LVTM and the proposed c-LVTM by 5-fold cross-validation. We transformed non-frontal face images to virtual frontal face images and then input the transformed images to a system that recognizes persons from the virtual frontal face images using a common subspace based face recognition algorithm, where the subspace for each face image was spanned by the slide window shifting extended sample set. The training images were precisely affine aligned using 15 landmark points and the testing images were roughly aligned using only 3 landmark points at left eye, right eye and mouth.

The image size was 32×32 in pixels and the face patch size was set to be 16×16 in pixels. The number of the cluster centers K was set to 4. The region of neighborhood searching ϵ was set to 5. The visual effects of the transformed virtual frontal face images using different methods are illustrated in Fig. 8. It can be seen that the generated virtual frontal face image using the proposed c-LVTM method has higher fidelity than that of other methods. This trend is further demonstrated in the following face recognition rate comparison which is illustrated in Fig. 9.

The recognition rate of the straightforward baseline method that using the non-frontal face images directly as input is much lower than that of using the virtually generated frontal face images as input, either using VTM, LVTM or the proposed c-LVTM. Furthermore, the recognition performance of the proposed c-LVTM outperforms the VTM and LVTM in two ways: 1) c-LVTM has a higher recognition rate than VTM and LVTM; 2) Though all methods have a rate decreasing trend as the pose angle increases, the proposed c-LVTM has a more robust property against pose angle degree. That is to say, as pose angle increases, the curve of rate-vs-angle for c-LVTM drops less drastically than that of VTM and LVTM. The recognition rate comparison results validate our assumption that learning both location specific and local 3D structure specific linear transforms can better capture the relationship between frontal and non-frontal patch pairs than just learning a single common linear transformation.

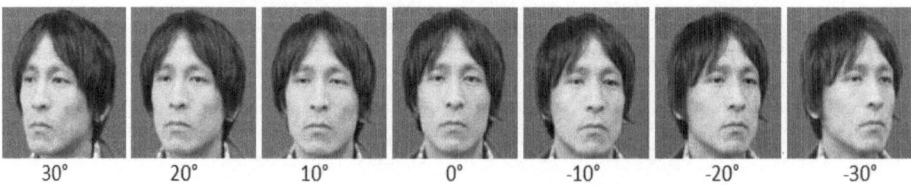

Fig. 7. The sample images of the multiple pose faces

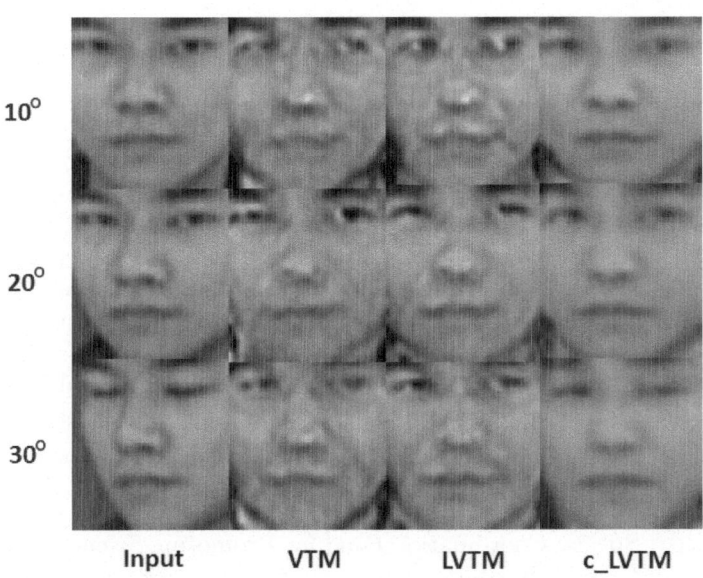

Fig. 8. The comparison of the visual effect of the transformed virtual frontal face image using different methods. It can be clearly seen that the virtual frontal face images generated using the proposed c-LVTM method have the highest visual fidelity.

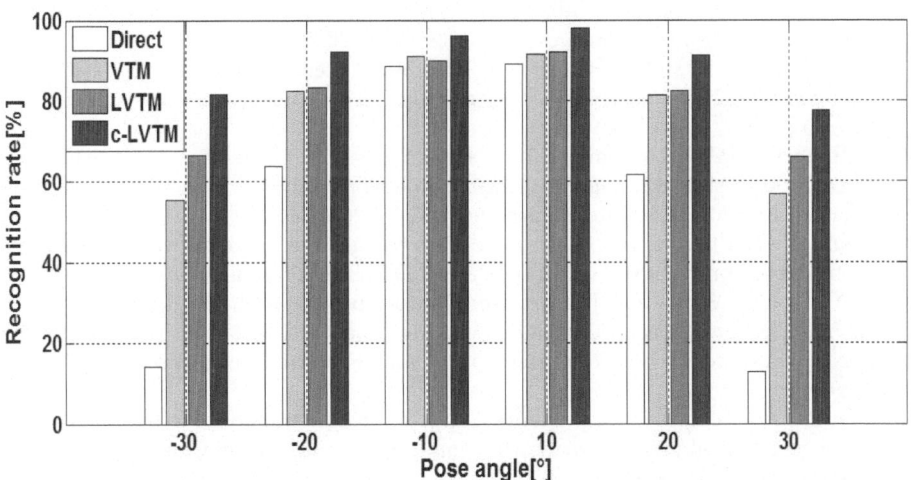

Fig. 9. Comparison of recognition rates across different angles. The input non-frontal face images are transformed using the VTM, LVTM and the proposed c-LVTM, respectively. The rate for the straightforward method of using the input non-frontal face images directly is also included for comparison.

5   Summary

In order to better exploit the underlying linear relationship between frontal and non-frontal pairs, this paper presented a framework for face recognition across pose based on virtual frontal view generation using the Local View Transition Model (LVTM) with local patches clustering. The proposed method further extended the LVTM by learning not only the local patch position specific transformations, but also the local 3D structure specific linear transforms. Experimental results showed the effectiveness of the proposed method. Although the main focus of this paper is on the problem of face recognition, the proposed framework for realistic virtual view generation is quite general. In the future, we would like to further investigate its performance evaluation not only on other more facial datasets, but also on databases in other domains such as multi-view object recognition or view invariant person identification using body images.

Acknowledgement. This work was supported by "R&D Program for Implementation of Anti-Crime and Anti-Terrorism Technologies for a Safe and Secure Society" Special Coordination Fund for Promoting Science and Technology of the Ministry of Education, Culture, Sports, Science and Technology, the Japanese Government. The face image dataset used in this work was provided by SOFTPIA JAPAN.

References

1. Zhao, W., Chellappa, R., Philips, P.J., Rosenfeld, A.: Face recognition: A literature survey. ACM Computer Survey 35(4), 399–459 (2003)
2. Turk, M.A., Pentland, A.P.: Face recognition using eigenfaces. In: Proc. 1991 IEEE Computer Society Conference on Computer Vision and Pattern Recognition, pp. 586–591 (1991)
3. Belhumeur, P.N., Hespanha, J.P., Kriegman, D.J.: Eigenfaces vs. fisherfaces: Recognition using class specific linear projection. IEEE Transactions on Pattern Analysis and Machine Intelligence 19(7), 711–720 (1997)
4. Blanz, V.G., Phillips, P.J., Vetter, T.: Face recognition based on frontal views generated from non-frontal images. In: Proc. 2005 IEEE Computer Society Conference on Computer Vision and Pattern Recognition, pp. 454–461 (2005)
5. Beymer, D.: Face recognition under varying pose. In: Proc. 1994 IEEE Computer Society Conference on Computer Vision and Pattern Recognition, pp. 756–761 (1994)
6. Utsumi, A., Tetsutani, N.: Adaptation of appearance model for human tracking using geometrical pixel value distribution. In: Proc. 6th Asian Conference on Computer Vision, pp. 794–799 (2004)
7. Kono, Y., Takahashi, T., Deguchi, D., Ide, I., Murase, H.: Frontal Face Generation from Multiple Low-Resolution Non-frontal Faces for Face Recognition. In: Koch, R., Huang, F. (eds.) ACCV 2010 Workshops, Part I. LNCS, vol. 6468, pp. 175–183. Springer, Heidelberg (2011)
8. Baker, S., Kanade, T.: Hallucinating faces. In: Proc. 2000 IEEE Conference on Automatic Face and Gesture Recognition, pp. 83–88 (2000)

9. Chai, X., Shan, S., Chen, X., Gao, W.: Locally linear regression for pose-invariant face recognition. IEEE Transactions on Image Processing 16(7), 1716–1725 (2007)
10. Beymer, D., Poggio, T.: Face recognition from one example view. In: Proc. 5th International Conference on Computer Vision, pp. 500–507 (1995)
11. Goesele, M., Curless, B., Seitz, S.M.: Multi-view stereo revisited. In: Proc. 2006 IEEE Computer Society Conference on Computer Vision and Pattern Recognition, pp. 2402–2409 (2006)

3D Facial Expression Synthesis
from a Single Image Using a Model Set

Zhixin Shu, Lei Huang, and Changping Liu

Institute of Automation, Chinese Academy of Sciences

Abstract. In this paper, we present a system for synthesizing 3D human face models containing different expressions from a single facial image. Given a frontal image of the target face with neutral expression, we first detect several key points denoting the shape of the face by Active Shape Model (ASM). Then we apply a RBF-based scattered data interpolation to reconstruct a 3D target face using a neutral expression 3D face model as reference. By analyzing a series of 3D expression face model, we segment the 3D reference model into regions automatically that each region is correspondent to a facial organ. From the expression set we construct a motion model for each facial action with respect to the target face in a local consistent manner. At last, the reconstructed 3D target face model with neutral expression and the facial action motion model are combined to generate 3D target face of various expressions. There are 3 contributions of our work: (1) We employ a set of registered 3D facial expression models as input, which enabled us to generate more complex and visual-realistic expressions than other parameter-based approaches and 2D image-based methods. (2) On the basis of a clustering-based segmentation, we developed a localized linear expression model, which make it possible for us to generate different facial expressions both locally and globally, thusly enlarge the space of synthesize output and break the limitation by the limited scale of the input expression model set. (3) A local space transform procedure is included that the output expression can fit distinct facial shapes regardless of the scarcity of variation of the facial shapes (fat or thin) in the input model set.

1 Introduction

Facial expression synthesis is significant to both the computer vision and the human perception research. In the view of the computer vision, the facial expression synthesis technology can help the face recognition systems to improve their stability and adaptability. On the other hand, synthesized facial expression is also useful for human perception research. Parameterized facial expressions help the researchers with the discovering of the human perceptual process of face and the construction of the perception models. In this paper, we present a new approach to synthesize a variety of facial expressions from a single input image.

Comparing with image, a 3D model can capture more details of head and face thus providing us wider space of details to percept the identity and the emotion.

J.-I. Park and J. Kim (Eds.): ACCV 2012 Workshops, Part II, LNCS 7729, pp. 272–283, 2013.

On the basis of face models, the expression synthesis and animation is an active research topic to many applications of video entertainment and human-computer interaction (HCI). A 3D face model can be obtained by 3D scanning using a laser scanner, reconstruction with structured light, or reconstruction from images, etc. Among them the image-based approach is the most appealing and challenging way because the image is easily obtained but provides less information than other approaches. There has been a lot of works on face modeling from images and model-based approach is the mainstream among most the state-of-the-art techniques. Blanz and Vetter [1] proposed 3D Morphable Model (3DMM), a system to create face models from a single image which provide high-quality result of face modeling. Their system uses a database of both geometry and texture. Kemelmacher and Basri [2] implement the shape-from-shading (SFS) approach within a model-based framework. They exploit the global properties of human faces as a constraint condition by using a reference face model.

On the basis of 3D face models, approaches of expression synthesis can be divided into 4 classes: interpolation, performance driven approach, muscle-based approach and parameterization approach. F. Pighin et al. [3] proposed a system of expression synthesis from photographs which can capture accurate geometry as well as textures, but painstaking model fitting process for each key frame is required and for each model the landmarks of the face need to be selected manually. Expression cloning [4][5][6] is another approach to assign expressions to face model but solving the correspondence between the models makes the problem complicated.

We start from a set of well-registered face models with different facial expressions. The input is a single image of a frontal target face. A neutral expression model of the target face is reconstructed by a model-based approach and we assign different expressions to the face by a motion model. There are three main contributions of our work. (1) We developed a simple clustering-based approach for face segmentation which divides the face model into regions. The region segmentation procedure effectively expanded the possible expression space thus enabled us to generate more facial expressions by local organ-wise manipulation; (2) Although almost everyone share an identical face structure, human faces may differ from each other in shape. Since our expression synthesis procedure is based on motion vectors, the motion direction and motion extent should vary with respect to different facial shapes. To solving this problem, we developed a local space approach to transfer the motion between models; (3) Localized linear facial motion model for expression synthesis is proposed that we can get different extents of each facial expression. Moreover, with the linear model, we can obtain smooth transition between expressions and generate natural facial expression animation.

Comparing with some other approaches, our method is superior in some aspects. Firstly, the 3D model-based paradigm can obtain better performance than most 2D image-based approaches. Since changes in facial action always lead to depth variation of face surface and depth variation will bring about changes of shadow. Since it is complex and time-consuming to take the unpredictable light condition of the input image into consideration, compute the shadow change within the change of different facial expressions is difficult in a 2D paradigm solely. It is easier to involve the shadow change in a 3D paradigm using the 3D display tool like OpenGL, etc. In addition, the expression change of the face model can be observed in wider range of view using a 3D model. Secondly, the localized linear expression model in our system enabled us to generate different facial expressions both locally and globally, which enlarge the space of synthesize output and break the limitation by the limited scale of the input expression model set. Thirdly, by using the local space transform, the output expression can fit distinct facial shapes regardless of the scarcity of variation of the facial shapes in the input model set.

The rest of the paper is organized as follow: Section 2 will give an overview of our system. Section 3 introduces the reconstruction part briefly. Section 4 introduces the clustering-based region segmentation. In section 5, the element of the synthesis part including the local transform and motion model will be introduced. The synthesis results of our system and an experimentation called " Expression Imitating" will be demonstrated in Section 6. The paper finishes with a conclusion and some related future works are suggested.

2   System Overview

As depicted in Fig 1, our system consists of two stages. The first stage is 3D face reconstruction and the second is 3D facial expression synthesis. In the face reconstruction part, we use a single image of a frontal human face and a 3D reference model as input, both of which containing neutral expression. We apply an active shape model (ASM) [7] to the image to localize the face as well as to obtain the major shape of the face and the organs. Based on the key points of the image face defined by ASM and their counterparts in the reference 3D model, which are predefined by manual selection, we reconstruct the 3D model of the target face through a RBF-based scattered data interpolation process. In the stage of facial expression synthesis, we start from analyzing a series of 3D face models, each of which represents a distinct expression. Including the neutral expression face model, also known as the reference, all of the input 3D faces are registered that each vertex in each model has its unique correspondent in any other model. At first, we segment the 3D models into regions using a clustering-based method so that we can implement better control of local face features and motions. From the reference models we extract the motion matrix for each facial action. After that, the motion matrix is transformed to fit the target face model by a local space approach which aims to keep the local features of the facial expressions. Based on the motion matrix, we generate a set of motion models to represent the expressions for each face region. Finally, the motion models are utilized to deform the 3D target face with neutral expression to generate facial expressions.

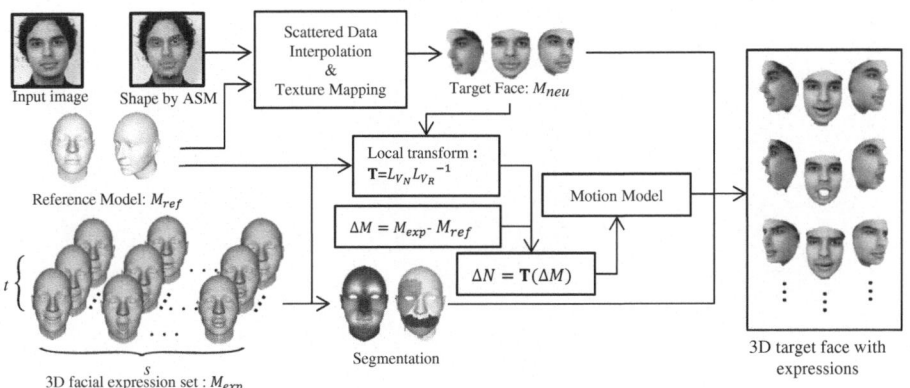

Fig. 1. System overview

3   3D Face Reconstruction

In order to obtain the major shape and the texture of the target face region, we apply ASM to the input image. The shape of the face is denoted by a vector S which consists of N key points: $S = (p_1, p_2, ..., p_N)$, where $p_i = (x_i, y_i), (i = 1, 2, ..., N)$.(Fig 2.(a)).There are three major functions of these points: (a) they define a 2D shape of the target face in the image; (b) they define the texture region of the target face; (c) some of these key points act as the clustering center for region segmentation.

To build a corresponding between the target face and the 3D model, N vertices in the model need to be found as the counterparts of the key points. We manually select the vertices in the reference model during the preparatory work. Given the key points of the target face in the image defined by the ASM and their counterparts in the model, we can construct a 3D model of the target face through a process of scattered data interpolation. Let S_{ref} denote the shape of the 3D reference face model: $S_{ref} = (q_1, q_2, ..., q_N)$ (Fig 2.(b)), where $q_i = (x_{ref_i}, y_{ref_i}, z_{ref_i}), (i = 1, 2, ..., N)$. In both the 2D image domain and the 3D model domain the origins are fixed to the key points at the tip of the nose.

The task of scattered data interpolation is to deform the model from M_{ref}, the 3D reference model, to M_{neu}, the neutral target face, according to the corresponding between S_{ref} and S. Our method is based on radial basis functions that has the form:

$$f(p) = c(p) + \sum_{i=1}^{N} \lambda_i \phi(\|p - p_i\|) \tag{1}$$

where $\phi(p)$ are radially symmetric basis functions, λ_i is the weight and $c(p)$ is a linear polynomial of the vertex coordinates. After calculating the coefficients [8], we can get the deformation of each point in the reference model through the

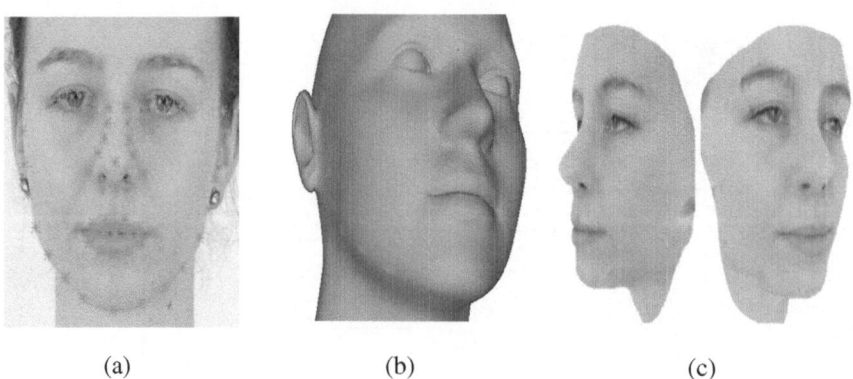

(a)   (b)   (c)

Fig. 2. Reconstruct the neutral-expression 3D face model from a single input image. (a) Obtain the facial shape in image by ASM (the green points denote the key points); (b) Shape of 3D reference model represented by 3D key points (denoted by green spots); (c) 3D target face obtained by interpolation and texture mapping.

interpolation function and thus obtain the shape of the 3D target face model with neutral expression (Fig 2.(c)).

4 Clustering-Based Facial Region Segmentation

There are two main reasons for us to conduct the region segmentation: (a) facial organs move differently in different expressions; (b) the regions can be manipulated independently so that more facial actions can be created by combination. While most of the previous works segment the model manually, we present a new method to segment the facial region automatically. In this section, a clustering based approach is described that we divide the face model into regions according to the position and activeness of each model vertex.

M_{exp} is a model set which contains various kinds of facial actions : (a) facial expressions like happy ,fear, disgust and sad ,etc. (b) local action of the face organs, i.e. stare, blink, pout, etc. (c) mouth action of different pronunciations. For each facial action, t models are included to represent different intensities (Figure 1). Assuming s kinds of facial actions are in the set, M_{exp} consists of $n = s \times t$ models: $M_{exp} = (M_{e1}, M_{e2}, ..., M_{en})$. Since each M_{ei} is well registered and scaled to M_{ref} we have a $3 \times m$ motion matrix for each model:

$$\Delta M_i = M_{ei} - M_{ref} = [\overrightarrow{\Delta v_{i_1}}, \overrightarrow{\Delta v_{i_2}}, ..., \overrightarrow{\Delta v_{i_m}}], (i = 1, 2, ..., n) \qquad (2)$$

$$\overrightarrow{\Delta v_{i_k}} = \overrightarrow{v_{ei_k}} - \overrightarrow{v_{ref_k}}, (k = 1, 2, ..., m) \qquad (3)$$

where m is the number of vertices in each model, $\overrightarrow{v_{ei_k}}$ is the kth vertex of M_{ei} , $\overrightarrow{v_{ref_k}}$ is the kth vertex of M_{ref} and $\overrightarrow{\Delta v_{i_k}}$ represents the motion of the kth vertex from M_{ref} to M_{ei} (Fig 3.(a)). The activeness of each model vertex among the expression model set M_{exp} can be evaluated with an m-dimensional vector

$$\overrightarrow{l} = (l_1, l_2, ..., l_m) \tag{4}$$

where $l_k = \sum_{i=1}^{n} |\overrightarrow{\Delta v_{i_k}}|, k = (1, 2, ..., m)$, $|*|$ is vector mode. In Fig 3.(b) we can see the activity distribution of the model vertices. A feature that manifest the information of both position and activeness of each vertex can be created as

$$\overrightarrow{f_k} = (\overrightarrow{v_{ref_k}}^{\mathrm{T}}, \mu l_k)^{\mathrm{T}}, k = (1, 2, ..., m) \tag{5}$$

where μ is a parameter controlling the effectiveness of activeness l_k . A clustering procedure is implemented to model vertices based on $\overrightarrow{f_k}$ and the result is adopted as the segmented regions of the face models (Fig 3.(c)). In experiment, a K-means clustering is adopted and the r initial clustering centers are selected manually at the positions of different organs.

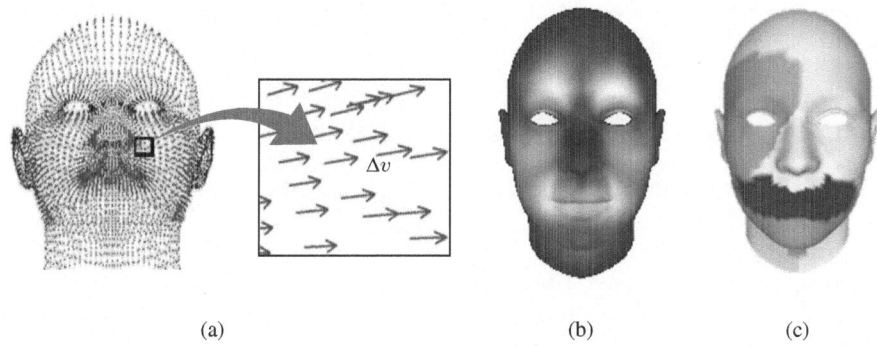

(a)   (b)   (c)

Fig. 3. Facial Region Segmentation. In (a) we illustrate the neutral expression model (blue points) and the motion vectors (green arrows); (b) is the activity distribution of the model vertices among the input expression set: the brighter color stand for higher vertex activity; (c) is the result of facial region segmentation.

5   Facial Expression Synthesis

The expression model set M_{exp} comprises s different facial actions and each action contains t models to represent different intensities: $M_{exp} = (M_1, M_2, , M_s)$ and $M_j = (M_{j_1}, M_{j_2}, M_{j_t}), (j = 1, 2, ..., s)$. M_j is the jth facial action in the set and M_{j_k} represents the model in action j with intensity k. Similarly, we have the motion matrix ΔM_i for each action j with intensity k which is represented as ΔM_{j_k}. In this section, we introduce the methods to obtain (a) ΔN_{j_k}, the motion matrix for the target face, which keep the local consistency with ΔM_{j_k} and (b) motion models for expression synthesis.

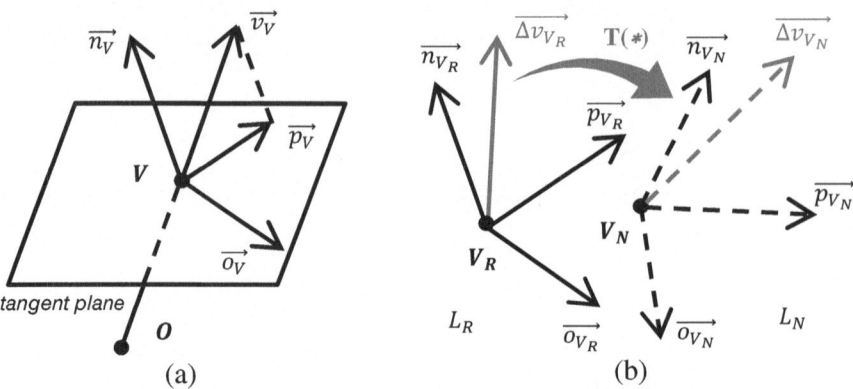

Fig. 4. Local space transform. (a) is a sketch of the local space. O is the global origin V is a vertex in the model. The tangent plane is the plane perpendicular to V's normal vector to the model surface. V's normal is defined by the arithmetic mean of the normal of the surface patches to which V belongs; (b) is a sketch of local space transform, which is designed to transfer a vector (colored in green) from a local space to another while keep the local coordinates.

5.1   Local Space Transform for Motion Vectors

According to the known motion matrix ΔM_{j_k} of M_{ref}, we can compute the correspondent motion matrix ΔN_{j_k} of M_{neu}.

At first, we build a local 3D Cartesian coordinate system (hereafter to be referred as the "local space") for each vertex V in both M_{neu} and M_{ref}. The orthogonal basis vectors of the local space are $\overrightarrow{n_V}$, $\overrightarrow{o_V}$ and $\overrightarrow{p_V}$ (Fig 4.(a)) that

$$\overrightarrow{o_V} = \overrightarrow{v_V} \times \overrightarrow{n_V} \qquad (6)$$

$$\overrightarrow{p_V} = \overrightarrow{n_V} \times \overrightarrow{o_V} \qquad (7)$$

$\overrightarrow{n_V}$ is the normal vector of the vertex V with respect to the model surface and $\overrightarrow{v_V}$ is a vector in the direction from the global origin O to V. $\overrightarrow{o_V}$ is the cross product of $\overrightarrow{v_V}$ and $\overrightarrow{n_V}$. $\overrightarrow{p_V}$ is the projection of the $\overrightarrow{v_V}$ on the tangent plane of V. In our approach, the vertex normal $\overrightarrow{n_V}$ is defined as the arithmetic mean of the normal vectors of the planes to which vertex V belongs on the surface.

For a vertex V_R in the M_{ref} the local space is represented as $L_{V_R} = (\overrightarrow{o_{V_R}}, \overrightarrow{p_{V_R}}, \overrightarrow{n_{V_R}})$ and for its counterpart V_N in M_{neu}, the local space is denoted by $L_{V_N} = (\overrightarrow{o_{V_N}}, \overrightarrow{p_{V_N}}, \overrightarrow{n_{V_N}})$. Considering a vector $\overrightarrow{\varepsilon}$ in global coordinate system, its local space representation in L_{V_R} is $\overrightarrow{\eta} = L_{V_R}^{-1} \overrightarrow{\varepsilon}$. Similarly, for a vector $\overrightarrow{\zeta}$ in L_{V_N}, its global form is $\overrightarrow{\mu} = L_{V_N} \overrightarrow{\zeta}$. Our goal is to transform $\overrightarrow{\Delta v_{V_R}}$, the motion vector of V_R, to $\overrightarrow{\Delta v_{V_N}}$, the motion vector of V_N while keep their local coordinates consistent(Fig 4.(b)). Thus we have:

$$\overrightarrow{\Delta v_{V_N}} = \mathbf{T}(\overrightarrow{\Delta v_{V_R}}) = L_{V_N} L_{V_R}^{-1} \overrightarrow{\Delta v_{V_R}} \tag{8}$$

ΔN_{j_k} is obtained by applying $\mathbf{T}(*)$ for each motion vector in ΔM_{j_k}. Noted that the global origin O is a vertex in the surface that neither $\overrightarrow{oO}$ nor $\overrightarrow{nO}$ exists, we assign identity matrix to both L_{O_R} and L_{O_N}.

5.2   Linear Motion Model for Expression Synthesis

In order to obtain expressions with more variation, a motion model is constructed for each facial action j based on the t motion matrix $\Delta N_{j_k}, k = 1, 2, ..., t$. The motion model has the form:

$$\Delta N_{j_l} = l \cdot A_j \tag{9}$$

where A_j is the coefficient matrix of the motion model (briefly as motion model) corresponding to the jth facial action. l is the intensity level of the facial action. For each facial action j, we compute the A_j on the basis of the ΔN_{j_l} with the corresponding $l = k$ by solving:

$$A_j = \underset{A_j}{\operatorname{argmin}} \sum_{k=1}^{t} \|\Delta N_{j_k} - k \cdot A_j\|^2 \tag{10}$$

The 3D facial expression is synthesized on the basis of the motion model and the reconstructed model of the target face:

$$E(\overrightarrow{\alpha}, \boldsymbol{\Gamma}) = M_{neu} + \sum_{i,j} (\alpha_i A_i + \gamma_{ij} R_{ij})$$

$$\overrightarrow{\alpha} = (\alpha_1, \alpha_2, ..., \alpha_s)$$

$$\boldsymbol{\Gamma} = \begin{bmatrix} \gamma_{11} & \cdots & \gamma_{1r} \\ \vdots & \ddots & \vdots \\ \gamma_{s1} & \cdots & \gamma_{sr} \end{bmatrix} \tag{11}$$

where E is the output of our system. M_{neu} is the neutral expression model of the target face that generated in the stage of 3D face reconstruction. $\overrightarrow{\alpha}$ is the facial action coefficients that α_i represents the intensity of the ith facial action in the model. $\boldsymbol{\Gamma}$ is the region control matrix that γ_{ij} represents the intensity of the jth region in the ith facial action. A_i is the motion model of facial action i. R_{ij} is the jth region of A_i.Different facial expressions can be obtained by designating corresponding α and $\boldsymbol{\Gamma}$ to (11).

6   Experimental Results

In our experiments, the input 3D facial expression set comprises 10 different facial actions ($s = 10$) including most of the basic facial expressions, e.g. smile, anger, blink, etc.For each facial action, 4 models ($t = 4$) corresponds to 4 different intensities are included. The face region is segmented into 5 parts ($r = 5$)

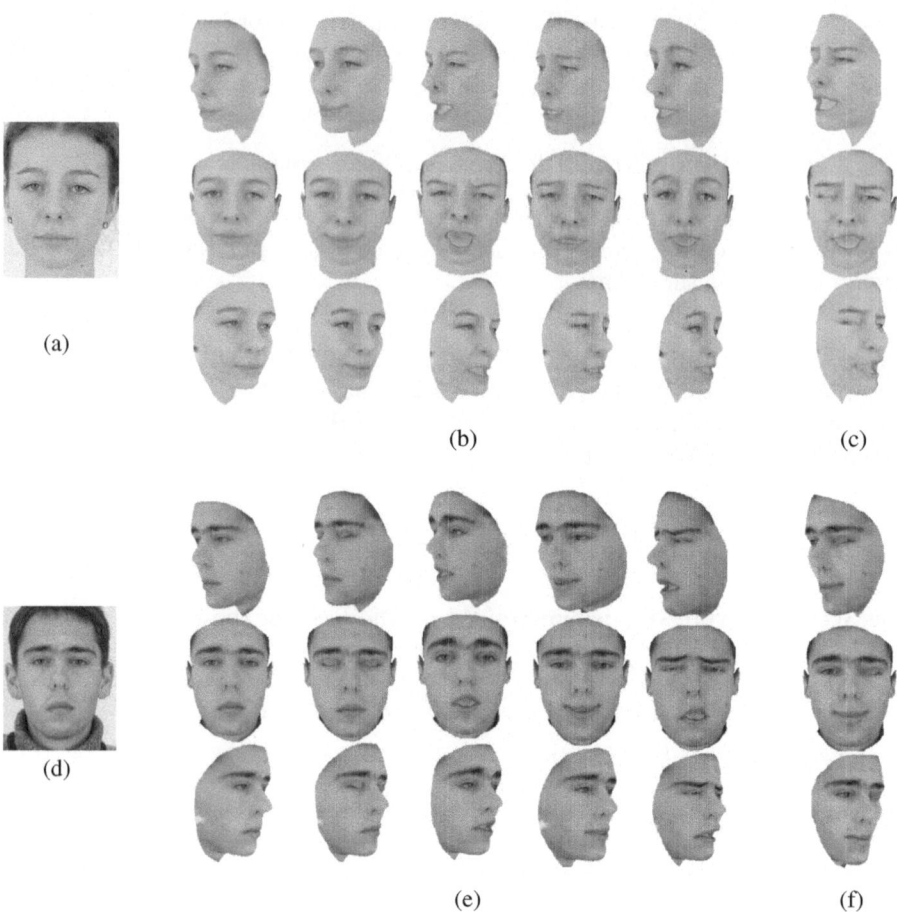

Fig. 5. Result of facial expression synthesis using images from the PUT database [9]. (a) and (d) are input images; (b) and (c) are part of the facial expressions synthesis results from image (a); (e) and (f) are part of the facial expressions synthesis results from image (d); (b) and (e) are results which presented the expressions existed in the input set while (c) and (f) are expressions that generated by region combination. Three rows of (b),(c),(e) and (f) corresponds to 3 distinct views. In (b) the expressions are neutral, smile, anger, sad, happy from left to right respectively; In (e) the expressions are neutral, blink, surprise, smile, nauseate from left to right respectively; In (c) the expression is generated by assigning the left eye with disgust, the right eye with blink, the mouth with both anger and happy. The new expression coincidentally deliver a feeling of pain, which is not included in the input expressions; In (f) the expression is generated by assigning the left eye with blink and mouth with smile, which subtly convey a feeling of frivolous.

Fig. 6. Result of "Expression Imitating". The odd columns are images of real expressions in different view angles. The 3D expression models are aligned at the right side of the corresponding images in the even columns. Images in the first row are input images and the target face with neutral expression. The parameters of the rest face models is set manually to "imitate" the image by their left. In order to facilitate the visual comparison, the model is adjusted to approximate the posture of the face in the image.

(Fig3.(c)): left eye, right eye, nose, mouth and the lower jaw. Our system has been tested in some public face databases. We conducted two experiments and both of which aim to show the effectiveness of our approach.

First, with a single input image which contains an upright frontal face with neutral expression, some certain facial expressions such as smile, blink, anger, etc. are synthesized. By assigning different values to $\vec{\alpha}$ the facial expression which is included in the input set can be easily obtained(see Fig 5.(b),(e)). On the basis of the segmented facial region, we can manipulate the facial expression locally by assigning different values to Γ (see Fig 5.(c),(f)).

Without the segmentation and local manipulation, the more complicated expressions will be unlikely to be synthesized only with the limited-numbered expressions in the input model set. By using the region segment result, we can see that some extra expressions can be simulated by regional expression combination. In Fig 5.(c), an expression conveying a feeling of pain, which is not included in the input set, has been successfully synthesized by combining the disgust left eye, the blink right eye, the anger mouth and the happy mouth in different intensity respectively.

In addition, to evaluate the ability of our facial expression synthesis system, we conduct an experiment called "Expression Imitating": a series of real facial expression pictures is involved to make a comparison to the synthesized facial expression. Our comparison procedure is based on the CMU-PIE database [10], which containing the images of some facial actions in different views. For each sample, the frontal face with neutral facial expression is adopted as input to reconstruct the target face M_{neu} . After that, we manually adjust the parameters of expression models to approximate the synthesized 3D expression model to a given image containing a certain facial expression. The result is shown in Fig 6, from which we can see that most of the expression in the database can be simulated by our system visually convincible.

From the second experiment, we can see the advantage of using a 3D model compare with the 2D image-based approach. A single 3D face model obtained by 3D face construction (the first row of column 2, 4, 6 in Fig 6) from the input image (the first row of column 1, 3, 5 in Fig 6) can generate the expressions in every possible view. In a 2D approach, the posture of the face in the image should be estimated firstly in order to get a suitable parameter. If the pose of the target changed, the parameters should be recalculated to fit the new view angle while the 3D approach need no extra computation for the parameters.

On the other hand, by carrying a lateral comparison of the synthesized result in Fig 6, we can see that the advantage of local space transform is explicit. The three target faces in Fig 6 have different facial shapes: the face in the first row has a rectangular shape; the face in the third row has a round shape and the face in the fifth row has a narrow chin. Using the local space transform, our motion model of expression which is derived from a monotonous facial shape can fit different facial shapes very well.

7   Conclusion and Future Works

We describe a facial expression synthesis system. Input of the system is a single frontal facial image. A model-based approach is utilized to reconstruct a 3D model for the face in the image. In addition, a new method of 3D face model segmentation is presented. Upon the face model segmentation and the reference model set, a new method of synthesizing facial expression by motion models is proposed. From the experiments we can see that our system works well for most common facial expressions. But there are still some work in the future to improve the system:(1) Some advanced reconstruction approaches could be introduced into our system to get a better 3D visual effect.(2) In the experimentation, we manually adjust the parameters of our system to fit the model to the input image, which remind us that some automated approaches could be introduced to uncover the pose and the expression of the face in a given image.

References

1. Blanz, V., Vetter, T.: A morphable model for the synthesis of 3d faces. In: Proceedings of the 26th Annual Conference on Computer Graphics and Interactive Techniques, pp. 187–194. ACM Press/Addison-Wesley Publishing Co. (1999)
2. Kemelmacher-Shlizerman, I., Basri, R.: 3d face reconstruction from a single image using a single reference face shape. IEEE Transactions on Pattern Analysis and Machine Intelligence 33, 394–405 (2011)
3. Pighin, F., Hecker, J., Lischinski, D., Szeliski, R., Salesin, D.: Synthesizing realistic facial expressions from photographs. In: ACM SIGGRAPH 2006 Courses, p. 19. ACM (2006)
4. Noh, J., Neumann, U.: Expression cloning. In: Proceedings of the 28th Annual Conference on Computer Graphics and Interactive Techniques, pp. 277–288. ACM (2001)
5. Pyun, H., Kim, Y., Chae, W., Kang, H., Shin, S.: An example-based approach for facial expression cloning. In: ACM SIGGRAPH Eurographics Symposium on Computer Animation, pp. 167–176 (2003)
6. Park, B., Chung, H., Nishita, T., Shin, S.: A feature-based approach to facial expression cloning. Computer Animation and Virtual Worlds 16, 291–303 (2005)
7. Cootes, T., Taylor, C., Cooper, D., Graham, J., et al.: Active shape models-their training and application. Computer Vision and Image Understanding 61, 38–59 (1995)
8. Xiong, P., Huang, L., Liu, C.: Initialization and pose alignment in active shape model. In: 2010 International Conference on Pattern Recognition, pp. 3971–3974. IEEE (2010)
9. Kasinski, A., Florek, A., Schmidt, A.: The put face database. Image Processing and Communications 13, 59–64 (2008)
10. Sim, T., Baker, S., Bsat, M.: The CMU pose, illumination, and expression (PIE) database of human faces. Carnegie Mellon University, The Robotics Institute (2001)

Face Hallucination on Personal Photo Albums

Yuan Ren Loke, Ping Tan, and Ashraf A. Kassim

Dept. of Electrical and Computer Engineering
National University of Singapore, Singapore

Abstract. This paper presents a new approach to generate a high quality facial image from a low resolution facial image, based on a large set of facial images belongs to the same person but varies in pose and expression. The input images are taken by low-end cameras or cameras from a long distance. The facial poses and expressions are not consistent and aligned. Firstly, using a low resolution facial image as a query image, a set of high resolution images with similar pose and expression is retrieved from the image examples by the proposed similarity measurement based on its shape and texture information of the query image. The selected images are then aligned with the query image and used as the candidates for the face hallucination. A Markov random field (MRF) model based on a new proposed color and edge constraints is introduced to find an optimum solution for the hallucination image. In the experiments, high textural details of hallucination images which are four to eight times larger than the original low resolution images were generated by the proposed face hallucination approach. The high resolution outputs of our method are significantly improved in quality compared to other image superresolution methods. Moreover, we also showed that our new approach is able to handle underexposure and noisy images.

1   Introduction

In digital photography era, most people have a lot of personal photos which recorded their memorable events in their personal computer. Since we are not professional photographers, some of them are not desirable. Those low quality images usually are taken by low resolution cameras such as webcams, inexpensive pocket cameras, mobile phones etc. In certain circumstance, the images have to be taken by cameras from a long distance. The subjects in these images usually are unclear due to their low resolution and poor quality of the lenses and camera sensors. Therefore, simple image interpolation approaches which enlarge the size of the images are not able to resolve the problem here. The textural details need to be enhanced but we only have a single image with unique pose and expression. However, there are a lot of other high quality images in the album. Although the expressions and poses in these images are not exactly same as the low quality image to be enhanced, these high resolution image still can be used as examples to improve the low quality images.

Image superresolution which aims to estimate a high resolution image from low resolution image(s), is a well-known ill-posed problem of long standing interest

J.-I. Park and J. Kim (Eds.): ACCV 2012 Workshops, Part II, LNCS 7729, pp. 284–295, 2013.

in image processing. It is similar to image restoration which aims to recover a good quality image from degraded images that are blurred and noisy. However, the objective of image superresolution technique is not only to recover good quality enhanced images, but also to increase size of the images. In addition, the problem of single image superresolution is more challenging.

2   Related Work

Freeman et. al. [1] proposed to perform a learning based image superresolution with a reference database with high resolution and low resolution image pairs. They used a MRF to model the relationship between the high resolution patches and low resolution patches and the relationship between adjacent high resolution patches. However, the quality of the outputs is often limited by the quality of the training patches. The results usually consist of some irregularities. Chang et. al. [15] proposed a neigbor embedding method to reconstructed high resolution image by its neighbors in the feature space. Fattal [2] proposed an image super-resolution approach to impose the edge constraint and conserve local intensities with respect to the low resolution input image. Yang et. al. [16] seek a sparse linear representation of low resolution input patch and use the coefficients of the representation to generate high resolution output. Sun et. al. [3] proposed a patch similarity measurement to select data with similar textural details to the input image for their image superresolution approach.

Single facial image superresolution is also known as face hallucination. It was first appeared in [4] by Baker et. al. It is a learning based facial image superresolution. The input image is a facial image for face hallucination. Instead of recovering all the unknown modules of image system such as motion, image alignment, noise etc., learning-based image superresolution approaches generally synthesize a high resolution image from low resolution image(s) with the help of an exemplar training set to learn a suitable model. In the learning approaches, facial features or facial models are used as the prior knowledge to preserve the facial structure. In [5,6], a MAP estimator is used to estimate the high resolution image that lies near to a PCA face sub-space. Mohammed et. al. [7] also used a PCA model and a non-parametric model for texture synthesis to generate a high resolution novel facial images. Jia and Kong [8] build a tensor model with different person, illumination, pose and expression to super-resolve the input face and synthesize a new facial expression of the face.

In [9], Joshi et. al. presented a data-driven based personal photo enhancement system. However, it is only applied on frontal images with same expression. They addressed that the image deblurring and color correction issue are required in pre-processing stage because images are obtained from different sources even thought the subject is the same person. In addition, these global correction methods are not able to enhance the image. A MRF local model with the prior images is used to enhance the image. In general, using an appropriate dataset as prior knowledge to model the input image is the key factor of the MRF model. It significantly improves the quality of the superresolution image.

Generally, data selection plays an important role in learning-based image superresolution especially face hallucination. Human being are more sensitive to the changes in the structure and textural details of facial images than other images such as building, trees, scenery images. In practice, a real-world low resolution image also consists of other image formation distortions such as motion blur, sensor noise etc. These problems have been addressed independently. Fergus et. al. [10] recovered blur kernels of the image with a gradient prior. Liu et. al. [11] estimated the noise based on a piecewise smooth image prior and the camera response functions.

In this paper, we propose to perform face hallucination on a single real-world low resolution facial image with a personal photo album which contains many high resolution images of the same person. The challenge of our problem is that the pose and expression of the facial images are not consistent. Moreover, the blurring filter, downsampling operator, and other image distortion are unknown in our inputs. In our approach, we first retrieve high resolution images with similar head pose and facial expression from the database based on the pose, face shape and textural details of eyes and mouth of the input face. The details is presented in the section 3. The selected examples are then used as candidates to generate a high resolution facial image by our proposed face hallucination approach in section 4.

3   Intelligent Image Selection

Having training images which are similar to the input image can narrow down the search space for estimating the superresolution image. It also can prevent from selecting inappropriate patches which generate the irregularities in the final output. In our proposed approach, there are three retrieval criteria based on pose, shape and texture. Firstly, the pose of input face is estimated based on its shape ratios. The input is classified into the corresponding pose category. Then, the shape and texture at each portion of face are estimated separately and represented as an expression descriptor. The training images with similar expression and pose are then selected based on the descriptor by our proposed similarity measurement.

(a)**Pose Discrimination:** Pose discrimination is used to find the facial images with similar pose. The method primarily relies on a general assumption that faces are a planar object. Each facial image is marked n facial feature points, p_i. Facial pose changes can be interpreted as the head rotation about three orthogonal axes. Since faces can be aligned by the feature points, the translation can be resolved by aligned the means of the feature points. $p'_i = p_i - \frac{1}{n} \sum_n p_i$. Changes in roll and pitch are very limited. Thus, we only need to measure the yaw changes here. To measure the similarity of yaw, a pose ratio is defined as $R_1 = \frac{l}{r}$ where l is the distance of the center of left eye to the nose tip and r is the distance of the center of right eye to the nose tip.

(b)**Shape Analysis:** Since faces are non-rigid objects, the shape structure of the faces is varied with the different facial expressions especially the region at eyes and mouth. Even though the pose of a face remains unchanged, the structure and textural details of the face can be changed significantly due to its expression. To select appropriate candidate images as training data, we need to retrieve images with similar expression. First, feature points at eye and mouth regions are extracted. The feature points at left eye region, right eye region and mouth region are denoted as $\mathbf{P}_i^l$, $\mathbf{P}_i^r$ and $\mathbf{P}_i^m$, respectively. Each set of the feature points is used to fit an ellipse by principal component analysis (PCA). The maximum spread is along the major axis of the ellipse, $\mathbf{a}$ which is also known as the first principal component direction. The minor axis of the ellipse, $\mathbf{b}$, is the second principal component direction. It is orthogonal to the first principal component direction. The two eigenvalues obtained from PCA are denoted as a and b. The ratio of them is defined as $D_S = \frac{b}{a}$. The shape ratio is scale invariant. Thus the ratio is not affected by the image resolution.

(c) **Color Constancy:** Since input images and training images are acquired from different cameras and under different lighting conditions, having a consistent color distribution of images is important in texture analysis. The color distributions of these images are very different. The skin color of the same person varies significantly according to the lighting conditions and camera sensors. To overcome this issue, a histogram equalization is applied on the input images. The images which have a similar color distribution of the training dataset are generated for the texture analysis. In our approach, the histogram equalization is applied on CIELAB color space. After input and training images are transformed from RGB to CIELAB color color space, their histograms in each color space are computed independently and normalized. The three normalized histograms from CIELAB color space can be interpreted as their color probability density functions (pdf). Assuming these color pdf are independent, the equalization transformation can be done independently. The input histogram and the reference histogram are denoted as $H(m)$, $H(n)$, respectively. Let the resolution of image denote as N and the scale is $[p_0, p_k]$, the equalization transformation $\mathfrak{T}(p)$ can be derived as $\mathfrak{T}(p) = \frac{p_k - p_0}{N} \int_{p_0}^{p} H(u) du + p_0$. It is noted that the equalization transformation is monotonic. Let $f(m)$ and $g(n)$ denote the equalized input histogram and the equalized reference histogram, respectively. $g^{-1}(n)$ denotes the inverse function of $g(n)$. Since $f(m)$ and $g(n)$ are equalized, $g^{-1}(f(m))$ transforms the input image to an image with similar histogram of the reference image.

(d)**Texture Analysis:** The objective of the facial texture analysis is to represent the local spatial information of input image and training images for similarity measurement. The Pyramid Histogram of Orientation Gradients (PHOG) proposed by Bosch et. al. [12] is used as our texture descriptor. Histogram of Orientated Gradients (HOG) [13] descriptor consists of a histogram of edge orientation gradients weighted by its corresponding magnitude within an image subregion quantized into K bins. First, the edge image is computed by Canny edge detector. Next, the gradient magnitude and gradient orientation are computed. The edge image, the gradient magnitude image and the gradient orientation image are divided into 2^l cells for l level. At each cell, the histogram of orientation gradients is computed. PHOG descriptor is concatenated HOG descriptor at each pyramid resolution level. The pyramid at level l has 2^l image subregions along each dimension or 4^l subregions in total. Each subregion is represented by a K-vector corresponding to the K bin of the histogram. Thus, the dimensionality of PHOG descriptor is $K \sum_{l=0}^{L} 4^l$ where L is the total number of the pyramid levels. Since the texture information at the center is more important than boundary region, we proposed an Extended PHOG (EPHOG) to emphasize the representation of the center subregion. At the center of the given images, another PHOG descriptor is applied on it. EPHOG is concatenated the original PHOG and the new PHOG descriptor which represents the center of the image.

3.1   Similarity Measurement

The image retrieval system aims to find a set of high resolution training images which are similar to input image in pose, shape and textural details. Firstly, the pose ratio, R_1 of input image is computed and then the corresponding training images are selected for the shape and texture analysis. To integrate the similarity of shape and texture descriptors, a similarity score is defined as

$$P(D_S^I, D_T^I|x) = P(D_S^I|x)P(D_T^I|x) \tag{1}$$

where the shape score is defined as $P(D_S^I|x) = 1 - \frac{(D_S^x - D_S^I)^2}{\max_{\{x\}}(D_S^x - D_S^I)^2}$. The D_S^x and D_S^I are the shape ratio. The interval of the shape score is $[0, 1]$. The texture score is defined as $P(D_T^I|x) = 1 - \frac{\chi^2(D_T^x, D_T^I)}{\max_{\{x\}} \chi^2(D_T^x, D_T^I)}$. D_T^x and D_T^I are the EPHOG descriptor vectors of the training image, x and input image, I, respectively. The χ^2 distance is used to measure the distance between two EPHOG decriptor vectors. The smaller χ^2 distance implies the more similar between two images. High resolution training images with high score are selected as the candidates for image superresolution. In facial images with different experessions, eyes and mouth arc two regions which vary significantly compared to other parts of the face. In our work, we select eyes and mouth regions as the query region of interest for texture analysis. Three sets of high resolution images are retrieved independently for image superresolution.

3.2   Image Alignment

Since the images are captured by different cameras, the images are unlikely aligned properly. Moreover, face is a non-rigid object. Expression changes distorts the structure of face significantly and make the alignment more difficult. To overcome this issue, we triangulate the feature points by Delaunay Triangulation and apply an affine warping on each corresponding triangle between high resolution training images and input image. The image alignment would not only ensure the input image is aligned with the training image, but also help to narrow down the search space in the image superresolution approach in the next section.

4   MRF Model for Face Hallucination

After the high resolution training images are selected and aligned, we need to match the patches in input image with the patches in these candidates subject to the constraint that the overlap region between two adjacent patches are smooth. Thus, a patch-based nonparametric MRF model is proposed to minimize the energy function.

$$E(x|\theta) = \sum_{s \in \mathcal{V}} \theta(x_s, p_s) + \sum_{(s,t) \in \mathcal{E}} \rho_{st}(x_s, x_t, p_s, p_t) \tag{2}$$

where set $\mathcal{V}$ denote the image patches obtained from the selected training data, p_s at coordinate x_s, θ_s denote the data penalty function and ρ_{st} denote the smoothness function of patch p_s and patch p_t and s and t denote the patch indices.

The input image is a poor quality low resolution image. It can be blurred, noisy, over- or underexposure image. In order to generating a visually-pleasing and aesthetically attractive facial image, a data penalty function based on the gradient and color information is proposed. In general, our eyes are more sensitive to certain colors than others. Using typical L2-norm to measure the difference between two color patches is not appropriate because our color perception is non-uniform. In addition, our eyes are also more sensitive to the region with large gradient changes such as edges. Moreover, edges are important components for preserving the structure of the face. Thus, we imposed a color cost and an edge cost into the penalty function, $\theta(x_s, p_s) = D_I(x_s, p_s)D_G(x_s, p_s)$.

The smoothness function, ρ_{st} is a constraint to ensure that the overlapping region of the adjacent patches must be as similar as possible. It is defined as $\rho_{st}(x_s, x_t, p_s, p_t) = D_I^{\Omega}(x_s, x_t, p_s, p_t)D_G^{\Omega}(x_s, x_t, p_s, p_t)$. ρ_{st} is also imposed a color cost, D_I^{Ω} and an edge cost D_G^{Ω} on the overlapping region, Ω.

4.1   Color Constraint and Edge Constraint

The color cost function, $D_I(x_s, p_s)$, is defined as

$$D_I(x_s, p_s) = 1 - \exp\left(-\lambda \Delta E_{CIE_{00}}(I(x_s, p_s), I_y)\right) \tag{3}$$

I_y is the patch extracted from the selected training images and $I(x_s, p_s)$ is the input patch to be optimized. $E_{CIE_{00}}$ is an extension of the L2-norm color difference function with five additional corrections on lightness, chroma, hue and chroma-hue interaction to resolve the perceptual uniformity issue defined by CIE [18]. The scale of D_I is [0, 1]. If two patches are similar in color, the color penalty is small. Similarly, the color smoothness function, D_I^{Ω} is defined as

$$D_I^{\Omega} = 1 - \exp\left(-\lambda \Delta E_{CIE_{00}}(I^{\Omega}(x_s, p_s), I^{\Omega}(x_t, p_t))\right) \tag{4}$$

where $I^{\Omega}(x_s, p_s)$ and $I^{\Omega}(x_t, p_t)$ are the overlapping regions of two adjacent patches.

Since we would like to preserve the edge information especially the strong edge information, an edge cost function is defined as

$$D_G = \begin{cases} 1 - \exp\left(-\lambda_1 \|G_x - G_y\|_2\right), & \|G_x\| > \varepsilon; \\ 1 - \exp\left(-\lambda_2 \|G_x - G_y\|_2\right), & \|G_x\| \leq \varepsilon; \end{cases} \quad (5)$$

where G_x and G_y are the gradient magnitude in input image I_x and training image, I_y, respectively. It is noted that $\lambda_1 \gg \lambda_2$ to preserve the strong edge. In addition, small $\|G_x\|$ is likely noise. The definition of edge smoothness function, D_G^Ω is similar to D_G. To minimize the energy function, $E(x|\theta)$, we use the sequential tree-reweighted message passing algorithm proposed in [14].

5   Experiments

Experiments for our face hallucination approach used five sets of high resolution training images. Each dataset consists of a person with different poses and expressions. Each dataset has 2000-5000 high resolution images. All these high resolution images have been labeled 38 feature points for alignment algorithm. Input images were captured from different sources such as mobile phone camera, web camera, low-end video camera etc. The size of the inputs ranges from 30×30 to 100×100. Due to the quality image, the low resolution input faces are manually labeled the 38 feature points. The superresolution images were enlarged the input images two times to eight times. Some training sample images are shown in Figure 1.

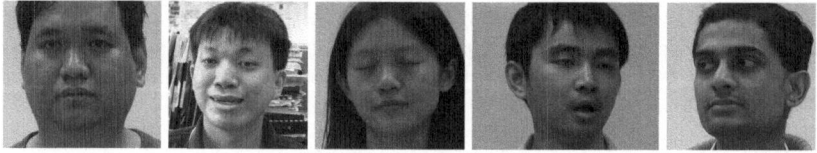

Fig. 1. High resolution training images with different expressions and poses extracted from each dataset

In the first experiment, our proposed image selection method was evaluated. Three measurements were compared in this experiment. The first approach applied the $L2$-norm to measure the texture similarity, $P(D_T^I|x)$. The second approach replaced $L2$-norm in the first approach to the χ^2 measurement. In the third approach, the shape ratio was integrated with texture analysis as mentioned in Section 3.1. Three training datasets were used in this experiment. 40 low resolution images with different poses and expressions were selected as query images for the performance evaluation. The high resolution training images in the top 20 rankings were retrieved from the databases. The retrieval rate (RR) is defined to evaluate the system. RR is defined as $RR(q) = \frac{H(q)}{N}$ where $H(q)$ is the number of ground truth images for a query image, q, found in the top N

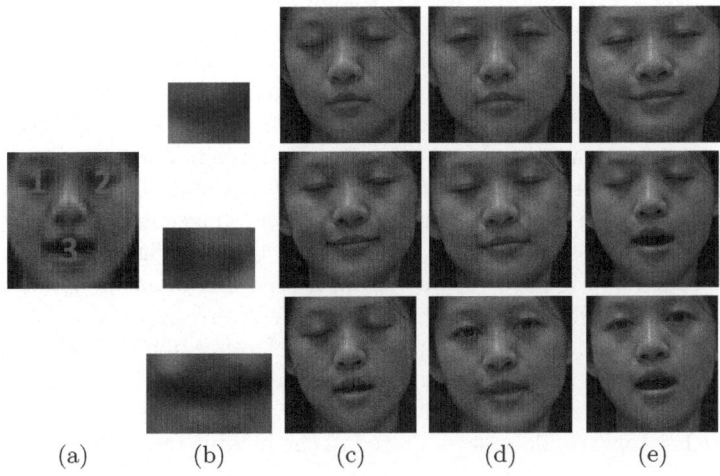

(a)   (b)   (c)   (d)   (e)

Fig. 2. (a)A low resolution query image (b)The region of interest (top to bottom: (1) left eye, (2)right eye, (3)mouth) with bicubic interpolation for image selection (c-e)The high resolution images retrieved from the training dataset

Table 1. Comparison results on the proposed similarity measurement

	$L2$-norm	χ^2	$\chi^2 + D_S$
Open Eye	88.33%	70.25%	**90.75%**
Closed Eye	52.25%	61.67%	**67.67%**
Open Mouth	10.00%	77.67%	**84.67%**
Closed Mouth	28.33%	81.50%	**91.50%**

retrieved images. In the experiment, the performance of $L2$-norm measurement was inconsistent compared to χ^2 measurement. The combination of shape and texture analysis boosted the accuracy of the retrieval system. The closed eye did not work well in all approaches because the extracted region is too small that the PHOG descriptor and shape ratio did not work properly. However, 67.67% of the retrieval rate was still acceptable for our image superresolution application. A query image and their corresponding retrieved high resolution images for different portions of face are shown in Figure 2. The results are shown in Table 1.

In the second experiment, we compared the difference between bicubic interpolation and our proposed method. We also evaluated the importance of our image selection method and alignment algorithm here. Three training datasets were used for the experiment. Low resolution images with different poses and expressions shown in Figure 3 (a) were used as the input images. Each input image was manually marked the 38 feature points due to the limitation of the feature detectors on low resolution images. The input image also applied the proposed color constancy method. Next, the input face was divided into three part, namely right eye region, left eye region and mouth region. The eyes and mouth region were used for texture analysis. 20 high resolution images were retrieved for each region of interest from the corresponding dataset by the proposed image selection algorithm. The retrieved images were then aligned with the low resolution input images by their corresponding 38 tracked feature points. The MRF based superresolution approach presented in section 4 was applied. The sample results with eight time magnification are shown in Figure 3 (e). Figure 3 (c) applied the proposed superresolution approach but the reference images are randomly selected and the alignment presented in section 3.2 is not applied. Figure 3 (d) applied the similar approach as Figure 3 (c) but the proposed alignment is applied. The intelligent image selection plays an important role in our approach. The proposed image selection method gave the appropriated reference images for our face hallucination approach to generate the high resolution image with the correct pose and expression. The superresolution images not only preserve the facial structure, but also have richer textural details compared to the bicubic interpolated images in in Figure 3(b).

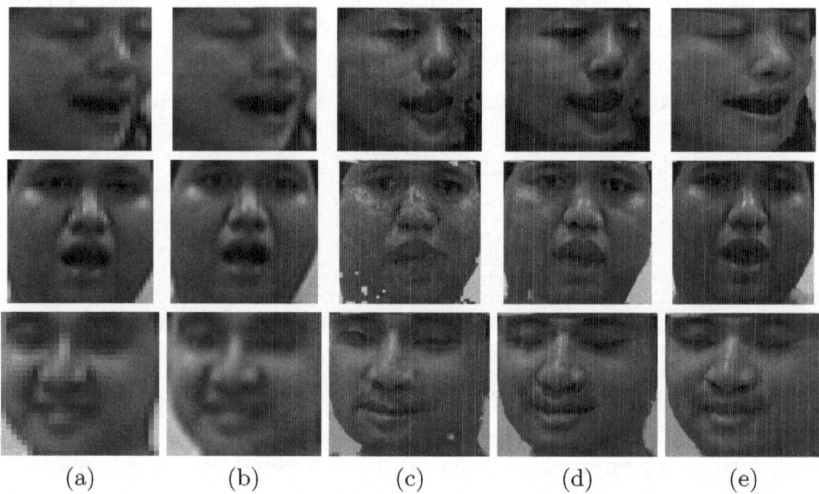

(a)   (b)   (c)   (d)   (e)

Fig. 3. (a)Input image (b)Bicubic interpolation (c)Superresolution image by randomly selected high resolution patches (d) Superresolution image by randomly selected the patches from the aligned images (e)Superresolution image by selected patches from the aligned images with similar pose and expression for ×8 magnification

In the third experiments, we compared the outputs of our method with the bicubic interpolation method, the VISTA approach [1], neighborhood embedding method [15], fast image upsampling method [17] and the sparse representation method [16]. Figure 4 and Figure 5 showed the comparison results for ×4 magnification and ×8 magnification respectively. The input images were captured by real-world low-end cameras which the blurring filter, down-sampling operator,

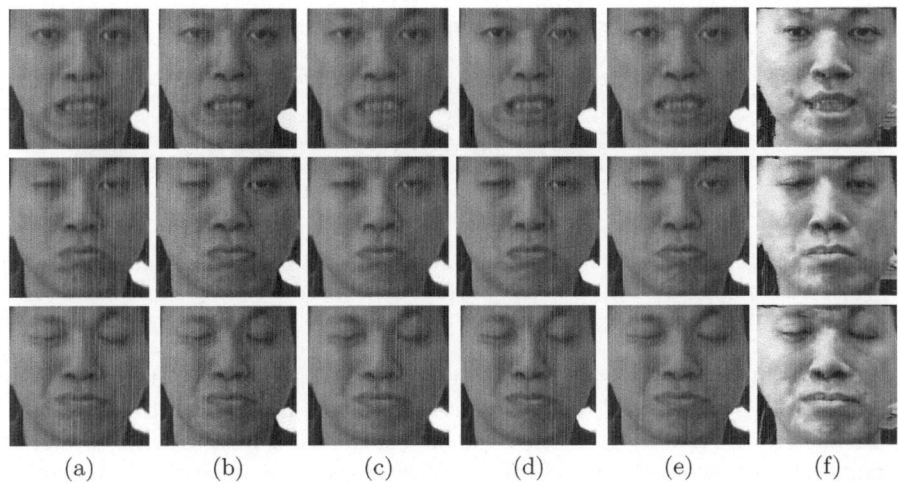

Fig. 4. (a) Bicubic interpolation (b)The VISTA approach [1] (c)Neighborhood embedding method [15] (d)Fast image upsampling method [17] (e) Sparse representation method [16] (f)Our method for ×4 magnification

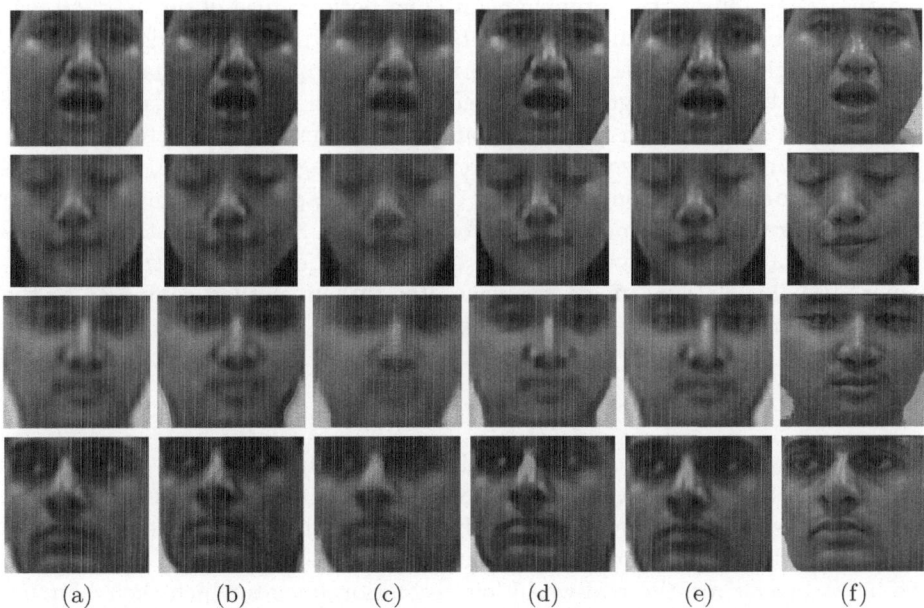

Fig. 5. (a) Bicubic interpolation (b)The VISTA approach [1] (c)Neighborhood embedding method [15] (d)Fast image upsampling method [17] (e) Sparse representation method [16] (f)Our method for ×8 magnification

other image distortions and noise are unknown. Those effects are also very difficult to predict from a single low resolution image with the limited information

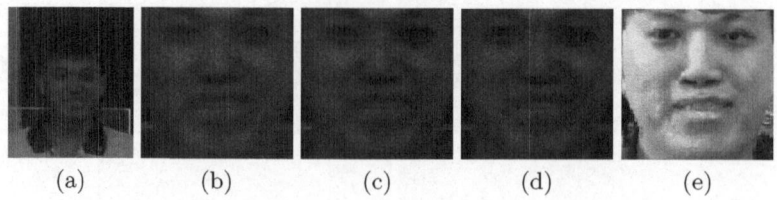

(a)   (b)   (c)   (d)   (e)

Fig. 6. (a) Input image (b)Bicubic interpolation (c)Neighborhood embedding method [15] (d) Sparse representation method [16] (e)Our method for ×2 magnification

about the cameras. The results by the VISTA approach showed in Figure 4 (b) and Figure 5 (b) consist of some artifacts. Using the implementations provided by Chang et. al. [15], we generated the superresolution images with 4x and 8x magnification shown in Figure 4 (c) and Figure 5 (c) respectively. Similarly, we also generated the corresponding superresolution images with 4x and 8x magnification by work in Shan et.al. [17] and Yang et. al. [16]. The results were shown in Figure 4 (d-e) and Figure 5 (d-e) respectively. Compare to our results shown in Figure 4 (f) and Figure 5 (f), our method significantly improved the quality and the resolution of the output images. The performance of our method was outstanding because our method do not require the prior knowledge about the image formation. Moreover, other superresolution approaches enhance interpolated images by adding high frequency details but our method replaces the poor quality input patches with the high quality patches from the selected images.

In the fourth experiment, an underexposure image with pepper noise was used for face hallucination. It is very commonly captured by an inexpensive pocket camera under the poor lighting environment. Since our method replaces the noisy image patches with the high quality patches from the training data, a more visually pleasing image was generated by our method compared to the other methods. The results are shown in Figure 6.

6   Conclusions

In this paper, we proposed a face hallucination algorithm based on a nonparametric MRF model with a novel color and edge constraints for personal photo albums. The challenge of the problem is that the facial poses and expressions in input images are not always same as the training images. In addition, the input images are the real-world low resolution images which their blurring filter, downsampling operator, other distortion effects and noise are unknown. In our approach, we proposed an image selection method based on texture and shape analysis to select the high resolution images with similar pose and expression from the high resolution training dataset. From the experiments, we showed that the retrieval rate is more than 84% in most of the cases. Our face hallucination approach is able to magnify the faces four to eight times. Moreover, it also worked on underexposure and noisy images.

References

1. Freeman, W.T., Pasztor, E.C., Carmichael, O.T.: Learning low-level vision. International Journal of Computer Vision 40(1), 25–47 (2000)
2. Fattal, R.: Image upsampling via imposed edge statistics. ACM Trans. on Graphics 26(3), 95 (2007)
3. Sun, J., Zhu, J.J., Tappen, M.F.: Context-constrained hallucination for image super-resolution. In: Proc. of CVPR, pp. 231–238 (2010)
4. Baker, S., Kanade, T.: Hallucinating faces. In: Fourth International Conference on Automatic Face and Gesture Recognition, pp. 83–89 (2000)
5. Capel, D., Zisserman, A.: Super-resolution from multiple views using learnt image models. In: Proc. of CVPR, vol. 2, pp. 627–634 (2001)
6. Liu, C., Shum, H.Y., Freeman, W.T.: Face hallucination: theory and practice. International Journal of Computer Vision 75(1), 115–134 (2007)
7. Mohammed, U., Prince, S.J.D., Kautz, J.: Visio-lization: generating novel facial images. ACM Trans. on Graphics 28(3), 57 (2009)
8. Jia, K., Gong, S.G.: Generalized face super-resolution. IEEE Trans. on Image Processing 17(6), 873–886 (2008)
9. Joshi, N., Matusik, W., Adelson, E.H., Kriegman, D.J.: Personal photo enhancement using example images. ACM Trans. on Graphics 29(2), 12 (2010)
10. Fergus, R., Singh, B., Hertzmann, A., Roweis, S.T., Freeman, W.T.: Removing camera shake from a single photograph. ACM Trans. on Graphics 25(3), 787–794 (2006)
11. Liu, C., Freeman, W.T., Szeliski, R., Kang, S.B.: Noise estimation from a single image. In: Proc. of CVPR, pp. 901–908 (2006)
12. Bosch, A., Zisserman, A., Munoz, X.: Representing shape with a spatial pyramid kernel. In: Proc. of the 6th ACM International Conference on Image and Video Retrieval, pp. 401–408 (2007)
13. Dalal, N., Triggs, B.: Histogram of oriented gradients for human detection. In: Proc. of CVPR, vol. 1, pp. 886–893 (2005)
14. Kolmogorov, V.: Convergent tree-reweighted message passing for energy minimization. IEEE Trans. on Pattern Anal. Mach. Intell. 28(10), 1568–1583 (2006)
15. Chang, H., Yeung, D.Y., Xiong, Y.: Super-resolution through neighbor embedding. In: Proc. of CVPR, vol. 1, pp. 275–282 (2004)
16. Yang, J., Wright, J., Huang, T.S., Ma, Y.: Image super-resolution via sparse representation. IEEE Trans. on Image Processing 19(11), 2861–2873 (2010)
17. Shan, Q., Li, Z., Jia, J., Tang, C.K.: Fast image/video upsampling. ACM Transactions on Graphics 27(5), 153 (2008)
18. Sharma, G., Wu, W., Dalal, E.N.: The CIEDE2000 color-difference formula: Implementation notes, supplementary test data, and mathematical observations. Color Research and Applications 30(1), 21–30 (2005)

Techniques for Mimicry and Identity Blending Using Morph Space PCA

Fintan Nagle[1,2], Harry Griffin[1], Alan Johnston[1,2], and Peter McOwan[2,3]

[1] University College London
[2] UCL CoMPLEX
[3] Queen Mary University of London

Abstract. We describe a face modelling tool allowing image representation in a high-dimensional morph space, compression to a small number of coefficients using PCA[1], and expression transfer between face models by projection of the source morph description (a parameterisation of complex facial motion) into the target morph space. This technique allows creation of an identity-blended avatar model whose high degree of realism enables diverse applications in visual psychophysics, stimulus generation for perceptual experiments, animation and affective computing.

1   Introduction

The human face is host to one of the highest-bandwidth channels of natural communication that can exist between two people. Facial expressions pass information from brain to brain through a sequence of processes: motor neuron spike trains, muscle activation, deformation of the facial surface, then via visible light to the observer's optic nerve and visual channels. Each medium encodes information in a different modality. In this work we focus on the development of novel technical methods to allow us to explore the expressor's and the observer's high-level neural codes and how they mediate identity perception.

Consider a high-resolution portrait video V of a distinctively recognisable dynamic expression sequence, and a second video M in which an actor is mimicking that expression. Intuitively, we can cognitively represent the *expression* and the two *identities* separately, because we can tell that the expression is constant and we can recognise the two people present.

In reality, identity and expression may not be easily dissociable. The brain's distributed face processing system[2] may code expressions in terms of faces it has already witnessed, or it may store them with reference to some average face, conflating expression representations with the identity of the average. Identity may also be coded from expression, as when we recognise someone from characteristic facial action. Previous efforts at face transfer, such as[3], have mainly focussed on transferring identity, not expression.

There is some evidence from primate studies[4] and clinically observed double dissociations[5] that processing pathways for identity and expression are separate. However, these two components are certainly not separable by simple

J.-I. Park and J. Kim (Eds.): ACCV 2012 Workshops, Part II, LNCS 7729, pp. 296–307, 2013.
© Springer-Verlag Berlin Heidelberg 2013

statistical techniques such as PCA. Although identity is responsible for more variance than expression, naive PCA conflates identity and expression in components of high variance[6].

Expression often involves motion[7–9]. The motion of the facial musculature induces textural changes in a 2D image due to deformation and lighting effects. Indeed, expression can be considered as change in a high-level representation of facial structure. This can be perceived either from static or moving images, and cognitive representations of identity may describe static or dynamic percepts. Attribute judgements from static and dynamic images may not be completely separate; it is possible that a static image might be used to seed a dynamic cognitive model. For example, viewing a static image of a person throwing a javelin can give a certain percept of acceleration and motion.

Our goal is the creation of diverse psychophysical stimuli and the design of experiments investigating whether discrimination and recognition are possible from motion alone. We describe two techniques aimed at retaining the dynamic characteristics of expression while manipulating facial form.

Firstly, we show how a PCA morph model can enable the projection or mimicry of source expressions onto a target face model. This allows the accuracy of natural mimicry to be measured by comparing the principal component loadings of an original video clip, to those of a mimicked clip, projected into the original actor's expression space. One can imagine quantifying mimicry improvement during learning to shed light on how the developing brain learns facial expressions through imitation[10].

Secondly, we leverage expression projection to create an identity-blended controllable morph model, or avatar, with an identity that is a composite of several sources. This enables the creation of identity-balanced stimuli.

The face space formalism[11] has proved an intuitive and useful guess at the brain's internal representation of facial characteristics. The naïve version[12] applies principal component analysis directly to raw image data in image space, producing a new coordinate frame whose axes correspond to the directions of greatest variation in the original data.

The PCA technique can be extended, firstly by including morphological data describing faces' shape and secondly by modifying the representation of each face so that it is relative to a mean. A series of portrait photographs can thus be used to create a coordinate frame which efficiently expresses the variation present in the input set. For ease of usage we will simply call this frame, face space, recognising the more generic use of this term elsewhere in the literature[11]. When the input image set comprises diverse configurations of one person's face, the resulting PCA space is effectively a controllable model of that face.

In the study of the perception of dynamic facial expression it would be very valuable to be able to separate facial motion from facial shape by mapping the motion onto an average face. However, generating an average dynamic avatar is non-trivial, requiring extensive software engineering efforts[13, 14], and 3D graphic techniques often suffer from a lack of photorealism leading to the uncanny valley effect[15].

One cannot simply average up multiple faces performing some action under instruction as the timing of the behaviour may differ radically between people, leading to temporal misalignment and temporal blur. We have developed a novel method to circumvent this problem. First we build individual expression spaces for multiple actors. We then project frames from a sequence into the multiple expression spaces and average over the result. The sequence of averages is then processed by PCA to provide a photorealistic mean expression space.

2   PCA Modelling of Morphed Faces

This section describes in detail the pre-processing and PCA operations necessary to transform a set of input images into a face model.

We begin with a set of n $h \times w$ input images I_i showing facial portraits of one person effecting different expressions. One image is chosen as the reference image I_r. Each image I_i is compared with the reference image using a gradient based motion estimation algorithm(the McGM[16, 17]) which produces, for each image, a full vector field (one vector per pixel) providing an estimate of the motion between the reference and target images. This is effectively a dense registration relation between I_r and each I_i. A vector from a point on I_r shows the new location of that point on I_i, and we refer to one such set of vectors as a warp field.

Each remaining image is represented by its difference from the reference image. I_r should show a neutral expression, with the eyes open and the mouth slightly open, showing the teeth and a small black area between them (so that the McGM can find a way to warp the reference image to reconstruct any dental or buccal features in the remaining images).

The multichannel gradient model (McGM[16, 17]) is a bio-inspired algorithm which calculates a basis set of spatio-temporal derivatives by convolving the image sequence with derivative of Gaussian filters, and then combines them to form derivatives of the Taylor expansion in space and time. Ratios of the resulting terms then yield robust estimates of image motion between the reference image and each additional frame. In practice, each pair (I_r and one I_i) are converted to greyscale and subsampled at several different resolutions before submission to the McGM. The resulting lower-resolution warp fields are combined into one field of the same size as I_i, which gives better results than single-scale motion analysis.

In practice it is useful to constrain the input to some degree. Rigid head movement (translation or rotation) should be kept to a minimum, either by recording protocol or image registration by face detection (good results have been obtained with the commercially available toolkit FaceAPI[18]), so that the warp fields represent primarily changes in expression and not head movement. The background should be a uniform colour so that it does not contribute to the final model. Exposure and white balance should be kept constant during recording. Once warp fields have been obtained for all images, the vectors at each pixel are averaged to give the mean warp field.

Each image I_i is now expressed as two components (a similar dissociation to Blanz and Vetter's separation of 3D shape and texture[19]):

- Texture T_i: an image showing the textural component of the face shown in I_i. Anatomical points on textures are aligned with each other and with the mean face, as T_i has been warped from I_i to align it with the mean face.
- Warp W_i: a warp field (full vector field the same size as T_i) showing how T_i should be warped in order to realign it with the expression shown in I_i; see Fig. 1(c). Operationally, the warp field is represented not by the relative displacement of each pixel (a true vector coding) but as the new position of each pixel (an absolute coding).

This representation decouples textural from configurational information, which is useful. This means each can be be manipulated separately, as when the warp field is amplified to caricature an expression[20].

Once each image I_i has been represented as a morph pair $\{T_i, W_i\}$, the data are nearly ready for principal component analysis. Two final preprocessing steps are performed: serialisation and mean relativisation.

As PCA operates on vectors of reals, not more complex data structures, each morph tuple is serialised into a morph vector. This is done by iterating column-wise across the elements of the texture's colour channel matrices (R, G and then B), followed by the x and y warp field matrices, and concatenating them into a vector.

The final vector is of length $w \times h \times (2 + 3)$, as each pixel in a $w \times h$ image is linked with 2 reals coding the x and y components of W_i and 3 reals coding the RGB colour channels of T_i. A morph vector contains all the information necessary to reconstruct a face image, which is done by deserialising the vector, displacing the texture pixels by the warp field, and interpolating. This process is termed morphing (after the special effect[21]) and gives a very high-fidelity reconstruction of I_i. The space of all possible morph vectors we refer to as morph space, and it is a superset of that of all possible images (image space) as every image can be exactly represented in morph space by a texture with zero warp field.

We thus obtain n morph vectors M_i. The mean morph M_m is found and subtracted from each M_i, giving mean-relative morph vectors R_i. This operation models the assumption that identity changes are cognitively encoded in terms of difference from a stored mean, although this property would of course be implemented differently in the neural substrate than in the model's source code. The relation $M_i = R_i + M_m$ splits each absolute morph vector (equivalent to a face image) into a constant part and a variable (relative) part. Fig. 1(b) illustrates the texture component of an example relative morph vector.

The relative morph vectors R_i are submitted directly to principal component analysis, which finds a new orthogonal coordinate frame such that, iteratively, each new axis encodes the maximum possible remaining variance in the data. We term the new frame a PCA space or expression space[1]; dimensionality is

---

[1] Face and expression spaces are mathematically identical, but expression is constant in the former and identity is constant in the latter.

  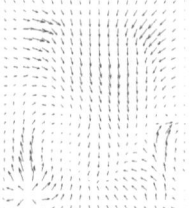

(a) One frame from a finished avatar.

(b) The (rescaled) relative texture of an I_i.

(c) Quiver plot showing every 5th vector of a warp field.

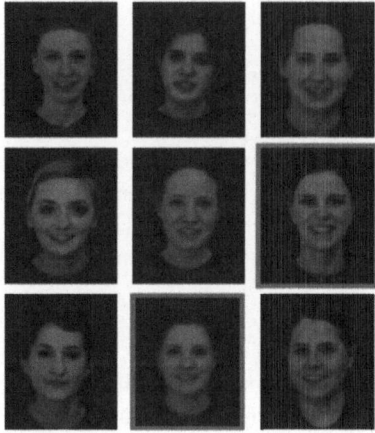 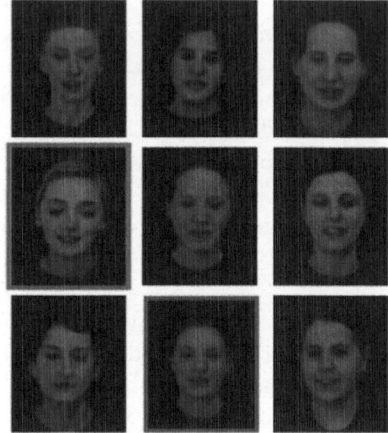

(d) Avatar generation. The red-bordered image is mimicked by seven morph models, producing the seven non-bordered images. These eight images with similar expression are then morph averaged, producing the blue-bordered image, which is an identity-blended frame of the final avatar.

(e) Another example of the generation of a blended-identity frame, this time showing a non-smiling expression with eyes closed.

Fig. 1. Components of a morph-modelled image and the avatar generation process

normally between 50 and 100, which allows for very accurate image reconstruction while retaining a high degree of compression (for 200×240 images, morph space dimensionality is $200 \times 240 \times 5 = 240,000$, so a 100-d expression space has a compression factor of 2400).

Faces can now be expressed in terms of their coordinates in expression space, which is to say their loadings on the principal component axes. Passage from expression space to morph space is done by multiplying by a matrix encoding the embedding of the expression space reference frame in relative morph space, which we term the expression space matrix P. Reconstruction of an image from loadings l involves

1. Projection of expression space coordinates into relative morph space: $R = P \times l$, where R is the relative morph space coordinate vector.
2. Addition of the morph mean to generate absolute morph space coordinates $M = R + M_m$
3. Image reconstruction by applying warp to texture and interpolating.

A full PCA face model consists of the expression space matrix P, the morph mean M_m and the variances of each principal component (used to generate faces in a realistic probability distribution).

3   Expression Projection

The twin decouplings of the morph space paradigm permit a very useful operation: the projection of an expression from one face model onto another. Consider two PCA models A and B generated from input images of two different people performing approximately the same sequence of expressions (or similar expressions in a different order).

One might imagine that if facial morphologies are similar, the warp fields for each expression will be similar. If lighting conditions and skin tones are similar, the texture components will also be similar. Once mean warp fields and textures have been found and subtracted, similarity of the relative warps and textures depends only on similarity of actual expressions, not underlying facial attributes (which are subtracted by the relativisation process).

The principal components (in other words, the orientation of expression space in image space), however, are not guaranteed to be similar, as PCA may alter the sign and ordering of components[1]. We therefore cannot rely on taking an image I_a of person A, calculating its PC-loadings using A, and passing these loadings to model B, to reconstruct an image of person B exhibiting a similar expression. This process would rely on the two expression spaces being similarly oriented in common image space (they do not need to have close origins, since the correct morph mean for each model is subtracted and added in each case). In practice, axis signing and ordering may be different, and simply transferring the loadings does not always result in effective expression mimicry.

A more robust mimicry method is the following:

1. Start with an image I_a of person A, along with a PCA space A for that individual and a PCA space B for person B. We have the morph means M_{ma} and M_{mb}.
2. Encode I_a as an absolute morph vector M_a and then a relative morph vector R_a with ($R_a = M_a - M_{ma}$).
3. Take the inner product of R_a with the PCA matrix for person B (P_b). This gives a set of PC-loadings in person B's expression space E_b. We have $L_b = R_a \times P_b$.

This step in effect calculates the best possible representation of the variable component of I_a (encoded by R_a) in E_b, the expression space for person B. Even if E_b is very differently oriented from E_a, R_a will still be sensibly reconstructable, even though its PC-loadings in E_b may be very different than in E_a. The requirement that E_a and E_b be similarly oriented is thus removed.

4. Generate a new relative morph vector R_b by multiplying the calculated loadings L_b by person B's PCA matrix: $R_b = P_b \times L_b$. This represents the reconstruction of R_a (the variable component of I_a) in E_b.

5. Add the mean morph vector M_{mb} of person B. This represents combining the constant component of model B with the variable component transferred from I_a. We are in effect applying the expression of I_a to the face of person B. We have $M_b = R_b + M_{mb}$.

6. Reconstruct an image I_b from M_b by applying the warp component to the texture component and interpolating.

This procedure is problematic when the source face is a very different shape from the target face, as projecting the source relative morph vector into the target PCA space will lead to an unnatural expression on the target (consider applying the warp of a smile exhibited on a wide, short face to a narrow, tall face). This mismatch can be avoided by transforming the warp and texture components of the morph vector before projection.

This is done by manually placing 3 keypoints on the morph mean images (which are generated during PCA modelling) for source and target identities, one at the centre of each eye and one at the centre of the philtrum. These define a triangle termed the faceframe. An affine transformation between source and target faceframes is defined and applied to the relative warp field and texture; this aligns them with the target faceframe, rendering it meaningful to add them to the target's morph mean. With 3 keypoints, only the eyes and philtrum are perfectly aligned, but the improvement is still substantial.

Note also that the projected vector is mean-relative. Adding the projection back on to the target's morph mean therefore only makes sense if the means describe a similar expression. For example, if the source's mean shows an open mouth and the target's mean a closed mouth, a warp field showing a closed mouth will contain much more motion if it is relative to the source. Relative to the target, it will be nearly zero, as the mean's mouth is already closed. Care must this be taken to ensure mean similarity. We found that including a diverse range of facial motion in the data worked well.

4   Avatar Generation

The projection process, as it allows expressions to be replicated across different identities, enables another useful technique: the fusion of PCA models of several different people into one PCA model depicting an artificial identity. Conceptually and aesthetically, the physical characteristics of this new face are a blend of those of its ingredient faces. Mathematically, the new face is generated by morphing together ingredient faces.

The naive mathematical implementation of blending is finding the mean in a representational space. As Galton found[22], this approach does not succeed if we average in image space, since not every point corresponds to an image. Direct averaging produces ghosted images which are textural but not anatomical hybrids. To generate sequences of realistic faces for mimicry animation we must therefore use morph space, in which many points in the subspace spanned by real data (input faces) correspond to anatomically plausible faces. Representing static images in morph space and finding the mean implements identity blending (like the classic morph special effect[21]) in static images. The same can be done for a PCA model by statically blending different expressions and then applying the PCA modelling procedure to the resulting images.

The following is a detailed description of our avatar generation process, as illustrated by Fig. 1(d).

1. We begin with k identities, each set S^k containing n^k portrait images of a particular person. These are subject to the same constraints on alignment and image characteristics as the single-identity PCA process described earlier.
2. A reference image r^k is chosen for each identity, subject to previously described constraints.
3. A PCA model is generated for each identity, as described. Each model brings with it a mean morph vector and an associated morph mean image.
4. Each morph mean image is displayed and 3 keypoints are manually placed, one at the centre of each eye and one at the centre of the philtrum. These define a triangle termed the faceframe f_i.
5. Each input image I_i^k is projected into $k-1$ other identities, producing k copies of the same expression, which form the set e_i^k. During projection, the morph vector is affinely transformed to bring its faceframe into line with that of the target identity.
6. Generated images are sharpened by convolving with an unsharp filter[23].
7. As images in the set e_i^k have been projected into different faceframes, they are not aligned, and so a second transformation is performed (this time on image data only, as the generated images have no warp component). Each image in each e_i^k can be transformed onto either a) a reference faceframe f_r (chosen arbitrarily from among the k models) or b) the faceframe corresponding to the source model for this expression (f_k).
8. Once images in each e_i^k have been aligned, they undergo the standard modelling process, but without the final step of PC analysis. In other words, they are each motion-compared with a reference image (we choose the projection onto the reference identity I_r); warp fields are generated and resourced to point from the mean warp; textures are reverse warped to align them onto the mean warp; fields and textures are assembled into morph vectors; and the mean morph vector is found. As each e_i^k contains images of the same expression across multiple identities, the mean morph image will also exhibit

that expression. Its identity, however, will have been morphed into a blend of the k original identities.

9. The standard PCA modelling process is run on the set of blended images from each e_i^k; as this only happens once, a new reference image can be chosen in order to obtain the best warp fields. Alternatively, the reference image of the reference model can be used. We obtain a PCA model with a wide range of emotional diversity and a common blended identity. Its final appearance is illustrated in Fig. 1(a).

5   Discussion

We have described a PCA modelling process allowing compression of face images into a small number of expression space coordinates (principal component loadings), projection of expressions from one face model to another, and generation of an identity-blended avatar.

The avatar generation technique outlined is robust across diverse facial expressions and variations in facial morphology. Realism can be reduced, however, when the McGM is not able to accurately describe facial deformation by warp fields in the initial PCA stage (due to unsuitable illumination or non-smooth facial features such as glasses, piercings or facial hair) or where face shapes are different enough to make alignment difficult during the two transformation stages. As affine transforms are only done based on three keypoints, corresponding facial features will not always be brought into alignment. This is only a major problem if identities vary greatly in head size.

We envisage that future work could compensate for this problem by automatically defining more keypoints using commercially available face recognition software and transforming by arbitrary warping instead of affine transformation, as long as anatomical realism is maintained. The addition of subexpression spaces which separately model individual features could allow constraints such as rigidity to be placed on specific features such as teeth and eyes. It would also allow comparison of the current (holistic) model with a local feature encoding scheme.

Mimicry also requires evaluation, which is only possible by human subjects. Meaningful expression transfer cannot be guaranteed without a full 3D model of the underlying facial musculature and skin deformation; a user evaluation of expression realism and accuracy of mimicry would be enlightening.

5.1   Applications of Mimicry

Our technique is nearly fully automated, requiring only one placement of three keypoints per identity (not per frame). We have also produced good results using completely automatic keypoint detection. Previous approaches required a large amount of manual warping[24] or manual feature coding[25]

Computational mimicry has two main uses: generation of face stimuli which imitate others, and characterisation of real mimicry by subjects. Given source morph vectors and a target PCA model, expressions can easily be projected from source to target. This can be done either for static faces or, by separately projecting each individual frame, for video sequences. Degrees of caricaturing can easily be applied. Such stimuli could form part of diverse psychophysical protocols, such as finding the detection threshold for erroneous mimicry or investigating to what extent caricaturing improves recognition. The ability to project expression means that it can be kept constant, making it easier to isolate effects due to changes in identity per se.

Natural mimicry can also be measured. Consider a protocol during which a subject is asked to mimic a portrait video sequence while a portrait recording is made. We can project the mimicry sequence from the subject into the PCA space of the stimulus, giving a sequence of PC loading vectors in the same expression space as the stimulus. This can be compared to the loading sequence of the mimicked video, either visually (by viewing the stimulus and its projection side-by-side) or using spatial distance measures or information theoretic measures such as mutual information or Shannon entropy. This would allow measurement of the distributed[2] face processing system's imitation accuracy.

The morph space PCA strategy is an intuitive, untrained way to measure mimicry in either human faces or artificial stimuli[26]; it could also be used to benchmark other algorithms that perform imitation. The morph vector framework presents opportunities for extension such as the inclusion of local feature characteristics or high-level data concerning illumination or head direction.

5.2   Generating Stimuli with the Avatar

The ability to generate a blended avatar renders it possible to generate realistic stimuli situated on the plausible side of the uncanny valley[15] (especially in terms of skin tone) but free from real-world identity. This avoids problems of privacy and ethics compliance (as the avatar is not a real person) and experimental bias (towards a particular identity). During the blending step, the output could be shifted away from the mean morph to change the appearance of the avatar, generating different identities for use in interactive software applications, games, or media.

The more source identities an avatar is built from, the more attractive the resulting PCA model will appear[27]. This could be leveraged in applications where positively-connoted faces can lend an advantage, such as animated user interface agents or virtual newsreaders[28].

In psychophysics, the avatar allows identity to be blended arbitrarily between different models. This allows generation of stimuli for experiments which explore identity perception.

The low computational demand of the PCA morph model also allows stimuli to be dynamically generated during an experiment. We can imagine guiding a subject to perform a search in face space (or expression space) by sequentially choosing between several different frames, the next set of frames being generated from coefficients in the direction indicated by the subject's choice.

5.3   Conclusion

Although identity and dynamic expression may not be dissociable or independently encoded in the face processing system, they are to a certain extent computationally dissociable, thanks to the encoding of dynamic expression as a high-level (PCA) mean-relative representation of motion. When the input data are constrained (in terms of lighting, head rigidity and image quality) such that a motion evaluator can generate good-quality warp fields, this dissociation allows motion and expression to be kept constant across multiple identities. In turn, this technique allows generation of sets of images with similar expressions which can be identity-blended. Both techniques enable the creation of useful psychophysical stimuli.

References

1. Jolliffe, I., MyiLibrary: Principal component analysis, vol. 2. Wiley Online Library (2002)
2. Haxby, J., Hoffman, E., Gobbini, M.: The distributed human neural system for face perception. Trends in Cognitive Sciences 4, 223–233 (2000)
3. Bitouk, D., Kumar, N., Dhillon, S., Belhumeur, P., Nayar, S.: Face swapping: automatically replacing faces in photographs. ACM Transactions on Graphics (TOG) 27, 39 (2008)
4. Hasselmo, M., Rolls, E., Baylis, G.: The role of expression and identity in the face-selective responses of neurons in the temporal visual cortex of the monkey. Behavioural Brain Research 32, 203–218 (1989)
5. Parry, F., Young, A., Shona, J., Saul, M., Moss, A.: Dissociable face processing impairments after brain injury. Journal of Clinical and Experimental Neuropsychology 13, 545–558 (1991)
6. Calder, A., Burton, A., Miller, P., Young, A., Akamatsu, S.: A principal component analysis of facial expressions. Vision Research 41, 1179–1208 (2001)
7. Rosenblum, M., Yacoob, Y., Davis, L.: Human expression recognition from motion using a radial basis function network architecture. IEEE Transactions on Neural Networks 7, 1121–1138 (1996)
8. Essa, I., Pentland, A.: Facial expression recognition using a dynamic model and motion energy. In: Proceedings of the Fifth International Conference on Computer Vision, pp. 360–367. IEEE (1995)
9. Bassili, J.: Facial motion in the perception of faces and of emotional expression. Journal of Experimental Psychology: Human Perception and Performance 4, 373 (1978)
10. Grossberg, S., Vladusich, T.: How do children learn to follow gaze, share joint attention, imitate their teachers, and use tools during social interactions? Neural Networks 23, 940–965 (2010)
11. Valentine, T.: Face-space models of face recognition. In: Computational, Geometric, and Process Perspectives on Facial Cognition: Contexts and Challenges, pp. 83–113 (2001)
12. Turk, M., Pentland, A.: Face recognition using eigenfaces. In: Proceedings of the IEEE Computer Society Conference on Computer Vision and Pattern Recognition, CVPR 1991, pp. 586–591. IEEE (1991)

13. Itti, L., Dhavale, N., Pighin, F.: Realistic avatar eye and head animation using a neurobiological model of visual attention. Technical report, DTIC Document (2003)
14. Rajan, V., Subramanian, S., Keenan, D., Johnson, A., Sandin, D., DeFanti, T.: A realistic video avatar system for networked virtual environments. In: Proceedings of IPT (2002)
15. Mori, M.: The uncanny valley. Energy 7, 33–35 (1970)
16. McOwan, P., Johnston, A.: The algorithms of natural vision: The multi-channel gradient model. In: First International Conference on Genetic Algorithms in Engineering Systems: Innovations and Applications, GALESIA (Conf. Publ. No. 414), pp. 319–324. IET (1995)
17. Anderson, K., McOwan, P.: Real-time Emotion Recognition Using Biologically Inspired Models. In: Kittler, J., Nixon, M.S. (eds.) AVBPA 2003. LNCS, vol. 2688, pp. 1056–1056. Springer, Heidelberg (2003)
18. Seeing Machines.: Faceapi product page (2012)
19. Blanz, V., Vetter, T.: A morphable model for the synthesis of 3d faces. In: Proceedings of the 26th Annual Conference on Computer Graphics and Interactive Techniques, pp. 187–194. ACM Press/Addison-Wesley Publishing Co. (1999)
20. Brennan, S.: Caricature generator: The dynamic exaggeration of faces by computer. Leonardo 18, 170–178 (1985)
21. Rotshtein, P., Henson, R., Treves, A., Driver, J., Dolan, R.: Morphing marilyn into maggie dissociates physical and identity face representations in the brain. Nature Neuroscience 8, 107–113 (2004)
22. Galton, F.: Composite portraits, made by combining those of many different persons into a single resultant figure. The Journal of the Anthropological Institute of Great Britain and Ireland 8, 132–144 (1879)
23. Polesel, A., Ramponi, G., Mathews, V.: Image enhancement via adaptive unsharp masking. IEEE Transactions on Image Processing 9, 505–510 (2000)
24. Liu, Z., Shan, Y., Zhang, Z.: Expressive expression mapping with ratio images. In: Proceedings of the 28th Annual Conference on Computer Graphics and Interactive Techniques, pp. 271–276. ACM (2001)
25. Theobald, B., Matthews, I., Cohn, J., Boker, S.: Real-time expression cloning using appearance models. In: Proceedings of the 9th International Conference on Multimodal Interfaces, pp. 134–139. ACM (2007)
26. Cook, R., Johnston, A., Heyes, C.: Self-recognition of avatar motion: how do I know it's me? Proceedings of the Royal Society B: Biological Sciences 279, 669–674 (2012)
27. Grammer, K., Thornhill, R.: Human (homo sapiens) facial attractiveness and sexual selection: the role of symmetry and averageness. Journal of Comparative Psychology 108, 233 (1994)
28. Müller, W., Spierling, U., Alexa, M., Rieger, T.: Face-to-face with your assistant. Realization issues of animated user interface agents for home appliances. Computers & Graphics 25, 593–600 (2001)

Facial and Vocal Cues in Perceptions of Trustworthiness

Elena Tsankova[1], Andrew J. Aubrey[2], Eva Krumhuber[1], Guido Möllering[1], Arvid Kappas[1], David Marshall[2], and Paul L. Rosin[2]

[1] School of Humanities and Social Sciences, Jacobs University Bremen, Campus Ring 1, 28759 Bremen, Germany
[2] School of Computer Science and Informatics, Cardiff University, Queens Buildings, 5 The Parade, Cardiff, CF24 3AA, United Kingdom

Abstract. The goal of the present research was to study the relative role of facial and acoustic cues in the formation of trustworthiness impressions. Furthermore, we investigated the relationship between perceived trustworthiness and perceivers' confidence in their judgments. 25 young adults watched a number of short clips in which the video and audio channel were digitally aligned to form five different combinations of actors' face and voice trustworthiness levels (neutral face + neutral voice, neutral face + trustworthy voice, neutral face + non-trustworthy voice, trustworthy face + neutral voice, and non-trustworthy face + neutral voice). Participants provided subjective ratings of the trustworthiness of the actor in each video, and indicated their level of confidence in each of those ratings. Results revealed a main effect of face-voice channel combination on trustworthiness ratings, and no significant effect of channel combination on confidence ratings. We conclude that there is a clear superiority effect of facial over acoustic cues in the formation of trustworthiness impressions, propose a method for future investigation of the judgmentconfidence link, and outline the practical implications of the experiment.

1 Introduction

Due to its evolutionary link to cooperation and survival, trust is a central aspect of human communication. Trust is typically established in the process of (continuous) interaction, but there is evidence that certain nonverbal information associated with trustworthiness can be processed already upon our very first encounter with an individual. For example, recent research by [1, 2] has shown that humans need no more than 100 ms to process facial features related to trustworthiness and generate an initial trustworthiness judgment. Such first impressions of trustworthiness, which emerge without the process of interpersonal interaction, are the focus of the current study. Here we investigate perceived trustworthiness, i.e., the subjectively estimated degree of another person's trustworthiness, rather than the actual trusting behavior in interaction. We also study how trustworthiness ratings are related to the person's own confidence in assessing the other's trustworthiness.

J.-I. Park and J. Kim (Eds.): ACCV 2012 Workshops, Part II, LNCS 7729, pp. 308–319, 2013.

In the above-mentioned research by Todorov and colleagues, perceived trustworthiness has been studied solely in the context of facial information. However, it has already been suggested that in addition to the visual channel, trustworthiness cues also travel via the auditory one [3]. The present study follows up on this and investigates the relative role of facial and vocal (paralinguistic, nonverbal) information in processing trustworthiness information. The study also looks at the way in which trustworthiness cues from one information channel (visual or auditory) bias perception of the information presented in the other channel. There is no prior empirical evidence on whether visual cues are more important than auditory cues for perceived trustworthiness, how they interact, and how people deal with inconsistent cues on different channels.

Not least because visual and auditory cues can be inconsistent, people may be able to form trustworthiness judgments quickly, but their confidence in those judgments may vary. Our study enables us to test whether consistent cues are associated with higher confidence ratings. More interestingly, though, by including confidence in one's assessment as a construct besides the perceived trustworthiness of the other, we can connect this psychological experiment to the more sociological-philosophical literature that emphasizes the leap of faith as an element in producing a state of trust [4–7]. Essentially, this literature claims that trust goes beyond the available information [8, 9]. In the context of our study this could mean that truly trusting individuals (i.e., those suspending uncertainty) generally give higher trustworthiness ratings, even when they are not fully confident about their assessment of the other (e.g., due to cue inconsistency). Alternatively, it could mean that highly confident individuals generally give higher trustworthiness ratings, even when they receive inconsistent cues. Both possibilities imply that the 'leap of faith' does not come as a result of the cognitive processing of trustworthiness information but is already part of this process. Our study explores whether the 'leap of faith' phenomenon can indeed be empirically observed and whether it is associated with certain characteristics of the visual and auditory cues or with certain types of study participants.

2 Predictions

The predictions we put forward in this study were twofold: first regarding the relationship between perceived trustworthiness and different combinations of visual and auditory cues; second about the relationship between confidence and perceived trustworthiness, and possible 'leap-of-faith' evidence. With respect to the relative role of facial and acoustic information in trustworthiness judgments, we took an explorative approach. We already knew from [10] that trustworthiness information is conveyed via facial dynamics. However, it is not so trivial to put forward a specific hypothesis about the degree to which visual and auditory information affect the perception of trustworthiness. There exists a large body of research on cross-modal integration in person identification ([11, 12] and many others), but not so much on face-voice information interaction in the case of person evaluation. But since trustworthiness has been shown to be closely related to emotional valence [13], we expected that previous findings on cross-modal integration in the case of emotion expression would apply to the perception of trustworthiness as well. [14] showed that both facial and vocal information play a role in the communication of emotion, but the influence of facial information is bigger. Based on their report one would expect that in the present study facial information will bias trustworthiness judgments in the following way: The trustworthy face + neutral voice combinations would receive the highest trustworthiness ratings, followed by the remaining combinations in the order neutral face + trustworthy voice, neutral face + neutral voice, neutral face + non-trustworthy voice, and non-trustworthy face + neutral voice.

With respect to the relationship between confidence and perceived trustworthiness, we did not expect a linear correlation but rather a bifurcation effect whereby individuals would generally give more extreme trustworthiness ratings (very high or very low) when they are confident in their judgments and would tend towards more moderate ratings when they are not confident in their judgments. It should be noted at this point that we expect overall lower confidence in cases where the visual and acoustic trustworthiness cues are inconsistent. For the 'leap of faith' argument, we would test whether high trustworthiness ratings occur in spite of lower confidence ratings or in spite of inconsistent cues and under which conditions this happens, e.g. specific patterns in terms of various cue combinations. We were not aware of any experimental study that has analyzed these effects.

3   Method

3.1   Participants

Data were collected from 25 (14 female) representatives of the student population of Jacobs University Bremen, age range $18 - 27$ years (M = 19.76, SD = 2.03), who volunteered to take part in this study in return for monetary compensation (5€per participant) and optional partial credit for one of two introductory methods courses. Informed consent was obtained from all participants prior to their respective individual testing sessions.

3.2   Stimuli

The stimulus material used to generate trustworthiness evaluations consisted of short audiovisual clips of various encoders saying the sentence "Hello, my name is Jo". This sentence was chosen because it is gender-neutral and does not contain any inherent trustworthiness cues. Thus, verbal content can be ignored and we can be relatively certain that trustworthiness cues in the audio channel are delivered via a acoustic route. Also, as it is an introductory phrase, this sentence adds a certain level of ecological validity to a situation of a first encounter with an unfamiliar person. Three types of recordings were obtained from 12 (six female) adult white Caucasian encoders. In one version encoders were instructed to sound "as neutral as possible", in the second version encoders were asked to sound trustworthy or "as if you are trying to convince somebody or if you want to make somebody believe you", and in the third they were asked to sound non-trustworthy or "as if you are trying to deceive somebody". It is important to note that none of the encoders came from the same student population as the target participant group. This measure was taken to prevent possible confounding effects of familiarity with the encoders as this could have potentially influenced trustworthiness evaluations. Female encoders wore no or minimal amounts of makeup and jewelry. There was variance with respect to facial hair in the case of male encoders, but that was considered a desirable phenomenon as it will allow us to study the effects of the presence or absence of facial hair on perceived trustworthiness.

The video and audio channels of each encoder's recordings were then digitally aligned (audio channel was warped to match video channel where necessary) so that, in addition to the non-warped neutral face + neutral voice, the following four combinations were obtained: neutral face + trustworthy voice, neutral face + non-trustworthy voice, trustworthy face + neutral voice, and non-trustworthy face + neutral voice. These combinations were digitally warped [15] to achieve perfect synchronization of lip movement and voice. The neutral face + neutral voice recording was included as it was available from the neutral recording version, without alignment or warping. Thus, we ended up with a total of 60 stimulus clips (five for each of the 12 encoders). Roughly, the duration of the clips was around two seconds each. Stimuli in this study were presented on a computer via MediaLab [16] using Philips Stereo Headphones SBC HP090.

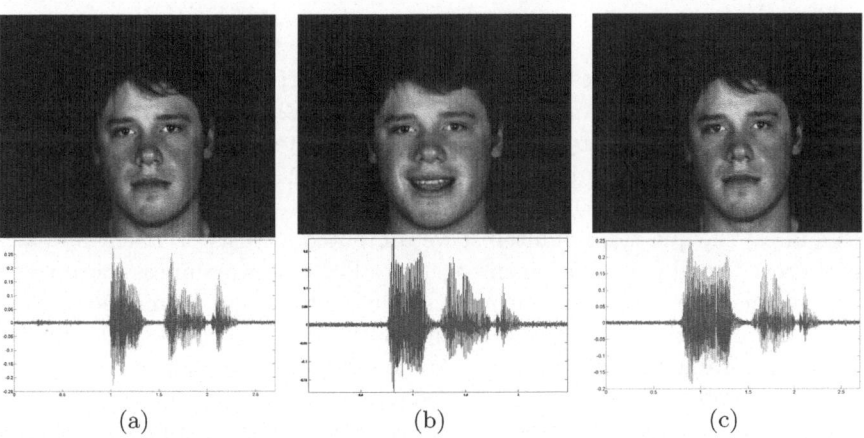

(a)   (b)   (c)

Fig. 1. Example of aligning (a) original neutral video and audio, with (b) original trustworthy video and audio. (c) neutral video with trustworthy audio. The trustworthy audio has been temporally aligned to the neutral video.

3.3   Procedure

Upon arrival at the lab participants were greeted by the experimenter and seated in the testing room, where they were asked to read and sign the informed consent form. The experimenter then gave a brief introduction to the study and left the room. Detailed instructions about the task were provided on the computer screen. Before the study began participants completed a short practice block to get used the task (the encoder in the practice clips was different from the ones in the actual study, to avoid participants becoming familiar with one of the target persons as this could have an impact on trustworthiness ratings later on). The experimental task required participants to watch each clip carefully, then evaluate the trustworthiness of the person in the clip, and finally provide a confidence rating of their trustworthiness evaluation. The exact phrasing of the trustworthiness and confidence questions was respectively *How trustworthy is this person?* and *How confident are you in this answer?*. Responses were always given sequentially on a seven-point Likert scale ranging from *not at all* to *very much*. The videos were shown in a random order on a within-subject basis. After having seen all videos, participants were asked whether they recognized any of the encoders. Participants were not able to correctly identify any of the actors and it was, therefore, not necessary to exclude any data due to familiarity effects between encoders and decoders.

At the end of the study, basic demographic information about age, gender, country of origin, and ethnicity was obtained from each participant. Upon completion of the experimental session participants were thanked and reimbursed for their help, and were given the opportunity to ask questions regarding the study.

4   Results

All ratings on the variables of interest were converted to standardized scores. Three participants who scored beyond the range ± 2.5 were treated as outliers and were not included in any further analyses reported here. The current report is, therefore, based on the data of 22 participants (12 female), age range 18 27 years (M = 19.82, SD = 2.15).

To address the questions of interest we computed a trustworthiness and a confidence score for all encoders taken together. The trustworthiness score was the mean average trustworthiness rating across all actors, and the confidence score was the mean average confidence rating. Both trustworthiness and confidence scores were obtained for all five video-audio stimulus combinations.

In the next step of the analyses we conducted a repeated measures analysis of variance (RM-ANOVA) on the trustworthiness scores with video-audio combination entered as a within-subject factor. The results indicated a main effect of channel combination on the mean trustworthiness ratings, $F(4, 18) = 27.70$, $p < .001, \eta_{p^2} = .86$ (see Figure 2). Pairwise comparisons showed that the trustworthy face + neutral voice condition was judged as being most trustworthy, with ratings significantly different from the other conditions ($ps < .01$). The conditions in which the trustworthiness of the voice was varied (e.g., neutral face + trustworthy voice and neutral face + non-trustworthy voice) did not differ significantly from each other and from the neutral face + neutral voice condition ($ps > .05$). These three conditions (neutral face + trustworthy voice, neutral face + non-trustworthy voice, and neutral face + neutral voice) all received higher trustworthiness ratings than the non-trustworthy face + neutral voice condition ($ps < .01$).

We further computed a RM-ANOVA on the confidence scores (again, with video-audio combination as a within-subject factor), but this analysis revealed no significant effect of channel combination on participants' confidence ratings, $p > .1$.

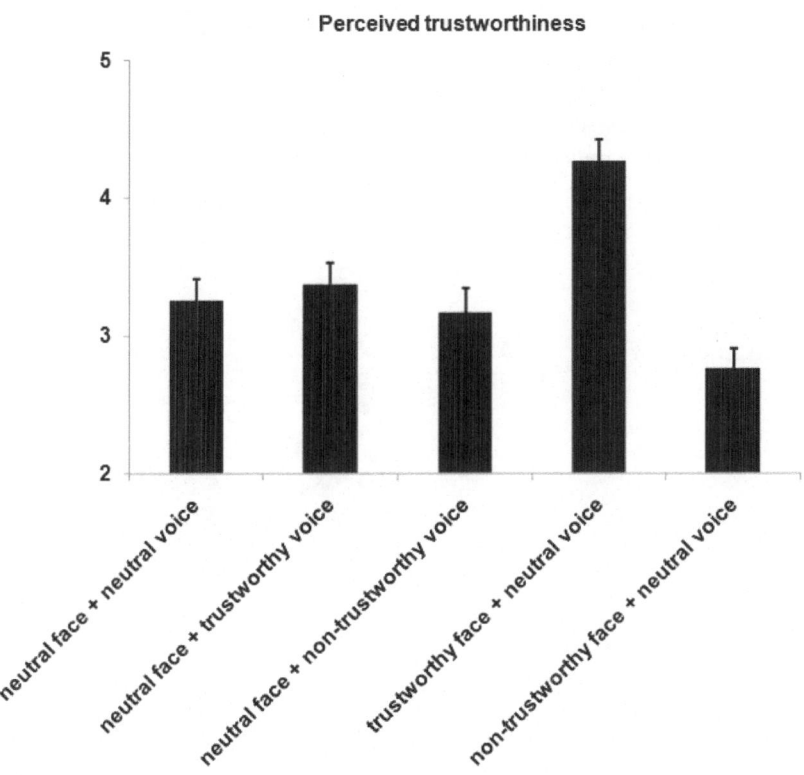

Fig. 2. Mean average trustworthiness ratings per channel combination. Error bars indicate 1 Standard Error.

With view to the 'leap-of-faith' hypothesis we plotted the trustworthiness scores against the confidence scores for each channel combination (see Figure 3). In this way we could obtain a more comprehensive view of our data distribution. Clearly, participants tended towards high confidence ratings and low to medium perceived trustworthiness ratings in all channel combinations. This data pattern posed a challenge to testing the predicted bifurcation model since there were no data points reflecting extremely low confidence or extremely high trustworthiness ratings. It did, however, seem likely that the relationship between perceived trustworthiness and confidence scores might follow a linear trend. We tested this possibility in a simple linear regression model, where confidence was entered as an independent variable and perceived trustworthiness as a dependent variable. The test was performed for all five channel combinations, but returned a significant result only in the case of the non-trustworthy face + neutral voice condition, $b = .42, t(20) = 2.07, p = .05$. In that case confidence score explained some of the variance in perceived trustworthiness, $R^2 = .18, F(1, 20) = 4.28, p = .05$.

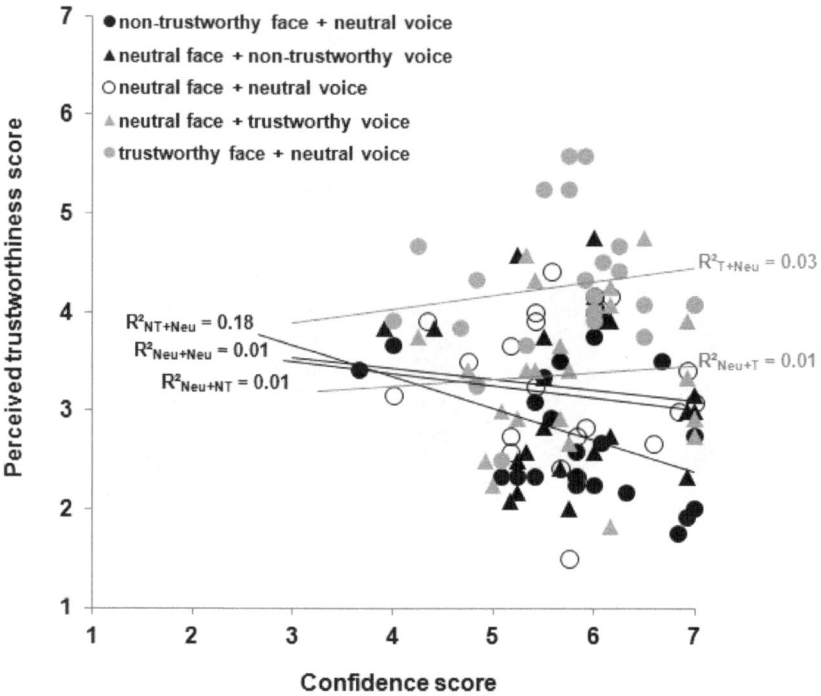

Fig. 3. Confidence scores plotted against perceived trustworthiness scores for all channel combinations

5   Discussion

With respect to the relative role of visual and auditory cues in explicit judgments of trustworthiness, our data clearly indicate a superiority effect of facial over acoustic information. Impression targets received highest trustworthiness ratings in the trustworthy face + neutral voice condition, and lowest ratings in the non-trustworthy face + neutral voice condition. From this, we can conclude that the trustworthiness information in this study was in fact conveyed via the facial cues. Our results are in line with Our results are in line with [14, 17], indicating a superior role of facial over vocal information in nonverbal communication, and thus provide further support for [13] of strong coupling between trustworthiness information and emotional valence.

We were interested in observing the relationship between confidence ratings and perceived trustworthiness ratings in order to assess whether higher confidence leads to more extreme trustworthiness ratings. We also hoped to find out whether lower confidence ratings might coincide with relatively high trustworthiness ratings that could be interpreted as being the result of a 'leap of faith'. Our results were not very clear in either respect. For the non-trustworthy face + neutral voice condition, we have evidence that some participants with relatively low confidence have given relatively high trustworthiness ratings and we could also see the expected effect that highly confident participants rated trustworthiness extremely low. Several explanations need to be considered. On the one hand, it is possible that what we conceptualized as a 'leap of faith', i.e. a form of suspending uncertainty, had already occurred when people were asked to indicate their level of confidence. This is supported by the fact that we did not press them in any way to give their ratings as quickly as possible. Participants could take their time and thus any 'leap-of-faith' effect would be contained in the ratings already. On the other hand, it is possible that initial trustworthiness ratings do not trigger a leap of faith yet. Only when it comes to interaction with the target and decisions to make one vulnerable or not are required, is there a need to suspend uncertainty. Thus our ratings would have been collected at a point before any observable 'leap of faith'. Finally, with a larger and more diverse sample, especially including more participants with low confidence in their trustworthiness ratings, it might have been possible to analyze why people with the same confidence rating give different trustworthiness scores for the same target (condition). A preliminary analysis of this kind may be possible, but the sample and model are a limiting factor in this. Further research can take our approach as inspiration for observing how people develop positive (or, negative) expectations in the face of uncertainty.

Future analysis of the current dataset could focus on the outliers and present case studies of specific response patterns. It is possible that the 'leap of faith', in the way defined earlier in this report, is only observed in participants who tend to give extreme responses and is, therefore lost with the exclusion of statistical outliers. The present study looks only at explicit judgments of trustworthiness, and it is possible that this creates a response bias in our participants i.e., when asked repeatedly to evaluate the trustworthiness level of targets they might become suspicious and assume that there is something non-trustworthy about the targets to begin with. Therefore, we propose another measure which could provide a more implicit measure of perceived trustworthiness that could be introduced in addition to the explicit one and could eliminate the bias reaction times [18]. The link between reaction times and confidence then could provide further empirical insights on the 'leap of faith' phenomenon.

6   Limitations

This study is a very first attempt at addressing the issues at hand, and as such naturally possesses a couple of limitations some related to the stimulus material, and some to the experimental design. To begin with, we recognize that the trustworthiness judgments we recorded might have been at least partially influenced by the facial geometry of the actors' faces, and thus are not solely the result of the nonverbal behavior manipulation. To eliminate this possibility, one might obtain trustworthiness judgments from a different participant sample based on still photographs of the actors' neutral facial expressions (e.g., screenshots from the original recordings). However, since no voice-based comparison stimulus exists, this would already imply a superior role of the face in the impression of trustworthiness. Therefore, in the current study we assumed natural variance of the actors' facial features' shape and configuration, much resembling real life. Given this variance, we considered it valid to assume that the trustworthiness judgments provided by our participants reflected the nonverbal behavior of the actors and any effects due to facial geometry cancelled each other out.

Further concerns might arise regarding the ecological validity of the recordings used in the present study. More precisely, it could be argued that actor portrayals represent an extreme or overexaggerated nature of the facial and vocal cues that would not have been observed in day-to-day communication. This is a common concern in research relying on actor databases, including emotion expression databases. However, there is no general consensus on what in fact constitutes "natural" behavior. Very often we intentionally try to convey a certain impression, thereby seeking to influence other people. For this we put extra emphasis on certain nonverbal behaviors. Other times, nonverbal behavior in the field is influenced by a wide range of additional factors such as mood or emotional state at that precise moment, our general frame of mind or the relationship with the communication partner. By applying standardized procedures in the laboratory, we can keep most of those factors constant (or rely on sufficient variance within the entire sample) and only manipulate the variables of interest. It is true that such conditions might result in an oversimplification of the studied phenomena, but we think that such powerful exemplars are a crucial first step in order to understand these phenomena.

One last issue related to the stimulus material touches upon the potential loss or warping of audio information during the channel alignment process, which could have resulted in the observation that the voice is less informative for judging trustworthiness. We do not exclude this possibility due to the novelty of the alignment algorithm. However, if this argument was true, then the neutral face + neutral voice condition should have received the highest trustworthiness judgments because there the stimuli did not undergo the alignment procedure and were most strongly reflecting real-life nonverbal behavior. Clearly, as can be seen in Figure 2, this was not the case. Instead, the neutral face + neutral voice condition was perceived as equally trustworthy to two other conditions where the voice had been warped. Therefore, it is safe to assume that no systematic bias in the trustworthiness judgments in this study occurred due to the channel alignment procedure.

We would like to emphasize the fact that since the present study was in itself a pilot one, its findings are yet inconclusive and need to be interpreted with care. Future investigation in the same direction would benefit from an experimental design covering four additional cue combinations (e.g., trustworthy face + trustworthy voice, trustworthy face + non-trustworthy voice, non-trustworthy face + trustworthy voice, and non-trustworthy face + non-trustworthy voice). In addition, one might be able to more clearly isolate the relative contributions of the face and voice in a between-subject design where two groups of perceivers are presented with single-channel cues (face only and voice only) and a third group is presented with the channel combination (face + voice). In fact, we are currently conducting a follow up study with the primary goal of addressing these issues. One also needs to keep in mind, however, the benefits of the simplistic design of the initial study. For example, by always keeping one cue channel constant (neutral with respect to trustworthiness) we were able to single out the exact contribution of one or the other channel. In combinations where both channels are different than neutral, the relative contribution of each modality is less clear, because the two channels produce combined effects on the resulting trustworthiness judgment. Such combinations are mostly interesting in the context of perceivers' sensitivity to cue consistency, and not so much along the lines of channels' relative contribution to the trustworthiness judgment (which was the purpose of the present study).

7   Conclusion

The study contributes new evidence on visual and acoustic influences on perceived trustworthiness, with particular emphasis on first impressions. Most trust research to date has investigated perceived trustworthiness in established relationships, but first impressions may determine who establishes relationships with whom. Hence prior research may be biased toward interactions that lasted beyond the initial encounter. Our research accounts for relationships that never formed and the relative role of visual and acoustic cues in this.

The present findings may have a wide range of practical implications both in the areas of everyday communication (e.g., sending a photo might be more powerful than giving a phone call) and computer animation (e.g., we see here that the generation of trustworthy virtual agent is primarily based on facial behavior rather than acoustic characteristics, [19]). Moreover, future research in the area with the expansion to microanalytic approaches has the potential for significant contributions to the computer-based recognition of trustworthy versus non-trustworthy facial behavior.

Finally, this study contributes exploratory insights into the possibility to observe the 'leap of faith' in an experimental set-up and to clarify to some extent whether the initial perception of another's trustworthiness, on trusting behavior later on, is based on "overdrawn information" [8]. Even though the results on the leap of faith in the narrower sense are inconclusive our study extends andpossibly challenges prior research by combining the question of perceived trustworthiness in the other with one's own confidence in making such an assessment. In short, we believe the current study constitutes a stimulating transdisciplinary approach to trust research, bridging experimental psychology methodology and sociological-philosophical research questions, and promising novel research designs.

Acknowledgements. This work has been conducted with the financial support of the Research Center CoWell at Jacobs University Bremen and the School of Psychology at Cardiff University. We thank Tony Manstead for his help with the recording of the actors.

References

1. Willis, J., Todorov, A.: First Impressions: Making up Your Mind after 100ms Exposure to a Face. Psychol. Sci., 592–598 (2006)
2. Ballew, C.C., Todorov, A.: Predicting Political Elections from Rapid and Unreflective Face Judgments. Proceedings of the National Academy of Sciences of the United States of America 104, 17948–17953 (2007)
3. Surawski, M., Ossoff, E.: The Effects of Physical and Vocal Attractiveness on Impression Formation of Politicians. Current Psychology 25, 15–27 (2006)
4. Giddens, A.: The Consequences of Modernity. Stanford University Press, Stanford (1990)
5. Giddens, A.: Modernity and Self-Identity. Polity Press, Cambridge (1991)
6. Möllering, G.: The Nature of Trust: From Georg Simmel to a Theory of Expectation, Interpretation and Suspension. Sociology 35, 403–420 (2001)
7. Möllering, G.: Trust: Reason, Routine, Reflexivity. Elsevier, Oxford (2006)
8. Luhmann, N.: Trust and Power. Two Works by Niklas Luhmann. Wiley, Chichester (1979)
9. Lewis, J.D., Weigert, A.: Trust as a Social Reality. Social Forces 63, 967–985 (1985)
10. Krumhuber, E., Manstead, A.S.R., Cosker, D., Marshall, D., Rosin, P., Kappas, A.: Facial Dynamics as Indicators of Trustworthiness and Cooperative Behavior. Emotion 7, 730–735 (2007)
11. Kamachi, M., Hill, H., Lander, K., Vatikiotis-Bateson, E.: 'Putting the Face to the Voice': Matching Identity Across Modality. Current Biology 13, 1709–1714 (2003)
12. Lachs, L., Pisoni, D.: Crossmodal Source Identification in Speech Perception. Ecological Psychology 16, 159–187 (2004)
13. Oosterhof, N.N., Todorov, A.: The Functional Basis of Face Evaluation. Proceedings of the National Academy of Sciences of the United States of America 105, 11087–11092 (2008)
14. Hess, U., Kappas, A., Scherer, K.R.: Multichannel Communication of Emotion: Synthetic Signal Production. In: Scherer, K.R. (ed.) Facets of Emotion: Recent Research, pp. 161–182. Erlbaum, Hillsdale (1988)
15. Aubrey, A., Kajic, V., Cingovska, I., Rosin, P., Marshall, D.: Mapping and Manipulating Facial Dynamics. In: Proc. IEEE Conf. Automatic Face and Gesture Recognition, pp. 639–645 (2011)

16. Jarvis, B.: MediaLab (Version 2008.1.33). Empirisoft Corporation, New York (2008)
17. Kappas, A., Hess, U., Scherer, K.R.: Voice and Emotion. In: Feldman, R.S., Rimé, B. (eds.) Fundamentals of Nonverbal Behavior, pp. 200–238. Cambridge University Press, Cambridge (1991)
18. Burns, C., Conchie, S.: Measuring Implicit Trust and Automatic Attitude Activation. In: Lyon, F., Möllering, G., Sanders, M. (eds.) Handbook of Research Methods on Trust, pp. 239–248. Edward Elgar, Cheltenham (2011)
19. Van Mulken, S., André, E., Muller, J.: An Empirical Study on the Trustworthiness of Life-Like Interface Agents. In: Bullinger, H., Ziegler, J. (eds.) Human Computer Interaction, pp. 152–156. Erlbaum, Mahwah (1999)

Disagreement-Based Multi-system Tracking

Quannan Li[1], Xinggang Wang[2], Wei Wang[3], Yuan Jiang[3],
Zhi-Hua Zhou[3], and Zhuowen Tu[1,4]

[1]Lab of Neuro Imaging, University of California, Los Angeles
[2]Huazhong University of Science and Technology
[3]National Key Laboratory for Novel Software Technology, Nanjing University
[4]Microsoft Research Asia

Abstract. In this paper, we tackle the tracking problem from a fusion
angle and propose a disagreement-based approach. While most existing
fusion-based tracking algorithms work on different features or parts, our
approach can be built on top of nearly any existing tracking systems
by exploiting their disagreements. In contrast to assuming multi-view
features or different training samples, we utilize existing well-developed
tracking algorithms, which themselves demonstrate intrinsic variations
due to their design differences. We present encouraging experimental
results as well as theoretical justification of our approach. On a set of
benchmark videos, large improvements (20% $\sim$ 40%) over the state-of-
the-art techniques have been observed.

1 Introduction

Object tracking has been a long standing problem in vision. Once a tracker gets
initialized, it starts to track the target in a video by performing two steps: (1)
making a prediction about the location of the target, and (2) updating its object
model (location, appearance, and shape) based on the prediction. This is in spirit
very similar to the bootstrapping and learning procedure in a learning algorithm.
With the recent success in detection-based tracking approaches, an increasing
amount of work has treated the tracking problem as a semi-supervised learning
problem [1–5]. Picking a target to track at the beginning provides supervised
data; the remaining of the frames for the tracker to explore do not contain label
information and thus is unsupervised. Due to the errors introduced in both the
prediction and model updating stage, nearly any tracker will eventually fail with
the errors being accumulated over the time.

Disagreement-based semi-supervised learning approaches [6], such as co-
training or tri-training [7, 8], provide a mechanism to allow classifiers trained
on different views or data samples to exploit unlabeled data. The learning process is a type of ensemble learning [9–11]. It involves multiple classifiers which
label the unlabeled data to update and improve each other [12]. From a different
angle, the use of multiple classifiers can be viewed as a fusion problem and it
has been shown that fusing complementary features in a tracking system often
leads to enhanced performances [13–15]. However, less efforts have been made

J.-I. Park and J. Kim (Eds.): ACCV 2012 Workshops, Part II, LNCS 7729, pp. 320–334, 2013.

in learning to fuse well-developed existing algorithms through semi-supervised learning; we will see later (in both theory and experiments) that a disagreement-based fusion significantly improves the performance over direct combination of features/systems [14, 16, 17].

In this disagreement-based multi-system tracking approach, we seek a balance between the current tracker and the level of agreements among other trackers. Our intuition is to find the location where the current tracker is confident but disagrees with other trackers, while other trackers reach a high degree of agreement. We provide both theoretical and experimental evidence to our approach and show much improved results over the state-of-the-art techniques on benchmark videos.

2   Related Work

A number of tracking methods have been proposed to perform fusion [13, 14, 18, 19, 16, 15, 17]. Different from [13, 17] where multiple parts were tracked and correlated, we deal with a single target. In [14, 16] multiple trackers were fused but these trackers represent different features and they were directly combined. In [18] the tracking approach was combined via the weighted combination of the PDFs. Different from [18], our method does not perform direct multiplication but seeks a balance between the PDF of one tracker and the degree of agreement by the other trackers; also, in our method, each tracker performs prediction separately maintaining certain independence and patches at the agreed positions can be recommended to update the other trackers. In [20], the tracking combination method is trained for specific scenarios. Different from [20], our method is based on the disagreement-based semi-supervised learning and do not require an off-line training process; also, it can be applied to general videos, and performs very well on a fairly large benchmark dataset. In [21], mutual information was used for the fusion. Here, the proposed fusion approach is based on the disagreements among the trackers. The most related work to our approach is [2], where the co-training idea was used to retrain classification-based trackers. However, [2] followed the standard co-training implementation using one specific type of classifier, SVM. In [19] several tracking algorithms were combined in a Bayesian framework whereas we here emphasize disagreement-based fusion through semi-supervised learning.

In disagreement-based semi-supervised learning, much of the work has been focused on using multi-view features [7] or different data samples [8]. The spirit of all such kind of approaches [7, 22, 8, 23] is to train multiple classifiers with disagreements, and then label the unlabeled instances for each other to update/improve the model. [24] provided PAC bounds with multi-view features, while [12] provided a sufficient condition for multi-view as well as single-view features. Recently, a sufficient and necessary condition was proved for disagreement-based semi-supervised learning, by establishing a connection between disagreement-based and graph-based approaches [25].

In this paper, we emphasize taking advantages of having various well-developed tracking algorithms. In the democratic co-learning framework [26], different algorithms are also used; however, their approach is for classification and a direct voting of all the methods is used. A main difference between tracking and classification is that there is no labeling information provided once the tracking process starts (it is a dynamic system), whereas most disagreement-based semi-supervised learning algorithms can still use labeled data in retraining. Notice that the existence of large disagreements among the classifiers is a premise for the learning or tracking process to continue [12], while the prediction is made by seeking the agreements among the classifiers. For example, in [22, 23], classifiers are learned so that they not only fit the supervised data well, but also themselves reach a high degree of agreement; the tri-training algorithm [8] uses confident and agreed data from two classifiers to help the third classifier. Our agreement is used in the prediction stage like the bootstrapping stage in [26], and we further emphasize the consistency with the information provided by the current tracker. Our work is also related to the active learning literature [27] but we do not have humans in the loop.

3   Disagreement-Based Tracking

The problem of making predictions in a tracking system has its own unique characteristic, and directly applying the standard co-training formulation [2] may not necessarily yield a good solution. Instead, we take advantages of having well-developed existing algorithms, *experts*, and combine them by exploiting the disagreements among the experts. The differences in the intrinsic design of the existing systems will naturally lead to a certain amount of biases/variations, a property the disagreement-based approaches requires [12].

3.1   Prediction of Single Tracker

In this section, we first clarify our notations for a single tracker. A tracker can be viewed as a learner denoted by $h^t = (A, f^t, X^t)$ since it always updates itself. Here, A is a specific tracking method e.g. mean-shift tracker [28], or particle filtering [29]; f^t is the underlying appearance model about the target at time t which can be represented by a discriminative model [5], generative model [4], or template matching [28]; X^t is the position of the target at time stamp t. Given a new image I^{t+1} at time stamp $t+1$, tracker h^t makes a prediction, X^{t+1}, about the position of the target and updates its underlying appearance model to f^{t+1}. We can view tracking a target of a tracker A as computing

$$q_A^{t+1}(x) \equiv p_A(y_x = +1 | I^{t+1}(x), f^t) \cdot p(X^{t+1} = x, X^t)$$
$$\text{and } \sum_x q_A^{t+1}(x) = 1 \tag{1}$$

Here $y_x = +1$ indicates the occurrence of target at location x and $I^{t+1}(x)$ is an image patch centered at x. Motion coherence is assumed that the prediction on the time stamp $t+1$ is smooth w.r.t. to the prediction on the time stamp t, for example, $p(X^{t+1} = x, X^t)$ can be a constant within a neighborhood of X_t and zero outside. This corresponds to the local search strategy adopted by most of the trackers.

Now that $q_A^{t+1}(x) \in [0,1]$ indicates how likely x_{t+1} is the correct position for the target. For an existing tracking algorithm, it may not strictly follow the formulation as in Eq. (1), but we still can use it so long as it outputs a probability map for the prediction.

3.2 Disagreement of Trackers

Suppose we have a set of existing trackers (experts) for making a prediction in a tracking system $S = \{h_i, i = 1..n\}$ with $n \geq 3$ being the number of trackers and each h_i is a tracker trained by tracking algorithm A_i. Given an input I^{t+1} at a time stamp $t + 1$, each tracker h_i computes a $q_i^{t+1}(x)$ to make a prediction of random variable X. Let $p^{t+1}(x)$ denote the ground probability map which indicates how likely x is the correct position, our general objective is to combine the probability maps by different trackers to obtain high probability modes in the "ground-truth" $p^{t+1}(x)$.

A direct way to fuse the multiple trackers is by linearly combining the probability maps together [15]. Here, we call it direct tracker fusion (DTF), which serves as a baseline algorithm:

$$\bar{q}^{t+1}(x) = \frac{1}{n} \sum_{i=1}^{n} q_i^{t+1}(x) \tag{2}$$

with the hope that $\bar{q}^{t+1}(x) \to p^{t+1}(x)$ as each tracker being unbiased and independent. Algorithms like [15] perform in this way with an adaption in the weighting parameters. The target location is retrieved by $\check{x}^{t+1} = \arg\max_x \bar{q}^{t+1}(x)$. In DTF, at each time, all trackers use the same prediction, $\check{x}^{t+1}$, and each tracker updates its appearance model to f_i^{t+1} based on $\check{x}^{t+1}$ separately and continues the tracking process. Fusing trackers leads to improvement over the original ones (see Section 3.2 and Section 4 for theoretical and empirical justification respectively).

However, predicting the position for the target w.r.t. Eq. (2) has a big drawback, i.e., the average performance $\bar{q}^{t+1}(x)$ of the n trackers may be degenerated by one bad tracker in the group. Here we give an example to illustrate this: suppose there are four trackers f_1, f_2, f_3, f_4 and two candidate positions x^* and x' at $t+1$, where x^* is the correct position for the target. The outputs of the four trackers on the two candidate positions are $q_1(x') = 0.9$, $q_2(x') = 0.4$, $q_3(x') = 0.4$, $q_4(x') = 0.4$ and $q_1(x^*) = 0.1$, $q_2(x^*) = 0.6$, $q_3(x^*) = 0.6$, $q_4(x^*) = 0.6$. The tracker f_1 is very confident but disagrees with other trackers and makes a wrong prediction. To some extent, this kind of tracker which is confident but disagrees with other trackers can be thought of as a outlier tracker. If we fuse the four trackers with direct tracker fusion (DTF), the position x' will be predicted as the position for the target according to $x = \arg\max_x \bar{q}^{t+1}(x)$. Unfortunately, we get a wrong position x' due to the outlier tracker f_1 at $t + 1$ although three trackers make correct prediction with confidence larger than 0.5.

Let ζ_{t+1} denote the probability mass on such an event that the average performance $\bar{q}^{t+1}(x)$ is degenerated by some outlier tracker f_i, i.e., the other $n-1$ trackers agree with each other and predict the correct position with high confidence while the DTF in Eq. (2) predict the position wrongly due to the outlier tracker f_i at $t + 1$. Next, we give the formulation for combining the multiple trackers based on their disagreement to avoid this kind of event for the purpose of robustness. Given n trackers, we still let each tracker perform prediction separately. If the current tracker is confident but disagrees with other trackers while other trackers reach a high degree of agreement, the current tracker is prone to be drifted to the agreed position of other trackers to reach more robust predictions. Our intuition is that we seek a balance between the generated distribution $q_i^{t+1}(x)$ of the current tracker and the degree of agreement by the other trackers as

$$Q_i^{t+1}(x) = (1 - \alpha)q_i^{t+1}(x) + \frac{\alpha}{n-1}[\sum_{j=1,j\neq i}^{n} q_j^{t+1}(x)] \cdot$$
$$\delta(\forall j \neq i, q_j^{t+1}(x) \geq TH)$$

$$(3)$$

and the specific location by the i-th tracker is $\tilde{x}_i^{t+1} = \arg\max_x Q_i^{t+1}(x)$. TH is a threshold corresponding to a confidence zone. α balances the importance of each tracker's own prediction and the influence from other trackers. The derivation of TH and α will be given in Section 3.2.

Note that the second term is non-zero only when all the other trackers have a high-degree agreement; this is different from the traditional fusion-based tracking [15] where weighted sum is performed; in addition, we emphasize that Eq. (3) focuses mainly on the places with high probability and it is not necessary to fit $p^{t+1}(x)$ at all xs as in the general statistical learning; our disagreement formulation in Eq. (3) can take advantage of this property.

Eq. (3) can be understood as the following: if the current tracker disagree with the other trackers while the other trackers are confident and agree with each other, the prediction of the current tracker will be influenced towards the agreed location (depending upon the overall probability map); otherwise, tracker h_i gives out a prediction as if there were no other trackers. In such a way, the trackers can keep relative independence and also enable *confident* interactions between each other. This makes our approach robust to outlier trackers. In addition, using the agreement of other trackers gives the overall system an ability to be self-aware of when the system starts to drift. This happens when all trackers have high entropy of $q_i^{t+1}(x)$ with large disagreement.

The overall output is then given by $x^{t+1} * = \arg\max_x \mathcal{Q}^{t+1}(x)$ and

$$\mathcal{Q}^{t+1}(x) = \frac{1}{n}\sum_{i=1}^{n} Q_i^{t+1}(x)$$

$$(4)$$

Note that $x^{t+1} *$ is the output of the overall system but it does not participate in the retraining of the individual trackers. The pseudo code of disagreement-based tracking is shown in Fig. 1. Tracking based on the disagreements among the trackers shows advantage over using a direct combination and we justify this point both theoretically and empirically in the following sections.

Given n trackers $\{h_i, i = 1..n\}$, each tracker $h_i = (A_i, f_i^t, X_i^t)$ adopts a specific tracking method A_i. At the time stamp $t = 0$, a target is manually identified located at X^0. All trackers start with the same X^0 and obtain their appearance model f_i^0. Given a new image I^{t+1} at time stamp $t + 1$,

- Each tracker h_i searches a local neighborhood around X_i^t and generate a probability map q_i^{t+1} using Eq. (1).
- Find modes of $\tilde{x}_i^{t+1}$ for $Q_i^{t+1}(x)$ as in Eq. (3). $\tilde{x}_i^{t+1}$ keeps a balance between the estimation of the current tracker and the level of agreements among other trackers.
- Assign $X_i^{t+1} = \tilde{x}_i^{t+1}$, sample patches around X_i^{t+1} and update the appearance model of each tracker to f_i^{t+1} using the embedded model updating/learning rule in A_i
- Based on $x^{t+1}{}^* = \arg\max_x \sum_i Q_i^{t+1}(x)$, report the $x^{t+1}{}^*$ as the tracking result for disagreement-based tracking (DBT).

Fig. 1. Pseudo code of disagreement-based tracking

Theoretical Justification. We first show that a linear combination of multiple trackers as in Eq. (2), direct tracker fusion (DTF), gains improvement over the individual systems. Let $p^{t+1}(x)$ denote the ground truth which indicates how likely x is the correct position, and let $q_i^{t+1}(x) \in [0, 1]$ be the output of algorithm A_i.

Lemma 1. *If we take an average of the predictions from all the experts: $\bar{q}^{t+1}(x) = \frac{1}{n} \sum_{i=1}^n q_i^{t+1}(x)$ as in Eq. (2), then the average is bounded in a PAC sense. We suppose that the n trackers are independent and unbiased: then $\bar{q}^{t+1}(x) \to p^{t+1}(x)$ as $n \to +\infty$.*

Proof. For any small $\epsilon > 0$, with Hoeffding inequality, we get that $P(|\bar{q}^{t+1}(x) - p^{t+1}(x)| \geq \epsilon) \leq 2\exp(-2n\epsilon^2)$. $\qquad\square$

Lemma 1 shows that $\bar{q}^{t+1}(x)$ can converge to the ground truth $p^{t+1}(x)$ exponentially. Let $error_i^{t+1}$ denote the error rate of the tracker f_i at $t + 1$, i.e., the probability that f_i predicts a wrong position for the target at $t + 1$, $error_{min}^{t+1} = \min_i\{error_i^{t+1}\}$ and $error_{max}^{t+1} = \max_i\{error_i^{t+1}\}$, we give the following theorem to show that fusing the multiple trackers according to Eq. (3) and Eq. (4) will improve the performance at least $\zeta_{t+1} - n(error_{max}^{t+1})^{n-1}$, contrasting to the direct tracker fusion (DTF) (ζ_{t+1} was defined in the previous section).

Theorem 1. *If we fuse the multiple trackers according to Eq. (3) and Eq. (4) where $\alpha \geq \frac{2}{3}$ and $TH \geq \frac{1}{2}$, contrasting to the direct tracker fusion in Eq. (2), the* *performance at $t+1$ can be improved at least $\zeta_{t+1} - n(error_{max}^{t+1})^{n-1}$, where ζ_{t+1} is the probability of the event that the average performance $\bar{q}^{t+1}(x)$ is degenerated by some outlier tracker.*

Proof. Let $\mathcal{X}^{t+1}$ denote the set of candidate positions at $t+1$. If there is some $x^* \in \mathcal{X}^{t+1}$, at which $q_i^{t+1}(x^*) \geq TH$ for all $i \in \{1, \ldots, n\}$, it is easy to find that such x^* is unique, since $TH \geq 0.5$ (Here we neglect the probability mass on the event that at $t+1$ there are two positions x and x' at which $q_i^{t+1}(x) = q_i^{t+1}(x') = \frac{1}{2}$). Considering Eq. (3) we get $\mathcal{Q}^{t+1}(x^*) = \frac{1}{n}\sum_{k=1}^{n} q_k^{t+1}(x^*)$, and x^* will be selected as the tracking result, no matter whether $\arg\max \mathcal{Q}^{t+1}(x)$ or $\arg\max \bar{q}^{t+1}$ is used. If for any $x \in \mathcal{X}^{t+1}$ there are less than $n-1$ trackers with $q_i^{t+1}(x) \geq TH$, then the second term of Eq. (3) is zero. So for any $x \in \mathcal{X}^{t+1}$, $\mathcal{Q}^{t+1}(x) = \frac{1-\alpha}{n}\sum_{k=1}^{n} q_k^{t+1}(x)$. Predicting the tracing result according to $\arg\max \mathcal{Q}^{t+1}(x)$ is equal to predicting according to $\arg\max \bar{q}^{t+1}$. Next we analyze the situation when there is some $\hat{x} \in \mathcal{X}^{t+1}$, at which $q_i^{t+1}(\hat{x}) < TH$ and $q_j^{t+1}(\hat{x}) \geq TH$ for all $j \neq i$. Obviously, such $\hat{x}$ is also unique, since $TH \geq 0.5$ and $n \geq 3$.

Case 1: $\hat{x}$ is the correct position for the target at $t+1$. We will show that even if $q_i^{t+1}(\hat{x})$ is very close to 0, i.e., tracker f_i is an outlier at $t+1$, it will not degenerate the fusion of the multiple trackers due to Eq. (3). We obtain

$$Q_i^{t+1}(\hat{x}) = (1-\alpha)q_i^{t+1}(\hat{x}) +$$
$$\frac{\alpha}{n-1}\sum_{j \neq i} q_j^{t+1}(\hat{x}) \geq (1-\alpha)q_i^{t+1}(\hat{x}) + \alpha \cdot TH \tag{5}$$

$$Q_{j,j\neq i}^{t+1}(\hat{x}) = (1-\alpha)q_j^{t+1}(\hat{x}) \geq (1-\alpha) \cdot TH$$

Thus,

$$\sum_{k=1}^{n} Q_k^{t+1}(\hat{x}) \geq (1-\alpha)q_i^{t+1}(\hat{x}) + \alpha \cdot TH +$$
$$(1-\alpha)(n-1)TH \tag{6}$$

For an incorrect position $x' \neq \hat{x}$, since $\sum_{x \in \mathcal{X}^{t+1}} q_k^{t+1}(x) = 1$, it is easy to see that

$$Q_i^{t+1}(x') = (1-\alpha)q_i^{t+1}(x') \leq (1-\alpha)(1 - q_i^{t+1}(\hat{x}))$$

$$Q_{j,j\neq i}^{t+1}(x') = (1-\alpha)q_j^{t+1}(x') \leq (1-\alpha)(1 - q_i^{t+1}(\hat{x}))$$
$$\leq (1-\alpha)(1 - TH) \tag{7}$$

Therefore,

$$\sum_{k=1}^{n} Q_k^{t+1}(x') \leq (1-\alpha)(1 - q_i^{t+1}(\hat{x}))$$
$$+ (1-\alpha)(n-1)(1 - TH) \tag{8}$$

We see that in general $\sum_{k=1}^{n} Q_k^{t+1}(\widehat{x}) \geq \sum_{k=1}^{n} Q_k^{t+1}(x')$ for $\alpha \geq \frac{2}{3}$ and $TH \geq \frac{1}{2}$. This makes the correct position more robust, i.e., the prediction of the disagreement-based tracking will never be influenced even if f_i is a outlier tracker. So the improvement is at least ζ_{t+1}.

Case 2: $\widehat{x}$ is not the correct position for the target at $t+1$. Since $q_j^{t+1}(\widehat{x}) \geq TH$ for all $j \neq i$ and $TH \geq \frac{1}{2}$, $n-1$ trackers predict the wrong position $\widehat{x}$ as the tracking result at $t+1$. Now we bound the probability of such event. Since $error_j^{t+1} \leq error_{max}^{t+1}$ and the multiple trackers are assumed to be independent, the probability mass on the event that $n-1$ trackers predict the position mistakenly is at most $n(error_{max}^{t+1})^{n-1}$. The worst situation is that the fusion according to Eq. (3) performs worse than the direct tracker fusion in **case 2** completely. We get Theorem 1 proved.   □

From Theorem 1 we know that the fusion will get benefit from Eq. (3) under the situation that one bad tracker degenerates the direct tracker fusion. When n (the number of the trackers) is large, it would be difficult for the remaining $n-1$ trackers to achieve some agreement (See the experiment "Non-Relax" in Section 4). In practice, we can relax this constraint, e.g., when two or more trackers achieve agreement, the agreement term would take effect.

Note that the performance of Eq. (2) and Eq. (3) depends on the correlation between the trackers. The correlation depends on two factors: (1) the intrinsic design of the trackers; (2) the training samples used to train the trackers. Two trackers with the same design trained on the same set of samples are highly correlated, and two different types of trackers trained on the same set of samples are more correlated than those trained on different set of samples. If the n trackers are the same, then using Eq. (3) shows no advantage over Eq. (2).

In summary, Lemma 1 suggests that fusing the multiple experts directly might gain exponential improvement, contrasting to the single tracker; Theorem 1 shows that our disagreement-based fusion method can provide more robustness to the tracking system, which motivates the use of Eq. (3) by keeping a balance between the current expert f_i and the agreement from the other experts.

4   Experiments

In the experiments, we make a comprehensive comparison between the performance of disagreement-based tracking, direct tracker fusion, and the individual trackers. Four trackers are used and the experiment is conducted on 11 commonly tested videos (listed in Table 1). The trackers used are MilTracker [5], the semi-supervised on-line boosting tracker (semiBoost)[3], Incremental Visual Tracker (IVT)[4], and Incremental Visual Tracker using edge information (IVTE).

Since the individual trackers perform prediction separately, the computational complexity of the proposed method only adds slight overhead over the individual ones with the multi-core processor and parallel computing. Compared with the large performance gain, this computational overhead is tolerable.

From the experiments, we observe that, statistically, each individual tracker gets significantly improved by using Eq. (3): the average center location error has been reduced by more than 12 pixels and the success rate has increased by $20\% \sim 40\%$. The result of DBT (Disagreement-Based Tracking) also outperforms the system by directly combining the original trackers, i.e., DTF, with 3.3 pixels reduction in center location error and 4.4% improvement in success rate. DBT also significantly outperforms PROST [30].

4.1 Implementation Details

While a wealthy body of tracking papers/systems have been reported, we found a few systems (with available source code) having decent performance on general videos. Here, we provide brief descriptions for these trackers we used with necessary changes made to them.

MilTracker adopts an online multiple instance learning algorithm to train a discriminative classifier. In order to handle the ambiguity of sampled patches, a bag of potentially positive image patches are extracted. MilTracker maintains a pool of Haar features and the online boosting mechanism is adopted.

SemiBoostTracker also adopts an online boosting mechanism and it formulates the update process in a semi-supervised fashion combined with a given prior. This helps to alleviate the drifting problem.

The IVT incrementally learns a low dimensional eigenspace representation to model the appearance changes of the object. In IVT model, the target is represented as a vector of gray-scale value, and the motion is modeled by an affine image warping. To propagate sample distributions over time, a particle filter framework is adopted. Since both MILTracker and SemiBoostTracker do not support affine transformation, we disabled the scaling and rotating ability of IVT. IVTE is similar to IVT, except that it uses level set as the feature.

The forms of the 4 tracking systems' outputs are rather different. MILTracker and SemiBoostTracker produce scores on local search regions; IVT and IVTE propagate probabilities via particles. In the experiment, we map the scores of MILTracker and SemiBoostTracker to the range $[0, 1]$ to produce probability maps q_{MIL} and q_{SBT} (The probability maps are normalized to make sure that $\sum_x q_A^{t+1}(x) = 1$). For IVT and IVTE, we keep the position entries (a_i, b_i) of the particle and use a parzen window approach to estimate the probability for prediction. For a point $x = (a, b)$ on the image, its probability is calculated as $q_{IVT}(x) = \sum_{i=1}^{M} w_i * \max\{0, 1 - \sqrt{(a - a_i)^2 + (b - b_i)^2}/L\}$. In our experiment, L is set to 15. A map q_{IVTE} is produced similarly as q_{IVT} for IVTE.

Based on q_{MIL}, q_{SBT}, q_{IVT}, and q_{IVTE} we respectively compute the corresponding Q_{MIL}, Q_{SBT}, Q_{IVT}, and Q_{IVTE} using Eq. (3) (Since 4 trackers are used, we relaxed Eq. (3) that when 2 trackers achieve confident agreement, the agreement term will take effect) and thus, each tracker makes its own prediction separately. As we have mentioned before, $\tilde{x}_i^{t+1}$ found by mean shift algorithm [28] can represent multiple points (modes). For each tracker, e.g., MILTracker, the one mode with the maximum value is reported as its prediction, significant modes found are used to retrain the tracker and as the seeds for further search at the next time stamp. For the results reported in our experiment, α is set to be 0.67 as suggested by the theoretical section. The threshold TH in Eq. (3) is set as $0.8/(\Lambda/3)$ (Λ is the size of the search window). For each tracker, 2 modes are kept.

4.2   Quantitative Results

The Average Center Location Error. The average center location error is a commonly used metric to measure the performance of tracking and is defined as the average error between the predicted locations to the ground truth. In Table 1, we summarize the results of the average center location error on all the 11 videos. It's clear that our disagreement based tracker outperforms the individual trackers, Direct tracker fusion, and the co-training scheme. For non-relax (using Eq. (3) directly without relaxation), it's less possible for the trackers to achieve some agreements, and the chance of interaction between trackers is reduced. Still the result is better than the individual trackers.

Table 1. Comparison of Average Center Location Error. Non-Relax indicates to use Eq. (3) directly without relaxation;Co-Training stands for the results using co-training method.

videos	MilT	IVT	IVTE	SBT	DTF	Co-Training	Non-Relax	DBT
Girl	31.9	25.2	18.1	19.3	20.6	39.8	23.3	**13.4**
CokeCan	20.5	55.3	11.0	14.9	9.3	49.0	7.9	**6.6**
Tiger1	**15.9**	71.9	56.6	20.9	37.7	64.1	49.0	31.2
Sylv	10.9	44.0	19.5	16	19.5	31.7	**7.3**	10.8
StatOcc	27.8	3.3	4.8	74.4	**2.5**	41.2	26.2	3.0
David	22.9	4.9	16.9	26.4	7.0	9.2	7.9	**4.1**
Cliffbar	12.0	31.4	78.6	29.9	27.1	45.6	16.7	**8.5**
Surfer	9.2	6.7	23.9	67.6	5.1	**4.7**	5.1	4.8
faceocc2	20.1	14.2	9.1	17	6.5	12.4	12.0	**6.1**
Indoor	17.2	30.2	193.5	116.5	4.7	61.9	10.7	**4.5**
faceocc	27.1	11.8	11.3	**6.8**	8.9	23.3	**7.6**	9.7
In all	20.8	21.8	22.3	37.3	11.5	32.7	15.8	**8.2**

The Success Rate. If the location error on one frame is less than a pre-specified threshold, the prediction is regarded as a successful prediction. The success rate is defined as the ratio of successful predictions over all the predictions. To compute the success rate, we set T as certain ratio of the average width of the target, i.e., $T = \beta * (w + h)/4$, where, w and h are the width and height of the target respectively. Conceptually, this is similar to the overlap score evaluation used in [30] and thus, success rate is conceptually similar to the tracked percentile in [30]. We observe that, the trackers can not track the target precisely at all times, if there is no overlap but the prediction of the tracker is not far from the target or within the search area of the tracker, it's often possible for the tracking process to recover. Our evaluation measurement can reflect such a phenomenon. We compare the success rate in Table 2 and from Table 2, the DBT achieves the highest success rate.

Table 2. Comparison of success rate when $\beta = 0.5$

MilT	IVT	IVTE	SBT	DTF	Non-Relax	DBT
0.596	0.733	0.753	0.592	0.862	0.84	0.900

Comparison of Probability Maps. In Fig. 2, we show how the probability maps are generated in Eq. (2) and Eq. (3) on a testing video, Tiger1. DTF drifts from the 30th frame; on this frame, the predictions of the trackers are rather diverse. In DTF, however, the individual tracker's prediction is not fully respected and it has to comply with the voted prediction; this is the primary reason for drifting and getting trapped; using Eq. (3) however leads to a more robust prediction. As we can see from the second figure, on the 30th frame, MILTracker and Semiboost tracker achieves certain agreement, but IVT and IVTE still complies with its own prediction since the agreement is weak. On the 35th frame, when MILTracker, IVT and Semiboost tracker achieve confident agreement, IVTE is pulled back to the agreed position and the four trackers merge again. The benefit of our disagreement-based tracking is obvious: the trackers then keep their relatively different traces and the risk of getting trapped is reduced.

(a)   (b)   (c)

Fig. 2. Illustration of the probability maps where four trackers (experts) are adopted (the figures have been scaled for visualization). (a) shows the results by DTF. (b) and (c) display the probability maps generated by disagreement-based tracking. The probability maps inside the dashed yellow rectangle are shown below the screen shots. Underneath each figure, from left to right, the probability maps are IVT, IVTE, MIL-Tracker, Semiboost tracker respectively (see the discussions about these trackers in the experiments). For DTF in (a), the fifth probability map is the combined map. For disagreement-based tracking in (b) and (c), the first rows shows the original probability maps, and the second row shows the Q_i^{t+1} computed by Eq. (3).

Comparison with Other Methods. PROST [30] is another fusion based tracking algorithm that adopts 3 trackers (a template model, an optical-flow based mean-shift tracker and an online random forest tracker). Table 3 and Table 4 compare the average location errors and the tracked percentage (computed using the overlap score in [30]) with PROST. From the two tables, we can observe that, our disagreement-based tracking outperforms PROST and achieves better tracked percentages on most of the videos.

Table 3. Comparison of average center location error with [30]

Method	Girl	tiger1	sylv	David	faceocc	faceocc2
[30]	19.0	**7.2**	**10.6**	15.3	**7.0**	17.2
Ours	**13.4**	31.2	10.8	**4.1**	9.7	**6.1**

Table 4. Comparison of tracked percentage with [30]

Method	Girl	tiger1	sylv	David	faceocc	faceocc2
[30]	89	**79**	73	80	**100**	82
Ours	**97**	30	**83**	**100**	**100**	**100**

The best experimental performance of Democratic Integration in [15] was achieved by using uniform qualities, which assigned equal weights to all the clues, and corresponded directly with DTF. In addition, we implemented the quality measure of normalized saliency and the performance was not as good as DBT: their average center location error is 13.8 with success rate 0.87.

We did not get the implementation of [2]. Nevertheless, we did experiment on some videos used in [2] and the results of DBT are better than [2] qualitatively (skipped here due to page limit). Moreover, we indeed implemented co-training and reported the result in Table 1 (average center location error 32.7), which is much worse than DBT.

Table 5. Performance by varying α and TH ($TH = R/(\Lambda/3)$)

R/α	0.8/0.3	0.8/0.67	0.8/0.85	0.7/0.67	0.9/0.67
Average Error	10.0	8.2	9.8	9.95	9.8
Success Rate	0.875	0.90	0.895	0.87	0.89

Robustness by Varying the Parameters. In table 5, we summarize the performance of disagreement-based tracking by varying the parameters α and TH, which are the two key parameters in Eq. (3). We can see from this table, by varying TH and α, the results (especially the success rate) do not change too much. This demonstrates the robustness of disagreement-based tracking. The average center location error has relatively larger change because on portions of the videos Tiger1 and Indoor, disagreement-based tracking gets distracted to positions distant from the targets. In such cases, the success rate does not vary too much, but the center location error is increased.

Screenshots of the Results. In Fig. 3, we compare the tracking results on the video Girl. This video undergoes several challenges: fast appearance change

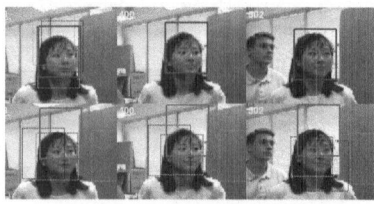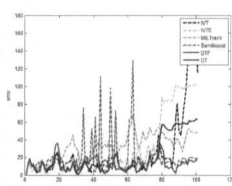

Fig. 3. Comparison of tracking results on the video Girl. The first row on the left shows the results of disagreement-based tracking; the second row on the left shows the results of 4 individual trackers and direct tracker fusion; the right plot shows the comparison of center location error (the number of the x axis denotes the number of predictions). The coloring scheme is: dotted green: IVT, dotted black: MILTracker, dotted blue: IVTE, dotted yellow: Semiboost tracker, solid Red: disagreement-based tracking, and solid magenta: Direct tracker fusion.

and occlusion. Although both MILTracker and IVTE can track the face of the girl successfully, the tracking process is not very stable. IVT drifts from the face at the 20th frame. From the 391th frame, direct tracker fusion also drifts and get trapped at the background of the images. As can be seen from both the screen shots and the error plot on the right of Fig. 3, we find that disagreement-based tracking tracks most robustly and accurately. In Fig. 3, we can also find a very nice property of democratic tracking that, the traces of the four trackers are similar but not exactly the same, thus, they can explore different spaces, recommend confident samples to other trackers, and thus avoid to be trapped at an incorrect position.

5 Conclusion

In this paper, we have introduced a disagreement-based tracking method which fuses multiple existing tracking systems in the following way that seeks a balance between the coherence of the current tracker and the degree of agreements among other trackers. In such a way, it enables the interaction between trackers and keeps the appealing characteristics of the trackers at the same time. As illustrated in the experiments, the balance complies with the characteristic of tracking. Disagreement-based tracking can be built on top of various existing well-developed tracking systems utilizing their intrinsic biases. Adopting several state-of-the-art tracking algorithms, our approach is able to improve each of them by a large margin on widely used benchmark videos in the literature

Acknowledgement. Zhuowen Tu and Quannan Li are supported by Office of Naval Research Award N000140910099, NSF CAREER award IIS-0844566 and NSF IIS-1216528; Wei Wang, Yuan Jiang and Zhi-Hua Zhou are supported by National Natural Science Fund of China 60975043.

References

1. Avidan, S.: Ensemble tracking. In: CVPR (2005)
2. Tang, F., Brennan, S., Zhao, Q., Tao, H.: Co-tracking using semi-supervised support vector machines. In: ICCV (2007)
3. Grabner, H., Leistner, C., Bischof, H.: Semi-supervised On-Line Boosting for Robust Tracking. In: Forsyth, D., Torr, P., Zisserman, A. (eds.) ECCV 2008, Part I. LNCS, vol. 5302, pp. 234–247. Springer, Heidelberg (2008)
4. Lim, J., Ross, D., Lin, R.-S., Yang, M.-H.: Incremental learning for visual tracking. Int'l J. of Comp. Vis. (2008)
5. Babenko, B., Yang, M.-H., Belongie, S.: Visual tracking with online multiple instance learning. In: CVPR (2009)
6. Zhou, Z.-H., Li, M.: Semi-supervised learning by disagreement. Knowledge and Information Systems 24, 415–439 (2010)
7. Blum, A., Mitchell, T.: Combining labeled and unlabeled data with co-training. In: COLT, pp. 92–100 (1998)
8. Zhou, Z.-H., Li, M.: Tri-training: Exploiting unlabeled data using three classifiers. IEEE Trans. on Know. and Data Eng. (2005)
9. Dietterich, T.G.: Ensemble Methods in Machine Learning. In: Kittler, J., Roli, F. (eds.) MCS 2000. LNCS, vol. 1857, pp. 1–15. Springer, Heidelberg (2000)
10. Kuncheva, L.I.: A theoretical study on six classifier fusion strategies. IEEE Tran. on PAMI 24, 281–286 (2002)
11. Bennett, K.P., Demiriz, A., Maclin, R.: Exploiting Unlabeled Data in Ensemble Methods. In: ACM SIGKDD 2002, Edmonton, Alberta, CA (2002)
12. Wang, W., Zhou, Z.-H.: Analyzing Co-training Style Algorithms. In: Kok, J.N., Koronacki, J., Lopez de Mantaras, R., Matwin, S., Mladenič, D., Skowron, A. (eds.) ECML 2007. LNCS (LNAI), vol. 4701, pp. 454–465. Springer, Heidelberg (2007)
13. Wu, Y., Huang, T.: Robust visual tracking by integrating multiple cues based on co-inference learning. In'l J. of Comp. Vis. (2004)
14. Siebel, N.T., Maybank, S.: Fusion of Multiple Tracking Algorithms for Robust People Tracking. In: Heyden, A., Sparr, G., Nielsen, M., Johansen, P. (eds.) ECCV 2002, Part IV. LNCS, vol. 2353, pp. 373–387. Springer, Heidelberg (2002)
15. Triesch, J., von der Naksvyrg, C.: Democratic integration: Self-organized integration of adaptive visual cues. Neural Computation (2001)
16. Spengler, M., Schiele, B.: Towards robust multi-cue integration of visual tracking. In: MVA (2003)
17. Kwon, J., Lee, K.M.: Visual tracking decomposition. In: CVPR (2010)
18. Leichter, I., Lindenbaum, M., Rivlin, E.: A general framework for combining visual trackers the 'black boxes' approach. Int'l J. of Comp. Vis. (2006)
19. Zhong, B., Yao, H., Chen, S., Ji, R.-R., Yuan, X., Liu, S., Gao, W.: Visual tracking via weakly supervised learning from multiple imperfect oracles. In: CVPR (2010)
20. Stenger, B., Woodley, T., Cipolla, R.: Learning to track with multiple observers. In: CVPR (2009)
21. Mundy, J.L., Chang, C.F.: Fusion of intensity, texture, and color in video tracking based on mutual information. In: AIPR (2010)
22. Collins, M., Singer, Y.: Unsupervised models for named entity classification. In: EMNLP (1999)
23. Leskes, B., Torenvliet, L.: The value of agreement a new boosting algorithm. J. of Comp. and Sys. Sci. (2008)

24. Dasgupta, S., Littman, M.L., McAllester, D.: Pac generalization bounds for co-training. In: NIPS (1999)
25. Wang, W., Zhou, Z.-H.: A new analysis of co-training. In: ICML, Haifa, Israel, pp. 1135–1142 (2010)
26. Zhou, Y., Goldman, S.: Democratic co-learning. In: Inte'l Conf. on Tools with Art. Intell. (2004)
27. Dasgupta, S.: Coarse sample complexity bounds for active learning. In: NIPS (2005)
28. Comaniciu, D., Ramesh, V., Meer, P.: Real-time tracking of non-rigid objects using mean shift. In: CVPR (2000)
29. Isard, M., Blake, A.: Condensation - conditional density propagation for visual tracking. Int'l J. on Comp. Vis. (1998)
30. Santner, J., Leistner, C., Saffari, A., Pock, T., Bischof, H.: Prost: Parallel robust online simple tracking. In: CVPR (2010)

Monocular Pedestrian Tracking
from a Moving Vehicle*

Zipei Fan, Zeliang Wang, Jinshi Cui, Franck Davoine**,
Huijing Zhao, and Hongbin Zha

Key Laboratory of Machine Perception(Ministry of Education), Peking University,
Peking, China

Abstract. Tracking of pedestrians from a moving vehicle equipped with
a monocular camera is still considered as a challenging problem in the
fields of both computer vision and robotics. In this paper, we address this
problem in a particle filter framework, which well incorporates different
cues from detector, dynamic model and target-specific tracking. In order
to eliminate the effect of ego-motion when predicting the movement of
pedestrians, we train one dynamic model for each driving behavior (moving forward, turning left/right) given a set of training trajectories. The
learnt dynamic model is then utilized to predict the future movement
of the pedestrian in the tracking process. We demonstrate our system
works robustly on challenging dataset with strong illumination changes.

1 Introduction

The research development in computer vision, especially on detection and tracking algorithms has generated a wide variety of applications in intelligent vehicle
systems [1]. However, difficulties such as illumination variat ion, partial and full
occlusion and scale change exist, and the effect of ego-motion nullifies several
simple but effective prediction strategies in static camera scenarios. Following
the development of detection algorithms [2], [3], tracking-by-detection is becoming a popular solution [4], [5], [6], [7], [8] to the difficulties mentioned above.
Such approaches combine information from both single frame and across frames
as well to recover trajectories.

In this paper, as illustrated on Fig. 2, we will adapt the multi-person tracking-
by-detection framework described in [4]. Because of the difficulty in eliminating
the effect of the vehicles ego-motion within monocular tracking systems, dynamic models for predicting target displacements is hard to design. In this paper, instead of estimating pedestrian location in the world coordination, we train
dynamic models by clustering trajectories of pedestrians under pre-selected driving behaviors (move forward, turn left, turn right) in a given training set. In the

---

* This work was supported by National Basic Research Program (973 Program) of
  China (No. 2011CB302202), NSFC Grant (No.91120004) and the NSFC-ANR Grant
  (No.61161130528).
** CNRS, LIAMA Sino-French IT Lab., Beijing, China.

J.-I. Park and J. Kim (Eds.): ACCV 2012 Workshops, Part II, LNCS 7729, pp. 335–346, 2013.
© Springer-Verlag Berlin Heidelberg 2013

(a)   (b)

Fig. 1. Some tracking results. Our algorithm tracks pedestrians well though occlusion, illumination change occurs

tracking process the prediction is done by choosing the dynamic model according the current driving behavior state.

The main contribution of the paper is to propose a novel approach to eliminate the effect of ego-motion of the camera, when the estimation of the camera motion and the exact location of the pedestrians in the world coordination are either unreliable or unavailable. We make the assumption that from the mounted camera's view, the pedestrians' trajectories follow particular pattern, which is hard to describe mathematically, while our approach of learning these patterns from training data would be a natural solution.

The rest of this paper is organized as follows: Section 2 introduces related work. Section 3.1 demonstrates the overview of our algorithm. Section 3.2 presents the tracking algorithm we propose. Section 3.3 presents the algorithm of labeling driving behavior. Section 3.4 details our design of dynamic model. Section 4 presents the analysis of experimental results. Section 5 concludes this paper.

2   Related Work

2.1   Pedestrian Detection and Tracking

The research of pedestrian detection has made a great progress recently [2], [9], which stimulates the research of combining the tracking and detection [4], [5], [6], [10], [3]. Such approaches fuse both the single and across frames' information and both general and specific model. These approaches perform well on several some quite challenging datasets. However for such approaches, the efficiency of detection must be taken into consideration. V. A. Prisacariu et al. [11] provides an excellent code of GPU-implementation of HOG detector, which reaches both high accuracy under complex environments and ideal time efficiency in our experiments.

Particle filter has become popular in the field of computer vision and robotics, especially for visual tracking [12], [4], [13], [14], [15]. A perfect introduction of particle filter algorithm can be found in [16]. It overcomes the limitation of Kalman filter, which is non-linear and multi-modal estimation. [12] proposes an Interactive MCMC-based particle filter visual algorithm, which successfully tracks the target in complex environment. In this paper, we utilize [12] as our tracker for each pedestrian, but when estimating the weight of each particle, we fuse more information from pedestrian detector and dynamic model.

2.2 Ego-Motion Estimation

Many interesting Visual Odometry [7], [17] or SLAM [18], [19] approaches exist, most of which are based on feature tracking and RANSAC-based filtering for outliers. In this paper, our goal is to distinguish the current driving behavior, instead of accurate estimation of ego-motion or mapping the surroundings. In this paper, we use [7] as an ego-motion estimator.

2.3 Dynamic Model

A serious problem for multi-person tracking is the dificulty in building a robust dynamic model, in another word, predicting targets future movement. In [10], stereo camera is used to estimate the depth of the scene, therefore the location of the pedestrians in 3-dimensional coordinate can be estimated and so does the ego-motion. However, in monocular application, this can hardly be done. However, recent research on trajectory analysis has provided a promising solution. In previous work, Trajectory analysis has successfully been applied in the field of video surveillance [8], [20], [21]. The analysis result can be used both as dynamic model and therefore enhance the tracking process and a classifier to distinguish the abnormal trajectories from normal ones. While on the moving platform, former approaches of trajectories analysis will fail, because the scene is changing, there will be a big difference of the trajectories of pedestrians when the car is moving forward and turning. So in this paper, instead of calculating the 3D location directly, we make prediction by the trajectory-clustering based dynamic model [8] in the training set under similar ego-motion. Here we consider three typical kinds of driving behavior that are moving forward, turning left and turning right.

3 Our Approach

3.1 Framework

As is shown in Fig. 2, we separate our tracking system into two parts: offline train ing and online tracking. In this figure, our contributions have been highlighted as red. In the training phase, we crop the behavior labels and trajectories (ext racted by our tracking algorith m without dynamic model) for the training video, and cluster the trajectories to build dynamic model for each driving behavior label. In the tracking phase, for each target in the test video, driving behavior label of current frame and its past trajectory is used to figure out which cluster it belongs to. And then, use that cluster to predict target's future movement.

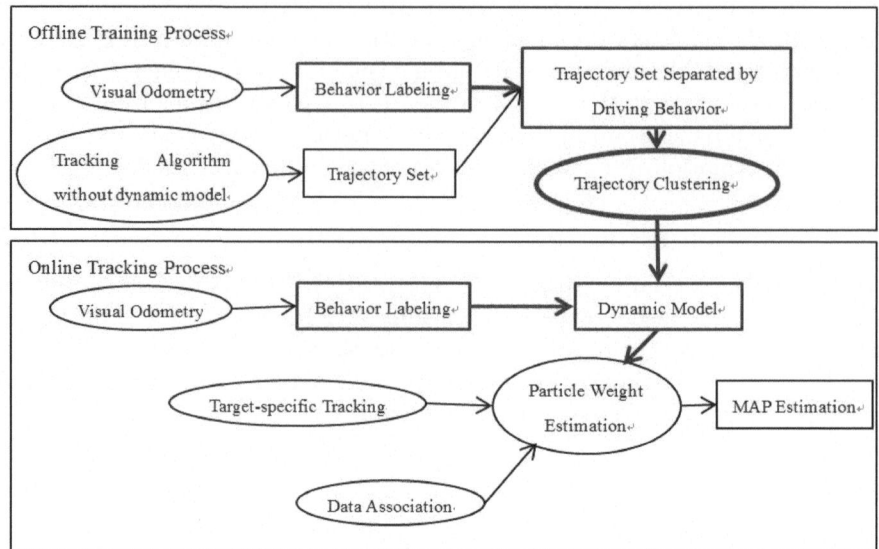

Fig. 2. Overview of our tracking system. Our contributions are highlighted in red.

3.2   Tracking

Our tracking algorithm fuses cues from four modules: visual odometry, data association, target-specific tracking, and dynamic model. The weight given to a proposed particle is estimated as a linear combination of three co mponents, similar but different from [4], as shown in (1),

$$w_{tr,p,l} = \alpha S_d(p,d) + \beta S_t(p,tr) + \gamma S_p(p,l) \tag{1}$$

where tr, p, l are the current tracker, particle, and label of the current driving behavior respectively, S_d, S_t, S_p are the importance terms given to detection, tracking and dynamic model respectively:

- Detection term, we apply greedy data association algorithm from [4] to match the detection results with trackers, and then penalizes the distance between the matched detection and the proposed position;
- Tracking term, estimated by the online learned observation model [12], which is a combination of 4 basic observation model selected from online updated template pool (including the hue, saturation, value and edge template from four most recent frames and the initial frame) by sparse principal component analysis;
- Dynamic model term, using the pre-trained dynamic model to make the prediction of the current movement, and penalizes the distance between the predicted and proposed positions of the particle.

In this paper, we use fasthog [11] as pedestrian detector and libviso2 [7] for visual odometry. Our contribution is mainly on dynamic model, we propose the integrated tracking system with trajectory clustering-based dynamic model, we cluster the trajectory separately on different driving behavior, so that our dynamic model can be adaptive to different driving behavior.

The parameters α, β, γ (set experimentally to 1.0, 1.5, 1.5 respectively) are controlling the contribution of each term.

3.3   Driving Behavior Labeling

Labeling the driving behavior is an important preparation for our dynamic model. We complete this part in two stages: First, we estimate the ego-motion of the vehicle using the monocular version of the library for visual odometry "libviso2" [7]. However, when used directly on our challenging video sequences, this stage suffers from continuous noise in the whole process and significant drift caused by severe illumination variations. Second, we filter out irregularities with a threshold on the acceleration and we label the current driving behavior by calculating the turning angle defined as:

$$\alpha_t \approx \sin \alpha_{t-N} = |\frac{s_{t-N} \times s_t}{|s_{t-N}| \cdot |s_t|}| \tag{2}$$

$$s_t = <x_t, y_t> - <x_{t-N}, y_{t-N}> \tag{3}$$

where $<x_t, y_t>$ is the location of the car in the world coordinate system and N is the interval of calculating the motion vector and we set it 5 to eliminate the noise. The we set a threshold to label the current driving behavior l_t. Experimentally, we set $\theta = 0.075$.

$$l_t = \begin{cases} \text{Move forward} & |\alpha_t| < \theta \\ \text{Turn left} & \alpha_t > \theta \\ \text{Turn right} & \alpha_t < -\theta \end{cases} \tag{4}$$

3.4   Dynamic Model

In our framework, dynamic models are used to propagate the hypotheses in the neighborhood of the target. Kalman filter for example, assumes a linear movement of the target. However linear assumption is improper in our case, because of the effects of perspective and ego-motion. So we design our non-linear dynamic model by clustering trajectories [8], [20]. Since "turning left" or "right" samples are less represented in the dataset than "moving forward" , clustering the trajectories separately for each driving behavior turns out to be a good choice.

To cluster the trajectories, we define the similarity between two trajectories as:

$$S(A, B) = d(A, B), +\delta(A, B) \tag{5}$$

where $d(A, B)$ and $f(A, B)$ are respectively the location and velocity Hausdorff distances [8] between A and B. δ controls the weight between $d(A, B)$

and $f(A, B)$ (experimentally δ is 0.25). After the similarity matrix is calculated, K-means based clustering (experimentally k is 5 for K-means clustering) is computed on the tree kinds of trajectories, for different driving behaviors labeled as Section 3.1 mentioned.

To predict the target's motion, we need to first identify the current set of trajectory clusters according to the current driving behavior, and then find out which cluster does the target belong to. In this paper, we define the likelihood that a trajectory A_T belongs to cluster k_T at time T as:

$$L(A_T, k_T) = \sum_{t=0}^{T} \frac{1}{2^T - t} log(p_c(T, k_T, A_T)) \tag{6}$$

$$p_c(T, k_T, A_T) = p_d(T, k_T, A_T) p_v(T, k_T, A_T) \tag{7}$$

$$p_v(T, k_T, A_T) = Gaussian(v_T^x, v_T^y; v_T^{px}, v_T^{py}, \sigma^2) \tag{8}$$

In (7), p_d is the density distribution of cluster k_T. In (8) v_T^x, v_T^y are velocities along the trajectory at time T, while v_T^{px}, v_T^{py} are the predicted velocities in the x and y coordinates. Eq.(8) means we estimate the current velocity score by a 2D Gaussian distribution centered on the predicted location with the perturbation of σ^2.

The cluster with highest matching score is selected as current dynamic model. To handle the irregular trajectories, we filter out those trajectories either are too short (experimentally the minimal length is 4) or do not belong to any of the clusters. If one cluster has been assigned to the trajectory, it will score every proposed particles in the current frame's particle filter weighting process, using the same approach as (7) (8) above.

4   Experimental Results

We evaluate our algorithm on the data recorded from a camera mounted on our experimental vehicle platform. The vehicle runs several rounds in our university during a winter sunny day for nearly 50 minutes (44306 frames), with strong illumination conditions and several people interacting. We separate the video into training and testing data, with 36100 frames and 8206 frames respectively. The tracking algorithm without prediction-by-clustering is processed on the training data to get the trajectories and train the dynamic models.

In this paper, we focus on evaluating our dynamic models, mainly in two aspects: first, we will show that our dynamic model successfully describes pedestrians' motion patterns under different types of ego-motion; second, quantitative comparision will evidence the usefulness of our dynamic models for tracking, comparing with the algorithm without our dynamic model.

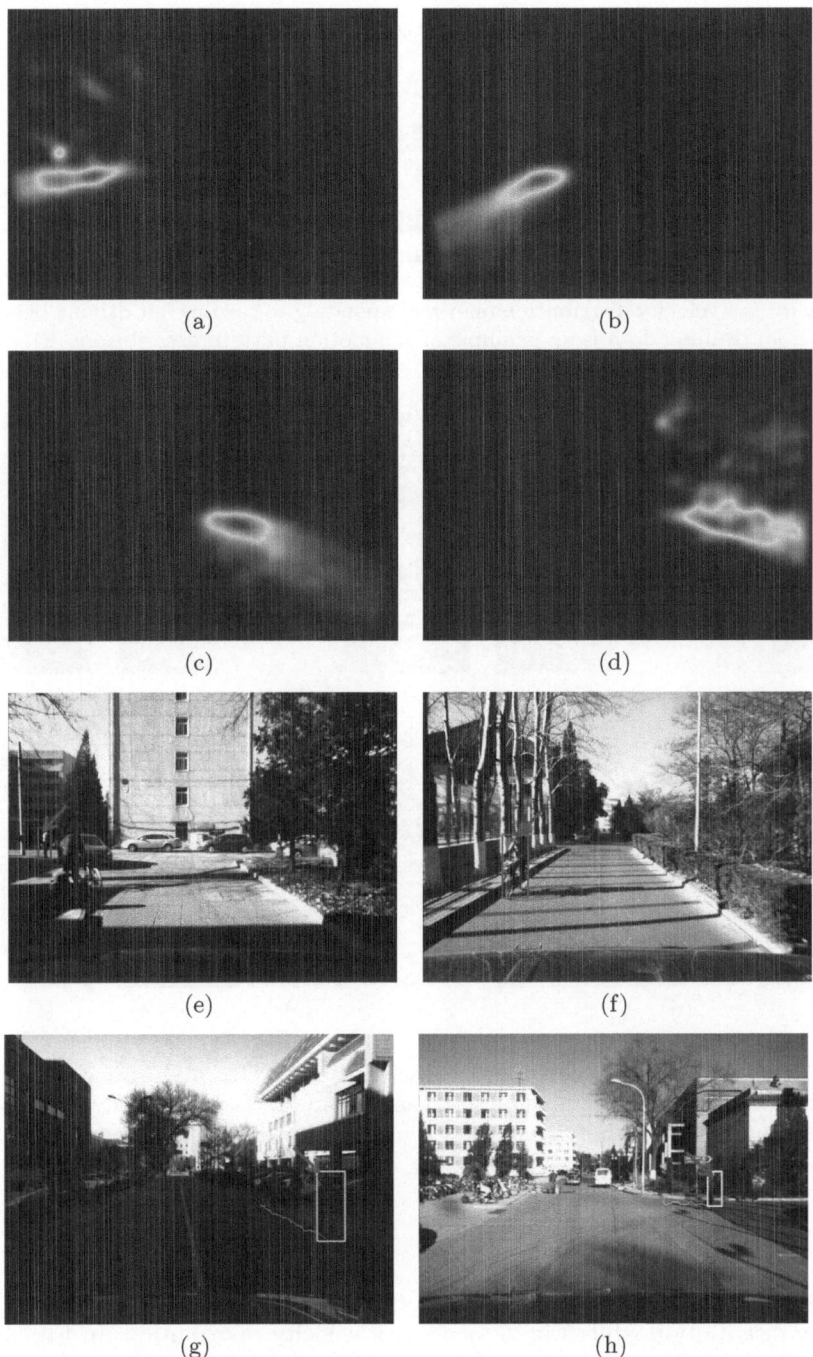

Fig. 3. Some of our clustering results. The first row are density distributions (in pseudo color) of four trajectory clusters when the car is moving forward, and the second row are corresponding pedestrians in the video.

Fig. 4. (a) is a velocity distribution map corresponding to turning left driving behavior, (b)(c) from training data is an example of the motion pattern describes in (a)

Fig. 5. comparison between tracking with dynamic model and without dynamic model. The two failure samples (a) and (b) are tracked without dynamic model, while (d) and (e) are with dynamic model. (c)(f) show the future movement predicted by our dynamic model.

4.1   Clustering Results

Fig. 3 and 6 show some of our trajectory clustering results. Fig. 3 focuses on density distribution while Fig. 6 focuses on velocity distribution. In Fig. 3, We can see that these four clusters describe the four different situations: (a) far from the left side of the vehicle, probably off-road; (b) near to the left side of the vehicle, they are probably on-road; (c) near to the right side of the vehicle; (d) far from the right side of the vehicle.

In Fig. 6, (a) is a velocity distribution map which describes one of the pedestrian motion pattern when the vehicle is turning left. The white line segment in (a) describes the estimated velocity vector from the camera's view. (b)(c) from training data is an example of the motion pattern described by (a), which is the pedestrian in front of the vehicle while the vehicle is turning left. Such motion pattern is non-linear and can be hardly described by dynamic model without eliminating the effects of ego-motion.

4.2   Tracking with Dynamic Model

In order to evaluate our tracking algorithm under different driving behaviors, we select 4 sequences from test dataset that includes the three driving behaviors (1, 4 are forward, 2 is turning left, 3 is turning right). Fig. 6 shows some results of these sequences. Table 1 illustrates the quantitative comparison between our algorithm with and without dynamic model. We can see that our dynamic model improves tracking under all the three driving conditions. Two examples in our test data shown in Fig.4 (a) and (d) are from tracking without our dynamic model, while (b) and (e) are from tracking with our dynamic model. (c) and (f) are the predicted next movement of the target by our dynamic model (the predicted motion vector is drew in the center of the bounding box and highlighted with a white arrow). For Seq. 4, the algorithm without dynamic model performs slightly better than with dynamic model, this is mainly because varying of the speed of vehicle may cause some drifts of movement prediction. From these above, we can see that our dynamic model can provide good prediction of target's future movement and successfully guide tracker to find its best location.

According to Table. 2, we can also find with our tracking method, the detection result can be improved greatly. With the tracking method, we can remove lots of false detection and improve the correct rate.

Table 1. Tracking results same to [1], "Objects" is the count of pedestian appearance in all frames. "# ids"is the number of pedestrians appeared in the sequence, W. and W.O. are short for with and without dynamic model(Section 4) respectively.

Seq.	Total Frame	Objects / # ids	True Positive W.	True Positive W.O.	False Positive W.	False Positive W.O.
1	248	684/22	518	511	28	41
2	126	514/10	304	302	33	35
3	238	697/11	332	298	23	90
4	163	460/9	333	337	16	12

Table 2. Comparison of detection result with/without tracking

	True Positive	Total	False Detection Rate
Detection without Tracking	316	679	53.5%
Detection with Tracking	431	441	2.3%

Fig. 6. Some of our tracking results.(a)(b)(c) are from seq.1, (d)(e)(f) are from seq.2, (g)(h)(i) are from seq.3, and (j)(k)(l) are from seq.4.

5   Conclusion

In this paper, we presented a multi-person tracking algorithm from moving platform using monocular only. We avoided calculating the exact 3D position of the targets and their absolute speed by a set of trajectories regarded as training set, instead we tried to eliminate the effect of ego-motion by finding similar trajectories under the same driving behavior. We showed the algorithm successfully predicts most of the targets future movement and indeed improve tracking results.

As a primary work of our far goal of intelligent vehicle with monocular camera only, many components of our work remain to be further improved and optimized. For instance, many pedestrians are missed by detector due to their abnormal pose (such as riding bicycle) or ill illumination condition. Knowing that the feature points on pedestrians are most likely to be outliers in visual odometry, we could properly use such information to enhance the detection process. Besides, manually filtering out false alarms and connecting broken trajectories in training set would also probably make result better. Besides, as can be seen from Table 1, the speed of ego-motion should also be included and a finer online learning dynamic model should also be taken into consideration in our future work.

References

1. Jungling, K., Arens, M.: Pedestrian tracking in infrared from moving vehicles. In: IEEE Intelligent Vehicles Symposium (2010)
2. Dalal, N., Triggs, B.: Histograms of oriented gradients for human detection. In: IEEE Conference on Computer Vision and Pattern Recognition, vol. 1, pp. 886–893 (2005)
3. Andriluka, M., Roth, S., Schiele, B.: People-tracking-by-detection and people-detection-by-tracking. In: IEEE Conf. on Computer Vision and Pattern Recognition, pp. 1–8. IEEE (2008)
4. Breitenstein, M.D., Reichlin, F., Leibe, B., Koller-Meier, E., Van Gool, L.: Online multiperson tracking-by-detection from a single, uncalibrated camera. IEEE Transactions on Pattern Analysis and Machine Intelligence 33, 1820–1833 (2011)
5. Song, X., Cui, J., Zha, H., Zhao, H.: Probabilistic detection-based particle filter for multi-target tracking. In: British Machine Vision Conference, BMVC (2008)
6. Zhao, T., Nevatia, R.: Tracking multiple humans in complex situations. IEEE Trans. Pattern Anal. Mach. Intell. 26, 1208–1221 (2004)
7. Libviso (2011), http://www.cvlibs.net/libviso2.html
8. Wang, X., Tieu, K., Grimson, E.: Learning Semantic Scene Models by Trajectory Analysis. In: Leonardis, A., Bischof, H., Pinz, A. (eds.) ECCV 2006, Part III. LNCS, vol. 3953, pp. 110–123. Springer, Heidelberg (2006)
9. Leibe, B., Seemann, E., Schiele, B.: Pedestrian detection in crowded scenes. In: Proc. IEEE Computer Society Conference on Computer Vision and Pattern Recognition (CVPR), vol. 1, pp. 878–885 (2005)
10. Ess, A., Leibe, B., Schindler, K., Van Gool, L.: A mobile vision system for robust multi-person tracking. In: IEEE Conf. on Computer Vision and Pattern Recognition. IEEE (2008)
11. Prisacariu, V., Reid, I.: fasthog-a real-time gpu implementation of hog. Department of Engineering Science, Oxford University, Tech. Rep. 2310 (2009)
12. Kwon, J., Lee, K.: Visual tracking decomposition. In: IEEE Conference on Computer Vision and Pattern Recognition, pp. 1269–1276. IEEE (2010)
13. Pérez, P., Hue, C., Vermaak, J., Gangnet, M.: Color-Based Probabilistic Tracking. In: Heyden, A., Sparr, G., Nielsen, M., Johansen, P. (eds.) ECCV 2002, Part I. LNCS, vol. 2350, pp. 661–675. Springer, Heidelberg (2002)
14. Isard, M., Blake, A.: ICONDENSATION: Unifying Low-Level and High-Level Tracking in a Stochastic Framework. In: Burkhardt, H.-J., Neumann, B. (eds.) ECCV 1998. LNCS, vol. 1406, pp. 893–908. Springer, Heidelberg (1998)

15. Khan, Z., Balch, T., Dellaert, F.: Mcmc-based particle filtering for tracking a variable number of interacting targets. IEEE Trans. Pattern Anal. Mach. Intell. 27, 1805–1819 (2005)
16. Arulampalam, M., Maskell, S., Gordon, N., Clapp, T.: A tutorial on particle filters for online nonlinear/non-gaussian bayesian tracking. IEEE Trans. on Signal Processing 50, 174–188 (2002)
17. Nistr, D., Naroditsky, O., Bergen, J.: Visual odometry. In: Proc. IEEE Computer Society Conference on Computer Vision and Pattern Recognition (CVPR), vol. 1, pp. 652–659 (2008)
18. Davison, A., Reid, I., Molton, N., Stasse, O.: Monoslam: Real-time single camera slam. IEEE Trans. Pattern Anal. Mach. Intell. 29, 1052–1067 (2007)
19. Tardif, J., Pavlidis, Y., Daniilidis, K.: Monocular visual odometry in urban environments using an omnidirectional camera. In: IEEE/RSJ International Conference on Intelligent Robts and Systems (IROS), pp. 2531–2538 (2008)
20. Song, X., Shao, X., Zhao, H., Cui, J., Shibasaki, R., Zha, H.: An online approach: Learning-semantic-scene-by-tracking and tracking-by-learning-semantic-scene. In: IEEE Conference on Computer Vision and Pattern Recognition, pp. 739–746. IEEE (2010)
21. Morris, B., Trivedi, M.: Learning trajectory patterns by clustering: Experimental studies and comparative evaluation. In: Proc. IEEE Computer Society Conference on Computer Vision and Pattern Recognition (CVPR), pp. 312–319 (2009)

Tracking the Untrackable:
How to Track When Your Object Is Featureless

Karel Lebeda[1], Jiri Matas[1], and Richard Bowden[2]

[1]Center for Machine Perception, Czech Technical University in Prague
[2]Centre for Vision, Speech and Signal Processing, University of Surrey

Abstract. We propose a novel approach to tracking objects by low-level line correspondences. In our implementation we show that this approach is usable even when tracking objects with lack of texture, exploiting situations, when feature-based trackers fails due to the aperture problem. Furthermore, we suggest an approach to failure detection and recovery to maintain long-term stability. This is achieved by remembering configurations which lead to good pose estimations and using them later for tracking corrections.

We carried out experiments on several sequences of different types. The proposed tracker proves itself as competitive or superior to state-of-the-art trackers in both standard and low-textured scenes.

1 Introduction

We present an approach to robustly track objects when they have limited or no visual features (such as distinctive texture). This is difficult as without consistent features many common assumptions used in tracking fail. We overcome this by using a novel formulation based on low level line correspondences which can operate with or without texture while avoiding the aperture problem.

Visual tracking is an active part of computer vision with a number of new approaches in recent years. The basic objective of tracking is, given a sequence of consecutive frames and the annotated pose of the object of interest in the first frame, to estimate the pose of this object in the rest of frames. Current techniques aim to learn the appearance of the tracked object [1, 2] or build a global model joining local trackers [3–5] to a robust frame.

Kalal *et al.*[1] proposed a method for on-line learning from positive and negative examples for tracking and detection (TLD, *track-learn-detect*). While positive samples arise from successful tracking, negative samples are found by contradictions. Other authors improve trackers by globally modelling the target. Matas and Vojir [4] joined local trackers (LK trackers [6]) to a *flock* and let each tracker converge to a feature good to track. Furthermore, they introduced new predictors of local tracker failure to cope with outliers. Similarly, Cehovin *et al.*[3] proposed a *coupled-layer visual model* in their LGT tracker, consisting of a local and global layer. While the local layer describes the target's local visual properties, the global layer encodes the target's global colour, motion and shape

J.-I. Park and J. Kim (Eds.): ACCV 2012 Workshops, Part II, LNCS 7729, pp. 347–359, 2013.

in a probabilistic manner. Dupac and Matas [5] used *zero shift points* for tracking – points with approximately even intensity function in their neighbourhood in all the directions. These points are tracked following the *shift field* and are connected hierarchically depending on the size of the neighbourhood.

These techniques works well when sufficient texture of the target object exists, which implies presence of features, which are good to track, e.g. blobs [7], Harris corners [8], or mentioned zero-shift points [5]. Unfortunately, real world scenes often contain objects without sufficient numbers of such features, or these lie on the boundary where the background affects their location and appearance. Without these features, even sophisticated methods like LGT or TLD often fail (see experimental evaluation). Conventional trackers often avoid edges because of the so called *aperture problem*, causing the edge points to be well defined only in one direction (perpendicular to the edge). However, with knowledge of this direction, a line correspondence can be established.

The edge features have been used in a number of previous articles, to solve a problem of tracking an object modelled by either 3D wireframe [9, 10] or by set of 2D edges [11, 12]. However, all of these approaches are based on the fitting of the *user-supplied* model to the image data. We are, on the contrary, trying to learn the object model online from the data, thus one of the challenges is to establish what features can be used to consistently track the object when no a priori information of the appearance is given.

Figure 1 illustrates the aperture problem when tracking a part of contour X_1 in frame f_1 to X_2 in f_2. True correspondences $\{a_2^*, b_2^*\}$ of points $\{a_1, b_1\}$ cannot be found directly as they are not uniquely defined in both parallel (to the edge) and perpendicular directions. When searching perpendicular to the edge, we find incorrect correspondences $\{a_2, b_2\}$ instead. If we assume a small movement between two consecutive frames, then we can expect these points to generate the same corresponding lines $\{k_2, l_2\}$ as true corresponding points. A point correspondence from the intersections of the lines (c_1, c_2) gives the true transformation regardless of the shift along the edge. The transformation

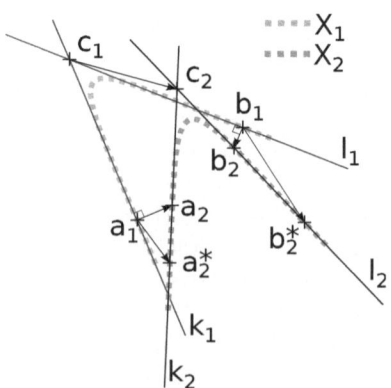

Fig. 1. Establishing edge correspondences (see text for detailed description)

between consecutive frames can be estimated directly from line correspondences, or from point correspondences of their intersections. Our approach is based on this principle.

The structure of this paper is as follows. We describe our tracking algorithm in section 2. In section 3 we address a question of points and lines and correspondences between frames; these correspondences are then used to estimate the inter-frame transformation in section 4. A long-term stability is addressed in section 5. The performance of our algorithm is experimentally evaluated in section 6. Finally, section 7 draws conclusions.

2   Algorithm Overview

The main objective of a tracking algorithm is to find the position of an object of interest in every frame of a video sequence. In other words, to find a transformation T_t from *model space* to the tracked area in every frame f_t. We estimate this transformation by transformation S_t of tracked areas of two consecutive frames f_{t-1} and f_t ($S_t = T_t \circ T_{t-1}^{-1}$). T_1 is supplied by the user in the form of an initial area annotated in the image space. In this work we restrict S_t to a *similarity transformation*.

When the frame f_1 is processed, the initial set of N_1 edge points $\{p_1^{[i]} | i \in \{1, ..., N_1\}\}$) is generated in the model space and transformed to the image space by the user-supplied T_1. Then lines $l_1^{[i]}$ defined by points are computed.

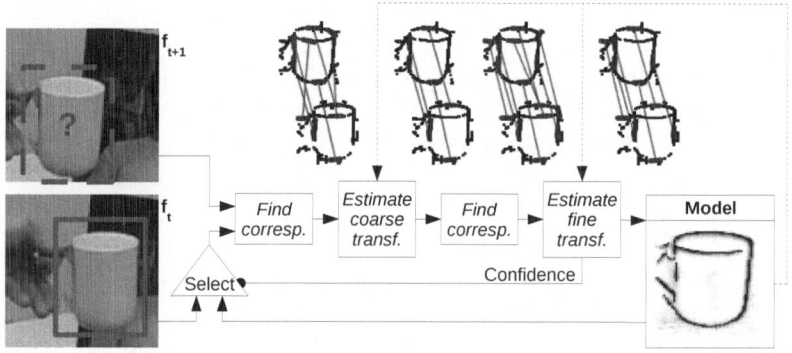

Fig. 2. Overview of the tracking algorithm

An overview of the iterative tracking procedure is outlined in Figure 2. Firstly, given two consecutive frames f_{t-1} and f_t, initial correspondences $(p_{t-1}^{[i]}; q_t^{[i]})$ are found. There are several ways to do this, we use a *guided edge search* (section 3.1) for coarse correspondences. A smooth movement is assumed and therefore points from the previous frame, moved by the transformation of the previous step $(S_{t-1}(p_{t-1}^{[i]}))$, can be used as an initial estimate of the new point locations. These correspondences are then used as input to a modified LO-RANSAC (section 4, [13, 14]) to find a coarse estimate of the transformation (S'_t and thus T'_t).

Using the coarse transformation estimate S'_t, new correspondences are computed by moving the points $p^{[i]}_{t-1}$ and employing an *unguided edge search* (see section 3.1). LO-RANSAC is then repeated with these correspondences to refine the transformation to S_t. S_t is usually more precise than S'_t and its number of inliers (which will be retained for the future computations) is higher.

As a measurement of the quality of the estimation, we introduce an *evidence score* E_t of the transformation S_t. E_t measures the fitness of points $p^{[i]}_{t-1}$ to the image of f_t and allows drift or tracking failures to be detected. In such a situation, the algorithm tries to recover and correct its pose.

We define a set of *inliers* of the resulting transformation as a subset of correspondences having an error smaller than or equal to a predefined *error threshold*:

$$\mathcal{I}_t = \left\{ \left(p^{[i]}_{t-1}; q^{[i]}_t \right) \middle| d \left(p^{[i]}_{t-1}, q^{[i]}_t | S_t, f_{t-1}, f_t \right) \leq \theta; \ i \in \{1, ..., N_{t-1}\} \right\} , \quad (1)$$

where d is a geometric error of corresponding lines defined by $p^{[i]}_{t-1}$ in frame f_{t-1} and $q^{[i]}_t$ in f_t. The points $q^{[i]}_t$ of inliers are retained for the next frame. To ensure a stable number of points we add a set of newly generated points to them. The new points are not cropped strictly to the tracked area and thus allow the model to grow slightly outside the original area.

3   Obtaining and Use of Correspondences

3.1   Search for Edge Correspondence

Unguided. Searching for the nearest strong edge is carried out in the direction of the gradient of image intensity [9]. Candidates for matching edge points are rated according to their magnitude of gradient and distance from the initial position by applying a Gaussian weighting. This process is iteratively repeated to convergence.

Guided. The guided searching of edges works in a different manner as we are not looking for a *strong edge* but for a *similar edge*. Instead of searching only in the direction of gradient, searches are also performed at angles shifted by $\frac{\pi}{20}$ and $\frac{\pi}{10}$ to both sides. The local gradient maxima are extracted and their similarities to the original edge are compared in terms of *change of gradient angle*, *change of appearance* and *spatial proximity*. This process has no iterations, correspondences are found in a single step.

3.2   Creation of Lines

Every line $l^{[i]}_t$ is computed from its defining point $p^{[i]}_t$ and orientation $\alpha^{[i]}_t$ (an angle of the image gradient). As angles of the normal vectors of lines are used in oriented evidence measurement (E_t, see section 4.1) and it is essential to distinguish angles with opposite orientation, normal vectors of lines must have angles in accordance with the image intensity gradient. Thus lines are "oriented".

3.3   Geometric Error of Line Correspondences

An aim of the algorithm is to find the "best" transformation S_t between two consecutive frames. The "best" usually means the one, which minimises some (robust) function of distance between projected and measured correspondences. But what does it mean for two lines l and l' to be close to each other?

Hartley [15] stated that distance (or geometric error) of lines has to be measured with respect to some point of interest. He suggested to use the distance between a line and line segment. This approach yields usable results. However, in our case it is necessary to calculate intersections between the lines and all four sides of the tracked quadrilateral and computational complexity is prohibitively large.

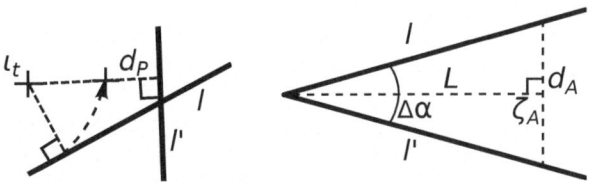

Fig. 3. Geometrical meaning of d_P and d_A

A faster approach is to see the distance as two independent components – difference of positions d_P and difference of angles d_A, with respect to a given *point of interest* (ι_t, e.g. centre of gravity of tracked area corners, as can be seen in Fig. 3). The error of position is defined as the difference between the perpendicular distances from the lines to ι_t, in normalised homogeneous coordinates as:

$$d_P = \left| \iota_t^T l \right| - \left| \iota_t^T l' \right| . \tag{2}$$

The error of angle is defined as the length of the shortest possible line segment with endpoints on the lines, going through point ζ_A, which is equidistant to the lines and its distance to their intersection is equal to L (L can be derived from the size of the tracked area, or set manually).

$$d_A = 2 \cdot L \cdot \tan \frac{\Delta\alpha}{2} , \tag{3}$$

where $\Delta\alpha = \alpha_t^{[i]} - \alpha_t^{[j]}$ is the angle between the lines, and finally

$$d = \sqrt{d_P^2 + d_A^2} . \tag{4}$$

This approach gives errors similar to Hartley's in significantly lower time (10-fold speed-up with correlation coefficient 0.9). It should be noted that d_P is strongly underestimated in the case of ι_t laying *between* the lines. However, as we are usually concerned with the distance of lines that are close to each other, this condition appears rarely (correspondences are incorrectly classified as inliers less than one percent of the time).

3.4   Transformation from Corresponding Lines

Generally, every line correspondence yields two equations regardless of an estimated transformation (linear in homogeneous coordinates), as well as point correspondences. Nevertheless, in the case of similarity, two lines are obviously not enough (scale ambiguity). The similarity transformation should be computed from at least three line correspondences. Line equations can be directly used, or alternatively equations from point correspondences of intersections.

In contrast to using points, an algebraic error of line correspondence is very different from the geometric error. Therefore the linear least squares solution is not viable. Hence the sum of squares of the geometric errors is minimised by a numerical iterative optimisation.

4   Frame-to-Frame Transformation

4.1   LO-RANSAC

The minimal sample is composed of three line correspondences. Intersections of the sampled lines are used for the computation of the hypothesis of S_t as a least squares solution (we have six linearly independent equations and only four degrees of freedom) from the point correspondences.

Standard (LO-)RANSAC use the number of inliers as a measurement of the quality of an estimated transformation. However, in the case of a cluttered background occupying a significant portion of the tracked area, the background-induced transformation may outweigh the correct one. Therefore we propose a different approach, measuring the quality of consistency of two frames, with respect to tracked points and to the evaluated transformation.

For every frame f_t, all the edges are detected by a Canny edge detector [16] and a distance transformation is performed. Evidence $e_t^{[i]}$ of a point $p_{t-1}^{[i]}$ in f_t is given by a modified oriented Chamfer distance [17, 18] of this projection as a product of inverse distance and an orientation weight:

$$e_t^{[i]} = e_{d;t}^{[i]} \cdot e_{A;t}^{[i]} = \frac{1}{1 + \left\| S_t(p_{t-1}^{[i]}) - c_t^{[i]} \right\|_2} \cdot \left(\frac{\cos(\alpha_{t-1}^{[i]} + \rho(S_t) - \alpha_t^{c[i]})}{2} + \frac{1}{2} \right) , \quad (5)$$

where $c_t^{[i]}$ is Canny's edge point in f_t nearest to $S_t(p_{t-1}^{[i]})$, $\alpha_{t-1}^{[i]}$ is the direction of gradient at point $p_{t-1}^{[i]}$ in f_{t-1}, $\alpha_t^{c[i]}$ is the direction of gradient at point $c_t^{[i]}$ in f_t and $\rho(S_t)$ is the rotation angle, given by transformation S_t.

Overall evidence of the transformation E_t is then computed as a mean evidence of all the points. To avoid situations when a solution is converging to a local optimum, representing impossible movements, a regularisation term is included as a multiplicative factor. This is a function of a scale change and of an overlap of old and new tracked areas.

$$E_t = \frac{1}{N_t} \sum_{i=1}^{N_t} e_t^{[i]} \cdot \min\left(\frac{\Gamma_t}{\Gamma_t'}, \frac{\Gamma_t'}{\Gamma_t} \right) \cdot \min\left(1, 2 \cdot overlap \right) , \quad (6)$$

where Γ_t is scale change of S_t and Γ_t' is an expected scale change.

4.2   On-Line Learning and Using Point Quality

For an area where the points had predicted a correct transformation in previous frames, it has a high probability of having good points in future frames. The image evidence $e_t^{[i]}$ is used as a point quality measurement. The *point quality field* Q_t is learned from these as follows. Q_1 in the first frame is taken directly from detected edge points $p_1^{[i]}$. At the end of processing each frame f_t, the Q_{t-1} is transformed by the estimated S_t and a forgetting factor employed by multiplying by a constant decay. The evidence $e_t^{[i]}$ of points $p_{t-1}^{[i]}$ projected to f_t is then added.

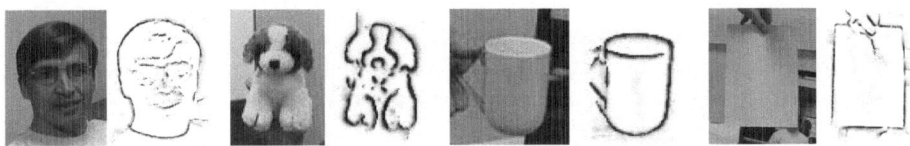

Fig. 4. Examples of images and learned point qualities

The resulting transformation obtained by LO-RANSAC is noisy despite the minimisation of geometric error in the LO step. This causes inaccuracies in estimated motion and thus drift. To remove this drift, the learned field Q_{t-1} is used. Points from the new frame are back-projected $(S_t^{-1}(q_t^{[i]}))$ and the fit is measured as a mean point quality at their positions. Parameters of the transformation are refined to maximise this fitting score by non-linear iterative optimisation.

5   Long-Term Relations

When a sudden decrease of evidence E_t is detected, confidence in the solution will be low and there may be strong drift or total loss of tracking. Then *correction* arises. The procedure of finding correspondences and the transformation is repeated with frame f_1 and initial model M_1 used instead of f_{t-1} and M_{t-1} and possibly with other frames and their models, if previously learned. A comparison of the fitness is carried out in terms of E_t and obtained inlier ratio. One of following situations appears (illustrated in Figure 5).

If the *current estimate* of S_t is the best solution (in terms of evidence score and inlier ratio), the current transformation is kept. Model M_{t-1} is transformed by estimated S_t and updated to get M_t. If the correcting model is a better fit than the current estimate then this *correcting transformation* is used rather than the current estimate and M_t is obtained from the correcting model (e.g. M_1). In this case, the assumption of a smooth movement has to be suppressed, as the correction transformation is not related to an actual movement of the tracked object but to a recovery from failure or drift. In the case, when both the current

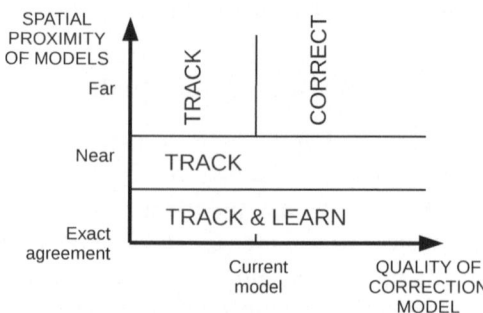

Fig. 5. Possible situations during correction, comparison of active (current) model M_{t-1} and the best correcting model

and the best correcting transformations yield similar movement of the tracked area, current estimate is kept. Optionally, this model (as proving itself as leading to the good estimate) may be learned for future corrections.

6   Experimental Evaluation

The performance of the trackers was evaluated on sequences tabulated in Tab. 1 and shown in Figures 6, 8, 10 and 12 (our results superimposed). The first two selected sequences are used for evaluation in a number of previous publications. The latter two are new, obtained specially for their lack of texture. Supplementary material includes videos of all the sequences. Original sequences and the improved ground truth points are made available to the wider community at the website: http://cmp.felk.cvut.cz/~lebedkar/sequences/.

To asses trackers' performance, the distance of the centre of the tracked area from its ground truth position was measured as well as an error in the scale estimation (size of the target object; the logarithm of ratio to the ground truth is shown in the graphs, 0 means no error at average). All the measurements are averaged over 20 runs. The results can be seen in Fig. 7, 9, 11 and 13.

The performance of the proposed FLOtrack (Feature-Less Object tracker) was compared to several recently published trackers, representing different approaches:

Table 1. Used Videosequences

Name	Resolution	Frames	Colour	Challenges	Prev. Used
DUDEK	720×480	1 145	grey	appearance change, occlusion, changing viewpoint	[2, 19]
DOG	320×240	1 353	grey	changes in scale, occlusion	[20, 21]
MUG	640×480	737	RGB	lack of texture, changes in scale, background	*new*
PAGE	640×480	539	RGB	lack of texture, changes in shape, background	*new*

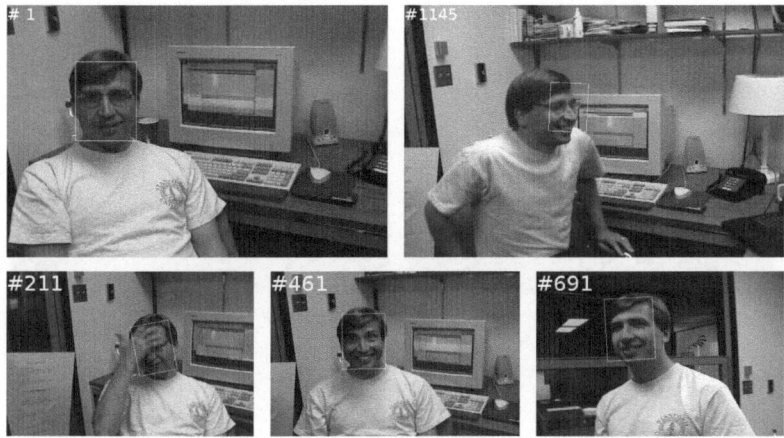

Fig. 6. Selected frames from DUDEK sequence

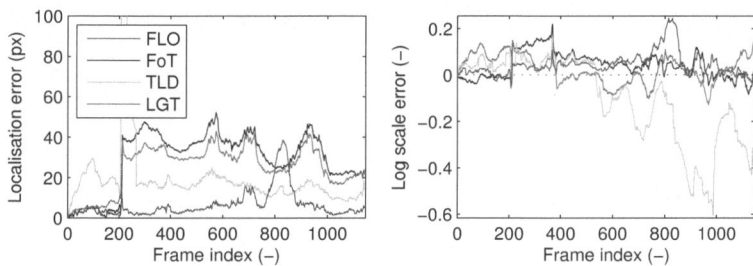

Fig. 7. DUDEK sequence evaluation

LGT by L. Cehovin [3], Flock of Trackers by T. Vojir (FoT [4]) and Z. Kalal's TLD [1]. The same settings were used for all the sequences.

6.1   Discussion

Dudek: The most challenging part is at about the 210th frame, when the face is occluded by the right hand. While FLOtrack's pose is corrected in several frames, TLD needs about 50 frames and other trackers never fully recover. FLO also experiences difficulties around frame 800. Here background points influence tracking and cause drift. Nevertheless, FLO recovers in every run.

Dog: In the range of frames between 700th and 1200th the challenges of this sequence are apparent. While FLO has no major problems and FoT experiences only light scale drift, the two others have severe problems, both in localisation and scale estimation.

Mug: Fig. 11 illustrates the inability of the LGT tracker on this scene. With no texture, the points simply drift off the mug and stay at the person's wrist. TLD consistently suffers from underestimation of the tracked area and sometimes loses

Fig. 8. Selected frames from DOG sequence

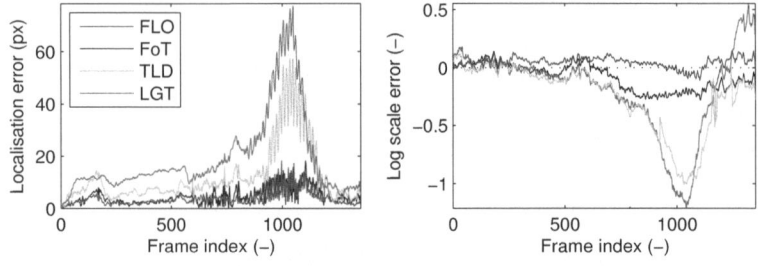

Fig. 9. DOG sequence evaluation

the object totally. FoT works well in this sequence, while FLO is comparable up to around frame 400 at which point FLO lost tracking in approximately half of the runs, resulting in a poor average score.

Page: Fig. 13 shows performance of the trackers for this sequence. LGT performs similarly to the mug sequence, all the points stabilise at person's hand and wrist. FoT uses only features from fingers and TLD often loses tracking and rarely re-detects even when paper returns to a pose similar to the starting one. FLO experiences difficulties, but still significantly outperforms all the others.

Additionally, we carried out qualitative tests of FLOtrack on further sequences with different challenges, such as a strong illumination change and a low reso-lution. The results are positive as FLO works well even in these conditions (see supplementary material).

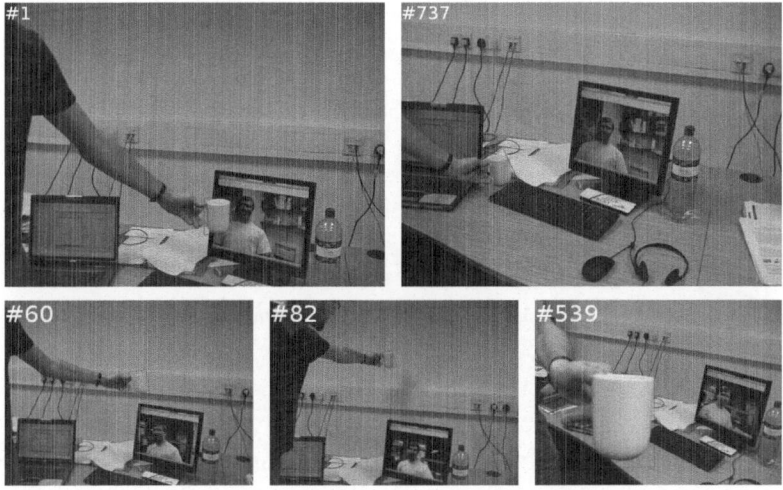

Fig. 10. Selected frames from MUG sequence

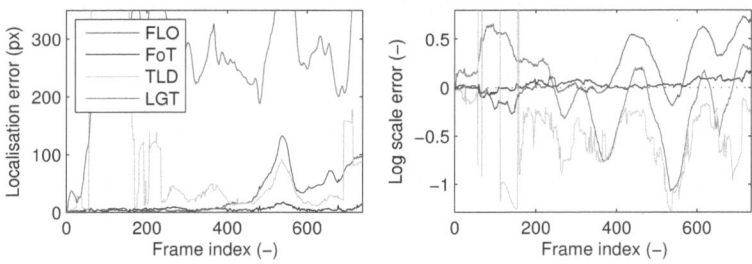

Fig. 11. MUG sequence evaluation

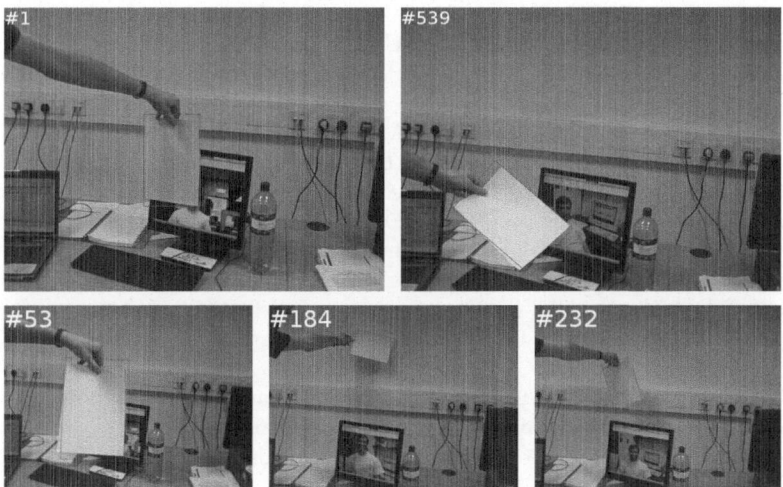

Fig. 12. Selected frames from PAGE sequence

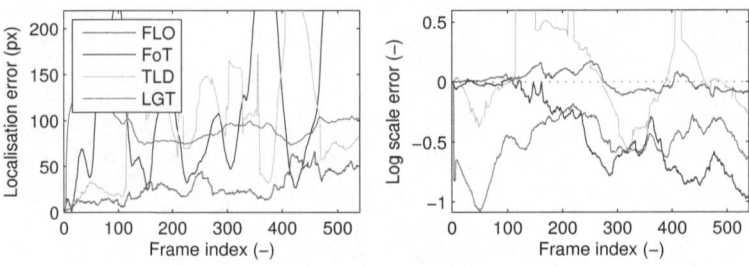

Fig. 13. PAGE sequence evaluation

7   Conclusion

We proposed and implemented a novel tracking algorithm based on low-level line correspondences with significantly lowered dependency on image features/texture. The tracker gives results competitive or superior to state-of-the-art trackers.

In future work, re-detection should be employed to upgrade to the long-term tracking. To increase stability, a memory holding the history of successful estimation should be increased from just remembering positions of good edgels (in fact learning the object contour) could be extended to hold configurations/combinations of complementary points (e.g. line triplets).

Acknowledgement. The authors were supported by the following projects: SGS11/125/OHK3/2T/13, GACR P103/12/2310 and EPSRC EP/I011811/1.

References

1. Kalal, Z., Matas, J., Mikolajczyk, K.: P-N learning: Bootstrapping binary classifiers by structural constraints. In: Proc. of CVPR, pp. 49–56 (2010)
2. Grabner, H., Grabner, M., Bischof, H.: Real-Time Tracking via On-line Boosting. In: Proc. of BMVC (2006)
3. Cehovin, L., Kristan, M., Leonardis, A.: An adaptive coupled-layer visual model for robust visual tracking. In: Proc. of ICCV (2011)
4. Matas, J., Vojir, T.: Robustifying the flock of trackers. In: Proc. of Computer Vision Winter Workshop, pp. 91–97 (2011)
5. Dupac, J., Matas, J.: Ultra-fast tracking based on zero-shift points. In: Proc. of ICASSP, pp. 1429–1432 (2011)
6. Lucas, B.D., Kanade, T.: An iterative image registration technique with an application to stereo vision. In: Proc. of Imaging Underst. Workshop, pp. 121–130 (1981)
7. Mikolajczyk, K., Schmid, C.: Scale and affine invariant interest point detectors. International Journal of Computer Vision 60, 63–86 (2004)
8. Harris, C., Stephens, M.: A combined corner and edge detector. In: Proc. of Alvey Vision Conference, pp. 147–151 (1988)
9. Harris, C., Stennett, C.: Rapid – a video rate object tracker. In: Proc. of BMVC (1990)

10. Drummond, T., Cipolla, R.: Real-time visual tracking of complex structures. IEEE Trans. PAMI 24, 932–946 (2002)
11. Tsin, Y., Genc, Y., Zhu, Y., Ramesh, V.: Learn to track edges. In: Proc. of ICCV, pp. 1–8 (2007)
12. Beveridge, J.R., Riseman, E.M.: How easy is matching 2D line models using local search? (1997)
13. Chum, O., Matas, J., Kittler, J.: Locally Optimized RANSAC. In: Michaelis, B., Krell, G. (eds.) DAGM 2003. LNCS, vol. 2781, pp. 236–243. Springer, Heidelberg (2003)
14. Fischler, M.A., Bolles, R.C.: Random sample consensus: A paradigm for model fitting with applications to image analysis and automated cartography. Communications of the ACM 24, 381–395 (1981)
15. Hartley, R.I.: Projective reconstruction from line correspondences. In: Proc. of CVPR, pp. 903–907 (1994)
16. Canny, J.: A computational approach to edge detection. IEEE Trans. PAMI 8, 679–698 (1986)
17. Olson, C.F., Huttenlocher, D.P.: Automatic target recognition by matching oriented edge pixels. IEEE Trans. Image Processing 6, 103–113 (1997)
18. Shotton, J., Blake, A., Cipolla, R.: Multiscale categorical object recognition using contour fragments. IEEE Trans. PAMI 30, 1270–1281 (2008)
19. Jepson, A.D., Fleet, D.J., El-Maraghi, T.F.: Robust online appearance models for visual tracking. IEEE Trans. PAMI 25, 1296–1311 (2003)
20. Ross, D., Lim, J., Yang, M.-H.: Adaptive Probabilistic Visual Tracking with Incremental Subspace Update. In: Pajdla, T., Matas, J. (eds.) ECCV 2004. LNCS, vol. 3022, pp. 470–482. Springer, Heidelberg (2004)
21. Chen, M., Pang, S.K., Cham, T.J., Goh, A.: Visual tracking with generative template model based on riemannian manifold of covariances. In: Proc. of Int. Conf. on Information Fusion, pp. 874–881 (2011)
22. Hartley, R.I., Zisserman, A.: Multiple View Geometry in Computer Vision. Cambridge University Press (2004)

A Robust Particle Tracker via Markov Chain Monte Carlo Posterior Sampling

Fasheng Wang[1,2], Mingyu Lu[1,*], and Liran Shen[1]

[1]School of Information Science and Technology, Dalian Maritime University
No.1 Linghai Road, Dalian 116026 China P.R.
[2]Department of Computer Science and Technology,
Dalian Neusoft Institute of Information
No.8 Software Park Road, Dalian 116023 China P.R.

Abstract. Particle Filters have grown to be a standard framework for visual tracking. This paper proposes a robust particle tracker based on Markov Chain Monte Carlo method, aiming at solving the thorny problems in visual tracking induced by object appearance change, occlusion, background clutter, and abrupt motion, etc. In this algorithm, we derive the posterior probability density function based on second order Markov assumption. The posterior probability density is the joint density of the previous two states. Additionally, a Markov Chain with certain length is used to approximate the posterior density, which consequently improves the searching ability of the proposed tracker. We compare our approach with several alternative tracking algorithms, and the experimental results demonstrate that our tracker is superior to others in dealing with various types of challenging scenarios.

1 Introduction

Visual tracking gains special attention in computer vision community due to its success in various real-world applications, such as intelligent visual surveillance, human-computer interaction, traffic monitoring or video indexing [1]. It has recently been addressed in real-world scenarios rather than a lab environment by many researchers, because it is much more challenging in complex real-world settings. Among the large amount of algorithms proposed in the literature, particle filter based tracking method has been applied with great success in solving various kinds of visual tracking problems. The basic idea behind such a method is to approximate the posterior probability density function recursively with a set of weighted particle (or sample) set evolving in the state space. The estimated object states can be as close to the optimal states as possible.

Isard [2] firstly used the particle filter, which was called CONDENSATION, to solve visual tracking problem. Thereafter, many researchers have been working on this topic. Jess Martnez-del-Rincn [3] proposed a Rao-Blackwellised particle filter (RBPF) based tracking algorithm, aiming at handling the uncertainties

---

* Corresponding author.

J.-I. Park and J. Kim (Eds.): ACCV 2012 Workshops, Part II, LNCS 7729, pp. 360–371, 2013.

induced by illumination change and short-time occlusions. He introduced a joint image characteristic-space tracking scheme which updates the model simultaneously to the object location, and the RBPF is used to avoid the curse of dimensionality. However, the algorithm can be failed in the face of long time full occlusions, or abrupt change of object position. Khan [4] adopted Markov random field (MRF) based motion prior model accompanied with Markov Chain Monte Carlo method within the particle filtering framework, to track multiple interacting ants. But the proposed tracking scheme only works well in lab environment with static background, and tends to fail in real-world scenarios. On the basis of Khans work, Cong [5] et al. proposed a new visual tracking algorithm within the particle filtering framework, which combined the color-based observation model with a detection confidence density obtained from the histograms o f oriented gradients (HOG) descriptor. The algorithm showed improved robustness in slight object occlusions with static background. However, the authors did not take into account the complex scenarios with severe appearance change and full occlusion, and the algorithm can only yield limited improvement. Choo [6] et al. proposed a hybrid Monte Carlo (HMC) method for 3D human tracking, which is much faster than the particle filter in 28D state space. But the authors did not use the algorithm in real-world applications, and it failed when several similar objects appeared in the scenario. Most of the algorithms in the literature are generally based on smooth motion assumption, that is, the target being tracked is moving with stable motion. However, in many real world applications, abrupt motions often appear due to camera switching, low frame rate video and uncertain object dynamics, which could cause the conventional tracking approach to fail since they violate the motion smoothness constraint. Kwon [7] et al. proposed a Wang-Landau Monte Carlo (WLMC) based tracking algorithm within the Bayesian filtering framework. The algorithm adopted a density of states (DOS) based prior distribution, and the images containing the object are divided into several subspaces with equal size. The tracker is guided to update the object state using the DOS computed online. Experimental results demonstrated that the tracker can handle different kinds of abrupt motions. But there is no rigorous convergence theory to support its convergence, and it could only achieve limited precision in statistical and physical applications [8]. Based on this work, Zhou [9] et al. proposed an adaptive stochastic approximation Monte Carlo (ASAMC) sampling based particle tracking algorithm. The authors constructed explicitly a DOS based trial distribution to replace the traditional filtering distribution as the proposal distribution, and all the samples are drawn from this trial distribution using the Metropolis-Hastings (MH) algorithm. The algorithm showed improved efficiency and accuracy in handling various abrupt motions. However, both the WLMC and the ASAMC based tracking algorithm are sensitive to background clutter and lighting condition changes. Moreover, it fails to track the object when the appearance changes drastically and the object is occluded persistently by obstacles. As is known to all, most of the particle filter based tracking algorithms in the literature are all based on first-order Markov assumption, that is, the current state at time t only depend on the state at time t-1. Although the assumption can simplify the expression of the posterior probability density function and the implementation of visual tracking, it suffers several disadvantages. On the one hand, the first-order assumption cannot accurately model the dynamics of the tracking objects. On the other hand, if the particles at time t-1 are lost or delayed, the performance of the algorithm would be severely affected.

In this paper, we address the tracking problems in different complex scenarios and proposed a new tracking algorithm based on Markov Chain Monte Carlo posterior sampling within the particle filtering framework. Firstly, the algorithm is based on second-order Markov assumption. We assume that the object state x_t depends on the states of the previous two time instant, x_{t-1} and x_{t-2}. This assumption can enhance the robustness of the tracker. Secondly, we adopt a Markov Chain with certain length to approximate the posterior probability density function discarding the traditional importance sampling based methods which tends to fail due to sample impoverishment. The Markov Chain is simulated using the MH algorithm; consequently the posterior density function is approximated using a set of unweighted sample sets $\{x_t^i\}_{i=1}^N$. This strategy can avoid the sample impoverishment problem suffered by traditional particle filter based trackers and avoid the local trap problem during visual tracking. The experimental results demonstrated that the proposed tracking algorithm showed improved robustness and accuracy in different types of challenging tracking scenarios.

2   Bayesian Tracking and Particle Filter

Suppose the object state x_t at time t is composed of position and scale information, $x_t=(x_t^p,x_t^s)$, where x_t^p is represented as the center of the object, and x_t^s is represented using the height and width of the object. The object tracking problem can be formulated as the Bayesian filtering problem, that is to say, recursively estimate the hidden state variable x_t, given a series of observations $z_{1:t}=\{z_1,z_2,...,z_t\}$ up to time t. The posterior probability density of state variable x_t is $p(x_{0:t}|z_{1:t})$, but the filtering density $p(x_t|z_{1:t})$ is often obtained which can be estimated by,

$$p(x_t|z_{1:t}) = cp(z_t|x_t) \int p(x_t|x_{t-1})p(x_{t-1}|z_{1:t-1}\mathrm{d}t \tag{1}$$

where $p(x_t|x_{t-1})$ is the system transition model describing the state evolution, $p(z_t|x_t)$ is the observation model, and c is a normalizing constant. After obtaining the posterior probability $p(x_t|z_{1:t})$, we can compute the Maximum a Posteriori (MAP) estimate over the sample set,

$$x_t^{MAP} = \mathrm{argmax} p(x_t^{(i)}|z_{1:t}), i = 1, ..., N \tag{2}$$

The estimation x_t^{MAP} calculated by (2) is considered as the best state estimates for the current time instant. But it is usually infeasible to calculate the integral in (1), especially for high dimensional state space. The particle filter is based on importance sampling, which uses a set of weighted samples $\{x_t^i, \omega_t^i\}_{i=1}^N$ to approximate the posterior probability distribution,

$$p(x_{0:t}|z_{1:t}) \approx \sum_{i=1}^N \omega_t^i \delta(x_{0:t} - x_{0:t}^i) \tag{3}$$

All the samples are drawn from a so called importance density $q()$, and the sample weights under first-order Markov assumption is updated recursively using,

$$\omega_t^i \propto \omega_{t-1}^i \frac{p(z_t|x_t^i)p(x_t^i|x_{t-1}^i)}{q(x_t^i|x_{t-1}^i, z_t)} \tag{4}$$

Details of the derivation of (4) are discussed in [10]. A main drawback of the conventional importance sampling based particle filter is the sample impoverishment problem, which can greatly deteriorate the performance of the tracking algorithm in real-world applications leading to local-trap problem especially in multiple objects tracking. Although many improvement strategies have been proposed in the literature, the applications of these are very limited in visual tracking.

3   Markov Chain Monte Carlo Sampling Based Particle Tracker

3.1   Second-Order Markov Assumption

The second-order Markov assumption assumes that the current state x_t depends on the previous two states x_{t-1}, x_{t-2}, so we can obtain

$$p(x_t|x_{0:t-1}) = p(x_t|x_{t-2:t-1}) \tag{5}$$

Based on this assumption, the posterior probability density $p(x_{0:t}|z_{1:t})$ is,

$$p(x_{0:t}|z_{1:t}) = \frac{p(z_t, x_{0:t}, z_{1:t-1})}{p(z_t, z_{1:t-1})} \tag{6}$$

$$= \frac{p(z_t|x_{0:t}, z_{1:t-1})p(x_{0:t}, z_{1:t-1})}{p(z_t, z_{1:t-1})} \tag{7}$$

$$= \frac{p(z_t|x_{0:t}, z_{1:t-1})p(x_{0:t}|z_{1:t-1})p(z_{1:t-1})}{p(z_t|z_{1:t-1})p(z_{1:t-1})} \tag{8}$$

$$= \frac{p(z_t|x_{0:t}, z_{1:t-1})p(x_{0:t}|z_{1:t-1})}{p(z_t|z_{1:t-1})} \tag{9}$$

$$= \frac{p(z_t|x_{0:t}, z_{1:t-1})p(x_t, x_{0:t-1}|z_{1:t-1})}{p(z_t|z_{1:t-1})} \tag{10}$$

$$= \frac{p(z_t|x_{0:t}, z_{1:t-1})p(x_t|x_{0:t-1}, z_{1:t-1})p(x_{0:t-1}|z_{1:t-1})}{p(z_t|z_{1:t-1})} \tag{11}$$

$$\propto p(z_t|x_{0:t}, z_{1:t-1})p(x_t|x_{0:t-1}, z_{1:t-1})p(x_{0:t-1}|z_{1:t-1}) \tag{12}$$

$$= p(z_t|x_t)p(x_t|x_{t-2:t-1})p(x_{0:t-1}|z_{1:t-1}) \tag{13}$$

In our tracking algorithm, we adopt the joint PDF $p(x_{t-1:t}|z_{1:t})$ under second-order Markov assumption. According to (13), the density $p(x_{t-1:t}|z_{1:t})$ can be formulated as:

$$p(x_{t-1:t}|z_{1:t}) \propto p(z_t|x_t)p(x_t|x_{t-2:t-1})p(x_{t-1}|z_{1:t-1}) \tag{14}$$

Our aim is to estimate the joint posterior probability density (14).

3.2   Markov Chain Posterior Sampling

The MCMC method constructs a Markov Chain in the state space to approximate the posterior distribution $p(x_{t-1:t}|z_{1:t})$ that converges to the stationary distribution $\psi(x)$. It is typically used to search the state space for the MAP estimates, or introduced into the particle filtering framework to suppress the sample impoverishment problem. Gilks [11] proposed the MCMC based improved particle filtering algorithm, the basic idea of which is to add a MCMC move step after the resampling process, which can guide the samples to move toward the more promising area. In this paper, we introduce the MCMC method within the particle filtering framework to construct a Markov Chain. Samples are drawn from this chain yielding an unweighted sample set $\{x_t^i\}_{i=1}^N$ which is used to approximate the joint posterior probability density $p(x_{t-1:t}|z_{1:t})$, that is,

$$p(x_{t-1:t}|z_{1:t}) = \frac{1}{N}\sum_{j=1}^{N}\delta(x_{t-1:t} - x_{t-1:t}^j) \tag{15}$$

In this way, the conventional importance sampling procedure can be avoided, while the MCMC method can enhance the searching ability of the particle filter in the state space. The classical algorithm to construct the Markov Chain is the MH algorithm [12]. For the given target posterior density $p(x_{t-1:t}|z_{1:t})$, the algorithm starts from a certain initial sample x_0. The samples are drawn from a so called proposal distribution in order to define the Markov Chain of time t. Suppose the state of the Markov Chain is x_k at the k_{th} iteration, the MH algorithm will generate a new sample x_{k+1} for the next iteration.

3.3   MCMC Posterior Sampling Based Particle Tracking

As mentioned above, we consider the joint posterior probability density $p(x_{t-1:t}|z_{1:t})$. By sampling from this target distribution at each iteration, a

Markov Chain is constructed. During this procedure, the states x_t and x_{t-1} will be updated simultaneously. At the proposal step, we propose a new joint state sample $\{x'_t, x'_{t-1}\}$ from the proposal distribution $Q()$.

$$\{x'_t, x'_{t-1}\} \sim Q(x_t, x_{t-1} | x_t^{k-1}, x_{t-1}^{k-1}) \tag{16}$$

Then, compute the acceptance rate of the proposed sample.

$$\psi_1 = \min\left(1, \frac{p(x'_t, x'_{t-1} | z_{1:t})}{Q(x'_t, x'_{t-1} | x_t^{k-1}, x_{t-1}^{k-1})} \frac{Q(x_t^{k-1}, x_{t-1}^{k-1} | x'_t, x'_{t-1})}{p(x_t^{k-1}, x_{t-1}^{k-1} | z_{1:t})}\right) \tag{17}$$

Using ψ_1, the sample is decided whether or not accepted at the acceptance step. If accepted, set $\{x_t^k, x_{t-1}^k\} = \{x'_t, x'_{t-1}\}$; otherwise, set $\{x_t^k, x_{t-1}^k\} = \{x_t^{k-1}, x_{t-1}^{k-1}\}$. In the following process, individual refinement steps are taken to sample the individual component of the joint state. Firstly, sample a state sample of x_{t-1} from the proposal distribution.

$$\{x'_{t-1}\} \sim Q(x_{t-1} | x_t^k, x_{t-1}^k) \tag{18}$$

Calculate the acceptance probability using

$$\psi_2 = \min\left(1, \frac{p(x'_{t-1} | x_t^k, z_{1:t})}{Q(x'_{t-1} | x_t^k, x_{t-1}^k)} \frac{Q(x_{t-1}^k | x_t^k, x'_{t-1})}{p(x_{t-1}^k | x_t^k, z_{1:t})}\right) \tag{19}$$

Decide whether or not accept the proposed sample x'_{t-1}. If accepted, set $\{x_{t-1}^k\} = \{x'_{t-1}\}$, otherwise, set $\{x_{t-1}^k\} = \{x_{t-1}^{k-1}\}$. Secondly, sample a state sample of x_t.

$$\{x'_t\} \sim Q(x_t | x_t^k, x_{t-1}^k) \tag{20}$$

The acceptance probability is computed using

$$\psi_3 = \min\left(1, \frac{p(x'_t | x_{t-1}^k, z_{1:t})}{Q(x'_t | x_t^k, x_{t-1}^k)} \frac{Q(x_t^k | x_t^k, x_{t-1}^k)}{p(x_t^k | x_{t-1}^k, z_{1:t})}\right) \tag{21}$$

Decide whether or not accept the proposed sample x'_t. If accepted, $\{x_t^k\} = \{x'_t\}$; otherwise, $\{x_t^k\} = \{x_t^{k-1}\}$. We can finally obtain the sample set $\{x_t^k\}_{k=1}^N$ which is used to compute the posterior probability density. The algorithm can be summarized as follow.

Algorithm 1: MCMC posterior sampling based particle tracking
  Input: Sample set of time t-1 $\{x_{t-1}^k\}_{k=1}^N$
  Output: Sample set of time t $\{x_t^k\}_{k=1}^N$, and state estimate $\hat{x}_t$
  FOR k=1,2,...,N,
  Step1: Propose $\{x'_t, x'_{t-1}\}$ using (16)
  Step2: Computer the acceptance probability ψ_1 using (17), and the accept $\{x'_t, x'_{t-1}\}$ with probability ψ_1.
  Step3: Refine x_{t-1}: Propose x'_{t-1} using (18).

Step4: Compute the MH acceptance probability ψ_2, then accept x'_{t-1} with probability ψ_2 (19).

Step5: Refine x_t: Propose x'_t using (20).

Step6: Compute the MH acceptance probability ψ_3 and accept x'_t with probability ψ_3 (21).

ENDFOR

Step7: Obtain the sample set $\{x^k_t\}^N_{k=1}$, and compute the state estimate $\hat{x}_t$

4   Implementation and Experiments

Firstly, we use rectangular area to represent the tracking object, which is defined by its spatial position center and the object scale, that is x=(x_0, y_0, s). As to the proposal distribution, we adopt the following as in [7].

$$Q(x'_t; x_t) = \begin{cases} Q_{AR}(x^{s'}_t ; x^s_t) & scale \\ Q_U(x^p_t) & position \end{cases} \quad (22)$$

We use a HSV color histogram based appearance model for the sake of handling illumination change. The likelihood function for the filtering distribution based on the HSV color histogram similarity is defined by the following equation.

$$p(z_t|x_t) = e^{-\xi d^2(H', H(x_t))} \quad (23)$$

where H' is the target reference model, $H(x_t)$ is the candidate model, ξ is a scaling parameter, and d is the Bhattacharyya distance over the HSV histogram which is defined by

$$d = \sqrt{1 - \zeta(H', H(x_t))} \quad (24)$$

where ζ is the Bhattacharyya similarity coefficient. As for the state transition dynamic models, we adopt a standard second order model (25) under second-order Markov assumption for the proposed algorithm, while a first order model (26) for the other tracking algorithms.

$$x_t = A_0 x_{t-1} + B_0 x_{t-2} + C_0 \nu_t \quad (25)$$

$$x_t = A_1 x_{t-1} + B_1 \nu_t \quad (26)$$

where A_0, B_0, A_1, B_1, C_0 are predefined coefficient matrix. For the sake of evaluating the performance of our proposed tracking algorithm, we use nine video sequences to test its performance and make comparison to the other alternatives: ASAMC [9], adaptive MCMC [13], WLMC [7], and HMC [6]. The video sequences are listed in Table 1.

4.1   Qualitative Results

For qualitative comparison purpose, we test our proposed tracker over different tracking scenarios and compare the tracking results with other algorithms. 300 samples are used by default.

Table 1. Video sequences used in experiments

Seq No.	Length(frames)	Description
1	813	"ChoiHongMan" CHM, Abrupt motion.
2	195	"Baby" Full and partial occlusions
3	539	"Hockey1" Background clutter, and partial occlusion.
4	101	"Hockey2" object interactions, similar moving object.
5	31	"Face1" Fast moving human face.
6	500	"Face2" object appearance and pose change.
7	113	"Soccer" object disappearing and re-entering.
8	50	"SeqMS" Frequently occluded human face.
9	220	"ChenNa" Illumination change.

Abrupt Motion-Fast Moving Object. The Face1 is used in this experiment to test the ability of the trackers of tracking a fast moving object with abrupt dynamic changes. Our tracker uses 300 samples while the others use 600. Figs. 1 displays the tracking results. The results show that our tracker shows better tracking accuracy than the other four trackers, although some trackers give acceptable results at certain frames. AMCMC shows the worst results.

Object Appearance Change. The goal of the experiment is to test the ability of handling object appearance change of different trackers. Face2 sequence is used in which a girl face is moving and rotating with appearance and pose

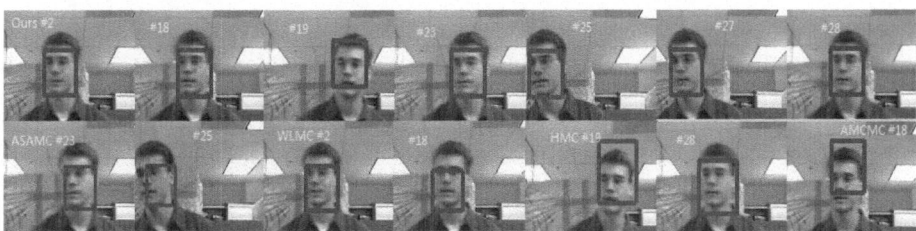

Fig. 1. Tracking results over the face1 sequence

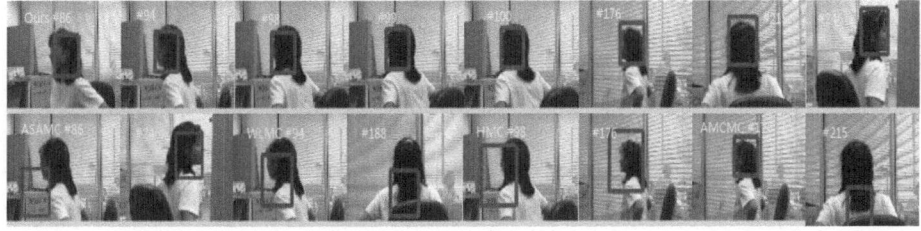

Fig. 2. Tracking results over the face2 sequence

changes in front of the camera. The tracking results are shown in Figs. 2. It is clearly shown that our tracker can accurately capture the object throughout the sequence. The first object rotation occurred between frame #86 and frame #100, and our tracker accurately tracked the object while other tracker failed. The second object rotation occurred between frame #176 and #246, our tracker can accurately track the object throughout the rotation process, and the other four frequently lose the object.

Background Clutter, Object Interaction and Occlusions. We use Hockey1 sequence for this experiment. In this sequence, the players wear the same sports suites moving with frequent occlusions and interactions, which make tracking a player more difficult. Sample frames of tracking results are shown in Figs. 3, which demonstrated that our proposed tracker could consistently track the object while other tracking algorithms failed frequently.

Fig. 3. Tracking results over the Hockey1 sequence

Illumination Change. This experiment aims to test the ability of handling illumination change. We use the ChenNa sequence for this tracking experiment. 100 samples are used for our tracker, while 300 for the others. We aim to track the girl's face in a scenario with illumination change. Tracking results are shown in Figs. 4. When illumination change occurred between frame #108 and #157, our tracker can accurately track the object. AMCMC algorithm shows acceptable results, but is worse than our tracker, while better than the other three algorithms.

4.2   Quantitative Results

Relative Position Error. In the quantitative comparison experiment, we compare the relative position error (RPE) of different trackers which is defined as:

$$\Delta p = \|(x, y) - (x_g, y_g)\|/s_g \tag{27}$$

where (x_g, y_g, s_g) is the ground truth state calibrated manually. We adopt this measurement for evaluating the tracking accuracy because it can facilitate comparing the tracking accuracy for the objects with different sizes [14]. As shown

Fig. 4. Tracking results over the ChenNa sequence

Table 2. RPEs of the algorithms over the 9 video sequences

Algorithms	Ours	ASAMC	WLMC	HMC	AMCMC
Face1	0.0240|0.0016	0.0670|0.0040	0.0330|0.0013	0.0280|0.0014	0.0650|0.0110
Face2	0.0028|0.0018	0.0082|0.0073	0.0055|0.0068	0.0068|0.0087	0.0070|0.0107
Hockey1	0.0070|0.0046	0.1821|0.0984	0.1450|0.1037	0.1491|0.1120	0.0948|0.0862
Hockey2	0.0127|0.0076	0.3511|0.1423	0.1840|0.1537	0.1871|0.1424	0.1010|0.0977
Soccer	0.0057|0.0053	0.1408|0.0819	0.0901|0.0815	0.1386|0.0968	0.1519|0.0725
Baby	0.0031|0.0022	0.0125|0.0106	0.0189|0.0135	0.0132|0.0096	0.0037|0.0030
CHM	0.0016|0.0010	0.0074|0.0022	0.0019|0.0013	0.0049|0.0079	0.0124|0.0111
SeqMS	0.0036|0.0023	0.0059|0.0056	0.0093|0.0074	0.0083|0.008	0.0215|0.0162
ChenNa	0.0014|0.0014	0.0116|0.008	0.0038|0.0019	0.0038|0.0022	0.0021|0.0015

Table 3. SRs of the algorithms over the 9 video sequences

Algorithms	Ours	ASAMC	WLMC	HMC	AMCMC
Face1	100%	83.87%	100%	100%	87.1%
Face2	100%	68.8%	87%	83.4%	84.8%
Hockey1	95.19%	0.74%	18.52%	19.26%	39.44%
Hockey2	82.18%	0%	22.77%	18.81%	37.62%
Soccer	80.53%	1.77%	30.97%	16.81%	11.5%
Baby	69.74%	23.08%	14.36%	21.54%	60.51%
CHM	94.1%	3.2%	88.19%	75.52%	42.31%
SeqMS	98%	94%	84%	78%	44%
ChenNa	97.73%	32.27%	82.73%	81.82%	97.27%
OverAll	93.05%	21.9%	62.96%	57.18%	55.15%

in Table 2, both the mean (left of each cell) and standard deviation (right of each cell) of our tracker over the nine sequences are consistently smaller than those of other trackers.

Success Rate. We define tracking to be lost when the distance between ground truth center and the estimated object center is larger than the calibrated radius of the object. The calibrated radius is defined as the smaller one of the half the width and half the height. This definition is different from that of [9], which defined tracking to be lost when the estimated center was not in the calibrated object area. The success rate (SR) is defined as the ratio between the successfully tracking frames and the total frames of the sequences. In this experiment, we calculate the overall success rate of different trackers over the nine sequences. We also compare the success rate of the trackers over a single sequence. Results are shown in Table 3. The results indicate that all the SRs of our proposed tracker over each of the sequences are higher than that of the other algorithms. The SRs of the other four algorithms fluctuate severely over different sequences which indicate that the robustness of the four trackers is worse than ours.

5   Conclusions

We have proposed a robust tracking algorithm within the particle filtering framework using the MCMC posterior sampling and second-order Markov assumption. In our tracking algorithm, we use a Markov Chain with certain length to simulate the approximated posterior probability density, which avoid the drawbacks of the traditional importance sampling based algorithm. The second-order Markov assumption can make better use of the history information and enhance the searching ability of our tracker. Experimental results have demonstrated that our proposed tracker can give stable and accurate tracking results in various tracking scenarios.

Acknowledgement. This work was supported by the National Natural Science Foundation of China (No. 60772063, 61073133, 61175053, 60973067, 61175096), the Innovation Group Project of China Education Ministry (No. 2011ZD010, Fundamental research fund of DMU (NO. 2012QN030), Foundation of Scientific Planning Project of Dalian City (No. 2011E15SF100), the Natural Science Foundation of Liaoning Education Ministry (No. L2011241, L2010043), and the Fundamental Research Funds for the Central Universities of P.R. China (No. 2011QN027).

References

1. Yilmaz, A., Javed, O., Shah, M.: Object Tracking: A Survey. ACM Computing Surveys 38, 1–45 (2006)
2. Isard, M., Blake, A.: CONDENSATION- conditional density propagation for visual tacking. IJCV 29, 5–28 (1998)
3. Martinez-del-Rincon, J., Orrite, C., Medrano, C.: Rao-Blackwellised particle filter for color-based tracking. PRL 32, 210–220 (2011)
4. Khan, Z., Balch, T., Dellaert, F.: MCMC-based particle filter for tracking a variable number of interacting targets. IEEE TPAMI 27, 1805–1819 (2005)
5. Cong, D.-N.-T., Septier, F., Garnier, C., Khoudour, L., Delignon, Y.: Robust Visual Tracking via MCMC-based Particle Filtering. In: ICASSP, Kyoto, Japan (2012)
6. Choo, K., Fleet, D.J.: People Tracking Using Hybrid Monte Carlo Filtering. In: ICCV, pp. 321–328 (2001)

7. Kwon, J., Lee, K.M.: Tracking of Abrupt Motion Using Wang-Landau Monte Carlo Estimation. In: Forsyth, D., Torr, P., Zisserman, A. (eds.) ECCV 2008, Part I. LNCS, vol. 5302, pp. 387–400. Springer, Heidelberg (2008)
8. Liang, F., Liu, C., Carroll, R.J.: Stochastic Approximation in Monte Carlo Computation. Journal of the American Statistical Association 102, 305–320 (2007)
9. Zhou, X., Lu, Y., Lu, J., Zhou, J.: Abrupt Motion Tracking Via Intensively Adaptive Markov-Chain Monte Carlo Sampling. IEEE TIP 21, 789–801 (2012)
10. Arulampalam, S., Maskell, S., Gordon, N., Clapp, T.: A tutorial on particle filters for on-line non-linear/non-Gaussian Bayesian tracking. IEEE TSP 50, 174–188 (2001)
11. Gilks, W., Berzuini, C.: Following a moving target: Monte Carlo inference for dynamic Bayesian models. Journal of Royal Statistic Society B 63, 127–146 (2001)
12. Geyer, C.J.: Introduction to Markov Chain Monte Carlo. In: Handbook of Markov Chain Monte Carlo. Chapman and Hall/CRC Press (2011)
13. Roberts, O., Rosenthal, S.: Examples of adaptive MCMC. Journal of Computational and Graphical Statistics 18, 349–367 (2009)
14. Yang, M., Fan, Z., Fan, J., Wu, Y.: Tracking nonstationary visual appearances by data-driven adaptation. IEEE TIP 18, 1633–1644 (2009)

A Framework for Inter-camera Association of Multi-target Trajectories by Invariant Target Models

Shahar Daliyot[1,2] and Nathan S. Netanyahu[1,3]

[1] Department of Computer Science, Bar-Ilan University, Ramat-Gan 52900, Israel
[2] Verint Systems Ltd., Herzliya 46733, Israel
[3] Center for Automation Research, University of Maryland,
College Park, MD 20742, USA

Abstract. We propose a novel framework for associating multi-target trajectories across multiple non-overlapping views (cameras) by constructing an invariant model per each observed target. Ideally, these models represent the targets in a unique manner. The models are constructed by generating synthetic images that simulate how targets would be seen from different viewpoints. Our framework does not require any training or other supervised phases. Also, we do not make use of spatiotemporal coordinates of trajectories, i.e., our framework seamlessly works with both overlapping and non-overlapping field-of-views (FOVs) as well as widely separated ones. Also, contrary to many other related works, we do not try to estimate the relationship between cameras that tends to be error prone in environments like airports or supermarkets where targets wander about different areas, stop at times, or turn back to their starting location. We show the results obtained by our framework on a rather challenging dataset. Also, we propose a black-box approach based on Support Vector Machine (SVM) for fusing multiple pertinent algorithms and demonstrate the added value of our framework with respect to some basic techniques.

1   Introduction

Tracking targets across multiple cameras, also known as the "handover" problem, is an important problem in computer vision, in general, and visual surveillance, in particular, where common applications are motivated by large surveillance systems installed in complex compounds. The goal is to maintain the identities of targets traveling across cameras, disappearing and later reappearing at a different location. The task of target tracking in a single view (single camera) has been studied quite extensively (see, e.g., [1,3,18]) and is not the focus of this work. We assume, in fact, that this fundamental building block is given. Tracking analysis of a single view produces a set of trajectories representing the target motions in a given FOV, independently from other views. Our goal is to associate trajectories belonging to the same target from multiple, independently-analyzed views.

To make aforementioned associations between trajectories, we compute a signature for each trajectory. This signature consists of models representing the

J.-I. Park and J. Kim (Eds.): ACCV 2012 Workshops, Part II, LNCS 7729, pp. 372–386, 2013.

target that originated the trajectory. Ideally, trajectory signatures belonging to the same target are more similar than those belonging to different targets. We obtain associations based on construction of invariant features. As opposed to common feature extraction techniques (e.g., SIFT [13]) that search for invariant features with respect to scale, rotation, etc., we explicitly promote invariance to any feature extracted from an observed target. We do that by generating synthetic images that simulate how the target would be seen from different viewpoints. This approach is different from various common approaches utilized in related work, whereby probability models evaluate inter-camera relationships to predict the associations between trajectories.

Our approach does not rely on spatiotemporal coordinates, color cues, and similar information that is usually exploited in other related work. Instead, by using our innovative scheme of invariant signature models constructed from a target's trajectory, we employ, essentially, an alternative technique for invariant feature extraction to finding trajectory association. Our solution can then be fused nicely with other, more common approaches. Indeed, we present also an SVM-based black-box fusion scheme that can easily be used to combine multiple pertinent algorithms and demonstrate its added value.

This paper is organized as follows. Section 2 provides an overview of related research in this area. Section 3 presents our framework for trajectory association. Section 4 presents our black-box approach for fusing a variety of pertinent algorithms. In Section 5 we present our experimental results. Section 6 makes concluding remarks.

2   Related Work

Recently, there has been a growing interest in tracking targets between blind regions, i.e., non-overlapping regions, which give rise to target disappearances at certain time intervals. In [4] a system for multi-camera tracking in "blind" regions is described. In order to match targets between different cameras, the authors assume that targets move in a fairly constant velocity. They utilize a known camera topology to construct a probabilistic prediction of where and when a target, disappearing from one view, will appear in a nearby view. This assumption does not hold in a supermarket-like environment where people wander about the different aisles, stop at times, or turn back to their starting location. In [9] a different approach is taken, whereby the camera topology and path probabilities are learned during a training phase. This method is constrained to a small number of cameras. It is quite cumbersome and even infeasible to have such a training phase when dealing with large scale premises with hundreds of cameras. On the other hand, an unsupervised method for learning the topology of a camera network is presented in [14]. The system learns "Entry" and "Exit" zones in each camera view according to a training dataset that contains a large number of sample trajectories. Then, a probabilistic graph representing links between these zones is learned based on the training dataset, using expectation maximization (EM) methods. The resulting graph is used to predict the location where a disappeared target might reappear. This unsupervised approach of learning a prediction model might suffer from poor accuracy in a large camera network with substantial amount of targets moving around, constantly disappearing and reappearing. The model might set high probabilities to links between "Entry" and "Exit" zones based on trajectories that are assumed to represent the same target but which in fact do not.

A very common approach of handling the problem of tracking across multiple disjoint views is based on color histogram comparisons (see, e.g., [8,10,11,15,17]). Issues such as differences in illumination, pose, and internal parameters between different cameras are discussed in [10]. The authors show that all brightness transfer functions, from a given camera to another, lie in a low-dimensional subspace which can be learned by supervised methods using labeled learning sets. In [11] the authors deal with the problem of tracking across multiple cameras by combining location prediction and appearance model matching. The location prediction is based on Kalman filters (KF), and the appearance model is constructed from multiple color distribution components, each of which is obtained by partitioning the blobs (representing the detected target) into their polar representation. This localization of color properties provides an advantage over a simple color model, whereby a single histogram represents the entire target.

The main task of our work is to find associations between trajectories of targets seen from several arbitrary viewpoints. Approaching the problem is often based on feature matching techniques, which have been studied extensively (see, e.g., [2,7,13,15,16]). The methods presented in these papers extract features that are invariant to affine transformations (i.e., scale, rotation, and shear), noise, illumination, etc. Such features can then be matched, regardless of the viewpoint of a target or an object. This insight serves as an inspiration for our target association scheme.

Fusion methods have also been introduced extensively in this domain. In [10] the authors combine space-time cues with an appearance matching scheme based on a Maximum Likelihood (ML) estimation framework. In order to overcome appearance changes between cameras, a brightness transfer function is learned during a training phase. Space-time cues are also learned during this training phase. The ML estimation framework is then used to assign correspondences between tracked targets. In [12] the authors combine three different feature types (the color histogram, covariance matrix, and Histogram of Oriented Gradients (HOG) features) using the Multiple Instance Learning (MIL) method. Similarity measures based on these features, among a set of automatically collected training samples, form a "feature pool". The MIL boosting algorithm is then applied to select discriminative features from this pool and their corresponding weighted coefficients. We propose a simple, yet effective black-box approach for fusing a variety of pertinent algorithms without any knowledge of the internal structure of the fused algorithms. This is in contrast to the above approaches whereby fusion is accomplished within the core of the algorithms, requiring the utilization of heuristics which makes it more complex.

3   Our Trajectory Association Framework

To associate trajectories across multiple views, we will compute a signature for each trajectory. These signatures will be used subsequently to compare between trajectories and obtain associations. Subsections 3.2–3.3 describe the process of signature computation and comparison.

3.1   Model Construction

We refer to the constructed model that describes, ideally, the target of a trajectory in a unique manner as an "invariant feature". The following paragraphs describe in detail the model construction.

ROI Selection. Contrary to the approach taken in many related works, we compute the descriptor over a specific region of interest (ROI), rather than the entire region where the target is captured. We will be interested specifically in regions containing a persons head. This is motivated by several reasons: (1) The head includes facial features that identify the target uniquely. (2) The head forms a fairly rigid surface relative to other moving parts of the human body like the arms. This helps when comparing descriptors computed from different viewpoints of the same target. (3) The head is the topmost body part, making it less prone to occlusions. (4) In practice, the region we use includes not only the head but also the upper body part that captures slightly areas from the neck and shoulders. This upper body part is still fairly rigid, yet often includes distinctive, informative characteristics about the target, e.g., texture of shirt, type of collar, tie, scarf, etc.

We extract the head region automatically from the bounding ellipse of the target in the frame obtained by a visual tracker. Calculating a bounding rectangle of the head region relative to the bounding ellipse is based on fixed coefficients learned a prior, as illustrated in Figure 1.

Descriptor Computation. Once a region is determined, a descriptor is computed from the image data in the region. We use a Histogram of Oriented Gradients (HOG) proposed in [7] to represent the data in the region. Namely, we compute a HOG feature by concatenating 8-bin orientation histograms calculated in $6{\times}6$ cells over the region, resulting in a row vector of size 288.

We note that the specific HOG feature we chose as a descriptor is not necessarily optimal for the task of trajectory association. Other types of descriptors could be used coherently with our framework. The type of descriptor used is not as important as the invariance required from the constructed model to changes in scale, rotation, viewpoint, and to some extent, also perspective distortions. This construction is the foundation of our framework as described henceforth.

Invariant Model Construction. Computing a single descriptor for the head region in the original frame (as captured by the camera) is not enough to match

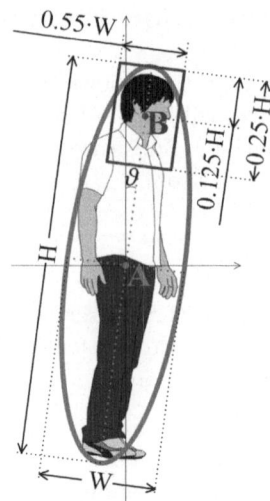

Fig. 1. Positioning of rectangle representing region of head relative to bounding ellipse of target; height and width of bounding rectangle are set with respect to lengths of major and minor axes of ellipse ($0.25H$ and $0.55W$, respectively); center of rectangle (point B) is located on major axis; rectangle (aligned with major and minor axes of ellipse) is tilted by an angle of ϑ

this instance of the head with other instances of the same target, taken from different viewpoints. Therefore, we simulate, as much as possible, how a given target would be seen from different viewpoints. We do that by applying various affine transformations to the ROI in the original frame and then computing the corresponding descriptors of these transformed regions. We realize that it is not possible to construct a full 3D model of a face given the data quality we deal with. Still, applying rather "small" affine transformations provides various simulated viewpoints of the head.

An affine transformation can be represented by the seven parameters: scale-x, scale-y, translation-x, translation-y, shear-x, shear-y and rotation. We denote these parameters by S_x, S_y, T_x, T_y, R_x, R_y, and ϑ, respectively. Each of the seven parameters can be expressed by a 3×3 transformation matrix of homogeneous coordinates. These parameters can be combined to form a variety of affine transformations by using matrix composition. Recall that the head region is captured by a rectangle, i.e., it can be represented by a 3×3 affine transformation matrix (with respect to a reference 1×1 rectangle centered at the origin) that is composed of several transformation matrices of homogeneous coordinates. Let G be that rectangular head region centered at point $C = (C_x, C_y)$ with width W, height H, and counter-clockwise rotation ϑ' with respect to C. The composition corresponding to G is given by

$$\begin{pmatrix} 1 & 0 & T_x \\ 0 & 1 & T_y \\ 0 & 0 & 1 \end{pmatrix} \begin{pmatrix} \cos\vartheta & -\sin\vartheta & 0 \\ \sin\vartheta & \cos\vartheta & 0 \\ 0 & 0 & 1 \end{pmatrix} \begin{pmatrix} 1 & R_x & 0 \\ 0 & 1 & 0 \\ 0 & 0 & 1 \end{pmatrix} \begin{pmatrix} 1 & 0 & 0 \\ R_y & 1 & 0 \\ 0 & 0 & 1 \end{pmatrix} \begin{pmatrix} S_x & 0 & 0 \\ 0 & S_y & 0 \\ 0 & 0 & 1 \end{pmatrix} \quad (1)$$

where

$$S_x = W \; ; \; S_y = H \; ; \; T_x = C_x \; ; \; T_y = C_y \; ; \; R_x = 0 \; ; \; R_y = 0 \; ; \; \vartheta = \vartheta' \qquad (2)$$

We compute $N \gg 1$ descriptors, corresponding essentially to N different affine transformations applied to the original head region. It is important that these transformations be diversified, albeit rather slightly, to avoid an intense distortion of the information contained in the relevant region. Having experimented with various values, we picked eventually $N = 1000$. (A larger number of transformations did not yield significantly better results.) We thus generate at random 1000 affine transformations as follows. Each affine transformation is associated with a 7-element list of the form $(\Delta S_x, \Delta S_y, \Delta T_x, \Delta T_y, \Delta R_x, \Delta R_y, \Delta\vartheta)$, where each element is chosen at random from the range $[-0.2, +0.2]$, i.e., each random generation corresponds to a perturbation of the given image by an affine transformation with the following parameters:

$$S_x = 1 + \Delta S_x \; ; \; S_y = 1 + \Delta S_y \; ; \; T_x = \Delta T_x \; ; \; T_y = \Delta T_y$$

$$R_x = \Delta R_x \; ; \; R_y = \Delta R_y \; ; \; \vartheta = \Delta\vartheta \qquad (3)$$

(with ϑ in radians). We end up with a list containing 1000 descriptors and a list containing corresponding representations of 1000 affine transformations. That is, for each i, $1 \le i \le 1000$, descriptor d_i corresponds to the set of coefficients of affine transformation t_i. Eventually, associations are determined by an iterative algorithm; at each iteration, given a descriptor d, it looks for the affine transformation t that "best fits" this descriptor. We have considered, among others, perspective transformations and non-linear transformations, but settled eventually for a simpler linear mapping according to the standard least squares approach. Let D be the domain of descriptors and let T be the domain of affine transformations. We want to determine a linear mapping $L : D \to T$. Since the output $t \in T$ consists of seven parameters, there is a need to determine seven linear mappings, $L_1, \ldots, L_7$, each of which corresponds to a single parameter of the ultimate affine transformation. Finding these linear mappings requires solving an overdetermined system of linear equations. Each of these seven linear mappings is essentially a list of 288 coefficients corresponding to the elements of a descriptor. The linear least squares method finds these coefficients, such that the error over the entire data (1000 instances in our case) is minimized. Once the seven linear mappings are determined, we concatenate them to form the final 288×7 matrix L. Given a descriptor d, the corresponding affine transformation t can be computed by:

$$[t]_{1\times7} = [d]_{1\times288} \cdot [L]_{288\times7} \qquad (4)$$

We can construct also the model that represents the head of a target in the frame using the data computed. The model consists of the following four items: (1) The vertical and horizontal gradient images calculated from the frame during the computation of the HOG feature. (2) *model-region*: the head region extracted from the bounding box of the target. (3) *model-descriptor*: a single descriptor computed for the *model-region*. (4) The linear mapping matrix L.

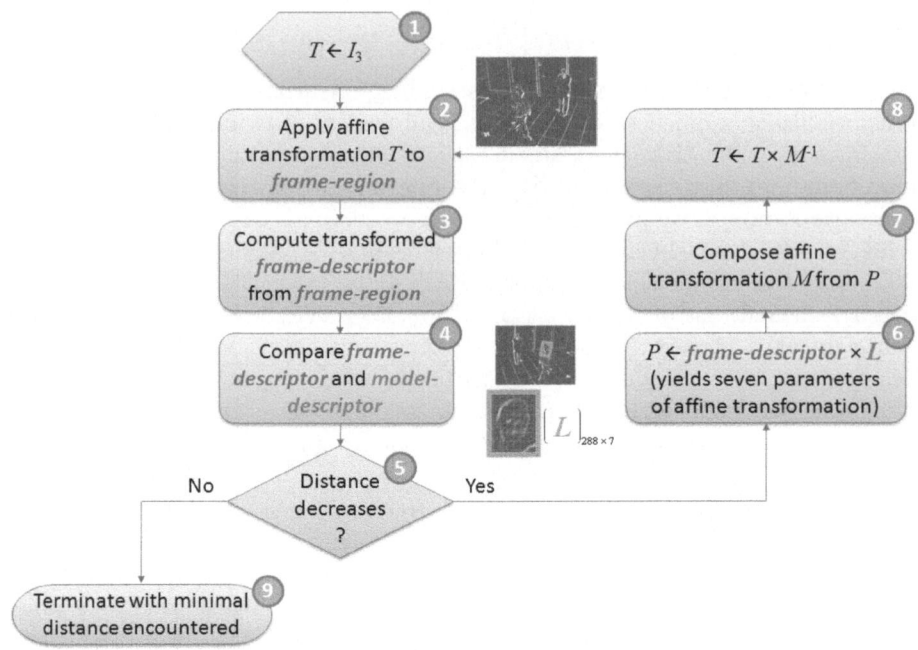

Fig. 2. Flowchart of iterative descriptor-to-model comparison

3.2   Signature Computation

Note that a model represents only a single target in a single frame. The purpose of a signature is to represent an entire trajectory, which spans usually multiple frames. A signature is a collection of models computed from frames of the trajectory. Through-out the frames of a trajectory, the target is captured from different viewpoints and in different poses. However, the target may look very similar in many frames. We want to keep only the frames in which the target looks different with respect to other frames of the trajectory. We start with an empty signature and append to it models from each frame that introduces new data not yet encountered. Starting with the first frame, the signature is still empty; hence, a model is always computed from this frame and added to the signature. Examining consecutive frames, the head in each frame is compared to models already appended to the signature in previous frames. If a head model is "similar enough" to any of the previous models, the frame in question is skipped. On the other hand, if the head is "distinct enough" from all previous models, a new model is computed for the current frame and appended to the signature.

Descriptor-to-Model Comparison. A fundamental element of the signature construction algorithm is the ability to compare a targets head in a given frame to models, which already reside in the signature. That is, to compute the distance between a given head and a model. Once we compute this distance, a simple threshold can be used to decide whether the head in question is "similar enough" to any of the models in the signature. We start by computing a single descriptor for the head region in a given frame. We will refer to this head region as *frame-region* and to the corresponding descriptor computed for that region as *frame-descriptor*. To compute the aforementioned distance, we use an iterative approach. The iterative descriptor-to-model comparison approximates the *frame-descriptor* to the *model-descriptor*. This is done by iteratively applying an (inverse) affine transformation to the *frame-region*, re-computing a descriptor for the resulting region and comparing it to the *model-descriptor*. Computing the (inverse) affine transformation that should be applied to the *frame-region* is done using the linear mapping matrix L and the descriptor obtained at each iteration. Figure 2 presents a flowchart of the iterative descriptor-to-model comparison. At the first iteration, the algorithm computes in Step 3 the *frame-descriptor* from the *frame-region*. Initially, the affine transformation T applied to the *frame-region* in Step 2 is the identity matrix, so that there is no effect on the region. In consecutive iterations, however, the *frame-region* will be transformed, yielding a so-called transformed *frame-descriptor*. In Step 4, the transformed *frame-descriptor* is compared to the *model-descriptor*, yielding a distance between the two. (The distance is the angle (in degrees) between the two unit vectors.) If the distance decreases, the algorithm continues to Step 6. Otherwise, the algorithm terminates, returning the minimal distance encountered so far. Steps 6-8 comprise the essence of our framework. First, the transformed *frame-descriptor* is multiplied by the linear regression matrix of the model, resulting in an affine transformation represented by a vector of seven parameters. Recall that the linear mapping matrix was calculated from corresponding pairs of affine transformations and descriptors. Intuitively, the product of an arbitrary descriptor and the linear regression matrix of a model yields an affine transformation that should approximate in some sense the descriptor of the model head to that arbitrary descriptor. Namely, the affine transformation obtained should approximate the *model-descriptor* to the *frame-descriptor*. However, since we want to approximate actually the *frame-descriptor* to the *model-descriptor*, we need to apply the inverse affine transformation to the *frame-region*. Steps 7 and 8 do exactly that. In Step 7 an affine transformation is composed out of the seven parameters obtained by the multiplication, and in Step 8 the inverse of this affine transformation is applied to the affine transformation T that process adjusts the (inverse) affine transformation applied to the *frame-region*, as long as the distance between the *frame-descriptor* and the *model-descriptor* continues to decrease.

Horizontal Flipping. We exploit the fact that a human head is usually symmetric along the horizontal axis to enhance a signature computation. In addition to comparing the descriptor representing a head region in a given frame to the models contained in the signature, we will compare also a descriptor derived from a horizontally-flipped image of the frame. The final distance between the head in a given frame and a head model is defined as the minimum distance between the *model-descriptor* and the two descriptors, i.e., the one computed for the original frame and that computed for the horizontally-flipped frame. This enables to reduce the number of models contained in the signature, thereby reducing computation time and space. Also, we exploit this flipping element, while comparing between signatures, to improve significantly the total matching rate (see next subsection).

3.3 Signature Comparison

Before presenting the algorithm for signature comparison, *CompareSignatures*, we define an algorithm for comparing models, *CompareModels*. In the previous subsection we presented an algorithm for comparing a single descriptor to a model, the iterative descriptor-to-model comparison. We later obtained an enhanced algorithm, *ModelWithHorzFlip*, using horizontal flipping. We utilized this enhanced algorithm for comparing also between models. Given two models, *model1* and *model2*, *CompareModels* compares the two models and returns a scalar representing the distance between them. The algorithm is symmetric, i.e., *model1* is compared to *model2* and vice versa. First the algorithm calls *CompDescToModelWithHorzFlip* with input information of the frame-region from *model1* and *model2* as the *model-descriptor*. This means that transformed descriptors are computed from the gradient images of *model1* and approximated to the *model-descriptor* contained in *model2* using the linear mapping matrix of *model2*. Then, the opposite comparison is performed, i.e., transformed descriptors computed from the gradient images of *model2* are approximated to the *model-descriptor* contained in *model1*. The distance between the two models is defined as the average between the values returned by the two comparisons performed.

We now sketch the *CompareSignatures* algorithm. Given two signatures, *sig1* and *sig2*, all models of *sig1* are compared to all models of *sig2* using the *CompareModels* algorithm. The minimum distance among all comparisons is returned as the distance between the two signatures. Final trajectory associations are obtained based on these distances by comparing them, for example, against a specified threshold.

4   Fusing Multiple Algorithms

The insight of establishing an improved outcome based on the fusion of multiple algorithms has long been recognized. Fusion methods have been introduced extensively in the domain of object matching across different views (see, e.g., [10,12]). We propose a simple, yet effective black-box approach for fusing a variety of matching algorithms. We let every algorithm run independently and return a distance (or grade) that is assigned to each pair of trajectories among a set of labeled training samples. Once the results from all algorithms are available, we train a Support Vector Machine (SVM) module [5] using the labeled training samples based on the distances obtained by the different algorithms. Samples of the SVM training set are compiled from all possible pairs of trajectories from the labeled training samples. Each sample has several features that are captured by the distances provided by the different algorithms for that specific sample. (Note that a sample represents a pair of trajectories; hence a distance or a grade is assigned to this pair by each algorithm.) The number of sample features is equal to the number of different algorithms that are fused. The sample class is assigned a value according to ground truth; if the two trajectories in the sample represent the same target, the class is set to 1, and otherwise to 0. The outcome of the training process is an SVM module that accepts a list of distances obtained by the different algorithms for a given pair of trajectories. The output of the SVM module is one of two possible classes, 1 and 0, as defined above. In order to assess the quality of the SVM module that was constructed, we used the well-known K-fold cross-validation technique (with $K = 10$). The accuracy was stable across all 10 repetitions, i.e., the variance between the 10 results was very low. This indicates that the module is well-balanced and coherent. Results are presented in the next section.

Using a labeled training set should not imply there is a need for a training phase or some supervised procedure each time the fusion is invoked with a new camera network. The labeled set is utilized to train an SVM module on the outputs of the fused algorithms. Theoretically, the same labeled dataset can be utilized repeatedly for different invocations and different algorithms. Of course, using a training set created from actual data of a given camera network is likely to yield better results. Note that the fusion is adapted in a modular manner upon the introduction of a new algorithm. One need only train a modified SVM module with the new additional data obtained by that algorithm, which is an unsupervised procedure that can be handled automatically.

The above is a "black-box" approach since no knowledge of the internal structure of the fused algorithms is needed. The only requirement from an additional fused algorithm is a coherent output of the distance (or grade) between all trajectory pairs. This is a rather elementary requirement from an algorithm whose purpose is to find associations between trajectories.

5   Experimental Results

All experiments were done using a 30-minute recorded scene of six cameras installed at the lobby of a facility according to the model shown in Figure 3. The views of the six cameras are shown in Figure 4. The scene includes both congested and sparse scenarios. A visual tracker was used to analyze separately the recorded video of each of the six cameras and produce trajectories of detected targets. Trajectories of 12 targets in the scene were manually labeled to form a dataset that serves as ground-truth. The 12 targets produced 98 trajectories in the scene. Each of the 98 trajectories was compared to all trajectories (including itself), yielding distances per each trajectory pair, i.e., ($\frac{1}{2} \times 98 \times 99 =$) 4851 entries altogether.

Figure 5 presents precision vs. recall results for our trajectory association framework before and after the horizontal flipping. (Precision and recall are determined

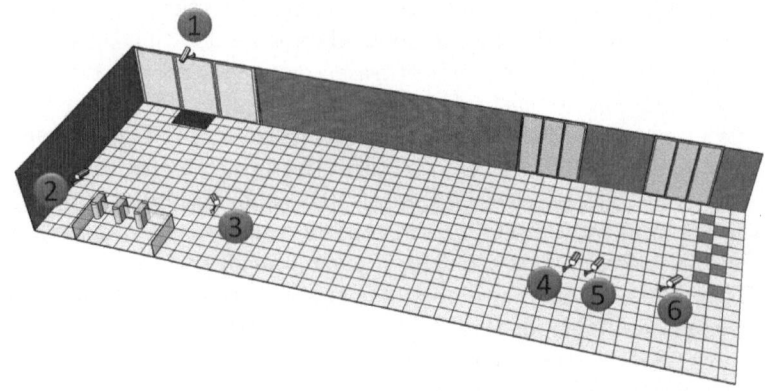

Fig. 3. Model of lobby showing positions of six cameras partially covering the area

by the number of "true-positives", "false-positives", and "false-negatives" returned by our algorithm over all trajectory pairs.) The optimal accuracy of 82% is quite impressive given the challenging dataset we experimented with. The accuracy is given by the F-measure (also known as the F_1-score), which is the harmonic mean of precision and recall, i.e.,

$$F\text{-measure} = 2 \times \tfrac{Precision \times Recall}{Precision + Recall} \qquad (5)$$

As mentioned in the Introduction, our algorithm for trajectory association does not consider spatiotemporal coordinates, color cues, or similar information that is exploited in other related work. Instead, it introduces a new type of information based on invariant features. This approach lends itself to further improvement by fusing it with other more common approaches. To validate this assumption, we implemented two additional basic algorithms and examined the results due to fusion of all three. The first additional algorithm is based on color histograms. That is, the distance between two trajectories is based on the Euclidean distance between the color histograms computed from the blobs of the relevant targets. Color histograms were constructed based on the three RGB channels of the frames. We allocate 16 bins to each of the three channels. We then concatenate them and normalize the resulting 48-bin histogram. The accuracy obtained by this algorithm is 76%. The second additional algorithm is based on spatiotemporal information. All trajectories from all cameras are projected into a common coordinate system and a grade is assigned to each pair of trajectories as follows. Two trajectories recorded at the same time frame but in distant locations are assigned a grade of −1, meaning they belong to different targets. This is based on the fact that a target cannot be in two different locations at the same time. A grade in the range $(0, 1]$ is assigned to two trajectories recorded at the same time frame that are in close vicinity. The exact grade is computed based on the spatial distance between the two trajectories. Finally, a grade of 0 is assigned to all other trajectory pairs, indicating the algorithm does not have sufficient

Camera 1   Camera 2   Camera 3

Camera 4   Camera 5   Camera 6

Fig. 4. Views of six cameras positioned according to model presented in Figure 3

information to obtain a reliable decision. The accuracy obtained by this algorithm for our dataset is 35%. This poor outcome is a result of the inability of the algorithm to compare between trajectories that are not adjacent in time or space. When executed only on a subset of the trajectories that do have adjacent spatiotemporal information it produces an impressive accuracy of 95%.

We fused our algorithmic framework with the color-based and spatiotemporal-based algorithms using our black-box technique. Fusing the three algorithms is pretty straightforward; we let every algorithm run independently and yield its own results. A sample (point in the graph) provided as input to the SVM module represents a pair of trajectories and consists of the distances/grades obtained by the three algorithms for that pair. A 10-fold cross validation procedure was executed to evaluate the accuracy of the SVM module. Figure 6 presents the final point classification according to the support vectors that were computed by the SVM module in one of the 10 iterations of the cross validation procedure. Precision, recall, and accuracy (F-measure) were calculated during each iteration and then averaged to yield the ultimate measures of the fusion. The bottom-line averages obtained for precision, recall, and accuracy are 94%, 83%, and 88%, respectively. Recall that the accuracies obtained when running the three algorithms independently were 35%, 76%, and 82% by the spatiotemporal-based algorithm (run on the entire dataset), the color-based algorithm, and our framework, respectively. Clearly, the bottom-line 88% accuracy obtained by the fusion comprises an improvement in comparison to each of the three independent algorithms. Evidently, the fact that each algorithm obtains trajectory associations based on a different type of information causes a significant mutual improvement.

Having established the added value of our framework relatively to the above basic approaches, it would be of interest, of course, to carry out a more comprehensive study with respect to various recent methods, e.g., [6,19].

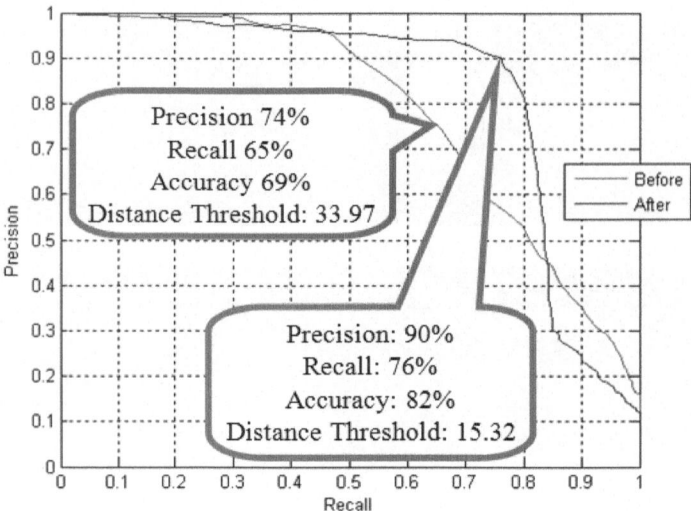

Fig. 5. Precision vs. recall before (red) and after (blue) horizontal flipping; note gain in accuracy of over 10%

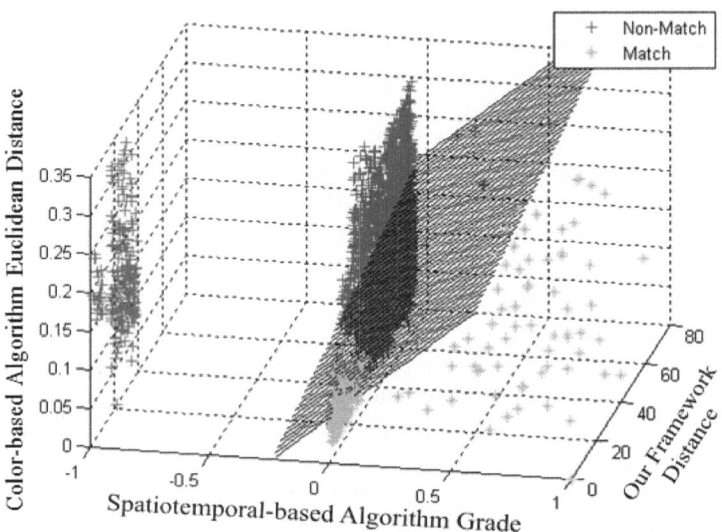

Fig. 6. Final classification of all pairs according to SVM module fusing our framework, color-based, and spatiotemporal-based algorithms: Pairs with positive classification, i.e., matches (cyan), pairs with negative classification, i.e., non-matches (magenta), and the planar separator computed by the SVM module

6 Conclusions

We have presented a novel basic framework for constructing an invariant descriptive model of a target for trajectory association across multiple views. Unlike many other related works, our system does not assume a constant velocity of moving targets and does not require a learning phase to construct a statistical model of target motions. Instead, we showed how to construct an invariant model for a target to represent it uniquely across multiple views. The model is constructed by generating synthetic target images that simulate how a target would be seen from different viewpoints. A regression analysis is then used to capture this simulated model in a compact and efficient manner. Next, an iterative convergence scheme is employed to compare between constructed models and find trajectory correspondences. We demonstrated the relatively high success rate of this approach, especially when fused with other, more common methods. Specifically, we examined fusion with two common techniques based on color histograms and spatiotemporal information. We showed that fusing the results from these common techniques with those of our framework yields an evident gain in accuracy. To carry out the fusion we introduced an effective black-box technique based on SVM. Our fusion technique cancels the need for sophisticated or heuristic methods to combine the outputs of multiple association algorithms.

The system was tested using real-world surveillance videos from a network of six cameras with nearly 100 trajectories. To the best of our knowledge, this is a much larger dataset than most datasets used in previous related work. Still, our system manages to yield a remarkable accuracy of close to 90%.

Acknowledgements. We would like to thank Dr. Shmuel Kiro for his fruitful technical discussions and suggestions and Oren Golan of Verint Systems for his consistent administrative support. Also, the support of the VULCAN Consortium under the Israeli Ministry of Industry, Trade and Labor is gratefully acknowledged.

References

1. Avidan, S.: Ensemble tracking. IEEE Transactions on Pattern Analysis and Machine Intelligence, 261–271 (2007)
2. Bay, H., Tuytelaars, T., Van Gool, L.: SURF: Speeded Up Robust Features. In: Leonardis, A., Bischof, H., Pinz, A. (eds.) ECCV 2006. LNCS, vol. 3951, pp. 404–417. Springer, Heidelberg (2006)
3. Changjiang, Y., Duraiswami, R., Davis, L.: Fast Multiple Object Tracking via a Hierarchical Particle Filter. In: Proceedings of the Tenth IEEE International Conference on Computer Vision, pp. 212–219 (2005)
4. Chilgunde, A., Kumar, P., Ranganath, S., Weimin, H.: Multi-Camera Target Tracking in Blind Regions of Cameras with Non-Overlapping Fields of View. In: Proceedings of the British Machine Vision Conference, pp. 397–406 (2004)
5. Cortes, C., Vapnik, V.: Support-Vector Networks. Machine Learning, 273–297 (1995)

6. D'Angelo, A., Dugelay, J.: People Re-idnetification in Camera Networks Based on Probabilistic Color Histograms. In: Proceedings of the SPIE Conference on Visual Information Processing and Communication, vol. 7882 (2011)
7. Dalal, N., Triggs, B.: Histograms of Oriented Gradients for Human Detection. In: Proceedings of the IEEE Conference on Computer Vision and Pattern Recognition, pp. 886–893 (2005)
8. Gilbert, A., Bowden, R.: Tracking Objects Across Cameras by Incrementally Learning Inter-camera Colour Calibration and Patterns of Activity. In: Leonardis, A., Bischof, H., Pinz, A. (eds.) ECCV 2006. LNCS, vol. 3952, pp. 125–136. Springer, Heidelberg (2006)
9. Javed, O., Rasheed, Z., Shafique, K., Shah, M.: Tracking across Multiple Cameras with Disjoint Views. In: Proceedings of the Ninth IEEE International Conference on Computer Vision, pp. 952–957 (2003)
10. Javed, O., Shafique, K., Rasheed, Z., Shah, M.: Modeling Inter-camera Space-time and Appearance Relationships for Tracking across Non-overlapping Views. Computer Vision and Image Understanding, 146–162 (2008)
11. Kang, J., Cohen, I., Medioni, G.: Continuous Tracking within and across Camera Streams. In: Proceedings of the IEEE Conference on Computer Vision and Pattern Recognition, pp. 267–272 (2003)
12. Kuo, C.-H., Huang, C., Nevatia, R.: Inter-camera Association of Multi-target Tracks by On-Line Learned Appearance Affinity Models. In: Daniilidis, K., Maragos, P., Paragios, N. (eds.) ECCV 2010, Part I. LNCS, vol. 6311, pp. 383–396. Springer, Heidelberg (2010)
13. Lowe, D.: Distinctive Image Features from Scale-Invariant Keypoints. International Journal of Computer Vision, 91–110 (2004)
14. Makris, D., Ellis, T.J., Black, J.: Bridging the Gaps Between Cameras. In: Proceedings of the IEEE Conference on Computer Vision and Pattern Recognition, pp. 205–210 (2004)
15. Mazzeo, P.L., Spagnolo, P., D'Orazio, T.: Object Tracking by Non-overlapping Distributed Camera Network. In: Blanc-Talon, J., Philips, W., Popescu, D., Scheunders, P. (eds.) ACIVS 2009. LNCS, vol. 5807, pp. 516–527. Springer, Heidelberg (2009)
16. Mikolajczyk, K., Schmid, C.: An Affine Invariant Interest Point Detector. In: Heyden, A., Sparr, G., Nielsen, M., Johansen, P. (eds.) ECCV 2002, Part I. LNCS, vol. 2350, pp. 128–142. Springer, Heidelberg (2002)
17. Montcalm, T., Boufama, B.: Object Inter-camera Tracking with Non-overlapping Views: A New Dynamic Approach. In: Proceedings of the Canadian Conference on Computer and Robot Vision, pp. 354–361 (2010)
18. Porikli, F., Tuzel, O., Meer, P.: Covariance Tracking Using Model Update Based on Means on Riemannian Manifolds. In: Proceedings of the IEEE Conference on Computer Vision and Pattern Recognition, pp. 728–735 (2006)
19. Zheng, W., Gong, S., Xiang, T.: Person Re-identification by Probabilistic Relative Distance Comparison. In: Proceedings of the IEEE Conference on Computer Vision and Pattern Recognition, pp. 649–656 (2011)

Colour Descriptors for Tracking in Spatial Augmented Reality

Thijs Kooi, Francois de Sorbier, and Hideo Saito

Keio University
Department of Information and Computer Science
3-14-1 Hiyoshi, Kohoku-ku, Yokohama, Japan
{tkooi,fdesorbi,saito}@hvrl.ics.keio.ac.jp

Abstract. Augmented Reality is an emerging research field, that aims for the composition of real and virtual imagery, by means of a camera and display device. Spatial augmented reality employs data projectors to augment the real world. In this setting, traditional tracking methods fall short due to the interference caused by the projector. Recent works assume a calibration process to model the projector and assume continuity in movement of the object being tracked. In this paper we present a tracking-by-detection system that does not require such a procedure and makes use of natural features represented by SIFT descriptors. We evaluate a set of photometric invariants that have previously been shown to improve the performance of object recognition, added to the descriptor to reduce the influence of the projector. We evaluate the descriptors based on precision-recall under projector distortion and the total system based on its tracking performance. Results show tracking is significantly more precise using one of the invariants.

1 Introduction

Augmented reality (AR) systems channel information from the real world through a sensor and in the process, *augment* the real data with some virtual content, deemed relevant for the application. The concept is most commonly applied to vision, in which case the displays are devices such as head-mounted displays, hand-held screens like tablets and smart phones or retinal displays, that project an image straight into the eye. Spatial Augmented Reality (SAR) [1] is a sub field of AR and makes use of data projectors as display devices. In this setting, the augmented information is no longer confined to some plane only visible to the user, but takes active part in its environment and generates a novel experience of the surroundings.

With the goal of AR to augment the real world with some virtual imagery, we are facing the challenge to properly align the two. Older systems use special hardware trackers or fiducial markers, which are invasive, expensive or difficult to use. Therefore, most recently developed AR systems employ natural features, acquired with a camera, for tracking. The most significant challenge in SAR, is that these methods are ineffective, due to the image projected onto the object being followed. This naturally results in two approaches: (1) treating the

J.-I. Park and J. Kim (Eds.): ACCV 2012 Workshops, Part II, LNCS 7729, pp. 387–399, 2013.

projector as interference or (2) modelling the projected image. The latter approach requires a-priori information about the reflectance properties of the surface and ambient illumination, which is obtained by means of some calibration procedure, as demonstrated by Audet et al. [2,3]. Here, tracking was done by template matching, which assumes some continuous motion between frames or an educated guess on the motion parameters. To enhance usability, a system omitting such a procedure and constraints would be preferable.

The work presented in this paper takes steps towards a *markerless, calibration-free* robust SAR system that does not require continuous motion or an initial estimate of the motion parameters. To this end, we make use of *tracking-by-detection*. The proper geometric invariance with respect to movement is obtained by using SIFT detection and description. An optimal photometric description is sought by evaluating several photometric invariant gradient representations in the SIFT descriptor and the remaining feature noise is filtered out by robust estimation in the form of RANSAC. We assume a planar, rigid surface using a static camera and projector, where the projected image has been aligned before the tracking commences. An illustration of an SAR setting is provided in figure 1. The contributions of this paper are twofold and can be summed up as follows: (1) we introduce a robust tracking system for SAR and (2) we underline the expedience of colour based information in local image descriptors, thereby confirming the results of previous work on this topic [4,5]. The rest of this paper is organised in the following way. Section 2 will cover some related work on projector-camera systems, SAR and colour feature description. In section 3, we will subsequently discuss the invariant descriptors used, followed by a brief exposition of the matching and tracking method in section 4. Section 5 will provide the experimental setup along with the acquired results and we will finish with a discussion and some future work in section 6.

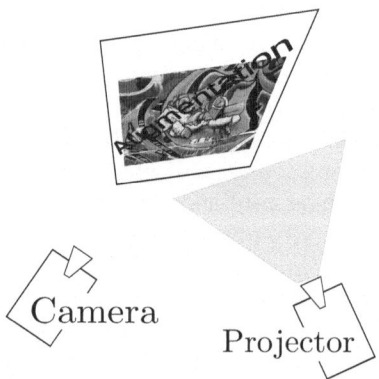

Fig. 1. Example of an SAR setting, where an image attached to a planar surface is augmented with some text

2   Related Work

SAR incorporates a projector-camera system, which has been the subject of some interesting research. Successful applications include the removal of shadows cast by the speaker during a presentation, by means of multiple projectors [6], combining multiple projectors to create large display walls [7], projection defocus analysis and correction [8,9], manipulating one object's colour and texture to make it look like another [10], and real-time correction of the geometry of the projected image according to the underlying geometry of the projection surface [11]. Audet [2,3] recently proposed an alignment strategy for SAR based on 4-point parametrisation, suggested by Baker et al. [12], template matching and Gauss-Newton optimisation of the homography parameters. A calibration procedure was employed to model the intrinsics of the projector and camera, colour mixing matrix and colour offset.

Even though colour can be a descriptive feature, it is still not common practice to add it to descriptors. Some work has been done, mostly in the field of object recognition. Funt and Finlayson [13] adapted a colour indexing scheme and showed that this improves previous results on object recognition. Geusebroek et al. [14] derived a set of photometric invariants, based on a previously presented Gaussian opponent model [15] and van Gemert et al. [4] showed that exploiting one of the invariants in an object recognition setting can improve performance. van de Wijer and Schmid [16] added colour and colour invariance properties to the SIFT [17] descriptor, which increased matching performance. van de Sande et al. [18] evaluate the performance of colour descriptors for object recognition and show performance can be increased.

Burghouts and Geusebroek [5] evaluated 5 SIFT based descriptors using greyscale and colour invariants, firmly put in a Gaussian spatio-spectral scale space framework and found an improvement in object recognition using colour. In the next section, we will briefly go over the derivation of the invariants, for a more rigorous treatment, the reader is referred to Geusebroek et al. [19,14,15]. A quick overview of the descriptors along with their invariance properties is provided in table 1.

3   Invariant Colour Descriptors

Since we are treating the projected image as interference, we are interested in image features that are as invariant as possible to distortion caused by the projector, yet descriptive enough for us to accurately track the surface. Any derivation of invariants relies on a physical model of light formation. For this, the Kubelka-Munk model can be applied [14], which models the reflected spectrum of a body, according to a material dependent reflection and absorption function. The resulting model of the spectrum at location $\mathbf{x}$ on the image plane is given by:

$$E(\lambda, \mathbf{x}) = e(\lambda, \mathbf{x})(1 - \varrho_f(\mathbf{x}))^2 R_\infty(\lambda, \mathbf{x}) + e(\lambda, \mathbf{x})\varrho_f(\mathbf{x}) \tag{1}$$

where $e(\lambda, \mathbf{x})$ denotes the illumination spectrum at $\mathbf{x}$, ϱ_f the Fresnel reflectance, λ the (visible) wavelength and R_∞ the material reflectance property. Koenderink

[20] showed that under several assumptions, the only reasonable function for probing the spatial structure of an image is the Gaussian. Geusebroek et al. [15] show an extension can be made with respect to the spectral dimension, giving a new model of colour measurement:

$$\hat{E}_{\lambda^k}^{\sigma_\lambda} = \int E(\lambda) G_{\lambda^k}(\lambda; \lambda_0, \sigma_\lambda) d\lambda \tag{2}$$

where $k \in \{0, 1, 2\}$, $G_{\lambda^k(\lambda_0, \sigma_\lambda)}$ indicates the kst Gaussian derivative centred around λ_0 with spectral bandwidth σ_λ, $\hat{E}$ indicates the spectral intensity (i.e., the greyscale channel), $\hat{E}_\lambda$ measures the 'yellow-blue' channel and $\hat{E}_{\lambda\lambda}$ describes the 'red-green' channel. The Gaussian opponent colour model can be computed from RGB values directly by the linear transformation [15]:

$$\begin{pmatrix} \hat{E} \\ \hat{E}_\lambda \\ \hat{E}_{\lambda\lambda} \end{pmatrix} = \begin{pmatrix} 0.06 & 0.63 & 0.27 \\ 0.30 & 0.04 & -0.35 \\ 0.34 & -0.60 & 0.17 \end{pmatrix} \begin{pmatrix} R \\ G \\ B \end{pmatrix} \tag{3}$$

For tracking the surface, we are interested in surface property R_∞ and should therefore look for transformations that isolate R_∞ from the other elements in the model. By making some assumptions on the environmental conditions, a set of invariants can be derived, that make use of the spatio-spectral scale space. Under equal energy but uneven illumination, we have that:

$$E(\lambda, \mathbf{x}) = i(\mathbf{x})\{\varrho_f(\mathbf{x}) + (1 - \varrho_f(\mathbf{x})^2 R_\infty(\lambda, \mathbf{x})\} \tag{4}$$

Where $i(\mathbf{x})$ denotes spatial intensity variation, since this is equal for all wavelengths. After taking first and second spectral derivative and adding the opponent colour model,

$$\hat{E}_\lambda = i(\mathbf{x})(1 - \varrho_f(\mathbf{x}))^2 \frac{\partial R_\infty(\lambda, \mathbf{x})}{\partial \lambda} \quad \hat{E}_{\lambda\lambda} = i(\mathbf{x})(1 - \varrho_f(\mathbf{x}))^2 \frac{\partial^2 R_\infty(\lambda, \mathbf{x})}{\partial \lambda^2} \tag{5}$$

we will see that their ratio $\hat{H} = \frac{\hat{E}_\lambda}{\hat{E}_{\lambda\lambda}}$ depends only on object reflectance $R_\infty(\lambda, \mathbf{x})$. Spatial derivation and application of the chain rule subsequently results in the invariant image gradient $\frac{\partial}{\partial j}\hat{H}$:

$$\hat{H}_j = \frac{\hat{E}_{\lambda\lambda}\hat{E}_{\lambda j} - \hat{E}_\lambda \hat{E}_{\lambda\lambda j}}{\hat{E}_\lambda^2 + \hat{E}_{\lambda\lambda}^2} \tag{6}$$

with $j \in \{x, y\}$, which is defined for $\hat{E}_\lambda^2 + \hat{E}_{\lambda\lambda}^2 > 0$.

If we further assume Fresnel reflectance is negligible, i.e., $\varrho_f(\mathbf{x}) \approx 0$, (1) reduces further to:

$$E(\lambda, \mathbf{x}) = i(\mathbf{x})R_\infty(\lambda, \mathbf{x}) \tag{7}$$

The surface property can be obtained by the ratio $\hat{C} = \frac{\hat{E}_\lambda}{\hat{E}}$ with spatial derivatives:

$$\hat{C}_{\lambda j} = \frac{\hat{E}_{\lambda j}\hat{E} - \hat{E}_\lambda E_j}{\hat{E}^2} \quad \hat{C}_{\lambda\lambda j} = \frac{\hat{E}_{\lambda\lambda j}\hat{E} - \hat{E}_{\lambda\lambda}E_j}{\hat{E}^2} \tag{8}$$

One more invariant can be derived, under yet again a stricter interpretation of the Kubelka-Munk model. If we consider diffuse reflectance, equal energy and uniform illumination with intensity i, (1) is reduced to:

$$E(\lambda, \mathbf{x}) = iR_\infty(\lambda, \mathbf{x}) \tag{9}$$

by taking the spatial derivative of (9) we get:

$$\hat{E}_j = \frac{\partial R_\infty(\lambda, \mathbf{x})}{\partial j} \tag{10}$$

If we let $\hat{W}_j = \frac{\hat{E}_j}{\hat{E}}$, we gain an expression that determines the change in object reflectance in the j direction, independent of illumination intensity. Similar properties can be defined for higher order spectral derivatives. Again employing the Gaussian colour model we get:

$$\hat{W}_j = \frac{\hat{E}_j}{\hat{E}} \quad \hat{W}_{\lambda,j} = \frac{\hat{E}_{\lambda,j}}{\hat{E}} \quad \hat{W}_{\lambda\lambda,j} = \frac{\hat{E}_{\lambda\lambda,j}}{\hat{E}} \tag{11}$$

Similar to Burghouts and Geusebroek, we added the photometric invariant gradients to the SIFT descriptor, along with the greyscale channel ($\hat{E}$), as used in the original version of SIFT. Addition of the intensity channel might seem contradictory, since we are trying to obtain invariance with respect to disturbing influences, but we found the combination between invariance and descriptiveness improved performance, thereby confirming their results.

Table 1. Photometric Invariants

Name	Given by	Invariance properties
E-grey	$\hat{E}$	None
E-colour	$\hat{E}, \hat{E}_\lambda, \hat{E}_{\lambda\lambda}$	None
W-colour	$\hat{W}, \hat{W}_\lambda, \hat{W}_{\lambda\lambda}$	Illumination intensity
C-colour	$\hat{C}_\lambda, \hat{C}_{\lambda\lambda}$	Object geometry and viewing angle, illumination intensity
H-colour	$\hat{H}$	Object geometry and viewing angle, illumination intensity, specular reflection

4   Tracking

The goal of the tracking stage is to compute the homography that maps points of the template to the points in each subsequent frame, by matching interest point locations between the two. With this homography, the projected image can be updated and aligned with the current orientation of the surface. An illustration is provided in figure 2. We performed matching by building a KD-tree on the interest points of the template, which reduces matching complexity and computing nearest and second nearest neighbours between the template

(a) Initial frame (template)   (b) Following frame, distorted augmentation   (c) Following frame, corrected augmentation

Fig. 2. Tracking problem. The goal is to compute the motion between the initial image and subsequent frames using interest point matching. The added text causes interference, rendering tracking more challenging.

points and frame points at every timestep. The nearest neighbour distance ratio (NNDR) between descriptor $\mathbf{d}$, its nearest $\mathbf{d}_1$ and second nearest neighbour $\mathbf{d}_2$

$$NNDR = \frac{||\mathbf{d} - \mathbf{d}_1||_2}{||\mathbf{d} - \mathbf{d}_2||_2} \tag{12}$$

as used by Mikolajczyk and Schmid [21] was applied, which they showed to improve matching performance. We employed the metric based on Euclidean distance in feature space and used RANSAC as a robust estimation technique for finding the homography between the set of matches.

We applied the *symmetric transfer function* [22] to compute the projection error. Not all putative matches will have equal confidence level and we can add this uncertainty into the optimisation by adding a weight term to the residual:

$$r_i = w_i\big(||\mathcal{M}_H \mathbf{x}_t^i - \mathbf{x}_{t+1}^i||_2 + ||\mathbf{x}_t^i - \mathcal{M}_H^{-1}\mathbf{x}_{t+1}^i||_2\big) \tag{13}$$

where w_i denotes a confidence weight of putative match $(\mathbf{x}_t^i, \mathbf{x}_{t+1}^i)$, which we computed by $1 - NNDR$ and $\mathcal{M}_H$ the current estimate of the homography. We applied a probabilistic setting to estimate the number of RANSAC iterations needed [23,24]. If we denote q to be the probability of sampling from the data a minimum sample set (MSS), containing no outliers, we can define the probability of sampling an MSS containing at least one outlier, for h consecutive iterations as $(1 - q)^h$. If we now define a probability threshold ϵ, such that $(1 - q)^h \leq \epsilon$ and invert the relation, we have that:

$$h \geq \left\lceil \frac{\log \epsilon}{\log(1 - q)} \right\rceil \tag{14}$$

By defining N_I to be the total number of inliers in the data, N the total number of points in the data and k the cardinality of the sample set, we compute q according to:

$$q = \frac{\binom{N_I}{k}}{\binom{N}{k}} = \frac{N_I!(N-k)!}{N!(N_I-k)!} = \prod_{i=0}^{k-1} \frac{N_I - i}{N - i} \approx \left(\frac{N_I}{N}\right)^k \quad (15)$$

By making a conservative approximation of the number of inliers $\hat{N}_I$ in the form of the maximum cardinality of the consensus set so far, we gain the following expression for h:

$$h = \left\lceil \frac{\log \epsilon}{\log \left(1 - (\hat{N}_I N^{-1})^k\right)} \right\rceil \quad (16)$$

Estimating the number of iterations rather then keeping them fixed improves performance and reduces computation time. As $\hat{N}_I$ increases, q and therefore h decrease, i.e., the better MSS we find, the less iterations we estimate to need and therefore the number of iterations is an indication of the estimated quality of the fit, evaluated by the residual functions defined before.

RANSAC attempts to minimise a loss function $\sum_i \rho(r_i)$ over all data, where in its original formulation $\rho(r)$ is defined as:

$$\rho_{RANSAC}(r) = \begin{cases} 1 & \text{if } r > \delta \\ 0 & \text{else} \end{cases} \quad (17)$$

with δ the outlier threshold. Without any extra computational costs, a more sensible loss function can be defined, based on the residual of the datum:

$$\rho_{MSAC}(r) = \begin{cases} r & \text{if } r \leq \delta \\ \delta & \text{else} \end{cases} \quad (18)$$

which is dubbed *MSAC* by Torr and Zissermann [23,24]. By combining this estimator with the confidence weighted matching, described earlier, the distance in feature space is propagated through to the learning phase of RANSAC and therefore directly taken into account in the final estimate of the homography.

5   Experiments and Results

This section provides details on the experimental setup and results after applying the presented method. Before proceeding, we would like to make clear the assumption made, when performing the experiments. The results reflect the following conditions. (1) The projector and camera are assumed to be static and the projected image is aligned with the surface at the first timestep. (2) The internal parameters of both camera and projector are static. (3) The projection surface is rigid, planar and displays a low amount of specular reflectance (a degree similar to ink printed on paper). (4) The image of the camera is sharp and sufficiently exposed and the image of the projector is taken to be sharp enough

to cause some interference. (5) There is no occlusion (6) there are no serious artefacts originating from the camera lens such as lens distortion, vignetting, etc.
and there is no image compression. During experimentation, we found it was
difficult to prevent clipping effects, using current hardware, since the exposure
correction of the camera is at least a frame behind. Because these are realistic
challenges, we have left them in the experiments. It is worth to note, however,
that these influences are known to inhibit performance of invariants.

5.1   Hardware and Implementation Details

The video's where recorded in 800x600, 24 BPP at 15 FPS using PPM file format
and test images where printed on standard A4 paper. We make use of a template which was extracted by sampling the first image from the image sequence,
which contains a projected image. Some functions in Marco Zuliani's RANSAC
toolbox where very helpful and an adaptation of the implementation of colour
SIFT, kindly provided by the authors [5], was used for the experiments. All parameter settings where held similar to those used by Lowe [17] and Burghouts
and Geusebroek [5], unless mentioned otherwise in the experiments. We used an
NNDR of 0.7, which performed best during initial testing. To filter out background points, we used a threshold to only consider points that actually moved
between frames, which was set to 3 pixels. Feature points where detected in $\hat{E}$,
using the method described by Lowe [17] and are equal for all descriptors.

5.2   Evaluation Strategy

To estimate the accuracy of the methods, we have used 2 different square test images of varying texture and colour. The first evaluation setting is an adaptation
of the setup used by Mikolajczyk and Schmid [21] and Burghouts and Geusebroek, testing only the effect of projector distortion in our case. The images
where attached to a rigid surface, which was held still throughout the process.
We used two different projector conditions in the form of video's taken from
the intro of TV shows. One causing mild interference and one causing strong
interference. This results in 4 video's with a total number of 151 frames. Interest
points in a frame with a projected image where computed and matched against
subsequent frames. For determining correspondences, the strategy proposed by
Schmid et al. [25] was applied, where the radius was set to 1 pixel in our setting.
Average over all frames where taken and precision and recall where computed
and interpolated by sweeping through the NNDR. Results are shown in figure 3.

The second setting tests the robustness of the tracking method with respect
to both geometric and photometric changes of the scene. Again, the images
where attached to a rigid surface, which this time was subjected to translation,
rotation and perspective movement. For each surface, this was done two times,
one time using slow movement and one time using very fast movement. We used
three different projector conditions: one where there is no projection at all and
two with the before mentioned video's, giving us a total of 12 video's, with in
total 1430 frames without any projection and 3740 frames with. Two frames

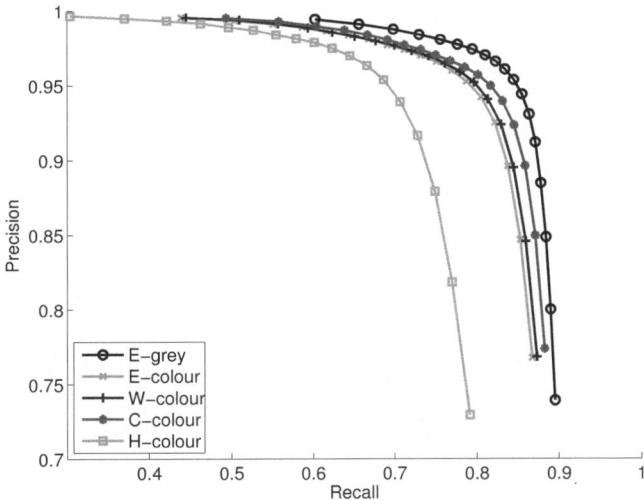

Fig. 3. Precision-Recall of Matching

of the different backgrounds and different projections are provided in figure 4. Please note that to keep conditions equal for all invariants, we did not explicitly update the projected image, but used the video's to emulate possible distortion, caused by the projector. To compensate for the random nature of RANSAC, each homography was estimated 5 times, each using a different random seed and treated as iid. The 4 corner points where annotated in every other frame, in each video and subsequently averaged, since they are typically not iid. This also compensates for small annotation errors. If we denote f to be the number of frames in a video, this leaves us with a total of $5\lceil \frac{f}{2} \rceil$ number of samples.

The following metrics provide insight in the performance. To begin with, we computed the Euclidean distance between the annotated points and their estimated

(a) Background 1, projection 1   (b) Background 2, projection 2

Fig. 4. Stills of test video's for tracking

position in each frame, by using the annotated corner points in the first frame and warping them according to each estimated homography. Secondly, we used the estimated number of RANSAC iterations as a mild indication of the estimated quality of the matches along with the number of inliers after finishing RANSAC. A large discrepancy between true and estimated quality of the homography would indicate a flaw in description or matching. A small number of high quality matches in combination with a high quality fit would be preferred, since the time complexity of RANSAC increases with the number of points.

5.3   Tracking Results

To set a benchmark and give an indication of the kind of error values the method should aim for, we first evaluated the tracking method without any projector interference. The mean and median Euclidean distance over all frames is provided in table 2, along with one standard deviation. A Kruskal-Wallis test, reveals no significant difference ($p = 0.33419$) in tracking performance based on the Euclidean distance. We subsequently ran the methods with projection. Same metrics along with the match rates, number of RANSAC iterations and number of inliers are given in table 3.

Table 2. Tracking results without projection

	Mean Eucl. Dist.	Median Eucl. Dist.
E-grey	16.98 +/- 13.25	16.0278
E-colour	16.51 +/- 4.99	16.0445
W-colour	16.69 +/- 8.88	16.0154
C-colour	16.36 +/- 4.87	15.9499
H-colour	18.46 +/- 22.75	16.0218

Table 3. Tracking results with projection

	E-grey	E-colour	W-colour	C-colour	H-colour
Nr. of Matches	210.03 +/- 110.95	113.83 +/- 83.39	119.66 +/- 91.19	138.22 +/- 88.95	97.98 +/- 68.36
Nr. of itt.	131.15 +/- 144.72	85.70 +/- 127.40	73.22 +/- 117.35	36.68 +/- 71.46	75.93 +/- 119.60
Nr. of inliers	138.67 +/- 110.40	90.45 +/- 82.84	97.71 +/- 90.26	116.99 +/- 89.46	79.10 +/- 67.75
Avg. Eucl. Dist.	94.39 +/- 1388.36	70.97 +/- 276.20	99.19 +/- 905.15	24.99 +/- 37.36	167.29 +/- 1212.93
Median Eucl. Dist.	19.092	20.1653	19.6647	18.9503	20.4581

Kruskal-Wallis tests reveal significant effects in all metrics. We subsequently computed a Wilcoxon rank sum on Euclidean distance between E-colour and E-grey, C-colour and E-grey and C-colour and E-colour, which all show significant effects ($p \ll 0.05$).

6   Discussion

Table 2 shows tracking is marginally though not significantly improved if there is no projector distortion. This is mainly reflected by the lower standard deviation of E-colour and C-colour, which represents a smaller amount of outliers. These outliers would produces strong quirks in the corrected image of the projector and less is therefore desirable.

It is interesting to see the discrepancy between the precision-recall curve and the tracking performance. Burghouts and Geusebroek also found that, when added to the SIFT descriptor, C-colour performs more or less equal to grey scale SIFT, when varying illumination colour, but outperforms it when varying viewpoint and illumination direction. It is therefore expected that this shows in the tracking without projection setting in our experiments. A different distance metric (Mahalanobis distance instead of Euclidean) and other minor variations in experimental setup and evaluation may have been the reason that this does not show significantly. Mikolajczyk and Schmid found that the NNDR not only improves matching performance, but also changes the shape of the curve slightly, which may have had some influence in the difference between previous work. It is known that clipping effect can severely impede the performance of invariants. We inspected both the stationary and tracking video's in the search for severe clipping effects, but even though it occurs, we did not find a big difference.

From the results in table 3 we can draw the following conclusions. The addition of colour clearly improves performance, except for H-colour. Similar conclusions where drawn by Burghouts and Geusebroek [5]. Lack of descriptiveness or instability due to non-linearity of the computation are probable causes. As far as colour goes, all arguments are in favour of C-colour. The number of matches is lower than E-grey, which improves speed of RANSAC and there are less iterations required to reach a consensus set with a high number of inliers. The median of E-grey is lower than E-colour but the average of E-colour is lower, which is caused by several outliers in the estimated position of the plane, using E-grey. The outliers are caused by difficult conditions. In general, all colour descriptors show a smaller standard deviation and are therefore better able to cope with these conditions.

7   Conclusion

In this paper we have presented a tracking-by-detection method for Spatial Augmented Reality (SAR) which, contrary to previous work on the topic, does not require a calibration procedure and does not assume continuous motion between frames. To this end, we have applied SIFT in combination with RANSAC for tracking and have evaluated a set of photometric invariants, added to the SIFT descriptor, from which the invariant C-colour significantly improved tracking performance under projector distortion. This confirms other work using the same invariant for object recognition.

Future work using the current assumptions will revolve mostly around optimisation. SIFT is known for its good performance, but a large amount of substantially faster methods have been developed. The extension to colour using gradient based descriptors is relatively straight forward: simply replace the image gradient or approximation thereof, by the invariants presented earlier. In this spirit, we made some attempts using SURF descriptors, but did not acquire similar performance. We suspect the better noise robustness of SIFT renders it more efficient in the current domain. Applying the method to pixel-ratio based detectors/descriptors such as FAST, randomised trees and FERNS is less obvious, but if this is done successfully, it will give a large boost in terms of computational speed.

Part of the assumptions made could be alleviated by allowing freedom of movement of the camera, thereby also tracking the projected image. Another extension would be compensating for radiometric variation and not only geometric change. Several mesh based methods have been developed and used in common AR settings, for augmenting non-rigid surfaces. A substantial challenge in applying this to SAR is that the projected image will be covariant with viewing angle. The use of multiple projectors and viewer tracking may provide a solution to this.

Acknowledgements. This work was partially supported by MEXT/JSPS Grant-in-Aid for Scientific Research (S) 24220004.

References

1. Bimber, O., Raskar, R.: Spatial augmented reality - Merging Real and Virtual Worlds. A. K. Peters, Ltd., Natick (2005)
2. Audet, S., Okutomi, M., Tanaka, M.: Direct image alignment of projector-camera systems with planar surfaces. In: CVPR, pp. 303–310. IEEE (2010)
3. Audet, S., Okutomi, M., Tanaka, M.: Augmenting moving planar surfaces robustly with video projection and direct image alignment. Virtual Reality, 1–12, 10.1007/s10055-012-0210-9
4. van Gemert, J.C., Burghouts, G.J., Seinstra, F., Geusebroek, J.M.: Color invariant object recognition using entropic graphs. International Journal of Imaging Systems and Technology 16, 146–153 (2006)
5. Burghouts, G.J., Geusebroek, J.M.: Performance evaluation of local colour invariants. Computer Vision and Image Understanding 113, 48–62 (2009)
6. Audet, S., Cooperstock, J.R.: Shadow removal in front projection environments using object tracking. In: Projector-Camera Systems, pp. 1–8 (2007)
7. Majumder, A., Brown, M.S.: Practical Multi-projector Display Design. A. K. Peters, Ltd., Natick (2007)
8. Zhang, L., Nayar, S.: Projection defocus analysis for scene capture and image display. ACM Transactions on Graphics 25, 907–915 (2006)

9. Oyamada, Y., Saito, H.: Blind deconvolution based projector defocus removing with uncalibrated projector-camera pair. In: IEEE International Workshop on Projector-Camera Systems, PROCAMS (2009)
10. Grossberg, M.D., Peri, H., Nayar, S.K., Belhumeur, P.N.: Making one object look like another: Controlling appearance using a projector-camera system. In: IEEE Computer Society Conference on Computer Vision and Pattern Recognition, vol. 1, pp. 452–459 (2004)
11. Johnson, T., Fuchs, H.: Real-time projector tracking on complex geometry using ordinary imagery. In: Projector-Camera Systems, p. 1 (2007)
12. Baker, S., Datta, A., Kanade, T.: Parameterizing homographies. Technical Report CMU-RI-TR-06-11, Robotics Institute, Pittsburgh, PA (2006)
13. Funt, B.V., Finlayson, G.D.: Color Constant Color Indexing. IEEE Transactions on Pattern Analysis and Machine Intelligence 17, 522–529 (1995)
14. Geusebroek, J.M., van den Boomgaard, R., Smeulders, A.W.M., Geerts, H.: Color invariance. IEEE Transactions on Pattern Analysis and Machine Intelligence 23, 1338–1350 (2001)
15. Geusebroek, J.-M., van den Boomgaard, R., Smeulders, A.W.M., Dev, A.: Color and Scale: The Spatial Structure of Color Images. In: Vernon, D. (ed.) ECCV 2000. LNCS, vol. 1842, pp. 331–341. Springer, Heidelberg (2000)
16. van de Weijer, J., Schmid, C.: Coloring Local Feature Extraction. In: Leonardis, A., Bischof, H., Pinz, A. (eds.) ECCV 2006. LNCS, vol. 3952, pp. 334–348. Springer, Heidelberg (2006)
17. Lowe, D.G.: Object recognition from local scale-invariant features. In: Proceedings of the International Conference on Computer Vision, ICCV 1999, vol. 2, pp. 1150–1157. IEEE Computer Society, Washington, DC (1999)
18. van de Sande, K.E.A., Gevers, T., Snoek, C.G.M.: Evaluating color descriptors for object and scene recognition. IEEE Transactions on Pattern Analysis and Machine Intelligence 32, 1582–1596 (2010)
19. Geusebroek, J.-M., Dev, A., van den Boomgaard, R., Smeulders, A.W.M., Cornelissen, F., Geerts, H.: Color Invariant Edge Detection. In: Nielsen, M., Johansen, P., Fogh Olsen, O., Weickert, J. (eds.) Scale-Space 1999. LNCS, vol. 1682, pp. 459–464. Springer, Heidelberg (1999)
20. Koenderink, J.J.: The structure of images. Biological Cybernetics 50, 363–370 (1984)
21. Mikolajczyk, K., Schmid, C.: A performance evaluation of local descriptors. IEEE Transactions on Pattern Analysis and Machine Intelligence 27, 1615–1630 (2005)
22. Hartley, A., Zisserman, A.: Multiple View Geometry in Computer Vision, 2nd edn. Cambridge University Press (2006)
23. Zisserman, A., Torr, P.H.S.: Robust parameterization and computation of the trifocal tensor. In: BMVC, Motion and Active Vision (1996)
24. Torr, P.H.S., Zisserman, A.: Robust computation and parametrization of multiple view relations. In: ICCV, pp. 727–732 (1998)
25. Schmid, C., Mohr, R., Bauckhage, C.: Evaluation of interest point detectors. International Journal of Computer Vision 37, 151–172 (2000)

Covariance Descriptor Multiple Object Tracking and Re-identification with Colorspace Evaluation

Andrés Romero, Michèle Gouiffés, and Lionel Lacassagne

Institut d'Éléctronique Fondamentale, UMR 8622, Université Paris-Sud XI,
Bâtiment 660, rue Noetzlin, Plateau du Moulon, 91400 Orsay

Abstract. This paper addresses the multi-target tracking problem with the help of a matching method where moving objects are detected in each frame, tracked when it is possible and matched by similarity of covariance matrices when difficulties arrive. Three contributions are proposed. First, a compact vector based on color invariants and Local Binary Patterns Variance is compared to more classical features vectors. To accelerate object re-identification, our second proposal is the use of a more efficient arrangement of the covariance matrices. Finally, a multiple-target algorithm with special attention in occlusion handling, merging and separation of the targets is analyzed. Our experiments show the relevance of the method, illustrating the trade-off that has to be made between distinctiveness, invariance and compactness of the features.

1 Introduction

Multiple objects tracking or matching is a classical task required in most surveillance systems. More than being useful for analyzing trajectories and behaviors in a mono-camera context, it is a challenging issue when objects have to be re-detected from a second camera under different set-ups, or at two very different times. The task faces many difficulties such as scale or appearance change, illumination variations or occlusion. Ideally, the representation of the target has to be chosen so as to be invariant and robust to such phenomena. Unfortunately, most color invariants, although robust against lighting changes, can reduce the separability between the targets and can lead to matching ambiguities. In addition, when targets have a non rigid motion or have low textural or structural contents, the gradient or corner-based methods, such as the classical KLT [1] or SIFT [2] are not appropriate.

Kernel-based methods like Mean-shift [3] are usually well adapted to such objects since they rely on a global statistical distribution. The price to pay is a decrease of discriminant power, therefore several attempts have enhanced the method by background subtraction [4], colorspace switch [5] and by using a spatio-colorimetric histogram [6]. Covariance Tracking [7] is an interesting alternative which employs a compact representation of the correlation between spatial and statistical features within the object window. High performances can

J.-I. Park and J. Kim (Eds.): ACCV 2012 Workshops, Part II, LNCS 7729, pp. 400–411, 2013.

be achieved even for low textured objects, since they are represented by a global model. However, the choice of the features is still an issue.

The aim of this paper is to develop a robust and fast solution for multiple target detection, labeling and re-identification. It evaluates the behavior of several sets of color and texture/gradient features for covariance matching. In addition, a strategy is proposed for multi-target matching. Note that, contrary to most tracking techniques [7] where the targets have a consistent trajectory, the present work focuses on matching in order to evaluate the descriptors in the context of large motion and object re-identification applications.

The continuation of the paper is structured as follows. Section 2 introduces the covariance matching and the descriptors. Then, Section 3 explains the principles of the objects handling, and how the occlusion, collision and separations are treated in a probabilistic context. To conclude, Section 4 compares the behavior of the features and evaluates the multi-target matching.

2   Covariance Descriptors

2.1   Principles

From each pixel in the observed image I_t of size $W \times H$ a feature vector is obtained by the mapping function ϕ, such that a $W \times H \times d$ dimensional feature image F is generated $F(x, y) = \phi(I, x, y)$, where local information represented by ϕ can be position, color, gradients, filter responses, etc. A rectangular region $\{\mathbf{z}_k\}_{k=1\cdots n}$ of n feature points is represented by the $d \times d$ matrix

$$\mathbf{C}_R = \frac{1}{n} \sum_{k=1}^{n} (\mathbf{z}_k - \mu)(\mathbf{z}_k - \mu)^T \tag{1}$$

where vector μ is the mean of the feature points inside the region. Targets are represented with covariance matrices $\mathbf{C}_R$ which preserve spatial and statistical information and allow to compare different sized regions. Tracking is performed searching for the most similar region in a list of candidate regions in I_t with the object's model in $t - 1$. However, direct arithmetic subtraction fails to compare covariance matrices because these type of matrices do not lie on the Euclidean space. The matching can be done applying the dissimilarity measure defined in [8] as the sum of squared logarithms of the generalized eigenvalues. Here, this distance is noted d_{cov}.

Adapting the model for changes of shape, size and appearance is also necessary. This is done by keeping a set of T previous covariance matrices $[\mathbf{C}_1 \cdots \mathbf{C}_T]$ where $\mathbf{C}_1$ denotes the current one, and by computing the mean covariance matrix through Riemannian geometry. A comprehensive explanation of the update mechanism can be found in [9].

2.2   Covariance Features

One of our objectives is to test the distinctiveness of several covariance matrices based on descriptors different both in size and nature. Classical features such as luminance I, image gradients (g_x, g_y), color RGB, and HSV models were tested as well as two color invariants that are worth of interest: the normalized (r, g, b), where r stands for $\frac{R}{R+G+B}$ (similar for G and B) because it offers a separation of luminance and color; then, invariant $L1$ from [10] is interesting because if offers a compact mixture of (r, g, b) and luminance by use of a color relevance measure. Finally, the LBP variance operator VAR_{LBP} is compared to the classical g_x and g_y. Specifically, the tested feature vectors combinations have the following generic form:

$$F_{\mathbf{A},\mathbf{B}} = \begin{bmatrix} x\ y\ \mathbf{A}\ \mathbf{B} \end{bmatrix} \tag{2}$$

where $\mathbf{A}$ is a color feature vector and $\mathbf{B}$ a texture descriptor. Five color features $\mathbf{A}$ are tested: Lum (a scalar luminance value), $[R\,G\,B]$, $[H\,S\,V]$, $[r\,g\,b]$, and $[L1\,V]$, as defined in [10] where $L1 = \max(r, g, b)$ and V is the luminance normalized in the object. Two texture descriptors are compared: $grads = [g_x\ g_y]$ is the gradient vector and LBP is the LBP variance value. Thus, ten feature vectors are compared.

2.3   Covariance Matching for Re-identification

Here, the *Mean Rimemannian Covariance* (**MRC**) matrices proposed by Bak et al. [11] are used to blend appearance information from multiple images.

Given a set of N covariance matrices $\{C_1, C_2, \cdots, C_N - 1\}$, the Karcher or Fréchet mean, is the value μ which minimizes the set of squared distances

$$\mu = arg\ \min_{C \in \mathcal{M}} \sum_{i=1}^{N} \rho^2(C, C_i) \tag{3}$$

For the case of covariance matrices, the value of μ is calculated iteratively, following the Newton gradient descent method for Riemannian manifolds. The approximate value of μ at step $t + 1$ is

$$\mu_{t+1} = exp_{\mu_t} \left[\frac{1}{N} \sum_{i=1}^{N} log_{\mu_t}(C_i) \right] \tag{4}$$

where, exp_{μ_t} and log_{μ_t} are specific operators uniquely defined on the Riemannian manifold. Equations (5) and (6) express how to calculate them.

$$Y = exp_X(W) = X^{\frac{1}{2}} exp(W) X^{\frac{1}{2}} \tag{5}$$

$$Y = log_X(W) = log(X^{-\frac{1}{2}} W X^{-\frac{1}{2}}) \tag{6}$$

Bak et. al [11] achieve great re-identification rates using a dense grid of of **MRC** matrices. For the case of human signatures, each image is scaled into a fixed size of 64×192 pixels where a grid of overlapping 16×16 pixel size cells is constructed. Neighboring cells are separated by 8 pixel steps. In total, 161 **MRC** are used to construct the human signature.

To reduce the re-identification computational cost we propose a different arrangement of **MRC** matrices. Images are re-scaled to 96×128 pixels, then, rings of concentric rectangles are formed around the image center with exponentially increasing areas allowing some area overlapping. The proposed pattern is inspired in FREAK and DAISY [12,13] but for rectangular covariance regions, in order to be easily accelerated by the integral images method.

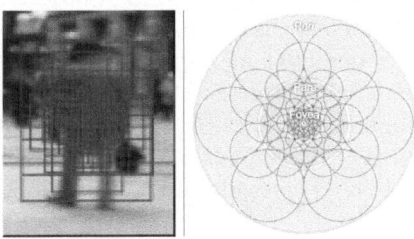

Fig. 1. Proposed **MRC** pattern and its resemblance to FREAK

In our configuration, a total of 42 **MRC** descriptors were employed mostly concentrated at the center while more variability is tolerated at the periphery. To further simplify things **MRC** descriptors are compared one to one in contrast to [11] where comparison is maid sliding one grid against the other.

3   Object Handling

In this section, we describe the multiple-tracking algorithm that we implemented together with the re-identification method (2.3). An approach similar to [14] was followed. A list is kept handling multiple levels of representation: blobs, people and groups. The algorithm receives an image I_t which corresponds to frame t. This image is introduced to the Sigma-Delta [15] background subtraction algorithm where data extracted from the set of previous frames $\{I_0, \cdots, I_{t-1}\}$ is employed to separate foreground and background into a binary image F_t.

Blobs are detected in F_t after applying the Light-Speed Labeling algorithm [16], signaling areas of the image of important change. Blobs of sufficient size are appended to an object list and small blobs are filtered out [1]. The accepted blobs inside the list are now considered objects to match/track exploiting their location, size, trajectory and appearance, modeled with the help of covariance descriptors.

In regular conditions, matching is done considering information of location, size and previous trajectory. Appearance information is used to confirm those estimations, a single low-dimensional (four to six features) covariance descriptor covering the whole object is preferred for simple situations. The complete set of

---

[1] The parameter used in the paper is 1500px for an image of 768×576.

descriptors described in section 4.1 is computed anyway as a preventive measure against faults such as object occlusions, target crossings and objects getting in and out from the scene.

When two or more existing objects become too spatially close, they are merged together to become a *group*. Groups are inserted into a separate list but in general, their location, trajectory and appearance are treated in the same way as any single object. Groups in contrast to single objects, are able to split into separate combinations of their composing original objects.

To each blob in F_t an identity is attributed, which comes from existing or newer objects/groups. As depicted in Fig.3, the identities can transit in five different states: *detected, tracked, occluded, collision* and *lost*. Consider a target blob B, and a set of N candidate objects $\{O_i\}_{i=0\cdots N-1}$, defined by their bounding boxes. For tracking purposes, the euclidean distance d_{bb} between the centers of B and O_i, denoted $\{d_{bb}(B, O_i)\}$, provides a first matching hint. Objects located far from B are filtered out by

$$d_{bb}(B, O_i) > K\max(W, H) \qquad (7)$$

where K is an adjustable factor and W, H correspond to the blob's width and height. Note that, when no assumptions can be made on the object location, for multiple-camera object re-identification for example, the location information of (7) is not taken into account.

Consider now a set $Z = \{O_j\}_{j=0\cdots M-1}$, formed by the objects which satisfy inequality (7) where $M \leq N$. A uniform probability $P(O_j) = 1/M$ is assigned to each object. These probabilities are updated considering the evidence provided by the set of distances $D_{bb} = \{d_{bb}(B, O_j)\}_{j=0\cdots M-1}$ as

$$P(O_j|D_{bb}) = \frac{d_{bb}^{-1}(B, O_j)}{\sum_j d_{bb}^{-1}(B, O_j)} \qquad (8)$$

Similarly, the set of covariance descriptor distances $D_{cov} = \{d_{cov}(B, O_j)\}_{j=0\cdots M-1}$ allows a second object probability update

$$P(O_j|D_{bb}D_{cov}) = \frac{exp\left(-d_{cov}(B, O_j)\right) P(O_j|D_{bb})}{\sum_j exp\left(-d_{cov}(B, O_j)\right) P(O_j|D_{bb})} \qquad (9)$$

the object O_j with the highest posterior probability is assigned to B if it surpasses a minimum threshold.

Groups are formed when multiple objects are merged into one blob, when their spatial distance $d_{bb}(O_i, O_j)$ is low (under a value which depends on the sizes of the two objects).

The covariance descriptors of each individual object are stored before grouping, and additional covariance descriptors are computed for each group, and matched as any other object. When the objects in a group cannot be matched individually with separated new candidate objects, then the matching of the whole group is performed. The individual objects are identified as group members and all of them are set in the state of *collision* sharing the same image location.

Figure 2, displays an example of a merge: at $t = 128$ a descriptor is calculated for the candidate fusion area (red dotted line). Next frame, (due to the closeness) only one blob is detected, and its covariance descriptor matches with the fusion area of previous frame. So, a group is created, it is tracked from frames $t = 130$ to $t = 135$, after this, each object is re-identified individually as described in section 4.1.

Fig. 2. Example of fusion and separation

Unmatched blobs are then compared with the objects considered in *collision* or *occluded* at $t − 1$.

Finally, covariance descriptors are regularly updated calculating their covariance mean (equation 3). To avoid model contamination, objects inside a group must not be updated (they are not reliable due to partial occlusion) until they are re-identified individually outside the group. The whole *object handling algorithm* is summarized in **Algorithm 1**.

4 Experiments and Results

Our experiments evaluated two different aspects: the re-identification success rate of the proposed **MRC** set for the different feature configurations (details in subsection 2.2), and the proposed multiple-target tracking algorithm.

4.1 Object Re-identification Experiments

To validate our method of re-identification we used the same performance measure of [11] and [17], which is the Cumulative Matching Characteristic (**CMC**) which represents the percentage of times the correct identity match is found in

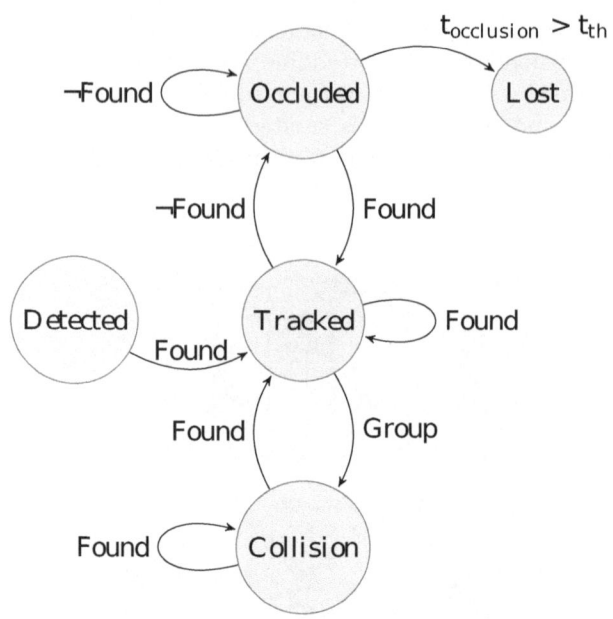

Fig. 3. Objects state transition

the top n matches. Tests were performed for the **ETHZ** [18] and **PETS'09 L1-Walking-Sequence 1** [19] datasets.

ETHZ dataset is composed by images from three different sequences (each one formed by 83, 35 and 28 individuals). Each individual, is captured by the same camera which suits just fine to our single-camera tracking objectives. For this experiment, eight different individuals from **PETS'09 L1-Walking-Sequence 1** were extracted taking discontinuous samples.

For each individual, 10 images were selected from the beginning ant the end and their **MRC** matrices (subsection 4.1) were calculated. The recognition rate was tested, taking random images and comparing against the registered signatures. Care was taken to avoid reusing any of the images occupied during signature calculation. Success is declared when the corresponding image identity is found inside the *top n* list.

To find out which is more discriminant, measurements were taken for the following feature configurations: 1)$F_{lum,LBP}$, 2)$F_{lum,grads}$, 3)$F_{RGB,grads}$, 4)$F_{L1,grads}$ and 5)$F_{HSV,grads}$. Figure 4 reports the results obtained for each sequence.

Except for the first sequence, $F_{RGB,grads}$ is the more powerful configuration to use, achieving good recognition percentages even for the rank-1 score. $F_{L1,grads}$ and $F_{HSV,grads}$ behave similar, because they explicitly separate luminance and colour and they show some resistance to low saturation conditions. On the other side, $F_{lum,grads}$ shows poor recognition performance being overtaken three times by $F_{lum,LBP}$ which has only 4 components.

Algorithm 1. Object handling algorithm
Input: Blobs list *blobs* and Object list *objList*
1 Get blobs covariance descriptors;
2 Match blobs - tracked objects;
3 Match blobs - candidate collisions;
4 Match blobs - occluded and collision objects;
5 Non-matched blobs create new objects inside *objList*;
6 Dissolve collisions with only one child;
7 Remove lost objects;
8 Update object states and models;
9 Detect candidate collisions;

The obtained re-identification rates are comparable to the ones reported in [11] employing a 75% less covariance matrices and fewer components inside them.

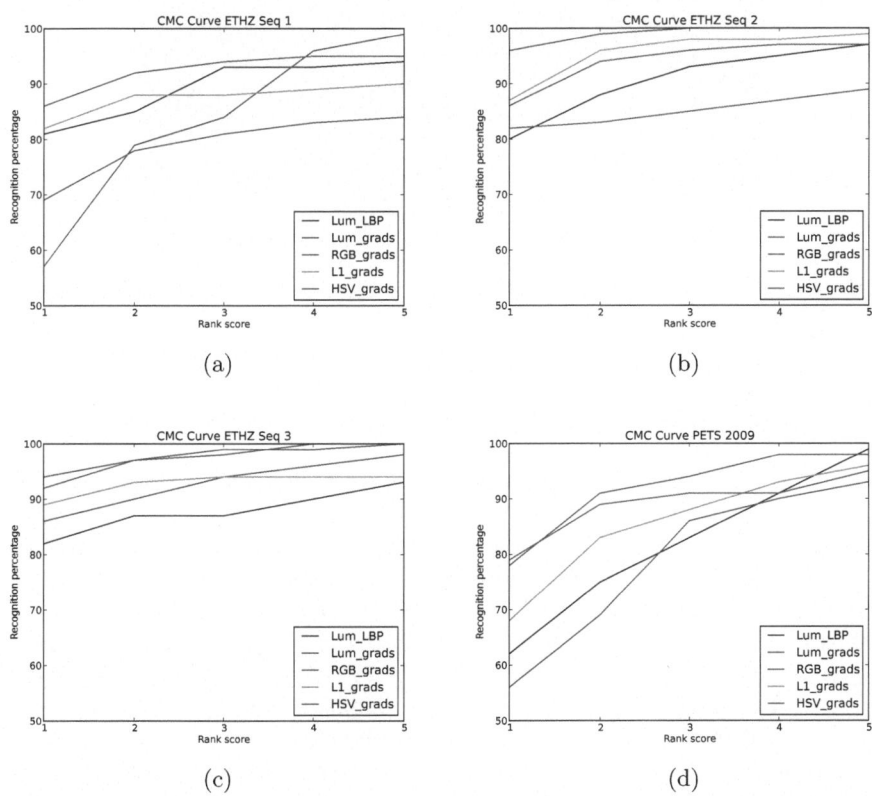

(a)   (b)

(c)   (d)

Fig. 4. Cumulative Matching Characteristic **CMC** curves. a) to c) for **ETHZ** sequences and d) for **PETS'09 L1-Walking-Sequence 1.**

Fig. 5. Some frames of PETS'09 showing the re-identification at different times and points of view

4.2 Multiple Object Tracking

Our tracking algorithm was tested on a randomly walking sparse crowd sequence from the **PETS'09 L1-Walking-Sequence 1 dataset** [19].

Feature vectors proposed in subsection 2.2 are evaluated considering Tracker's Purity (**TP**) [20], which is the ratio of frames a tracker ϵ_i correctly identifies a target $n_{i,j}$ to the total number of frames the tracker exists n_i: $\mathbf{TP} = \frac{n_{i,j}}{n_i}$.

Feature vector combinations which lead to the finest **TP** results were: 1) $F_{RGB,grads}$, 2) $F_{L1,LBP}$, 3) $F_{lum,grads}$, 4) $F_{L1,grads}$, 5) $F_{RGB,LBP}$ and 6) $F_{lum,LBP}$. **TP**s for these combinations are displayed in Figure 6. The points on the circle of radius **TP** $= 1$ are related to objects which are always correctly identified in the sequence. The more area is covered, the more often the targets are correctly identified. Mean **TP** values for these combinations are shown in Table 1. Obviously, tracker purity **TP** increases when using color since it provides relevant information that improves distinctiveness. Note that, although $F_{L1,LBP}$ is less distinctive than $F_{RGB,grads}$, it has two advantages: the covariance matrix

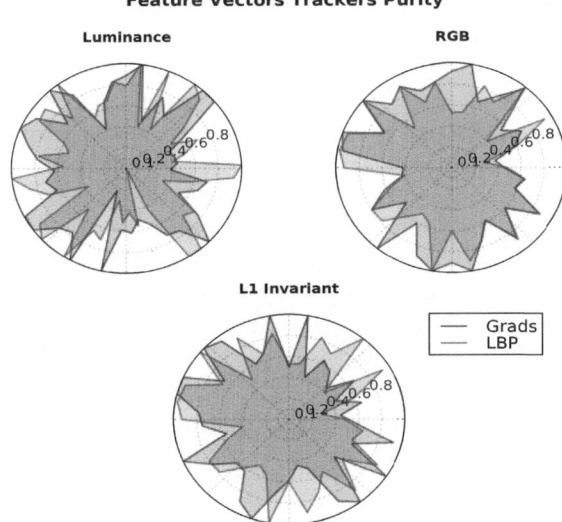

Fig. 6. Trackers Purity **TP** radar plots for different feature vectors

Table 1. Mean **TP** and Frames per Second (FPS) associated with the proposed Feature Vectors

Space	*Grads*		*LBP*	
	Mean **TP**	FPS	Mean **TP**	FPS
Lum	0.72455	22.887	0.68523	23.628
RGB	0.75046	14.574	0.70516	18.284
L1	0.72209	23.082	0.72823	23.290

is more compact (5×5 instead of 7×7) therefore the matching is more rapid, and it offers a better invariance against illumination variations as shown in [10]. The features vectors based on HSV and (r, g, b) are not convincing, since for low saturation the hue is ill-defined and (r, g, b) is not distinctive enough. $F_{lum,LBP}$ degrades severely in comparison to $F_{lum,grads}$, while in the case of the mixture $L1$, the use of VAR_{LBP} does not alter the performances.

5   Conclusions

We have proposed mainly three things in this article: 1) a reduced set of **MRC** matrices which achieves similar to state of the art performances, 2) the incorporation of the VAR_{LBP} operator which produces smaller matrices and 3) a tracking algorithm which blends localization, trajectory and appearance information providing it with re-identification capabilities.

Here, we have evaluated several descriptors. Note that the use of the invariant $L1$ [10], especially $F_{L1,LBP}$ allows to maintain a performance similar to RGB

while being more compact. The use of $L1$ to match images from different cameras and suffering drastic changes of color illumination, will be subject of further investigation.

The proposed multiple-target matching has shown encouraging results. Indeed, although there is intentionally no constraint on the temporal consistency of the trajectories, most objects are correctly matched due to the good distinctiveness of the chosen covariance features. Single crossings between two targets are handled fine regardless of the chosen feature vector combination. Still there are some targets non-consistently identified throughout the sequence (those with low **TP**). This is due to some issues not considered by the model. For example, some problems occur when several targets are crossing each other while some of them experiment background occlusion.

Acknowledgments. This research is supported by the European Project ITEA2 SPY^2.

References

1. Tomasi, C., Kanade, T.: Detection and tracking of point features. Order A Journal on the Theory of Ordered Sets and its Applications 7597, 22 (1991)
2. Lowe, D.G.: Distinctive Image Features from Scale-Invariant Keypoints. International Journal of Computer Vision 60, 91–110 (2004)
3. Comaniciu, D., Meer, P.: Mean shift: A robust approach toward feature space analysis. IEEE Trans. Pattern Anal. Mach. Intell. 24, 603–619 (2002)
4. Jeyakar, J., Babu, R.V., Ramakrishnan, K.R.: Robust object tracking using local kernels and background information. In: ICIP, pp. V:49–V:52 (2007)
5. Laguzet, F., Gouiffés, M., Lacassagne, L., Etiemble, D.: Automatic color space switching for robust tracking. In: ICSIPA, pp. 295–300. IEEE (2011)
6. Birchfield, S.T., Rangarajan, S.: Spatiograms versus histograms for region-based tracking. In: CVPR, pp. II:1158–II:1163 (2005)
7. Porikli, F., Tuzel, O., Meer, P.: Covariance Tracking using Model Update Based on Lie Algebra. In: IEEE CVPR, vol. 1, pp. 728–735 (2006)
8. Förstner, W., Moonen, B.: A metric for covariance matrices. In: Qua Vadis Geodesia, pp. 113–128 (1999)
9. Ojala, T., Pietikainen, M., Harwood, D.: Performance evaluation of texture measures with classification based on kullback discrimination of distributions. In: ICPR, pp. A:582–A:585 (1994)
10. Romero, A., Gouiffés, M., Lacassagne, L.: Feature points tracking adaptive to saturation. In: ICSIPA, pp. 277–282. IEEE (2011)
11. Bak, S., Corvee, E., Bremond, F., Thonnat, M.: Multiple-shot Human Re-Identification by Mean Riemannian Covariance Grid. In: Advanced Video and Signal-Based Surveillance, Klagenfurt, Autriche (2011)
12. Alahi, A., Ortiz, R., Vandergheynst, P.: FREAK: Fast Retina Keypoint. In: IEEE Conference on Computer Vision and Pattern Recognition (2012)
13. Tola, E., Lepetit, V., Fua, P.: DAISY: An efficient dense descriptor applied to wide-baseline stereo. IEEE Trans. Pattern Anal. Mach. Intell. 32, 815–830 (2010)

---

^{2} Surveillance imProved sYstem http://www.ppsl.asso.fr/spy.php

14. McKenna, S.J., Jabri, S., Duric, Z., Rosenfeld, A., Wechsler, H.: Tracking groups of people. Computer Vision and Image Understanding 80, 42–56 (2000)
15. Lacassagne, L., Manzanera, A., Dupret, A.: Motion detection: Fast and robust algorithms for embedded systems. In: 2009 16th IEEE International Conference on Image Processing (ICIP), pp. 3265–3268 (2009)
16. Lacassagne, L., Zavidovique, B.: Light speed labeling for risc architectures. In: 2009 16th IEEE International Conference on Image Processing (ICIP), pp. 3245–3248 (2009)
17. Gray, D., Brennan, S., Tao, H.: Evaluating appearance models for recognition, reacquisition, and tracking. In: 10th IEEE International Workshop on Performance Evaluation of Tracking and Surveillance, PETS (2007)
18. Schwartz, W.R., Davis, L.S.: Learning Discriminative Appearance-Based Models Using Partial Least Squares. In: Brazilian Symposium on Computer Graphics and Image Processing (2009)
19. Ferryman, J., Shahrokni, A.: Pets2009: Dataset and challenge. In: 2009 Twelfth IEEE International Workshop on Performance Evaluation of Tracking and Surveillance (PETS-Winter), pp. 1–6 (2009)
20. Smith, K., Gatica-perez, D., Marc Odobez, J., Ba, S.: Evaluating multi-object tracking. In: Workshop on Empirical Evaluation Methods in Computer Vision (2005)

Iterative Hypothesis Testing for Multi-object Tracking with Noisy/Missing Appearance Features

Amit Kumar K.C., Damien Delannay, Laurent Jacques,
and Christophe De Vleeschouwer

ICTEAM Institute, Université catholique de Louvain, Louvain-la-Neuve, Belgium[**]
{amit.kc,laurent.jacques,christophe.devleeschouwer}@uclouvain.be,
damien.delannay@keemotion.com

Abstract. This paper assumes prior detections of multiple targets at each time instant, and uses a graph-based approach to connect those detections across time, based on their position and appearance estimates. In contrast to most earlier works in the field, our framework has been designed to exploit the appearance features, even when they are only sporadically available, or affected by a non-stationary noise, along the sequence of detections. This is done by implementing an iterative hypothesis testing strategy to progressively aggregate the detections into short trajectories, named tracklets. Specifically, each iteration considers a node, named key-node, and investigates how to link this key-node with other nodes in its neighbourhood, under the assumption that the target appearance is defined by the key-node appearance estimate. This is done through shortest path computation in a temporal neighbourhood of the key-node. The approach is conservative in that it only aggregates the shortest paths that are sufficiently better compared to alternative paths. It is also multi-scale in that the size of the investigated neighbourhood is increased proportionally to the number of detections already aggregated into the key-node. The multi-scale and iterative nature of the process makes it both computationally efficient and effective. Experimental validations are performed extensively on a 15 minutes long real-life basketball dataset, captured by 7 cameras, and also on PETS'09 dataset.

1 Introduction and Overview

Multi-object tracking is a fundamental issue in computer vision. It supports high-level semantic scene analysis in numerous and various applications. Vehicle trajectories are, for example, collected to control traffic monitoring solutions [1]. People displacement analysis is important to improve the security of public spaces [2], or to understand sport actions [3], for example.

Due to the recent improvement in object detection, many detection-based approaches have been proposed to handle the multi-target tracking problem.

[**] Part of the work is supported by the Belgian NSF and by the WIST3 Walloon Region project SPORTIC.

J.-I. Park and J. Kim (Eds.): ACCV 2012 Workshops, Part II, LNCS 7729, pp. 412–426, 2013.
© Springer-Verlag Berlin Heidelberg 2013

In such approaches, plausible object locations are first estimated in each individual frame,together with some features characterizing the appearances of the detected objects. Afterwards, graph-based solutions are generally envisioned to match the detections in consecutive frames. For example, in [4], authors explicitly model the spatial layout and mutual occlusion constraints, and use linear programming relaxation to solve the multi object tracking problem. The K-shortest paths approach has been investigated in [5], while the authors in [6] use min-cost flow network and greedy approach to estimate the number of tracks, as well as their birth and death states. In [7], the problem is formulated as a global maximum a posteriori estimation over a directed acyclic hyper graph. Similarly, [8] casts the problem into finding maximum weighted independent sets of a graph.

As a main limitation, most of these methods implicitly assume that the appearance features are known with the same level of reliability all along the time. Therefore, they exploit them in a uniform manner with time. In practice, however, most features are subject to non-stationary noise, and their ability to discriminate objects varies with the scene context. For example, colour histograms appear to be quite noisy in presence of occlusions, and object positions do not help to disambiguate a clutter of detections. In some other cases, highly discriminant appearance features are only available sporadically (and under certain configurations only). For example, in sports, a number on a jersey is visible only when facing the camera. In such time-varying observation process, the task of tracking of objects, while taking into account the position and all the available appearance features, is non-trivial.

To illustrate this, we now explain the limitations of conventional graph-based approaches when dealing with sporadic/noisy features. In short, graph-based approaches assign a node to each detection. Edges are then defined to connect the nodes, and each edge gets a cost that reflects the *dissimilarity* between the two nodes it connects. Afterwards, a (K-)shortest-path algorithm is generally applied to find the trajectories of the (K) targets. The approach has proven to be effective in scenarios for which the features are collected with same level of accuracy for each detection. In contrast, the approach is not appropriate in cases for which appearance features are noisy or missing at some time instants. This is because the (in)consistency of the appearances observed along a path cannot any more be measured simply based on the accumulation of the appearance (dis)similarities measured between pairs of consecutive nodes, since the relevance of those dissimilarity measurement directly depends on the availability and reliability of the corresponding appearance estimates. Hence, most graph-based methods fail to properly exploit noisy or sporadic appearance cues.

To address the above limitations, our paper introduces a new paradigm to aggregate detections into objects trajectories. Similar to numerous previous works, it adopts a graph-based formalism, but fundamentally goes from a paradigm that builds on comparisons of appearances between consecutive nodes towards a new paradigm that investigates the graph under some target appearance hypothesis.

Each iteration of the algorithm works as follows. A node, named key-node, is selected to define a target appearance hypothesis. Given this hypothesis, a shortest-path computation algorithm is considered to investigate how to aggregate the key-node with its temporal neighbours in the graph, while promoting the nodes that share this target appearance hypothesis. The process is repeated iteratively, each node becoming a key-node at some step of the algorithm, and each tracklet defining the nodes of the updated graph to be used in subsequent iterations of the algorithm. To limit the computational complexity while giving the opportunity to build long trajectories, we adopt a multi-scale strategy to define the size of the temporal neighbourhood to investigate around a key-node. Specifically, the size of the observation window is made proportional to the size of the key-node, i.e. to the number of detections that have been aggregated into the key-node during the earlier steps of the algorithm. In addition, to avoid misleading the overall multi-object tracking process due to a wrong aggregation decision, *e.g.*, caused by some inappropriate appearance hypothesis, the shortest-path connecting the key-node to the extremity of its observation neighbourhood is only validated when it is sufficiently shorter than the alternative paths connecting each one of its extremities to the opposite extremity of the observation window.

The advantages of our proposed approach can be summarized as follows. Primarily, it naturally favours the aggregation of detections that share a similar appearance, even if those detections are not adjacent in time. It can also naturally account for different levels of reliability in the observation process, typically by giving more credit to the reliable appearance measurements when defining the cost associated to the discrepancy between the target appearance hypothesis and a node appearance estimate. Hence, the algorithm becomes able to effectively exploit sporadic or noisy features, which is a significant step forward compared to the state-of-the-art. As a second advantage, the multi-scale and progressive nature of the algorithm not only mitigates its computational load, but also helps in selecting long term matching based on more reliable tracklet appearance estimations, which directly benefits to the overall accuracy of the algorithm.

To the best of our knowledge, the only graph-based previous work exploiting sporadically available appearance cues has been presented in [9]. In this work, the authors assume a discrete set of N possible appearances, and end up in a K-shortest paths computation on a N-layered graph, K being the number of targets, and N corresponding to the number of appearance hypothesis. Our method is more flexible than [9], in the sense that it does not require prior knowledge of the discrete set of possible appearances, as required in [9]. Our approach naturally adapts to the observation of new appearances in the scene, and can handle an arbitrary number of appearances. In addition, our approach is computationally efficient due of its iterative and multi-scale nature. In particular, it avoids the (slow) linear programming optimization required in [9]. Moreover, we avoid the computational burden associated to the construction of a N-layer graph by embedding the hypothesis testing within an iterative local aggregation framework.

The rest of the paper is organized as follows. Section 2 defines the graph terminology. The tracking algorithm is explained in Section 3. Section 4 discusses the experimental results and demonstrates the approach on a real-life basketball dataset.

2   Graph Formalism and Notations

As an input, the algorithm receives the set of candidate targets detected independently at each time instant, as described in [10]. Apart from the detection time t and the location $\boldsymbol{x}$, the detector computes K appearance features $\boldsymbol{f}_i$ ($1 \leq i \leq K$) for a target. Since a feature might be noisy or even missing, the detector outputs a confidence value $c_i \in [0, 1]$ for each feature ($c_i = 0$ standing for a missing feature). A detection $\mathbf{d}$ is therefore characterized by the vector

$$\boldsymbol{d} = (t, \boldsymbol{x}, \mathcal{F}, \boldsymbol{c}),$$

where $\mathcal{F} = \{\boldsymbol{f}_1, \cdots, \boldsymbol{f}_K\}$ and $\boldsymbol{c} = (c_1, \cdots, c_K)$. The set of detections at a given time t is denoted as $\mathcal{D}^t$. As introduced in Section 1, the proposed algorithm adopts a graph-based formalism to progressively aggregate the detections into *tracklets* using a graph-based formalism. We define a graph $\mathcal{G} = (\mathcal{V}, \mathcal{E}, \mathcal{W})$ by:

- a set of nodes, with each node corresponding to a tracklet, *i.e.*, $\mathcal{V} = \{v_k : 1 \leq k \leq \#\mathcal{V}\}$,
- a set of edges, $\mathcal{E} \subset \mathcal{V} \times \mathcal{V}$, defining the connectivity between the nodes in $\mathcal{V}$,
- and a set of weights, $\mathcal{W} : \mathcal{E} \to \mathbb{R}^+$, weighting these nodes and edges.

Initially, individual detections define the nodes of the graph. Detections are then aggregated into tracklets, which define the nodes of the updated graph. The proposed iterative aggregation process is presented in details in Section 3, including the definition of cost and edges between nodes. Here, we only introduce the associated terminology. Formally, along the aggregation process, a tracklet v is defined to be collection of chained detections, *i.e.*, $v = \left(\boldsymbol{d}^{(1)}, \boldsymbol{d}^{(2)}, \cdots, \boldsymbol{d}^{(N)}\right)$, N being the length of the tracklet, also denoted as $\mathcal{L}(v) = N$. Notice that the chain is ordered, in the sense that the detection times $t_{d^{(i)}}$, $0 < i \leq N$, are such that $t_v^{(s)} = t_{d^{(1)}} < t_{d^{(2)}} < \cdots < t_{d^{(N)}} = t_v^{(e)}$, with $t_v^{(s)}$ and $t_v^{(e)}$ respectively denoting the starting and ending time of the tracklet. Figure 1 depicts how the tracklets are gathered into a graph in the proposed framework.

Notice that pairs of tracklets are connected only between their extremities, in such a way that each connection maintains the increasing ordering of the detection times composing the two tracklets. The weight $\mathcal{W}(u, v)$ is introduced to denote the linking cost between two nodes $u, v \in \mathcal{V}$. It is formally defined in Section 3. In short, it typically decreases with the likelihood that nodes u and v correspond to the same physical target. In addition, we introduce the *inner cost* $\mathcal{W}(v, v)$ of a node v to denote the cost of traversing tracklet v from its starting time to its ending time. It typically depends on the length $\mathcal{L}(v)$ of the node, *i.e.*, $\mathcal{W}(v, v) = \rho \mathcal{L}(v)$, and is introduced to avoid that long nodes create 'short-cuts' in the graph.

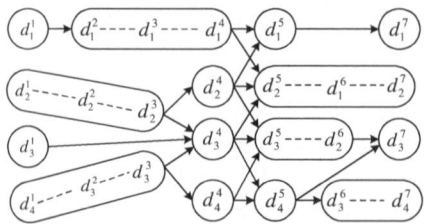

Fig. 1. Our graph formalism. The kth detection at time t is denoted by d_k^t. Some of them are aggregated into tracklets. Each node corresponds to a tracklet. An edge that connects two nodes u and v has a cost $\mathcal{W}(u, v)$.

In the sequel, we use two more graph notations. Since we consider a recursive algorithm that incorporates new detections at each time instant t, the graph is continuously incremented with time, and is denoted $\mathcal{G}^t$ at a given time t. Second, $\mathcal{G}^t_{[t_1, t_2]}$ represents a graph formed by selecting in $\mathcal{G}^t = (\mathcal{V}^t, \mathcal{E}^t, \mathcal{W}^t)$ the tracklets $v \in \mathcal{V}^t$ having at least one extreme time component inside the temporal window $[t_1, t_2]$. The connectivity $\mathcal{E}^t$ and the weight $\mathcal{W}^t$ are restricted accordingly from these selected tracklets in order to form $\mathcal{E}^t_{[t_1, t_2]}$ and $\mathcal{W}^t_{[t_1, t_2]}$.

3   Iterative Hypothesis Testing

The global flow of our proposed iterative aggregation algorithm is presented in Algorithm 1. We observe that the graph manipulated by our method is continuously incremented by the new detections, encountered as time is evolving. Once all the detections computed at time t have been connected to the previous graph, the algorithm iteratively investigates how each node of the graph can be aggregated with its neighbours. This investigation starts with a node, named key-node, and is performed either in the forward or in the backward direction. As controlled by the *dir* flag in Algorithm 1, the direction of investigation changes at each time instant to propagate the appearance hypothesis associated to the key-node both towards the future and the past of this key-node, thereby making the global process symmetric with respect to time.

The remainder of the section details the behaviour of the two functions introduced in Algorithm 1, namely *IncrementGraph* and *IterativeAggregation*. The incrementation of the graph with time is presented in Section 3.1. The core of our proposed multi-scale and iterative aggregation strategy is then explained in Section 3.2.

3.1   Incrementing the Graph with Time

At time $t = 1$, the graph is just a set of detections at that instant, *i.e.*, $\mathcal{G}^1 = (\mathcal{D}^1, \emptyset)$. At time $t > 1$, the graph is obtained by adding new detections to the so-called previous graph, $\mathcal{G}^{t-1}$, resulting from earlier steps of the algorithm, up to time $t - 1$.

Algorithm 1. Recursive aggregation of tracklets with time

Input: Set of detections, $\mathcal{D}^t$
Output: Graph at time t, $\mathcal{G}^t$
Procedure:
dir $\leftarrow$ +1 {/* Note: dir=1 means forward and -1 means backward aggregation */}
for each time instant t **do**
   if $t = 1$ **then**
    $\mathcal{G}^t = (\mathcal{D}^t, \emptyset)$
   else
    $\mathcal{G}^t \leftarrow$ IncrementGraph($\mathcal{G}^{t-1}, \mathcal{D}^t$) {/* See Section 3.1 */}
    $\mathcal{G}^t \leftarrow$ IterativeAggregation($\mathcal{G}^t$, dir) {/* See Section 3.2 */}
    dir $\leftarrow$ -dir
   end if
end for

The procedure is referred to as **IncrementGraph** in Algorithm 1. It connects the detections $\mathcal{D}^t$ at time t to the graph $\mathcal{G}^{t-1}$. All nodes ending later than time $t - \tau_{\max}$ are linked to all the current detections. We set $\tau_{\max} = 120$. This allows to investigate connections up to 6 seconds (at the frame rate of 20 fps) in the past, which is sufficient to make the algorithm robust to most practical occurrence of missed detections. We present the effect of this parameter on the performance in the supplementary material. The set of links (or edges), connecting the detections computed at t, is denoted $\mathcal{E}^t$. Those edges are directed and "time-forwarded". Therefore, the graph $\mathcal{G}^t$ is directed and acyclic (DAG), and permits only causal traversals. Nevertheless, the graph can be globally reversed in order to allow anti-causal paths for processing purposes (see Section 3.2). Each edge $e \in \mathcal{E}^t$ is characterized by a cost which measures the distance between two nodes. As the appearance features are often unreliable or even unavailable for most elements in $\mathcal{D}^t$, the distance between a node $v \in \mathcal{V}^{t-1}$ and a new detection $d \in \mathcal{D}^t$, is computed based only on the position and motion related parameters. Specifically, the cost $\mathcal{W}^t(v, d)$ is defined as

$$\mathcal{W}^t(v, d) = \begin{cases} \left[1 + \gamma \times (t - t_v^{(e)} - 1)\right] g_{\text{sp}}(v, d) & \text{if } t - t_v^{(e)} \leq \tau_{\max}, \\ \infty & \text{else,} \end{cases} \quad (1)$$

with the metric g_{sp} measuring the distance between detection d and the predicted position of the object corresponding to node v. It is defined as:

$$g_{\text{sp}}(v, d) = \left\| \boldsymbol{x}_d^{(s)} - \boldsymbol{x}_v^{(e)} - \dot{\boldsymbol{x}}_v^{(e)}(t - t_v^{(e)}) \right\|_2, \quad (2)$$

where the term $\dot{\boldsymbol{x}}_v^{(e)}$ is the velocity, at the end of tracklet v. It is zero for unit length tracklets, and is computed from the last 2 detections of the tracklet otherwise. The factor $\gamma > 0$, typically set to 3, introduces penalty for missed detections.

Track Creation, Deletion and Duplicate Detections. Here, we provide a brief remark on how a track is created and deleted. Each node corresponds to a track and we create node as soon as a novel detection is introduced. It is removed either if it is aggregated with other nodes, or if its length is too small compared to its distance to the current time, i.e., $\mathcal{L}(v) \ll t - t_v^{(e)}$. In the latter case, we consider it as false positive.

In a typical tracking scenario, duplicate detections arise which might result in two tracklets that partly overlap in time; despite they correspond to the same object. Such time overlap prevents the desired aggregation of the tracklets in a graph that is only composed of forward links. To mitigate this, we introduce backward links in such a way that the graph is still directed and acyclic. They make the aggregation of overlapped tracklets possible.

3.2 Iterative Tracklet Aggregation

Once the graph has been incremented with novel detections, the objective is to aggregate the nodes that correspond to the same physical object. To exploit appearance cues that are noisy, or only available sporadically. Therefore, we cannot rely on conventional propagation of appearance similarity measures between consecutive nodes. Instead, we promote a novel aggregation paradigm, founded on iterative hypothesis testing process.

Overview of the Contribution. In this approach, each iteration selects a node, named key-node, and studies how to aggregate this key-node with its forward or backward neighbourhood, under the assumption that the observed key-node appearance defines the reference appearance of the tracked object. Given this hypothesis, paths that go through nodes that do (not) share the key-node appearance are promoted (penalized). This is done simply by decreasing (increasing) the cost to go through a node of the graph when the appearance of that node is similar (different) to that of the key-node. Hence, all appearance cues, even the sparse or inaccurate one, can be exploited to drive the selection of aggregated paths within the graph. Since the process is repeated with each node being the key-node, all observed appearance hypotheses are examined.

The hypothesis testing approach is complemented by a multi-scale strategy. It consists of defining the size of the neighbourhood to be proportional to the length of the key-node. The advantages are twofold. First, the approach makes sense from the tracking efficiency point of view, since longer nodes are more likely to have accumulated reliable and accurate knowledge about their appearance, which should be exploited to connect them with other nodes with the same appearance, even if they are far away. Second, with respect to the computational complexity, the fact that long time frame neighbourhoods are only investigated when detections already got the opportunity to be aggregated into tracklets of sufficient length reduces the actual number of nodes to be considered when dealing with large neighbourhood windows. Moreover, our approach adopts an extreme caution while aggregating the nodes in the graph. A plausible aggregation computed during an iteration will only be validated and integrated to the graph structure for subsequent iterations if it is reliable enough.

We now detail the practical implementation of the the algorithm, which is presented in Algorithm 2.

Multiscale Iterative Hypothesis Testing. Formally, the key-node is denoted v_{key}. It is selected at each iteration among the set of nodes, $\mathcal{R}^t$, that have not yet been investigated at time t. The aggregation of the key-node with its neighbours is then investigated in an observation window that precedes or follows the key-node, depending on the sign of the *dir* flag. The size of the observation window is proportional to the length of the key-node. As explained earlier, this proportional definition allows traversal in different time-scales. We use Δ to denote the observation window interval. Hence, $\Delta = [t_{v_{\text{key}}}^{(e)}, t_{v_{\text{key}}}^{(e)} + \kappa\,\mathcal{L}(v_{\text{key}})]$ in the forward mode ($dir = 1$), or $\Delta = [t_{v_{\text{key}}}^{(s)} - \kappa\,\mathcal{L}(v_{\text{key}}), t_{v_{\text{key}}}^{(s)}]$ in the backward mode ($dir = -1$), where κ is the window proportionality constant. We use $\kappa = 5$. Detailed results on varying κ are provided on the supplementary material.

Given the key-node v_{key} and its observation window Δ, we define the graph $\mathcal{G}_\Delta$ to investigate how the key-node can be aggregated with its neighbours to define an appearance-consistent path under the assumption that the tracked object appearance is defined to be the key-node appearance. The function is named **GraphHypothesis** because it returns the graph that is used to test the key-node appearance hypothesis. The graph $\mathcal{G}_\Delta$ is directly derived from the graph $\mathcal{G}^t$, by cutting $\mathcal{G}^t$ according to the limits of the observation window, and updating the inner costs of the nodes within the window to reflect the hypothesis made about the target appearance. In short, the inner cost $\mathcal{W}(v,v)$ of a node $v \in \mathcal{V}_\Delta$ is increased (decreased) if it has a different (similar) appearance than the one of the key-node.

The inference of the tracklet features based on the features observed in the set of detections directly depends on the characteristics of the features observation process. If, for example, the observation process is affected by outliers, a RANSAC approach could help in capturing the right appearance model. On the other hand, if the observations are independent and affected by Gaussian noise, then a weighted average provides an appropriate inference. In this paper, we use a weighted average for the tracklet appearance as an example of practical implementation. Then, the average i^{th} feature of a node v is computed as:

$$\overline{\boldsymbol{f}}_i^{(v)} = \frac{1}{C_i} \sum_{t=1}^{\mathcal{L}(v)} c_{i,v}^{(t)} \boldsymbol{f}_{i,v}^{(t)}, \tag{3}$$

where $C_i = \sum_{i=1}^{\mathcal{L}(v)} c_{i,v}^{(t)}$. In particular, $\overline{\boldsymbol{f}}_i^{(\text{ref})}$ denotes the average i^{th} feature of the key-node, used as an hypothesis reference. Let $D(v)$ denote the value by which the inner cost of node v is incremented due to its dissimilarity with the appearance of the key-node. Then,

$$D(v) = N \sum_{i=1}^{K} \underbrace{\left(\alpha_i \left\| \overline{\boldsymbol{f}}_i^{(\text{ref})} - \overline{\boldsymbol{f}}_i^{(v)} \right\|_1 + (1 - \alpha_i) w_i^{(\text{fix})} \right)}_{:= w_i^{(v)}} \tag{4}$$

Algorithm 2. IterativeAggregation

Input: Graph at time t, $\mathcal{G}^t$; Direction of aggregation, dir
Output: Updated graph at time t, $\mathcal{G}^t$
Procedure:

    Initialize: $\mathcal{R}^t \leftarrow \mathcal{V}^t$ {/* $\mathcal{R}^t$ is the set of nodes that are yet to be scheduled for hypothesis testing. */}
    while $\mathcal{R}^t \neq \emptyset$ **do**
        $v_{\text{key}} \leftarrow \text{Schedule}(\mathcal{R}^t)$
        $\Delta \leftarrow$ Limits of the observation window
        $\mathcal{G}_\Delta \leftarrow \text{GraphHypothesis}(\mathcal{G}^t, \Delta, v_{\text{key}})$
        $v_{\text{agg}} \leftarrow \text{Aggregate}(v_{\text{key}}, \mathcal{G}_\Delta)$
        if $v_{\text{agg}} \neq v_{\text{key}}$ **then**
            $\mathcal{G}^t \leftarrow \text{Simplify}(\mathcal{G}^t, v_{\text{agg}})$
        end if
        $\mathcal{R}^t \leftarrow \mathcal{R}^t \setminus v_{\text{agg}}$
    end while

Aggregate: *Refer to Figure 2 for the illustration.*

    Input: Key-node, v_{key}; Windowed graph for testing key-node hypothesis, $\mathcal{G}_\Delta$
    Output: Set of nodes that can be aggregated, v_{agg}
    Procedure:
    $(S_b, S_{\text{sb}}) \leftarrow$ Best and second best shortest paths from v_{key} to the other extremity of $\mathcal{G}_\Delta$
    if $\text{Cost}(S_b) \ll \text{Cost}(S_{\text{sb}})$ **then** {/* Here $\ll$ signifies that path S_b is sufficiently better than path S_{sb}. */}
        $\mathcal{G}_\Delta^- \leftarrow \text{ReverseDirection}(\mathcal{G}_\Delta)$
        $(S_{b'}, S_{\text{sb}'}) \leftarrow$ Best and second best shortest paths from v_b to the other extremity of $\mathcal{G}_\Delta^-$
        if $\text{Cost}(S_{b'}) \ll \text{Cost}(S_{\text{sb}'})$ **then**
            if $v_{b'} = v_{\text{key}}$ **then**
                $v_{\text{agg}} \leftarrow S_b$
            end if
        end if
    else
        $v_{\text{agg}} \leftarrow v_{\text{key}}$
    end if

In Algorithm 2, the function **Schedule** selects a node for hypothesis testing that has not yet been scheduled. Different scheduling mechanisms can be envisioned. However, in this paper, we select the nodes that are "sufficiently long" and "not so far from t". That is, we schedule the nodes in decreasing order of $\mathcal{L}(v)/\max\{1, t - t_v^{(e)}\}$. Since long nodes are more likely to have accumulated sufficient appearance features, they get more priorities. Similarly, by exploring nodes that are recent in time, we prevent the fast growth of the graph.

The parameter α_i is introduced to give less weight to the appearance features that are computed on short tracks as they are prone to error. L_1 norm is chosen for the computational reasons, but could be replaced by other metrics like Bhattacharyya distance, etc.

$$\alpha_i = \begin{cases} 0 & \text{if } C_i \leq C_{\min}, \\ 1 & \text{if } C_i \geq C_{\max}, \\ \frac{C_i - C_{\min}}{C_{\max} - C_{\min}} & \text{otherwise.} \end{cases} \tag{5}$$

where $C_{\min}$ and $C_{\max}$ are the limits to define if the feature is considered reliable or not. We set $C_{\min} = 20$ and $C_{\max} = 100$. An analysis of effect of these values on the performance is presented in the supplementary material. When $\alpha_i \to 1$, $w_i^{(v)} \to \left\| \overline{\boldsymbol{f}}_i^{(\text{ref})} - \overline{\boldsymbol{f}}_i^{(v)} \right\|_1$ and when $\alpha_i \to 0$, $w_i^{(v)} \to w_i^{(\text{fix})}$. The term $w_i^{(\text{fix})}$ is introduced so that a node, which definitely looks similar to the key-node $(D(v) \approx 0)$, is favoured compared to a node for which no appearance features is available $\left(D(v) \approx N \sum_i w_i^{(\text{fix})} \right)$. It corresponds to the noise level, affecting the feature. Empirically, we set $w_i^{(\text{fix})} = 5$ for all $1 \leq i \leq K$ and is related to the unit of the detection. After the inner costs of the nodes have been incremented by $D(v)$, a shortest path algorithm is applied. For this, the DAG shortest path algorithm is preferred because of the inherent directed and acyclic nature of the graph. The cost of a path is defined to be the sum of costs of the edges and the inner costs of the nodes along it, and is given by the function **Cost** in the algorithm.

Even though it seems that updating the costs requires additional scanning of the graph, it is mitigated by the concept of *visitors* in the shortest path algorithm. The visitors allow to update the costs of the nodes or edges "in place" by invoking various events.

Path Ambiguity Estimation and Validation. Having the cost of edges been defined in order to take the displacement as well as the appearance into consideration, the shortest path S_b, which connects the key-node to a set of nodes within the window, reasonably corresponds to a single physical object (same appearance, and coherent motion) and that could thus be aggregated into a single node. To limit the risk of connecting nodes that correspond to two distinct objects, we check the level of ambiguity of the shortest path by comparing its cost to the costs of a set of paths that could constitute reasonable alternative to connect the extremities of the shortest path to the opposite extremity of the observation window. Figure 2 illustrates this process.

This validation process is run in two steps. In the first step, the shortest S_b and the second shortest S_{sb} paths are considered. Moreover, the ends of the best and second-best paths are denoted as v_b and v_{sb} respectively. The shortest path S_b is considered being sufficiently better than S_{sb} only if several conditions are met: (i) $\text{Cost}(S_b) < K_1$, (ii) $\text{Cost}(S_b)/\text{Cost}(S_{\text{sb}}) < K_2$, and (iii) $\text{Cost}(S_{\text{sb}}) > K_3$. The thresholds K_1 and K_3 vary linearly with Δ. We set $K_2 = 1/3$. The sensitivity on varying K_2 is detailed in the supplementary material.

If all conditions are met, the second step of the validation process is considered. For this, the graph is reversed by flipping the direction of all the edges of $\mathcal{G}_\Delta$. It is mentioned as **ReverseDirection** in the algorithm. The shortest $(S_{b'})$ and second shortest $(S_{\mathrm{sb}'})$ paths linking v_b with the opposite extremity of the observation window are then computed. If $S_{b'}$ leads to the original key-node, *i.e.*, if $v_{b'} = v_{\mathrm{key}}$, and if a similar set of conditions hold for $S_{b'}$ and $S_{\mathrm{sb}'}$, then the path S_b is considered to be *unambiguous*, and is replaced by a single node in the path. This procedure is called **Simplify** in the algorithm. It updates the appearance features of the node as in Equation 3 and also the motion parameters. It keeps only the edges connecting the extremities of the aggregated path to the rest of the graph. Other connections involving intermediate nodes are removed. Since all the nodes along S_b are aggregated into a single node (and thus the intermediate nodes are removed from the graph), it resembles the greedy matching of the nodes and is thus suboptimal.

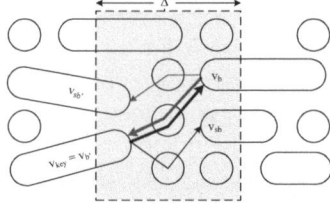

Fig. 2. Illustration of the **Aggregate** function in Algorithm 2. Within the window, the best (thick arrow) and the second best (thin arrow) paths are searched. Blue and red arrows represent forward and backward directions respectively.

4   Evaluation

The proposed algorithm has been evaluated on the APIDIS dataset [11]. The dataset has been generated by 7 cameras distributed around a basket-ball game. The candidate detections on the ground plane at each frame are independently computed, as described in [10]. The ID and the position of players have been manually defined at every second of a 15 minutes period. This provides the reference ground truth used in our evaluation.

In the remainder of the section, we first present the appearance features that are considered to support the tracking algorithm. We then discuss the multi-object tracking evaluation metrics, followed by the results and comparison with other approaches.

4.1   Appearance Features

In the APIDIS dataset, the jersey colour and the digit printed on it $(K = 2)$, are computed for each candidate detection. Specifically, a rectangular box is positioned at the expected height of the player shirt. The first feature is the average colour, which is computed as the average blue component divided by the sum of average red and green components, over the silhouette of the player within the rectangular box. However, depending on whether the detection is close (far) from the camera, and also whether the detection is visible (occluded) in each camera view, the measurement is considered (discarded). The second feature is a set of candidate shirt digits, obtained by running a digit-recognition algorithm in the same rectangular region. These features are computed for all available cameras. Finally, the confidence of the feature is computed as the ratio of number of cameras, for which features are computed, to the total number of available cameras. A sample of such a measurement is shown in Figure 3 for a specific player (digit=8, colour=0.25). It can be easily observed that the appearance characteristics are noisy and sometimes missing in our scenario. Indeed, these features cannot always be reliably measured for each frame because of occlusions, illumination change, unfavourable shirt orientation with respect to the camera, etc.

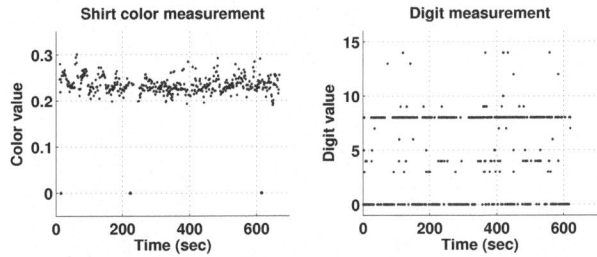

Fig. 3. Shirt colour and digit measurement along time for a player (digit=8, colour=0.25). Zero values indicate that no measurements are available.

4.2 Performance Metrics

We evaluate the performance of the proposed tracking algorithm based on the well-known CLEAR-MOT metrics [12]. A standard metric for evaluating object trackers is the Multiple Object Tracking Accuracy (MOTA), defined as:

$$\text{MOTA} = 1 - \frac{\sum_t (\text{m}_t + \text{fp}_t + \text{re}_t + \text{sw}_t)}{\sum_t g_t}, \tag{6}$$

where g_t is the number of ground-truth objects present at time t; $\text{m}_t, \text{fp}_t, \text{re}_t$ and sw_t are the number of misses, false positives, track reinitializations and track switches. We write, $\text{FP} = \sum_t \text{fp}_t, \text{RE} = \sum_t \text{re}_t, \text{SW} = \sum_t \text{sw}_t, \text{MS} = \sum_t \text{m}_t, \text{GT} = \sum_t g_t$. A switching error occurs when the tracker starts following another object, whereas a reinitialization error occurs when the tracker fails to track the object at some time and a new track is assigned for the same object later on. The error due to switching is more problematic as it might lead to significant errors in higher level interpretation of the scene. An error due to a miss means that the tracker does not have any estimate for the corresponding ground truth. Similarly, a false positive represents that the tracker outputs some estimate for which no ground truth position is available.

4.3   Results

Figure 4 compares the performance obtained by the proposed algorithm when different set of appearance features are exploited. The results are obtained by enabling (or disabling) certain features in the algorithm and running on the 15 minutes long video sequence. There are all together 7460 ground truth positions, *i.e.*, GT=7460. As we can see, the switches and re-initializations are reduced substantially. However, the false positives increase slightly. When we incorporate only the digit feature, it can be seen that digit features, even though they are highly sparse, can disambiguate some tracks. However, the improvements are not as ample as those from the colour feature. It can be justified due to the fact that colour feature is available more often than the digit feature, and hence some tracks might not have accumulated sufficient digit information. When both features are used, not only the switches are reduced but also the gaps between tracklets are bridged (thereby, reducing the re-initializations and misses). To compute the tracking effectiveness, the authors of [9] applied their method on the same APIDIS dataset. With their notion of MOTA, we achieve 94.5%, whereas they achieve 92.8%.

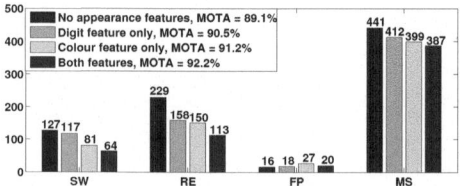

Fig. 4. Components of MOTA metric (15 min video) for different cases. Indeed, exploitation of color and digit features help to reduce the errors.

We compare our results with other approaches namely [13,14,5,7]. These results have been adopted from [7] which have been computed just for 500 frames. Therefore, we also ran our algorithm on the same 500 frames. The results are shown in Table 1. As the results in [7] were computed only with colour, we also used just the colour feature. From Table 1, we can see that the proposed method outperforms several popular state-of-the-art methods, in scenarios relying on target appearance estimation, but also in scenarios that only exploit detections positions as input features. We explain the benefit observed in this latter case by the progressive and conservative nature of our proposed aggregation process. In case of matching ambiguities, instead of validating the shortest path (and possibly committing an error), the algorithm waits for more observations. When validating other unambiguous paths, their motion parameters are estimated more accurately which, in turn, can benefit in resolving the ambiguities.

In addition, our approach is applied on camera view 1 of the PETS'09 S2.L1 dataset. First, targets are detected by using the deformable parts model [15]. Then, 8-bin histograms are computed on RGB channels separately, which are

Table 1. MOTA metric, computed on 500 frames of the APIDIS dataset

Method	MOTA
Track-before-detect [13]	0.614
Trajectory association [14]	0.781
K-shortest path [5]	0.586
Second order with exclusion [7]	0.735
Baseline: proposed method (position only)	**0.828**
Proposed method (position+colour feature)	**0.864**

then stacked to obtain a 24-bin feature vector. The feature is considered reliable only if the overlap between the bounding boxes is less than 20%. It takes only 42 seconds in MATLAB (3GHz 4-core CPU, 4 GB RAM), to estimate the feature and process all 795 frames. It results in (MOTA, MOTP)=(0.83, 0.74), which is competitive to several other tracking algorithms like Berclaz et al.[5] (0.82, 0.56), Breitenstein et al.[16] (0.75, 0.60) and Andriyenko et al. [17] (0.89, 0.56). These performance metrics are extracted from [17].

5   Conclusion and Future Perspectives

The paper proposed a framework for matching of detections while exploiting partial appearance features. It proceeds with hypothesis testing in an iterative framework which considers the input data at different time scales. The iterative principle helps in aggregating the appearance observations on tracklets by computing unambiguous local paths, thereby creating nodes with more reliable appearance cues. It also reduces the size of the graphs, and thus the complexity, handled by successive iterations of the algorithm. The multi-scale aspect allows matching decisions to be taken at different time horizons and is elegantly embedded in the framework. Future work will focus on the generalized inference of tracklet appearance and also on investigating different scheduling mechanisms for selecting key-node.

References

1. Ocakli, M., Dermirekler, M.: Video tracker system for traffic monitoring and analysis. In: IEEE Signal Processing and Communication Applications (2007)
2. Piciarelli, C., Micheloni, C., Foresti, G.: Trajectory based anomalous event detection. IEEE Transactions on Circuits and Systems for Video Technology (2008)
3. Alahi, A., Boursier, Y., Jacques, L., Vandergheynst, P.: Sports players detection and tracking with a mixed network of planar and omnidirectional cameras. In: ICDSC, Como, Italy (2009)
4. Jiang, H., Fels, S., Little, J.: A linear programming approach for multiple object tracking. In: CVPR (2007)
5. Berclaz, J., Fleuret, F., Turetken, E., Fua, P.: Multiple object tracking using k-shortest paths optimization. PAMI 33, 1806–1819 (2011)

6. Pirsivash, H., Ramanan, D., Fowlkes, C.C.: Globally-optimal greedy algorithms for tracking a variable number of objects. In: CVPR (2011)
7. Russell, C., Setti, F., Agapito, L.: Efficient second order multi-target tracking with exclusion constraints. In: BMVC (2011)
8. Brendel, W., Amer, M., Todorovic, S.: Multiobject tracking as maximum weight independent set. In: CVPR, pp. 1273–1280 (2011)
9. Shitrit, H.B., Berclaz, J., Fleuret, F., Fua, P.: Tracking multiple people under global appearance constraints. In: ICCV (2011)
10. Delannay, D., Danhier, N., Vleeschouwer, C.D.: Detection and recognition of sports(wo)men from multiple views. In: ICDSC, Como, Italy (2009)
11. http://www.apidis.org/dataset/
12. Bernardin, K., Stiefelhagen, R.: Evaluating multiple object tracking performance: the clear mot metrics. J. Image Video Process. 2008, 1:1–1:10 (2008)
13. Taj, M., Cavallaro, A.: Multi-camera track-before-detect. In: ICDSC, Como, Italy (2009)
14. Anjum, N., Cavallaro, A.: Trajectory association and fusion across partially overlapping cameras. In: AVSS, pp. 201–206 (2009)
15. Felzenszwalb, P.F., Girshick, R.B., McAllester, D., Ramanan, D.: Object detection with discriminatively trained part based models. PAMI 32, 1627–1645 (2010)
16. Breitenstein, M.D., Reichlin, F., Leibe, B., Koller-Meier, E., Van Gool, L.: Online multiperson tracking-by-detection from a single, uncalibrated camera. PAMI 33 (2011)
17. Andriyenko, A., Schindler, K., Roth, S.: Discrete-continuous optimization for multi-target tracking. In: CVPR (2012)

Novel Adaptive Eye Detection and Tracking for Challenging Lighting Conditions

Mahdi Rezaei and Reinhard Klette

The *.enpeda..* Project, Tamaki Campus
The University of Auckland, New Zealand

Abstract. The paper develops a novel technique that significantly improves the performance of Haar-like feature-based object detectors in terms of speed, detection rate under difficult lighting conditions, and reduced number of false-positives. The method is implemented and validated for driver monitoring under very dark, very bright, and normal conditions. The framework includes a fast adaptive detector designed to cope with rapid lighting variations, as well as an implementation of a Kalman filter for reducing the search region and indirect support of eye monitoring and tracking. The proposed methodology effectively works under low-light conditions without using infrared illumination or any other extra lighting support. Experimental results, performance evaluation, and comparing a standard Haar-like detector with the proposed adaptive eye detector, show noticeable improvements.

1 Introduction

Since the early 2000s, researchers such as Viola and Jones [13], Jesorsky et al. [7], or Hsu et al. [6] made important progress in model- and learning-based object detection methods. Despite of general improvements in detection methods, face and especially eye detection under non-ideal lighting conditions still requires further improvements. Even in recent efforts such as [10,12,15], limited verification tests have been applied for normal situations only. Driver-behaviour monitoring is an example for a challenging environment for eye analysis, where the light source is not uniform, symmetric, or the light intensity may rapidly and repeatedly change (e.g. due to entering a tunnel, shadow, turning into very bright light, or even sun strike). Although recent techniques for frontal face detection under normal lighting conditions are quite robust and precise [9,15,14,18], sophisticated and sensitive tasks such as driver eye-status monitoring (open, closed) and gaze analysis, are still far away from being solved accurately.

Among related work, there are publications on single and multi-classifier approaches for the addressed area of applications. Brandt et al. [2] designed a coarse-to-fine approach for face and eye detection using Haar wavelets to measure driver blinking with satisfactory results under ideal conditions. Majumdar [10] introduced a hybrid approach to detect facial features using Haar-like classifiers in HSV colour space but tested it on a very limited number of frontal faces only. Zhua and Ji [17] introduced robust eye detection and tracking under

J.-I. Park and J. Kim (Eds.): ACCV 2012 Workshops, Part II, LNCS 7729, pp. 427–440, 2013.

variable lighting conditions; however, their method is not effective without support of IR illumination. Research results in the field often suffer from a lack of verification and performance analysis on a wide-range of video or image data.

In this paper, we pursue four goals, (1) to improve noisy measurements of a Haar-classifier by a more stable solution for detecting and localizing features in the image plane, (2) to reduce the total computational cost by minimizing the search region based on a Kalman-filter face tracker, therefore indirectly reducing the "eye status" detection cost, (3) to minimize false detections by having a limited operational area within the image plane, and most importantly, (4) to overcome issues of eye-detection failures due to sophisticated lighting conditions, by introducing a novel technique, an *adaptive Haar detector*.

Section 2 outlines the main idea. Section 3 discusses our adaptive detection method. Section 4 informs about the implementation of the tracking module along with technical considerations for face and eye tracking. Section 5 discusses details of experimental and validation results with a performance analysis, comparing our method with the Viola-Jones method. Section 6 concludes.

2   Main Idea and Brief Methodology

Figure 1 illustrates the overall structure of our approach. Using Haar-feature based classifiers, two possible options are considered. In Option 1 we quantify the region of interest (ROI_1) as 100% for each classifier, which means that the whole area of an input image needs to be searched, from top left to the bottom right. Generally, such a full search needs to be repeated three times via three individual Haar classifiers in order to detect "face", "open eyes", or "closed eye" status. However, utilizing ground truth information for eye localization on two standard databases of FERET [3] and YALE [16], we estimated the eye location in a range of 0.55-0.75 as shown in Fig. 2. In case of head tilt, the eye location may vary in a range of 0.35-0.95 on one side of the face. Therefore, assuming an already detected face, an eye classifier can perform on regions A and B only (ROI_2), which are only 5.2% of the input image. If face detection fails, then a second full search in the image plane is required for eye classification (ROI_3), as we do not have any prior estimation for face location. This causes a total search cost of 200%. If the open-eye classifier detects at least one eye in segment A within the last five frames, then the closed-eye classifier is called to look at region C (ROI_4). This region covers about 3% of an VGA image. In brief, assuming an already detected face in the first stage, we have a maximum search area of 108.2% (ROI_1 + ROI_2 + ROI_4) for complete eye status analysis, while a face detection failure leads to a search cost of 203% (ROI_1 + ROI_3 + ROI_4).

Assessing 1,000 recorded frames and ground truth faces, we measured the mean size of the detected faces as 185×185 pixels which covers only 11% of a VGA image plane. Based on this idea we plan for a partial search that potentially defines a limited search region as Option 2 instead of Option 1.

As shown in Fig. 1, implementing a face tracker reduces the search region for face detection, thus a faster eye analyses through the tracked face region. Later we discuss that using a tracking solution as Option 2, the total search cost can be reduced to around 34.6% ($ROI_{1\_1} + ROI_2 + ROI_4$) which is at least 6 times faster than Option 1.

In addition to an optimized search policy, we require proper face localization from the first step, followed by robust eye status detectors to be applicable for all lighting conditions. Figure 3 shows examples that point to a need for robust eye detectors to be adaptive under extremely challenging lighting conditions. This requirement is considered in the left feedback cycle in Fig. 1, called classifier tuning and adaptation phase, which introduces adaptive Haar-like eye detection to overcome the weakness of a standard Viola-Jones detector [13].

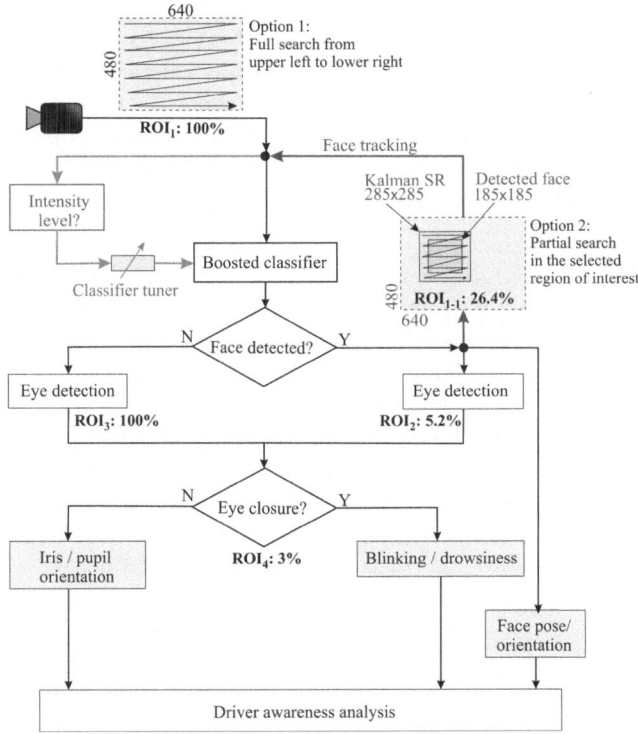

Fig. 1. Driver awareness monitoring: brief flowchart

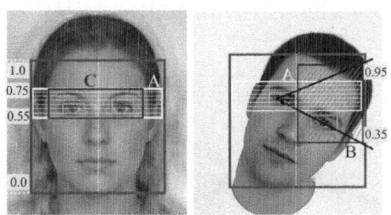

Fig. 2. Optimized ROIs for eye tracking, after successful face detection

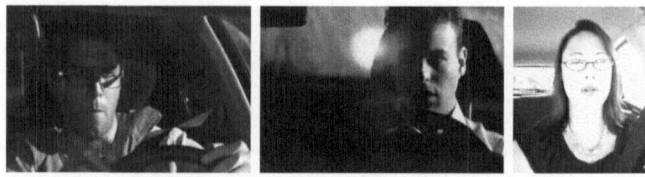

Fig. 3. Examples of "unusual" lighting conditions while driving, causing difficulties for driver's eye monitoring (images in the public domain)

3 Adaptive Classification and Detection

The adaptation module is detailed in three sub-sections, addressing the recognition of weakness of a Viola-Jones detector [13], statistical analysis of intensity changes around the eye region, and dynamic parameter adaptation to overcome inefficiency of Haar-feature based detectors under non-ideal conditions.

3.1 Weakness of Viola-Jones for Challenging Lighting Conditions

We tried five well-recognized and publicly available [11] Haar classifiers developed by Castrillon, Lienhart, Yu, and Hameed, in our nominated application, driver monitoring. Although they are quite robust for non-challenging and normal lighting scenes, we realized that due to frequent shadows and artificial lighting in day and night, those Haar-like classifiers are likely to fail. The situation becomes even more complicated when a part of the driver's face is brighter than the other part (due to light falling in through a side-window), making eye status detection extremely difficult. We also compiled and trained our own classifier based on a large dataset of +12,000 positive images from YALE [16], FERET [3], BioID [1], and FTD [5]. Applying an AdaBoost machine learning technique [4], we obtained best results for our trained Haar-like classifier with the following parameters:

- Positive images of size 21×21 pixel
- Number of weak classifiers (stages): 15
- Minimum expected hit rate for each stage: 99.8%
- Maximum acceptable false alarm at each stage: 40%
- Trimming threshold: 0.95

After training and creation of the classifier, we had to utilize the classifier in the real-world with parameters which are normally similar to the trained parameters. Main parameters are:

- Initial search window size (SWS) which should normally be the same as the scale size of positive images (i.e. 21×21 as above)
- Scale factor (SF) to increase the SWS in each subsequent search iteration (e.g. 1.2 which means 20% increase in window size for each search iteration)

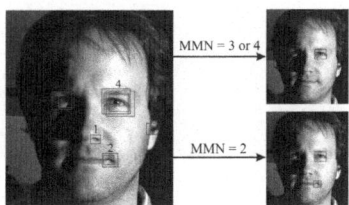

Fig. 4. Left: Initial results of eye detection. Right: Finally detected eyes after trimming by different MNN factors.

– Minimum expected number of detected neighbours (MNN) which is needed to confirm an object, when there are multiple object candidates in a small region (e.g. 3)

In general, a smaller SF means a more detailed search in each iteration, but it also causes higher computational costs.

Decreasing MNN, causes increase of detection rate; however, it increase false detection rate as well. Larger values for MNN lead to more strictness to confirm a candidate for face or eye, thus a reduced detection rate.

Figure 4 shows potential issues related to the MNN parameter in Viola-Jones techniques. The figure shows 10 initial eye candidates before trimming by the MNN parameter. Detections are distributed in 5 regions, each region shows 1 to 4 overlapping candidates. In order to minimize this problem, it is common to assign a trade-off value for the MNN parameter to gain the best possible results. Figure 4, top right, shows one missed detection with MMN equals 3 or 4, and Fig. 4, bottom right, shows one false detection with MMN equals 2. MMN equals 1 causes 3 false detections, and any MMN greater than 4 will lead to no detection at all; so there is no optimum MNN value for this example.

We conclude that although we can define trade-off values for SWS, SF, and MNN to obtain the optimum detection rate for "ideal" video sequences, however, a diversity of changes in light intensity over the target object can still significantly affect the performance of the given classifier in terms of TP and FP rates.

3.2   Hybrid Intensity Averaging

To cope with the above mentioned issues, we propose that Haar-classifier parameters have to be adaptive, varying with time depending on lighting changes. Figures 5 and 6 illustrate that we cannot measure illumination changes by simple intensity averaging over the input frame: In the driving application, there can be strong back-light from the back windshield, white pixel values around the driver's cheek or forehead (due to light reflections), or dark shadows on the driver's face. All these conditions may negatively affect the overall mean intensity measurement. Analysing various recorded sequences, we realized that pixel intensities around eyes can change independently from surrounding regions in the input sequence. Focusing on a detected face region, Fig. 5, right, shows very dark and very bright spots in two sample faces (in a grey-level range of 0-35

Fig. 5. Mean intensity measurement for eye detection, by excluding *very bright* and *very dark* regions. Green: thresholding range 210-255. Blue: thresholding range 0-35.

and 210-255, respectively). It also shows a considerable illumination difference for the left and right side of a driver's face. Apart from the iris intensity (for dark or light eyes), the surrounding eye intensities play a very crucial role in eye detection. Thus, proper classifier parameter adjustment based on proper intensity measurement in the region surrounding an eye can guaranty robust eye detection. Following this consideration, we defined white rectangles around eyes (Fig. 5, right) which can not only provide a good approximation of both vertical and horizontal light intensities around the eyes, but they are also very marginally influenced by green or blue (very bright or very dark) regions.

Considering an already detected face, and expected eye regions A and C based on Fig. 2, we can geometrically define C_r, F_r, C_l, and F_l as being the optimum regions in order to gain an accurate estimation of light intensity around the eyes (see Fig. 6). We also consider independent classifier parameters for the left and right half of the face, as each half of the face may receive different and non-uniform light exposures.

Performing a further analytical step, Fig. 5, right, shows that a few small green or blue segments (extreme dark or light segments) have entered the regions of white rectangles. This can affect the actual mean intensity calculations in the C or F regions. Thus, in order to reduce the affect of this kind of noise into our measurements, we apply a hybrid averaging by combining *mean* and *mode* (Mo) of pixel intensities as follows:

$$I_r(\alpha) = \frac{1}{2} \cdot \left[\alpha \cdot \text{Mo}(C_r) + \frac{(1-\alpha)}{m} \sum_{i=1}^{m} C_{r_i} + \alpha \cdot \text{Mo}(F_r) + \frac{(1-\alpha)}{n} \sum_{j=1}^{n} F_{r_j} \right] \quad (1)$$

where $I_r(\alpha)$ is the hybrid intensity value of the *right eye* region of the face, m and n are the total numbers of pixels in C_r and F_r regions; C_r and F_r are in $[0, 255]$ point to cheek and forehead light intensity, respectively.

An α-value of 0.66 assumes a double importance of *mode* intensity measurement compared to *mean* intensity; Integration of *mode* reduces the impact of eye iris colour (i.e. blue segments) as well as of very bright pixels (i.e. green segments) for our adaptive intensity measurement. Similarly, we can calculate $I_l(\alpha)$ as hybrid intensity value of *left eye* region.

3.3   Parameter Adaptation

The final step of the detection phase is *classifier parameter adjustment* based on the measured I_r and I_l values, to make our classifier adaptive for every

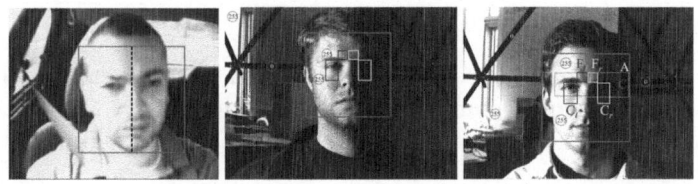

Fig. 6. Selected regions to sum up the mean intensity around eyes. Images Source: Yale database.

Table 1. Optimum parameters for 10 selected intensity levels in terms of *Search Window Size*, *Scale Factor*, and *Minimum Number of Neighbours*. FS: detected *Face Size*.

Light Intensity	*SWS*	*SF*	*MNN*
0	FS/5.0	1.10	1
20	FS/4.5	1.12	2
50	FS/3.5	1.30	9
75	FS/4.0	1.15	7
90	FS/4.0	1.30	10
120	FS/4.2	1.25	16
155	FS/5.0	1.35	15
190	FS/4.5	1.30	14
220	FS/4.6	1.25	9
255	FS/4.0	1.35	7

single input frame. Now we need to find optimum parameters (SWS, SF, MNN) for all the intensity ranges between 0 to 255, which is a highly time-consuming practical tasks. Instead, we defined optimum parameters for 10 selected intensities, followed by a data interpolation method to extend those parameters to all intensity ranges.

Table 1 shows optimum parameter values for 10 data points obtained from 20 recorded videos in different weather and lighting conditions. These factors are adjusted to lead to the highest rate of true positives for each of the 10 given intensities. The parameter values in Table 1 show a non-linear behaviour over intensity changes; therefore we apply non-linear *cubic interpolation* and *Lagrange interpolation* to extend adapted values to an intensity range from 0 to 255.

4   Tracking and Search Minimization

This section pursues the goals of minimizing the search region for eye status detection, time efficiency, less computational cost, more precise detections, and lower rate of false detection.

4.1   Tracking Considerations

A simple tracking around the previously detected face can easily fail due to a fast change in both face size and moving trajectory (Fig. 7). Therefore we need to perform a dynamic and intelligent tracking strategy to minimize the search region. We apply a Kalman filter.

Figure 8 shows the brief structure of Kalman filter [8] including time update and measurement steps, where $\hat{x}_k^-$ and $\hat{x}_k^+$ are priori and posteriori states estimated for centre of the detected face, z_k is a *Haar-classifier measurement vector*, A is an $n \times n$ matrix referred to as *state transition matrix* which transforms the previous state at time step $k-1$ into the current state at time step k, B is an $n \times l$ matrix referred to as *control input transition matrix* which relates to optional control parameter $u \in \Re^l$, w_k is process noise which is assumed to be Gaussian and white, and H is an $m \times n$ matrix called *measurement transition matrix* which relates to state to the measurement. P_k^- and P_k^+ are the *priori and posteriori estimation error covariance* based on predicted and measured values (by Haar-classifier), and K_k is the *Kalman gain*.

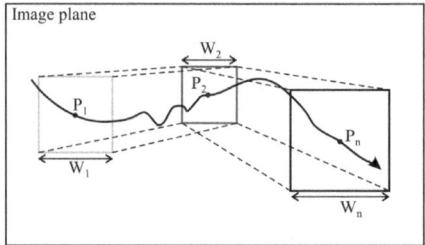

Fig. 7. Sample movement trajectory: simultaneous changes in face position and size

4.2   Filter Modelling and Implementation

There are many different motion equations such as linear acceleration, circular acceleration, or Newton mechanics; however, for a driver's face movements we simply consider a linear dynamic system and assume constant acceleration in a short Δt time frame. We implemented and modelled the filter with the following elements. First, the state vector is defined as $x_t = \begin{bmatrix} x & y & w & v_x & v_y & a_x & a_y \end{bmatrix}^T$. Also based on the theory of motion we have that

$$p(t+1) = p(t) + v(t)\Delta t + a(t)\frac{\Delta t^2}{2} \qquad (2)$$

$$v(t+1) = v(t) + a(t)\Delta t \qquad (3)$$

where p, v, a, Δt are position, velocity, acceleration, and time difference between input images, respectively. Δt is the average processing time which is the time between starting the process of face detection on a given frame at time k until end of the eye detection process and accepting the next frame at time $k+1$.

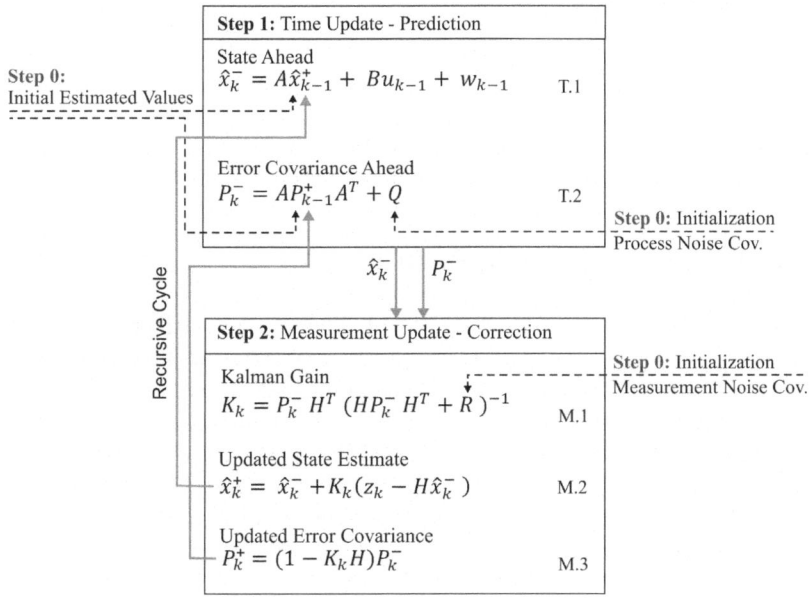

Fig. 8. Kalman filter in a nutshell

Relying on the definition of state transition in equation T.1, we modelled the transition matrix A as a 7×7 matrix. Thus, the next state is estimated as below:

$$
\begin{pmatrix} x_k \\ y_k \\ w_k \\ v_{x_k} \\ v_{y_k} \\ a_{x_k} \\ a_{y_k} \end{pmatrix} = \begin{pmatrix} 1 & 0 & 0 & \Delta t & 0 & \frac{1}{2}\Delta t^2 & 0 \\ 0 & 1 & 0 & 0 & \Delta t & 0 & \frac{1}{2}\Delta t^2 \\ 0 & 0 & 1 & 0 & \Delta t & 0 & 0 \\ 0 & 0 & 0 & 1 & 0 & \Delta t & 0 \\ 0 & 0 & 0 & 0 & 1 & 0 & \Delta t \\ 0 & 0 & 0 & 0 & 0 & 1 & 0 \\ 0 & 0 & 0 & 0 & 0 & 0 & 1 \end{pmatrix} \cdot \begin{pmatrix} x_{k-1} \\ y_{k-1} \\ w_{k-1} \\ v_{x_{k-1}} \\ v_{y_{k-1}} \\ a_{x_{k-1}} \\ a_{y_{k-1}} \end{pmatrix} + \begin{pmatrix} rand(2e-3) \\ rand(2e-3) \\ . \\ . \\ . \\ . \\ rand(2e-3) \end{pmatrix}
$$

Before running the filter in the real world, we need to have a few initializations including $\hat{x}_0^+$, $\hat{P}_0^+$, R, Q, and H. R directly depends on measurement accuracy of our camera along with the accuracy of our face/eye detection algorithm. Comparing the ground truth information and the result of our face detection method, we determined the variance of the measurement noise in our system as $R = rand(1e-4)$. The determination of Q is generally more difficult than of R, and needs to be tuned manually. A good tuning of R and Q stabilizes K_k and P_k very quickly after a few iteration of filter recursion. We got the best system stability when we set the process noise to $Q = rand(2e-3)$.

We take $H = 1$ because the measurement is composed of only the state value and some noise. Matrix B is omitted as there is no external control for driver face movement. For $\hat{x}_0^+$ and $\hat{P}_0^+$ we assumed the initial position of a face at position $x = 0$ and $y = 0$, with an initial speed of zero, as well as posteriori error covariance of 0. We also considered Δt as being between 33 and $170ms$, based on computation cost and a processing rate between 6 to $30Hz$.

5   Experiments and Validation

A grey-level VGA camera at a distance of $60cm$ to the driver seat is used to take continuous recording of the driver seat area. Figures 9-11 show results of eye status detection before and after implementation of the adaptation module. The images have been selected from 20 recorded video sequences with extremely varying lighting. Table 2 provides details of TP and FP detection rates performed on 2 sample videos (5 minutes each) and 2 face datasets (2,000 images each).

Figure 12 illustrates partial face tracking results and error indices in x-coordinates, for 450 recorded sequences while driving. Using adaptive Haar-like detectors, we rarely faced detection failure for more than 5 continued frames; however, we intentionally deactivated the detection module for up to 15 continued frames to see the tracking robustness (grey bars, Fig. 12). Results show promising tracking without any tracking divergence. Similar results were obtained for y-coordinates and face size tracking. Comparing ground truth and tracking results, the average error index was ± 0.04. Adding a safety margin of 4% around the tracking results, we can recursively define an optimized rectangular search region for the adaptive Haar-classifier instead of a blind search in the whole image plane. We define $P1_k = (x1_{k-1} - 0.04 \times 640, y1_{k-1} - 0.04 \times 480)$ and $P2_k = (x2_{k-1} + 0.04 \times 640 + 1.4w, y2_{k-1} + 0.04 \times 480 + 1.4w)$ as the optimized search region, where $P1_k$ and $P2_k$ point to the upper-left and lower-right corners of search regions at time k; pairs of $(x1_{k-1}, y1_{k-1})$ and $(x2_{k-1}, y2_{k-1})$ are upper-left and lower-right coordinates of predicted faces at time $k - 1$; w is the predicted face width at time $k - 1$.

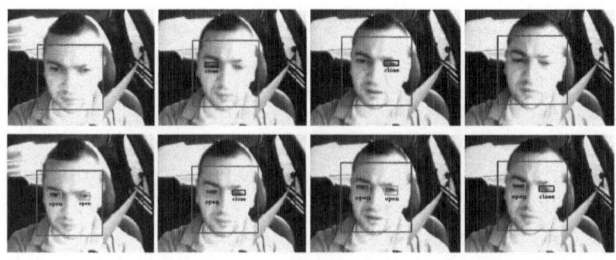

Fig. 9. Video 1: Face and eye detection under sun-strike; standard classifier (top) vs. adaptive classifier (bottom)

Fig. 10. Video 2: Face detection with sunglasses under light-varying conditions; standard classifier (top) vs. adaptive classifier (bottom)

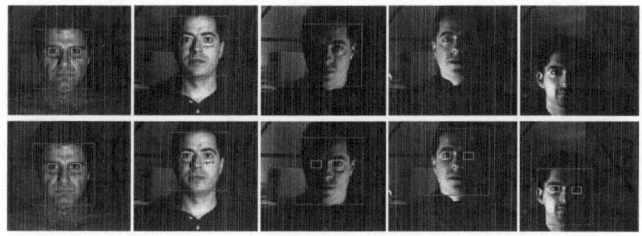

Fig. 11. Face and eye detection under difficult lighting conditions; standard classifier (top) vs. adaptive classifier (bottom). Image source: Yale database.

Table 2. Performance analysis for *standard* and *adaptive classifier*

	Standard V-J Classifier						Proposed Adaptive Classifier					
	Face		Open		Closed		Face		Open		Closed	
	TP	FP	TP	FP	TP	FP	TP	FP	TP	FP	TP	FP
Video 1	97.5	0.01	82	3.2	86	4.4	99.3	0	96.1	0.27	95.7	0.32
Video 2	81.1	1.02	-	0.5	-	0.32	94.6	0.01	-	0.01	-	0
Yale DB	86.3	0.05	79.4	0.1	-	0.07	98.8	0.02	97.3	0	-	0.01
Closed Eye DB	92.2	0.06	87.5	3.7	84.2	3.9	99.5	0.02	99.2	0.4	96.2	0.18

Figure 13.*a* shows good tracking after 5 frames, and 13.*b* shows perfect tracking after 10 frames. Figure 13.c displays failed face detection due to face occlusion while steering; however, successful face tracking (yellow frame) lead to proper eye detection. Figure 13.d shows another good "match" of detection and tracking with accurate closed-eye detection. Figure 14 shows very good results for eye detection and tracking at night when having sharp back-lights, strong shades, and very dark conditions.

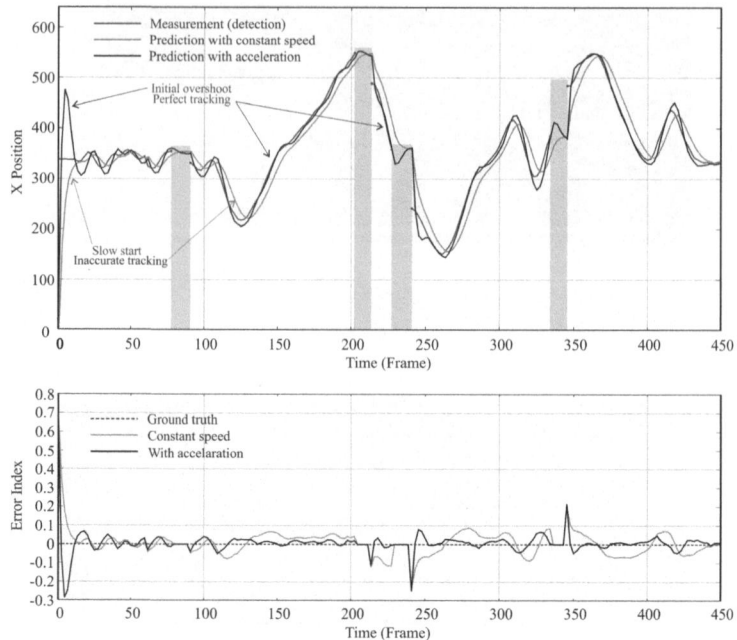

Fig. 12. Tracking results with and without accelerations. Grey blocks: No detections, but still good tracking results.

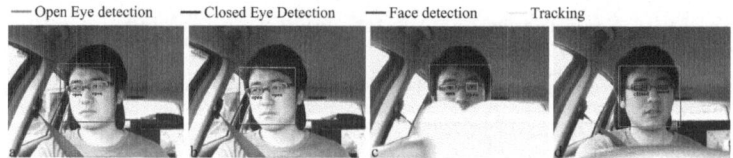

Fig. 13. *Face, open eye,* and *closed eye* detection and tracking while driving in daylight.

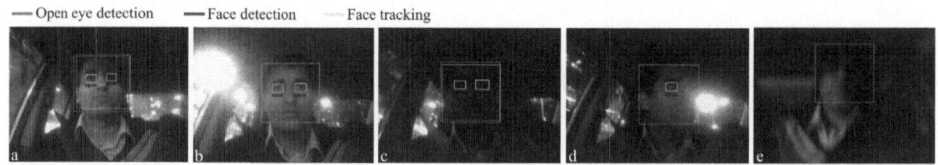

Fig. 14. *Face, open eye,* and *closed eye* detection and tracking at night under difficult lighting. (e) detection failure due to motion blur; however still robust tracking.

6   Conclusions

The paper introduced a fast and effective detection framework that enables accurate facial feature tracking under difficult lighting, performing clearly better than standard Viola-Jones classifiers. While the *tracking module* minimized the search region, the *adaptive module* focused on the given limited region to adapt the SWS, SF, MNN parameters depending on pixel by pixel intensity changes in eye-pair surrounding areas. Both modules recursively boosted each other; the face tracking module provides an optimum search area for eye detection, and in turn, the adaptive eye detection module provides geometrical face location estimation to support a Kalman tracker in case of tracking failure. We gained six times faster processing and more accurate results using only a low-resolution VGA camera, without any application of IR light, or any preprocessing or illumination normalization techniques. Our solution is suggested to be an amendment for any kind of a Haar-like classifier, to improve in challenging environments. For future work we suggest integration of pre-processing techniques and a comparison with illumination invariant methods such as SIFT or LBP.

References

1. BioID Database (2012),
   http://www.bioid.com/downloads/software/bioid-face-database
2. Brandt, T., Stemmer, R., Rakotonirainy, A.: Affordable visual driver monitoring system for fatigue and monotony. Systems Man Cybernetics 7, 6451–6457 (2004)
3. FERET Face Database (2012), http://www.itl.nist.gov/iad/humanid/feret/
4. Freund, Y., Schapire, R.E.: A decision-theoretic generalization of on-line learning and an application to boosting. J. Computer System Sciences 55, 119–139 (1997)
5. Face of tomorrow database (2012), http://www.faceoftomorrow.com/posters.asp
6. Hsu, R., Abdel-Mottaleb, M., Jain, A.K.: Face detection in color images. IEEE Trans. Pattern Analysis and Machine Intelligence 24, 696–706 (2002)
7. Jesorsky, O., Kirchberg, K.J., Frischholz, R.W.: Robust Face Detection Using the Hausdorff Distance. In: Bigun, J., Smeraldi, F. (eds.) AVBPA 2001. LNCS, vol. 2091, pp. 90–95. Springer, Heidelberg (2001)
8. Kalman, R.E.: A new approach to linear filtering and prediction problems. J. Basic Engineering 82, 35–45 (1960)
9. Li, S.Z., Jian, A.K.: Handbook of Face Recognition. Springer, London (2011)
10. Majumdar, A.: Automatic and robust detection of facial features in frontal face images. In: Proc. Int. Conf. Computer Modelling Simulation, pp. 331–336 (2011)
11. World recognized Haar classifier contributors for face and eye detection (2012),
    http://opencv.willowgarage.com/wiki/Contributors
12. Tsao, W., Lee, A.J.T., Liu, Y., Chang, T., Lin, H.: A data mining approach to face detection. Pattern Recognition 43, 1039–1049 (2010)
13. Viola, P., Jones, M.: Rapid object detection using a boosted cascade of simple features. In: Proc. CVPR, pp. 511–518 (2001)
14. Wu, Y., Ai, X.: Face detection in color images using AdaBoost algorithm based on skin color information. In: Proc. Int. Workshop on Knowledge Discovery Data Mining, pp. 339–342 (2004)

15. Wu, J., Brubaker, S.C., Mullin, M.D., Rehg, J.M.: Fast asymmetric learning for cascade face detection. IEEE Trans. Pattern Analysis and Machine Intelligence 30, 369–382 (2008)
16. YALE Face Database (2012),
    http://cvc.yale.edu/projects/yalefacesB/yalefacesB.html
17. Zhu, Z., Ji, Q.: Robust real-time eye detection and tracking under variable lighting conditions and various face orientations. J. Computer Vision Image Understanding 98, 124–154 (2005)
18. Zhang, X., Gao, Y.: Face recognition across pose: A review. Pattern Recognition 42, 2876–2896 (2009)

Obstacles Extraction Using a Moving Camera

Shaohua Qian[1], Joo Kooi Tan[1], Hyoungseop Kim[1], Seiji Ishikawa[1],
and Takashi Morie[2]

[1] Department of Mechanical & Control Engineering, Kyushu Institute of Technology
[2] School of Brain Science & Engineering, Kyushu Institute of Technology

Abstract. A method of automatic obstacles detection is proposed which
employs a camera mounted on a vehicle. Although various methods of
obstacles detection have already been reported, they normally detect
moving objects such as pedestrians and bicycles. In this paper, a method
is proposed for detecting obstacles on a road, irrespective of moving or
static, by the employment of the background modeling and the road
region classification. The background modeling is often used to detect
moving objects when a camera is static. In this paper, we apply it to
the moving camera case to get foreground images. Then we extract the
road region using SVM. In this road region, we carry out region classification. Then we can delete all the things which are not obstacles in
the foreground images using the result of the region classification. In the
performed experiments, it is shown that the proposed method is able to
extract the shapes of both static and moving obstacles in a frontal view
from a car.

1 Introduction

In recent years, autonomous collision avoidance systems have been researched
and developed for realizing safe driving using cameras and computers. These
systems are designed to warn the drivers the presence of obstacles on the road
and help them take a necessary action in advance. In these systems, the ability
to detect obstacles is essential. We know that vision is important for the safe
driving. Given the vision information around the car, drivers can use this vision
information to help them take proper actions. Therefore, a vision-based obstacle
detection system is the mainstream of current researches in the intelligent vehicle
technology.

The existing obstacles detection methods are separated into three categories
[1]: (i) The method using a monocular static camera. This method detects obstacles based on the optical flows which are inconsistent with the main movement
direction of vehicles [2,3]. This method is sensitive to camera motion. It cannot detect static obstacles, and it can only be used to detect moving obstacles.
(ii) The method using a monocular moving camera. This method detects obstacles based on searching for their features, such as shape [4,5] and symmetry
[6]. This method can only be used to detect one kind of specific object, such
as pedestrian detection or vehicle detection. (iii) The method based on stereo

J.-I. Park and J. Kim (Eds.): ACCV 2012 Workshops, Part II, LNCS 7729, pp. 441–453, 2013.
© Springer-Verlag Berlin Heidelberg 2013

vision [7,8]. Scene images are captured using two or more cameras from different angles simultaneously, and then the obstacles are detected through matching. This method needs huge calculation cost.

In this paper, we propose an obstacles detection method using a vehicle-mounted monocular camera. This camera records the road environment in front of a vehicle when the vehicle is moving, and the computer deals with these captured images to realize obstacles detection. The output of this method is the shape of obstacles. After having obtained the obstacle information, the drivers can react quickly and make the corresponding actions accurately to prevent car accidents. Here correct obstacles are defined as arbitrary objects which protrude from the ground plane in the road region, including static and moving objects. Road marks in the road region (e.g. zebra crossings) and objects outside the road region are considered as incorrect obstacles.

When a car is moving forward, the stationary objects in the visual are considered as the background, and the foreground can be obtained based on the background model. (e.g. because the road is almost no texture, the road can be considered to be static in the visual.) In this premise condition, the foreground image which has obtained from background modeling contains the shape of the obstacles and the edges of moving buildings. In order to extract the shape of the obstacles in the foreground image, the following operations are employed. Firstly, extraction of the road region in the input image using the SVM. The features of road are described by texture features and color features. Secondly, non-road region in the result of the road region detection are classified as noise region and obstacles region. After region classification, we have three kinds of regions, noise region, obstacles region and road region. Finally, all objects inside the noise region in the foreground image are considered as noises and can be deleted using the noise region. Road marks in the foreground image are considered as incorrect obstacles inside the road region and can be deleted using the road region. The shape of the obstacles (e.g. pedestrians) in the foreground image is extracted using the obstacles region.

The most significant differences between the proposed method and the existing obstacles detection methods are as follows. In the first place, the proposed method detects arbitrary objects, irrespective of static objects or moving objects, which may pose a threat to safe driving on the road, not just detect specific objects. This is helpful, because the objects which have fallen on the road from a car are dangerous to driving. Most of the existent methods, however, concentrate only on detecting moving objects such as pedestrians, bicycles and cars. In the second place, the output of the proposed method is the shape of obstacles. Currently most of the existing obstacles detection methods cannot extract the shapes of obstacles. They only use a rectangle which surrounds an obstacle to represent a detected obstacle. This shape information is important for obstacles recognition and classification. If the detected obstacles are judged as a pedestrian, we can use the shape to carry out his/her motion recognition. In the third place, the proposed method can be applied for speeds up to 40 km/h which is usually the speed limit within the city.

2   Outline of the Proposed Method

Outline of the proposed method is given in the following: Given a video image sequence taken by a moving camera, in section 3, in order to obtain the shape of obstacles, we employ Gaussian Mixture Model for reconstructing the background and getting the corresponding foreground images. The obstacles are to be included in the foreground images. In section 4, because some obstacles outside the road region and noises are also included in the foreground images, we detect the road region using Support Vector Machines to delete these incorrect obstacles and noises. Section 5 describes region classification using the result of road region detection and extraction of obstacles on the road. Experimental results are shown in section 6 and the results are evaluated in section 7. Finally the paper is concluded in section 8.

3   Background Modeling

A typical method of detecting obstacles is background modeling, by which we can get the shape of an object directly. Among numerous background modeling methods, we employ Gaussian Mixture Model (GMM) [9] in the present study, because GMM is robust to illumination change. The GMM models the gray value of each pixel on an image as a mixture of K Gaussian distributions. Different Gaussian distributions in the mixture represent different pixel values. The background models are determined by the parameters of these Gaussian distributions. Then we regard those pixels which do not match the background distributions as foreground pixels. In order to deal with illumination change, the mixture models are updated frame by frame.

The GMM is robust when employed in a fixed camera case. But, in this research, the employed camera is moving, since it is mounted on a car. In order to employ GMM in a moving camera case, we need to construct a virtual scene using the real scene. We then employ the GMM in this virtual scene in reconstructing the background.

3.1   Virtual Scene Construction

In this research, since a camera is mounted on a vehicle, when the vehicle is moving, the camera is as well moving. This is the real situation. In this scene, if we assume that the camera is static, we see that buildings, the road and static objects are moving according to the relative motion. But, since the road has almost no texture, we can assume that the road is static. So we obtain a virtual scene in which the camera is static; the road area which is classified as the background is static; objects (including static and moving objects) and pedestrians on the road, buildings, road marks and zebra crossings which are classified as the foreground are moving. After getting this virtual scene, we employ the GMM to this scene to reconstruct the background.

3.2   Gaussian Mixture Model

The history of a particular pixel is a time series of pixel values, i.e. the pixel values of a particular pixel in the image sequence over time. At any time t $(t=1,2,\ldots,T)$, the history of a particular pixel (x_0, y_0) is given by

$$\{X_1,\ldots,X_T\} = \{I(x_0,y_0,t) : 1 \leqslant t \leqslant T\} \tag{1}$$

where I is the gray value of pixel (x_0, y_0).

For the history of a particular pixel, $\{X_1,\ldots,X_T\}$, K Gaussian distributions are used to model these pixel values. The probability of the pixel value X_t is

$$P(X_t) = \sum_{k=1}^{K} w_{k,t} * \eta(X_t, \mu_{k,t}, \sigma_{k,t}^2) \tag{2}$$

where K is the number of Gaussian distributions in the mixture, $w_{k,t}$ is the weight of the kth Gaussian distribution at time t; $\mu_{k,t}$ is the mean value of the kth Gaussian distribution at time t; $\sigma_{k,t}^2$ is the covariance of the kth Gaussian distribution at time t, and η is a Gaussian probability density function defined by

$$\eta(X_t, \mu_{k,t}, \sigma_{k,t}^2) = \frac{1}{\sqrt{2\pi}\sigma} exp\{-\frac{1}{2}\frac{(X_t - \mu_{k,t})^2}{\sigma_{k,t}^2}\} \tag{3}$$

For each pixel in the image, K Gaussian distributions are used to model the history of this pixel. These K Gaussian distributions are ordered by the value of $w_{k,t}/\sigma_{k,t}^2$. After the ordering, the distributions which represent background distributions most likely are on the top, whereas the distributions which represent background distributions less likely are on the bottom. Then among these K distributions, we consider the first B distributions as background models, where

$$B = argmin_b(\sum_{k=1}^{b} w_k > T_B) \tag{4}$$

where T_B is a threshold. Every pixel value, X_t, is checked against these B Gaussian distributions; if it matches them, this pixel is a background pixel, otherwise a foreground pixel. According to this, we can extract the foreground image from the input image. The match is defined by

$$\frac{|X_t - \mu_{k,t-1}|}{\sigma_{t-1}} \leq T_{gauss} \tag{5}$$

where T_{gauss} is a threshold.

4   Road Region Detection

Because the camera is moving, the foreground which obtained from the background modeling often contains a lot of noises. These noises are mostly caused

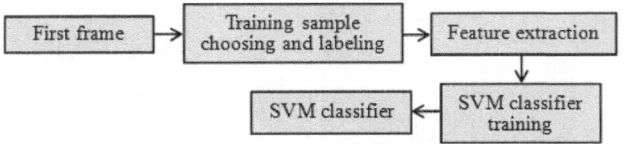

Fig. 1. Outline of the SVM classifier training

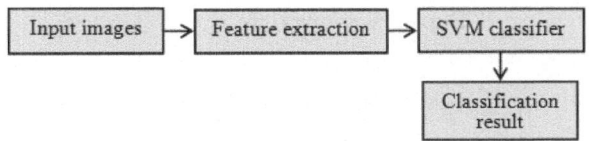

Fig. 2. Outline of the test

by the objects outside the road region. In order to delete these noises, we need to detect the road region. In this paper, we detect a road region using the Support Vector Machines (SVM) [10,11].

This method includes two steps: training and test. Fig. 1 shows an outline of the SVM classifier training. In the first frame of the input video, a small sample of pixels is labeled by a human supervisor as road class or non-road class. For each pixel in this sample, we extract a feature vector by feature extraction. These feature vectors are considered as the training data of SVM. Then, these training data are used to train a SVM classifier.

Fig. 2 gives an outline of the test (classification) using the SVM classifier. In others frames of the input video, a feature vector is extracted for each pixel using the same feature extraction method. These feature vectors are considered as the testing data. Then, each pixel in the input video is classified as the road pixel or non-road pixel using the trained SVM classifier and the testing data.

4.1   Feature Extraction

In this paper, the features we used are color features and texture features. Color features are three features in HSV color space. For texture features, five Haralick texture features [12] are used as follows:

$$Energy = \sum_i \sum_j \{p(i,j)\}^2 \tag{6}$$

$$Entropy = -\sum_i \sum_j p(i,j)log\{p(i,j)\} \tag{7}$$

$$Contrast = \sum_{n=0}^{N_g-1} n^2 \{\sum_{i=1}^{N_g} \sum_{j=1}^{N_g} p(i,j)\}, |i-j| = n \tag{8}$$

Fig. 3. Training data selection

$$Inverse \quad difference \quad moment = \sum_i \sum_j \frac{1}{1+(i-j)^2} p(i,j) \qquad (9)$$

$$Correlation = \frac{\sum_i \sum_j (ij)p(i,j) - \mu_x\mu_y}{\sigma_x\sigma_y} \qquad (10)$$

Here $p(i,j)$ is an element value in co-occurrence matrix. μ_x, μ_y and σ_x, σ_y are the mean values and covariance calculated from a co-occurrence matrix. A feature vector is combined by color features and texture features as follows;

$$F_{i,j} = [f_{t_1(i,j)}, f_{t_2(i,j)}, f_{t_3(i,j)}, f_{t_4(i,j)}, f_{t_5(i,j)}, f_{c_1(i,j)}, f_{c_2(i,j)}, f_{c_3(i,j)},] \qquad (11)$$

where $f_{t_n(i,j)}$ is the nth Haralick texture at the point (i,j), $f_{c_n(i,j)}$ is the nth value in HSV color space at the point (i,j).

4.2   Training Database Initialization

The first frame in the input video is used as the training image of SVM, the other frames in the input video are used as the testing images.

In the first frame, the training data is selected and labeled by a human. Two rectangle windows are used by a supervisor to select the training data on the image as shown in Fig. 3. A green window is placed in the road region. The pixels located in this green window are labeled as positive samples. A red window is placed outside the road region. The pixels located in this green window are labeled as negative samples. These two samples constitute the training database.

5   Obstacles Extraction

In order to extract the shape of obstacles in the foreground images, we need to delete two kinds of things: road marks and noises outside the road. For road marks, we can use the result of the road region detection to delete them. For the noises outside the road; when we use the result of the road region detection to delete noises, it also deletes the obstacles outside of the road. In order to solve this problem (reserving the obstacles and deleting the noise outside the road), we need to divide the non-road region into obstacles region and noise region. Then we use this noise region to delete the noise outside the road in the foreground image.

(a)   (b)   (c)   (d)

Fig. 4. (a) The foreground image, (b) the result of road region detection,(c) road region template image, (d) the result of region classification

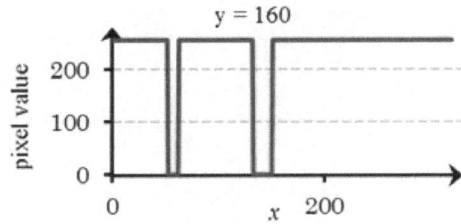

Fig. 5. Pixel values distribution of the 160th row in the road region image

5.1   Region Classification

Here we want to divide the non-road region into obstacles region and noise region.

Figure 4 (b) and (c) show the result of road region detection and the corresponding road region template image. In this road region template image, *black* pixels are road pixels, whereas *white* pixels are non-road pixels. If we check the pixels of one particular row in this image, we get a curve as shown in Fig. 5. Based on this curve, we consider *white* regions (high values) which have two adjacent *black* regions (low values) both on the left and right sides as the obstacles region. We check each row in the road region template image to carry out region classification. Figure 4 (d) shows the result of the region classification.

5.2   Post-processing

In the foreground image, we check each black pixel's position in the template. If it is located in the road region (*black*) or noise region (*white*), we set it *white*. By this operation, road marks and other noises are respectively deleted. Then we carry out erosion operation and region growing as the post-processing.

6   Experiments and Results

This section presents some experimental results. The camera which is fixed in front of the vice driver's seat records the road conditions in front of the car when the car moves forward. Because correct obstacles are defined as arbitrary objects

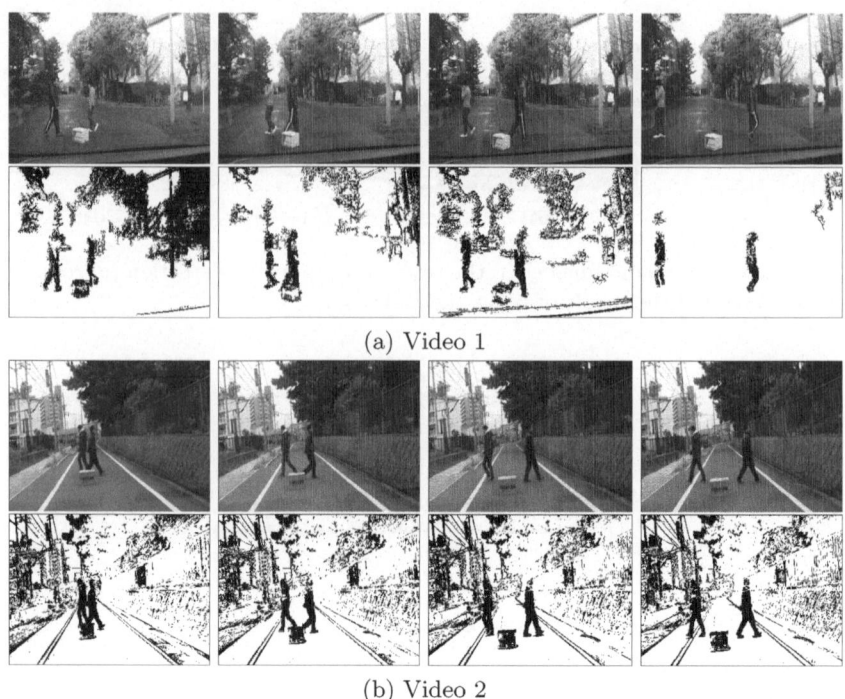

(a) Video 1

(b) Video 2

Fig. 6. The results of background modeling. First row: Input images, second row: Foreground images.

which protrude from the ground plane in the road region, including static and moving objects. Road marks in the road region (e.g. zebra crossings) and objects outside the road region are considered as incorrect obstacles. According to this definition, correct objects in video 1 and video 2 are two pedestrians and a box.

In the first place, we reconstruct the background model using a Gaussian mixture model in the input images. Fig. 6 show the input images and corresponding foreground images. Because the camera is moving, the foreground images as shown in Fig. 6 contain a lot of noises. These noises are mostly caused by the objects outside the road region. In order to delete these noises, we need to detect the road region.

In the second place, we detect the road region in the input images using the Support Vector Machines. Fig. 7 shows the results of two road region detection methods: the method using SVM (first row) and the method using motion compensation (second row) [13-15]. In these resultant images, the purple color region means the road region. The detection method using motion compensation needs to calibrate the camera and calculate the motion parameters of the car; due to the precision error of these calculations the detection results are worse than the detection method using SVM.

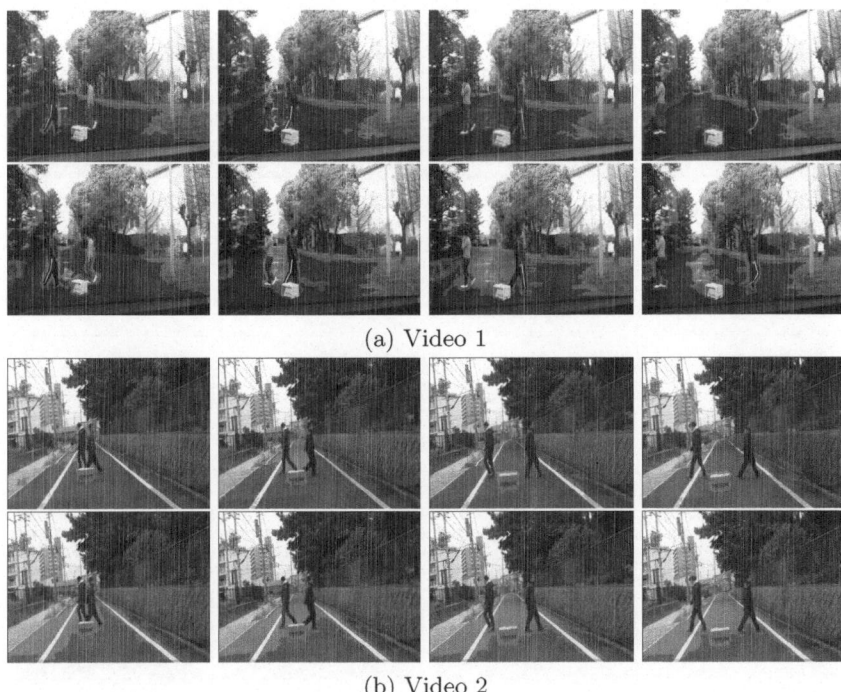

(a) Video 1

(b) Video 2

Fig. 7. The results of road region detection. First row: the method using SVM, second row: the method using motion compensation.

In the third place, we carry out region classification in the road template image, and then delete road marks and noises in the foreground image using the result of the region classification. In Fig.8, the second row shows the results of obstacles detection in which the road detection is done using SVM, the third row shows the results of obstacles detection in which the road is detected using motion compensation (comparative experiment). Since the road region detection accuracy has been improved, the results of obstacles detection are improved. Compared with the results of comparative experiment, the shapes of obstacles of the proposed method are more complete.

7   Evaluation

In order to evaluate the effectiveness of the proposed obstacles detection method, we compare the result of obstacles detection (shown in Fig. 9 (b)) with the Ground Truth (shown in Fig. 9 (a)). In the resultant image of comparison (shown in Figure 9 (c)), the red area means the overlap part of (a) and (b), this part is called *True Positive*: Blue means the part which is included in (b) but not in (a)

(a) Video 1

(b) Video 2

Fig. 8. The results of obstacles detection. First row: input images, second row: the result using SVM (the proposed method), third row: the result using motion compensation (comparative experiment).

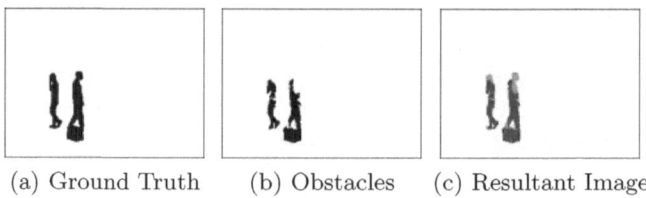

(a) Ground Truth   (b) Obstacles   (c) Resultant Image

Fig. 9. Images employed for evaluation

and this part is called *False Positive*: Green means the part which is included in (a) but not in (b) and this part is called *True Negative*. We calculate 'recall' using the following formula;

$$recall = \frac{TP}{GT} \times 100[\%] \tag{12}$$

TP is the area of *True Positive*; *GT* is the area of obstacles in Ground Truth. If *recall* is larger than 0.5, we consider this object has been extracted. Then we calculate 'Recall' and 'Precision' using the following formulas;

$$Recall = \frac{N_{TP}}{N_{GT}} \times 100[\%] \tag{13}$$

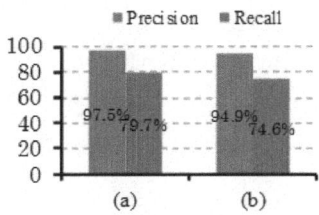

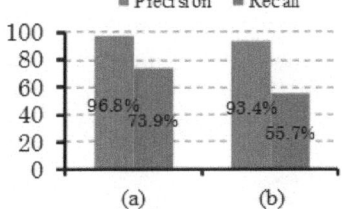

Fig. 10. The result of evaluation (video 1)   **Fig. 11.** The result of evaluation (video 2)

$$Precision = \frac{N_{TP}}{N_{TP} + N_{FP}} \times 100[\%] \tag{14}$$

where N_{TP} is the number of correct objects in the resultant images; N_{GT} is the number of objects in the *GT* images; N_{FP} is the number of incorrect objects in the resultant images. The results of evaluation are shown in Fig. 10-11. (a) is the evaluation of the results of obstacles detection in which the road detection method using SVM (our method); (b) is the evaluation of the results of obstacles detection in which the road detection method using motion compensation was employed (comparative experiment). From the result of evaluation, we know that the result of the proposed obstacles detection method is better.

8   Conclusions

In this paper, we proposed an obstacles detection method using a video taken by a vehicle-mounted monocular camera.

The proposed method employed the GMM in the moving camera scene. GMM is an effective background modeling method which was often used in the static camera scene. This expansion is important to the industrial applications of a vehicle-mounted camera based obstacle detection system.

The proposed method compared two road region detection methods. According to the results, the detection method using SVM is better than the method using motion compensation, which is also showed in the resultant images of obstacles detection. This is caused by the precision error of camera calibration and motion parameters calculation in the road region detection method using motion compensation.

The proposed method can detect arbitrary objects which may pose a threat to safe driving on the road, not just to detect specific objects. The proposed method can also detect both static and moving objects simultaneously.

The output of the proposed method is the shape of obstacles. Currently most of the existing obstacles detection methods cannot extract the shape of obstacles. They only use a rectangle which surrounds an obstacle to represent a detected obstacle. We understand that extraction of the shape of an obstacle is important for obstacle recognition. If obstacles are pedestrians, we can also use the shape to carry out human motion recognition.

Acknowledgement. This work was supported by JSPS KAKENHI (22510177) which is greatly acknowledged.

References

1. Demonceaux, C., Kachi-Akkouche, D.: Robust obstacle detection with monocular vision based on motion analysis. In: Proc. IEEE Intelligent Vehicles Symposium, pp. 527–532 (2004)
2. Kruger, W., Enkelmann, W., Rossle, S.: Real-time estimation and tracking of optical flow vectors for obstacle detection. In: Proc. IEEE Intelligent Vehicles Symposium, pp. 304–309 (1995)
3. Lefaix, G., Marchand, E., Bouthemy, P.: Motion-based obstacle detection and tracking for car driving assistance. In: Proc. Pattern Recognition, pp. 74–77 (2002)
4. Broggi, A., Bertozzi, M., Fascioli, A., Sechi, M.: Shape-based pedestrian detection. In: Proc. IEEE Intelligent Vehicles Symposium, pp. 215–220 (2000)
5. Lutzeler, M., Dickmanns, E.D.: Road recognition with marveye. In: Proc. IEEE Intelligent Vehicles Symposium, pp. 341–346 (1998)
6. Kuehnle, A.: Symmetry-based vehicle location for AHS. In: Proc. SPIE-Transportation Sensors and Controls: Collision Avoidance, Traffic Management, and ITS, vol. 2902, pp. 9–27 (1998)
7. Bertozzi, M., Broggi, A.: Gold: a parallel real-time stereo vision system for generic obstacle and lane detection. In: Proc. IEEE Image Processing, 62–81 (1998)
8. Labayrade, R., Aubert, D.: Robust and fast stereovision based obstacles detection for driving safety assistance. In: Proc. Machine Vision Applications, pp. 624–627 (2004)
9. Stauffer, C., Grimson, W.E.L.: Adaptive back-ground mixture models for real-time tracking. In: Proc. Conference on Computer Vision and Pattern Recognition, vol. 2, pp. 246–252 (1995)
10. Zhou, S., Gong, J., Xiong, G., Chen, H., Iagnemma, K.: Road Detection Using Support Vector Machine based on Online Learning and Evaluation. In: Proc. IEEE Intelligent Vehicles Symposium, pp. 21–24 (2010)

11. Zhou, S., Iagnemma, K.: Self-supervised Learning Method for Unstructured Road Detection using Fuzzy Support Vector Machines. In: Proc. ICRA, pp. 1183–1189 (2010)
12. Haralick, R.M.: Statistical and Structural Approaches to Texture. Proc. IEEE 67, 786–804 (1979)
13. Yamaguchi, K., Kato, T., Ninomiya, Y.: Vehicle ego-motion estimation and moving object detection using a monocular camera. In: Proc. International Conference on Pattern Recognition, pp. 610–613 (2006)
14. Yamaguchi, K., Watanabe, A., Naito, T.: Road region estimation using a sequence of monocular images. In: Proc. International Conference on Pattern Recognition, pp. 1–4 (2008)
15. Yamaguchi, K., Kato, T., Ninomiya, Y.: Ego-motion estimation using a vehicle mounted monocular camera. Proc. The Institute of Electrical Engineers of Japan 129(12), 2213–2220 (2009)

Scene Text Detection and Tracking for a Camera-Equipped Wearable Reading Assistant for the Blind

Faustin Pégeot[1] and Hideaki Goto[2]

[1] Graduate School of Information Sciences,
Tohoku University, Sendai, Japan
[2] Cyberscience Center,
Tohoku University, Sendai, Japan

Abstract. Visually impaired people suffer daily from their disability to read textual information. One of the most anticipated blind-assistive devices is a system equipped with a wearable camera capable of finding the textual information in natural scenes and translating it into sound through a speech synthesizer. To avoid duplicate readings, the device should be able to recognize text areas with the same content, and group them to obtain a single result. Scene text detection and tracking methods attract a lot of interest for these purposes. However, this field is still challenging and methods of scene text detection and tracking are yet to be perfected. This paper proposes a scene text tracking system capable of finding text regions and tracking them in video frames captured by a wearable camera. By combining a text detection method with a feature point tracker, we obtain a robust text tracker which produces much less false positive text images at 2.9 times faster speed compared with the conventional method.

1 Introduction

There are several hundreds of millions people who suffer from visual deficiency. These people cannot access the text information surrounding us, and this has a huge impact on their quality of life. Lots of improvements have been made in the recent years to help them in their daily lives, with alternative forms of information using sound or touch. On the other hand, the development of text reading devices is still in an early stage. A wearable camera capable of finding text regions in natural scenes, reading them and translating them into speech is probably one of the most anticipated blind-assistive devices.

To develop such a device, text tracking has been found necessary [1] [2]. Indeed, recognizing all the text strings in every frame woud be computationally impractical since text recognition engines are resource-consuming, especially for east-asian languages such as Japanese and Chinese, with thousands of characters. Moreover, the user of the system does not want to hear the same text over and over. So, we should utilize text tracking to associate the same text string across multiple frames.

J.-I. Park and J. Kim (Eds.): ACCV 2012 Workshops, Part II, LNCS 7729, pp. 454–463, 2013.

Other related devices include systems for robots with character recognition capability [3]. Some promising results have been found for retrieval of information in videos, such as subtitles or movie credits [2]. In these applications, the motion of text is assumed to be simple and steady. However, because the movement of a camera worn by a visually deficient person is basically unpredictable, the precision decreases very much [4]. Furthermore, the tracking accuracy is made worse by the errors due to the text detection phase preceeding the text region tracking.

This paper presents a wearable camera system that detects, tracks and extracts text areas. Since text detection is prone to errors, we apply a robust object tracking technique directly to the input video sequence before the text detection to achieve higher overall performances. In Section 2, we present an overview of the system, the global algorithm, as well as the text detection and localization methods used. Section 3 describes the text tracking algorithm. Section 4 shows experimental results and evaluations.

2   Outline of the System

In this section, we will expose the global algorithm used for the system, before a more thorough description of the text detection and localization methods used.

The prototype of the device consists of a head-mount color camera (640 × 480px) and a laptop PC. In the complete process, the text captured by the camera is detected and tracked using the algorithm described in Sec. 2.1. The algorithm returns text images that are binarized and recognized by an OCR engine. The final text string is then read through a speech synthesizer. In this paper, we focus on text detection and text tracking parts only.

2.1   Algorithm Overview

Conventional methods [2] [3] [4] generally use text detection every frame, and, using various methods, compare the results from frame to frame in order to evaluate the motion of the text areas. Our approach is a bit different. In order to predict the motion of the text areas, the proposed algorithm does not rely on the results of text detection, but uses an object tracking method. The goal is to reduce the impact of a flawed text detection on the text tracking system.

We apply text detection, track the detected text areas for several frames, and apply text detection again. Then, we compare the results of tracking with the results of text detection. We call this step text area matching.

In this paper, we will call F_{int} the number of frames between two text detections. The method follows the following pattern, as illustrated in Fig. 1:

- First, perform text detection to obtain locations of text areas
- Track the text areas for F_{int} frames.
- Perform text detection again.
- Match previously tracked text areas with newly detected text areas.

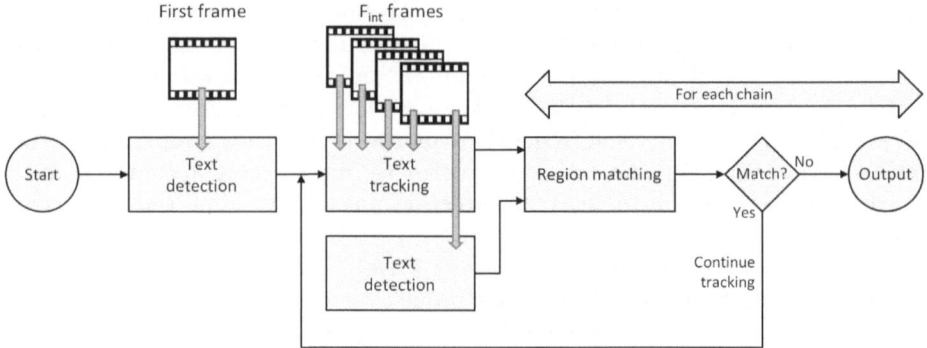

Fig. 1. The outline of the system

During the matching step, we not only associate text areas that match together, but also find if text areas appear or disappear from the screen. Note that the text tracking in our system is based on a well-developed feature point tracking.

2.2   Text Detection

The purpose of text detection is to highlight the parts of the image containing characters. Scene text detection is still a challenge for researchers, and at the present time, no methods of scene text detection can achieve both a good precision and a good recall in general cases. Neural-network based methods [5] give good results but require a machine learning phase.

For our purpose, we need a text detection method with a good recall rate, i.e. we want most of the text areas to be detected, because it would otherwise impede the performances of text tracking. A speed close to real-time is good but not necessary since the text detection is not applied for every frame.

While not being the most precise, DCT-based text detection proposed in [6] gives the best recall rate.

To describe the process simply, DCT feature value method divides the image into 8×8 pixel blocks and computes the DCT feature value for each of them. DCT feature value is a real number used as a measure of the stroke complexity of the pixel block. The blocks with high feature values will be considered as text, while the ones with low values will be discarded. Using a thresholding method explained more in detail in the next section, we determine which blocks are selected as text blocks. An exemple of output is shown in Fig. 2.

2.3   Text Localization

Associated with text detection, text localization defines clearly the boundaries of the text areas. For this purpose, we first obtain a threshold using Fisher's linear discriminant. Pixel blocks with feature values greater than the threshold are colored white, the other black. Then, white pixel blocks are grouped and

 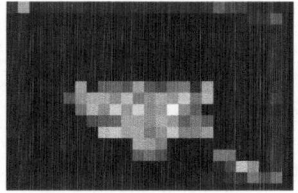 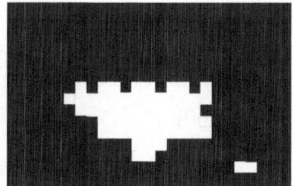

Fig. 2. Example of text detection output using DCT feature value [6]

bounded together in a rectangular area using a Connected Components analysis. If the area of the bounding rectangle is less than 400px, it is discarded, as it is more likely to be noise than an actual text region.

However, the bounding rectangle method has a limitation. By using connected components, there is a high probability, for instance with edges of signboards, often recognized by mistake as text, that the final bounding rectangle may include more than the intended text region. Fig. 3 illustrates this problem. The edge of the signboard has been mistakenly detected as text, and the rectangle on the left image does not enclose the text properly.

To solve this problem, we add another part to the localization procedure. The borders of the bounding rectangle are shrinked one after another to fit better with the real boundaries of the text region.

We call A_0 the area of the initial bounding rectangle, and A_0^t the area of pixel blocks colored white after the thresholding. The ratio of text in the rectangle is defined as:

$$R_0 = \frac{A_0^t}{A_0}$$

Then, we shrink the bounding rectangle from one pixel block, first the uppermost line, then in the other directions consecutively. We call, similarly, A_1, A_1^t and $R_1 = \frac{A_1^t}{A_1}$ the parameters for the shrinked rectangle. To measure if the shrinked rectangle fits better the text region, we compute:

$$F = \frac{R_1 - R_0}{R_0}$$

 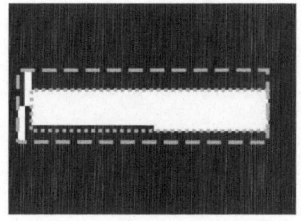 

Fig. 3. Shrinking text regions

If the improvement factor $F > \omega$, with ω a threshold set as 5%, the rectangle is shrinked. The process continues with another shrinked box until no further improvement is possible. In the middle image of Fig. 3, we show an example of the shrinking process. We can see the text region first localized using Connected Components, drawn with a dashed line, and the final, shrinked text region, drawn with a dotted line. The result is seen in the right image. As we can see, the boundaries fit better the text region.

3   Text Tracking Using Feature Points

We now need to track the detected text areas. For that purpose, we use local feature points scattered in the image. The positions of these points are tracked from frame to frame. The general motion of a text area is computed using the data of the motion of the points. We chose a sparse method (feature points) over a global method (optical flow) because of several reasons. First, we do not need the optical flow for the whole image. Second, by choosing and selecting feature points, we ensure a good-quality tracking for text areas.

3.1   Features Selection and Tracking

We compared several types of features detectors to choose the one most suitable for text tracking. We chose 3 sample images containing text in complicated background, applied different feature detectors, and counted the portion of features in text for each of them. Results are displayed in Fig. 4.

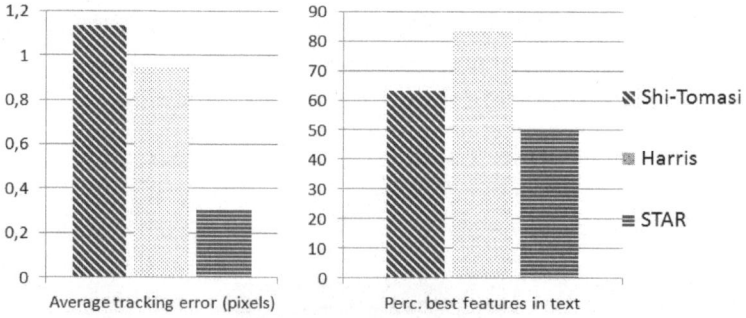

Fig. 4. Results of feature selection

The portion was calculated by counting how many, out of the best 100 features, are part of a text area. Using this measure, Harris corner detection is the most suitable for our application, with more than 80% of the best features detected located in text areas.

In our experiments we used the standard pyramidal Lucas-Kanade tracker algorithm [7]. Since the performances of the tracker are very dependent on the nature of the feature points, we also compared the tracking error of the feature points on a set of frames. The average tracking error for Harris corner detection is not the best, but sufficient for our application.

3.2   Motion Extrapolation

Once every feature point has been tracked, their motion is stored in motion vectors. Each feature point P^i is assigned a motion vector (V_x^i, V_y^i) that describes their motion between two frames, in pixels.

Fig. 5. Feature points in the same text area and their motion vectors

We group together all the feature points inside the same text area. Motion extrapolation now consists in describing the geometric transformation of the text area knowing the motion of a few points inside, as illustrated in Fig. 5. Indeed, the text area can follow any arbitrary motion, for instance zoom, translation, rotation, or any combination of these. The general problem considering all types of deformation leads to a complex system. To simplify it, we assume simpler motions, in order to reduce the number of unknowns. For example, if we allow only translations, zoom and rotations, the system is reduced to 4 unknowns, the translation vector, the dilatation factor and the angle of rotation.

For our model, we further simplified this model by considering the motion to be purely translational, removing all deformation and rotation degrees. The simplification leads to a system with two unknowns, V_x and V_y, which are the two components of the global motion vector of the text area. We assume the text area contains n feature points, with $n > 0$, which leaves us to a set of $2n$ equations. Then, using the least-squares principle to solve the over-determined system:

$$V_x = \frac{1}{n} \sum_{i=1}^{n} V_x^i \quad , \quad V_y = \frac{1}{n} \sum_{i=1}^{n} V_y^i$$

where n is the number of features in the considered text area and $(V_x^i, V_y^i), i = 1..n$ their motion vectors.

If the video quality is low, for instance because of motion blur or poor illumination, the feature points can not all be tracked with good precision. However, if enough points are being tracked correctly, the average motion for the text area

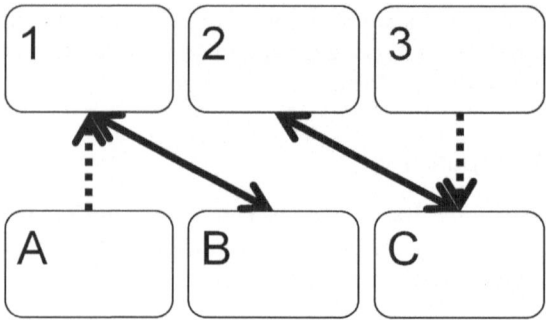

Fig. 6. Nearest neighbor matching process

is more accurate. The features selection method described in section 3.1 ensures that there is always a sufficient number of points for a precise tracking. So, even if a few feature points fail to be tracked correctly, the tracking system is still robust.

3.3 Text Area Matching

We want to apply text detection the least as possible since it is prone to errors. However, text detection is still necessary at repeated intervals for two reasons. First, it can be used to verify and correct the position of the text areas if the tracking fails. Second, it is used to detect when new text areas appear, or if previous text areas disappear, from the camera's field of view.

We apply text detection once every F_{int} frames, where F_{int} is the frame interval between two detection steps. After the text detection step n, we have two sets of text areas: the text areas detected during step $n-1$ and then tracked, and the ones detected during step n. Now, we need to match these two sets of text areas.

The text areas from step $n-1$ have already been tracked. Consequently, we know their positions within the tracking errors, and we can match the text areas directly using the distances between them.

We use a Nearest-Neighbor (NN) search method for this step:

- Each newly detected text area is matched to the nearest previous tracked text area in distance. The distance between the areas' centers must be less than 25 pixels (a large evaluation of the tracking and detection errors coupled).
  - Identically, previous text areas are matched to their nearest neighbours.
  - When text areas both match together, the tracking is continued.
  - When a new text area has no match, it forms a new chain.
  - When an old text area has no match, the tracking is terminated.

Fig. 6 illustrates the process. The positions of areas 1, 2 and 3 have been evaluated using the tracking process described in sec. 3.2. In parallel, the positions of areas A, B and C have been obtained using text detection. After NN-matching,

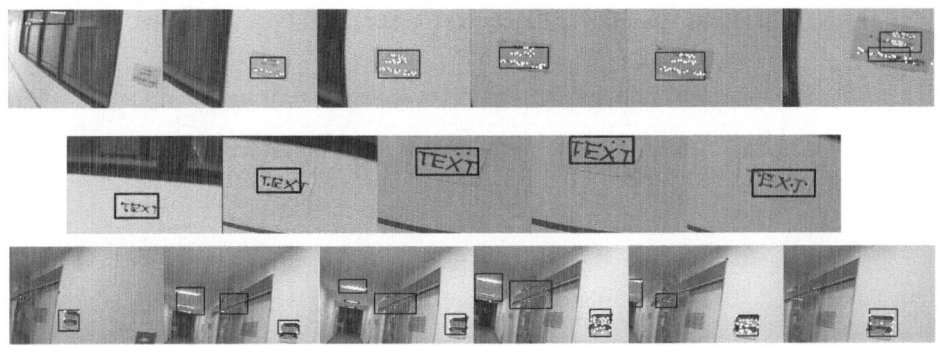

Fig. 7. Results of text area tracking

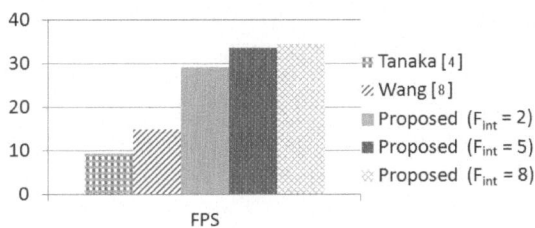

Fig. 8. Speed of text tracking

we conclude that, in this frame, the area 3, which found no match, has disappeared. Similarly, the area A which found no match has just appeared and will be tracked until the next detection step.

Once tracking is terminated, we choose to ouput the text area only if it was tracked succesfully for more than T_f frames. This is motivated by the fact that even though scene text detection has a lot of false positive results, only real text is repeatedly detected under various conditions, such as different illuminations, etc. We set the threshold for the number of frames $T_f = 15$ frames.

4   Experimental Results and Discussion

The program was run on a standard desktop computer equipped with an Intel Core 2 Duo (2.9GHz). The method was tested on three different videos, where a person is walking a corridor with several signboards, at various speeds and under different lighting conditions. The videos have in total 3,700 frames and 30 signboards.

We tested the method for frame intervals $F_{int} = 2, 5,$ and 8. As shown in Fig. 8, the system runs at 33.5 fps for $F_{int} = 5$. The obtained speed is 2.9 times faster than the conventional method [8], and is even faster than needed for real-time text tracking. This gives extra computational time to incorporate more specific and precise methods for further implementations.

Table 1. Number of candidate text areas (DCT Feature value)

	Scene 1	Scene 2	Scene 3
Text detection for every frame	1351	1813	2886
Proposed method (F_{int}=2)	621	887	1352
Proposed method (F_{int}=5)	227	360	504
Proposed method (F_{int}=8)	136	239	348

Table 2. Number of output results after tracking

	Scene 1	Scene 2	Scene 3
Number of text areas (GT)	8	12	10
Proposed method (F_{int}=2)	57	90	31
Proposed method (F_{int}=5)	13	25	17
Proposed method (F_{int}=8)	12	21	13

Fig. 7 shows several result examples. The first and second lines are demonstrations of our system on simple examples. The third line shows a few cases of mistakenly detected areas (false positives) in case of more complicated backgrounds.

Table 1 shows the number of candidate text areas detected (includes false positive detections). Naturally, the number decreases as we increase the frame interval. Basically, this means that by using a greater frame interval, we can reduce the number of false positive text detections.

Table 2 shows the number of areas after tracking. A lot of candidate regions, most of which are unwanted false positive text regions, disappear because they failed to be tracked for more than the pre-defined threshold $T_f = 15$ frames. Thus, the proposed method can effectively reduce the false positive output.

The on-screen time of each text signboard was measured, as well as the duration it was tracked. Table 3 presents these tracking performances. For example, if a text area is visible for 100 frames and is tracked for a total of 70 frames, its tracking time is 70%. As we can see, the performances drop when the number of frames between two text detections increases, especially for Scene 2, where the camera motion is the fastest among the three. The feature point tracker fails if the camera motion is too fast. Although the proposed method gives superior results compared with the conventional methods, some further improvements in the tracking speed would be necessary. The performance of the camera system is also crucial, as we need less motion blur when the user is moving the camera.

Finally, the average number of text image outputs per signboard is 2.5. This is a great improvement compared to the conventional method [8], with about 8 text images per signboard. For comparison, using text detection without text tracking results in an average 50 text images per signboard.

Table 3. Tracking performances

Tracking time (%)	Scene 1	Scene 2	Scene 3
Proposed method ($F_{int}=2$)	60.1	42.0	58.6
Proposed method ($F_{int}=5$)	58.6	40.8	56.1
Proposed method ($F_{int}=8$)	51.5	25.7	48.3

5   Conclusions

Scene text tracking has various applications, one of them being blind people assistance. Unfortunately, there was no method precise, reliable and fast enough for practical applications in scene text detection in video. By combining a feature point-based area tracking to the text detection method, we managed to obtain a result of 2.5 outputs per text image, which is about a third of the conventional method [8], at a three-times faster frame rate.

Further improvements could be added to improve this result. For example, the same signboard could once disappear from the camera view and come into it again in the real application. Our future work includes an improvement of the text tracking so it can deal with such broken traces.

References

1. Lyu, M., Song, J., Cai, M.: A comprehensive method for multilingual video text detection, localization, and extraction. IEEE Transactions on Circuits and Systems for Video Technology 15, 243–255 (2005)
2. Jiang, H., Liu, G., Qian, X., Nan, N., Guo, D., Li, Z., Sun, L.: A fast and effective text tracking in compressed video. In: Tenth IEEE International Symposium on Multimedia, ISM 2008, pp. 136–141 (2008)
3. Létourneau, D., Michaud, F., Valin, J.M.: Autonomous mobile robot that can read. EURASIP J. Appl. Signal Process., 2650–2662 (2004)
4. Tanaka, M., Goto, H.: Text-tracking wearable camera system for visually-impaired people. In: 19th International Conference on Pattern Recognition (2008)
5. Lienhart, R., Wernicke, A.: Localizing and segmenting text in images and videos. IEEE Transactions on Circuits and Systems for Video Technology 12, 256–268 (2002)
6. Goto, H.: Redefining the DCT based feature for scene text detection. International Journal on Document Analysis and Recognition 11, 1–8 (2008)
7. Bouguet, J.Y.: Pyramidal implementation of the Lucas Kanade feature tracker. Technical report, OpenCV Document, Intel Microprocessor Research Labs (2000)
8. Wang, B., Goto, H.: Scene text detection and tracking for wearable camera system. IEICE Technical report, PRMU2010-156, pp. 47–52 (2011)

Object Tracking across Non-overlapping Cameras Using Adaptive Models

Xiaotang Chen, Kaiqi Huang, and Tieniu Tan

National Laboratory of Pattern Recognition,
Institute of Automation, Chinese Academy of Sciences
{xtchen,kqhuang,tnt}@nlpr.ia.ac.cn

Abstract. In this paper, we propose a novel approach to track multiple objects across non-overlapping cameras, which aims at giving each object a unique label during its appearance in the whole multi-camera system. We formulate the problem of the multiclass object recognition as a binary classification problem based on an AdaBoost classifier. As the illumination variance, viewpoint changes, and camera characteristic changes vary with camera pairs, appearance changes of objects across different camera pairs generally follow different patterns. Based on this fact, we use a categorical variable indicating the entry/exit cameras as a feature to deal with different patterns of appearance changes across cameras. For each labeled object, an adaptive model describing the intraclass similarity is computed and integrated into a sequence based matching framework, depending on which the final matching decisions are made. Multiple experiments are performed on different datasets. Experimental results demonstrate the effectiveness of the proposed method.

1 Introduction

In recent years, multi-camera visual surveillance systems have attracted much attention and been widely used in applications such as continuously tracking interested objects, early warning of abnormal events, etc. Using multiple cameras, the area of surveillance is expanded and the occlusion problem which is very challenging for single cameras can be solved to a certain degree. To continuously track objects in wide areas, the key problem is to establish correspondences of different object sequences observed under multiple cameras with either overlapping or non-overlapping views. To solve this problem, the overlapping multi-camera tracking systems generally require calibration of cameras and make use of geometric constrains (e.g. homography constrains) [1, 2]. While tracking objects across non-overlapping views is more challenging, for the observations under different cameras are widely separated in both time and space because of blind areas between cameras. In this paper, we focus on object tracking across multiple non-overlapping cameras.

Because few of spatio-temporal cues can be used in tracking multiple objects across non-overlapping cameras, the tracking decisions depend heavily on object matching. However, the appearance changes greatly across cameras due to

J.-I. Park and J. Kim (Eds.): ACCV 2012 Workshops, Part II, LNCS 7729, pp. 464–477, 2013.

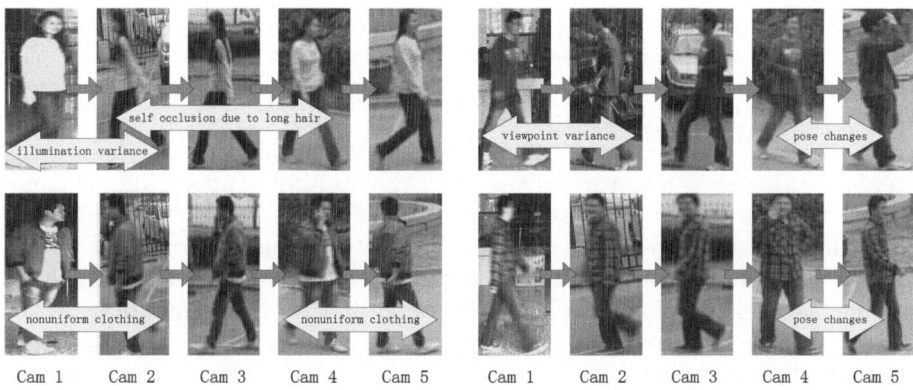

Fig. 1. Observations under multiple cameras with non-overlapping views. Each person walks along the same path from Cam 1 to Cam 5.

many factors, such as illumination variance, viewpoint changes, pose changes, nonuniform clothing, self-occlusions, different camera characteristics, as shown in Figure 1. The appearance changes of different persons across the same camera pair (e.g. from Cam 1 to Cam 2, or from Cam 2 to Cam 3) generally have the same pattern. For example, the appearances suffer from illumination variance and viewpoint changes between Cam 1 and Cam 2, while they almost remain the same between Cam 2 and Cam 3. In other words, the pattern of appearance changes varies according to the entry/exit cameras (the camera the object exits from and the camera it enters into). Based on this fact, various methods have been proposed over the last few years to deal with appearance changes across cameras. Methods [3–5] present brightness transfer functions or color transfer approaches to transform color between each pair of entry/exit cameras. In [6], camera transfer functions are applied to transform not only color but also other features. In this paper, we also deal with the appearance changes differently according to entry/exit cameras. Unlike previous work, we take entry/exit cameras as a categorical variable (or a feature) in the AdaBoost algorithm rather than transform the appearance from one camera to another.

Many methods [7–11] take the problem of object matching across cameras as object re-identification. Method [7] extracts color and texture histogram features to represent the objects, and uses machine learning tools to learn the similarity between any two objects. D'Angelo, A. *et al.* [8] propose a probabilistic color histogram based on a fuzzy K-Nearest Neighbors classifier. Xiaotang, C. *et al.* [9] present a direction-based stochastic matching method using directional cues of objects. These object re-identification methods always return a best match in a dataset for a given object, assuming the given object exists in the dataset. However, for a real multi-camera object tracking system, the number of moving objects is uncertain and an object with a new identity may appear at any time, in which case the identity of this new object can never be defined as anyone in the dataset, thus, defining each newcomer as the best match found by object re-identification methods can not always work in the real multi-camera object tracking systems. Obviously, directly setting a fixed threshold of the output similarities for all the objects is not proper. To solve this problem, the proposed method learns an adaptive model for each object to help drawing the final matching decisions.

Overall, the contributions in this paper lie in two aspects: (1) taking the entry/exit cameras as a feature to deal with the problem of appearance changes varying with entry/exit cameras; (2) using an adaptive model for each object in tracking objects across cameras instead of setting common rules for all the objects. Note that AdaBoost is used as a black-box machine learning tool. Other machine learning tools can also be used coherently with our framework.

Figure 2 shows the flowchart of the proposed system. Firstly, a blob sequence is initialized for an object when it comes into view, then after the initialization is completed, the blob sequence is matched against other objects which have left the views in a sequence based matching framework in order to determine whether this newcomer has been previously tracked or not. Once its identity is specified, it is continuously tracked under the single camera until it leaves. Finally, the dataset is updated by this labeled object.

This paper is organized as follows. In Section 2, the strategy of single camera object tracking used in our multi-camera system is presented. Section 3 describes a novel approach of tracking objects across non-overlapping cameras. Experimental results and conclusions are given in Section 4 and Section 5 respectively.

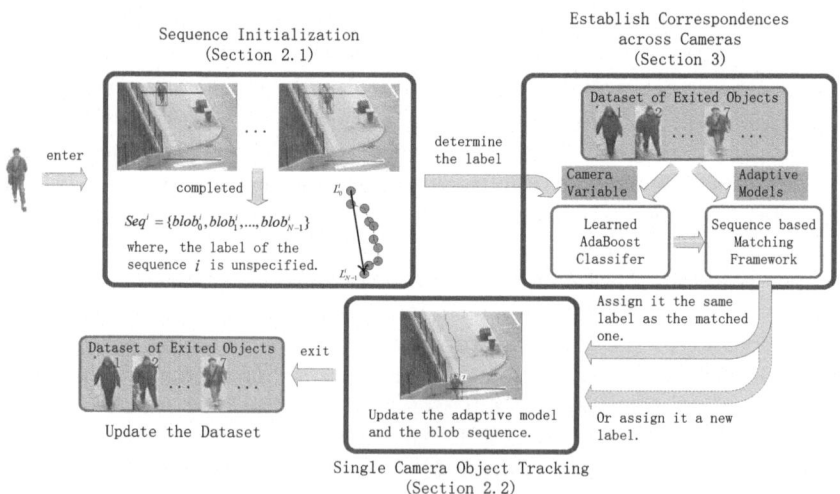

Fig. 2. The flowchart of our method

2   Single Camera Object Tracking

The task of single camera object tracking is to establish correspondences of blobs in successive frames under single cameras.

2.1   Sequence Initialization

In this paper, we use entry/exit lines to mark the locations of entry/exit zones. Each object entering the field of view of a camera must cross an entry/exit line, which initializes a sequence of blobs, as shown in Figure 2. If the diagonal line (from upper left to lower right) of the bounding box of a blob intersects an entry/exit line, then we believe this blob, denoted by $blob_n$, is detected in the entry/exit zones. The correspondences between $blob_n$ and uninitialized blob sequences $\{unSeq^i | unSeq^i = \{blob_0^i, blob_1^i, ...blob_{n-1}^i\}, n < N\}$ are established by finding the nearest location:

$$i^* = \arg\min_i \left\{ Dis(L_n, L_{n-1}^i) | Dis(L_n, L_{n-1}^i) < D_T \right\} \tag{1}$$

where i is the unspecified label of the blob sequence and D_T is a given threshold. L_n and L_{n-1}^i denote the locations of $blob_n$ and the latest blob in $unSeq^i$. When a match is found, $blob_n$ is used to extend $unSeq^{i^*}$. Otherwise, a new $unSeq^i$ is established. As long as the sequence length is up to N, the initialization of this sequence is completed. To be clear, $unSeq^i$ refers to the blob sequence which has not finished initialization yet. Once its initialization is completed, a label (or identity) is determined and assigned to it by establishing correspondences across cameras (described in Section 3). Then $unSeq^i$ becomes Seq^i, which is an initialized blob sequence with a specified label.

2.2   Establishing Correspondences under Single Cameras

For a blob detected in non-entry/exit zones, denoted by $blob_N$, we match it against initialized blob sequences $\{Seq^i\}$, where the label i has been specified, under the same camera using only directional cues. In Figure 3, each blob sequence has a direction vector, from the earliest location L_0 to the latest location L_{N-1}, roughly giving the moving direction of the corresponding object. Under a reasonable assumption that the moving direction of the same object does not change greatly in successive frames, $blob_N$ is assigned to Seq^{i^*} as:

$$i^* = \arg\min_i \left\{ \theta_i | \theta_i = \angle \left(\overrightarrow{L_0^i L_{N-1}^i}, \overrightarrow{L_0^i L_N} \right), \theta_i < \theta_T \right\} \tag{2}$$

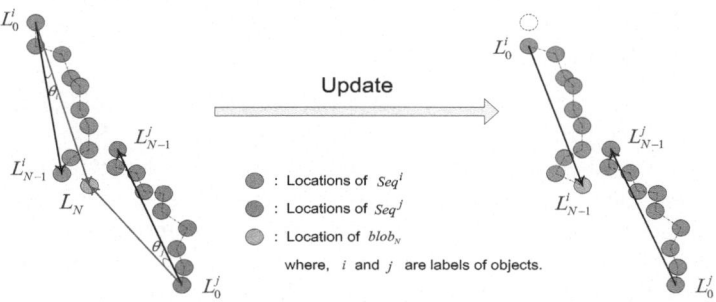

Fig. 3. Establishing correspondences under single cameras

where L_N is the location of $blob_N$. θ_T is a given threshold. Then, $blob_N$ is used to update Seq^{i^*} and its direction vector, as shown in Figure 3. The blobs detected in entry/exit zones are dealt with in the same way as the blobs in non-entry/exit zones first. Then it is matched against $\{unSeq^i\}$ as mentioned in Section 2.1, if the first round of match fails.

Establishing correspondences under single cameras is done depending only on the directional cues, not the appearance cues, making it efficient and fast when applied to tracking multiple objects in a large-scale multi-camera system.

3   Establishing Correspondences across Cameras

Once a blob sequence completes initialization, we match it against labeled objects having left the connected entry/exit zones to define the identification (or label) of this blob sequence.

3.1   Learning the Adaptive Model

For each blob, we only deal with the area occupied by the object. Several appearance-based features are explored to represent the blob. They are described in the following:

MCSH. In RGB color space, a major color spectrum histogram (MCSH) is computed for the blob, and the similarity is measured using [12]. The number of major colors is 50. The threshold of color distance is set to 0.06.

Histogram. The color histogram (H) of the blob is used, and normalized by the size of the object in pixels. Distance is measured using histogram intersection.

Major Colors. To incorporate color spatial information into the representation, the blob is partitioned into three parts using the method in [13], corresponding to the head, the upper body, and the lower body. A major color is clustered using K-means for the upper body (MU) and lower body (ML) respectively. Euclidean Distance is used to measure the distance between two colors.

Spatiogram. Spatiogram (S) [14] is also applied for the purpose of describing the spatial distribution of color. The similarity between two spatiograms is computed as the weighted sum of the similarity between two histograms.

LBP. LBP [15] is used as a texture feature, and comparison is done by the Bhattacharyya distance.

As mentioned above, the similarity vector between $blob^i$ and $blob^j$ is measured as $\{Sim^{ij}_{MCSH},\ Sim^{ij}_H, Sim^{ij}_{MU}, Sim^{ij}_{ML}, Sim^{ij}_S, Sim^{ij}_{LBP}\}$, or Sim^{ij} for short, where i and j are labels of the corresponding objects. To deal with different patterns of appearance changes across cameras, we use a categorical variable $E^i E^j$ in addition, indicating the camera E^i whose view $blob^i$ enters into and the camera E^j which $blob^j$ exits from. In the training process, the vector $[Sim^{ij}, E^i E^j]$ computed from two blobs corresponding to the same object (namely, $i = j$) is taken as a positive sample, while the negative sample set corresponds to different objects. Thus, the multiclass object recognition problem is formulated as a binary classification problem. The AdaBoost classifier is used to solve this problem with decision trees as weak classifiers. The output of AdaBoost is a weighted sum over the learned weak classifiers. The larger the value the more similar the two blobs. Different with other AdaBoost algorithms, we use learned adaptive thresholds for different objects to deal with the outputs instead of using a sign function.

An adaptive model for each labeled object is learned by calculating and updating the adaptive threshold. It is calculated based on the similarities of two blobs in the initialized blob sequence and updated over time, thus it indicates the intraclass similarity of the corresponding object to a certain degree. Given the initialized blob sequence $\{blob_0^i, blob_1^i, \cdots, blob_{N-1}^i\}$ of $Object^i$ (namely Seq^i), the adaptive threshold is computed as:

$$Thre^i = \frac{1}{N-1} \sum_{n=0}^{N-2} f\left([Sim(blob_n^i, blob_{n+1}^i), E^i E^i]\right) \tag{3}$$

where $f(*)$ is the learned AdaBoost classifier.

When $Object^i$ is continuously tracked under Camera E^i, we update Seq^i, as described in Sec.2.2, as well as $Thre^i$:

$$Thre^i \leftarrow \frac{f\left([Sim(blob_{N-1}^i, blob_N), E^i E^i]\right) + Thre^i}{2} \tag{4}$$

where $blob_N$ is the blob which updates Seq^i.

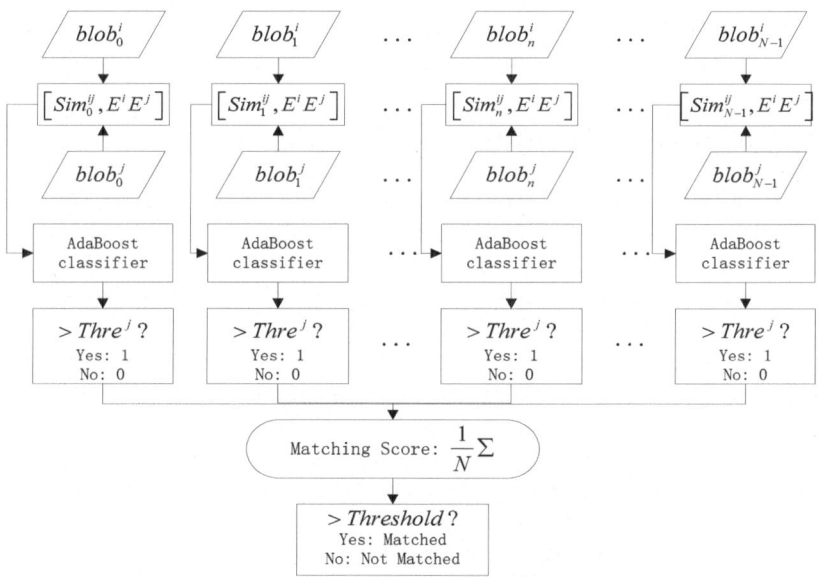

Fig. 4. Sequence based matching framework

3.2 Object Matching across Cameras

Following the work in [16], the weighted cross correlated model is used to estimate the topology of a multi-camera network. And MCSH [12] is applied to this
model, by which connectivity and average transition time between two entry/exit
zones are learned.

Given Seq^i with an unspecified label i, which completes initialization at time
t, we match it against labeled objects (i.e. $Object^j$) leaving the connected en-
try/exit zones during the time window $[t - T_{Z^i Z^j} - T, t - T_{Z^i Z^j} + T]$, where
Z^i and Z^j denotes the entry/exit zone where Seq^i detected and the entry/exit
zone which $Object^j$ exits from respectively. T allows a small fluctuation around
the average transition time $T_{Z^i Z^j}$.

Figure 4 shows the sequence based matching framework. Firstly, a feature
vector is extracted from each blob. Secondly, a similarity vector is calculated
and the camera variable $E^i E^j$ is obtained as well. Finally, a decision is made
according to the outputs of AdaBoost, where $Thre^j$ is the adaptive threshold
of $Object^j$. If no one successfully matches Seq^i, then $Object^i$ is identified to be
a new object and a new label is assigned to it. Otherwise, it retains the same
label as the one with a maximal matching score. This framework is based on
the assumption that if $blob_n^i$ and $blob_n^j$ belong to the same object, the similarity
between them agrees with the similarity between two blobs of $Object^i$ or $Object^j$,
as the appearance changes across cameras have been considered by using $E^i E^j$.
$Thre^j$ has been updated over time along with Seq^j, so it is relatively stable and
represents the similarity between every two blobs of $Object^j$. $Thre^i$ of Seq^i can
also be calculated using Eq.3, however, it is relatively unstable as it has not been
updated over time and could be greatly influenced by noise (false detections).
Thus, we choose $Thre^j$ for comparison rather than $Thre^i$.

4 Experimental Results and Analysis

The first experiment is conducted to show the performance of the proposed object matching method using the AdaBoost classifier based on only color and
texture features, in which samples are not representations of single blobs, but
similarity measures between them. To compare with other methods, the experiment is based on the VIPeR dataset [17].

The dataset contains 1264 images of 632 pedestrians and each pedestrian has
two images seen from different views. Since no camera information is given by
this dataset, we train the AdaBoost classifier without using the categorical variable $E^i E^j$. As done in [7], we randomly split the dataset into two halves: the
training set and the testing set. We show the average of the results on multiple random train-test sets and compare it to 5 different benchmark methods[1],
Template (sum-of-squared distances matching), Histogram, Hand Localized Histogram [18], Principal Axis Histogram [2], and ELF 200 [7]. The results are
presented using cumulative matching characteristic (CMC) curves, as shown in

---

[1] Results of other methods are from the work [7].

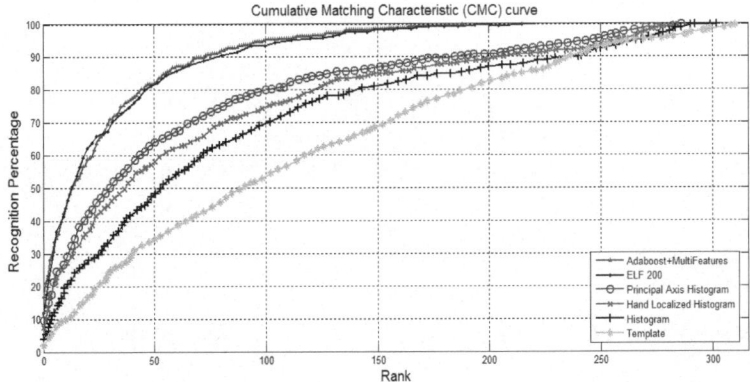

Fig. 5. CMC curves of different methods

Figure 5. It indicates that the performance of our method matches or exceeds other methods, and the rank 1 matching rate is about 16.77%. Figure 6 shows some examples of the matching results using our method. However, this experiment can not fully reveal the performance of our method, for the camera variable $E^i E^j$ is not used which provides important information and can greatly improve the performance of object recognition across cameras.

To the best of our knowledge, there is no public dataset providing images of pedestrians observed under multiple cameras (no less than three cameras). Thus, to demonstrate the effectiveness of the camera variable $E^i E^j$ for automated recognition, we conduct the second experiment based on a dataset[2] which consists of manually labeled pedestrians from off-line videos. This dataset contains 585 images of 39 pedestrians seen from three cameras with non-overlapping views, among which 25 pedestrians walk across cameras. Each pedestrian has 9 images under a single view. Some examples of the dataset is shown in Figure 9. As the samples are similarity measures between any two images from the same or different cameras, we collect 2358 positive samples and 13203 negative samples for the training set, and 1251 positive samples and 10530 negative samples for the testing set. The results are presented using receiver operator characteristic (ROC) curves in Figure 7. Our method using the camera variable $E^i E^j$ outperforms the one that does not using it, demonstrating the effectiveness of the camera variable.

To demonstrate the performance of our method on object tracking across cameras, we conduct the third experiment based on a non-overlapping multi-camera system both indoors and outdoors. Figure 8 (a) shows the layout of the system. The learned connections and average transition time between entry/exit zones across cameras are shown in Figure 8 (b).

The dataset used to train the AdaBoost classifier is the same dataset in the second experiment. The proposed object tracking algorithm is tested on videos[3]

---

[2] This dataset is available on the website: http://www.datatang.com/Member/76804

[3] The videos are available on the website: http://www.datatang.com/Member/76804

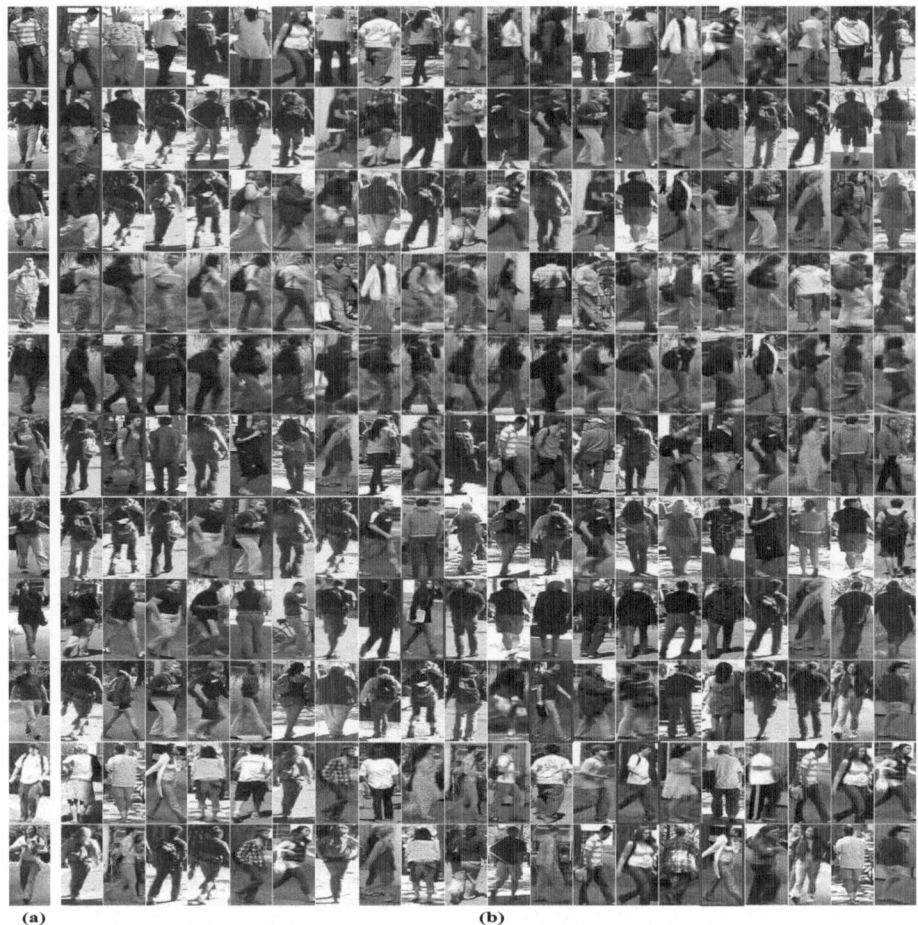

(a)   (b)

Fig. 6. Examples of the matching results. (a) Reference image; (b) Top 20 results (sorted left to right). The correct matches are circled by red lines.

containing 39 pedestrians and 10 of them walk across cameras. In this case, we only consider pedestrians rather than cyclists or vehicles. D_T and θ_T is set to 10 and 1.0 (radian) respectively. The sequence length N is set to 15, and the *Threshold* in the sequence based matching framework is set to 0.5. Some examples of tracking results are shown in Figure 10. The tracking accuracy is about 94.9% ($\frac{37}{39}$). All the pedestrians who transfer across cameras are correctly recognized and retain unique labels in the whole videos. In the view of Camera 3, two pedestrians exchange their labels while one is leaving and the other is

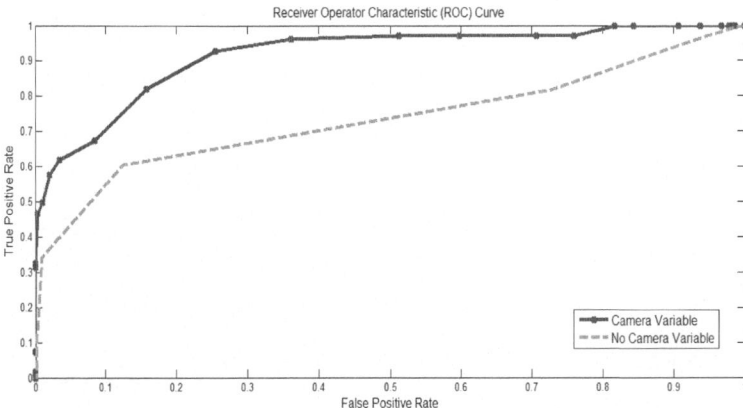

Fig. 7. The receiver operator characteristic curve showing the effectiveness of the categorical variable $E^i E^j$

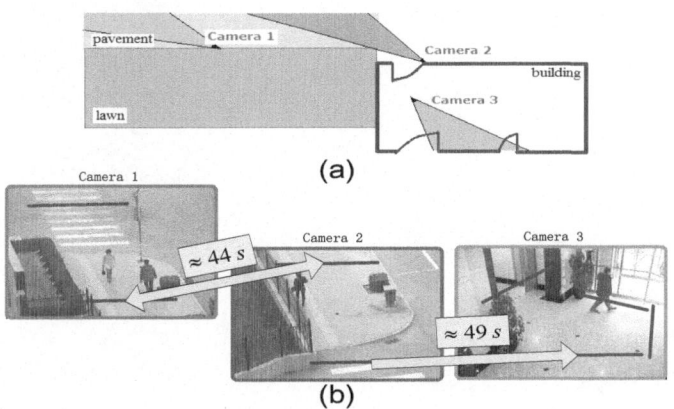

Fig. 8. (a) The layout of the multi-camera system, (b) Learned spatio-temporal cues across cameras

Fig. 9. Some examples used for training. Each column is the same pedestrian in different entry/exit cameras (Column 1-4: from Cam 2 to Cam 3; Column 5-8: from Cam 1 to Cam 2; Column 9-12: from Cam 2 to Cam 1).

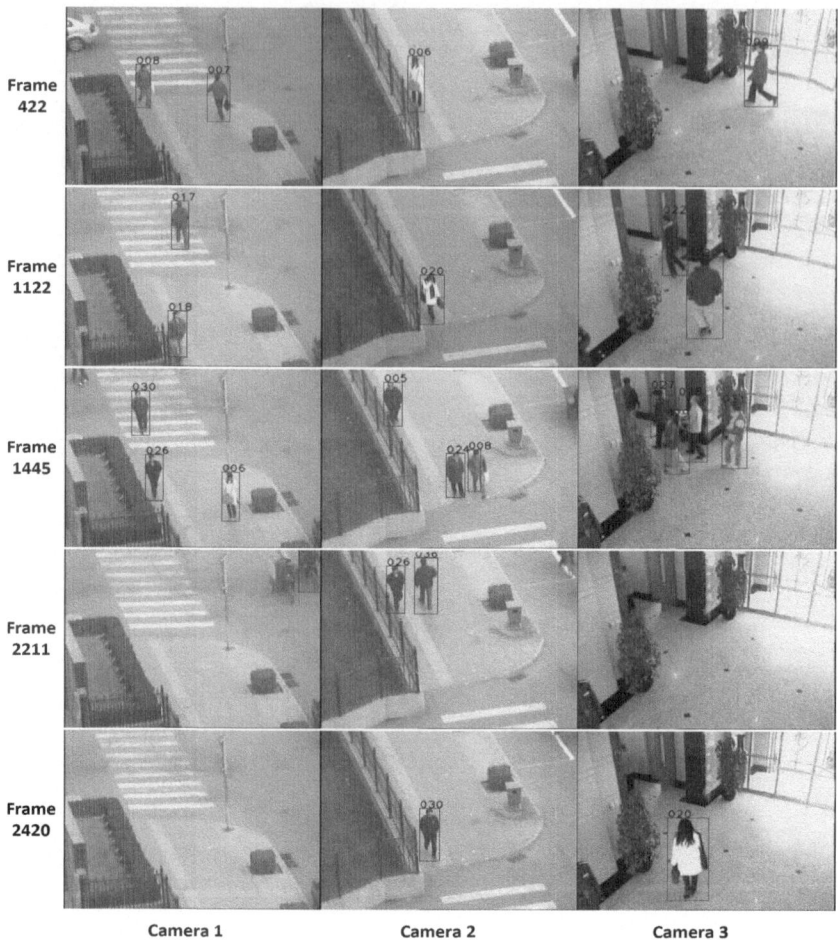

Fig. 10. Examples of tracking results. Note that all pedestrians retain unique labels.

entering almost at the same time and in the same place. This failure can be avoided by combining appearance cues with directional cues in the single camera tracking.

Table 1 shows some outputs of the AdaBoost classifier in the sequence based matching framework, demonstrating the effectiveness of adaptive models. Using the proposed adaptive model, Person 26 (enter) and Person 30 (exit) are impossible to be recognized as the same person due to a very small matching score, while Person 8 (enter) and Person 8 (exit) are successfully matched, although the similarity scores between Person 26 (enter) and Person 30 (exit) are generally larger than that between Person 8 (enter) and Person 8 (exit).

Table 1. Outputs of the AdaBoost classifier in the sequence based matching framework

index	Person 8 (enter) frame 1330-1344 vs. Person 8 (exit) frame 557-571	Person 26 (enter) frame 2187-2201 vs. Person 26 (exit) frame 1511-1525	Person 26 (enter) frame 2187-2201 vs. Person 30 (exit) frame 1621-1635
	$Thre^8 = -15.1423$	$Thre^{26} = -8.0389$	$Thre^{30} = 0.3742$
1	-15.6932	**-7.7642**	-7.2057
2	**-9.3708**	**-1.5252**	-7.2057
3	**-9.0034**	**-1.5252**	-5.4428
4	**-9.0034**	**1.4108**	-7.2057
5	**-9.0034**	**2.0720**	-7.2057
6	**4.3071**	**-5.7348**	-7.2057
7	**4.3071**	**-1.3836**	-3.5291
8	**-8.2373**	**-5.7348**	-7.2057
9	**-8.2373**	**-1.3836**	-8.2373
10	**-8.2373**	**-1.5252**	-9.2772
11	**-8.2373**	**-1.3836**	-8.2373
12	**-8.2373**	**-1.5252**	**4.3071**
13	**-8.2373**	**2.0720**	-7.2057
14	**-8.2373**	**2.0720**	**2.2356**
15	**-8.2373**	**0.8478**	-9.2772
Matching Score	0.93	1.00	0.13

5   Conclusions

In this paper, we have presented a solution to the problem of object tracking across non-overlapping cameras. Unlike previous work, our method deals with the appearance changes differently according to different entry/exit cameras by using a camera variable. Adaptive models are learned and integrated into the sequence based matching framework to draw final conclusions. Experiments have demonstrated that using the camera variable improves the performance, adaptive models are necessary and effective, and the proposed method performs well in tracking multiple objects across non-overlapping cameras. To extend our work, appearance cues can be exploited and integrated into the single camera object tracking, and more distinctive features can be applied to represent the objects when matching across cameras. Future work will focus on object tracking in large-scale multi-camera systems.

Acknowledgement. This work is funded by National Natural Science Foundation of China (Grant No. 61175007), the National Key Technology R&D Program (Grant No. 2012BAH07B01), the National Basic Research Program of China (Grant No. 2012CB316302).

References

1. Khan, S., Shah, M.: Consistent labeling of tracked objects in multiple cameras with overlapping fields of view. IEEE Transactions on Pattern Analysis and Machine Intelligence 25, 1355–1360 (2003)
2. Hu, W., Hu, M., Zhou, X., Tan, T., Lou, J., Maybank, S.: Principal axis-based correspondence between multiple cameras for people tracking. IEEE Transactions on Pattern Analysis and Machine Intelligence 28, 663–671 (2006)
3. Javed, O., Shafique, K., Shah, M.: Appearance modeling for tracking in multiple non-overlapping cameras. In: IEEE Computer Society Conference on Computer Vision and Pattern Recognition, pp. 26–33 (2005)
4. Prosser, B., Gong, S., Xiang, T.: Multi-camera matching using bi-directional cumulative brightness transfer functions. In: Proceedings of the British Machine Vision Conference (2008)
5. Jeong, K., Jaynes, C.: Object matching in disjoint cameras using a color transfer approach. Machine Vision and Applications 19, 443–455 (2008)
6. Montcalm, T., Boufama, B.: Object inter-camera tracking with non-overlapping views: a new dynamic approach. In: Canadian Conference Computer and Robot Vision, pp. 355–361 (2010)
7. Gray, D., Tao, H.: Viewpoint Invariant Pedestrian Recognition with an Ensemble of Localized Features. In: Forsyth, D., Torr, P., Zisserman, A. (eds.) ECCV 2008, Part I. LNCS, vol. 5302, pp. 262–275. Springer, Heidelberg (2008)
8. D'Angelo, A., Dugelay, J.: People re-identification in camera networks based on probabilistic color histograms. In: Visual Information Processing and Communication, SPIE Electronic Imaging, vol. 7882 (2011)
9. Chen, X., Huang, K., Tan, T.: Direction-based stochastic matching for pedestrian recognition in non-overlapping cameras. In: 18th IEEE International Conference on Image Processing, pp. 2065–2068 (2011)
10. Cai, Y., Huang, K., Tan, T.: Human Appearance Matching Across Multiple Non-overlapping Cameras. In: 19th International Conference on Pattern Recognition, pp. 1–4 (2008)
11. Cai, Y., Huang, K., Tan, T.: Matching Tracking Sequences Across Widely Separated Cameras. In: IEEE International Conference on Image Processing, pp. 765–768 (2008)
12. Piccardi, M., Cheng, E.D.: Multi-frame moving object track matching based on an incremental major color spectrum histogram matching algorithm. In: IEEE Computer Society Conference on Computer Vision and Pattern Recognition - Workshops, p. 19 (2005)
13. Hu, M., Hu, W., Tan, T.: Tracking people through occlusions. In: International Conference on Pattern Recognition, pp. 724–727 (2004)
14. Birchfield, S., Rangarajan, S.: Spatiograms versus histograms for region-based tracking. In: IEEE Computer Society Conference on Computer Vision and Pattern Recognition, pp. 1158–1163 (2005)

15. Ojala, T., Pietikäinen, M., Harwood, D.: A comparative study of texture measures with classification based on featured distributions. Pattern Recognition 29, 51–59 (1996)
16. Niu, C., Grimson, E.: Recovering non-overlapping network topology using far-field vehicle tracking data. In: International Conference on Pattern Recognition, pp. 944–949 (2006)
17. VIPeR dataset, http://vision.soe.ucsc.edu/?q=node/178
18. Park, U., Jain, A.K., Kitahara, I., Kogure, K., Hagita, N.: ViSE: visual search engine using multiple networked cameras. In: International Conference on Pattern Recognition, pp. 1204–1207 (2006)

Combining Fast Extracted Edge Descriptors and Feature Sharing for Rapid Object Detection

Yali Li, Fei He, Wenhao Lu, and Shengjin Wang

State Key Laboratory of Intelligent Technology and Systems
Department of Electronic Engineering, Tsinghua University, Beijing, P.R. China
wgsgj@tsinghua.edu.cn, liyali@ocrserv.ee.tsinghua.edu.cn

Abstract. We mainly focus on feature sharing problem for object detection in cluttered scenes. The contributions are two-fold. First, a novel kind of edge/contour descriptors is presented and they serve as the basic features for sharing. Compared with HOGs (histograms of oriented gradients), the descriptors show the approximately equivalent efficiency while much less computational lost. Second, to exploit feature sharing techniques for object detection, a mathematical representation of shared features for "sliding-window" based object detection methods is given. Also with the newly defined shared features, a learning framework based on Real-Adaboost algorithm and a reusing framework based on look-up table are proposed. Experimental results show both the efficiency of proposed features and feature sharing method.

1 Introduction

Object detection is important for a wide range of applications, such as video surveillance and content-based image and video retrieval. Much progress has been achieved in object detection and some techniques are mature enough for practical applications. However, the problem is still challenging. It is mainly because that even for objects of the same class, their appearances can vary a lot when illuminations and viewpoints changes. Also reducing the computational cost is also an important issue.

Many efforts have been devoted to the research of object detection methods and much progress has been achieved. Among them, various kinds of features are proposed for training classifiers. Haar-like features [1] which can be computed fast through integral images have been proven efficient in face detection. Lienhart et al. [2] introduce rotated Haar-like features and propose rotated summed area table (RSAT) for fast extraction of rotated features. Besides, HOG descriptors [3] which compute edge intensity in different orientations are the most common-used features in object detection. Edgelets [4] which extract the contour of objects are typical features for object detection with AdaBoost classifiers. In sum, Haar-like features focus on gray-scale information so they do not perform well for car and human detection, while the computational costs of HOGs and Edgelets are high.

Generally we can divide current object detection methods into two categories as part-model based approaches [5] [6] [7] [8] and sliding-window based approaches [1] [3] [9] [10] [11]. Part-model based approaches implicitly or explicitly

J.-I. Park and J. Kim (Eds.): ACCV 2012 Workshops, Part II, LNCS 7729, pp. 478–490, 2013.
© Springer-Verlag Berlin Heidelberg 2013

divide the object into parts and build up models for these parts. Object positions can be predicted through part locations with the mutual relationship between the locations of objects and parts. Typical methods like pLSA (probabilistic Latent Semantic Analysis) [5], Implicit Shape Model (ISM) [6], Hough forests [7] achieve good results. The object detection system presented by Felzenszwalb et al. [8] combines coarse-scale object models and fine-scale part models has shown excellent performance on Pascal Challenge Test [17] [18]. Different from these, sliding-window based approaches combine features with trained discriminative classifiers for object detection. Exhaustive search is a characteristic of these approaches since all candidate sub-windows in test images are scanned and discriminated whether they are object regions or not. To reduce the computational cost, several techniques are applied for features sharing, such as tree-structured classifiers [9], feature-centric evaluation [10]. Torralba et al. [11] propose a systemized feature sharing scheme for multi-class multi-view object detection which can make the runtime increases approximately logarithmically with the number of classes.

In this paper, we concentrate on the feature sharing problems in object detection. To achieve this, we firstly develop fast computed descriptors to extract the edges/ contours of objects. The descriptors are used to train the discriminative classifiers for detection. Furthermore, we find that when training parallel discriminative classifiers for multiple task object detection(i.e. multi-view object detection, multiple objects detection), the common characteristics among different object classes or different views of one object are often ignored. To utilize these common characteristics, we present a novel definition of shared features and propose the features sharing methods to reduce the needed feature number for detection. Experimental results validate the efficiency of proposed features and the performance of sharing feature method.

The remainder part of this paper is organized as follows. Section 2 introduces the proposed edge/contour descriptors. The shared features learning framework is given in Section 3. Section 4 presents the experimental results on existing database and the conclusion is given in Section 5.

2   Local Edge/Contour Descriptors

Based on the observation that edges/contours are one of the most salient patterns for object detection, we present a novel kind of features which can precisely represent the intensities and orientations of edges/contours here. Firstly, four edge operators are used to acquire the edge maps in four directions as horizontal, vertical, left tilted and right tilted (see Fig. 1(a)). Since the descriptors are defined by the rectangles in edge maps, we name them as Integral Oriented Gradient Intensity features, abbreviated as IOGI descriptors. The features are categorized into two types and the details are introduced as follows.

The IOGI features of type-I are defined by adjacent rectangles in single edge maps. As illustrated in Fig. 1(b), two restrictions on the distribution of rectangles are imposed. First, the directions of sharing sides for adjacent rectangles

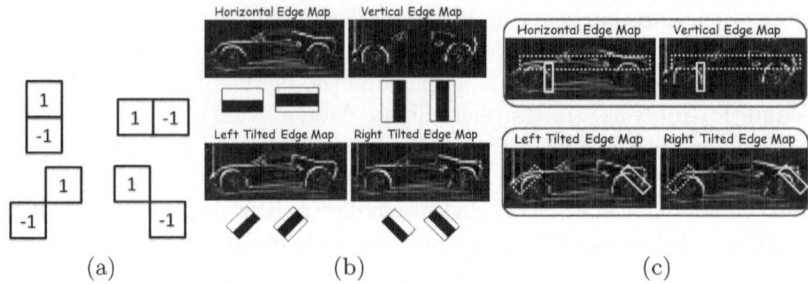

Fig. 1. The definition for Edge/Contour features. (a)Edge operators. (b)Type-I IOGI features. (c) Type-II IOGI features.

should be consistent with the directions of edge maps. For example, in horizontal edge maps, the adjacent sides of white and black rectangles should be horizontal. Second, the number of white and black rectangles might be different but the pixels number inside the white and black rectangles are the same. Several examples of rectangles for different edge maps are shown in Fig. 1(b). It is like restricted prototypes of Haar-like features in edge maps however the calculation is different. Suppose the sum of the pixel values inside the black and white rectangles are II_1 and II_2, respectively, the final feature value for IOGI descriptors of type-I is as eq.(1):

$$(II_1 - II_2) / (II_1 + II_2).$$ (1)

The feature value is set to zero when both II_1 and II_2 are zero. The feature values of IOGI features of type-I range from -1 to 1 and they can describe the positions and intensity of edges. That is, if the white rectangle lies in strong-edge region and the black rectangle located in the smooth region, the values of IOGI features are prone to achieve 1. Reversely the feature value tends to be -1.

Different from type-I IOGI descriptors, type-II features are defined by two orthogonal edge maps. As shown in Fig. 1(c), there are four different combinations of rectangles. Suppose the sums of the pixel values inside the rectangles on two edge maps are II_1 and II_2, respectively, the calculation of type-II IOGI descriptors is the same as type-I. The IOGI features of type-II can represent main directions of edges. That is, if inside the rectangles the horizontal edges are much stronger than vertical edges, the absolute value of IOGI features is prone to 1. Analogous to Haar-like features [2], both type-I and type-II IOGI features can be fast extracted by integral images or tilted integral images.

3   Features Sharing

3.1   The Definition of Shared Features

Here we take the toy problem to illustrate the importance of feature sharing. If we want to detect circles, rectangles and ellipses together with trained classifiers, a straight-forward way is to independently train three classifiers with several circle, rectangle and ellipse samples. As indicated in Fig. 2(a), there might be some similar "parts" between circles, rectangles and ellipses (red and blue rectangles). We can represent these similar "parts" with shared features and apply feature sharing techniques to improve the performance of detectors. To mathematically define the shared features, we denote the common features as $(\mathbf{p}, \psi)$, where $\mathbf{p}$ denotes the location of features and ψ denotes the operation of feature extraction. For example, for IOGI descriptors, ψ represents the calculation of the sum of pixel values inside the rectangles and get the final features value via eq.(1).

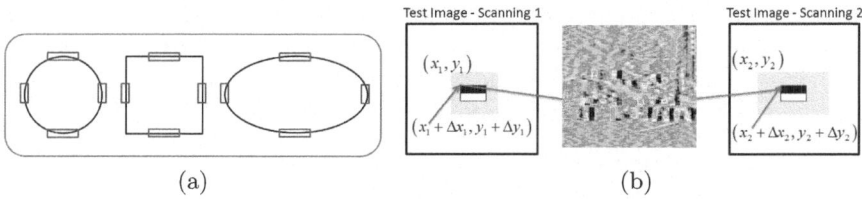

(a)   (b)

Fig. 2. The definition of shared features. (a)A toy problem. (b)Sharing extracted feature values when exhaustively testing all candidate sub-windows.

When detecting multiple objects in images with "sliding window" based methods, all the possible locations of objects are scanned and tested by several combined classifiers. That is to say, on each fixed scale several times of scanning should be done and several features need to be extracted. As illustrated in Fig. 2(b), if the feature values at all candidate locations are extracted and stored in the feature-value maps before classifier testing, for different classifiers in the parallel structure the feature extraction step can be replaced by getting "pixel values" from the feature-value maps. Then if K classifiers use the features which can be denoted as $(\mathbf{p}, \psi)$ and furthermore have the same ψ, the features extraction step only needs to be done once. Different from the definition of shared features in [11], the features for sharing defined here only are required to having the same ψ while not $\mathbf{p}, \psi$. Therefore our feature sharing methods also do not require that all the train samples are normalized into the same resolution. The shared features definition here removes the limitation in locations of features as well as the same normalized sample size.

3.2 The Criterion to Find Weak Hypotheses of Shared Features Learning

The shared features learning scheme here is based on Real-Adaboost(RAB) algorithm which fits an additive logistic regression model by stage-wise and approximation optimization of the loss function as eq.(2):

$$J(F) = E\left[\exp\left(-yF(\mathbf{x})\right)\right].$$
(2)

where $F(\mathbf{x})$ is the strong classifier of Adaboost and y refers to the label of samples. To solve multi-view multi-object detection problem with K parallel classifiers, for each sample a label $k \in \{\pm 1, \pm 2, \cdots, \pm K\}$ is assigned. For positive samples, $k > 0$. Reversely $k < 0$ for negative samples. The loss function can be extended into eq.(3):

$$J(F) = \sum_k I(y = k)E\left[\exp\left(-sgn(k)F(\mathbf{x})\right)\right]. \tag{3}$$

where $I(y = k)$ is the indication function and it is 1 only if the label of the sample is equal to k. It can be proven that $J(F)$ is the upper bound of classification error ratio with one scale factor. Traditional RAB greedily minimizes the upper bound of training error as eq.(2) by minimizing Z_t on each round. Analogously, given the loss function as eq.(3), the weak learned parallel classifiers should try to find the weak hypothesis h_t to minimize Z_t as eq.(4):

$$Z_t = \sum_k \sum_i I(y_i = k)D(i)exp\left[-sgn(k)h_t(\mathbf{x})\right]. \tag{4}$$

where $sgn(k)$ indicates the sign of k. Using domain-partitioning hypothesis, we let

$$W_k^j = \sum_{i:x_i \in X_j \cap y_i = k} D(i). \tag{5}$$

which is the weighted fraction of examples fall in block j with label k. Then we can get the weak hypotheses h_t and minimum optimization Z_t at each round of boosting as:

$$h_t^j = 0.5\ln\left(\sum_{k>0} W_k^j / \sum_{k<0} W_k^j\right), Z_t = 2\sum_j \sqrt{\left(\sum_{k>0} W_k^j\right)\left(\sum_{k>0} W_{-k}^j\right)}. \tag{6}$$

Up to now we have got the mathematical criterion for shared feature learning among parallel classifiers. That is, for each round iteration in the boosting learning, we need to find Z_{min} and the weak classifier represented by h^j. That is the mapping relation between feature values and corresponding weak hypotheses. The remaining issue is to find the optimal features combinations.

3.3   Finding the Shared Features Combinations

In the above sub-section we have got the minimum optimization function Z_t for each iteration of boosting algorithm. To simplify the discussion, here we omit the subscript t and view Z as a function with variables of different features for corresponding classifiers. According to the definition of shared features in 4.1, we denote feature for kth class as $(\mathbf{p}_k, \psi_k)$ and set ψ_k to equal. That is,

$$Z\left(\mathbf{p}_1, \mathbf{p}_2, \cdots, \mathbf{p}_K, \psi\right). \tag{7}$$

If the normalized sample size for kth classifier is $W_k \times H_k$, the size of candidate shared features need be smaller than the samples size of all K classifiers. Suppose the feature size is $w \times h$, then size of shared features should satisfy eq.(8):

$$w \leq \min(W_1, \cdots, W_K), h \leq \min(H_1, \cdots, H_K). \tag{8}$$

Moreover, the range of $\mathbf{p}_k$ can be determined by the definition of ψ and the samples size $W_k \times H_k$. That is,

$$0 \leq x_k \leq W_K - w, 0 \leq y_k \leq H_K - h. \tag{9}$$

By eq.(8) we can get a discrete set of ψs. When ψ is fixed, the feasible variable range of $\mathbf{p}_k$ can be got by eq.(9). We need to traverse all possible ψs and find the best combinations of $\mathbf{p}_k$ for each ψ. The one with minimum Z as eq.(7) is chosen as the solution. We find that to get the best combination of $\mathbf{p}_k$ with fixed ψ is a combinatorial optimization problem. A straightforward way is to traverse all the possible solutions but the time complexity is exponential. Here we introduce a genetic programming based method to find the combinations of $\mathbf{p}_k$ here. It can get a sub-optimum solution for this problem which is enough for many applications. The method proceeds as follows.

(1) Initialization: The important sampling method is used to generate the initial population. With Cauchy inequality, we can get the result as shown in eq.(10):

$$Z = 2 \sum_j \sqrt{\left(\sum_{k>0} W_k^j\right)\left(\sum_{k>0} W_{-k}^j\right)} \geq 2 \sum_{k>0} \sum_j \sqrt{W_k^j W_{-k}^j} = \sum_{k>0} Z_k. \tag{10}$$

where $Z_k = 2 \sum_j \sqrt{W_k^j W_{-k}^j}$ is the minimum criterion for each classifier in the parallel structure. By adding the variables into eq.(7), we can get:

$$Z(\mathbf{p}_1, \mathbf{p}_2, \cdots, \mathbf{p}_K, \psi) \geq \sum_{k>0} Z(\mathbf{p}_k, \psi). \tag{11}$$

That is to say, those features which can better discriminate positive and negative samples in single classifier are also prone to show higher performance when used for sharing. Based on this, we apply important sampling techniques to generate the initial population. We let the $\mathbf{p}_k$ with smaller $Z(\mathbf{p}_k, \psi)$ have more possibility to be chosen as solution bit to form the individuals in the initial generation.

(2) Scoring and ranking: The optimal criterion as eq.(6) is reversed to serve as the fitness function. That is, if an individual gets lower value of Z, it has higher fitness score. For each generation, we rank all the individuals by their corresponding Z value in ascending order. The individuals with higher Z values are wiped out while those with lower Z are kept to generate the next generation.

(3) Producing the next generation: The children individuals of the next generation are produced by operations as retaining, crossover and mutation. For retaining, the individual in parent generation is replicated as children individual. For crossover operation, it has three steps: (i) randomly choose two different

individuals in parent generation; (ii) randomly choose several crossover element bits; (iii) exchange the elements in these bits to generate two children individuals. Eq.(12) illustrates an example of generating two examples using crossover operation.

$$
\left.\begin{array}{l}
(\mathbf{p}_1^{i_1}, \cdots, \mathbf{p}_m^{i_m}, \mathbf{p}_{m+1}^{i_{m+1}}, \cdots, \mathbf{p}_K^{i_K}) \\
(\mathbf{p}_1^{j_1}, \cdots, \mathbf{p}_m^{j_m}, \mathbf{p}_{m+1}^{j_{m+1}}, \cdots, \mathbf{p}_K^{j_K})
\end{array}\right\} \rightarrow \left\{\begin{array}{l}
(\mathbf{p}_1^{i_1}, \cdots, \mathbf{p}_m^{i_m}, \mathbf{p}_{m+1}^{j_{m+1}}, \cdots, \mathbf{p}_K^{j_K}) \\
(\mathbf{p}_1^{j_1}, \cdots, \mathbf{p}_m^{j_m}, \mathbf{p}_{m+1}^{i_{m+1}}, \cdots, \mathbf{p}_K^{i_K})
\end{array}\right. \tag{12}
$$

In mutation operation, several elements (can be one) are chosen to be changed, as shown in eq.(13):

$$
(\mathbf{p}_1^{i_1}, \cdots, \mathbf{p}_n^{i_n}, \cdots, \mathbf{p}_K^{i_K}) \rightarrow (\mathbf{p}_1^{i_1}, \cdots, \mathbf{p}_n'^{i_n}, \cdots, \mathbf{p}_K^{i_K}). \tag{13}
$$

Considering that the gene-good parent individuals are prone to generate good children individuals, here we apply the proportion of choice techniques to speed up the optimization process. That means the parents with higher fitness have higher opportunity to be chosen to generate individuals of the next generation.

3.4   Shared Features Using

When using shared features in detection stage, for each shared feature a look-up map is set up for search and reference. The feature values are stored in the map with locations as indexes and flag bits are attached to help identify whether the feature value has been computed and stored. If firstly used, the feature value is extracted and stored and the corresponding flag bit is set true. When next used, the feature value can be immediately acquired by looking up in the map without repetitive extraction. As illustrated in Fig. 3, the extracted feature values are stored in white boxes and the black boxes indicate that the features have not been extracted. The more times a feature is shared in detection stage, the more time is saved. Essentially, feature sharing is a space for time technique. To save the required memory, the hash table based can be applied.

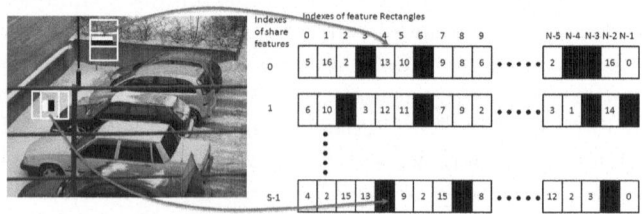

Fig. 3. The using of shared features when detecting objects

4   Experiments and Results

Both the efficiency of the proposed IOGI features and feature sharing techniques is evaluated. Experiments are divided into two parts as single task object detection to test the performance of IOGI features and multiple task object detection to test the performance of feature sharing methods. The PC setting is Core(TM) 2 Duo 3.16GHz CPU with the RAM of 3.48GB.

4.1   IOGI Features

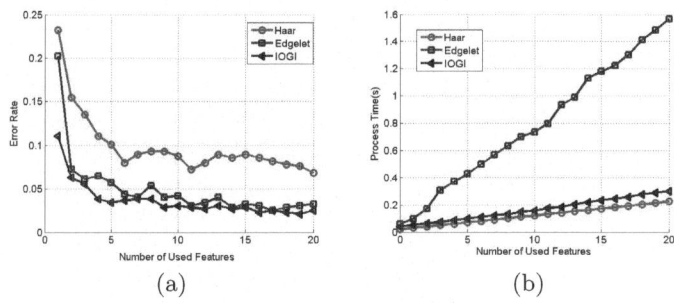

(a)   (b)

Fig. 4. The performance of different features. (a) Error curves on UIUC car dataset; (b) Computational efficiency of different types of features.

(a)   (b)

Fig. 5. Some detection results.(a)Examples of detected results in UIUC car set. (b)Examples of detected results in TUD motorbike set.

The train set of UIUC car dataset [12] consists of 550 side-view car images and 500 non-car images. We split both the car and non-car images into two parts. The first half are for training and the second half are for testing. Both the car and non-car sample images are resized into 75×30. Since there are a large amount of IOGI features, Real-Adaboost algorithm with look-up table (LUT) is applied to select the most appropriate feature subset and construct a strong classifier to discriminate between positive and negative samples. The error rates on test set with different number of features are recorded as measurement. The extraction for Haar-like features [1] and Edgelet features [4] are re-implemented and the error rate curves under the same conditions are also recorded for comparison. From curves in Fig. 4(a) we can see that with the same number of features, the error rates of the proposed IOGI features are lower than those of Haar-like features and Edgelet features. It also means that the IOGI features have better discriminative efficiency. Besides, we also test the computational efficiency of proposed IOGI features here. The cost time of the above boosted classifiers on a 320×240 video is recorded. Each frame from the image sequences is scaled from its original size down to 75×30 with the ratio 1.26 and the slide pixels for shifting sub-windows in both horizontal and vertical directions are 2. The processing time of classifiers with different types of features is shown in Fig. 4(b). It can be seen from the figure that the proposed IOGI features show equal computational efficiency with Haar-like features, which is much less than that of Edgelet features. Besides, we also construct a car detector and a motorbike detector with cascaded-Adaboost algorithm based on IOGI features. 550 car images from the train set of UIUC car dataset [12] are used as positive samples to train car detector. And 826 side-view motorbike images from the Caltech motorbike dataset [13] are used to train the motorbike detector. For both detectors, thousands of background images are collected to serve as negative samples. Three sets as UIUC single scale car set, UIUC multi-scale car set [12] and TUD motorbike set [13] are used to test the performance of trained detectors. The UIUC single scale car set has 170 images containing 200 car instances with the scale as 100×40 and the multi-scale set consists of 108 images with 139 cars with scale ranging from 89×36 to 212×85. Besides, the TUD motorbike dataset consists of 115 images with 125 motorbike instances under various scales. When detection, post-process techniques as clustering are applied to get the final detection results and only when the intersection between a detection response and a ground truth box is larger than 50% of their union, it can be viewed as a successful detection. The Equal Error Rates (EER) are listed in Table 4.1 to serve as the measurements. The best achieved EERs on three sets are 99.5%, 97.5% and 90%, respectively.

Table 1. Category detection rates for public datasets

Methods	UIUC Single Scale Car	UIUC Multi Scale Car	TUD motorbikes
[6]	97.5%	95.0%	87.0%
[15]	98.5%	98.2%	89.3%
Ours	99.5%	97.5%	90.4%

Several detection results are shown in Fig. 5 and red rectangles indicate the results. It needs to be noted that the other two methods in Table 4.1 are based on HOG descriptors. From the table we can see that the performance of IOGI based detectors show equivalent performance when compared with HOG based detectors. Combined with Fig. 4(b) we can see that the proposed IOGI features perform slightly better with HOG descriptors while show equal computational efficiency with Haar-like features.

4.2   Features Sharing

To evaluate the efficiency of features sharing, parallel-structured classifiers for multi-class object detection are trained with proposed IOGI descriptors. Several positive and negative samples belong to different classes are collected. They are randomly split into two parts. One is for training. Detection rates and false alarm rates are evaluated on the whole set. We set the number of used features the same as the classes number and the ROC curves are plotted (black lines). Besides, the ROC curves of boosting without feature sharing as non-shared boost are also plotted for comparison (red lines). Also when the sample sizes are the same for all K classes, the ROC curve of feature sharing method proposed by Torralba et al. [11] as Joint-Boost is also given (blue lines). Three experiments as follows are done.

Multi-View Face: We choose five views of faces as left profile, left half-profile, frontal, right half-profile and right profile. 300 positive samples are collected for each classes. Negative samples of the same number are randomly cropped from background images. Among them, 100 positive/negative samples are for training. The sample size are 24×24. The example positive samples and the ROC curves are shown in Fig. 6(a).

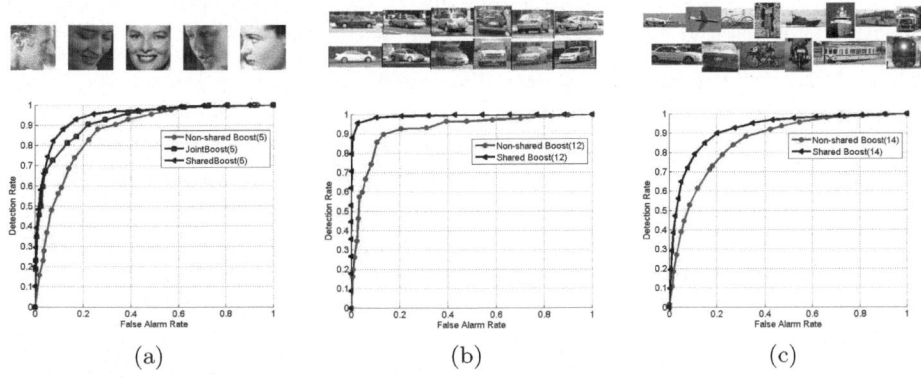

(a)   (b)   (c)

Fig. 6. ROC curves on test sets of the share-trained classifiers and single-trained classifiers. (a)multi-view face detection. (b)multi-view car detection. (c) Pascal 2007 outdoor objects.

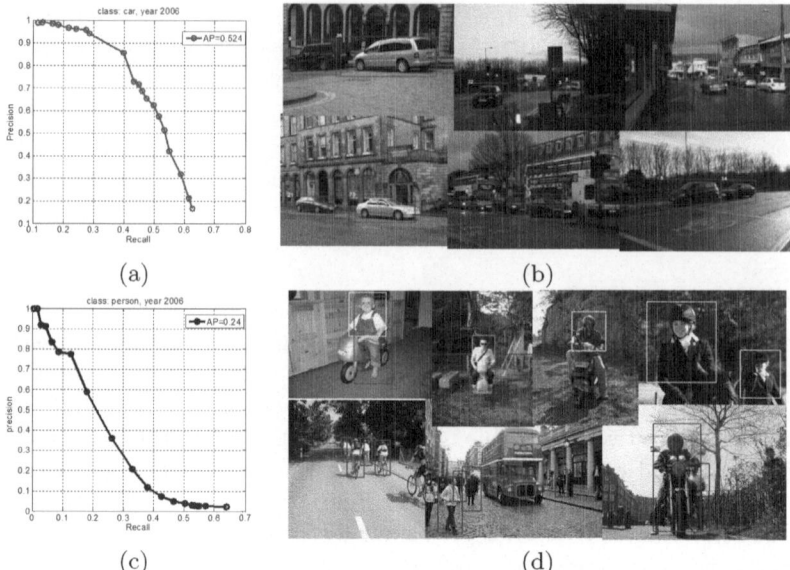

Fig. 7. Performance on Pascal 2006 dataset. (a)PR curve on car set. (b)Detection results on car test set. (c)PR curve on person set. (b)Detection results on person test set.

Multi-view Car: The experiments are done on multi-view car dataset [19]. It contains totally 1279 complete car images under seven different views. With flipping, car examples of 12 views from 0 330 degrees at intervals of 30 degrees can be obtained. The first 50 positive samples for each view are used as positive train samples. Besides, 150 sub-images are randomly cropped from background images without cars and the first 50 ones are also for training. The example positive samples and the ROC curves are shown in Fig. 6(b).

Pascal 2007 Outdoor Objects: The outdoor subset of Pascal 2007 VOC dataset [18] contains annotated examples of seven categories as aeroplane, bicycle, boat, bus, car, motorbike, train. Considering about the view changes, for each class samples are split into two parts according to the aspect ratio. In total 14 classes of objects are obtained. The samples in train set are used for training the classifiers and the ones in both train and validation set are for testing. The same number of train/test negative samples are cropped from other images which do not contain the objects. Positive samples and ROC curves are given in Fig. 6(c).

The performance of classifiers can be evaluated with the area under ROC curve (AUC). Since proposed feature sharing method can learn the shared parts among multiple classes and reduce the used features number, the AUC of shared boost are obviously higher than that of non-shared boost. It also means that with the same number of used features, shared boost can achieve better performance than common boost. From Fig. 6(a) we can also find that the AUC of our proposed shared boost is also higher than that of Joint-Boost. That is because our definition provides a much larger set of features for sharing. Therefore the performance is also better.

The performance of proposed feature sharing techniques is also evaluated in the car and person set of PASCAL VOC 2006 [17]. The car detector is trained with labeled samples from Leibe's multi-view car dataset [19]. We divide the car samples into two parts by the aspect ratio and resize the sample sizes into 45x30 and 50x20, respectively. For each part, a 15-layer cascaded-Adaboost classifier is trained and the feature sharing techniques are applied among the first 2 layers of two parallel classifiers. Similarly the person detector is trained with upright body and head-shoulder sample images. Train samples are collected from the train set of Inria pedestrian database [3] and the train set of Pascal 2006. The resized sample size for upright body images is 24×60 while that for head-shoulder images is 32×32. For both detectors, post-process techniques as clustering are applied to get the final detection results. We use the neighbors number around the detection bounding box and the output of the boosted classifiers to get the confidence of detection response. Only if the intersection between a detection response and a ground truth box is larger than 50% of their union, the detection can be viewed as a successful one. The precision/recall curves are plotted in Fig. 7(a)(c) and several detection results are shown in Fig. 7(b)(d). From Fig. 7 we can see that the average precision of our car detector and person detector are 0.524 and 0.24, respectively. Since we do not apply part models when trained the detectors, we compare the experimental results with the latent SVM based detectors [8] with only root filters. The trained person detector performs slightly better on Pascal 2006 dataset. While on the car set, the performance of our car detector fall in between the latent-SVM based detectors with 1 root filter and 2 root filters. It needs to be noted that our detectors show competitive computational efficiency. The cost time of trained multi-view car detector is less than 500 ms when dealing with a test image with the resolution as 640×480. That is much lower than most of the existing methods.

5   Conclusions

We exploit techniques to speed up object detection in this article. First the IOGI descriptors which can efficiently extract the edges/contours of objects as well as fast computed are presented. After that, feature sharing techniques for Adaboost algorithm based sliding-window object detection methods are given. The techniques can be used to reduce used features number for multiple task detection. Experiments are done to evaluate the efficiency of proposed techniques. The future work will be to combine part-based models to improve the performance of detectors.

References

1. Viola, P., Jones, M.: Rapid object detection using a boosted cascade of simple features. In: CVPR (2001)
2. Lienhart, Maydt, J.: An Extended Set of Haar-like Features for Rapid Object Detection. In: ICPR 2002 (2002)
3. Dalal, N., Trigges, B.: Histograms of Oriented Gradients for Human Detection. In: ICCV (2005)
4. Wu, B., Nevatia, R.: Detection and Tracking of Multiple, Partially Occluded Humans by Bayesian Combination of Edgelet Based Part Detectors. IJCV 75(2), 247–266 (2007)
5. Li, F.-F., Perona, P.: A bayesian hierarchical model for learning natural scene categories. In: CVPR 2005 (2005)
6. Leibe, B., Leonardis, A., Schiele, B.: Robust object detection with interleaved categorization and segmentation. In: IJCV (2008)
7. Gall, J., Lempitsky, V.: Class-specific hough forests for object detection. In: CVPR (2009)
8. Felzenszwalb, P.F., Girshick, R.B., McAllester, D.A., Ramanan, D.: Object detection with discriminatively trained part-based models. IEEE Trans. Pattern Anal. Mach. Intell. 32(9), 1627–1645 (2010)
9. Wu, B., Nevatia, R.: Cluster boosted tree classifier for multi-view, multi-pose object detection. In: ICCV (2007)
10. Schneiderman, H.: Feature-centric evaluation for efficient cascaded object detection. In: CVPR 2004 (2004)
11. Torralba, A., Murphy, K.P., Freeman, W.T.: Sharing visual features for muticlass and multiview object detection. IEEE Trans. Pattern Anal. Mach. Intell. 29(5), 854–869 (2007)
12. Agarwal, S., Roth, D.: Learning a Sparse Representation for Object Detection. In: Heyden, A., Sparr, G., Nielsen, M., Johansen, P. (eds.) ECCV 2002, Part IV. LNCS, vol. 2353, pp. 113–127. Springer, Heidelberg (2002)
13. Fergus, R., Perona, P., Zisserman, A.: Object class recognition by unsupervised scale-invariant learning. In: CVPR 2003 (2003)
14. Fritz, M., Leibe, B., Caputo, B., Schiele, B.: Integrating representative and discriminant models for object category detection. In: ICCV 2005 (2005)
15. Villamizar, M., Moreno-Noguer, F., Andrade-Cetto, J., Sanfeliu, A.: Efficient rotation invariant object detection using boosted Random Ferns. In: CVPR 2010 (2010)
16. Papageorgiou, C., Poggio, T.: A trainable system for object detection. Int'l J. Computer Vision 38(1), 15–33 (2000)
17. Everingham, M., Zisserman, A., Williams, C., Van Gool, L.: The PASCAL Visual Object Classes Challenge 2006 (VOC 2006) Results (2006), http://www.pascal-network.org/challenges/VOC/voc2006/
18. Everingham, M., Van Gool, L., Williams, C.K.I., Winn, J., Zisserman, A.: The PASCAL Visual Object Classes Challenge 2007 (VOC 2007) Results (2007), http://www.pascal-network.org/challenges/VOC/voc2007/
19. Leibe, B., Cornelis, N., Cornelis, K., Van Gool, L.: Dynamic 3d scene analysis from a moving vehicle. In: CVPR 2007 (2007)

Motion Segmentation by Velocity Clustering with Estimation of Subspace Dimension

Liangjing Ding[1], Adrian Barbu[2], and Anke Meyer-Baese[1]

[1] Department of Scientific Computing, Florida State University
[2] Department of Statistics, Florida State University

Abstract. The performance of clustering based motion segmentation methods depends on the dimension of the subspace where the point trajectories are projected. This paper presents a strategy for estimating the best subspace dimension using a novel clustering error measure. For each obtained segmentation, the proposed measure estimates the average least square error between the point trajectories and synthetic trajectories generated based on the motion models from the segmentation. The second contribution of this paper is the use of the velocity vector instead of the traditional trajectory vector for segmentation. The evaluation on the Hopkins 155 video benchmark database shows that the proposed method is competitive with current state-of-the-art methods both in terms of overall performance and computational speed.

1 Introduction

The task of motion segmentation is to label a set of tracked feature points from several moving objects into different groups based on their motions. This is an important step in many computer vision problems, such as robotics, inspection, video surveillance, etc. Motion segmentation has been studied mostly in the case of the affine camera model, under which the vectors of feature points from each rigid motion lie in a subspace of dimension four or less [1], thus the motion segmentation problem can be posed as a subspace separation problem. The main difficulty in subspace separation is that it is usually hard to determine the number of subspaces and their dimension. For example, tracked feature points from a static background might lie on a 2-dimensional subspace, while points from other motions might lie on subspaces of dimension 3 or 4. Moreover, practical motion scenes usually exhibit partially dependent motions, such as when two objects have the same rotational but different translational motion relative to the camera [2], or for articulated motions [3].

Many methods [4], [5], [6], [7], [8] project the feature trajectories onto a smaller dimensional space and perform clustering on the projected points. This approach not only provides computational advantages, but also imposes some sort of a spatial prior on the point trajectories.

Unlike earlier attempts to find a best projection dimension for subspace separation, this paper proposes to perform subspace separation for all possible dimensions. Based on this idea, this paper proposes a motion segmentation approach

J.-I. Park and J. Kim (Eds.): ACCV 2012 Workshops, Part II, LNCS 7729, pp. 491–505, 2013.

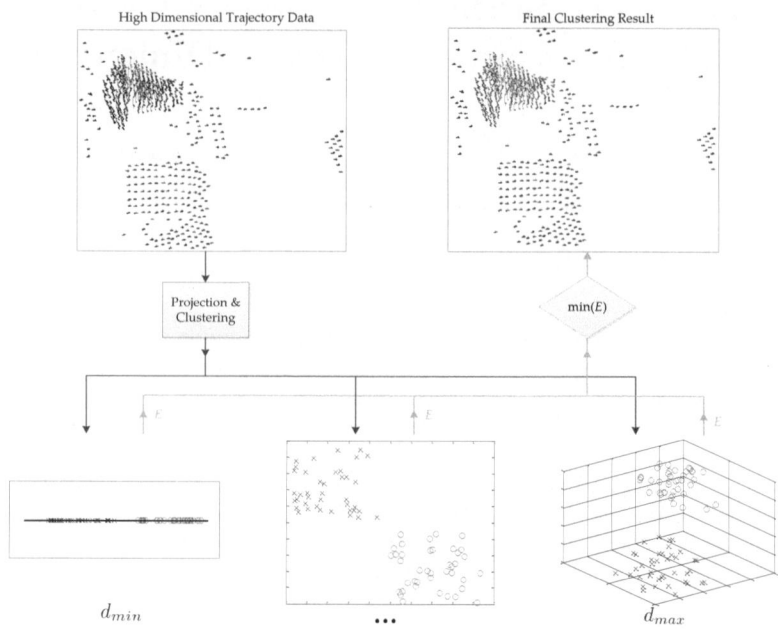

Fig. 1. Illustration of the process of selection of the best result after performing spectral clustering in spaces of dimensions in the range $[d_{min}, d_{max}]$. The selection is based on a clustering error measure described in Section 3.4.

which performs spectral clustering in many dimensions, and then carefully selects the result with the best separability using a novel clustering error measure.

Related Work. Early works of multiframe 3-D motion segmentation based on matrix factorization [9], [10] find the segmentation by thresholding the entries of a similarity matrix built from the factorization of the matrix of data points. However, the thresholding process is very sensitive to noise and such methods are only provably correct when the subspaces are independent. The Generalized Principal Component Analysis (GPCA) [7] is an algebraic method for subspace separation which could deal with dependent motions, but it is not robust to data contaminated by outliers and noise. Some statistical methods, such as Agglomerative Lossy Compression (ALC) [6], RANSAC [11], Multi-Stage Learning (MSL) [2], etc, can handle noise in the data, but their assumptions about the distribution of the noise are not optimal. In recent years, spectral clustering has become a widely used method in motion segmentation. Based on the fact that a point and its k-nearest neighbors (k-NNs) often belong to the same subspace, Local Subspace Affinity (LSA) [8], Spectral Local Best-fit Flats (SLBF) [12], Locally Linear Manifold Clustering (LLMC) [13] use the angle or distance between a point and the subspace fitted through the point and its k-NNs to construct the affinity measure for spectral clustering. However, the neighbors of a point could belong to different spaces, especially when close to the intersection of two subspaces. Also, the selected neighbors may not span the underlying subspace. The spectral clustering (SC) method [5], which uses the angular information between trajectories as affinity, is simple and efficient, but its criterion to select the best subspace dimension is noise-sensitive. More recently, some approaches such as Spectral Curvature Clustering (SCC) [14], Sparse Subspace Clustering (SSC) [4], and Low-Rank Representation (LRR) [15], use the so-called *sparsity* information as the affinity measure. Optimization is always involved in these methods, which makes them computationally expensive.

Our Contributions. In this work, we provide two main contributions. First, we use the velocity vector as a preprocessing step to reduce the influence of the errors accumulated during feature point tracking. This step proves to be very important for improving performance. Second, we present a method for estimating the optimal projection dimension for spectral clustering. We use the angular information between the points proposed in SC [5] to build the affinity matrix. Compared to the SC algorithm, the proposed method presents a different strategy for selecting the best subspace dimension. The SC finds the best dimension before performing the spectral clustering, and the dimension is determined by the so called *relative gap* which is related to the eigenvalues of a Lapacian matrix L . However, when the noise level is large, the relative gap is not very effective. Instead, our method performs spectral clustering after projecting to each of the possible dimensions in a range $[d_{min}, d_{max}]$, and then selects the best result based on a novel clustering error measure. The advantage of the proposed strategy is that the performance is much more robust to data corrupted by noise. Moreover, the complexity of the resulting algorithm remains low as long as the number of motions is small. When applied to the motion segmentation data from the Hopkins155 database [16], the proposed method is competitive with the current state-of-the-art methods both in terms of segmentation accuracy and computational speed.

2   Mathematical Background

Recent works on motion segmentation [5], [14], [4], [15] usually considered the affine camera model. The affine camera model assumes an affine projection model, which generalizes orthographic, weak-perspective and paraperspective projection. Under the affine camera model, a point on the image plane (x, y) is related to the real world point (X, Y, Z) by

$$\begin{bmatrix} x \\ y \end{bmatrix} = \underbrace{K \begin{bmatrix} 1\ 0\ 0\ 0 \\ 0\ 1\ 0\ 0 \\ 0\ 0\ 0\ 1 \end{bmatrix} \begin{bmatrix} R\ t \\ 0^T\ 1 \end{bmatrix}}_{A \in \mathbb{R}^{2 \times 4}} \begin{bmatrix} X \\ Y \\ Z \\ 1 \end{bmatrix}, \tag{1}$$

where A is the affine motion matrix, which is determined by the camera calibration matrix $K \in \mathbb{R}^{2 \times 3}$ and the relative orientation of the image plane with respect to the world coordinates $(R, t) \in SE(3)$.

Let $t = (x^1, y^1, x^2, y^2, \ldots, x^F, y^F)^T$ be a trajectory of a tracked feature point in F frames. Given P trajectories undergoing the same rigid motion, the *measurement matrix* $W = [t_1, t_2, \ldots, t_P]$ is constructed. From equation (1), W can be decomposed into a *motion matrix* $M \in \mathbb{R}^{2F \times 4}$ and a *structure matrix* $S \in \mathbb{R}^{4 \times P}$ as

$$W = MS$$

$$
\begin{bmatrix}
x_1^1 & x_2^1 & \cdots & x_P^1 \\
y_1^1 & y_2^1 & \cdots & y_P^1 \\
\vdots & \vdots & \ddots & \vdots \\
x_1^F & x_2^F & \cdots & x_P^F \\
y_1^F & y_2^F & \cdots & y_P^F
\end{bmatrix}
=
\begin{bmatrix}
A^1 \\
\vdots \\
A^F
\end{bmatrix}
\begin{bmatrix}
X_1 & \cdots & X_P \\
Y_1 & \cdots & Y_P \\
Z_1 & \cdots & Z_P \\
1 & \cdots & 1
\end{bmatrix}.
$$

where A^f is the affine motion matrix at frame f. It implies that $\text{rank}(W) \leq 4$. In other words, under the affine camera model, the 2-D trajectories of a set of 3-D points from a rigidly moving object reside in a subspace of dimension at most 4. Also, it is worth noting that the rows of each A^f involve linear combinations of the first two rows of the rotation matrix R^f, hence $\text{rank}(W) \geq \text{rank}(A^f) = 2$.

Additionally, the entries of the last row of the structure matrix S are identically 1. It is easy to derive the orthographic camera model [1]. Define the registered trajectories as

$$\tilde{t}_i = t_i - \frac{\sum_{i=1}^{P} t_i}{P},$$

then the registered measurement matrix

$$\tilde{W} = [\tilde{t}_1, \tilde{t}_2, \ldots, \tilde{t}_P] \tag{2}$$

is at most rank 3. This means that the trajectories are in a 3-D affine subspace within the 4-D space.

3   Motion Segmentation by Spectral Clustering

This paper only focuses on the problem of segmentation of tracked feature point trajectories. The goal is to find labels for all trajectories, to group them according to their corresponding motions. Also, we assume that the number of different motions is already known.

3.1   Noise Reduction Using Velocity Vectors

Methods for reducing the noise level in the trajectory data is an area that did not receive enough attention in previous work. Noise is an inevitable by-product of feature tracking. Tracking errors are introduced with each new frame, due to factors such as aliasing, non-constant brightness, lack of texture, occlusion, and so on. These errors tend to accumulate and the total tracking error tends to grow as the number of frames increases.

In order to reduce the effect of the accumulated error in the motion segmentation, we use the velocity vector to characterize the trajectories, which is defined by

$$[x^1 - x^2, y^1 - y^2, \ldots x^{F-1} - x^F, y^{F-1} - y^F, x^i, y^i]^T, i \in [1, \cdots, F] \quad (3)$$

With the exception of the last two rows, the entries of the other rows are replaced with the corresponding velocities. In the last two rows, the feature locations of the i-th frame are kept. The selection of i is not crucial. In this paper, we use $i = F$ but we could as well use $i = 1$ for example. The advantage is that the velocities in each frame contain only the tracking error from the previous frame to the current frame, and not the error accumulated from the starting frame. A similar velocity has been used to measure the distance between trajectories for motion segmentation in [17].

It is easy to see that when the measurement matrix W' is built from the velocity vectors, no information is lost since the original measurement matrix W can be recovered from W' by simple row operations. Because of this, the ranks of W and W' are the same. In other words, the subspace clustering problem has not been changed. However, even though the velocity matrix differs from the original measurement matrix only by row operations, the subspace projections are different because these row operations cannot be represented by a rotation matrix.

Table 1. The SSE and variance of the distances from the projected points to the fitted subspaces in 3D for a synthetic experiment. The projected points were generated from trajectories with different signal-to-noise ratio (SNR).

	SSE	Variance
Distance Vector (No noise added)	0	0
Velocity Vector (No noise added)	0	0
Distance Vector (SNR = 10)	0.256e-5	0.0011e-5
Velocity Vector (SNR = 10)	0.106e-5	0.0004e-5
Distance Vector (SNR = 5)	1.058e-5	0.005e-5
Velocity Vector (SNR = 5)	0.208e-5	0.001e-5

The noise reduction effect of using the velocity vector can be well observed in a synthetic experiment. For this purpose, 242 synthetic trajectories of length 20 were generated for two different motions, perfectly following the affine camera model. The starting feature points were randomly chosen in the first frame, and different levels of Gaussian tracking errors were introduced based on the displacement of feature points. If denote the tracker as f, and noise as n, to a point p_i in frame i, the tracked point in the next frame would be $p_{i+1} = f(p_i) + n$.

The trajectories were projected to a 3D subspace by truncated SVD. A plane was fitted in a least squares sense to the projected points of each motion. The sum of squared error (SSE) and variance of the distances from projected points to the fitted planes are shown in table 1. One could see that by using velocity

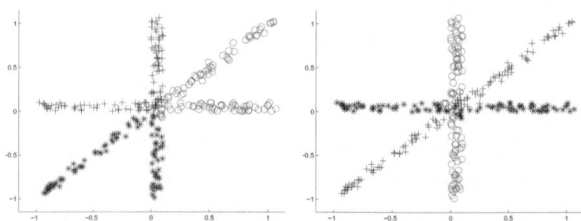

Fig. 2. Spectral clustering of lines with a distance-based affinity mixes points from different subspaces (left), while the angle-based affinity (4) separates them very well (right)

vectors the noise is reduced, and the reduction is greater when the tracking errors are larger. Since the projected points obtained by velocity clustering are closer to satisfying the planarity assumption, it should be expected that the segmentation results would also be better.

3.2 Spectral Clustering of Subspaces

Spectral clustering [18], [19] is a popular technique for solving motion segmentation problems [20], [21], [8], [12], [13], [4], [15], [14]. One challenge in applying spectral clustering is the construction of a good affinity matrix. Two points that lie in two different subspaces and are near the intersection of the subspaces may be close to each other. Conversely, a pair of points in the same subspace could be far from each other. As a consequence, one cannot use the typical distance-based affinity.

SC [5] proposes an affinity measure based on the angle between two vectors, defined by

$$A_{ij} = (\frac{x_i^T x_j}{\|x_i\|_2 \|x_j\|_2})^{2\alpha}, i \neq j, \alpha \in \mathbb{N} \tag{4}$$

where x_i, x_j are two vectors. The parameter $\alpha > 1$ is used to increase the separation and should be tuned according to the noise level. It has been proved [5] that the proposed affinity measure (4) guarantees that each point x_i has a higher connection with its own group than the others. Figure 2 shows the power of angle-based affinity over the distance-based affinity in clustering 1D subspaces in 2D. This paper also uses the angular information to build the affinity matrix. While SC [5] suggests to set $\alpha = 4$ for motion segmentation, in the experiments of section 5, we find that $\alpha = 2$ could produce better results for our algorithm.

3.3 Best Subspace Dimension

Most motion segmentation methods usually require the projection to a low dimensional space where the clustering is performed. The dimension of this projection space has a large impact on the speed and accuracy of the final result. GPCA [7] suggests to project trajectories onto a 5-dimensional space. However, five dimensions are not sufficient to complex scenes, such as scenes with articulated or nonrigid motions. Motivated by compressive sensing [22], ALC [6] chooses to use the sparsity-preserving dimension

$$d_{sp} = \min_{d \geq 2D \log(2F/d)} d$$

for D-dimensional subspaces (with $D = 4$ for motion segmentation). SC [5] wants the intersection of different subspaces to have minimal dimension and proposes to set dimension $d = kD+1$, where k is the number of motions and $1 \leq D \leq 4$ for motion segmentation; the d used for clustering is searched in range $[k+1, 4k+1]$ by some relative gap.

The main difficulty for selecting the best subspace dimension is that the dimension of one affine subspace is not fixed. If one tries to find the correct dimension by setting a threshold of noise, this scheme will not work well because different scenes may have different thresholds.

The search strategy in SC [5] is innovative, but the range of possible dimensions that are searched is a parameter that needs to be tuned. Moreover, the criterion to select the best dimension in SC [5] is related to the noise level, and is not optimal in some scenarios, as we will see in experiments.

In this paper, we don't look for the best subspace dimension directly. Instead, we employ an exhaustive strategy. Since the best dimension is unknown and hard to determine, our method performs clustering after projecting to spaces of all possible dimensions, then the best result is chosen by a clustering measure. Based on this idea, finding the best subspace dimension is not necessary in this paper. What we need to do is to find a bound on the possible dimensions.

The dimension of one affine subspace S is not fixed but is bounded by

$$2 \leq \dim(S) \leq 4.$$

If there are k linear affine subspaces in general position embedded in space S_k, we would expect

$$2k \leq \dim(S_k) \leq 4k.$$

This is the range of space dimensions that will be used in our method. The best dimension will be determined using the clustering error measure defined in the next section.

3.4   Motion Error Measure

When the spectral clustering is performed in the selected spaces, a number of results will be obtained. A question is raised naturally: how to select the best one? In this paper we investigate two types of estimators of the segmentation error, both based on a RMSE error measure for each trajectory.

The Tomasi-Kanade factorization [1] allows us to write the registered matrix $\tilde{W}$ in equation (2) as

$$\tilde{W} = \tilde{M}\tilde{S}$$

where $\tilde{M}$ is a $2F \times 3$ matrix and $\tilde{S}$ is a $3 \times P$ matrix. There is an inherent ambiguity in $\tilde{M}$ and $\tilde{S}$ but we will show that it is irrelevant for the error measure.

Any registered trajectory $\tilde{t}$ in $\tilde{W}$ will have a corresponding point $\tilde{P} \in \mathbb{R}^3$ obtained by least squares:

$$\tilde{P} = \underset{\tilde{P}}{\operatorname{argmin}} \|\tilde{t} - \tilde{M}\tilde{P}\|^2.$$

We define the RMSE error of $\tilde{t}$ as

$$\mathrm{RMSE}_{\tilde{W}}(\tilde{t}) = \sqrt{\frac{\min_{\tilde{P}} \|\tilde{t} - \tilde{M}\tilde{P}\|^2}{F}} \tag{5}$$

The RMSE error is measured in pixels and can be viewed as the tracking error for one trajectory.

Remark 1. *The $RMSE_{\tilde{W}}(\tilde{t})$ is invariant to the choice of $\tilde{M}$ and $\tilde{S}$ in the decomposition $\tilde{W} = \tilde{M}\tilde{S}$.*

Proof. To any 3×3 invertible matrix A, $\tilde{W} = \tilde{M}AA^{-1}\tilde{S}$. It can be easily verified that $\mathrm{RMSE}_{\tilde{W}}(\tilde{t})$ in equation (5) does not change when $\tilde{M}$ is multiplied by an invertible matrix A. Moreover, any decomposition $\tilde{W} = \tilde{M}'\tilde{S}'$ has $\tilde{M}' = \tilde{M}A$ for some invertible matrix A. ∎

Given a labeling L of the trajectories, obtain for each label l the registered measurement matrix $\tilde{W}^l$ containing all trajectories with label l. Based on $\tilde{W}^l$ we define two types of estimators of the segmentation error.

The first type is just the sum of the RMSE errors of all registered trajectories based on their corresponding motion matrices

$$E(L) = \sum_{l=1}^{k} \sum_{i, L(i)=l} \mathrm{RMSE}_{\tilde{W}^l}(\tilde{t}_i). \tag{6}$$

The second type makes the contribution of each registered trajectory comparing to a threshold τ

$$E_\tau(L) = \sum_{l=1}^{k} \sum_{i, L(i)=l} I(\mathrm{RMSE}_{\tilde{W}^l}(\tilde{t}_i) \geq \tau). \tag{7}$$

where $I(\cdot)$ is the indicator function taking on value 1 if its argument is true or 0 otherwise.

In a perfect segmentation result, each trajectory would have a small RMSE error because of the affine camera model, resulting in a small clustering error $E(L)$ and $E_\tau(L)$.

A number of segmentation results can be obtained by projecting the original trajectories to spaces of different dimensions and performing clustering in those spaces. The problem is how to select from the obtained segmentations the one with the smallest error. For that we can use an estimator that correlates well with the segmentation error.

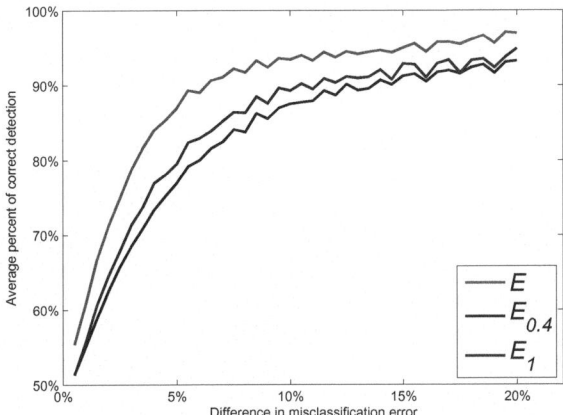

Fig. 3. The average percentage of times the proposed error estimators find the better segmentation out of two segmentations vs. their difference in misclassification errors for sequences with 3 motions in the Hopkins 155 dataset. If one segmentation is much better than the other, it will be found most of the time.

We propose to use the measures $E(L)$ and $E_\tau(L)$ to rank the obtained segmentations. We evaluated the capability of these error measures to find the better one out of two segmentations on the sequences with 3 motions in the Hopkins 155 dataset (See section 5). It is expected that when one segmentation is much better than the other (i.e. the error difference is large), the better segmentation should be found more often. Different segmentations were obtained in this way: for one sequence, a random set of $p\%$ of trajectories ($p \leq 50$ is a random number) were assigned random labels, while the labels of the remaining trajectories were untouched. 500 sets of segmentation were generated for each sequence (25000 segmentations in total for all sequences). At last, we calculated the difference in misclassification error and the average correct detection rate shown in Figure 3. One can see that if one segmentation is much better than the other, it will be found most of the time. Also, $E(L)$ always outperforms the other two estimators. Thus in this paper, $E(L)$ is adopted.

4   Complete Procedure

Dimension Reduction. Dimension reduction seems to be a standard procedure for motion segmentation by spectral clustering in [5], [6], [7]. It can improve the computational tractability without adversely affecting the quality of the segmentation, since in general the projection onto an arbitrary d-dimensional space preserves the multi-subspace structure of data lying on subspaces with dimensionality strictly less than d. There are two different strategies in dimension reduction: the random sampling [14], [4] and the truncated SVD [9], [5], [2]. This paper uses the latter method for dimension reduction from $W' \in \mathbb{R}^{2F \times P}$ to $X = [x_1, ..., x_P]^T \in \mathbb{R}^{D \times P}$ in our framework, where D is the dimension of the subspace. The truncated SVD is related to the factorization-based methods [23],

[24], which use the SVD, $W = U \Sigma V^T$, to obtain a shape interaction matrix $Q = VV^T$. In order to deal with the noise and dependencies, we use the truncated SVD of the velocity measurement matrix, $W' \approx U_D \Sigma_D V_D^T$.

Details of Spectral Clustering. After the projection for dimension reduction, the spectral clustering method is applied to obtain the clustering result.

The affinity matrix is constructed using the angular affinity metric in equation (4). In fact, the affinity matrix can be easily calculated as $Q = (\tilde{V}_D \tilde{V}_D^T)^{2\alpha}$, where $\tilde{V}_D$ is the V_D with normalized rows. This normalization ensures that only the angular information is taken into account.

From the affinity matrix, the corresponding Laplacian matrix L is obtained. Then the k largest eigenvectors of L are found, where k is the number of clusters. A matrix A is formed by stacking the k eigenvectors in columns. Finally, the segmentation of the trajectories follows by applying K-means clustering to the rows of $\tilde{A}$, which is obtained by normalizing the rows of A .

Selection of the Best Result. According to section 3.3, to ensure that the best result is not missed, an exhaustive search strategy is employed. Let $d_{min} = 2k$ and $d_{max} = 4k$ be the minimal and maximal subspace dimensions, motion segmentation is performed in spaces with all dimensions D in the range $D \in [d_{min}, d_{max}]$. Then the best result is selected among all results based on the smallest clustering error (6) or (7). The whole procedure is illustrated in Figure 1 and described in Algorithm 1.

Algorithm 1. Velocity Clustering with Estimation of Subspace Dimension

---

Input: The measurement matrix $W = [t_1, t_2, \ldots, t_P] \in \mathbb{R}^{2F \times P}$ whose columns are point trajectories, and the number of clusters k.
Preprocessing: Build the velocity measurement matrix W' by row transformations of W given by eq. (3).
for $D = d_{min}$ **to** d_{max} **do**
  1. Perform SVD: $W' = U \Sigma V^T$
  2. Build the N-by-D data matrix

$$X_D = [v_1, \ldots, v_D]$$

where v_i is the i-th column of V.
  3. Apply spectral clustering to the N points in X_D using the affinity measure (4).
  4. Compute the clustering error E_D of the segmentation result using eq. (6).
end for
Output: The segmentation result with the smallest error E_D.

---

5   Experiments

The Hopkins 155 Dataset [16] has been created with the goal of providing an extensive benchmark for testing feature based motion segmentation algorithms.

(a) Checkerboard   (b) Traffic   (c) Articulated

Fig. 4. Sample images from some sequences of three categories in the Hopkins 155 database with ground truth superimposed

It contains video sequences along with some feature points extracted and tracked in all the frames. The ground-truth segmentation is also provided for evaluation purposes. Based on the content of the video and the type of motion, the 155 sequences can be categorized into three main groups: *checkerboard*, *traffic* and *articulated*. Figure 4 shows sample frames from three videos of the Hopkins 155 database with the feature points superimposed. The sequences contain degenerate and non-degenerate motions, independent and partially dependent motions, nonrigid motions, etc. Since the trajectories were obtained by an automatic tracker, they could be considered as slightly corrupted by noise.

We have tested our algorithm on the image sequences from the Hopkins 155 database, as well as several other state-of-the-art algorithms: ALC [6], SC [5] and SSC [4]. For each algorithm on each sequence, we recorded the misclassification rate defined as

$$\text{Misclassification Rate} = \frac{\#\ \text{of misclassified points}}{\text{total}\ \#\ \text{of points}} \quad (8)$$

The parameter setting in our method are $\alpha = 2$, $d_{min} = 2k$, $d_{max} = 4k$, and the locations of the last frame are kept to build the velocity matrix. The results on sequences with 2, 3 motions and the whole dataset are presented in Table 2 and compared with the three state-of-the-art and baseline methods. We also show in the table the results of the algorithm with fixed subspace dimension $D = 4k$ as well as results without using the velocity preprocessing step.

One could see that by using the velocity for clustering the misclassification error decreases by about 0.8% while by using the clustering error measure to decide the best segmentation the error decreases from 4.91% to 0.99%. Thus the clustering error measure has a large impact in the spectral clustering performance while the velocity clustering has a smaller but also important impact.

Compared to other motion segmentation algorithms, our approach outperforms for the 3 motion sequences and for all the sequences combined and is outperformed on the two motion sequences by SC [5] and SSC [4]. We achieve an overall misclassification error of 1.10% for 3 motions, around half of the best reported result (SC [5]); an overall error of 0.96% for 2 motions, coming close

Table 2. Misclassification rate (in percent) for sequences of full trajectories in the Hopkins 155 dataset (Subscript $4k$ means using fixed dimension $4k$ instead of dimension search, and superscript $*$ means not using velocity for clustering)

Method	ALC	SC	SSC	Our Method$_{4k}^*$	Our Method*	Our Method$_{4k}$	Our Method
Checkerboard (2 motion)							
Average	1.55	0.85	1.12	2.07	1.38	1.38	**0.67**
Median	0.29	0.00	0.00	0.30	0.00	0.00	0.00
Traffic (2 motion)							
Average	1.59	0.90	**0.02**	6.87	1.35	8.25	0.99
Median	1.17	0.00	0.00	1.33	0.30	1.09	0.22
Articulated (2 motion)							
Average	10.70	1.71	**0.62**	6.02	2.56	2.46	2.94
Median	0.95	0.00	0.00	0.99	0.88	0.88	0.88
All (2 motion)							
Average	2.40	0.94	**0.82**	3.67	1.48	3.25	0.96
Median	0.43	0.00	0.00	0.51	0.00	0.00	0.00
Checkerboard (3 motion)							
Average	5.20	2.15	2.97	4.38	1.06	2.28	**0.74**
Median	0.67	0.47	0.27	1.37	0.58	0.51	0.21
Traffic (3 motion)							
Average	7.75	1.35	**0.58**	27.80	8.22	19.21	1.13
Median	0.49	0.19	0.00	32.27	1.42	28.28	0.21
Articulated (3 motion)							
Average	21.08	4.26	**1.42**	6.18	6.18	18.95	5.65
Median	21.08	4.26	0.00	6.18	6.18	18.95	5.65
All (3 motion)							
Average	6.69	2.11	2.45	9.17	2.78	6.62	**1.10**
Median	0.67	0.37	0.20	1.99	0.67	0.85	0.22
All sequences combined							
Average	3.37	1.20	1.24	4.91	1.78	4.01	**0.99**
Median	0.49	0.00	0.00	0.57	0.00	0.24	0.00

to the best performing SSC [4]; and an overall error of 0.99% for the whole database, which is better than the other methods. Our method always obtains good results for checkerboard sequences which have the most complicated scenes (including both translation and rotation motions) in the dataset.

The performance on the articulated sequences with 3 motions is worse than the SC, possibly because these sequences don't obey the rigid motion model and thus the RMSE measure might not be accurate. On the other hand, when the motions follow the rigid model, the RMSE measure helps obtain very good results. This is clearly visible in the three motion checkerboard sequences, where our algorithm obtains errors less than half of the other algorithms.

From the cumulative distributions in Figure 5, we see that for 2 motions, our method is comparable to the best method SSC; and for 3 motions, our method outperforms all others. Moreover, the largest error of our method for 3 motions is about 10%, while that of the other methods is around 40%.

Table 3. Average computing time for sequences in the Hopkins 155 database

	ALC	SC	SSC	Our Method
2 motions	7.85m	0.53s	2.27m	0.72s
3 motions	16.77m	1.34s	4.08m	1.81s

Table 3 shows that the average computing time per sequence (obtained on a 2.66GHz Core 2 Duo computer with Matlab on Linux) for sequences with 2 motions is less than 1 second, while that for sequences with 3 motions is less than 2 seconds. In comparison to other methods, our method is much faster than ALC and SSC, but slightly slower than SC.

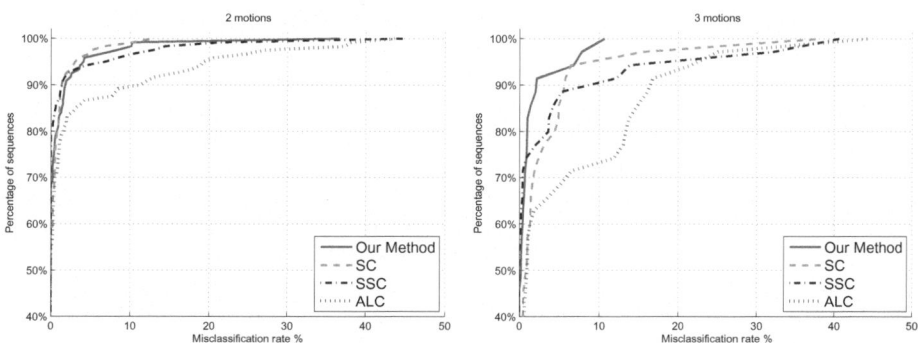

Fig. 5. The cumulative distribution of the misclassification rate for two and three motions in the Hopkins 155 database

6 Conclusion and Future Work

In this paper, we presented a method for segmenting moving objects using spectral clustering. The method uses the velocity vectors as the input for clustering, which is more robust to accumulated errors, and then applies spectral clustering in all possible subspace dimensions. The final segmentation is selected from the obtained results using a novel clustering error measure. Our evaluation on the Hopkins 155 database shows that the method is competitive with current state-of-the-art methods, both in terms of overall performance and computational speed. The algorithm has been shown to be robust to different types of scenes and motions present in the Hopkins 155 database, while remaining very efficient in computation time.

Future work will study how to extend the method to deal with incomplete trajectories, and hopefully, treat complete and incomplete trajectories equally.

References

1. Tomasi, C., Kanade, T.: Shape and motion from image streams under orthography: a factorization method. International Journal of Computer Vision 9, 137–154 (1992)
2. Sugaya, Y., Kanatani, K.: Geometric Structure of Degeneracy for Multi-body Motion Segmentation. In: Comaniciu, D., Mester, R., Kanatani, K., Suter, D. (eds.) SMVP 2004. LNCS, vol. 3247, pp. 13–25. Springer, Heidelberg (2004)
3. Yan, J., Pollefeys, M.: A factorization approach to articulated motion recovery. In: IEEE Conference on Computer Vision and Pattern Recognition, vol. II, pp. 815–821 (2005)
4. Elhamifar, E., Vidal, R.: Sparse subspace clustering. In: IEEE Conference on Computer Vision and Pattern Recognition (2009)
5. Lauer, F., Schnörr, C.: Spectral clustering of linear subspaces for motion segmentation. In: IEEE International Conference on Computer Vision (2009)
6. Rao, S., Tron, R., Vidal, R., Ma, Y.: Motion segmentation in the presence of outlying, incomplete, or corrupted trajectories. IEEE Transactions on Pattern Analysis and Machine Intelligence 32, 1832–1845 (2010)
7. Vidal, R., Ma, Y., Sastry, S.: Generalized principal component analysis. IEEE Transactions on Pattern Analysis and Machine Intelligence 27, 1945–1959 (2005)
8. Yan, J., Pollefeys, M.: A General Framework for Motion Segmentation: Independent, Articulated, Rigid, Non-rigid, Degenerate and Non-degenerate. In: Leonardis, A., Bischof, H., Pinz, A. (eds.) ECCV 2006. LNCS, vol. 3954, pp. 94–106. Springer, Heidelberg (2006)
9. Costeira, J., Kanade, T.: A multibody factorization method for independently moving objects. International Journal of Computer Vision 29, 159–179 (1998)
10. Gear, C.W.: Multibody grouping from motion images. International Journal of Computer Vision 29, 133–150 (1998)
11. Fischler, M.A., Bolles, R.C.: RANSAC random sample consensus: A paradigm for model fitting with applications to image analysis and automated cartography. Communications of the ACM 26, 381–395 (1981)
12. Zhang, T., Szlam, A., Wang, Y., Lerman, G.: Hybrid linear modeling via local best-fit flats. In: IEEE Conference on Computer Vision and Pattern Recognition, pp. 1927–1934 (2010)
13. Goh, A., Vidal, R.: Segmenting motions of different types by unsupervised manifold clustering. In: IEEE Conference on Computer Vision and Pattern Recognition (2007)
14. Chen, G., Lerman, G.: Spectral curvature clustering. International Journal of Computer Vision 81, 317–330 (2009)
15. Liu, G., Lin, Z., Yu, Y.: Robust subspace segmentation by low-rank representation. In: International Conference on Machine Learning (2010)
16. Tron, R., Vidal, R.: A benchmark for the comparison of 3-d motion segmentation algorithms. In: 2007 IEEE Conference on Computer Vision and Pattern Recognition, pp. 1–8. IEEE (2007)
17. Brox, T., Malik, J.: Object Segmentation by Long Term Analysis of Point Trajectories. In: Daniilidis, K., Maragos, P., Paragios, N. (eds.) ECCV 2010, Part V. LNCS, vol. 6315, pp. 282–295. Springer, Heidelberg (2010)
18. Shi, J., Malik, J.: Normalized cuts and image segmentation. IEEE Transactions on Pattern Analysis and Machine Intelligence 22 (2000)

19. Ng, A.Y., Jordan, M.I., Weiss, Y.: On spectral clustering: analysis and an algorithm. In: Advances in Neural Information Processing Systems 14, vol. 2 (2001)
20. Wang, H., Culverhouse, P.: Robust motion segmentation by spectral clustering. In: Proc. of the British Machine Vision Conference, pp. 639–648 (2003)
21. Park, J., Zha, H., Kasturi, R.: Spectral Clustering for Robust Motion Segmentation. In: Pajdla, T., Matas, J. (eds.) ECCV 2004. LNCS, vol. 3024, pp. 390–401. Springer, Heidelberg (2004)
22. Donoho, D., Tanner, J.: Counting faces of randomly-projected polytopes when the projection radically lowers dimension. American Mathematical Society 22, 1–53 (2009)
23. Costeira, J., Kanade, T.: A multi-body factorization method for motion analysis. In: Proceedings of the 5th International Conference on Computer Vision, pp. 1071–1076 (1995)
24. Kanatani, K.: Motion segmentation by subspace separation and model selection. In: Proceedings of the 8th IEEE International Conference on Computer Vision, vol. 2, pp. 586–591 (2001)

Beyond Spatial Pyramid Matching: Spatial Soft Voting for Image Classification

Toshihiko Yamasaki and Tsuhan Chen

The University of Tokyo / Cornell University

Abstract. Recently, spatial partitioning approaches such as spatial pyramid matching (SPM) are commonly used in image classification to collect the global and local features of the images. They divide the input image into small sub-regions (typically in a hierarchical manner) and generate a feature vector for each of them. Although the codes for the descriptors are assigned softly in modern image feature representation techniques, the codes must fall into only a single sub-region when forming the feature vector. In other words, the soft code assignment is used in the descriptor space but the codes are still "hard" voted from the view point of the image space. This paper proposes a spatial soft voting method, in which the existence of the codes are expressed by a Gaussian function and the maps of the existence are sampled to form a feature vector. The generated feature vectors are "soft" both in the descriptor space and the image space. In addition, extra computational cost as compared to SPM is negligibly small. The concept of the spatial soft voting is general and can be applied to most hard spatial partitioning approaches.

1 Introduction

A bag-of-features (BoF) model [1, 2] is one of the most successful approaches for efficient and effective feature representation for image classification and retrieval. The basic idea of the BoF model consists of three steps: (1) generating appearance descriptors extracted from local patches, (2) assigning codes to them, and (3) binning/pooling the codes.

A lot of engineering efforts have been done to improve the performance of each stage mentioned above. One of the most significant progress in the second stage was the "soft voting" [3–6]. In so-called "hard voting" approaches, only a single visual word was assigned to each descriptor. Therefore, two descriptors which were very close to each other in the descriptor space could be assigned different codes when they were near the cluster boundary. The distance from the descriptor and the code word was also discarded. In [3, 4], soft codes were assigned depending on the distance from the descriptors to the code words. In [5], a sparse code (Sc) representation was employed. Then, the locality-constrained linear coding (LLC) [6] was developed to ensure that only the code words that were closer to the descriptor would be picked up.

One of the milestones for the third stage was the spatial pyramid matching (SPM) [7]. The SPM divided the images into increasingly fine sub-regions and

J.-I. Park and J. Kim (Eds.): ACCV 2012 Workshops, Part II, LNCS 7729, pp. 506–519, 2013.

generated histograms of local descriptors in each sub-region. In this manner, approximate global geometric correspondence was considered. Since this approach was very powerful, a lot of further investigations have been done such as optimal partitioning and so on.

From the view point of "softness," however, the SPM and its descendants are still "hard": all the descriptors must fall into one of the sub-regions in each layer. No matter how many sub-regions the image is partitioned into, one local feature can contribute to only a single sub-region. In this paper, a spatial soft voting (SSV) method, which achieves the "soft voting" in the image space, is proposed. Namely, the existence of a descriptor is broadcasted to its neighboring pixels whose influence (weight) is represented as a Gaussian function of the image-space distance from the local feature point. The proposed algorithm can improve the performance of most of the algorithms that use SPM. In addition, the SSV is compatible with the conventional soft voting in the feature space. The combination of both the soft voting in the image space and that in the feature space can outperform the state-of-the-art algorithms for image classification.

The rest of this paper is organized as follows. In Sec. 2 we will talk about some related works. Sec. 3 describes our proposed method followed by experimental results in Sec. 4. Finally, Sec. 5 concludes our paper. Some more detailed experiments are presented in the Appendix.

2   Related Work

In this section, we review soft code assignment and SPM-related techniques.

In the BoF representation, k-means clustering is often used to vector quantize the local descriptors and to assign discrete indices to the local descriptors. This enables us to achieve compact representation of an image in the form of a histogram. In the vector quantization (VQ) based method [1], only a single code index was assigned to each local descriptor and then a histogram of the indices were generated. This assignment is called "hard voting." The histogram can be regarded as the probability density over the code words if the histogram is normalized. Such hard assignment of the code indices leads to errors because of variability in the feature descriptor. Even if descriptors are close to each other in the feature space, they might fall into different bins if they were near the cluster boundaries. The hard voting does not consider the distance from the local descriptor to the code word, either. To solve this hard voting problem, Ref. [8] used Gaussian Mixture Model (GMM), in which the set of key points were represented directly as a probability density function, over which a kernel was defined. [3] used an Gaussian function of the distance from the descriptor to the code words to calculate weights to the code words. So, the assigned weights could be considered as the probabilities of belonging to the code words, which were calculated by a kernel function. In the code word uncertainty (UNC) model [4], the weights were further normalized by the sum of the distances. In [5], a sparse code representation was employed in conjunction with max pooling. The sparse coding had advantages in capturing salient patterns of local descriptors into a codebook and better representation of the descriptors with much less quantization error. Then, the locality-constrained linear coding (LLC) [6] was developed to ensure that descriptors that are closer in the feature space are assigned similar code words whereas the sparse coding does not hold such assumption. Besides, the computational complexity for the LLC was smaller than [5]. The performance of the sparse coding based method was improved by two-layer sparse coding [9], label consistent K-SVD [10] which enforce discriminability in sparse codes during the dictionary learning process, non-negative sparse coding with low-rank and sparse matrix decomposition [11], and affine sparse codes with AdaBoost [12].

The SPM is one of the key elements in state-of-the-art image classification. In the simplest BoF representation [2], a histogram of code indices was generated using descriptors in a whole image, thus the spatial information of the descriptors in the image space was lost. To overcome this problem, a spatial pyramid matching (SPM) technique has been proposed in [7]. The SPM partitioned the image into increasingly fine-grid sub-regions and extract histograms of local descriptors in each sub-region. The global feature was captured in the coarse-grid histograms and local spatial information was described in the finer-grid histograms. After the success of the SPM method, a lot of variations have been proposed in terms of the number of pyramid levels, how to partition the image, weights to the histograms, pooling methods, and so on. In [5–7], 1×1, 2×2, 4×4 (, and 8×8 in [7]) spatial pyramids were used while [13–16] used 1×1, 2×2, 1×3. Ref. [7] assigned weights to each histogram based on the pyramid match kernel [17]. Simply concatenating the histograms was also frequently used [5, 6]. Instead of concatenating the histograms, Ref. [18] proposed a discriminative SPR, which formed the image feature as a weighted sum of semi-local features over all pyramid levels. It has also been demonstrated that the max pooling [5, 6] outperformed conventional average pooling. In the max pooling, peaky response of certain local descriptor to the codebook was well maintained, whereas such characteristics would be smoothed and washed out in the average pooling. Average pooing and max pooling can be regarded as scaled l_1-norm and l_∞-norm, respectively. Recently, geometric l_p-norm feature pooling (GLP) was presented [19], which learns the location-dependent optimal p considering spatial distribution patterns of LLC responses across different classes. GLP worked well only if all the images were aligned and intra-class spatial variances were small.

There are some more image-space-aware BoF representations in addition to [19]. Spatial weighting for BoF was proposed in [20]. This approach required ground-truth segmentation information during the training phase to learn the position of the object from a point of view relative to the training features, which is not practical. Instead of partitioning the input image in a pyramid manner, Cao et al. proposed a linear and circular ordered BoF model [21]. Local features of an image were projected to different directions or points to generate a series of ordered bag-of-features, based on which different families of spatial BoFs were designed to capture the invariance of object translation, rotation, and scaling. Feature context [22] was a technique inspired by the shape context [23] in the shape retrieval community. Ref. [24] considered co-occurrence of descriptor

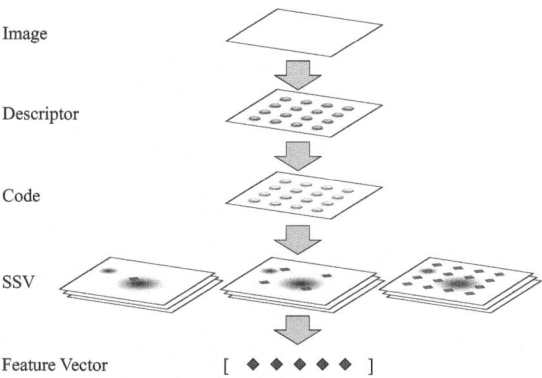

Image

Descriptor

Code

SSV

Feature Vector   [◆ ◆ ◆ ◆ ◆]

Fig. 1. Overview of the image classification flowchart. Here we replace the SPM building block with our proposed SSV method.

pairs in the image space. Ref. [25] considered the mean and variance of the occurrences of each code word by Fisher vector representation instead of using the SPM. In addition, Ref. [25] focused on more compact feature representation rather than classification performance improvement. Ref. [26] did the clustering of the descriptors in each sub-region and used the codebook as features. Ref. [27] proposed a method to learn the optimal partitioning.

Although there are some image-space-aware BoF representations as listed above, most of them are based on learning of the spatial distribution of the code words. In our proposed SSV, on the other hand, the existence of the code words are represented by a Gaussian function and the pooling is conducted in a "soft" manner. All the Gaussian functions are independent of each other and therefore no spatial statistics is included in our approach.

3 Spatial Soft Voting

The spatial partitioning can be considered as quantization of the image space. Therefore, from the view point of "softness", the SPM and its descendants are spatially "hard." Therefore, this paper introduces "softness" to the image space in feature vector generation.

3.1 Classification Procedure

Fig. 1 shows the pipeline of the proposed image classification procedure. Different from conventional approaches, the spatial pyramid representation is replaced with our spatial soft voting (SSV). After local feature descriptors are extracted, a soft voting method encodes the descriptors by assigning a few code indices with weight values. In this paper, Sc [5] and LLC [6] are employed as examples.

After the soft codes are assigned to all the descriptors, the "existence" map of the code words is generated. The existence is calculated as a function of the weights assigned to the descriptors and spatial distance from them as described in 3.2. Such map is generated for each code word. Then, the existence values are sampled on a regular grid from all the maps and they are concatenated to form a feature vector. When sampling, a concept of max pooling [5, 6] is employed.

Finally, the generated feature vectors are fed to a machine learning algorithm to classify them into one of the pre-determined image categories.

The proposed SSV algorithm is compatible with most of the conventional SPM-based algorithms. It can work well with most of local feature descriptors, soft code assignment methods, and machine learning algorithms.

3.2   Spatial Map Generation

The local descriptors are extracted either on a regular grid, on random points, or on detected key points. Then, each local descriptor is softly encoded. Up to this point, the procedure is same as most of the conventional methods.

In our proposed SSV method, a map of existence of each local descriptor for each code word is generated. Let $\mathbf{X}$ be a set of descriptors in a D-dimensional feature space, i.e. $\mathbf{X} = [\mathbf{x}_1, \cdots, \mathbf{x}_M]^T \in \Re^{M \times D}$. The weights for the code words can be obtained as cluster membership indicators $\mathbf{C} = [\mathbf{c}_1, \cdots, \mathbf{c}_M]^T$ in the soft voting process. In the Sc, $\mathbf{C}$ is obtained by the following optimization problem

$$\min_{\mathbf{C}} \sum_{i=1}^{M} ||\mathbf{x}_i - \mathbf{Bc}_i||^2 + \lambda|\mathbf{c}_i|, \quad \text{subject to } |c_{ij}| < 1, \forall j = 1, 2, \cdots, K$$

where $\mathbf{B}$ is the codebook, c_{ij} is the weight for the jth elements in $\mathbf{c}_i$, and K is the codebook size. $\mathbf{C}$ in LLC is calculated as

$$\min_{\mathbf{C}} \sum_{i=1}^{M} ||\mathbf{x}_i - \mathbf{Bc}_i||^2 + \lambda||\mathbf{a}_i \odot \mathbf{c}_i||^2 \tag{1}$$

where $\odot$ represents the element-wise multiplication, and $\mathbf{a}_i \in \Re^D$ is the locality adapter that tries to assign as close code word to the descriptor as possible.

At this stage, each local descriptor $\mathbf{d}_i$ consists of three components: a descriptor $\mathbf{x}_i$, a code $\mathbf{c}_i$, and its location in the image (p_i, q_i). In order to make the dimension of the generated feature vectors consistent and compact, only the location information is normalized to $n \times n$. Namely, the descriptors and the codes are untouched but only the coordinates are normalized as follows:

$$p_i' = p_i \times n/(\text{width of the image}), \quad q_i' = q_i \times n/(\text{height of the image}) \tag{2}$$

Then, the existence values are sampled on a regular grid: (s, t), $s, t = 0.5, \cdots, n - 0.5$. The map for the jth element in the ith descriptor is calculated as follows.

$$P_{ij}(s, t) = c_{ij} \exp\left(-\frac{(s - p_i')^2 + (t - q_i')^2}{\sigma}\right), \quad s, t = 0.5, \cdots, n - 0.5 \tag{3}$$

Eq. 3 is a function of three parameters: the code (c_{ij}), the distance from the sampling point, and σ. Therefore, it does not necessarily mean that the descriptors closer to the sampling points have larger score. A toy example of the existence map generation for the ith code is illustrated in Fig. 2.

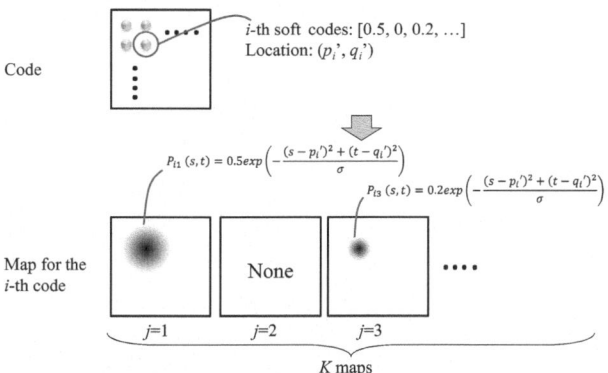

Fig. 2. Toy example of the existence map generation. The soft code is used as weights to the Gaussian function. The map is generated for all the elements in the codes.

3.3 Max Existence Sampling

Inspired by the success of the max pooling in the SPM framework [5, 6], "max existence sampling" is employed in this paper. In the max pooling, the maximum response to each code word among the descriptors in the sub-region is sampled as shown in the following:

$$c_{out} = \max(c_{in_1}, ..., c_{in_{Nsub}}) \tag{4}$$

where max functions in a row-wise manner, returning a vector the same size as c_{in}. N_{sub} is the number of descriptors in the sub-region. In our approach, the existence of the jth element for the location of (s,t) is sampled as in the following.

$$P_j^{out}(s,t) = \max(P_{1j}(s,t), ..., P_{Mj}(s,t)), \quad j = 1, \cdots, K \tag{5}$$

The average existence sampling is also possible. When $\sigma = \infty$ and $n = 1$, the result is exactly the same as those in [5, 6] whose spatial pyramid is only 1×1.

The feature vector is generated by concatenating the sampled existence values for all j. Therefore, the dimension of the feature vector would become $n \times n \times K$ for the $n \times n$ map configuration.

3.4 Grabbing Both Global and Local Features

The more sampling points and the smaller σ we have, the more locality-conscious the generated feature vectors would become. On the other hand, if the number of the sampling points is less and σ is larger, the feature vectors grab the global information. It is also possible to embed both global and local spatial information into a feature vector as in SPM [7] by increasing the number of the sampling points in a hierarchical manner.

Table 1. Comparison between SPM and SSV using the Caltech-101 dataset. The results with (*) were obtained by executing the source codes downloaded from the authors' sites.

	SPM (hard)	SSV (soft)
-Sc [5]		
$1\times1, 2\times2, 4\times4$	*71.8±1.14	74.6±0.82
4×4	-	74.6±0.82
8×8	-	75.1±0.58
-LLC [6]		
$1\times1, 2\times2, 4\times4$	*72.6±0.64	73.6±1.10
4×4	-	73.8±0.41
8×8	-	74.6±0.59

4   Experiments and Results

In this section, experimental results are shown based on four widely used datasets. We employed the ScSPM [5] and LLCSPM [6] for comparison. Note that Sc and LLC are different soft code assignment techniques because LLC is not sparse. The reason for choosing [5, 6] is that the source code is available on the authors' project site[1]. So anyone can reproduce the results in [5, 6] and those in this paper. The source code of this paper is available on our project page. In addition, we can guarantee that all the configurations other than the SSV part is the same.

Our descriptor extraction followed the one in [5]. Namely, the SIFT descriptors extracted from 16×16 pixel patches were densely sampled from each image on a grid with step size of 6 pixels. The images were all prepossessed into gray scale. Note that the local descriptor used in [6] was HoG [28] but in this paper SIFT is employed because the source code mentioned above also assumed SIFT. A simple linear SVM was used as the classifier.

Important to note is that the scope of this paper is to show the superiority of our SSV over the SPM. If different feature descriptors or different machine learning algorithms are employed, the performance could be made better but it is far beyond the scope of this paper.

4.1   Caltech-101 Dataset

The Caltech-101 dataset [29] contains 9,145 images in 101(+1 for background) classes including a variety of objects with large variance in shape. On the other hand, the position, orientation, and size of the objects are roughly aligned. The number of images per category was from 31 to 800. The images were resized to be no larger than 300×300 pixels with preserved aspect ratio. We mostly followed a common testing procedure as in [6, 10, 12, 30, 31]: 30 samples were

---

[1] http://www.ifp.illinois.edu/~jyang29/resources.html

Table 2. Comparison between SPM and SSV using Caltech-256. The results with (*) were obtained by executing the source codes downloaded from the authors' sites.

	SPM (hard)	SSV (soft)
-Sc [5]		
$1\times1, 2\times2, 4\times4$	*34.0±0.17	39.5±0.24
4×4	-	39.1±0.41
-LLC [6]		
$1\times1, 2\times2, 4\times4$	*35.8±0.15	37.0±0.25
4×4	-	37.1±0.24
8×8	-	37.5±0.21

randomly picked up from each class and the rest of them were used for testing. This process was repeated 5 times and the performance was measured by using average accuracy over 102 classes (102 accuracy values were averaged). The codebook sizes were set as 1,024 for Sc and 2,048 for LLC, respectively.

The performance comparison is shown in Table 1. Note here is that the results of ScSPM and LLCSPM were obtained by executing the source codes downloaded from the authors' project site. Therefore, the numbers are different from those in their papers [5, 6]. It is observed that the SSV-based methods always work better than the SPM-based ones even with the two different soft code assignment methods. The performance is increased by 2.8∼3.3% for Sc and 1.0∼2.0% for LLC. Please also notice that the dimension of the feature vector for the 4×4 grid is only 16/21 (=76%) of the original SPM while achieving better results. Some recent works [11, 12, 19, 22] outperform our ScSSV and LLCSSV but as we discussed already, ScSPM and LLCSPM are used as our baselines to show the validity of our SSV. [11, 12] proposed soft code assignment methods and they can be combined with our SSV. And please be aware that the proposed SSV already outperforms recent descriptor distribution aware methods such as [24] that considered the spatial relation of the descriptors, [18] that considered spatial distribution of descriptors by assigning weights to sub-regions, [32] that used probabilistic model for the SPM, and [9] that introduced another sparse coding in the pooling stage. Please see Appendix A for more detailed comparison with the previous works.

4.2   Caltech-256 Dataset

The Caltech-256 dataset holds 30,608 images in 256 categories with much higher intra-class variability and higher object location variability compared with Caltech-101. Each category contains at least 80 images. Same as in 4.1, all the images were resized so that they are no larger than 300×300 with aspect ratio being preserved. We trained a codebook with 4,096 bases. The number of training data was 30 for each object class. The improvement from SPM-based methods is 5.1∼5.5% for Sc and 1.2∼1.7% for LLC as demonstrated in Table 2. The improvement rates are

Table 3. Comparison using PASCAL VOC 2011 dataset. (*)The results of LLCSPM were obtained by executing the source codes downloaded from the authors' sites.

	SPM (hard)	SSV (soft)
-LLC [6]		
$1 \times 1, 2 \times 2, 4 \times 4$	*29.9±0.53	31.2±0.76
4×4	-	31.2±0.10
8×8	-	31.3±0.29

Table 4. Comparison using 15-Scene Categories dataset. (*)The results of ScSPM and LLCSPM were obtained by executing the source codes downloaded from the authors' sites.

	SPM (hard)	SSV (soft)
-Sc [5]		
$1 \times 1, 2 \times 2, 4 \times 4$	*80.3±0.41	81.5±0.55
4×4	-	81.6±0.40
-LLC [6]		
$1 \times 1, 2 \times 2, 4 \times 4$	*80.9±0.66	81.3±0.80
4×4	-	81.3±0.70

larger than those for the Caltech-101 data set because the baseline performance (original ScSPM and LLCSPM) is low and there is more room for the improvement.

4.3 PASCAL VOC 2011 Dataset

The PASCAL VOC 2011 dataset [33] holds 14,961 images in 20 categories with much higher intra-class variability, higher object location variability, and less bias compared with Caltech-101 and Caltech-256. All the images were resized so that they are no larger than 300×300 with aspect ratio being preserved. We trained a codebook with 2,048 bases. The number of training data was 30 for each object class. As shown in Table 3, the performance of SSV is still better than SPM even with such high intra-class and object location variability. The performance improvement is 1.3~1.4%.

4.4 15-Scene Categories Dataset

The 15-Scene Categories dataset [7] holds 2,285 images in 15 categories. All the images were resized so that they are no larger than 300×300 with aspect ratio being preserved. We trained a codebook with 1,024 bases for ScSPM and 2,048 bases for LLCSPM. The number of training data was 100 for each object class. As shown in Table 4, the performance of SSV is still better than SPM even with such high intra-class and object location variability. The performance improvement is 1.3% for ScSPM and 0.4% for LLCSPM.

Table 5. Comparison among different sampling strategies and σ's

σ	0.1	0.2	0.5	1.0	2.0	5.0
(1) 1×1	46.1	**47.9**	46.4	44.4	43.4	42.5
(2) 2×2	56.8	62.5	**65.7**	64.8	59.7	50.2
(3) 4×4	63.9	69.6	**73.8**	**73.8**	71.8	65.5
(4) 8×8	60.2	66.1	71.6	73.7	**74.6**	73.3
(1)-(3)	72.2	**73.6**	73.3	71.0	66.6	58.6
(1)-(4)	72.2	72.9	**74.0**	**74.0**	72.5	67.8
(1)-(3)$_{opt}$	73.6 $(\sigma_1, \sigma_2, \sigma_3) = (0.2, 0.5, 0.5)$					
(1)-(4)$_{opt}$	74.4 $(\sigma_1, \sigma_2, \sigma_3, \sigma_4) = (0.2, 0.5, 0.5, 2.0)$					

Table 6. Comparison between max sampling and average sampling using Caltech-101

Algorithms	Max	Average
LLCSSV		
1×1, 2×2, 4×4	73.6±1.10	68.0±0.41
4×4	73.8±0.41	69.2±0.75
8×8	74.6±0.59	69.8±0.62

4.5   Experiments Revisit with Caltech-101 Dataset

Sampling Density, Hierarchical Sampling, and σ. Table 5 compares the classification performance as a function of the sampling density and σ. It can be observed that the optimal σ is within a narrow range of 0.2~2.0: the optimal values do not vary a lot because the size of the existence map is decided by the number of the sampling point, which makes the impact of the σ value almost the same regardless of the sampling density. The table shows the more sampling point we have, the larger σ's yield better results. It is also demonstrated that as the sampling density is increased, the classification accuracy is improved as well. The performance of the hierarchical sampling is almost the same as those of 4×4 and 8×8 sampling. Choosing the best σ in each hierarchical level does not necessarily enhance the accuracy. We also confirmed that the optimal σ's are independent of the dataset. Hereafter, the σ's are fixed at 0.5 for 4×4, 2.0 for 8×8, and 0.2 for 1×1, 2×2, 4×4.

Existence Sampling Methods. We also compared the max sampling (eq. 5) with the average sampling. More specifically, the existence map is generated by averaging all the existence values at a certain point:

$$P_j^{out}(s,t) = \text{average}(P_{j1}(s,t), ..., P_{jM}(s,t)) \tag{6}$$

As shown in Table 6, the max sampling is better than the average sampling. When the max sampling is employed, just like the max pooling [5, 6], peaky responses of certain descriptors can be well embedded to the generated feature

Table 7. Comparison of soft code handling using Caltech-101 dataset

Weight calculation	as is	sqrt	square	all one
LLCSSV				
1×1, 2×2, 4×4	73.6	73.1	65.9	71.6
4×4	73.8	73.1	66.4	71.7
8×8	74.6	74.0	67.7	72.9

Table 8. Comparison of SIFT sampling point density using Caltech-101 dataset

Sampling grid for SIFT	LLCSPM (1×1, 2×2, 4×4)	LLCSSV (4×4)
4	73.8±1.21	74.8±0.64
6	72.6±0.64	73.8±0.41
8	70.0±1.45	71.6±0.78

vector. Due to the same reason, normalizing the sampled existence values over the image does not work well.

Soft Code Handling. In our paper, the soft codes assigned to the descriptors are used as the weight values for the existence calculation as shown in eq. 3. We compared it with three different weight decision strategies: sqrt ($\sqrt{|c_{ij}|}$), square (c_{ij}^2), all one ($c_{ij} = 1$ if $c_{ij} > 0$, 0 otherwise). Simply using the soft codes as weights is the best as shown in Table 7.

Grid Spacing for SIFT Description. The classification performance is improved when the sampling point for the SIFT descriptor extraction is denser. Even in such cases, SSV methods outperform SPM methods by 1.0∼1.6% as shown in Table 8.

Codebook Size. We further evaluated the performance with respect to codebook training. We tried three sizes: 1,024, 2,048, and 4,096. The results are shown in Table 9. It is shown that the accuracy increases when the codebook size is increased from 1,024 to 2,048 but saturates after that. It should also be noted that increasing the codebook size affects the computational efficiency both in code assignment and machine learning.

4.6   Discussion

The computational complexity is almost the same as those of SPM-based methods [5, 6]. The extra cost for the SSV is calculating the existence for all the descriptors for all the sampling points and taking the maximum, which is much cheaper than the soft code assignment. The dimension of the feature vectors is the number of sampling points times K, which is similar to the other approaches whose dimension is the number of sub-regions times K.

Table 9. Dependency on the codebook size

Codebook size	1,024	2,048	4,096
LLCSPM	*71.5	*72.6	*72.7
LLCSSV			
1×1, 2×2, 4×4	71.0	73.6	74.2
4×4	70.3	73.8	73.9
8×8	72.0	74.6	74.1

Table 10. Classification rate (%) comparison on Caltech-101. The results with (*) were obtained by executing their source code.

Algorithms\ train	5	10	15	20	25	30
Kernel Codebook [4]	-	-	-	-	-	64.14±1.18
ScSPM [5]	-	-	67.0±0.45	-	-	73.2±0.54
*ScSPM [5]	-	-	-	-	-	71.8±1.14
LLCSPM [6]	51.15	59.77	65.43	67.74	70.16	73.44
*LLCSPM [6]	*49.8	*60.4	*65.1	*68.6	*70.8	*72.6
Code Relation [24]	-	-	66.88	-	-	74.25
D-SP [18]	-	-	-	-	-	67.21±0.67
LC-KSVD2 [10]	54.00	63.10	67.70	70.50	72.30	73.60
RLDA [32]	-	-	67.4±0.5	-	-	73.7±0.8
Hie Sc [9]	-	-	-	-	-	74.00
LR-Sc+SPM [11]	-	-	69.58± 0.97	-	-	75.68± 0.89
RBC+FC [22]	-	-	69.63±0.84	-	-	77.09±0.74
l_p-norm FP [19]	59.35	-	70.34	-	-	82.60
DA-Sc [12]	66.13	73.09	78.38	82.36	83.20	83.28
Field Learning [27]	-	-	-	-	-	75.3±0.70
Ours, ScSSV						
1×1, 2×2, 4×4	53.2±0.51	63.2±0.52	67.6±0.75	70.2±0.48	71.8±0.49	74.6±0.82
4×4	53.5±0.33	63.2±0.67	67.6±0.44	70.5±0.65	72.1±0.32	74.6±0.82
8×8	54.6±0.54	64.3±0.50	68.7±0.34	71.6±0.32	72.9±0.47	75.1±0.58
Ours, LLCSSV						
1×1, 2×2, 4×4	51.3±0.24	61.8±0.48	66.6±0.25	69.4±0.38	71.8±0.31	73.6±0.53
4×4	50.5±0.38	61.4±0.51	66.2±0.37	69.4±0.32	71.7±0.51	73.8±0.41
8×8	51.6±0.22	62.6±0.31	67.5±0.41	70.5±0.78	72.6±0.38	74.6±0.59

5   Conclusions and Future Work

In this paper we proposed a simple but promising image-space-aware soft feature generation method called spatial soft voting (SSV) for image classification. In stead of forcing the codes of the descriptors to be assigned to one of the sub-regions as in conventional approaches, the existence values of the codes were calculated in the image space and they were sampled to form feature vectors. The hierarchical sampling was also possible to grab both global and local features of images. The validity of the proposed algorithm has been verified by the experiments. Since the soft code assignment with SPM is very popular in state-of-the-art image classification, we believe the suggested SSV will help SPM-based systems perform better. One of the largest advantages of the proposed method is that it is applicable to most feature description and machine learning algorithms as long as the spatial pyramid framework is employed.

References

1. Sivic, J., Zisserman, A.: Video google: A text retrieval approach to object matching in videos. In: ICCV, pp. 1470–1477 (2003), BoF
2. Csurka, G., Dance, C., Fan, L., Willamowski, J., Bray, C.: Visual categorization with bags of keypoints. In: Workshop on Statistical Learning in Computer Vision, ECCV (2004)
3. Philbin, J., Chum, O., Isard, M., Sivic, J., Zisserman, A.: Lost in quantization: improving particular object retrieval in large scale image databases. In: CVPR (2008)
4. van Gemert, J.C., Geusebroek, J.-M., Veenman, C.J., Smeulders, A.W.M.: Kernel Codebooks for Scene Categorization. In: Forsyth, D., Torr, P., Zisserman, A. (eds.) ECCV 2008, Part III. LNCS, vol. 5304, pp. 696–709. Springer, Heidelberg (2008)
5. Yang, J., Yu, K., Gong, Y., Huang, T.: Linear spatial pyramid matching using sparse coding for image classification. In: CVPR (2009)
6. Wang, J., Yang, J., Yu, K., Lv, F., Huang, T., Gong, Y.: Locality constrained linear coding for image classification. In: CVPR (2010)
7. Lazebnik, S., Schmid, C., Ponce, J.: Beyond bags of features: Spatial pyramid matching for recognizing natural scene categories. In: CVPR (2006)
8. Farquhar, J., Szedmak, S., Meng, H., Shawe-Taylor, J.: Improving "bag-of-keypoints" image categorisation. Technical report, University of Southampton (2005)
9. Yu, K., Lin, Y., Lafferty, J.: Learning image representations from the pixel level via hierarchical sparse coding. In: CVPR (2011)
10. Jiang, Z., Lin, Z., Davis, L.S.: Learning a discriminative dictionary for sparse coding via label consistent k-svd. In: CVPR (2011)
11. Zhang, C., Liu, J., Tian, Q., Xu, C., Lu, H., Ma, S.: Image classification by non-negative sparse coding, low-rank and sparse decomposition. In: CVPR (2011)
12. Kulkarni, N., Li, B.: Discriminative affine sparse codes for image classification. In: CVPR (2011)
13. Marszalek, M., Schmid, C., Harzallah, H., van de Weijer, J.: Learning object representations for visual object class recognition. In: Visual Recog. Challange Workshop (2007)
14. Perronnin, F., Sánchez, J., Mensink, T.: Improving the Fisher Kernel for Large-Scale Image Classification. In: Daniilidis, K., Maragos, P., Paragios, N. (eds.) ECCV 2010, Part IV. LNCS, vol. 6314, pp. 143–156. Springer, Heidelberg (2010)
15. Yang, J., Yu, K., Huang, T.: Efficient Highly Over-Complete Sparse Coding Using a Mixture Model. In: Daniilidis, K., Maragos, P., Paragios, N. (eds.) ECCV 2010, Part V. LNCS, vol. 6315, pp. 113–126. Springer, Heidelberg (2010)
16. Zhou, X., Yu, K., Zhang, T., Huang, T.S.: Image Classification Using Super-Vector Coding of Local Image Descriptors. In: Daniilidis, K., Maragos, P., Paragios, N. (eds.) ECCV 2010, Part V. LNCS, vol. 6315, pp. 141–154. Springer, Heidelberg (2010)

17. Grauman, K., Darrell, T.: The pyramid match kernel: discriminative classification with sets of image features. In: ICCV, vol. 2, pp. 1458–1465 (2005), Pyramid match kernel, base of SPM
18. Harada, T., Ushiku, Y., Yamashita, Y., Kuniyoshi, Y.: Discriminative spatial pyramid. In: CVPR (2011)
19. Feng, J., Ni, B., Tian, Q., Yan, S.: Geometric l_p-norm feature pooling for image classification. In: CVPR (2011)
20. Marszaek, M., Schmid, C.: Spatial weighting for bag-of-features. In: CVPR (2006), Spatial relation but segmentation is required
21. Cao, Y., Wang, C., Li, Z., Zhang, L., Zhang, L.: Spatial bag-of-features. In: CVPR (2010)
22. Wang, X., Bai, X., Liu, W., Latecki, L.J.: Feature context for image classification and object detection. In: CVPR (2011)
23. Belongie, S., Malik, J., Puzicha, J.: Shape matching and object recognition using shape contexts. IEEE TPAMI 24, 509–522 (2002)
24. Huang, Y., Huang, K., Wang, C., Tan, T.: Exploring relations of visual codes for image classification. In: CVPR (2011)
25. Krapacy, J., Verbeek, J., Jurie, F.: Modeling spatial layout with fisher vectors for image categorization. In: ICCV (2011)
26. Boureau, Y., Roux, N.L., Bach, F., Ponce, J., LeCun, Y.: Ask the locals: Multi-way local pooling for image recognition. In: ICCV (2011)
27. Jia, Y., Huang, C.: Beyond spatial pyramids: Receptive field learning for pooled image features. In: CVPR (2012)
28. Dalal, N., Triggs, B.: Histograms of oriented gradients for human detection. In: CVPR (2005)
29. Li, F.-F., Fergus, R., Perona, P.: Learning generative visual models from few training examples: an incremental bayesian approach tested on 101 object categories. In: CVPR Workshop on Generative-Model Based Vision (2004)
30. Zhang, H., Berg, A., Maire, M., Malik, J.: Svm-knn: Discriminative nearest neighbor classification for visual category recognition. In: CVPR (2006)
31. Griffin, G., Holub, A., Perona, P.: Caltech-256 object category dataset. Technical Report 7694, California Institute of Technology (2007)
32. Karayev, S., Fritz, M., Fidler, S., Darrell, T.: A probabilistic model for recursive factorized image features. In: CVPR (2011)
33. Everingham, M., Van Gool, L., Williams, C.K.I., Winn, J., Zisserman, A.: The PASCAL Visual Object Classes Challenge 2011 (VOC 2011) Results

A   Comparison Using Caltech-101

Detailed comparison of the image classification performance is shown in Table 10. Our SSV-based methods outperform the SPM with Sc [5] and LLC [6]. It also outperforms most of the previous works and comparable to most of the state-of-the-arts.

Efficient Geometric Re-ranking
for Mobile Visual Search

Junwu Luo and Bo Lang

State Key Laboratory of Software Development Environment
Dept. of Computer Science and Engineering, Beihang University
Beijing, China
{luojunwu,langbo}@nlsde.buaa.edu.cn

Abstract. The state-of-the-art mobile visual search approaches are based on the bag-of-visual-word (BoW). As BoW representation ignores geometric relationship among the local features, a full geometric constraint like RANSAC is usually used as a post-processing step to re-rank the matched images, which has been shown to greatly improve the precision but at high computational cost. In this paper we present a novel and efficient geometric re-ranking method. Our basic idea is that the true matching local features should be not only in a similar spatial context, but also have a consistent spatial relationship, thus we simultaneously introduce *context similarity* and *spatial similarity* to describe the geometric consistency. By incorporating these two geometric constraints, the co-occurring visual words in the same spatial context can be regarded as a "visual phrase" and significantly improve the discriminative power than single visual word. To evaluate our approach, we perform experiments on Star5k and ImageNet100k dataset. The comparison with the BoW method and Soft-assignment method highlights the effectiveness of our approach in both accuracy and speed.

1 Introduction

With the development of mobile internet and the fast proliferation of smart phones, a new class of augmented reality applications [1–3] based on CBIR technology, named mobile visual search, are becoming increasingly popular and prevalent. Users can use camera-phone to take photos of objects they are interested in and then recognize the objects by visual search engine. Such applications can be used for identifying products, artworks, print media, landmarks etc.

Today, the state-of-the-art algorithm for mobile visual search is referred to as bag of visual words (BoW), this idea is borrowed from text retrieval. For image, robust local feature descriptors are extracted from image database and then clustered into a vocabulary tree, whose leaf nodes are treated as visual words. Next, each local feature is mapped to the visual word closest to it under the vocabulary tree and then an image can be represented as a bag of visual words. Further, taking visual words as keys, images in the database are indexed by inverted files for fast retrieval.

J.-I. Park and J. Kim (Eds.): ACCV 2012 Workshops, Part II, LNCS 7729, pp. 520–532, 2013.

While BoW has two major issues: First, hard quantization makes many matching features that should be quantized to the same visual word end up in different words due to quantization error. Second, BoW representation ignores the geometric relationship among local features, thus it brings false feature matches and reduces the accuracy. Considering this, Soft-assignment [4] has been proposed to quantize each feature to more than one visual word. The greedy N-best paths scheme [5] also reduces the quantization error but increases the query time. Furthermore, several approaches about Geometric Verification (GV) have been proposed recently. In [6], a geometric transformation of the location of features in query image and the database images is estimated by Hough scheme and RANSAC-based method, but it is computationally expensive and prohibits the list of candidate images for verification to a small number, then some groups in [7, 8] have investigated different ways to speed up this GV process. To exploit the geometric information without explicitly estimating the affine transformation, week geometric verification is proposed. The authors Jegou et al. [9] use the angle and scale information of features to check the geometric consistency. In [10], the location information of features is used to define the relative geometric ordering relationship between two bundled features. The authors in [11] propose a fast geometric re-ranking method by the observation that the matching feature pairs should have consistent distance ratios, orientation difference and scale difference.

For mobile visual search, a significant rotation between the query image and the database image easily occurs as the camera angle is not fixed. Thereby a robust geometric similarity should be rotation invariant besides scale invariant. In view of this, this paper presents a novel and efficient geometric verification method. Our basic idea is that two true matching local features should be not only in a similar spatial context, but also have a consistent spatial relationship, thus we simultaneously introduce *context similarity* and *spatial similarity* to describe the geometric consistency. We show in experiments that our geometric verification schemes can achieve higher retrieval accuracy than BoW method and Soft-assignment method, while only introducing a modest time penalty.

The paper is organized as follows. Section 2 gives a detailed introduction of our geometric verification method. In Section 3, the index and score method is presented. Section 4 discusses our experiment results. Finally we conclude in Section 5.

2   Proposed Geometric Verification Method

Like text retrieval, the context of a word refers to the surrounding text of this word (e.g. a sentence or a paragraph). Analogically, we can define the context of a local feature as a spatial region where it falls inside, thus local features representing the same object tend to have a consistent spatial context. As illustrated in Fig. 1, the true matching local features should be not only in a similar spatial context (e.g. density and shape of the ellipse region), but also have a consistent spatial relative relationship (e.g. position). Based on this observation, we simultaneously introduce *context similarity* and *spatial similarity* to describe the geometric consistency. By incorporating these two geometric constraints to the matching pairs, the co-occurring visual words in the same spatial context can be regarded as a "visual phrase" and significantly improve the discriminative power than single visual word.

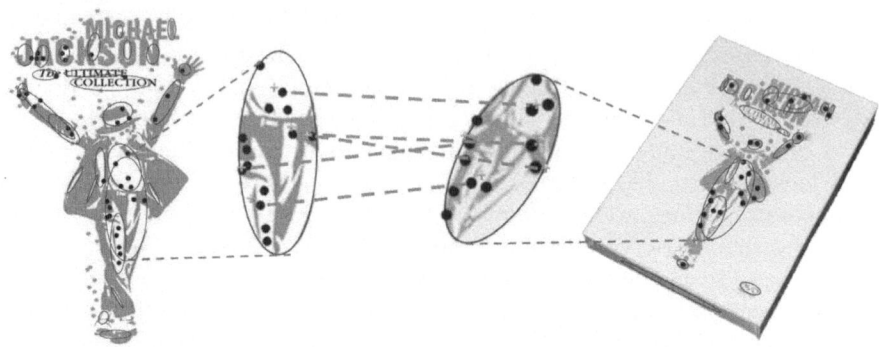

Fig. 1. An illustration of spatial context and spatial relationship of the matching pairs. Left: a clean database image. Right: a real-world image captured by phone. Dot marks local feature, red ellipse marks MSER region, cyan dotted line marks a pair of matched local features.

For a local feature, we define its spatial context as a MSER region [12] where it falls inside, not simply take the circular or grid region as the context, for we think the MSER region is more semantic relevant to a specified object and can narrow down the semantic gap to a certain extent.

Denote the local feature by $p = (x, y, w)$, where (x, y) is the location and w is the visual word. Denote the MSER region by ellipse $r = (x, y, a, b, \alpha)$, where (x, y) is the center, (a, b) is the length of major and minor axis, and α is the angle between X-axis and the major-axis.

For image I, we denote the set of local features by $P = \{p_i\}$ and the set of MSER regions by $R = \{r_i\}$, then we can group the local features by MSER region, it defines as follow:

$$g_i = \{p_j | p_j \propto r_i, p_j \in P\} \tag{1}$$

where $p_j \propto r_i$ means the local feature p_j falls inside the region r_i. Empty g_i is discarded. Note that one local feature may fall into multiple regions or may not fall into any region.

If two local features are quantized the same visual word, then we say these two local features are a matched pair. Furthermore, we think the two corresponding regions where the two local features fall inside are a matched pair. Now, we define a matching score $S(r_i, r_j)$ for a pair of matched region, which consists of a context similarity $CS(r_i, r_j)$ and a spatial similarity $SS(r_i, r_j)$:

$$S(r_i, r_j) = \alpha \cdot CS(r_i, r_j) + \beta \cdot SS(r_i, r_j) \tag{2}$$

where α and β are weighting parameters, whose sum equals 1.

2.1   Context Similarity

As shown in Fig. 1, for two true matched regions, not only should they contain a similar number of local features (we say their density are very close), but also their shape should be similar. Thus we can define the context similarity between two MSER regions as follows:

$$CS(r_i, r_j) = Density_{sim}(r_i, r_j) + Shape_{sim}(r_i, r_j) \tag{3}$$

where $Density_{sim}(r_i, r_j)$ is the density similarity, and $Shape_{sim}(r_i, r_j)$ is the shape similarity. For density similarity, it defines as:

$$Density_{sim}(r_i, r_j) = \frac{min(Density(r_i), Density(r_j))}{max(Density(r_i), Density(r_j))} \tag{4}$$

where $Density(r_i)$ and $Density(r_j)$ are respectively the number of local features fall inside region r_i and r_j.

For shape similarity, we calculate it based on the "characteristic triangle" of ellipse. As shown in Fig. 2(a), the vertex of minor axis and the two foci form the characteristic triangle ΔPF_1F_2. If the corresponding characteristic triangles of two ellipses are similar, then we think these two ellipses are similar, so we can calculate the shape similarity by the characteristic triangle similarity.

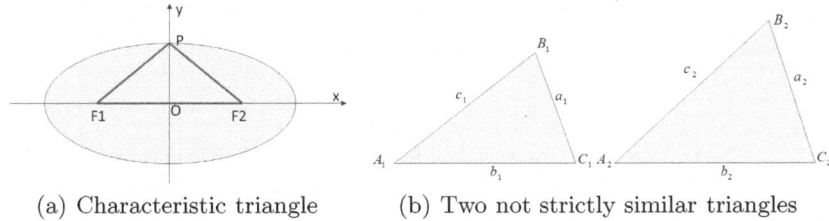

(a) Characteristic triangle   (b) Two not strictly similar triangles

Fig. 2. Shape similarity between two MSER region defined by their characteristic triangle similarity. Left: characteristic triangle of an ellipse. Right:two similar but not strictly similar triangles.

As shown Fig. 2(b), two similar but not strictly similar triangles $\Delta A_1B_1C_1$ and $\Delta A_2B_2C_2$, we define their similarity as:

$$T_{sim}(\Delta A_1B_1C_1, \Delta A_2B_2C_2) = \frac{1}{(1 + |\frac{a_1}{d_1} - \frac{a_2}{d_2}| + |\frac{b_1}{d_1} - \frac{b_2}{d_2}| + |\frac{c_1}{d_1} - \frac{c_2}{d_2}|)} \tag{5}$$

where a_1, b_1, c_1 and a_2, b_2, c_2 are respectively the edges of $\Delta A_1B_1C_1$ and $\Delta A_2B_2C_2$, d_1 and d_2 are respectively the perimeters of these two triangles. If two triangles are strictly similar, the value of T_{sim} equals 1, otherwise it is closer to 0.

2.2   Spatial Similarity

For mobile search scenario, a significant rotation between the query image and the database image easily occurs, so a robust spatial similarity should be rotation invariant besides scale invariant. Here we propose two effective schemes to define the spatial similarity based on the location information of local features between two matched MSER regions.

For two matched local features $p = (x_1, y_1, w)$ and $q = (x_2, y_2, w)$, we denote their matched pair by $m = ((x_1, y_1), (x_2, y_2), w)$. Then for two matched MSER regions r_i and r_j, we can obtain a set of matched local feature pairs:

$$M = \{m_k\} = \{((x_{k1}, y_{k1}), (x_{k2}, y_{k2}), w_k) | (x_{k1}, y_{k1}) \propto r_i, (x_{k2}, y_{k2}) \propto r_j\} \quad (6)$$

Angle Consistence. As shown in Fig. 3, it is an illustration of two matched MSER regions and their matched local feature pairs. For two matched local feature pairs $m_i = \{(x_{i1}, y_{i1}), (x_{i2}, y_{i2}), w_i\}$ and $m_j = \{(x_{j1}, y_{j1}), (x_{j2}, y_{j2}), w_j\}$, if they are both the true matched pair, then the angle α_1 (before matching) and α_2 (after matching) should be similar, this conclusion is established though it exists a significant rotation or scale change between query image and database image.

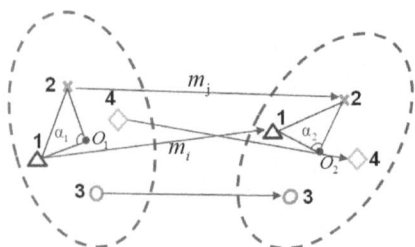

Fig. 3. The illustration of two matched MSER regions and their matched local feature pairs. There exists a rotation between these two MSER regions, but the angle α_1 (before matching) and α_2 (after matching) should be similar.

Based on this assumption, we firstly define an indicator function that measures the angle consistence between two matches m_i and m_j:

$$\delta(m_i, m_j) = \begin{cases} 1 & if |1 - \alpha_2/\alpha_1| < \delta_D, \\ 0 & otherwise. \end{cases} \quad (7)$$

where δ_D is a predefined threshold, it represents the allowable error range for angle change. Then, we can define the spatial similarity between two MSER regions as follows:

$$SS(r_i, r_j) = \frac{\Sigma_{i \neq j, m_i, m_j \in M} \delta(m_i, m_j)}{C_{|M|}^2} \quad (8)$$

where M is the set of matched local feature pairs between region r_i and r_j, $C^2_{|M|}$ is the combinations of M.

Circle Distance Consistence. As illustration in Fig. 4, we start from the vertical orientation and divide the MSER region into several sectors with the same angle in clockwise, and then number them with $1, 2, \cdots, N$. The circle distance between two sectors is defined as the relative difference of their number. For two true matched MSER regions, all the matched local feature pairs should have a consistent circle distance.

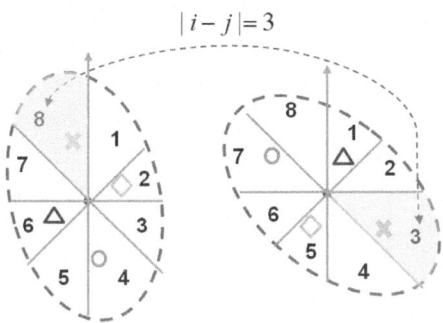

Fig. 4. The illustration of circle distance between two sectors from two matched MSER regions

Firstly, we calculate the circle distance d_i for each matched local features pair $m_i = \{(x_{i1}, y_{i1}), (x_{i2}, y_{i2}), w_i\}$, it is obvious that d_i satisfy $0 \le d_i \le N-1$. After that we can obtain a set of circle distances $S = \{d_i\}$, and then split S into N subset by the value of circle distance:

$$S_k = \{d_{ki} | d_{ki} = k, d_{ki} \in S\} \quad (0 \le k \le N-1) \tag{9}$$

Based on statistics theory, we can use the k corresponding to the maximum size of subset S_k as the region circle distance. Then the spatial similarity between two MSER regions can be defined as follows:

$$SS(r_i, r_j) = \frac{\max_k |S_k|}{|M|} \tag{10}$$

where M is the set of matched local feature pairs between region r_i and r_j.

3   Index and Score

For efficient large-scale indexing and retrieval, we use an inverted index [13]. As shown in Fig. 5, the structure of our index is composed of two parts: 1) the global inverted index using the term frequency information for traditional text retrieval. Each visual word has an entry in the index which contains the list of images where the visual word appears; 2) the local inverted index for each image using spatial information for geometric verification. Each visual word has an entry containing the list of local features which are assigned to this visual word, and then each local feature has an entry containing the list of MSER regions where the local feature falls inside. (Note that, in one image, multiple local features may be quantized to a visual word and a local feature also may belong to multiple regions)

For each local feature, we use 20 bits to record the location information: respectively 10 bits for X-coordinate and Y-coordinate (In our experiments, all the images are resized to no larger than 640×640). For each MSER region, we record the central point coordinates (20 bits), length of major and minor axis (20 bits), the angle between X-axis and the major-axis in degree (9 bits).

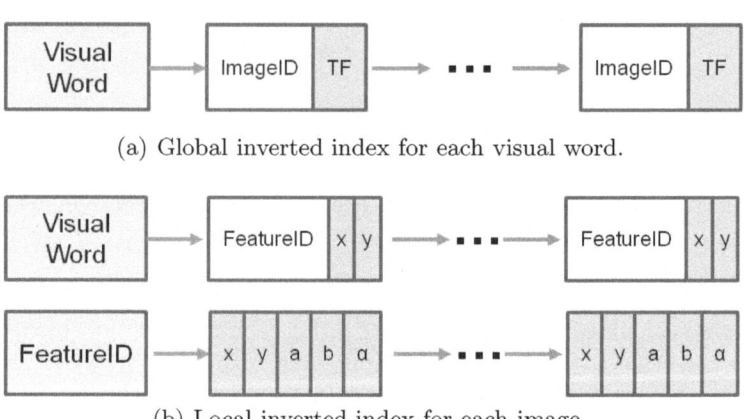

(a) Global inverted index for each visual word.

(b) Local inverted index for each image.

Fig. 5. Inverted index structure

After indexing, we firstly use the TF-IDF score [14] to measure the similarity between the query image Q and database image I, and rank the matched images. Then we re-rank them incorporating geometric information. The new scoring formula is as follows:

$$Score(Q, I) = \alpha \cdot Score_{tfidf}(Q, I) + \beta \cdot Score_{GV}(Q, I), \alpha + \beta = 1 \qquad (11)$$

where $Score_{tfidf}(Q, I)$ is the standard TF-IDF score, and $Score_{GV}(Q, I)$ is the geometric verification score which can be calculated by Algorithm 1 based on the local inverted index shown in Fig. 5(b). Note that we should normalize these two scores respectively before merging. By this way, we have augmented the bag-of-visual-word model with geometric constraints and improve the discriminative power of visual word.

Algorithm 1. Calculate Geometric Verification Score

Input: query image Q and candidate image I
Output: geometric verification score
1: Initialize the MSER region pairs set $T = \emptyset$.
2: Generate the common visual words set Ω between Q and I.
3: **foreach** visual word $w_i \in \Omega$ **do**
4:     $P_i^Q \leftarrow$ local features set quantized to visual word w_i in image Q.
5:     $P_i^I \leftarrow$ local features set quantized to visual word w_i in image I.
6:     **foreach** local feature $p_m^Q \in P_i^Q$ **do**
7:         **foreach** local feature $p_n^I \in P_i^I$ **do**
8:             $R_m^Q \leftarrow$ MSER regions set where local feature p_m^Q falls inside in image Q.
9:             $R_n^I \leftarrow$ MSER regions set where local feature p_n^I falls inside in image I.
10:            Add MSER region pair (r_i, r_j) to T where $r_i \in R_m^Q, r_j \in R_n^I$.
11:        **end for**
12:    **end for**
13: **end for**
14: Initialize $Score_{GV}(Q, I) = 0$
15: **foreach** MSER region pair $(r_i, r_j) \in T$ **do**
16:    $CS(r_i, r_j) \leftarrow$ context similarity between r_i and r_j.
17:    $SS(r_i, r_j) \leftarrow$ spatial similarity between r_i and r_j.
18:    $S(r_i, r_j) \leftarrow \alpha \cdot CS(r_i, r_j) + \beta \cdot SS(r_i, r_j)$, where $\alpha + \beta = 1$.
19:    $Score_{GV}(Q, I) \leftarrow Score_{GV}(Q, I) + S(r_i, r_j)$.
20: **end for**
21: **return** $Score_{GV}(Q, I)$.

---

4   Results and Discussion

A series of experiments are conducted to evaluate our proposed geometric re-ranking method for mobile visual search. Firstly, we explain the datasets and evaluation criteria, and then compare the overall performance with the state-of-the-art methods on different datasets.

The common settings for all experiments are summarized here. All the images in the dataset are resized to no larger than 640×640. For each image, Scale Invariant Feature Transform (SIFT) [15], one of the most popular and robust local features, is extracted and quantized based on vocabulary tree as [16] did. The vocabulary tree with depth 5 and branch factor 10 is built by hierarchical K-Means clustering.

4.1   Datasets and Evaluation Criteria

In the following, we list the datasets and evaluation criteria in our experiments. The statistics are summarized in Table 1.

Star5k is collected by [17], and contains $5,613$ images from posters and CD covers of 107 singers. The 50 query images are captured by mobile camera in

Fig. 6. Example images from the Star5k dataset. Top: a clean database image, Bottom: a real-world image with various distortions captured by mobile camera.

real environment, which contain much more variations like complex background and poor light conditions. Fig. 6 has shown some examples from this dataset.

ImageNet100k is an image dataset crawled from internet according to the WordNet hierarchy [18]. It includes $111,489$ images of 8 categories, e.g. animal, plant, food, activity. We employ this dataset as distracters to test performance and scalability of our methods. We have ensured there is no overlap with Star5k dataset.

ILSVRC30k is a dataset of $30K$ images provided by ImageNet Large Scale Visual Recognition Challenge 2010 (ILSVRC2010), which covers large varieties of images, e.g. scenes, objects. We use this dataset to train the vocabulary tree in this paper. It also has no overlap with the other datasets mentioned above.

Table 1. Statistics of the image datasets in the experiments

Dataset	# images	# queries	# descriptors
Star5k	5,613	50	1,852,649
ImageNet100k	111,489	N/A	31,923,123
ILSVRC30k	30,000	N/A	9,127,190

For mobile visual search, we concern if the original image is retrieved from the dataset and ranked on the top, namely the rank of the correct answer. Thus we use mean reciprocal rank (MRR) [19] to evaluate the performance. The MRR is defined as follows:

$$Score = \frac{1}{n} \sum_i \frac{1}{rank_i} \qquad (12)$$

where n is the number of all the queries, and $rank_i$ is the position of the original image in the retrieval result list. In this paper, we only consider the top 10 results to satisfy the mobile search scenario.

4.2   Evaluation

Comparisons. We compare the performance of the following six approaches: (1) BoW, the standard bag of visual words method; (2) BoW+a-GV, geometric verification with angle consistence based on BoW; (3) BoW+cd-GV, geometric verification with circle distance consistence based on BoW; (4) Soft, Soft-assignment, and the number of words is set to 3, under which the best performance is achieved. (5) Soft+a-GV, geometric verification with angle consistence based on Soft; (6) Soft+cd-GV, geometric verification with circle distance consistence based on Soft. In our implementation, the spatial score weight β is set to be 0.3, and the number of MSER regions sector N in GV is set to be 8 (see below for experiment study of the effect of varying β and N) . We employ different numbers of images from ImageNet100k as distracters, i.e., the first 30k, 60k and the entire set of 100k images.

Table 2. Performance comparison of different approaches on various datasets

	Star5k	Star5k+30k	Star5k+60k	Star5k+100k
BoW	0.884	0.792	0.765	0.740
BoW+a-GV	0.916(+3.2%)	0.853(+6.1%)	0.818(+5.3%)	0.795(+5.5%)
BoW+cd-GV	0.904(+2.0%)	0.835(+4.3%)	0.808(+4.3%)	0.795(+5.5%)
Soft	0.909	0.857	0.846	0.828
Soft+a-GV	0.933(+2.4%)	0.900(+4.3%)	0.881(+3.5%)	0.875(+4.7%)
Soft+cd-GV	0.924(+1.5%)	0.902(+4.5%)	0.882(+3.6%)	0.869(+4.1%)

Table 2 compares the above six approaches using MRR, leading to three major observations:

(1) Compared to BoW and Soft, our geometric verification methods can improve the values of MRR. This is because BoW and Soft ignore the geometric relationship among local features and bring false feature matches. It indicates the importance of post-processing based on geometric information.

(2) BoW+a-GV achieves slightly better performance than BoW+cd-GV. The reason is that angle consistence has stronger geometric constraints than circle distance consistence.

(3) We use ImageNet100k as distracters to test scalability. Results shown our methods always boost the performance over BoW and Soft when scaling up the number of distracters. It demonstrates that our methods have good generalization capability and scalability on large scale dataset.

Impact of Spatial Score Weight β. The value of β in Eq.11 determines the weight of spatial score in the new scoring method. We evaluate the performance of BoW+a-GV method using different β from 0 to 0.5 on Star5k dataset. As shown in Table 3, The most effective value of β is around 0.3.

Table 3. Performance comparison of different values of β

(α, β)	$(0.5, 0.5)$	$(0.6, 0.4)$	$\mathbf{(0.7, 0.3)}$	$(0.8, 0.2)$	$(0.9, 0.1)$	$(1.0, 0.0)$
$MRR@10$	0.832	0.893	**0.916**	0.904	0.884	0.884

Impact of Number of MSER Region Sectors N. As introduced in Section 2.2, a MSER region is divided into several sectors to incorporate more spatial information. We evaluate the performance of BoW+cd-GV method using different N from 2^0 to 2^5 on Star5k dataset. As shown in Table 4, the performance is improved with more sectors in the partition. However, the improvement is not significant when N is large. Therefore, we selected $N = 8$ in the experiments.

Table 4. Performance comparison of different sector number of MSER region in GV

N	1	2	4	**8**	16	32
$MRR@10$	0.886	0.894	0.899	**0.904**	0.904	0.902

Runtime. We perform our experiments on a desktop with dual Core CPU of $2.13GHz$ and $16G$ memory. The SIFT and MSER feature extraction time is not included. All the algorithms are implemented by Java. Fig. 7 shows the average query time for one image using Star5k+ImageNet100k dataset. We can see that our geometric re-ranking methods take a little more time than BoW and Soft (about $30 \sim 100$ms), while achieving a higher retrieval accuracy.

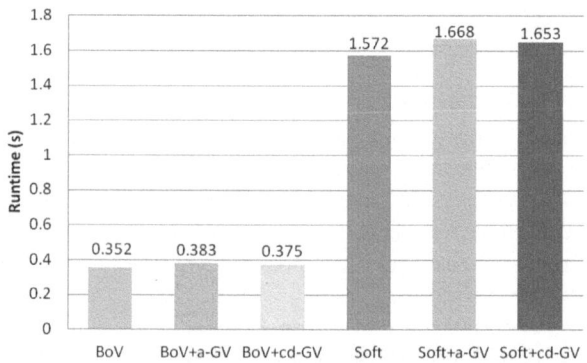

Fig. 7. Average query time comparison of different approaches

5   Conclusion

In this paper, we propose a novel and efficient geometric re-ranking method for mobile visual search. By regarding the MSER region as the spatial context of local features, we simultaneously introduce context similarity and spatial similarity to describe the geometric consistency. It significantly improves the features discriminative power. The experiments on Star5k and ImageNet100k dataset validate the effectiveness of our methods in both accuracy and speed.

There is still much room for future improvement, e.g., exploring a better strategy to define the spatial context of local features at the semantic level to achieve the goal of narrowing down semantic gap. Another direction is to consider the co-occurrence probability of visual words based on the spatial context, and boost their weight in standard bag of visual words method.

References

1. Google: Goggles, http://www.google.com/mobile/goggles/
2. Nokia: Point and Find, http://www.pointandfind.nokia.com
3. Amazon: SnapTell, http://www.snaptell.com
4. Philbin, J., Chum, O., Isard, M., Sivic, J., Zisserman, A.: Lost in quantization: Improving particular object retrieval in large scale image databases. In: IEEE Conference on Computer Vision and Pattern Recognition, CVPR 2008, pp. 1–8. IEEE (2008)
5. Schindler, G., Brown, M., Szeliski, R.: City-scale location recognition. In: IEEE Conference on Computer Vision and Pattern Recognition, CVPR 2007, pp. 1–7. IEEE (2007)
6. Philbin, J., Chum, O., Isard, M., Sivic, J., Zisserman, A.: Object retrieval with large vocabularies and fast spatial matching. In: IEEE Conference on Computer Vision and Pattern Recognition, CVPR 2007, pp. 1–8. IEEE (2007)
7. Chum, O., Matas, J., Kittler, J.: Locally Optimized RANSAC. In: Michaelis, B., Krell, G. (eds.) DAGM 2003. LNCS, vol. 2781, pp. 236–243. Springer, Heidelberg (2003)
8. Chum, O., Werner, T., Matas, J.: Epipolar geometry estimation via RANSAC benefits from the oriented epipolar constraint. In: Proceedings of the 17th International Conference on Pattern Recognition, ICPR 2004, vol. 1, pp. 112–115. IEEE (2004)
9. Jegou, H., Douze, M., Schmid, C.: Hamming Embedding and Weak Geometric Consistency for Large Scale Image Search. In: Forsyth, D., Torr, P., Zisserman, A. (eds.) ECCV 2008, Part I. LNCS, vol. 5302, pp. 304–317. Springer, Heidelberg (2008)
10. Wu, Z., Ke, Q., Isard, M., Sun, J.: Bundling features for large scale partial-duplicate web image search. In: IEEE Conference on Computer Vision and Pattern Recognition, CVPR 2009, pp. 25–32. IEEE (2009)
11. Tsai, S.S., Chen, D., Takacs, G., Chandrasekhar, V., Vedantham, R., Grzeszczuk, R., Girod, B.: Fast geometric re-ranking for image-based retrieval. In: 2010 17th IEEE International Conference on Image Processing (ICIP), pp. 1029–1032. IEEE (2010)
12. Matas, J., Chum, O., Urban, M., Pajdla, T.: Robust wide-baseline stereo from maximally stable extremal regions. Image and Vision Computing 22, 761–767 (2004)
13. Manning, C.D., Raghavan, P., Schutze, H.: Introduction to information retrieval, vol. 1. Cambridge University Press, Cambridge (2008)
14. Salton, G., Buckley, C.: Term-weighting approaches in automatic text retrieval. Information Processing & Management 24, 513–523 (1988)

15. Lowe, D.G.: Distinctive image features from scale-invariant keypoints. International Journal of Computer Vision 60, 91–110 (2004)
16. Nister, D., Stewenius, H.: Scalable recognition with a vocabulary tree. In: 2006 IEEE Computer Society Conference on Computer Vision and Pattern Recognition, vol. 2, pp. 2161–2168. IEEE (2006)
17. Liu, X., Lou, Y., Yu, A.W., Lang, B.: Search by mobile image based on visual and spatial consistency. In: 2011 IEEE International Conference on Multimedia and Expo (ICME), pp. 1–6. IEEE (2011)
18. Deng, J., Dong, W., Socher, R., Li, L.J., Li, K., Fei-Fei, L.: Imagenet: A large-scale hierarchical image database. In: IEEE Conference on Computer Vision and Pattern Recognition, CVPR 2009, pp. 248–255. IEEE (2009)
19. Voorhees, E.M.: The TREC-8 question answering track report. In: Proceedings of TREC, vol. 8, pp. 77–82 (1999)

Intelligent Photographing Interface with On-Device Aesthetic Quality Assessment

Kuo-Yen Lo[1], Keng-Hao Liu[2], and Chu-Song Chen[1,2]

[1] Research Center for Information Technology Innovation, and
[2] Institute of Information Science, Academia Sinica, Taipei, Taiwan
{kylo,keng3,song}@iis.sinica.edu.tw

Abstract. This paper proposes a efficient method for instant photo aesthetics quality assessment that can be implemented on general portable devices. The classification performance is guaranteed to 0.89 on benchmark photo database. We also port our method onto a middle-level tablet computer to execute instantly and we find it reaches good acceptable efficiency. Moreover, an aesthetic information display to present the aesthetics evaluation results to users is introduced.

1 Introduction

Photo aesthetic quality assessment aims to classify the photographs into high or low quality automatically. Tong et al. [1] attempted to classify photographs into those taken by professionals or home users using low-level features derived from computer vision techniques. Datta et al. [2] also employed a set of low-level features then followed by a classifier to achieve photo quality assessment. Ke et al. [3] designed more semantic features based on the perceptual factors that present the difference between high and low quality photos to increase the performance. These works are the earliest representatives in this topic.

Later, Luo et al. [4] proposed regional features to improve assessment results by utilizing subject region detection methods. This work was refined by Luo et al. [5] by improving existing features with a variety of subject detection algorithms, such as super-pixel segmentation, layout and human detection. Dhar et al. [6] introduced a high-level attributes layer to make the subject-based framework more integrated. Those works tended to use more high-complexity and describable features to imitate the photography rules. The contribution is indeed significant but the computational overhead is increased rapidly.

Other works [7,8] adopted bottom-up strategies to acquire more improvement than conventional rule-specific method. They extracted the hidden composition relations of image by using bag-of-words since many aesthetic factors cannot be simply described by common photography rules. Despite those works set another benchmark in this topic, they also suffered from the issue of computation efficiency. Furthermore, those bottom-up features are usually not describable so that they cannot provide direct feedback to users. Recently, a few web-based applications using the techniques of photo aesthetics were proposed.

J.-I. Park and J. Kim (Eds.): ACCV 2012 Workshops, Part II, LNCS 7729, pp. 533–544, 2013.
© Springer-Verlag Berlin Heidelberg 2013

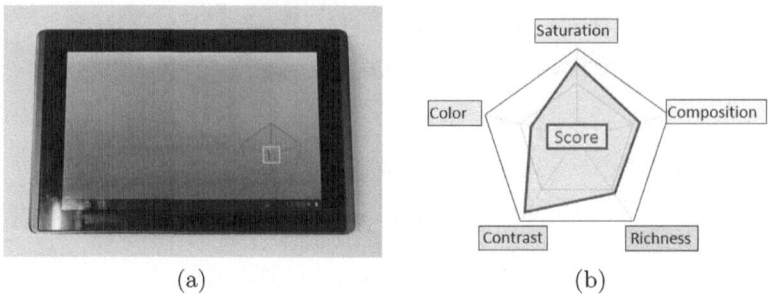

(a)   (b)

Fig. 1. The overview of our proposed mobile application. (a) Tablet device with our instant photo aesthetics system (b) The proposed five aesthetic perspectives of photography.

Datta et al. [9] created a web-based aesthetic quality inference system (AC-QUINE) which automatically rates the quality of photos that uploaded from web users. Yao et al. [10] proposed an on-site composition and aesthetics feedback system that can provide aesthetic score and retrieved exemplars for mobile consumers. Both works are considered as off-line applications since they require supports of network transitivity, computational ability and storage of server.

As the rapid growth of mobile commodity shown in the market, more and more people use them to take life photos. Compared to existing off-line applications, creating an autonomous photo aesthetics quality assessment system on mobile devices is urgently demanded. With the advantage of programming flexibility and considerable computational ability of mobile devices, it is possible to design a near instant aesthetics evaluation system which can assist users to take photos. However, those well-crafted works [10][5] cannot be run on such devices because of the huge computational complexity that portable devices cannot afford. Besides, using a bunch of aesthetics features without appropriate manipulation can not give describable feedback to users. Therefore, the key to reach this goal is to adopt more efficient and describable techniques for implementation on mobile devices. For instance, Vazquez et al. [11] proposed an assisted photography method for people with visual impairments to improve picture composition by using efficient methods.

In this paper, we design an efficient way of running instant photo quality assessment on popular portable devices. We first design a set of aesthetics features that are describable, discriminative, and computationally efficient. Then those features are clustered hierarchically into five groups that are semantically independent to display describable aesthetic information to users. Finally, the proposed scheme is implemented and visualized in on-line (near real-time, or instant) aesthetic assessment system through the live view screen of mobile device as a strong aid for creation of photos.

This paper is organized as follows: Section 2 describes the feature extraction for our assessment work. Section 3 evaluates the feature performance and corresponding runtime on two different platforms. Section 4 introduces aesthetic information display method. Finally, the conclusion is drawn in Section 5.

(a) CP=0.79   (b) CP=0.4

Fig. 2. Five dominant colors generated by proposed CP feature in flower scenes (a) High quality photo (b) Low quality photo.

2 Aesthetic Feature Extraction

This section introduces several efficient aesthetic features for our assessment system. They follow the rules of photography. Our principle of designing features mainly follows instance-based (data driven) rule, where the features are created by analyzing the training database which contains large amount of high/low quality-labeled photos. The rule-specific approach is also adopted to make assessment work more effective. In addition, we do not utilize any computation consuming techniques, such as subject detection or image segmentation. Only low-cost features are considered in our system to mitigate the computational burden of mobile devices.

2.1 Color Combination

A good combination of colors within an image is directly related to visual appearance and attractiveness. We call such combination as Color Palette (CP). To make our method efficient, we simply consider the color distribution of an image. The main issue is to find few dominant colors such that they occur frequently in the image and are dissimilar to each other.

We divide each channel of the HSV color space into 16 bins to construct a $16^3 = 4096$ bins color histogram. The center of each bin in the HSV space is called a candidate color. Our first goal is to find several key colors dominating the entire color distribution from 4096 candidate colors. First, we approximate the color distribution of the image by the histogram built on the candidate colors, $\mathbf{H} = \{h(i) \mid i = 1, ..., 4096\}$, where $h(i)$ is the number of pixels associated with the i-th bin in the image. Denote $C_i \in R^3$ to be the i-th candidate color, and we treat $h(i)$ as its weight. Let $\mathbf{D} = \{C_i, h(i) \mid h(i) > 0, i = 1, ..., 4096\}$ be the dataset consisting of the weighted samples. We then apply weighted k-means algorithm to $\mathbf{D}$ and obtain N cluster centers. Note that the clustering process is performed in only three-dimensional space and so it is very efficient to compute.

Despite the N colors associated to the cluster centers can be employed as the dominant colors, they could suffer from the problem that these centers are not the colors appearing in the image since they are averages of candidate colors. In practice, we seek to find nearby candidate colors with high weights instead, which should be more representable. For each cluster j $(j=1,...,N)$ we find the j-th dominant color by

$$\arg\max_{C_i \in cluster\ j} \alpha h(i) + (1 - \alpha)\|C_i - V_j\|^{-1}. \tag{1}$$

where V_j is the center of cluster j, and $\alpha \geq 0$ is a parameter balancing between the high-weight requirement and the closeness to the cluster center. The number of dominant colors is set as $N=5$ and α is set as $10/\max(h(i))$ in our implementation. Fig.1 shows two examples of the dominant colors obtained by our method.

Once the dominant colors are obtained, an image is reduced to a 5×3 (channels)=15-d vector. To conduct a feature for aesthetic-value assessment based on color information, we utilize an instance-based approach instead of using rule-based approaches such as color-harmony [12]. Our empirical study finds that the former often performs better as more details can be utilized. For any imput image, we find its k-Nearest Neighbors (kNN) among the training photos in the 15-d space via Hungarian Algorithm [13]. Let n_H and n_L be the numbers of high and low-quality neighbors found by kNN with $k=25$, respectively, where $k=n_H + n_L$, we then construct the CP feature by their difference, $f_1 = (n_H - n_L)/2k + 0.5$. In our work, the training set typically contains thousands of photos for each label. Since kNN is only performed in 15-d space, it is still very efficient to compute.

2.2   Composition

A proposed efficient composition feature is called Edge Composition (EC). We follow instance-based learning principle to measure image composition measured from training images, instead of using traditional rule-specific methods (e.g., rule of thirds or visual balance). Such a design not only directly reflect the composition properties of image but also requires less complexity. This feature is operated on H, S, and V channels individually. We first calculate the average of edge-intensity maps of thousands of high (low) quality training photos to build a high (low) quality edge template. The edge-intensity is measured by laplacian filter. Since the edges in an image could reflect object boundaries, it assumes that the spatial pattern of edges will benefit to the assessment of photos with salient objects. Let the L1 distance between the input image and the high/low quality edge templates be d_H and d_L respectively. The value $d_H - d_L$ for the three channels then serve as the EC features f_2 to f_4. Those values are proportional to the composition of high-quality photos.

Fig. 3. Examples of photos containing subjects and normalized subject/background-prior maps obtained by EC feature (H, S, and V channels). (a) Photo examples. (b) Subject-prior maps. (c) Background-prior maps.

2.3   Contrast

Contrast is considered as important aesthetic factor in rules of photography. It measures the dynamic range of photos. There are two types of contrast feature we use: Histogram Contrast and Spatial Contrast.

Histogram Contrast. Histogram Contrast (HC) calculated the width of dominant range in color histograms of image. We follow [3] to compute them as the widths of 98% mass of both RGB-mixed and gray-level histograms (f_5, f_6). In general, high-quality photos have higher contrast values in common.

Spatial Contrast. The general contrast feature merely measures the range of color histograms, where the information of spatial contents is ignored. We propose a new feature called Spatial Contrast (SC) which measures the related contrast between subject and background regions in the spatial domain.

To obtain SC, we first need prior information of subject region and background region. We directly employ the edge template obtained by EC feature as a subject-prior map ($\mathbf{MAP}_{sub}$) to save computation power. The background-prior map ($\mathbf{MAP}_{bkg}$) can be further calculated by the subtraction from subject-prior map. Both maps have fixed size and are normalized to sum-to-one. Fig. 3(b-c) show the examples of prior maps. For any query image $\mathbf{I}$ with the same size of prior maps, we create three 16-bin histograms for H, S, and V channels to present color distribution of subject regions, denoted as $\mathbf{H}_{sub}^H$, $\mathbf{H}_{sub}^S$, $\mathbf{H}_{sub}^V$, and other three present color distribution of background regions, denoted as $\mathbf{H}_{bkg}^H$,

$\mathbf{H}_{bkg}^{S}$, $\mathbf{H}_{bkg}^{V}$, by using prior maps. To create those histograms, the basic unit of count (vote) for each color bin is deployed by the weighting values on corresponding spatial locations of $\mathbf{MAP}_{sub}$ and $\mathbf{MAP}_{bkg}$, instead of by one. More specifically, let $\mathbf{h}_{A}(k)$ presents the weight of k-th bin of histogram $\mathbf{H}_{A}$ in a certain channel, where $A \in \{sub, bkg\}$ and k=1,2,...,16. Thus, the $\mathbf{h}_{A}(k)$ can be obtained by:

$$\mathbf{h}_{A}(k) = \sum_{i,j} \mathbf{MAP}_{A}(i,j), \qquad (2)$$

where (i,j) denotes the spatial locations of the pixels contributing to k-th bin. Therefore, the histogram $\mathbf{H}_{A}$ not only retains statistics of color distribution but also takes the spatial properties into account. Once six histograms are obtained, the SC feature is defined by $f_7 = \mathbf{dist.}(\mathbf{H}_{sub}^{H} - \mathbf{H}_{bkg}^{H})$, $f_8 = \mathbf{dist.}(\mathbf{H}_{sub}^{S} - \mathbf{H}_{bkg}^{S})$, and $f_9 = \mathbf{dist.}(\mathbf{H}_{sub}^{V} - \mathbf{H}_{bkg}^{V})$ respectively, where $\mathbf{dist.}$ denotes L2 norm between any two 16-d histograms. The SC feature measures the relative difference between subject region and background region. In general, high quality photos usually possess higher SC.

2.4   Richness

People usually feel more pleasant to those images containing richness of spatial contents. A feature that can measure the variability of image content is further required. Two features are proposed to measure the richness in both spatial and color aspects.

Spatial Richness. Since the human eye used to scan image in horizontal or vertical way, the image with severe changes in both directions usually attracts more humans attention. We propose a simple feature, called Spatial Richness (SR), to measure such properties. We segment the image into 6 stripes uniformly in both vertical and horizontal directions, and compute the sum of differences of edge-intensity maps of all the adjacent stripes for H, S, and V channels. Features f_{10} to f_{12} are thus generated.

Color Richness. A good low-level statistic can also contribute to aesthetics prediction. We design a feature called Color Richness (CR) that assumes that high-quality photos always have more colors with higher hue counts than low-quality ones [3]. To obtain CR, we directly make use of the 16-bin HSV histograms obtained from CP feature. Then we calculate the number of bins with corresponding counts higher than a given threshold for each histogram. Features f_{13} to f_{15} are presented as the HSV counts for H, S, and V channels respectively.

2.5   Average Saturation

Average Saturation (AS) is considered as another indispensable statistic in computational aesthetics for high-quality photos [2]. It calculates the average value of saturation channel in HSV color space for an image (f_{16}). Basically, photos with higher saturation usually produce more aesthetic feeling.

It should be noted that we do not employ the averages of H and V channels as features duo to the reasons: The average of Hue is generally not related to aesthetic quality because it just corresponds to human's preference. The average of Value is also not a factor to aesthetics since the exposure status is automatically balanced by camera devices in most of cases. Therefore, only the average of Saturation can be useful statistic to express aesthetic emotion of image.

3   Feature Evaluations and Experiments

After developing the describable features for photo aesthetics assessment, it is necessary to evaluate the classification performance of those features and corresponding on-device computation time. In this section we introduce the photo database and validation methods used in this paper. The performance of proposed aesthetics features will be demonstrated in detail, and the corresponding efficiency report of on-device computation will be also presented.

3.1   Database and Setting

We choose the publicly available photo database provided by CUHK [5] for experiments. It consists 7 categories of photos and each photo in the database has been assigned as high-quality or low-quality label. We only use those photos with obvious subjects inside the scene, referred to as Animal, Plant, Static categories of photos, which totally contain 2078 high-quality photos and 7573 low-quality photos, for our implementation. Half of them are selected as training photos and the rest as testing ones.

Once feature extraction is done, a binary classifier can be learned by using Support Vector Machine (SVM) based on high/low-quality training photos to evaluate the classification performance for testing photos. In our setting, the random partition repeats 10 times and the averaged results are reported. The performance index we use is Area Under the ROC Curve (AUC) since it is a better measure for unbalanced datasets. In general, higher AUC presents higher classification ability.

3.2   Feature Performance

To balance between performance and computational efficiency, we assign a small scale 240x180 as working resolution in feature extraction for each image. Table. 1 tabulates the classification performance and computational consumption of our proposed features. First row "AUC" shows the AUC value of each individual aesthetics feature, referred to as Edge Composition (EC), Color Palette CP), Histogram Contrast (HC), Spatial Contrast (SC), Spatial Richness (SR), Color Richness (CR), and Average Saturation (AS) respectively. To evaluate the performance of single feature, we trained a individual SVM classifier for each feature

Table 1. The classification performance (AUC) of proposed aesthetic features

	Edge Comp.	Color Palette	Spatial Contrast	Hist. Contrast	Spatial Richness	Color Richness	Avg. Saturation	Overall (16-d)
AUC	0.80	0.77	0.67	0.67	0.79	0.70	0.64	0.89

separately to obtain its own AUC result. From Table. 1, it can be observed that those "well-designed" features, such as EC, CP, and SR, have better classification abilities while other low-statistical features, such HC and AS, just give fair results. We owe this to the reason that those well-designed features adopted instance-based principle to extract statistical information from training samples so that the values they generated are much accordance with the characteristics of datasets. Nevertheless, other simple features designed by traditional rule-specific approach have fair performance but could help to promote overall classification rate. The final column in row AUC shows the classification result using all 7 features (16-d) simultaneously. It reached considerable classification accuracy 0.89. Furthermore, it is noticed that the CP feature performed the best among all features we used. It suggests that the color combination is indeed one of the crucial factor to determine photo's quality. However, such a superior feature usually requires relatively higher computational resource to implement. It is obviously a trade-off between effectiveness and efficiency.

3.3 On-Device Computation Time

Table. 2 tabulates the required computation time for each feature implemented on PC (Win7 64-bit, Intel Core i5 CPU at 3.4MHz, 16GB RAM, Matlab) and middle-level mobile device (i.e. Tablet PC) (ASUS Transformer TF101, Android 3.2, NV Tegra2 CPU at 1.0MHz, 1GB RAM) respectively. There are two observations: 1.Those features with higher computational complexity require much computation time. For example, the CP feature contains complex techniques such as histogram construction, weighted k-means, and kNN algorithms, and Hungarian matching so that it requires much more time to process than other features. 2.PC runs faster than mobile device does for all features. The PC requires averaged 82ms to run whole 16-d features per image while the mobile needs 288ms. This result matches our expectation because PC has more advantages on computation power in terms of higher CPU clock and larger capacity of memories. Even though the mobile device has such inherent disadvantages, it can run whole features instantly in near 2.3fps including all system loading factors.

Besides, we found both systems own different properties on computing the same features. For instance, CP feature needs close processing time on both platforms, but EC feature requires greatly different processing time. This implies that to make the aesthetics evaluation system working more efficiently, appropriate optimization on programming is further required.

Table 2. The average computation time (millisecond) of each proposed aesthetics feature on different platforms.

	Edge Comp.	Color Palette	Spatial Contrast	Hist. Contrast	Spatial Richness	Color Richness	Avg. Saturation	Overall (16-d)
On PC	11	40	12	3	2	14	0.1	82
On Mobile	160	54	22	10	25	16	1	288

4   Aesthetic Information Display

For building a good aesthetics system, it is necessary to create an interface to display aesthetic information once the aesthetic prediction is calculated. Opposite to most previous works where the assessment results are simply obtained by feature extraction followed by a trained classifier, additional describable, referable and useful-feedback information that can suggest users to re-frame current scene is indispensable for an instant photo aesthetic assessment system.

We simply utilize hierarchical structure to insert an additional layer that constitutes five dominant aesthetics indices for such a purpose. Finally, we design a simple interface to display such information on portable devices. The details are going to be described as follows.

4.1   Hierarchical Structure for Aesthetics Evaluation

In order to reveal the "hidden aesthetic issues" when evaluating an image, the diagram of our aesthetics assessment system is designed as a three-layer hierarchical structure. Fig. 4 shows the diagram of hierarchical structure of proposed system. From Layer 1 (named as Feature layer) to Later 2 (named as Aesthetic index layer), we first group our 16-d aesthetic feature set to five dominant groups according to their attributes. They are Composition, Saturation, Color,

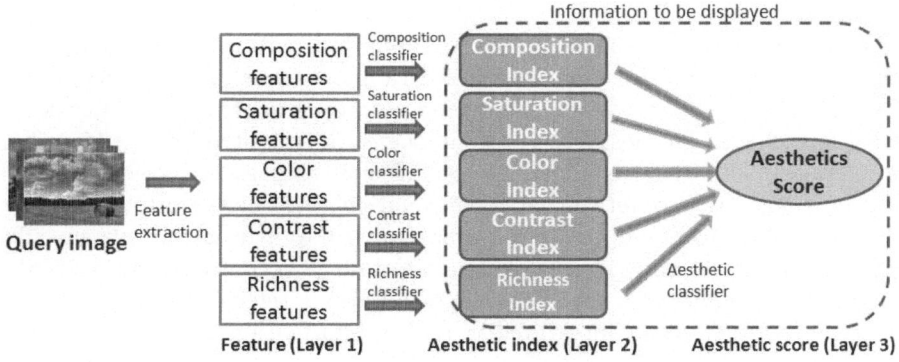

Fig. 4. The hierarchical structure of proposed aesthetics system

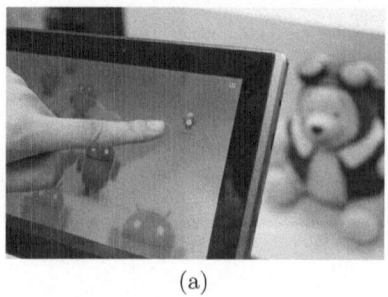
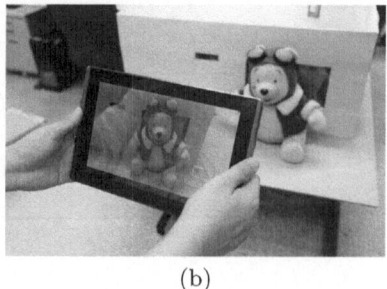

(a)   (b)

Fig. 5. Demonstration of proposed instant photo aesthetic quality assessment system. (a) Mobile program (c) Real-life scenario.

Contrast, and Richness respectively, which are considered as five common aesthetic indices in photography. The Composition index presents the degree of good composition by using EC feature (f_2 to f_4). The Saturation index measures saturation degree of image by simply using Avg. saturation feature (f_{16}. The Color index presents the degree of color combination of image using CP feature (f_1). The Contrast index evaluates both intensity contrast (histogram-based) and spatial contrast by considering HC and SC features simultaneously (f_5, f_6, f_7 to f_9). Finally, The Richness index detects the richness of image content in both spatial and color aspects by taking SR and CR features (f_{10} to f_{12}, and f_{13} to f_{15}).

For each group, we then train an independent SVM classifier (via probability model) with training database to measure the degree of each index then it is normalized to [0,1] according to corresponding index range. (The index range are the lowest and largest values of the distribution of a certain aesthetics index in the database. Before acquiring this range, we removed top and bottom 3% samples to avoid the influence of extremely good or bad samples.) These main purpose of aesthetics indices is to inform users the current sub aesthetic degree of five general perspective in photography. Users can improve overall aesthetics quality of current photo scene by means of adjusting those ones which are lower than a certain threshold.

Since each image has been represented as a 5-d feature vector in Layer 2, from Layer 2 to Layer 3 (named as Aesthetic score layer), another SVM classifier can be trained to evaluate the overall aesthetics quality (aesthetics score) for a query image. The aesthetics score indicates the quality of image for a simple reference. This score is generally high if those indices in Layer 2 are high enough. It is worthy to notice that the evaluation performance of photo quality will not be significant improved if Layer 2 is removed from our system. The classification rate of using hierarchical maintains 0.88 according to our extended experiment. It implies that the components in Layer 2, referred to as five aesthetics indices, could preserve integrated statistical information for photo aesthetics assessment.

(a)

(b)

Fig. 6. Assessment examples of proposed instant photo aesthetic quality assessment system. (a) Scenes with different poses (b) Scenes with different background.

4.2   Interface Design on Display

An ideal photo quality assessment system not only needs precise evaluation ability, but also requires a good presentation to users. To display the aesthetics information on the screen of mobile device, we design a pentagon graph to display the information of aesthetics indices and overall score introduced in Section 4.1. The illustration of pentagon graph is shown in Fig. 1(a). The outside pentagon defines five aesthetics indices while the inner pentagon indicates corresponding index values. Figs. 5(a-b) demonstrate the display interface of our aesthetics system where the pentagon graph is set in the lower right corner on the screen. Fig. 6 further demonstrates the assessment results of proposed system implementing on the scenes with different scenarios. By means of such layout design for aesthetics presentation, user can easily understand the current status of photo aesthetics when taking photos, and would have intention to improve those indices with low values to obtain higher aesthetics score.

5   Conclusions and Future Work

This paper introduces an on-device photographing interface with aesthetic quality feedback on mobile devices. We first experimentally show that our proposed aesthetic features could reach great classification performance. Later, the implementation on a modern tablet computer with considerable response speed is achieved. The experiments demonstrate that the computational ability of middle-level mobile device is sufficient to implement image processing techniques in instant manner. In future works, we will investigate more discriminative features with better evaluation accuracy, explore the ways to optimize the system overall efficiency, and create an interactive system coupled with more useful feedback for advanced photographing interface.

Acknowledgement. This work was supported in part by National Science Council, Taiwan, under the grants NSC 101-2221-E-001-015-MY2.

References

1. Tong, H., Li, M., Zhang, H.-J., He, J., Zhang, C.: Classification of Digital Photos Taken by Photographers or Home Users. In: Aizawa, K., Nakamura, Y., Satoh, S. (eds.) PCM 2004. LNCS, vol. 3331, pp. 198–205. Springer, Heidelberg (2004)
2. Datta, R., Joshi, D., Li, J., Wang, J.Z.: Studying Aesthetics in Photographic Images Using a Computational Approach. In: Leonardis, A., Bischof, H., Pinz, A. (eds.) ECCV 2006, Part III. LNCS, vol. 3953, pp. 288–301. Springer, Heidelberg (2006)
3. Ke, Y., Tang, X., Jing, F.: The design of high-level features for photo quality assessment. In: CVPR (2006)
4. Luo, Y., Tang, X.: Photo and Video Quality Evaluation: Focusing on the Subject. In: Forsyth, D., Torr, P., Zisserman, A. (eds.) ECCV 2008, Part III. LNCS, vol. 5304, pp. 386–399. Springer, Heidelberg (2008)
5. Luo, W., Wang, X., Tang, X.: Content-based photo quality assessment. In: ICCV (2011)
6. Dhar, S., Ordonez, V., Berg, T.: High level describable attributes for predicting aesthetics and interestingness. In: CVPR (2011)
7. Marchesotti, L., Perronnin, F., Larlusa, D., Csurka, G.: Assessing the aesthetic quality of photographs using generic image descriptors. In: ICCV (2011)
8. Su, H.H., Chen, T.W., Kao, C.C., Hsu, W.H., Chien, S.Y.: Scenic photo quality assessment with bag of aesthetics-preserving features. In: ACM Multimedia (2011)
9. Datta, R., Wang, J.Z.: Acquine: Aesthetic quality inference engine real-time automatic rating of photo aesthetics. In: ACM MIR (2010)
10. Yao, L., Suryannarayan, P., Qiao, M., Wang, J.Z., Li, J.: Oscar: On-site composition and aesthetics feedback through exemplars for photographers. IJCV (2011)
11. Vazquez, M., Steinfeld, A.: An assisted photography method for street scenes. In: WACV (2011)
12. Desnoyer, M., Wettergreen, D.: Aesthetics image classification for autonomous agents. In: ICPR (2010)
13. Burkard, R., Dell'Amico, M., Martello, S.: Assignment Problems. SIAM (2009)

Camera Pose Estimation of a Smartphone at a Field without Interest Points

Ruiko Miyano, Takuya Inoue, Takuya Minagawa,
Yuko Uematsu, and Hideo Saito

Department of Information and Computer Science, Keio University
{rui,inoue,takuya,yu-ko,saito}@hvrl.ics.keio.ac.jp

Abstract. An Augmented Reality (AR) system on mobile phones has recently attracted attention because smartphones have increasingly been popular. For an AR system, we have to know a camera pose of a smartphone. A sensor-based method is one of the most popular ways to estimate the camera pose, but it cannot estimate an accurate pose. A vision-based method is another way to estimate the camera pose, but it is not suitable to a scene with few interest points such as a sports field. In this paper, we propose a novel method of a camera pose estimation for a scene without interest points by combining a sensor-based and a vision-based approach. In our proposed method, we use an acceleration and a magnetic sensor to roughly estimate a camera pose, then search the accurate pose by matching a captured image with a set of reference images. Our experiments show that our proposed method is accurate and fast enough to apply a real-time AR system.

1   Introduction

An Augmented Reality (AR) technology which projects virtual annotations onto a camera image has attracted attention in recent years. Especially an AR system which is using mobile devices has increased. Takacs et al. built an outdoor AR system which makes annotations of building information for mobile phones [1]. Yovcheva et al. explored recent researches in order to develop an AR system for tourism using smartphones [2]. As described in these papers, a system using mobile devices has advantages of being able to utilize devices such as a camera, a sensor, and a GPS. Moreover smartphones are suitable for an AR system used by ordinary people because they have become widespread in these days.

To project annotations onto the right position, a camera pose of a smartphone must be estimated. Some researchers tried to estimate a camera pose of a smartphone for an indoor AR navigation system [3, 4]. They achieved their goal by employing two approaches. The first approach is a sensor-based approach which uses sensors such as an acceleration and a magnetic, and the second approach is a vision-based approach which uses images captured with the camera. A vision-based approach is used to estimate the accurate camera pose by extracting local features. However a vision-based approach cannot be applied to a situation with few local features such as a sports field.

J.-I. Park and J. Kim (Eds.): ACCV 2012 Workshops, Part II, LNCS 7729, pp. 545–555, 2013.

The contribution of this paper is to propose a method that estimates a camera pose of a smartphone at a field without interest points. We have achieved our goal by combining a sensor-based and a vision-based approach which does not use interest points. Our method has been experimented on a soccer field and evaluated regarding the processing time and the accuracy.

2   Related Research

Researches about a camera pose estimation are extremely important to develop an AR system.

Many AR services which use a sensor-based approach are developed in recent days [5–7]. In these services, a GPS and an electronic compass are used to obtain a position and a camera pose of a mobile device. Tokusho and Feiner introduced an AR street view system using a GPS sensor and a digital pedometer [8]. A sensor-based approach has advantage that a position and a pose of devices can be obtained without complex processing. However there is problem that smartphones are susceptible to noise.

On the other hand, many researchers have focused on a vision-based approach for a robust camera pose estimation. For example, Kato and Billinghurst proposed a marker based camera pose estimation method [9], and Klein and Murray proposed a local feature based camera localization techniques [10]. Klein and Murray also reduced computational cost of a camera pose estimation in order to apply it to a mobile application. They demonstrated an AR system on mobile phones by reducing the number of local features used to estimate a camera pose [11]. These methods sometimes do not work well when a sufficient number of feature points can not be detected to estimate the camera pose. Chen et al. proposed a framework for recognizing scenes using a panorama image [12]. In this research, input image sequences are matched to parts of a panorama image based on template matching. Since this method does not use interest points, we have tried to adopt the idea of using a panorama image in order to a camera pose estimation.

Furthermore Atzori et al. proposed an indoor navigation system based on both approaches [4]. An initial position of a camera is obtained from 2D barcode. A camera position is updated by sensor information. And an accurate pose is estimated by comparing SURF [13] between a current image and reference images. However this system requires barcode and textures with local features.

To address these problems, we propose to combine a sensor-based and a vision-based approach which does not use interest points.

3   Proposed Method

In this paper, we define three coordinate systems: $C^W(X^W, Y^W, Z^W)$ for a target field, $C^C(X^C, Y^C, Z^C)$ for a camera and $C^I(u, v)$ for an image captured by a smartphone (Fig. 1). We assume that a camera is not translated, but only

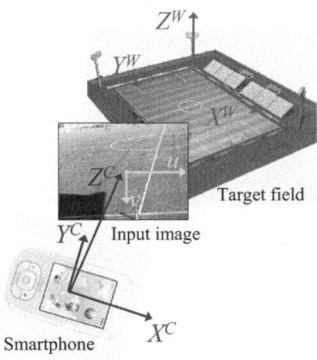

Fig. 1. Definition of the coordinate system

rotated. Because, for example, users in spectators' stands do not change their seat positions. Hence our final goal is to estimate the rotation from C^C to C^W.

Fig. 2 shows the overview of our proposed method. An initial estimate of the camera pose is provided from sensors such as an acceleration and a magnetic. Then the pose is refined by comparing a captured image with reference images created by panorama images like the research of Chen et al. [12].

In preprocessing, two types of information are obtained from a smartphone. The first information are images captured by a smartphone and the second information are camera angles calculated from sensors. Then panorama images are generated to create reference images in online processing.

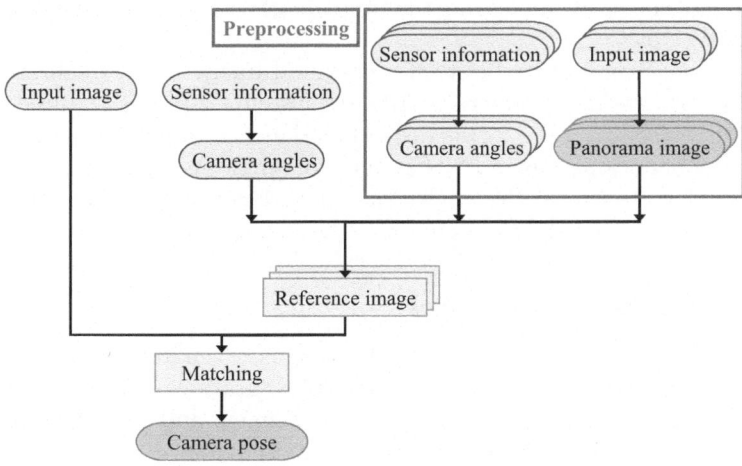

Fig. 2. Overview of the proposed method

(a) Left side   (b) Right side

Fig. 3. Panorama images

In online processing, a smartphone captures an image and a camera angle every frame. However this camera angle is not suitable to be directly used because of noise. Therefore an accurate camera angle is refined by a vision-based approach. To do that, a captured image is compared with reference images generated from panorama images and camera angles. Finally an accurate camera pose is obtained from the most similar reference image.

3.1 Preprocessing

In preprocessing, panorama images are generated for image matching in online processing. The smartphone must be moved to capture images and sensor information of the whole target field. Three angles such as a pan, a tilt, and a roll can be obtained from sensor information. Pan, tilt, and roll angles of a camera mean the rotation around Y^C, X^C, and Z^C axes in Fig. 1, respectively.

If one panorama image is generated from all captured images, there will be distortion of the target field in the panorama image. Therefore captured images are classified into several groups according to pan angles. Then panorama images of every groups $G = \{P^1, P^2, \cdots P^i, \cdots\}$ are generated by image mosaicing [14, 15]. Fig. 3 shows examples of panorama images created by Image Composite Editor (ICE) [15].

To know relative relation between a captured image and a panorama image P^i in online processing, a camera coordinate system C_{center}^{Ci} is required (Fig. 4). $C_{center}^{Ci}(X_{center}^{Ci}, Y_{center}^{Ci}, Z_{center}^{Ci})$ is defined so that the camera is pointing to the center of the panorama image P^i. The rotation angle $(p_{center}^i, t_{center}^i, r_{center}^i)$ of C_{center}^{Ci} is calculated from the angles that have been used to generate P^i: the medium angle between the minimum and the maximum one in these angles.

3.2 Creating Reference Images

In online processing, an accurate camera pose is estimated. Reference images are created from panorama images. Then an accurate camera pose is selected based on image matching between a captured image and reference images. In this section, we explain how to create reference images.

Reference images are created through three steps. First, smartphone gets an image and a camera angle. Second, variation ranges are added to the angle

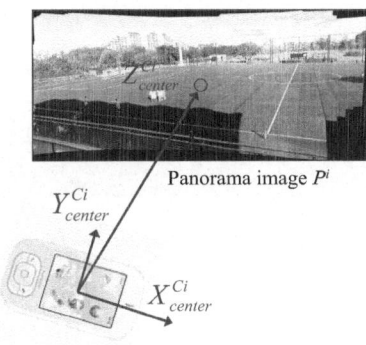

Y^{Ci}_{center}

Panorama image P^i

X^{Ci}_{center}

Fig. 4. Camera coordinate system regarding a panorama image

from sensors to handle errors of sensors. Finally reference images are created by clipping parts of a panorama image at the angles which include variation ranges. An image similar to a captured image is expected to be in these reference images.

A smartphone captures an image and angles such as a pan $p_{current}$, a tilt $t_{current}$, and a roll $r_{current}$ every frame. The panorama image P^i that is the nearest to the current pose is selected according to a pan angle $p_{current}$ in order to create reference images. A rotation angle from camera coordinate system C^{Ci}_{center} to the current pose is calculated by Eq. 1.

$$(\theta, \phi, \psi) = (p_{current} - p^i_{center}, t_{current} - t^i_{center}, r_{current} - r^i_{center}) \qquad (1)$$

The accurate pose is assumed to be around the pose obtained from sensors. Therefore we define three variation ranges: Δp as a pan angle, Δt as a tilt angle and Δr as a roll angle. And they are added to Eq. 1 like Eq. 2. Multiple reference images are created based on Eq. 2. We define Q^j and $D = \{Q^1, Q^2, \cdots, Q^j, \cdots, Q^N\}$ as a reference image and a group of reference images, respectively. N means the total number of reference images. That is to say, if the number of Δp, Δt and Δr are n_p, n_t and n_r, variable N is $n_p \times n_t \times n_r$.

$$(\theta, \phi, \psi)_j = (p_{current} - p^i_{center} + \Delta p, t_{current} - t^i_{center} + \Delta t, r_{current} - r^i_{center} + \Delta r) \quad (2)$$

To create a reference image Q^j from the panorama image P^i, the coordinates of Q^j's corners in P^i are required. First, the coordinates of Q^j in C^{Ci}_{center} are calculated from the camera angle $(\theta, \phi, \psi)_j$. Then coordinates of Q^j in P^i are calculated by projecting coordinates in C^{Ci}_{center} and Q^j is clipped.

The coordinates of the image plane Q^j in C^{Ci}_{center} are calculated by rotating an image plane I^{Ci} in Fig. 5(a). I^{Ci} means an image when the camera is capturing the center of the panorama image P^i. Coordinates which represent four corners of I^{Ci} are described as $\boldsymbol{a}^{Ci} = (-\frac{w}{2}, -\frac{h}{2}, f)$, $\boldsymbol{b}^{Ci} = (\frac{w}{2}, -\frac{h}{2}, f)$, $\boldsymbol{c}^{Ci} = (\frac{w}{2}, \frac{h}{2}, f)$ and $\boldsymbol{d}^{Ci} = (-\frac{w}{2}, \frac{h}{2}, f)$ in C^{Ci}_{center}. Variable w and h mean width and height of a captured image.

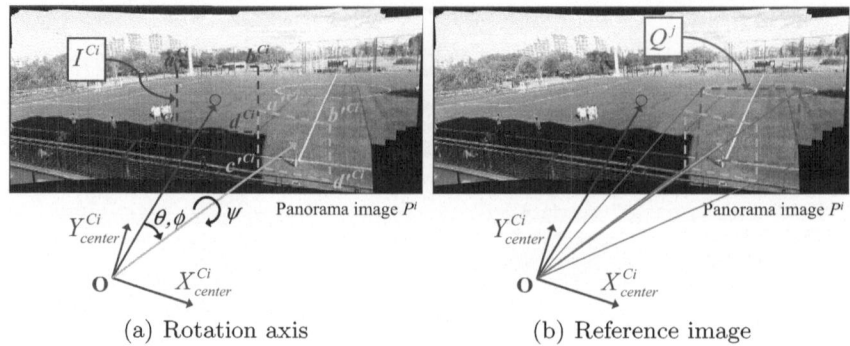

(a) Rotation axis   (b) Reference image

Fig. 5. Rotation about camera coordinate system C^{Ci}_{center}

To rotate I^{Ci}, the vector from the origin of C^{Ci}_{center} to the center point of I^{Ci} should be calculated. This vector is represented as $(0, 0, f)$ in C^{Ci}_{center}. Variable f means a focal length of the camera. This vector is rotated using θ and ϕ at first (Eq. 3). Then it is used as an axis to rotate four corners by ψ (Eq. 4). A coordinate a'^{Ci}, b'^{Ci}, c'^{Ci} and d'^{Ci} in Fig. 5(a) denote the coordinates of Q^j in C^{Ci}_{center}.

$$\begin{pmatrix} X' \\ Y' \\ Z' \end{pmatrix} = \begin{pmatrix} 1 & 0 & 0 \\ 0 & \cos\phi & -\sin\phi \\ 0 & \sin\phi & \cos\phi \end{pmatrix} \begin{pmatrix} \cos\theta & 0 & -\sin\theta \\ 0 & 1 & 0 \\ \sin\theta & 0 & \cos\theta \end{pmatrix} \begin{pmatrix} 0 \\ 0 \\ f \end{pmatrix} \tag{3}$$

$$\mathbf{R} = \cos\psi \begin{pmatrix} 1 & 0 & 0 \\ 0 & 1 & 0 \\ 0 & 0 & 1 \end{pmatrix} + (1 - \cos\psi) \begin{pmatrix} X'^2 & X'Y' & Z'X' \\ X'Y' & Y'^2 & Y'Z' \\ Z'X' & Y'Z' & Y'^2 \end{pmatrix} + \sin\psi \begin{pmatrix} 0 & -Z' & Y' \\ Z' & 0 & -X' \\ -Y' & X' & 0 \end{pmatrix} \tag{4}$$

Four corners in P^i can be calculated by projecting four corners in C^{Ci}_{center} like Fig. 5(b). For example, if coordinates in C^{Ci}_{center} is (X, Y, Z), coordinate (x, y) in P^i is calculated using Eq. 5.

$$(x, y) = (X \frac{f}{Z}, Y \frac{f}{Z}) \tag{5}$$

Thus four corners in P^i are obtained, the reference image Q^j can be created by clipping this region.

3.3 Image Matching

The camera pose is estimated using reference images created in section 3.2.

Every reference image Q^j is linked to a camera angle $(\theta, \phi, \psi)_j$ like Fig. 6. This angle is represented by Eq. 2.

A captured image is compared with each reference image using SSD (Sum of Squared Differences). Then the most similar image Q^g which has the highest score of SSD is selected from all reference images. A camera angles $(\theta, \phi, \psi)_g$ corresponding to the image Q^g is the accurate camera pose.

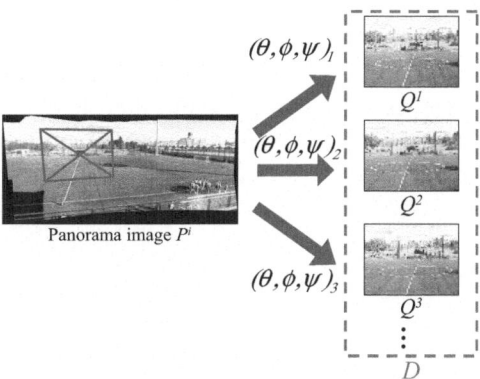

Fig. 6. Reference images

4   Experimental Results

We carried out two experiments to evaluate our proposed method. One is an experiment which measures the processing time for estimating the camera pose (Section 4.1). The other is an experiment which evaluates the accuracy of a camera pose estimation (Section 4.2).

We developed a camera pose estimation system using a smartphone and a server PC. The smartphone captured both images and sensor information, and the server PC estimated a camera pose of the smartphone. Captured data were transferred via TCP/IP connection over wireless LAN.

Here is our experiment environment.

- Smartphone
  - OS: Android 2.3
  - CPU: Samsung Exynos 4210 Orion Dual-core 1.2GHz
  - RAM: 1.00 GB
- Server PC
  - OS: Windows 7 Professional 64 bit
  - CPU: Intel Xeon 2.67GHz
  - RAM: 4.00 GB
- Smartphone camera
  - Resolution (pixel): 320   240
- Panorama images
  - Number: 3
  - Resolution (pixel): 1076×485, 1397×560 and 1171×545
- Variation ranges
  - $\Delta p \in \{-4.0, -3.0, \cdots, 3.0, 4.0\}$
  - $\Delta t \in \{-1.0, 0.0, 1.0\}$
  - $\Delta r \in \{-2.0, -1.5, \cdots, 1.5, 2.0\}$

The number of Δp, Δt, and Δr were $n_p = 9$, $n_t = 3$, and $n_r = 9$, respectively. The variable N in section 3.2, the total number of reference images, was $n_p \times n_t \times n_r = 243$.

4.1   Processing Time

In this experiment, we show the processing time to estimate the camera pose. The processing time includes time to select a panorama image, to create reference images, to calculate SSD and to obtain the camera pose of the smartphone. Table 1 shows an average and a variance of the processing time using 1307 frames.

Table 1. Processing time

	Average	Variance
Processing Time (ms)	127.4	1.8

From this result, it seems that our proposed method can be used in a real-time processing.

4.2   Accuracy

In this experiment, we projected a center line of a soccer field onto a smartphone image to evaluate the accuracy of our proposed method.

The homography matrix between a soccer field (X^W, Y^W) and a smartphone image (u, v) is computed from a camera pose of a smartphone. Therefore we evaluate this matrix in order to evaluate the camera pose. For the accuracy evaluation of this matrix, the center line is projected and the projection error is calculated.

To project the center line onto a smartphone image, a homography matrix $\mathbf{H_{t \to i}}$ between the soccer field and a smartphone image is required. The homography matrix between a panorama image and each reference image can be calculated in section 3.2. Then the homography matrix $\mathbf{H_{P^i \to i}}$ between a panorama

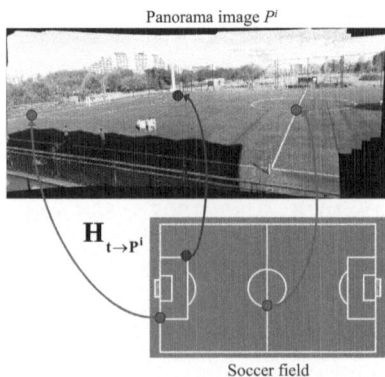

Fig. 7. Corresponding points between panorama image and soccer field

$$\mathbf{H}_{t \to P^i} \qquad \mathbf{H}_{P^i \to i}$$

Soccer field   Panorama image P^i Smartphone input image

Fig. 8. Homography matrices that relates smartphone image and soccer field

image and a smartphone image is selected in section 3.3. The homography matrix $\mathbf{H}_{t \to P^i}$ between the soccer field and a panorama image is calculated in preprocessing by manually inputting corresponding points like Fig. 7. Therefore $\mathbf{H}_{t \to i}$ can be calculated by Eq. 6 like Fig. 8.

$$\mathbf{H}_{t \to i} = \mathbf{H}_{p \to i} \cdot \mathbf{H}_{t \to p} \tag{6}$$

We project the center line using homography matrix $\mathbf{H}_{t \to i}$. Fig. 9 shows result images. From these results, it seems our proposed method is efficient no matter where the camera is capturing.

We also calculated the projection error to compare two methods, which are using only sensor information and using both sensor information and a captured image. 5 points on the center line are projected onto a smartphone image. These points are manually compared with ground truth points. We define these distance as projection errors. Table 2 shows an average of the projection errors of 50 frames. Fig. 10 shows comparison of result images between two methods. It seems that the combination of sensor information and an image is effective to estimate the camera pose. It costs about 10 fps to project the center line to a smartphone image via TCP/IP connection, which is enough to apply to a real-time processing.

Fig. 9. Result images

Table 2. Projection error

	Only sensor	Sensor & image
Projection error (pixel)	19.97	8.69

(a) Only sensor   (b) Sensor & image   (c) Only sensor   (d) Sensor & image

Fig. 10. Comparison between two methods

Fig. 11. Examples of a sports AR system

5   Conclusion and Future Work

We proposed a camera pose estimation method for a smartphone without interest points.

We achieved our goal by combining a sensor-based and a vision-based approach which does not use interest points. A rough camera pose is estimated using sensors, then an accurate camera pose is calculated by matching a captured image with a set of reference images.

Two experiments were carried out to validate our proposed method regarding the processing time and the accuracy. We confirmed that our proposed method can accurately estimate the camera pose and it is fast enough to apply a real-time AR system.

However our method is effective only when the smartphone is not translated, but only rotated. As future works, we expand our proposed method to deal with any camera motion using the pedometer.

Moreover we plan to develop a mobile AR system using our proposed method, for example an AR system for watching sports like Fig. 11. There are many researches about sports player detection, recognition and tracking [16–18]. Players' positions and identities can be analyzed using these researches. Annotations of players are projected by the homography matrix between a smartphone image and the sports field. Thus we plan to apply our method to a sports AR system.

Acknowledgement. This research is supported by National Institute of Information and Communications Technology, Japan.

References

1. Takacs, G., Chandrasekhar, V., Gelfand, N., Xiong, Y., Chen, W.C., Bismpigiannis, T., Grzeszczuk, R., Pulli, K., Girod, B.: Outdoors augmented reality on mobile phone using loxel-based visual feature organization. In: ACM International Conference on Multimedia Retrieval, ICMR (2008)
2. Yovcheva, Z., Buhalis, D., Gatzidis, C.: Overview of smartphone augmented reality applications for tourism. e-Review of Tourism Research (eRTR) 10, 63–66 (2012)
3. Kang, J.: Technique of tangible user interfaces for smartphone. In: International Conference on Information and Computer Applications, ICICA (2012)
4. Atzori, L., Dessi, T., Popescu, V.: Indoor navigation system using image and sensor data processing on a smartphone. In: International Conference on Optimization of Electrical and Electronic Equipment, OPTIM (2012)
5. Tonchidot: Sekai camera (2009), http://sekaicamera.com/
6. Layar: Layar (2009), http://www.layar.com/
7. mTrip: mtrip (2009), http://www.mtrip.com/
8. Tokusho, Y., Feiner, S.: Prototyping an outdoor mobile augmented reality street view application. In: IEEE International Symposium on Mixed and Augmented Reality, ISMAR (2009)
9. Kato, H., Billinghurst, M.: Marker tracking and hmd calibration for a video-based augmented reality conferencing system. In: International Workshop on Augmented Reality, IWAR (1999)
10. Klein, G., Murray, D.: Parallel tracking and mapping for small ar workspaces. In: IEEE International Symposium on Mixed and Augmented Reality, ISMAR (2007)
11. Klein, G., Murray, D.: Parallel tracking and mapping on a camera phone. In: IEEE International Symposium on Mixed and Augmented Reality, ISMAR (2009)
12. Chen, C.S., Hsieh, W.T., Chen, J.H.: Panoramic appearance-based recognition of video contents using matching graphs. IEEE Transactions on Systems, Man, and Cybernetics 34, 179–199 (2004)
13. Bay, H., Ess, A., Tuytelaars, T., Van Gool, L.: Surf: Speeded up robust features. Computer Vision and Image Understanding (CVIU) 110, 346–359 (2008)
14. Levin, A., Zomet, A., Peleg, S., Weiss, Y.: Seamless Image Stitching in the Gradient Domain. In: Pajdla, T., Matas, J. (eds.) ECCV 2004. LNCS, vol. 3024, pp. 377–389. Springer, Heidelberg (2004)
15. Microsoft-Research: Image composite editor (2011), http://research.microsoft.com/en-us/um/redmond/groups/ivm/ice/
16. Delannay, D., Danhier, N., Vleeschouwer, C.D.: Detection and recognition of sports(wo)men from multiple views. In: Third ACM/IEEE International Conference on Distributed Smart Cameras, ICDSC (2009)
17. Kasuya, N., Kitahara, I., Kameda, Y., Ohta, Y.: Real-time soccer player tracking method by utilizing shadow regions. In: 18th International Conference on Multimedia (2010)
18. Shitrit, H.B., Berclaz, J., Fleuret, F., Fua, P.: Tracking multiple people under global appearance constraints. In: International Conference on Computer Vision, ICCV (2011)

Hierarchical Scan-Line Dynamic Programming for Optical Flow Using Semi-Global Matching

Simon Hermann and Reinhard Klette

The .enpeda.. Project, Department of Computer Science
The University of Auckland, New Zealand

Abstract. Dense and robust optical flow estimation is still a major challenge in low-level computer vision. In recent years, mainly variational methods contributed to the progress in this field. One reason for their success is their suitability to be embedded into hierarchical schemes, which makes them capable of handling large pixel displacements. Matching-based regularization techniques, like dynamic programming or belief propagation concepts, can also lead to accurate optical flow fields. However, results are limited to short- or mid-scale optical flow vectors, because these techniques are usually not combined with coarse-to-fine strategies. This paper introduces fSGM, a novel algorithm that is based on scan-line dynamic programming. It uses the cost integration strategy of semi-global matching, a concept well known in the area of stereo matching. The major novelty of fSGM is that it embeds the scan-line dynamic programming approach into a hierarchical scheme, which allows it to handle large pixel displacements with an accuracy comparable to variational methods. We prove the exceptional performance of fSGM by comparing it to current state-of-the-art methods on the KITTI Vision Benchmark Suite.

1 Introduction

The objective of optical flow algorithms is to estimate a vector field that describes the 2D pixel displacement between two consecutive frames of an image sequence. The observed 2D motion in the image plane represents the projected 3D object motion of a real-world scene. Currently the most successful optical flow algorithms use variational calculus to minimize a global error function. In order to handle large displacements, variational optical flow methods are embedded into hierarchical schemes, refining an optimal prior solution successively at subsequent levels. See for example Brox et al. [1,2], Zach et al. [22], and Werlberger et al. [20].

Discrete optimization techniques for optical flow estimation have been published before. For example Quénot [15], Sun [17], Felzenszwalb and Huttenlocher [3], Gong and Yang [6], Lempitsky et al. [11], and Lei and Yang [12]. None of them however report on large-scale optical flow results (i.e. flow vectors around 100 pixels). The main reason is that correspondence costs for the finite set of all

J.-I. Park and J. Kim (Eds.): ACCV 2012 Workshops, Part II, LNCS 7729, pp. 556–567, 2013.
© Springer-Verlag Berlin Heidelberg 2013

potential pixel displacements need to be calculated. This set increases quadratically, when the search distance is extended, which limits the maximum amount of optical flow displacements that can be handled within reasonable run-time.

In this paper we present fSGM, a novel algorithm that is based on the *scanline dynamic programming principle* (scan-line DP). It regularizes correspondence costs via cost accumulation and integrates accumulated costs following the concept of *semi-global matching* (SGM), as proposed by Hirschmüller [8] for the task of stereo matching. The major novelty of fSGM is that it embeds scan-line DP into a hierarchical scheme. As a result, even large optical flow displacements can be estimated with similar accuracy as variational methods and within reasonable run-time.

Lei and Yang [12] also propose a coarse-to-fine strategy for a discrete optimization technique. However, they employ the coarse-to-fine concept in their region-tree-based refinement method "to accommodate the sampling inefficiency problem" and not to overcome large pixel displacements. Gong and Yang [6] use scan-line DP to calculate optical flow, and this is probably the closest approach to ours. However, their method is limited to a maximum displacement of 25 pixels.

This paper is structured as follows. Section 2 recalls the regularization process as known from SGM for stereo analysis. Section 3 provides the outline of the proposed novel method for optical flow calculation. Evaluation results are discussed in Section 4, followed by conclusions in Section 5.

2   Semi-Global Matching for Stereo

The SGM concept [8] generalizes single-line dynamic stereo matching [14] into a multi-line integration strategy. It can be described as two-step approach: First, the cost of pixel correspondences is established for all possible *disparity labels* d (or simply *disparities*) in the defined search space $\mathbb{D} = \{0, \ldots, d_{max}\}$ of non-negative integers. Second, these calculated costs are regularized along scan-lines that run across the image domain by employing an accumulative dynamic programming scheme. Accumulated regularization costs from multiple scan-lines with different directions are then integrated. Optimal disparities are selected based on a winner-takes-all cost evaluation.

2.1   Cost Regularization

The cost regularization procedure implements cost accumulation along an oriented scan-line, which is a 1D linear path identified by a direction vector $\mathbf{a}$. The cost $L_{\mathbf{a}}$, defined for a pixel location p and a disparity d, is accumulated between image border and p. Consider the segment $p_0, p_1, \ldots, p_n$ of the path defined by $\mathbf{a}$, with p_0 on the image border, and $p_n = p$. The cost at pixel p_i, for disparity $d \in \mathbb{D}$, on the scan-line defined by $\mathbf{a}$ is for $i = 1, 2, \ldots, n$ recursively defined as

$$L_{\mathbf{a}}(p_i, d) = C(p_i, \chi(d)) + \mathcal{M}_i - \min_{\eta \in \mathbb{D}} L_{\mathbf{a}}(p_{i-1}, \eta) \tag{1}$$

In this definition, we have

$$\mathcal{M}_i = \min \begin{cases} L_{\mathbf{a}}(p_{i-1}, d) \\ L_{\mathbf{a}}(p_{i-1}, d-1) + P_1 \\ L_{\mathbf{a}}(p_{i-1}, d+1) + P_1 \\ \min_{\eta \in \mathbb{D}} L_{\mathbf{a}}(p_{i-1}, \eta) + P_2 \end{cases} \tag{2}$$

where $C(p, \chi(d))$ is the data cost when matching p for disparity d. The function

$$\chi(d) = \begin{cases} (-d, 0), & \text{if the left image is chosen as base image} \\ (d, 0), & \text{if the right image is chosen as base image} \end{cases} \tag{3}$$

defines the sign of the column offset as a 2D displacement of the corresponding pixel. Regularization penalties P_1 and P_2 enforce piecewise disparity consistency along a scan-line. In this paper we use eight uniformly distributed path directions $\mathbf{a}$ for accumulation (i.e. to the right, left, up, down, and the four in-between angles).

2.2   Penalty Adjustment

Penalties P_1 and $P_2^\star$ are given as external parameters. They implement the Potts model as follows. For a solution d, a constant penalty cost of P_2 is assigned to all disparity labels $g \neq d$ starting at p. By penalizing all labels equally, disparity jumps at depth discontinuities are preserved. In order to model smooth transitions of non-fronto parallel surfaces, a smaller penalty $P_1 < P_2$ is assigned to all labels g with $|g - d| = 1$ which lie within the immediate disparity neighborhood of d. P_2 is constant for all labels, but it is locally adjusted for each pixel p_i as follows:

$$P_2(p_i) = \max \left\{ \frac{P_2^\star}{|I(p_{i-1}) - I(p_i)|}, P_1 + \delta \right\} \tag{4}$$

where $\delta > 0$. This adjustment links the regularization procedure with the underlying image data since the magnitude of the forward difference in direction $\mathbf{a}$ scales the penalty at each p_i. The rationale behind this is to improve performance at depth discontinuities as they are more likely to occur at intensity edges. Another motivation is to reduce the streaking effect which is inherent to scan-line optimizations.

3   Optical Flow with Semi-Global Matching

The previous section describes the accumulation procedure of semi-global matching for the stereo case. SGM is assumed to operate on rectified image pairs. Thus, in order to calculate the data cost for a disparity d at (i, j), the disparity value itself defines the column offset that needs to be added to (or subtracted from) the pixel location of the base image B to find the corresponding pixel in the match image M [i.e. $B(i, j) = M(i \pm d, j)$]. As already mentioned above, this stereo correspondence problem is defined for a 1D search range.

For unconstrained optical flow estimation, however, the corresponding task is set within a 2D search domain since any 2D displacement is potentially possible. Next, we outline how to embed the SGM concept into a coarse-to-fine approach in order to robustly solve mid- and large-scale optical flow displacements. We refer to it as fSGM, where 'f' is short for 'flow'.

3.1   1D Stereo to 2D Optical Flow Search Space

We describe the SGM extension from the 1D stereo to the 2D optical flow search space by using a bijective discrete mapping

$$\phi : \mathbb{D} \longrightarrow \mathbb{O} \subset \mathbb{Z}^2, \text{ with } \phi(d) = (\Delta u, \Delta v) \tag{5}$$

that translates a disparity label into a unique 2D pixel offset $\phi(d)$. The offset domain $\mathbb{O} \subset \mathbb{Z}^2$ is defined by a positive integer f_m specifying the maximum possible discrete flow, such that

$$\mathbb{O} = \{(\Delta u, \Delta v) \mid |\Delta u| \leq f_m \wedge |\Delta v| \leq f_m\} \tag{6}$$

where the offset $(\Delta u, \Delta v)$ describes a pixel-accurate flow estimate. The inverse mapping is:

$$\phi^{-1} : \mathbb{O} \longrightarrow \mathbb{D}, \text{ with } \phi^{-1}(\Delta u, \Delta v) = d \tag{7}$$

We now adjust Eqn. (1) as follows:

$$L_{\mathbf{a}}(p_i, d) = C(p_i, \phi(d)) + \mathcal{M}_i - \min_{\eta \in \mathbb{D}} L_{\mathbf{a}}(p_{i-1}, \eta) \tag{8}$$

with

$$\mathcal{M}_i = \min_{\eta \in \mathbb{D}} \{L_{\mathbf{a}}(p_{i-1}, d) + P_\kappa ||\phi(\eta) - \phi(d)||_1\} \tag{9}$$

In the Potts model, see Eqn. (2), a step function is utilized for the cost regularization summand $\mathcal{M}_i$. The effect is that piecewise constant solutions are enforced. This model is sufficient for the stereo case but for the optical flow a linear function is more appropriate. The reason is that optical flow vectors are not piecewise constant but vary within small pixel distances. In Eqn. (9), $||\phi(\eta) - \phi(d)||_1$ refers to the L_1 distance of two disparity values within the offset domain $\mathbb{O}$. P_κ is the penalty factor that scales the slope of the linear function.

3.2   Penalty Adjustment

Although optical flow vectors only tend to have small variations even for different static objects, where one is occluding the other, they tend to 'jump' at occlusion edges if there is either a significant depth discontinuity, or an independently moving object passing through the scene.

In such a case, the full linear model tends to over-regularize. Therefore, a truncation of the linear model is more appropriate. Flow results η that lie within a certain vicinity of the reference solution d are penalized depending on their relative distance within the offset domain $\mathbb{O}$. Candidates outside this neighborhood all are treated equally, as in the Potts Model, to maintain flow discontinuities. The implementation of the truncated linear function follows Felzenszwalb and Huttenlocher [3], for $d = 0, \ldots, d_{max}$, and is implemented as follows:

$$L_{\mathbf{a}}(p_i, d) = \max\{L_{\mathbf{a}}^{\Gamma}(p_i, d), L_{\mathbf{a}}^{\Lambda}(p_i, d) + P_\tau P_\kappa\} - \min_{\eta \in \mathbb{D}} L_{\mathbf{a}}(p_{i-1}, \eta) \qquad (10)$$

with

$$L_{\mathbf{a}}^{\Lambda}(p_i, d) = C(p_i, \phi(d)) + L_{\mathbf{a}}(p_{i-1}, d) \qquad (11)$$

where P_τ refers to the truncation factor. The term L^{Γ} is calculated for every d in a forward and a backward pass as follows:

$$L_{\mathbf{a}}^{\Upsilon}(p_i, \phi^{-1}(\Delta u, \Delta v)) = \min \begin{cases} L_{\mathbf{a}}^{\Lambda}(p_i, \phi^{-1}(\Delta u, \Delta v)) \\ L_{\mathbf{a}}^{\Lambda}(p_i, \phi^{-1}(\Delta u - 1, \Delta v)) + P_\kappa \\ L_{\mathbf{a}}^{\Lambda}(p_i, \phi^{-1}(\Delta u, \Delta v - 1)) + P_\kappa \end{cases} \qquad (12)$$

for Δu and Δv running from $-f_m + 1, \ldots, f_m$. The backward pass reads:

$$L_{\mathbf{a}}^{\Gamma}(p_i, \phi^{-1}(\Delta u, \Delta v)) = \min \begin{cases} L_{\mathbf{a}}^{\Upsilon}(p_i, \phi^{-1}(\Delta u, \Delta v)) \\ L_{\mathbf{a}}^{\Upsilon}(p_i, \phi^{-1}(\Delta u + 1, \Delta v)) + P_\kappa \\ L_{\mathbf{a}}^{\Upsilon}(p_i, \phi^{-1}(\Delta u, \Delta v + 1)) + P_\kappa \end{cases} \qquad (13)$$

with Δu and Δv running backwards from $f_m - 1, \ldots, -f_m$.

The implementation handles the quadratic formulation of the regularization model with linear run-time complexity. However, the actual run-time of the algorithm is still doubled when compared to a solution implementing the Potts model because one extra pass through the offset domain $\mathbb{O}$ is required.

With the exception of using a truncated linear function instead of a step function, there is in fact no significant difference to the stereo regularization process. The only adaption is that data costs are calculated for a 2D search space and not for 1D column offsets. This is possible because the accumulation procedure only regularizes data costs that correspond to a unique label d. The interpretation of label d is unimportant. The only requirement is that the data costs at label d of neighboring pixels correspond to the same solution.

The optimal label d_{opt} is identified by a winner-takes-all approach, as in the stereo case. In other words, the label with the minimum aggregated cost is selected and mapped via $\phi(d)$ to the corresponding 2D optical flow result.

3.3   Coarse-to-Fine Scheme for Mid-Scale Optical Flow

In the previous section we outlined how to use the SGM integration process, originally designed for the 1D stereo case, for optical flow calculation. Threshold f_m defines the maximum flow that can be calculated.

First we note that by increasing f_m to $f_m + 1$, we add $8 \cdot (f_m + 1)$ new pixel positions that also need to be considered for a possible correspondence. To maintain a reasonable run-time of the algorithm on current hardware, the search space is limited to a value of $f_m = 7$. This results in a maximum of $d_{max} = 225$ labels, which is insufficient for the number of labels required for optical flow calculations. Therefore, we need to embed fSGM into a coarse-to-fine approach.

The following concept is adapted from the coarse-to-fine scheme that Gehrig et al. [4] proposed for the stereo case. This concept can be applied for 'mid-scale optical flow' which we consider as a 2D displacement of up to 20 pixels. We generate image pyramids $\mathcal{P}_{t_0}$ and $\mathcal{P}_{t_1}$ of the input images I_{t_0} and I_{t_1} and run fSGM instances in parallel on each pyramid level l, with $l = 0, ..., l_{max}$. Flow results of each level are filtered such that only valid flow vectors are kept. The remaining vectors are then scaled up and merged with the next higher resolution level. In cases where flow vectors from level l and $l - 1$ fall on the same pixel location, the result from level $l - 1$ is favoured, assuming a higher accuracy for the higher resolution at $l - 1$.

Filtering is performed as follows. First, we segment the displacement field into homogeneous flow regions. To be precise, we label a valid optical flow displacement (u_p, v_p) at pixel p with an invalid label d_{inv} if there is at least one valid displacement (u_q, v_q) at a pixel q being 8-adjacent to p such that $||(u_p, v_p) - (u_q, v_q)||_2 > \gamma$, for a given threshold γ.

In other words, if any two 8-adjacent flow vectors vary by more than γ then both pixels are invalidated. We refer to the result of this process as being a *homogeneous flow map*, homogenized by threshold γ. Using flood-fill segmentation we find the largest 8-connected homogeneous flow region and take this as the valid flow result at the current level. Figure 1 in Section 4.2 shows some results using this concept for mid-scale optical flow.

Problems arise if frame rates are not sufficiently high such that displacement vectors are larger than 20 pixels. In these cases the mid-scale approach becomes too inaccurate. Therefore, the standard coarse-to-fine concept should be considered which is based on the principle that flow vectors of level l are used to initialize the estimation process at level $l - 1$.

So far, the regularization process handles a 2D search space instead of a 1D search space, but it assumes that data costs of labels refer to the same optical flow solution. This assumption is now violated, because data costs are calculated around initial flow vectors which need to be added to the 2D pixel offset $\phi(d)$ of label d. In other words, a regularization process for large-scale optical flow needs to accommodate the situation in which two neighboring pixels are initialized by different flow vectors. In that case data costs for label d at pixel p_i refers to a different solution than label d at pixel p_{i-1}. This problem is addressed in the following section.

3.4   Regularization with Initial Flow Results

For the following adaptation of the regularization process, we consider it to be embedded into a standard coarse-to-fine approach. The upscaled optical flow vector at pixel p_i from previous pyramid level $l+1$ is referred to as $(u_{p_i}^{l+1}, v_{p_i}^{l+1})$.

We recall from the previous section that in case that initial flow vectors of neighboring pixels are different, a label d does not represent an identical solution for both pixels anymore. Depending on the initial flow differences, the solution for label d at pixel p_i is either located at another label d' at p_{i-1}, or there is simply no data cost calculated at p_{i-1} for the solution represented by d. However, we assume that changes in optical flow between neighboring pixels are either small or large (at flow discontinuities).

For the first case we define a mapping ϑ that establishes a correspondence between labels of neighboring pixels which represent the same solution. For the latter case, this correspondence is not required as changes relate to flow discontinuities.

Before defining a correspondence mapping ϑ, $\mathbb{D}$ is extended to $\mathbb{D}^{\mathrm{inv}}$ by adding a unique integer value d_{inv} that lies outside the domain $\mathbb{D}$. This integer identifies a label that cannot be mapped. Now we define the mapping as

$$\Theta : \mathbb{Z}^2 \times \mathbb{D} \longrightarrow \mathbb{D}^{\mathrm{inv}}, \text{ with } \vartheta(x,y,d) = \begin{cases} \phi^{-1}(x,y), & \text{if } (x,y) \in \mathbb{O} \\ d_{\mathrm{inv}} & \text{otherwise} \end{cases} \qquad (14)$$

The regularization process is now described by

$$L_{\mathbf{a}}(p_i, d) = \max\{L_{\mathbf{a}}^{\Gamma}(p_i, d), L_{\mathbf{a}}^{\Lambda}(p_i, d) + P_{\tau}P_{\kappa}\} - \min_{\eta \in \mathbb{D}} L_{\mathbf{a}}(p_{i-1}, \eta) \qquad (15)$$

with

$$L_{\mathbf{a}}^{\Lambda}(p_i, d) = C(p_i, (u_{p_i}^{l+1}, v_{p_i}^{l+1})) + \phi(d) + \qquad (16)$$
$$L_{\mathbf{a}}(p_{i-1}, \vartheta(u_{p_i}^{l+1} - u_{p_{i-1}}^{l+1}, v_{p_i}^{l+1} - v_{p_{i-1}}^{l+1}, d)) \qquad (17)$$

In cases where a label has no corresponding solution at the previous pixel we define a default cost with $L_{\mathbf{a}}(p, d_{\mathrm{inv}}) = c_{def}$, with $d \in \mathbb{D}$. The result of the regularization process at level l corresponds to the optimal 2D translation w.r.t. the initial flow results. In other words, $(u_{p_i}^{l+1}, v_{p_i}^{l+1}) + \phi(d_{opt})$ is the solution at pyramid level l.

4   Evaluation and Discussion

The previous section described a scan-line dynamic programming implementation that can be embedded into a coarse-to-fine scheme. Next, we specify the algorithm's configuration that we use for our evaluation.

4.1   Algorithm Configuration

To calculate the data cost between pixels we use the *census cost function*. This function has been identified [9] to be very descriptive and robust, especially under strong illumination variations. Since this is a crucial feature for real-world applications the function is increasingly applied for both, stereo [16] and optical flow estimation methods [20].

The census cost is based on the *census transform* as introduced by Zabi and Woodfill [21]. A binary *signature* vector is assigned to each pixel position $p = (i, j)$ of the base and match image. It is calculated based on the ordinal characteristic of the intensity $I_p = I_{(i, j)}$ of an image I in relation to intensities within a defined neighborhood. This transform is performed once on the base and the match image prior to cost calculations. A signature is stored as a bit string in an integer matrix of the same dimensions as the given image. The signature sequence is generated as follows:

$$\text{census}_{\text{sig}}(I_{(i, j)}) = \left\{ \Upsilon\big[I_{(i, j)} \geq I_{(i+x, j+y)}\big] \right\}_{(x, y) \in \mathcal{N}} \tag{18}$$

where $\Upsilon[\cdot]$ returns 1 if true, and 0 otherwise; $\mathcal{N}$ denotes a neighborhood with respect to the origin. In our implementation we use a 11×11 window. The actual *census cost* is the Hamming distance of two signature vectors and can be calculated very efficiently [18].

The data costs are calculated for the domain $\mathbb{O}$ specified by parameter $f_m = 7$. The regularization function is configured with $P_\kappa = 12$, and $P_\tau = 6$. Image pyramids with 15 levels are employed for the coarse-to-fine process. A factor of $\zeta = 0.8$ is used for down scaling. Outliers at each level are filtered by a median filter. For mid-scale optical flow we employ pyramids with only three levels and with a scaling factor of $\zeta = 0.5$. The flow map segmentation process uses $\gamma = 3$ and does not require median filtering.

4.2   Mid-Scale Optical Flow

Figure 1 shows results of our mid-scale optical flow algorithm on a dataset provided for the currently running HCI Bosch Robust Vision Challenge.[1] The top two image rows [13] show optical flow in the mid-scale range with good results. The bottom row shows a frame from a sequence recorded on a motorway and is dominantly defined by large-scale optical flow. This example shows inaccuracies of calculated values on the road for the mid-scale optical flow method. Clearly the mid-scale approach lacks accuracy but has the advantage that all pyramid levels can be processed in parallel.

4.3   Large-Scale Optical Flow

Geiger et al. [5] recently introduced The KITTI Vision Benchmark Suite. It currently features 195 testing and 194 training stereo pairs with semi-dense ground

---

[1] http://hci.iwr.uni-heidelberg.de/Static/challenge2012/

Fig. 1. Mid-scale fSGM results on images of the HCI Bosch Challenge

truth generated by a Velodyne HDL-64E laser range-finder. The images were taken from recorded sequences of real-world driving scenarios and can be used to evaluate stereo and optical flow methods. The algorithms, evaluated on this dataset have to deal with realistic illumination conditions, a high image resolution of 1240 × 376 pixels and large pixel displacements.

We submitted our results to this benchmark to be ranked against current state-of-the-art optical flow algorithms. Table 1 shows the four top ranked algorithms along with baseline results of the highly recognized work by Brox et al. [2] (LDOF) and Zach et al. [22] (DB-TV-L1). This list refers to the state on 10 October 2012.

Table 1. Ranking of the KITTI flow benchmark on 10 October 2012

Rank	Method	Out-Noc	Avg-Noc	Setting	Runtime
1	PCBP-Flow	5.88 %	1.7 px	ms	180 s
2	fSGM	11.03 %	3.2 px	-	60 s
3	TGV2CENSUS	11.14 %	2.9 px	-	4 s
4	GC-BM-Bino	18.93 %	5.0 px	ms	1.3 s
9	LDOF	21.86 %	5.5 px	-	60 s
12	DB-TV-L1	30.75 %	7.8 px	-	16 s

The algorithms that are listed in Table 1 need to be distinguished between methods operating on monocular image sequences and those utilizing additional information of the stereo pair. The latter are identified by the setting *ms*. The table shows the reference evaluation error index (Out-Noc). For this index the evaluation is performed on a 100% dense optical flow map and flow vectors at all pixels with available ground truth are considered to be correct if they do not deviate by more than a spatial distance of 3 pixels from the ground truth. If a method does not provide 100% optical flow density, such as GC-BM-Bino [10], a simple background interpolation technique is applied prior to evaluation. We summarize that our method fSGM is currently ranked second w.r.t. the reference index closely followed by TVG2CENSUS [19].

The algorithms PCBP-Flow and GC-BM-Bino belong to the category of algorithms employing stereo information. They both use the *motion-stereo* constraint (setting ms). At the time of writing, all methods using additional stereo information are either anonymous or are still to appear. However, according to information from KITTI and the provided method descriptions the constraint can be characterized by the following two features. First, by using epipolar geometry from the stereo pair, the 2D optical flow search space can be reduced to a 1D search space. Second, because this constraint is employed, the algorithms are not capable of handling independently moving objects. Therefore, they may be hard to apply in applications such as driver assistance systems. fSGM as well as TVG2CENSUS on the other hand do handle general motion patterns.

Our fSGM implementation is C++ based. It is executed on an Intel Core2Duo, and has currently a run-time of 60 seconds. The main reason for this comparatively slow run-time is that we do not utilize any parallel hardware processing (such as multiple CPU cores, any SSE optimization, or GPU) and that we still need very fine-scaled image pyramids to gain high-quality results.

Figure 2 shows a strong and a weak example of fSGM performing on the KITTI testing dataset. Despite the already exceptional performance for a scan-line DP algorithm on the KITTI benchmark, there are still many possibilities left for further improvement.

Future Research. We aim at the development of a robust optical flow treatment at image borders and occlusions. Another focus of our research is on the improvement of the run-time of fSGM while keeping its performance.

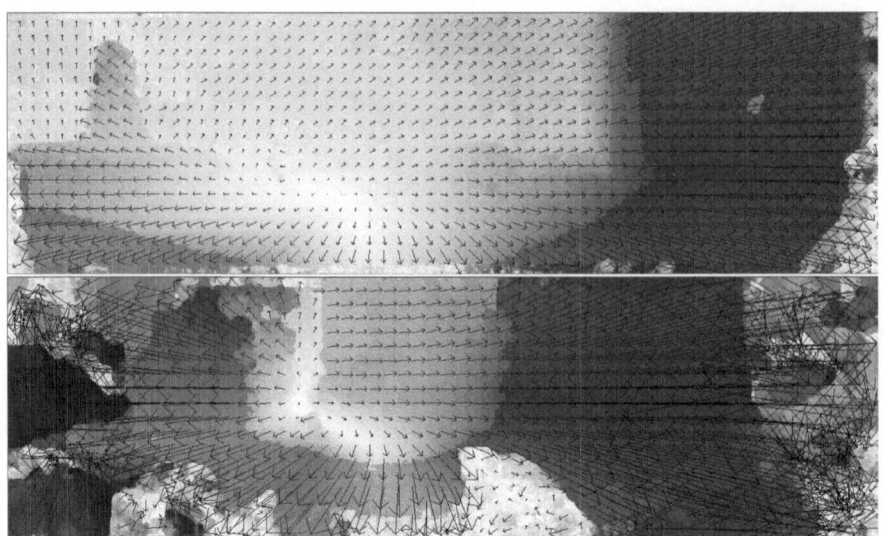

Fig. 2. fSGM results for frame 163 (strong performance) and frame 166 (weak performance) of the KITTI testing dataset

5   Conclusions

This paper presented fSGM as a novel technique and algorithm to calculate optical flow. It is based on a scan-line dynamic programming concept that follows the cost integration strategy of semi-global matching. The significant novelty is that fSGM is the first scan-line DP algorithm for optical flow that is successfully embedded into a coarse-to-fine approach. This enables it to handle large pixel displacements with a performance that is usually only known from variational methods. fSGM currently ranks second on the KITTI Vision Benchmark Suite.

References

1. Brox, T., Bruhn, A., Papenberg, N., Weickert, J.: High Accuracy Optical Flow Estimation Based on a Theory for Warping. In: Pajdla, T., Matas, J. (eds.) ECCV 2004. LNCS, vol. 3024, pp. 25–36. Springer, Heidelberg (2004)
2. Brox, T., Malik, J.: Large displacement optical flow: descriptor matching in variational motion estimation. IEEE Trans. Pattern Analysis Machine Intelligence 33, 500–513 (2011)
3. Felzenszwalb, P.F., Huttenlocher, D.P.: Efficient belief propagation for early vision. Int. J. of Computer Vision (IJCV) 70, 41–54 (2006)
4. Gehrig, S.K., Eberli, F., Meyer, T.: A Real-Time Low-Power Stereo Vision Engine Using Semi-Global Matching. In: Fritz, M., Schiele, B., Piater, J.H. (eds.) ICVS 2009. LNCS, vol. 5815, pp. 134–143. Springer, Heidelberg (2009)
5. Geiger, A., Lenz, P., Urtasun, R.: Are we ready for autonomous driving? The KITTI Vision Benchmark Suite. In: Proc. Computer Vision Pattern Recognition, CVPR (2012)

6. Gong, M., Yang, Y.-H.: Estimate large motion using the reliability-based motion estimation algorithm. In: Proc. Int. Joint Conf. Artificial Intelligence (IJCAI), vol. 2, pp. 1120–1126 (1985)
7. Hermann, S., Morales, S., Vaudrey, T., Klette, R.: Illumination Invariant Cost Functions in Semi-Global Matching. In: Koch, R., Huang, F. (eds.) ACCV 2010 Workshops, Part II. LNCS, vol. 6469, pp. 245–254. Springer, Heidelberg (2011)
8. Hirschmüller, H.: Accurate and efficient stereo processing by semi-global matching and mutual information. In: Proc. IEEE Int. Conf. Computer Vision Pattern Recognition (CVPR), vol. 2, pp. 807–814 (2005)
9. Hirschmüller, H., Scharstein, D.: Evaluation of stereo matching costs on images with radiometric differences. IEEE Trans. Pattern Analysis Machine Intelligence 31, 1582–1599 (2009)
10. Kitt, B., Lategahn, H.: Trinocular optical flow estimation for intelligent vehicle applications. In: Proc. IEEE Int. Conf. Intelligent Transportation Systems (2012) (to appear)
11. Lempitsky, V., Roth, S., Rother, C.: FusionFlow: Discrete-continuous optimization for optical flow estimation. In: Proc. IEEE Int. Conf. Computer Vision Pattern Recognition, CVPR (2008)
12. Lei, C., Yang, Y.-H.: Optical flow estimation on coarse-to-fine region-trees using discrete optimization. In: Proc. Int. Conf. Computer Vision (ICCV), pp. 1562–1569 (2009)
13. Meister, S., Jähne, B., Kondermann, D.: Outdoor stereo camera system for the generation of real-world benchmark data sets. Optical Engineering 51, paper 021107, 6 pages (2012)
14. Ohta, Y., Kanade, T.: Stereo by two-level dynamic programming. In: Proc. Int. Joint Conf. Artificial Intelligence (IJCAI), vol. 2, pp. 1120–1126 (1985)
15. Quénot, G.M.: Computation of optical flow using dynamic programming. In: Proc. IAPR Workshop Machine Vision Appl. (MVA), vol. 3, pp. 249–252 (1996)
16. Ranftl, R., Gehrig, S., Pock, T., Bischof, H.: Pushing the limits of stereo using variational stereo estimation. In: Proc. IEEE Intelligent Vehicles Symposium (IV), pp. 401–407 (2012)
17. Sun, C.: Fast optical flow using 3D shortest path techniques. Image Vision Computing 20, 981–991 (2002)
18. Warren, H.S.: Hacker's Delight, pp. 65–72. Addison-Wesley Longman, New York (2002)
19. Werlberger, M.: Convex approaches for high performance video processing, PhD thesis (2012)
20. Werlberger, M., Trobin, W., Pock, T., Wedel, A., Cremers, D., Bischof, H.: Anisotropic Huber-L1 optical flow. In: Proc. British Machine Vision Conference (BMVC), pp. 1–11 (2009)
21. Zabih, R., Woodfill, J.: Non-parametric Local Transform for Computing Visual Correspondence. In: Eklundh, J.-O. (ed.) ECCV 1994, Part II. LNCS, vol. 801, Springer, Heidelberg (1994)
22. Zach, C., Pock, T., Bischof, H.: A Duality Based Approach for Realtime TV-L^1 Optical Flow. In: Hamprecht, F.A., Schnörr, C., Jähne, B. (eds.) DAGM 2007. LNCS, vol. 4713, pp. 214–223. Springer, Heidelberg (2007)

Novel Multi-view Synthesis from a Stereo Image Pair for 3D Display on Mobile Phone

Chen-Hao Wei, Chen-Kuo Chiang, Yu-Wei Sun,
Mei-Huei Lin, and Shang-Hong Lai

National Tsing Hua University, Hsinchu, Taiwan 30013, R.O.C.

Abstract. In this paper we present a novel view synthesis method for mobile platform. A disparity-based view interpolation is proposed to synthesize the virtual view from a left and right image pair. This makes to full use of the available information of both images to give accurate interpolation results and greatly decreases the pixel number in the disoccusion region, thus reducing the errors introduced by the patch-based image inpainting. Two boundary refining schemes considering gradient and color coherence and applying directional filters on boundaries are proposed to improve the synthesized results. Experimental results show that the proposed method is effective and suitable for mobile environment.

1   Introduction

More and more 3D display applications can be found in the high-tech products, including 3D LCD/LED displays, 3D laptops, 3D cameras and mobile phone devices. Nowadays, advanced 3D display technologies allow people to experience realistic 3D scenes without wearing 3D glasses. People watch 3D contents from different viewpoints by changing the viewing directions. This requires multi-view contents rather than two views for stereoscopic displays.

However, most stereoscopic cameras can only capture two views. It is necessary to convert stereoscopic contents from two views to multiple views for free viewpoint system. The major challenge to create such content is that the regions covered by the foreground objects in the original view may be disoccluded when views direction changes. Thus, filling the disocclusion regions is very critical to achieve high-quality view synthesis. Conventional system produces virtual views based on the three steps: preprocessing of depth image, image warping and hole filling. Since the disocclusion region is likely to be large, some researcher focus on the hole filling methods based on image inpainting techniques [1,2] to fill in the disocclusion regions.

Considering the limited resources on mobile devices and the high complexity of the previous methods, a novel view synthesis method is proposed using a binocular image for autostereoscopic displays. In contrast to the conventional method [3] that virtual views are interpolated after filling the disocclusion regions from warped images, we propose to interpolate the virtual views by warped images prior to the inpainting step in a disparity-based manner. After interpolation, patch-based image inpainting is applied to fill holes. This greatly reduces

J.-I. Park and J. Kim (Eds.): ACCV 2012 Workshops, Part II, LNCS 7729, pp. 568–579, 2013.

the number of hole pixels, thus improving the synthesized quality and execution time for mobile environment. Two edge smoothing schemes are proposed to reduce the artifacts around object boundaries of the synthesized views.

2   Related Work

Image inpainting is a technique widely utilized to recover the disocclusion regions. Recent exemplar-based approaches are commonly used in image inpainting. In [4], a fast algorithm was presented to propagate texture and structure in a small patch. The success of structure propagation was dependent on the order in which the filling proceeds. The confidence value in the synthesized pixel values was propagated in a manner similar to the propagation of information in inpainting. Shen et al. [5] selected similar patches to fill in the unknown regions with gradient maps. In that case, the image can be reconstructed from gradients by solving a Poisson equation. Overall, inpaint-based methods fill each hole by selecting the patch from a large candidate set. This requires high computational power and large memory when the number of holes is large.

Contrary to the greedy methods, some approaches formulate the image completion as discrete global optimization problems. In [6], image completion was automatically solved using an efficient BP algorithm. In [7], the image was completed with manually added structure information. Huang et al. [8] improved the hole filling method in [6] by adding the structure information into the global optimization formulation and solved the optimization problem with a two-step BP. In their method, only a single image was considered for the completion. Above methods need user interaction or solve optimization of objective function which are not suitable for mobile platform.

3   Disparity-Based View Interpolation

Since we have rightmost and leftmost views with their corresponding disparity maps, we can interpolate the virtual views by using images that are warped to the new view. In autostereoscopic display, image warping is generally degenerated to one-dimensional displacement along the horizontal scanline based on the assumption that the human eyes are in parallel to the screen at the same horizontal line when watching the display. Thus, we simplify the description of image warping to one-dimensional displacement along the horizontal here.

Assume position of the leftmost view $v=0$ and the position of rightmost view $v=1$. The desired intermediate view k lies between 0 and 1 which is the relative position of the leftmost view. Denote I_L, D_L, I_R and D_R the color and disparity images of the leftmost and rightmost views. Let I_0 and I_1 be the warped images from leftmost view and rightmost view. The relation between the position x' and the original position x can be described by the following equation [3]:

$$s = k \times \left(d_0 + \left(D_L(x,y) - g_0 \right) \frac{d_1 - d_0}{g_1 - g_0} \right) \tag{1}$$

$$x' = x + round(s) \tag{2}$$

where k is the relative position of the desired view, $(d_0; d_1)$ is the desired disparity range in pixel unit on the display, and $(g_0; g_1)$ is the input disparity range. The value of the position (x',y) in the warped image I_0 can be defined as follows:

$$a = x' - s \tag{3}$$

$$p = floor(a + 1) \ and \ q = floor(a) \tag{4}$$

$$I_0(x', y) = (a - q) \times I_L(p, y) + (p - a) \times I_L(q, y) \tag{5}$$

Similarly, the warped image I_1 can be derived by using I_R. Denoted D_0 and D_1 the warped disparity maps of I_0 and I_1, respectively. D_0 can be obtained in the way that I_0 warps by Eq.5. After two views I_L and I_R are warped to the desired virtual view I_0 and I_1, these two warped images are used for view interpolation. Intuitively, I_0 and I_1 can be linearly combined to obtain the desired view I_k. However, artifacts are introduced in the synthesized view very often. The inconsistency occurs when the depth $D_0(x, y)$ is very different from $D_1(x, y)$. The difference of depth information indicates that these two pixels of the same position are occluding and occluded pixels one another. In other words, interpolating these pixels from two warped images directly makes the pixel in the virtual view either transparent to the background or inconsistent between foreground objects. In this sense, a disparity-based view interpolation is introduced. We do not apply linear interpolation for both pixels. Instead, by considering the relation of depth, only the pixel in the front contributes to the synthesized pixel $I_k(x, y)$. The desired view I_k is obtained as follows:

$$I_k(x, y) = \begin{cases} I_0(x, y) & if \ D_0(x, y) - D_1(x, y) > D_{th} \\ I_1(x, y) & if \ D_1(x, y) - D_0(x, y) > D_{th} \\ I_0 \times (1 - k) + I_1 \times k & otherwise \end{cases} \tag{6}$$

where D_{th} is a threshold for difference of depth. When the difference of depth is below the threshold, two pixels are used for interpolation. Otherwise, only the occluding pixel contributes to $I_{k(x,y)}$.

4   Boundary Refinement

The results of view synthesis usually contain artifacts in object boundaries. This is mainly caused by the inaccurate depth information used by forward warping. The original left and right views are warped to the desired view in the first place. If the depth information are not accurate enough around object boundaries, forward warping warps the real object boundaries further away from their actual positions, making the object boundaries in the synthesized view very aliasing. In addition, errors are often introduced by the process of disocclusion region

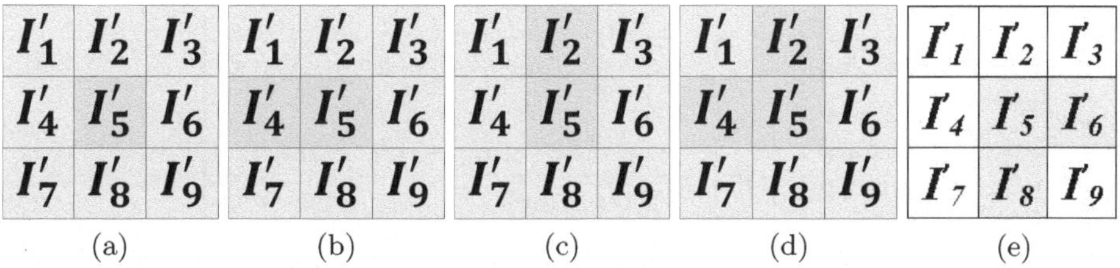

I'_1 I'_2 I'_3	I'_1 I'_2 I'_3	I'_1 I'_2 I'_3	I'_1 I'_2 I'_3	I_1 I_2 I_3
I'_4 I'_5 I'_6	I'_4 I'_5 I'_6	I'_4 I'_5 I'_6	I'_4 I'_5 I'_6	I_4 I_5 I_6
I'_7 I'_8 I'_9	I'_7 I'_8 I'_9	I'_7 I'_8 I'_9	I'_7 I'_8 I'_9	I_7 I_8 I_9
(a)	(b)	(c)	(d)	(e)

Fig. 1. Four cases of left and up neighboring pixels. (a) I'_5 is a edge pixel and its left and up neighboring pixels are not. (b) The left neighboring pixel I'_4 of I'_5 is a edge pixel. (c) The up neighboring pixel I'_2 is a edge pixel. (d) Both I'_2 and I'_4 are edge pixels. (e) A simple example of object boundary that contains 3 edge pixels: I'_5, I'_6 and I'_8 and their neighboring pixels.

recovering. These situations can be found in many literatures of view synthesis work [9,10]. Therefore, two boundary refinement are proposed to improve the quality for view synthesis.

4.1 Gradient and Color Coherence

To refine the artifacts around boundaries, intuitively average filters can be applied to smooth the artifacts. However, the boundary edges become blurred as well. Here, gradient and color coherence of corresponding pixels are considered along the object boundary in the disocclusion region between the refined and the original images. Denote I, I' the original image and the refined image, respectively. The gradient and color of object boundaries of I' should be similar to that of I. To simplify the calculation of gradient, we consider the gradient along x-direction and y-direction by the difference between current pixel and its left and up neighboring pixels. Take a example in Fig. 1 (e), there are three edge pixels: I'_4, I'_5 and I'_7. The gradient coherence of I'_4 is defined as:

$$
\begin{aligned}
I'_5 - I'_4 &= (I_5 - \Delta d_5) - (I_4 - \Delta d_4) \\
I'_5 - I'_2 &= (I_5 - \Delta d_5) - (I_2 - \Delta d_2)
\end{aligned}
\tag{7}
$$

where Δd_n represents the disparity of pixel index n. The color coherence of I'_5 is simply defined as:

$$
I'_5 = I_5 - \Delta d_5
\tag{8}
$$

Considering all pixels along edges with their gradient and color coherence, we can construct the following linear system $Ax = b$:

$$
\begin{bmatrix} 1 & 0 & 0 \\ 1 & 0 & 0 \\ 1 & 0 & 0 \\ -1 & 1 & 0 \\ 0 & 1 & 0 \\ 0 & 1 & 0 \\ 0 & 0 & 1 \\ -1 & 0 & 1 \\ 0 & 0 & 1 \end{bmatrix} \begin{bmatrix} I_5' \\ I_6' \\ I_8' \end{bmatrix} = \begin{bmatrix} (I_5 - \Delta d_5) - (I_4 - \Delta d_4) + I_4' \\ (I_5 - \Delta d_5) - (I_2 - \Delta d_2) + I_2' \\ I_5 - \Delta d_5 \\ (I_6 - \Delta d_6) - (I_5 - \Delta d_5) \\ (I_6 - \Delta d_6) - (I_3 - \Delta d_3) + I_3' \\ I_6 - \Delta d_6 \\ (I_8 - \Delta d_8) - (I_7 - \Delta d_7) + I_7' \\ (I_8 - \Delta d_8) - (I_5 - \Delta d_5) \\ I_8 - \Delta d_8 \end{bmatrix} \tag{9}
$$

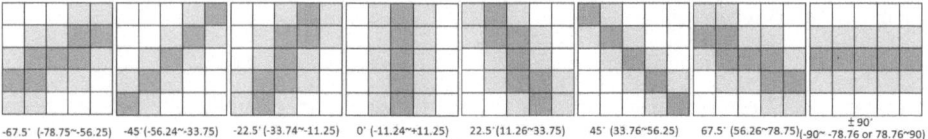

-67.5˚ (-78.75~-56.25)   -45˚(-56.24~-33.75)   -22.5˚(-33.74~-11.25)   0˚ (-11.24~+11.25)   22.5˚(11.26~33.75)   45˚ (33.76~56.25)   67.5˚ (56.26~78.75)   ±90˚ (-90~ -78.76 or 78.76~90)

Fig. 2. Eight quantized orientations of gradient. Note that the orientation of gradient is perpendicular to edge orientation.

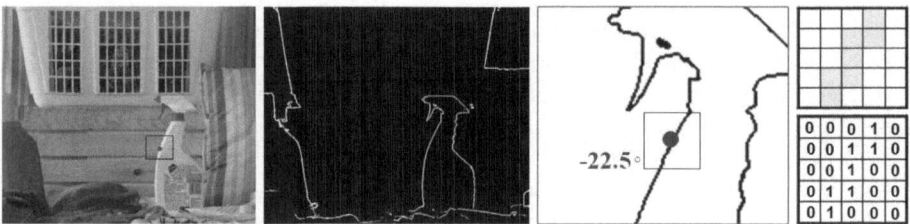

Fig. 3. Applying directional filter on object boundary. From left to right: The first is the synthesized image. A concise edge map can be obtained by depth map. The third is the orientation of gradient and the last depicts the corresponding filter.

To build the matrix A automatically, the neighboring pixels of the current edge pixel are examined to see whether they are also edge pixels or not. Four cases are depicted in Fig. 1 (a) to (d). According these four cases, 1 and −1 are placed into the matrix A and the value is calculated by the known variables for the corresponding entry in matrix b. When the linear system is large, the execution time to find the solution increases dramatically. For fast approximation on mobile platform, the pixel number is counted along edges. When it is over a threshold k, the edge is cut into separate lines to reduce the size of linear system.

4.2   Directional Filter

Using spatial filters to reduce the image artifacts is simple and fast. To make use of the merits for mobile platform, an alternative is to take edge directions into account so that the edge structure can be preserved when spatial filters are applied to object boundaries. Firstly, for every pixel along edges, orientation of gradient is calculated. It is not likely to have a filter for every possible orientation. In practice, the orientations are quantized into several bins, as depicted in Fig. 2. Then, directional filters are applied according to the quantized orientation.

The size of each filter is 5x5. They can be considered as directional filters. It is different from the mean filter since only pixels along the edge orientation (which are represented by the gray grids) contribute to smoothing. Note that each pixel in the conventional averaging filter is equally weighted. In the proposed directional filters, we can set the dark gray pixel to 1 and all the others to 0. A second choice is to apply extended directional filter. Each dark pixel is extended one pixel to its left and right (represented by light gray pixels). Then, different weighting can be set. In our implementation, 1 and 0.5 are given to the dark gray pixels and their extended part, respectively.

Once the synthesized virtual view is obtained, directional filters can be applied to refine the object boundary. However, applying edge detection directly may produce too many edge points. In this work, edge detection is applied on warped depth map to find the discontinuity between objects and background without details inside objects. In Fig.3, we can notice that the edge map contains only boundaries of main objects. Then, the gradient is calculated for each edge point to obtain the direction of one edge. After quantized the edge direction, the corresponding directional filter is chosen to refine the object boundary. A sample example is given in Fig.3.

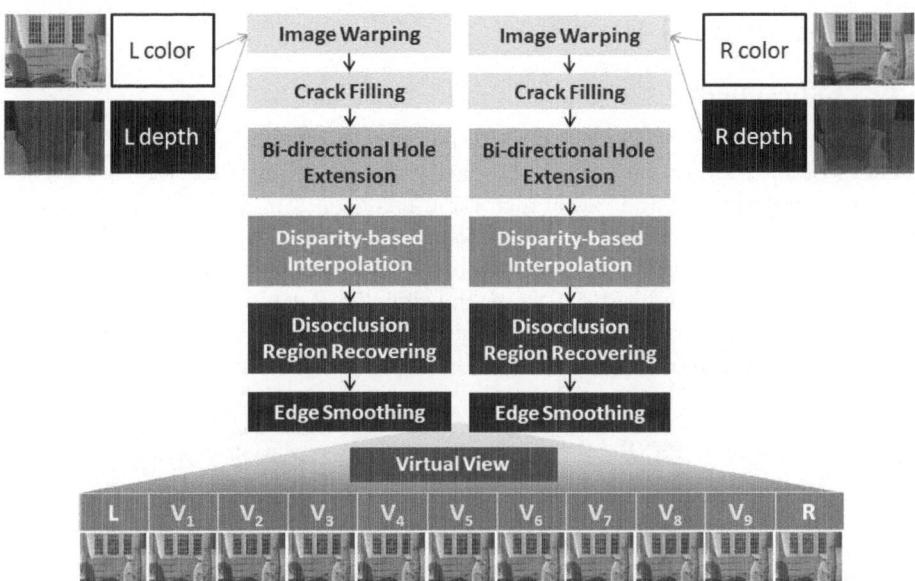

Fig. 4. Flow chart of proposed system framework

5 System Framework

The input of our system contains two color images and their corresponding depth maps of leftmost and right most views. It aims to synthesize virtually in-between views. The first step is image warping based on the depth information. According to Eq. 1, each pixel is warp to the new position in the desired view.

After warping, it contains internal empty regions in the warped images due to the depth discontinuities. These regions can be classified into two types, image cracks and disocclusion regions. Image cracks are generally caused by noises or digital numerical precision, whereas disocclusion regions come from the sharp depth discontinuity. Image cracks are very thin lines (usually 1-pixel wide). Here, the mean filter is first applied to fill the cracks. We observe that some mixed pixels appeared around the boundaries of the disoccusion regions usually cause the color inconsistency during view synthesis. To remove the errors, we extended the holes boundaries in bi-directions by using image dilation. Then, disparity-based interpolation is applied to generate virtual views. This step can cover large number of image holes since most information in the disocclusion regions in one view can be found from the other view based on disparity information. For those remaining large holes, exemplar-based image inpaint method [3] is exploited to recover those regions. The first step is to calculate the filling priority for every patch which contains holes. The second step is to search the best k matches among a set of candidate patches. Last, fill the unknown pixels in the current patch in a weighted manner of these k candidates.

6 Experimental Results

6.1 Experimental Setting

The experiments are conducted on Middlebury Stereo Datasets 2005 [11] which consists of 7 views for each topic. Disparity maps are provided for views 1 and 5. Images are rectified and radial distortion has been removed. The resolution of images provided is 671x555. We chose four topics in our experiment: Baby, Art, Laundry and Bowling. We have shown our evaluation based on Windows 7 and mobile phone with Android platform. Specifically, it runs on a Intel QuadCore Q9300 2.5GHz CPU with 2G RAM. As our mobile environment, we used HTC EVO 3D with 1.2GHz dual core and 1G RAM.

Table 1. PSNR (dB) of different view interpolation methods

Data	Baseline Method	Basis Interpolation Method	Proposed Method
Art	28.5405	28.4744	28.4570
Baby	35.0441	34.8765	35.0476
Bowling	27.5040	27.1295	28.3395
Laundry	28.0766	27.6523	28.5151

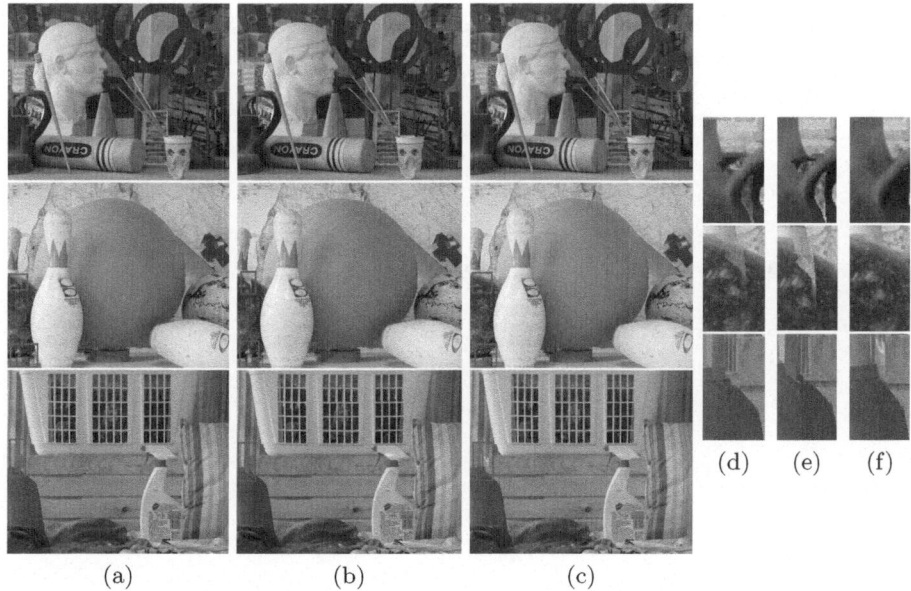

(d)   (e)   (f)

(a)   (b)   (c)

Fig. 5. View interpolation results of different methods. Column (a)(d), (b)(e) and (c)(f) are results from baseline interpolation, basis method and the proposed method, respectively.

Table 2. PSNR comparison of the unrefined image and refined by gradient and color coherence and directional filters

PSNR(dB)	Unrefined	Gradient and Color Coherence	Directional Filter
Art	28.46	28.51	30.54
Baby	35.05	35.32	35.34
Bowling	28.34	28.50	28.56
Laundry	28.52	28.64	29.48

Table 3. Time comparison of the coherence method, fast coherence and directional filters

Time(s)	Gradient and Color Coherence	Fast Coherence Method	Directional Filter
Art	61.26	1.90	1.48
Baby	26.33	1.29	0.99
Bowling	63.25	1.51	1.33
Laundry	98.05	2.22	1.22

6.2 Results of View Interpolation

Taking *view*1 and *view*5 of seven views from **Middlebury** dataset as leftmost and rightmost view, *view*3 can be regarded as the ground truth of the synthesized virtual view in the middle. Two view interpolation methods are compared

Table 4. PSNR comparison of different view synthesis methods

PSNR(dB)	DIBR	VSRS	Proposed
Art	17.34	20.45	28.46
Baby	23.15	24.22	35.05
Bowling	20.68	21.92	28.34
Laundry	18.24	21.18	28.52

Table 5. Execute time of each component on Android phone

Component	Time(s)
Warping	0.506653
Crack filling	0.105804
Bi-directional hole extension	0.055237
Interpolation	0.138886
Disocclusion region recovering	8.340515
Edge smoothing	0.424774

Fig. 6. The results of boundary refinement using gradient and color coherence and directional filters. (a)(g) Results of coherence method. (b)(h) Input: part of Laundry and Bowling. (c)(i) Results of coherence method. (d)(j) Results of filter method. (e)(k) Ground truth. (f)(l) Results of filter method.

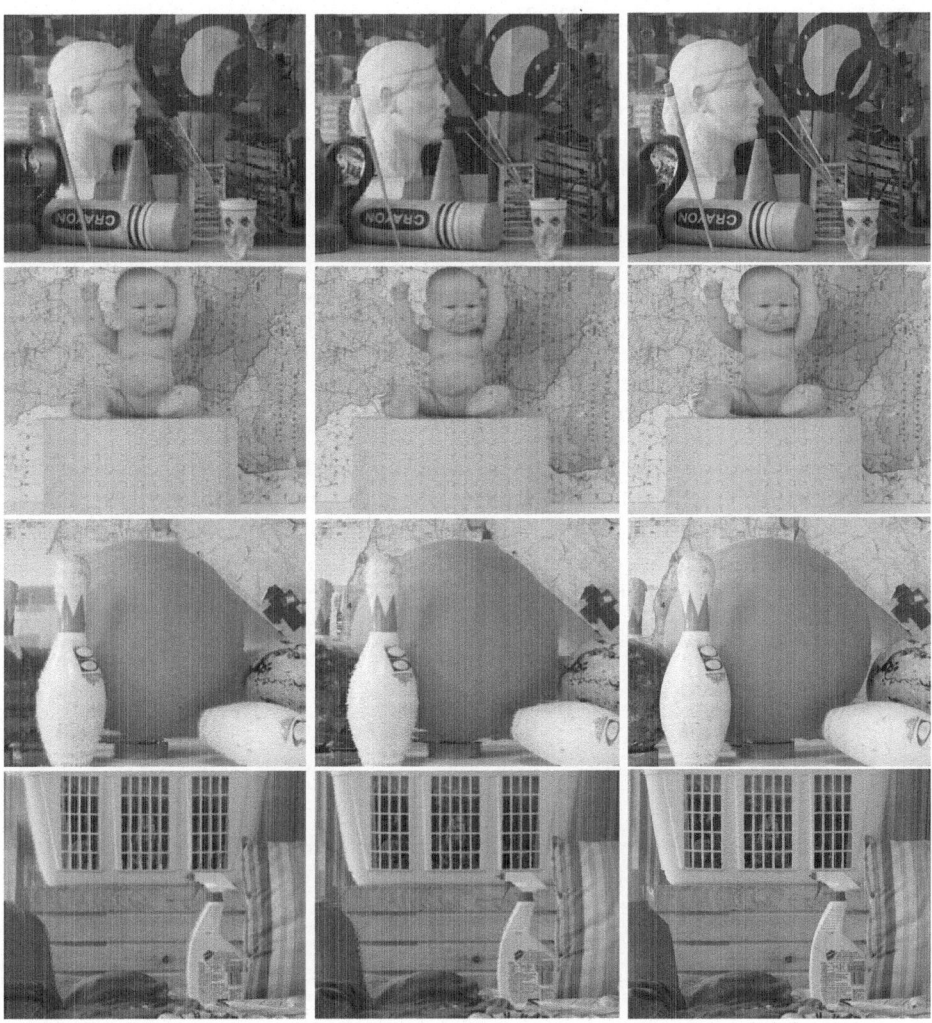

Fig. 7. View synthesis results of different methods. The first column to the third column are the results of DIBR, VSRS and the proposed method, respectively.

to the proposed disparity-based view interpolation. Direct interpolation of two views are used as the baseline method. Interpolation by using leftmost view as a basis image and find available information from the rightmost view [3] are referred as basis interpolation method. Fig. 5 shows the results of different interpolation method. We can notice that without considering the depth information, the foreground objects are likely to be affected by background pixels, making themselves transparent or containing errors. The PSNR comparison is listed in Table 1. The proposed method has higher PSNR in all cases.

6.3   Results of Boundary Refinement

We compared the method considering gradient and color coherence and the directional filter method for boundary refinement. The results are shown in Fig. 6. We can see that both methods improve the original synthesized images with aliasing around object boundaries. The PSNR comparison is given in Table 2. The results are very comparable of both methods. The execution time is listed Table 3. To solve large linear system, the time complexity is rather high. By exploiting fast method, the execution time of both methods are close. Considering the results of PSNR and time of directional filter method are better, we used such method in our system framework in the next subsection.

6.4   Overall Performance

We compared the proposed view synthesis method to the view synthesis reference software ($VSRS$) [12] and depth image based rendering ($DIBR$) [13]. Fig. 7 shows the synthesized results from different methods. The PSNR comparison is given in Table 4. We can see the proposed method outperforms significantly to other methods. The execution time on HTC EVO mobile platform is listed in Table 5. We can note that the proposed disparity-based interpolation and the edge smoothing component via directional filters are very fast. Both components runs under 0.5 seconds.

7   Conclusion

In this paper, an novel view synthesis system is presented with a disparity-based interpolation method and two boundary refinement schemes for mobile phone. By our experimental results, the PSNR of the synthesized view is significantly improved by the proposed method. Overall, our method is effective, simple and suitable for mobile platform.

References

1. Bertalmio, M., Sapiro, G., Caselles, V., Ballester, C.: Image inpainting. In: Proceedings of the 27th Annual Conference on Computer Graphics and Interactive Techniques, SIGGRAPH 2000, pp. 417–424. ACM Press/Addison-Wesley Publishing Co., New York (2000)
2. Oh, K.J., Yea, S., Ho, Y.S.: Hole filling method using depth based in-painting for view synthesis in free viewpoint television and 3-d video. In: Proceedings of the 27th Conference on Picture Coding Symposium, PCS 2009, pp. 233–236. IEEE Press, Piscataway (2009)
3. Lin, S.J., Cheng, C.M., Lai, S.H.: Spatio-temporally Consistent Multi-view Video Synthesis for Autostereoscopic Displays. In: Muneesawang, P., Wu, F., Kumazawa, I., Roeksabutr, A., Liao, M., Tang, X. (eds.) PCM 2009. LNCS, vol. 5879, pp. 532–542. Springer, Heidelberg (2009)

4. Criminisi, A., Pérez, P., Toyama, K.: Object removal by exemplar-based inpainting. In: CVPR (2), pp. 721–728 (2003)
5. Shen, J., Jin, X., Zhou, C.: Gradient Based Image Completion by Solving Poisson Equation. In: Ho, Y.-S., Kim, H.-J. (eds.) PCM 2005, Part I. LNCS, vol. 3767, pp. 257–268. Springer, Heidelberg (2005)
6. Komodakis, N.: Image completion using global optimization. In: Proceedings of the 2006 IEEE Computer Society Conference on Computer Vision and Pattern Recognition, vol. 1, pp. 442–452. IEEE Computer Society, Washington, DC (2006)
7. Sun, J., Yuan, L., Jia, J., Shum, H.Y.: Image completion with structure propagation. ACM Trans. Graph. 24, 861–868 (2005)
8. Ting, H., Chen, S., Liu, J., Tang, X.: Image inpainting by global structure and texture propagation. In: Proceedings of the 15th International Conference on Multimedia, MULTIMEDIA 2007, pp. 517–520. ACM, New York (2007)
9. Xi, M., Wang, L.H., Yang, Q.Q., Li, D.X., Zhang, M.: Multiview virtual image synthesis for auto-stereoscopic display based on two views. In: International Conference on Systems and Informatics (ICSAI), pp. 1966–1970 (2012)
10. Ahn, I., Kim, C.: Depth-based disocclusion filling for virtual view synthesis. In: International Conference on Multimedia Expo (ICME) (2012)
11. Scharstein, D., Pal, C.: Learning conditional random fields for stereo. In: CVPR (2007)
12. ISO/IEC JTC1/SC29/WG11: View synthesis software manual, release 3.5. In: MPEG (2009)
13. Fehn, C.: Depth-image-based rendering (dibr), compression, and transmission for a new approach on 3d-tv. In: Proceedings of SPIE, vol. 5291, pp. 93–104 (2004)

An Accurate Method for Line Detection and Manhattan Frame Estimation

Ron Tal and James H. Elder

York University
{rontal,jelder}@yorku.ca

Abstract. We address the problem of estimating the rotation of a camera relative to the canonical frame of an urban scene, from a single image. Solutions generally rely on the so-called 'Manhattan World' assumption [1] that the major structures in the scene conform to three orthogonal principal directions. This can be expressed as a generative model in which the dense gradient map of the image is explained by a mixture of the three principal directions and a background process [2]. It has recently been shown that using sparse oriented edges rather than the dense gradient map leads to substantial gains in both accuracy and speed[3]. Here we explore whether further gains can be made by basing inference on even sparser extended lines. Standard Houghing techniques suffer from quantization errors and noise that make line extraction unreliable. Here we introduce a probabilistic line extraction technique that eliminates these problems through two innovations. First, we accurately propagate edge uncertainty from the image to the Hough map through a bivariate normal kernel that uses natural image statistics, resulting in a non-stationary 'soft-voting' technique. Second, we eliminate multiple responses to the same line by updating the Hough map dynamically as each line is extracted. We evaluate the method on a standard benchmark dataset [3], showing that the resulting line representation supports reliable estimation of the Manhattan frame, bettering the accuracy of previous edge-based methods by a factor of 2 and the gradient-based Manhattan World method by a factor of 5.

1 Introduction

The problem of single-view 3D reconstruction is of great practical interest, with potential applications in 3D mapping, 2D-to-3D film conversion, and robot navigation. The problem is in general ill-posed, but indoor scenes and outdoor urban scenes contain many regularities that can be exploited. One of these is the so-called 'Manhattan World' assumption [1] that the oriented structure of the scene is dominated by three mutually orthogonal 3D directions comprising the Manhattan frame. This constraint has been used in a number of interesting recent algorithms for single-view 3D reconstruction [4,5], and can in principle be used to recover, up to a scale factor, the overall 3D structure of a scene conforming to the Manhattan assumption. Thus accurate and reliable estimation of the Manhattan frame, i.e., the 3D rotation of the Manhattan scene relative to the

J.-I. Park and J. Kim (Eds.): ACCV 2012 Workshops, Part II, LNCS 7729, pp. 580–593, 2013.
© Springer-Verlag Berlin Heidelberg 2013

camera, is a crucial problem in single-view 3D reconstruction, and is the problem we address here.

Coughlan and Yuille [2] devised a generative framework for the problem in which the dense gradient map is explained by a mixture of the three principal Manhattan directions and a random background process. A maximum-likelihood technique is then used to estimate the rotation of the Manhattan frame relative to the camera. Later, Denis et al. [3] adapted this framework to use sparse oriented edges rather than the dense gradient map. In using edges, they reduced the size of the input space by a factor of 10, making the algorithm much more efficient. Further, they argued that by eliminating redundancy, the naive Bayes model on which the framework is based becomes more accurate.

In this work, we take the next logical step and explore whether accuracy can be further improved by using a relatively small number of accurately extracted lines. This additional reduction in redundancy further improves the accuracy of the naive Bayes model. Moreover, since the Manhattan structure in the scene is typically organized as line segments rather than randomly positioned edges, this approach should improve the discrimination between Manhattan and non-Manhattan structures, and lead to more accurate estimation.

Our contribution is twofold:(1) An improved line extraction framework that uses more accurate, adaptive probabilistic kernels, and (2) a maximum likelihood method for estimating the pose of the camera with respect to the Manhattan frame. We evaluate the performance of our contributions on a standard benchmark dataset [3] and show that our method performs better than leading methods from the literature.

The standard method for extended line extraction is to histogram edges into Hough maps. However, this standard Hough approach is sensitive to bin resolution and suffers from noise and quantization error. To overcome these problems, we introduce a probabilistic line extraction technique that relies upon two innovations. First, we introduce a novel kernel-based voting scheme that accurately propagates observation uncertainty from the image domain, using natural statistics, into Hough space. Second, we eliminate multiple redundant detections by selecting lines in a greedy fashion and probabilistically subtracting the evidence for detected lines from the Hough map.

To relate these lines to the Manhattan frame, we adapt the Manhattan World mixture model to use line observations, replacing the image-based observation model relating gradient position and orientation to vanishing point locations, with an observation model on the Gauss sphere. Finally, we adapt the maximum likelihood technique used by Denis et al. [3] to our new observation model.

2   Overview of the Hough Transform

The standard Hough transform [6,7] uses the normal parametrization of a line in polar coordinates,

$$\rho = x cos(\theta) + y sin(\theta), \tag{1}$$

where ρ represents the signed distance of the line from the origin (the principal point) and $\theta \in (0, \pi]$ represents the angle of the normal relative to the image

Fig. 1. The parametric representation of a line in the Hough transform

frame (ρ is positive if the point on the line closest to the origin lies in the lower half of the image, negative if it lies in the upper half: Fig. 1). The Hough transform involves the computation of a histogram $H(\rho, \theta)$ representing the image evidence for all possible lines passing through the image. Detection of an edge in the image can be taken as evidence for the existence of a line passing through the edge. If we neglect noise in the estimated position (x, y) of the edge, then the edge is consistent with the one-dimensional sinusoidal family of lines satisfying (1), and bins of $H(\rho, \theta)$ intersected by this sinusoid are incremented [7]. However, edge detection normally provides an estimate of edge orientation as well as its position, and it has long been recognized [8,9] that this orientation information can be used to limit the family of possible lines to a section of the sinusoid.

If edges were not noisy, the standard Hough transform would work well. Each oriented edge would map to a single bin in the Hough map, and all edges associated with the same line would map to the same bin. Due to noise in both edge position and orientation, edges associated with the same line will in practice map to different bins. This means that peaks in the Hough map will in general be displaced from the correct (ρ, θ) values, some peaks will be missing, and there will be many false peaks that do not correspond to real lines in the image. Further, the exact nature of these errors depends strongly on the histogram resolution: coarse resolutions lead to mislocalization and missed lines (due to merging of neighbouring lines), while fine resolutions lead to many false positives.

These problems have motivated research in probabilistic Houghing methods that map each edge to a smooth local distribution over the Hough parameters. The Hough map thus becomes a summation of overlapping smooth kernels, resulting in fewer false peaks. Since kernel dimensions are defined in absolute (ρ, θ) units, the number of false positives becomes independent of histogram resolution, permitting fine resolutions that allow precise localization. See [10] for a good survey of early probabilistic methods.

Stephens[11] modeled the parameters (position and orientation) of each edge as independent when conditioned upon line parameters. The probability of each edge given a hypothesized line is modeled as a mixture of a normal distribution, reflecting generation of the edge by the line, and a uniform distribution, reflecting generation by some independent background process. A probabilistic Hough map is then computed by summing the log-likelihood over all edge observations. The proposed method is extremely computationally intensive, as it requires that the log likelihood be computed for every edge observation at every cell of the Hough map. No results on real data are reported.

Using only edge position, Kiryati & Bruckstein [12] reported a probabilistic Houghing method based on a bivariate normal model of edge position error, but demonstrated only limited results (detection of a single line from a single image). Li and Xie [13] detectd line endpoints as seed features and then paired these with randomly-selected edge points, using first-order error analysis to propagate uncertainty into the Hough map as a bivariate normal kernel. They also reported only limited, qualitative results.

Rather than using random edge pairs, Fernandes and Oliveira [14] formed local clusters of approximately collinear edges, computed a least-squares fit of a line to the cluster, estimated the uncertainty in the slope-intercept parameters of the line in the image, and propagated this uncertainty to the (ρ, θ) variables of the Hough map using first-order linear uncertainty propagation [15]. Each cluster thus contributes an oriented Gaussian kernel to the Hough map. The authors reported qualitative results on a wide array of images.

Barinova et al.[16] adopted an approach somewhat similar to Stephens'[11], modeling edge observations as conditionally independent, but explicitly introducing hidden association variables that tie edge observations to hypothesized lines. While they used both edge orientation and location information to generate soft votes in the Hough map, they employed a heuristically-derived fixed kernel rather than explicitly modelling error propagation from the image to the Hough domain. Perhaps the most interesting component of their approach is a greedy method for selecting associations once the initial Hough map is constructed. This results in a dynamically changing Hough map that avoids heuristic post-processing such as histogram smoothing or non-maxima suppression. This principle is implemented using probabilistic vote subtraction.

The probabilistic line detection technique we propose is in part inspired by these prior efforts. We use a kernel-voting technique to generate a smooth Hough map. As in [13,14], we employ first-order error propagation to determine the kernel in the Hough domain, and use a kernel subtraction technique related to the method of [16] to eliminate false positives without the need for post-processing steps like non-maximum suppression and smoothing.

However, our method also has several advantages in terms of effectiveness and efficiency. First, unlike [12], we use both edge position *and* orientation information, as both provide valuable constraints. Second, we avoid seed identification, clustering and line-fitting pre-processing steps [13,14], which increase the complexity of implementation and the number of parameters that must be tuned. Rather we propagate each observed edge independently, using learned image statistics and first-order error propagation. Unlike the fixed kernel of Barinova et al.[16], this leads to a voting kernel that varies dramatically over Hough space, better capturing the variation in uncertainty with edge parameters. We show that these kernels can easily be precomputed and stored, making the voting process efficient. Unlike [11,16] we use a normal model rather than a mixture model for each conditional observation, and accumulate the sum of the likelihood as opposed to the log likelihood, so that the Hough map encodes the estimated length of hypothesized lines.

3   Proposed Framework

3.1   Propagating Observation Uncertainty to the Hough Domain

Edge detectors provide both position and orientation information. Thus, it makes sense to model the uncertainty of each observation with respect to both. That way, each edge can contribute a single bivariate normal (BVN) kernel to the Hough map (Fig. 2).

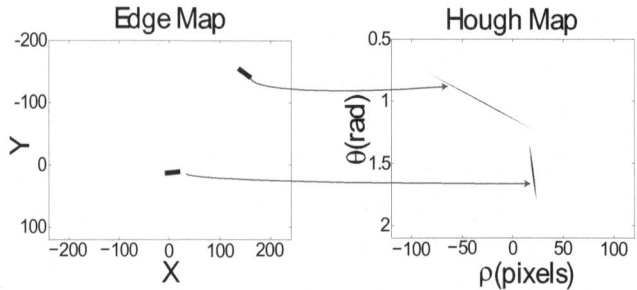

Fig. 2. Each edge votes according to a BVN kernel

We detect edges using a multiscale edge detector[17] that provides accurate sub-pixel edge position and orientation estimates, though our analysis is applicable to any edge detection algorithm such as Canny[18]. We model the uncertainty of edge observation by examining the statistics of image edges with respect to ground-truth lines[3]. Fig. 3 shows empirical densities for the displacement and orientation deviation of edges within one pixel of each line. We model each as a mixture of a normal distribution generated by the line and a uniform distribution generated by a background process. We assume isotropic displacement error, so that the data in Fig. 3 provide sufficient statistics to estimate the three key uncertainties σ_x, σ_y and σ_θ, i.e., the space constants of the normal models for horizontal and vertical displacement and angular error. The maximum likelihood estimates are: $\sigma_x = \sigma_y = 0.49$ pixels and $\sigma_\theta = 5.3$ deg. For comparison, we have also computed the same statistics for the popular Canny edge detector (MATLAB implementation with default parameters) and the space constants are: $\sigma_x = \sigma_y = 1.3$ pixels and $\sigma_\theta = 3.9$ deg.

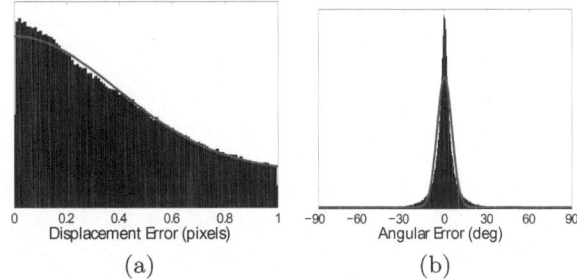

Fig. 3. Edge observation uncertainty with respect to ground-truth lines. Red curves show fit of two-component normal+uniform mixture model. (a) Distance of edge elements from ground-truth line. (b) Angular deviation of edge element from ground-truth line.

Approximating the deviations in these three dimensions as independent, we define the covariance of uncertainty of edge observations in the image domain as

$$C_I = \begin{bmatrix} \sigma_x^2 & 0 & 0 \\ 0 & \sigma_x^2 & 0 \\ 0 & 0 & \sigma_\theta^2 \end{bmatrix}, \tag{2}$$

where we use σ_x for both horizontal and vertical displacement. Using linear propagation of uncertainty[15], the covariance of the corresponding parameters in the Hough domain can be computed as

$$C_h(x, y, \theta) = \nabla P_h C_I \nabla P_h^T \tag{3}$$

where ∇P_h is the Jacobian of the parameter vector with respect to the observation vector

$$\nabla P_h = \begin{bmatrix} \dfrac{\partial \rho}{\partial x} & \dfrac{\partial \rho}{\partial y} & \dfrac{\partial \rho}{\partial \theta} \\[2ex] \dfrac{\partial \theta}{\partial x} & \dfrac{\partial \theta}{\partial y} & \dfrac{\partial \theta}{\partial \theta} \end{bmatrix} = \begin{bmatrix} \cos(\theta) & \sin(\theta) & \dfrac{\partial \rho}{\partial \theta} \\[2ex] 0 & 0 & 1 \end{bmatrix}$$

and

$$\frac{\partial \rho}{\partial \theta} = -x \sin(\theta) + y \cos(\theta). \tag{4}$$

Thus, (3) becomes

$$C_h(x, y, \theta) = \begin{bmatrix} \sigma_x^2 + \left(\dfrac{\partial \rho}{\partial \theta}\right)^2 \sigma_\theta^2 & \dfrac{\partial \rho}{\partial \theta} \sigma_\theta^2 \\[3ex] \dfrac{\partial \rho}{\partial \theta} \sigma_\theta^2 & \sigma_\theta^2 \end{bmatrix}. \tag{5}$$

The vote each edge observation contributes at each location in the Hough map is described by the BVN distribution

$$\text{Vote}_i(\rho, \theta | x_i, y_i, \theta_i) = \frac{1}{2\pi |C_h|} \exp\left(-\frac{1}{2}\begin{bmatrix}\rho - \rho_i \\ \theta - \theta_i\end{bmatrix}^T C_h^{-1} \begin{bmatrix}\rho - \rho_i \\ \theta - \theta_i\end{bmatrix}\right), \qquad (6)$$

where ρ_i and θ_i correspond to a direct mapping of the ith edge onto Hough space. The Hough map $H(\rho, \theta)$ is the sum of these votes at each location due to all observations

$$H(\rho, \theta) = \sum_i \text{Vote}_i(\rho, \theta | x_i, y_i, \theta_i). \qquad (7)$$

Note that while σ_x and σ_θ are constants, $\partial\rho/\partial\theta$ varies over the image: for our images, $\partial\rho/\partial\theta$ ranges from roughly -400 to 400. (Note that this dramatic variation underlines the importance of modelling the first-order error propagation rather than using a fixed kernel.) We sample $\partial\rho/\partial\theta$ in increments of 1 over this range, computing and storing the voting kernel given by (6) for each value, truncating at 3σ. Then when constructing the Hough map, for each edge observation (x_i, y_i, θ_i), $\partial\rho/\partial\theta$ is computed, the nearest stored kernel is selected, shifted to the corresponding (ρ_i, θ_i) location in the Hough map, and added to $H(\rho, \theta)$.

3.2   Peak Selection

After a Hough map is constructed, the peaks that correspond to dominant line features need to be identified. Ideally, image lines correspond to the modes of H. However, due to the limited sample of edges in the image, spurious local maxima will occur despite the kernel voting scheme, especially within the neighborhood of true peaks. Typically, this problem is addressed by either post-hoc smoothing [19] or non-maxima suppression. However, both approaches involve the ad-hoc selection of parameters, and in practice we find neither method works well.

Inspired by the recent dynamic updating approach of [16], we propose a simple, iterative, greedy technique, in which the kernel contributions of edges that can be associated with a detected line are subtracted from the Hough map. At each iteration, the global maximum of H is determined and is added to the list of selected lines. All edges that fall within a 3σ bound in terms of both distance from the line and angular deviation are identified. The probabilistic kernels for all of the identified edges are then subtracted from H (Fig. 4). This procedure is repeated until $\max(H)$ is smaller than a specified threshold which is a fraction of the global maximum of the Hough map. Using 5-fold cross-validation on the training images, we found that a threshold of 0.25, resulting in detection of 36.5 lines per image on average, yields minimum error for our Manhattan frame estimation application.

3.3   Manhattan Frame Estimation

We adapt the probabilistic mixture model used by Coughlan and Yuille [1,2] and Denis et al.[3] to observations consisting of extended lines. Each observed

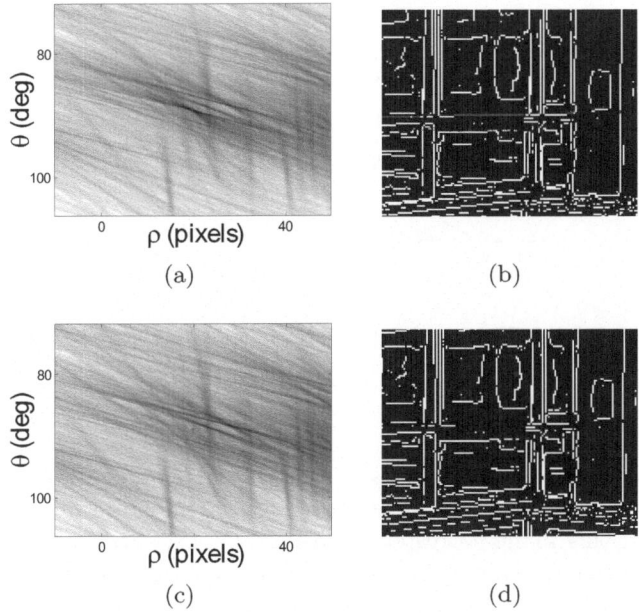

(a)   (b)

(c)   (d)

Fig. 4. Iterative peak detection with vote removal (a) Hough map with global maxima shown in red. (b) Corresponding edge map. Edges associated with the detected peak are shown in red. (c) Updated Hough map after kernel subtraction (d) Residual edge map.

line ℓ_i is assumed to be generated by a latent Manhattan variable m_{ℓ_i} that can assume values representing four classes of structure: vertical, horizontal(1), horizontal(2), background. If we knew the Euler angles Ψ describing the rotation of the camera with respect to the Manhattan frame of the scene, we could compute the probability of observing the line as a mixture of the four possible causes

$$p(\ell_i|\Psi) \;=\; \sum_{m_{\ell_i}} p(\ell_i|\Psi, m_{\ell_i}) p(m_{\ell_i}) \tag{8}$$

where $p(m_{\ell_i})$ is the prior probability for the latent variable m_{ℓ_i} and $p(\ell_i|\Psi, m_{\ell_i})$ is the probability of a line conditioned on m_{ℓ_i}. Next, we provide a statistical model for $p(\ell_i|\Psi)$.

Previous gradient- and edge-based methods [2,3] used an error model that relates localized, oriented image features to a set of vanishing points. Lines, however, are infinite in length and so cannot be related to vanishing points in the same way. Instead, it is natural to employ a Gauss sphere representation [20] (Fig. 5(a)). A line in the image plane, together with the optical centre of the camera, form an interpretation-plane that can be represented by its normal vector ℓ. In a noise-free world, all lines conforming to the same 3D orientation will produce interpretation plane normals that are coplanar. Thus, the error of

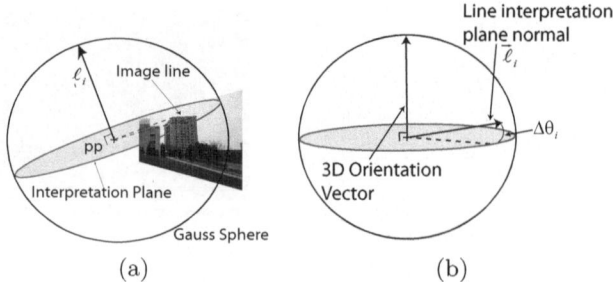

(a)   (b)

Fig. 5. Modeling uncertainty on the unit circle. (a) A line detected in the image can be represented in the Gauss sphere by its interpretation plane normal. (b) Error model for a line in the Gauss sphere.

an estimated line with respect to its true 3D orientation is given by the angular deviation of the interpretation-plane normal for the line from this common plane. The normal of this plane corresponds to the 3D orientation of the lines. Denis et al. [3] found that a Gauss-sphere error model was less accurate than an image model for edge primitives. However, line primitives have the potential to be much more accurate, as they integrate over many edges, and we find that for lines, the Gauss-sphere model works well.

We develop a statistical model for $p(\ell_i|\Psi, m_{\ell_i})$ by considering hand-labeled ground-truth lines from the database[3]. The error $\Delta\theta_i$ is the angular deviation of the interpretation plane normal for a line ℓ from the plane normal to the 3D Manhattan orientation vector (Fig. 5(b)). A histogram of $\Delta\theta_i$ indicates that a Laplace model for $p(\ell_i|\Psi, m_{\ell_i})$ is suitable (Fig. 6). Thus,

$$p(\ell_i|\Psi, m_{\ell_i}) = \frac{1}{2b}e^{\frac{-|\Delta\theta m_{\ell_i}|}{b}} \tag{9}$$

when m_{ℓ_i} is not background, where $b = 0.80$ deg for horizontal lines and $b = 0.57$ deg for vertical lines.

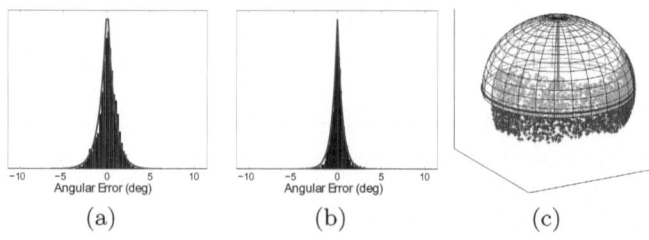

(a)   (b)   (c)

Fig. 6. Distribution of line interpretation plane normals with respect to their associated (a) horizontal and (b) vertical Manhattan direction. Maximum likelihood Laplace model shown in red. (c) Sample of uniform distribution of lines observable in the image.

We make the assumption that lines produced by the background process are uniformly distributed over the Gauss Sphere. However, to be observable, a section of the line must fall within the subtense of the image. This leads to a distribution that is roughly uniform over a rectangular region of the angular coordinates of the Gauss Sphere (Fig. 6(c)). For the camera parameters associated with the images in the YorkUrbanDB database, this leads to the uniform likelihood $p(\ell_i | m_{\ell_i} = B) = \dfrac{1}{0.68\pi}$.

The association priors $p(m_{\ell_i})$ may depend on the line extraction algorithm and its parameters, and thus cannot be determined using the ground-truth data. The association priors are therefore learned independently for each line extraction algorithm and parameter setting, by maximizing the likelihood of the mixture model over the training data.

The next stage is to estimate the 3D orientation of the Manhattan frame given a set of line observations $\mathbb{L} = \{\ell_i\}_{1...N}$ and a model for their association. The likelihood of the relative rotation of the Manhattan frame with respect to the camera is given by the product of the probabilities of observing all lines

$$p(\mathbb{L}|\Psi) = \prod_i p(\ell_i|\Psi) \tag{10}$$

so that

$$\hat{\Psi} = \arg\max_{\Psi} \sum_i \log p(\ell_i|\Psi). \tag{11}$$

Kosecka and Zhang [21] and Schindler and Dellaert [22] have used the EM algorithm to estimate $\hat{\Psi}$. This works well when 3D orientations are independent. However, satisfying the Manhattan constraint (orthogonality of the 3 Manhattan directions) makes the M-step a constrained optimization with no closed-form solution. Thus applying EM requires gradient descent on each M-step, increasing computational cost and the risk of missing the global optimum. For these reasons, Denis et al.[3] found empirically that a search based on the standard iterative BFGS quasi-Newton algorithm[23] performs better than EM on the Manhattan problem, and we apply the same method here.

4   Results

4.1   Line Detection

We evaluate our extended line detection algorithm on the standard, publicly-available YorkUrbanDB test dataset [3]. We define a correct detection as an extended line passing within 3 pixels of both endpoints of a ground truth line segment in the dataset. We compute a 1:1 greedy bipartite match between lines and ground truth segments: additional lines matching the same segment are considered false positives. Fig. 7 shows that precision-recall performance for our method compares favourably to a recent state-of-the-art method for line detection [16]. Although one may be tempted to ask whether this improved accuracy

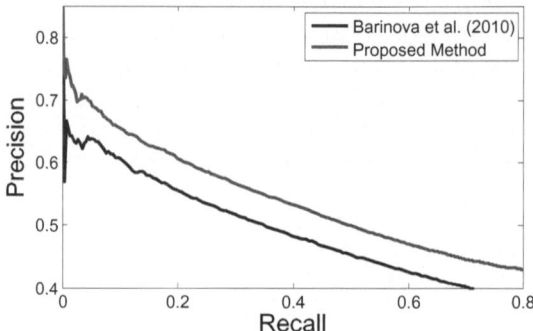

Fig. 7. Quantitative evaluation on the York UrbanDB test dataset (51 images) showing Precision recall plot for the proposed extended line detection method vs. Barinova et al.[16].

should be attributed to the improved localization and detection of edges, we find that when supplying our line detection algorithm with edge estimates obtained from the Canny edge detector used by [16], our proposed method still outperform theirs. Our experiments show that in fact the superior performance of our method is due primarily to its lower susceptibility to the multiple response problem, likely deriving from a faithful representation of the underlying statistics and more accurate propagation of uncertainty to the Hough domain.

4.2   Manhattan Frame Estimation

We evaluate our probabilistic line-based method for Manhattan Frame estimation on the YorkUrbanDB benchmark database [3], comparing against previous gradient-based [2] and edge-based [3] methods, as well as a standard non-probabilistic Hough method implemented in-house. We note that while algorithms for vanishing point detection have been published more recently (e.g., [24,25]), these have not specifically addressed the problem of estimating the Manhattan frame, and our own attempts to adapt these methods to this problem have led to poor results (average error of 19 deg).

The parameters for the line-based method were selected using a process of cross-validation, in which the training set was divided using a random 50-50 split: the Hough transform parameters were tuned and the association priors were learned on one subset and evaluated on the other. This process was repeated over 20 trials, and the configuration that produced the lowest mean error on the training data was used for final evaluation on the test set. With optimized parameters, our probabilistic line-based method uses 36.5 lines per image, on average, to estimate the Manhattan frame.

We measure performance by the angle between the estimated Manhattan frame and ground truth. Evaluation on the test set (Fig. 8) shows that our probabilistic Hough method, incorporating accurate non-stationary kernels and

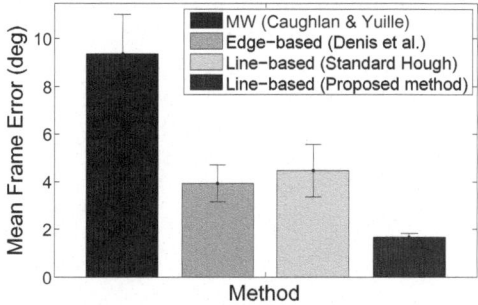

Fig. 8. Mean error of estimated camera pose relative to ground truth

Fig. 9. Examples of automatically detected lines and association with Manhattan frame. All lines used to estimate the Manhattan frame are shown. Lines are colour coded according to the most probable Manhattan cause (red: vertical, green & blue: horizontal, yellow: background).

dynamic kernel subtraction, increases accuracy by a factor of more than 2 over prior methods, achieving an average frame error of 1.7 degrees. We provide a wide range of images to demonstrate the effectiveness of our contribution in both accurate line detection and robust Manhattan frame estimation in Fig. 9.

At roughly 8 s/image, our probabilistic line detection method is much faster than the original gradient-based Manhattan World method [2] (22 s/image), and comparable to the edge-based method of Denis et al. [3] (5 s/image). (All times based upon a 3.0 GHz Intel Core Duo CPU.) Running time is dominated by the time required to add and subtract kernels from the Hough map: these operations could be greatly accelerated if mapped to the GPU.

5   Conclusion

We have developed a novel probabilistic, dynamic line-based method for estimating the Manhattan frame of a scene, and shown it to be more than twice as accurate as the state-of-the-art, using a standard publicly-available benchmark dataset. Importantly, we have shown that standard line extraction methods are insufficient: achieving this improvement depends crucially on novel methods for a) accurate propagation of uncertainty in edge detection to the Hough domain, and b) dynamic updating of the Hough map to reflect inferred edge-line associations.

The superiority of our line-based method over the prior edge-based state-of-the-art [3] likely reflects a) greater accuracy of the naive Bayes mixture model for the less redundant and noisy line primitives and b) inferential leverage deriving from the tendency for orientation energy to be organized more as lines in the Manhattan structures and more as texture in the background.

References

1. Coughlan, J.M., Yuille, A.L.: Manhattan world: Compass direction from a single image by Bayesian inference. In: International Conference on Computer Vision (1999)
2. Coughlan, J.M., Yuille, A.L.: Manhattan world: Orientation and outlier detection by Bayesian inference. Neural Computation 15, 1063–1088 (2003)
3. Denis, P., Elder, J.H., Estrada, F.J.: Efficient Edge-Based Methods for Estimating Manhattan Frames in Urban Imagery. In: Forsyth, D., Torr, P., Zisserman, A. (eds.) ECCV 2008, Part II. LNCS, vol. 5303, pp. 197–210. Springer, Heidelberg (2008)
4. Yu, S.X., Zhang, H., Malik, J.: Inferring spatial layout from a single image via depth-ordered grouping. In: CVPR Workshop (2008)
5. Hedau, V., Hoiem, D., Forsyth, D.: Recovering the spatial layout of cluttered rooms. In: ICCV 2009, pp. 1849–1856 (2009)
6. Hough, P.V.C.: Method and Means for Recognizing Complex Patterns. U. S. Patent 3, 069, 654 (1962)
7. Duda, R.O., Hart, P.E.: Use of the Hough transformation to detect lines and curves in pictures. Communications of the ACM 1, 11–15 (1972)
8. O'Gorman, F., Clowes, M.: Finding picture edges through collinearity of feature points. IEEE Transactions on Computers C-25, 449–456 (1976)
9. Ballard, D.: Generalizing the hough transform to detect arbitrary shapes. Pattern Recognition 13, 111–122 (1981)

10. Klviinen, H., Hirvonen, P., Xu, L., Oja, E.: Probabilistic and non-probabilistic Hough transforms: overview and comparisons. Image and Vision Computing 13, 239–252 (1995)
11. Stephens, R.: Probabilistic approach to the Hough transform. Image and Vision Computing 9, 66–71 (1991)
12. Kiryati, N., Bruckstein, A.M.: Heteroscedastic Hough transform (HtHT): an efficient method for robust line fitting in the 'errors in the variables' problems. Comput. Vis. Image Underst. 78, 69–83 (2000)
13. Li, Q., Xie, Y.: Randomised Hough transform with error propagation for line and circle detection. Pattern Analysis and Applications 6, 55–64 (2003)
14. Fernandes, L.A.F., Oliveira, M.M.: Real-time line detection through and improved Hough transform voting scheme. Pattern Recognition 41, 299–314 (2008)
15. Bevington, P.R.: Data Reduction and Error Analysis for the Physical Sciences. McGraw-Hill (1969)
16. Barinova, O., Lempitsky, V., Kohli, P.: On detection of multiple object instances using Hough transforms, pp. 2233–2240 (2010)
17. Elder, J.H., Zucker, S.W.: Local scale control for edge detection and blur estimation. Transactions on Pattern Analysis and Machine Intelligence 20, 699–716 (1998)
18. Canny, J.: A computational approach to edge detection. IEEE Transactions on Pattern Analysis and Machine Intelligence PAMI-8, 679–698 (1986)
19. Bishop, C.M.: Pattern Recognition and Machine Learning, 1st edn. Springer (2006)
20. Barnard, S.T.: Interpreting perspective images. Artificial Intelligence 21, 435–462 (1983)
21. Košecká, J., Zhang, W.: Video Compass. In: Heyden, A., Sparr, G., Nielsen, M., Johansen, P. (eds.) ECCV 2002, Part IV. LNCS, vol. 2353, pp. 476–490. Springer, Heidelberg (2002)
22. Schindler, G., Dellaert, F.: Atlanta world: An expectation maximization framework for simultaneous low-level edge grouping and camera calibration in complex manmade environments. In: CVPR 2004, pp. 203–209 (2004)
23. Avriel, M.: Nonlinear Programming: Analysis and Methods. Prentice Hall (2003)
24. Tardif, J.P.: Non-iterative approach for fast and accurate vanishing point detection. In: ICCV 2009 (2009)
25. Barinova, O., Lempitsky, V., Tretiak, E., Kohli, P.: Geometric Image Parsing in Man-Made Environments. In: Daniilidis, K., Maragos, P., Paragios, N. (eds.) ECCV 2010, Part II. LNCS, vol. 6312, pp. 57–70. Springer, Heidelberg (2010)

Hierarchical Stereo Matching
Based on Image Bit-Plane Slicing

Huei-Yung Lin and Pin-Zhi Lin

Department of Electrical Engineering
National Chung Cheng University
Chiayi 621, Taiwan

Abstract. We propose a new stereo matching framework based on image bit-plane slicing. A pair of image sequences with various intensity quantization levels constructed by taking different bit-rate of the images is used for hierarchical stereo matching. The basic idea is to use the low bit-rate image pairs to compute rough disparity maps. The hierarchical matching strategy is then performed iteratively to update the low confident disparities with the information provided by extra image bit-planes. Since the disparity computation is carried out on a need-to-know basis, the proposed technique is suitable for remote processing of the images acquired by a mobile camera. Our method provides a hierarchical matching framework and can be combined with the existing stereo matching algorithms. Experiments on Middlebury datasets show that our technique gives good results compared to the conventional full bit-rate matching.

1   Introduction

Stereo matching is considered as one of the most challenging and unsolved problems in computer vision. Due to the broad applicability to many application domains such as multimedia, 3D display and robotics, it has attracted the researchers' attentions for over several decades. In the past few years, a fairly large amount of computational algorithms has been proposed to cope with the stereo matching problem. The objective is usually to obtain an accurate disparity map from a pair of images. Based on the well-studied camera and scene geometry, the 3D structure of the scene can then be derived using the image formation parameters. In the early development of stereo algorithms, the block matching techniques exploiting local constraints have been extensively investigated [1–3]. Recent advances on stereo matching, in contrast, are mostly based on the energy minimization framework for global optimization [4]. Some well-known techniques include graph cuts [5, 6], belief propagation [7], and dynamic programming [8, 9], etc. Those methods and the variations have been shown to provide significant disparity estimates on the Middlebury stereo evaluation datasets [10].

In addition to the methodologies for matching cost computation and disparity optimization, there also exist some frameworks such as multi-resolution stereo matching. This hierarchical stereo matching technique can be implemented using image pyramids to improve the disparity estimates based on the coarse-to-fine

J.-I. Park and J. Kim (Eds.): ACCV 2012 Workshops, Part II, LNCS 7729, pp. 594–605, 2013.

paradigm [11, 12]. It can also be adopted to generate the aggregation window with a pyramid shape for better depth discontinuity handling and hardware-accelerated computation [13]. Although the hierarchical matching strategy has been investigated in terms of spatial image resolution change, no related research has been done based on the modification of image intensities.

In this work we address the feasibility of hierarchical stereo matching in the intensity domain. Similar to the conventional image pyramids, a series of images with less and less information encoded is constructed hierarchically. The lower level image in the hierarchy contains more detailed information present in the higher level. However, the information reduced is in terms of the representation for image intensity, instead of the image resolution. By creating an image pyramid in the intensity domain, the low bit-rate stereo matching is possible with a smaller memory allocation for each image.

To implement the intensity-based hierarchical matching technique, a variate bit-rate image processing approach similar to bit-plane slicing is proposed. The stereo matching algorithm is first carried out on a low bit-rate image pair. If the resulting disparity meets some quality requirements, the stereo matching process is terminated. Otherwise, an extra bit-plane is provided for the image pair to perform higher bit-rate stereo matching. This process continues until the disparity estimate is satisfactory or the finest level in the hierarchy is reached. The proposed technique can thus be used to reduce the data usage or transmission overhead, especially for the applications with the images captured by a mobile camera and transmitted for remote processing.

To the best of the author's knowledge, the idea of intensity-based hierarchical matching technique has never been presented before. The proposed variate bit-rate matching scheme can be considered as a framework and used to combine with the existing stereo matching algorithms. It is shown in this paper that the quality of disparity estimates does not drop significantly when the incomplete information with low intensity quantization is provided. Thus, the trade-off between performance and data bit-rate for stereo matching is worth investigating.

2   Variate Bit-Rate Matching

Given a pair of images for stereo matching, most existing algorithms take the image intensity values as raw data input for disparity computation. For a general grayscale image with 256 quantization levels, each pixel is represented by an 8-bit integer. Stereo matching is carried out on the images with full 8-bit intensity depth. However, as the number of bits per pixel increases for high quality image representation, it might be unnecessary or impractical to perform stereo matching using the full intensity depth. The investigation on how to use the reduced intensity depth for stereo matching has become an important issue. More precisely, if an n-bit stereo image pair is given, is it possible to derive a comparable or better disparity map using only the m-bit information where $m < n$?

To deal with the problem of stereo matching on the image pairs obtained from various quantization levels, we propose a hierarchical stereo matching technique based on a variate bit-rate image representation. For an n-bit grayscale image, the intensity value of any pixel can be represented by

$$I = \sum_{i=1}^{n} a_i \cdot 2^{i-1} \tag{1}$$

using bit-plane decomposition, where a_i is the i-th bit of the intensity value. If the image is to be approximated with an intensity depth less than n bits, then the higher order bits should be used since they contain more significant information. Based on these facts, the variate bit-rate representation for a k-bit image approximation is given by

$$I(k) = \sum_{i=n-k+1}^{n} a_i \cdot 2^{i-1} \tag{2}$$

where $k = 1, 2, \ldots, n$. An image represented by $I(k)$ can be thought as the one with 2^k quantization levels on the intensity value.

Using the image representation given by Eq. (2), a variate bit-rate stereo matching technique can be carried out for disparity computation from a pair of n-bit images. Instead of the full intensity depth, we can use only the k most significant bits of the n-bit image pair for stereo matching. If the derived disparity estimate does not meet some predefined criteria, then stereo matching is performed again on either the entire or certain parts of the image using $I(k+1)$, the $k + 1$ most significant bits of the pixels. This process may continue until the disparity is satisfactory. It is clear that the worst case happens when $k = n$ and the entire image is processed. In this case, the result is identical to the one computed from the original n-bit image pair.

One major advantage of the proposed variate bit-rate stereo matching technique is its capability on bitwise processing of the image data. Once the image pair of bit-plane n is obtained, stereo matching can be performed immediately for disparity computation. The quality of disparity estimates can then be gradually improved by providing lower order bit-planes "upon request". Thus, the data rate for disparity computation in terms of the number of bits per pixel is generally less than n when processing the n-bit images.

The framework of hierarchical stereo matching using the variate bit-rate image representation does not involve any specific stereo matching algorithms. To take the advantage of hierarchical matching exclusively on the pixels with increased intensity depth (bit-rate), it is preferable to adopt the algorithms which are capable of regional processing. However, as shown in the experiments, the global matching algorithms provide good disparity estimates for low bit-rate image pairs, and can be used to initialize the variate bit-rate hierarchical matching.

3   The Algorithm

Given a pair of n-bit images, I_l and I_r, the corresponding k-bit image representation, $I_l(k)$ and $I_r(k)$, can be obtained from Eq. (2), where $k = 1, 2, \ldots, n$.

Let the image pair with k-bit quantization levels be denoted by $S(k)$, then the associated disparity map $D(k)$, which consists of the disparity $d(x, y, k)$ for each pixel (x, y), can be computed by stereo matching algorithms using the two k-bit images. Since the objective is the low data rate disparity computation, it is natural to use the smallest k to perform stereo matching on $S(k)$, provided that the quality of $D(k)$ is satisfactory.

For variate bit-rate stereo matching, the quality of disparity estimate obtained from $S(k)$ is defined as follows. First, the disparity error rate of an image region R is given by

$$E(D(k)) = \frac{1}{R} \sum_{x,y} \|d(x, y, k) - d(x, y, 0)\| \tag{3}$$

where the pixel $(x, y) \in R$, and $D(0)$ represents the ground truth disparity with $E(D(0)) = 0$. The quality of disparity estimate can then be written as the reciprocal of disparity error rate, i.e., $Q(D(k)) = E(D(k))^{-1}$.

In general, the more intensity quantization level implies the less stereo mismatch. Thus, $D(n)$ can be used for quality assessment by writing Eq. (3) as

$$E(D(k)) = \frac{1}{R} \sum_{R} \|D(k) - D(n)\| \tag{4}$$

if the ground truth disparity is not available. In this case, when all of the n bits per pixel are used we have $E(D(n)) = 0$ as expected.

The variate bit-rate stereo matching is to find the smallest $k(x, y)$ for each pixel $(x, y) \in R$ such that Eq. (4) is equal to zero or less than a quality threshold. More precisely, we will solve for the required intensity depth

$$k(x, y) = \arg \min_i \{\|d(x, y, i) - d(x, y, n)\| \leq t\} \tag{5}$$

with the disparity error $t \geq 0$ for each pixel (x, y). Since $d(x, y, n)$ cannot be computed until the last image bit-plane is available, $k(x, y)$ is estimated using an iterative technique by comparing the disparity estimates obtained from $S(k)$ and $S(k - 1)$, where $k = 2, 3, 4, \cdots$. If the disparity difference for a pixel (x, y) is less than a threshold, i.e.,

$$\|d(x, y, k) - d(x, y, k - 1)\| \leq t, \tag{6}$$

then $d(x, y, k)$ is selected. Otherwise, $k(x, y)$ is increased until the condition (6) holds. Mathematically, Eq. (5) can be modified as

$$k(x, y) = \arg \min_i \{\|d(x, y, i) - d(x, y, i - 1)\| \leq t\} \tag{7}$$

with $i = 2, 3, \cdots, n$.

It should be noted that Eq. (7) is derived pixelwise and the error disparity threshold t takes on the integer values $0, 1, 2, \cdots$. If the spatial proximity is considered for disparity computation, then the variate bit-rate representation can be obtained regionally as

$$k = \arg \min_i \{\frac{1}{R} \|D(i) - D(i - 1)\| \leq t\} \tag{8}$$

where $D(\cdot)$ takes on the region R and $t \geq 0$.

To increase the bit-rate on specific pixels of the image pair for iterative stereo matching, a binary mask is created according to the disparity difference given by Eq. (6). That is, the stereo matching on the k-bit image pair $S(k)$ is carried out only on the regions assigned by the binary mask $B(k)$. Depending on the tolerance of the error disparity threshold t, $B(k)$ might contain many small isolated regions. In order to preserve the local depth continuity of the scene, they are eliminated using morphological erosion and dilation operations. The associated disparities on the regions are updated with the disparity of the surrounding pixels given by the latest disparity estimate $D(k - 1)$ obtained from the stereo image pair $S(k - 1)$.

In addition to provide the k-th bit information to the reference image (the left image in our case) for stereo matching, it is also required to assign the search ranges for all pixels covered by the binary mask $B(k)$ in the left image. The search range is used to decide the regions in the right image to have the k-th bit information for stereo matching. Similar to the mask $B(k)$ used for the reference image, a binary mask $M(k)$ is generated to update the k-th bit information for the right image. Since stereo matching is carried out along the image scanlines for rectified image pairs, $M(k)$ is created by taking the union of possible matches for all of the regions in $B(k)$. More specifically, suppose $B(k)$ consists of m connected components, i.e.,

$$B(k) = \bigcup_{i=1}^{m} B_i(k) \tag{9}$$

and the maximum disparity in the region $B_i(k)$ obtained from the disparity map $D(k - 1)$ is d_i. Then $M(k)$ is given by the union of m regions

$$M(k) = \bigcup_{i=1}^{m} M_i(k) = \bigcup_{i=1}^{m} \{ \langle \bigcup_{j=0}^{d_i} B_i(k) \rangle_j \} \tag{10}$$

where $\langle \cdot \rangle_j$ indicates the translation of a set by j pixels along the image scanlines.

The proposed hierarchical stereo matching technique is carried out iteratively by providing more information using the next image bit-plane. It is clear that the binary mask for the left (reference) image possesses the property $B(1) \supset B(2) \supset B(3) \supset \cdots$. Moreover, we have $M(i) \supset M(i + 1)$ for the right image since the correspondence search region $M_i(k + 1)$ in the right image $I_r(k + 1)$ is constrained by the maximum disparity obtained from performing stereo matching on the region $M_i(k)$. Thus, it is guaranteed that, while the disparity estimate is gradually improved, the amount of extra bits to be added to the image pair is decreased for each iteration.

Now, suppose the additional information from bit-plane k is added to form the image pair $S(k)$. The extra-bit is updated on the regions $B(k)$ and $M(k)$ for the left and right images, respectively. Each pixel of the images contains various bits of data representation and has different levels of intensity quantization. To derive the disparity map $D(k)$ from the image pair $S(k)$, two different matching schemes, performing on the bit-rate updated pixels only or on the whole image, are adopted. For the first case, stereo matching is carried out only on the newly updated region of the left image given by $B(k)$. Thus, as the image bit-rate increases, the computational cost of each iteration decreases for this *partial matching* scheme. For the second case, even only part of the image is updated with new pixel values, stereo matching is performed on the whole image which contains mixed bit-rate pixels. This *full matching* scheme is similar to the conventional stereo matching method and has a fixed computational cost for all iterations. Since the image region without extra-bit update can still gain some information from the bit-rate update region $B(k)$ while evaluating the cost function using a local matching window, the disparity estimates using partial and full matching schemes will be slightly different. As shown in the experiments, the full matching scheme provides better results at the cost of more computation involved.

One important issue for the hierarchical variate bit-rate matching technique is how to select the lowest bit-rate and use the associated image pair for initial stereo matching. It is clear that the images with very low bit-rate, such as binary images or images with low intensity quantization levels, are not informative for stereo matching. If no additional information is available, an initial disparity estimate can be derived using $S(k)$ with $k = n/2$ for an n-bit image pair. Eq. (7) is then carried out to improve the disparity quality with higher bit-rate image representation. As shown in the experiments, there is a significant improvement in terms of bad matching pixels on 3-bit image pairs. Thus, a 4-bit stereo image pair can be used to derive a suitable initial disparity map and identify the regions for further disparity computation.

Since the hierarchical stereo matching is carried out iteratively on different bit-rate image pairs, it is possible to perform different matching algorithms for each disparity computation. More specifically, global matching algorithms can be used derive the initial disparity maps followed by local matching algorithms for regional improvement. This hybrid stereo matching strategy takes the low bit-rate images for global matching and high bit-rate images for local matching. Consequently, the final disparity result can be improved with a reduced average image bit-rate for overall computation.

4   Implementation and Discussion

In this section, we describe the implementation and experiments of the stereo matching technique based on bit-plane slicing. The hierarchical matching framework is served as a basis for conventional stereo matching approaches. Thus, two local algorithms– *sum of absolute difference* (SAD) and *normalized cross correlation* (NCC), and two global algorithms– *graph cut* (GC) [5] and *belief propagation* (BP) [7] are used to test the effectiveness of our method. Four standard Middlebury stereo datasets, Tsukuba, Venus, Teddy and Cones are used in our experiments [14]. The evaluation details on the percentage of bad matching pixels and the average image bit-rate versus the pixel bit-rate are presented using Tsukuba dataset.

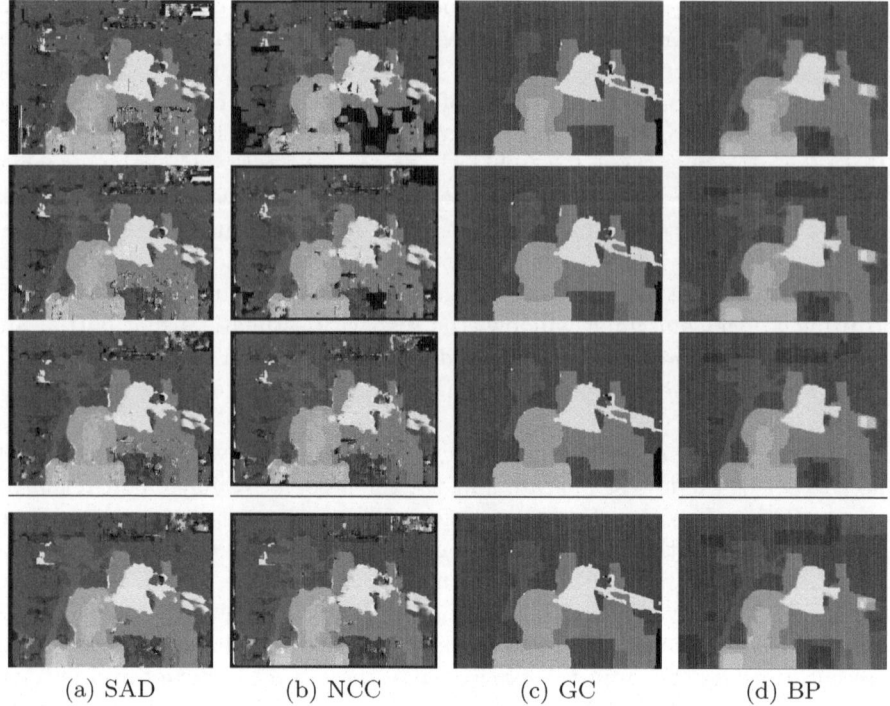

| (a) SAD | (b) NCC | (c) GC | (d) BP |

Fig. 1. The disparity maps computed using different initial bit-rate image pairs with four matching algorithms, SAD, NCC, GC and BP. The top three rows: the image pairs with 2 and 3 bpp, 3 and 4 bpp, and 4 and 5 bpp, are used for iterative stereo matching. The last row: the disparity maps computed using the full 8-bit image pairs.

To perform the hierarchical stereo matching using image bit-plane slicing, the initial low bit-rate image pairs $S(k)$ and $S(k+1)$ are first selected. Stereo matching is then carried out iteratively by incorporating higher bit-plane information until the highest bit-rate image pair is used. The selection of the initial bit-rate will affect not only the final disparity quality, but also the total amount of data used for disparity computation. Compare the ground truth disparity with the results from the fixed bit-rate matching, it is seen that rough disparity maps can be obtained using low bit-rate image pairs. Thus, we will now evaluate the performance of hierarchical stereo matching starting from the initial bit-rate of 2, 3 and 4 for the local and global matching algorithms.

Figure 1 shows the results obtained using various initial bit-rates for SAD, NCC, GC and BP algorithms. The disparity estimates as shown in the top three rows are computed using the full matching scheme with the initial stereo image pairs of 2 and 3 bpp (bits per pixel), 3 and 4 bpp, and 4 and 5 bpp, respectively. Since stereo matching is carried out iteratively until the 8-th bit information is used, the above results are derived after 5, 4, and 3 iterations, respectively.

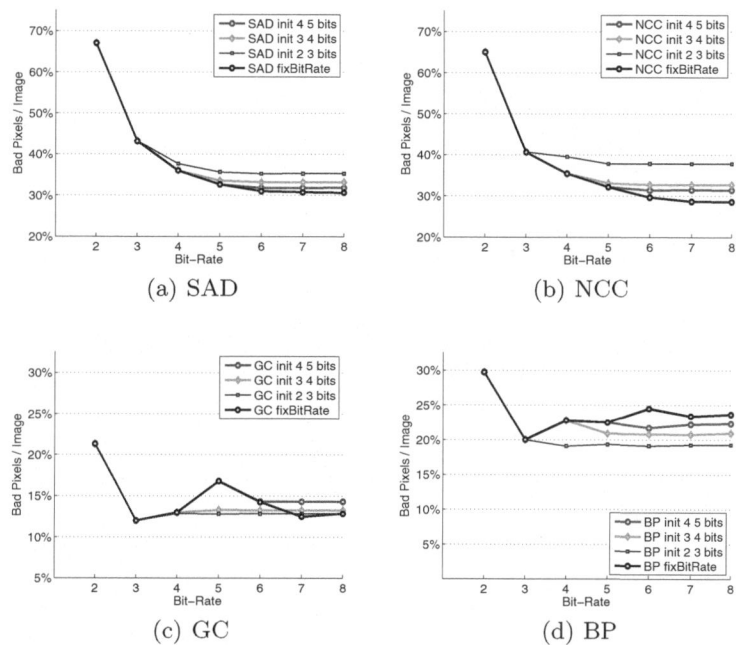

Fig. 2. The percentage of bad matching pixels versus the pixel bit-rate for Tsukuba dataset using SAD, NCC, GC and BP algorithms. Different curves represent the iterative stereo matching using different initial bit-rate image pairs. For the local methods SAD and NCC, high initial bit-rates imply better final disparity estimates after the iterations. But this is not the case for the global methods GC and BP.

Figure 2 shows the trend of bad matching rates with the initial bit-rates of 2, 3, 4 bpp for different matching algorithms. For the local methods SAD and NCC, high initial bit-rates imply better final disparity estimates after the iterations (see Figures 2(a) and 2(b)). But this is not the case for the global methods such as GC and BP. As shown in Figures 2(c) and 2(d), higher initial bit-rates for iterative stereo matching do not necessarily give good initial disparity estimates and always result in higher bad matching rates after all iterations. Moreover, the fixed bit-rate matching (the black circles in Figure 2) always gives the lowest bad matching rate for SAD and NCC, but it is not the case for GC and BP. For the BP algorithm, the conventional fixed bit-rate matching even gives the worst matching result.

The average image bit-rates used for hierarchical stereo matching are shown in Figure 3. Although high initial bit-rates for iterative stereo matching results in high average image bit-rates after iterations, the amounts of data used for computation are reduced by 20% – 30% for SAD and NCC, and 50% – 70% for GC and BP. As shown in Figures 2(a), 2(b), 3(a) and 3(b), the data rate is inversely proportional to the error matching rate for SAD and NCC.

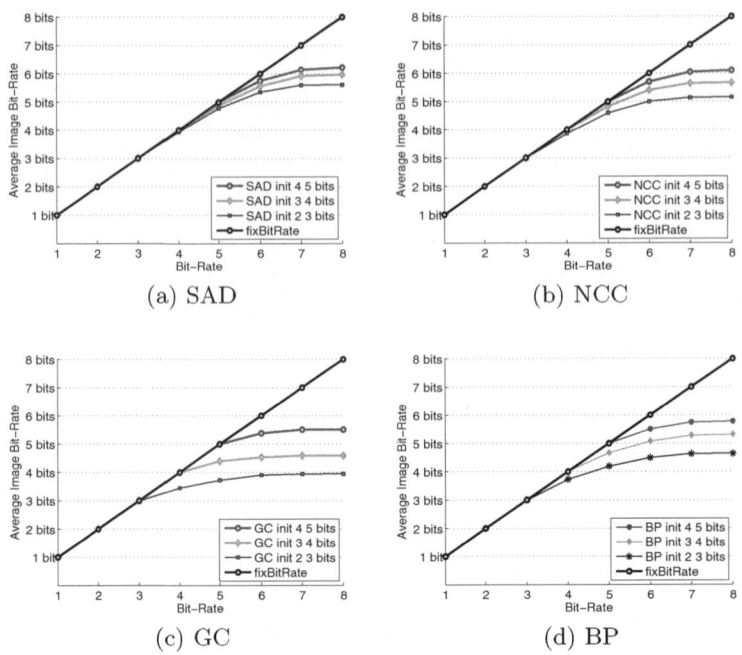

Fig. 3. The average image bit-rate versus the pixel bit-rate (intensity quantization level) for Tsukuba dataset using SAD, NCC, GC and BP algorithms. Different curves represent the iterative stereo matching using different initial bit-rate image pairs. Although high initial bit-rates for iterative stereo matching results in high average image bit-rates after iterations, the amounts of data used for computation are reduced by 20% – 30% for SAD and NCC, and 50% – 70% for GC and BP.

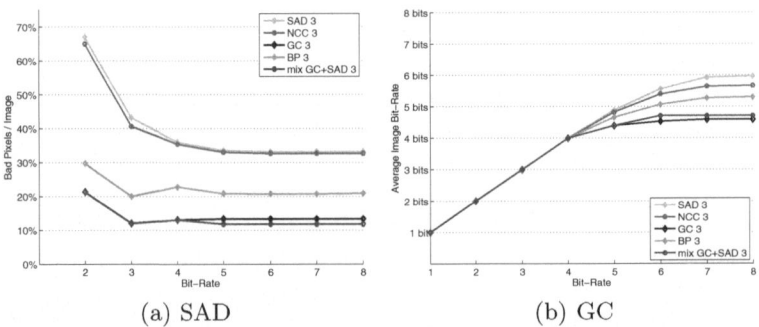

Fig. 4. Performance evaluation of Tsukuba dataset using different SAD, NCC, GC, BP and hybrid (GC+SAD) matching algorithms. The image pairs of 3 bpp and 4 bpp are used for initial disparity computation. The hybrid method with mixed GC and SAD results in the lowest bad matching rate of 11.8% at the cost of increased average image bit-rate (4.71 bpp).

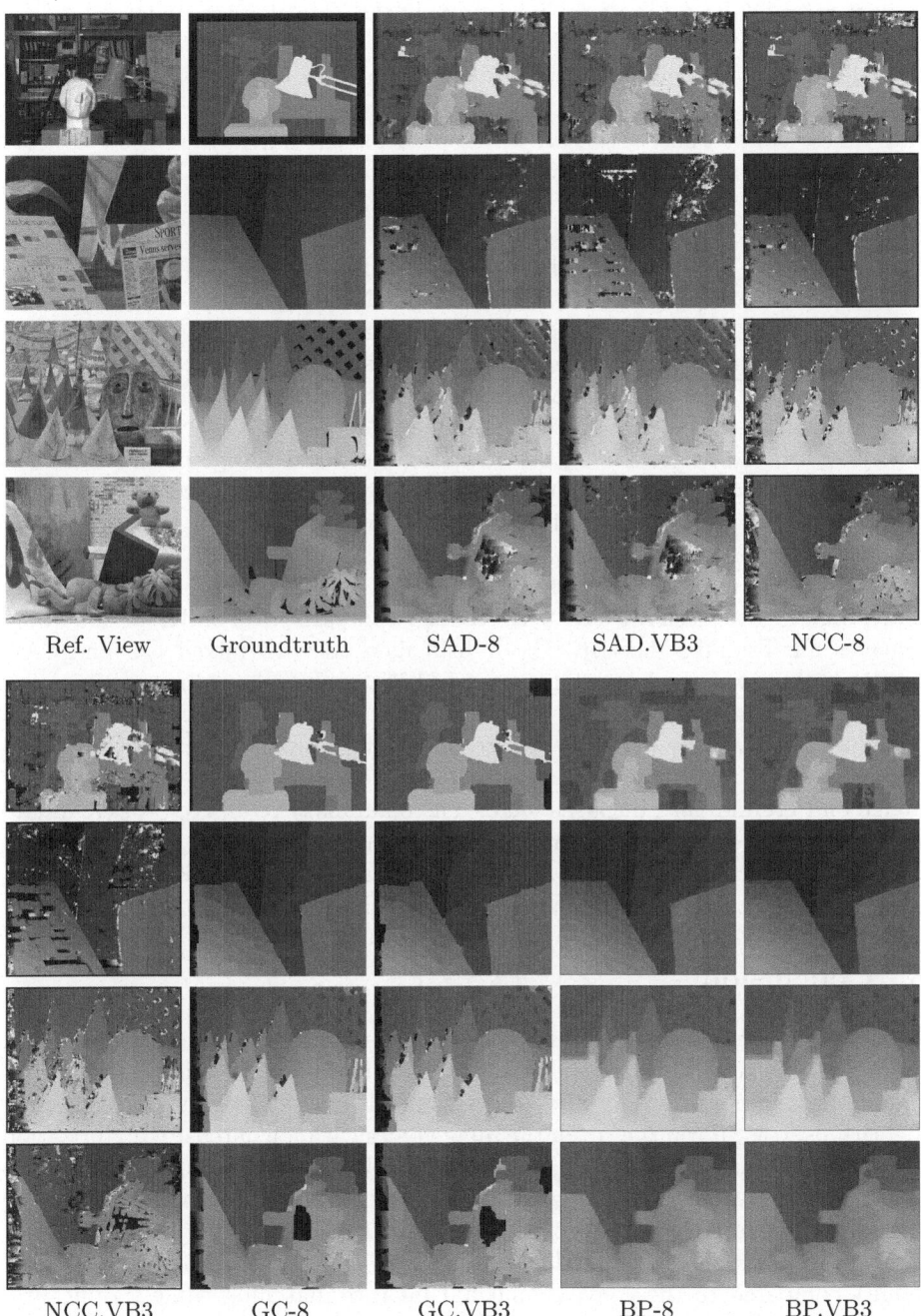

Fig. 5. The results of Tsukuba, Venus, Cones and Teddy datasets computed using the conventional stereo matching algorithms and with the proposed hierarchical stereo matching technique

However, for GC and BP algorithms, starting the iterative stereo matching from low initial bit-rate image pairs not only ends up with the low data rate, but also provides the low bad matching rate. This fact indicates that the global matching algorithms such as GC and BP can achieve fairly good disparity using low bit-rate image pairs. Thus, they can be adopted for the early stages (low bit-rate matching) of a hybrid stereo matching technique.

From the performance evaluation of hierarchical stereo matching with different algorithms (as shown in Figures 2 and 3), we have the following observations: (1) The overall performance for the global methods (GC, BP) is better than the local methods (SAD, NCC). (2) SAD and NCC algorithms always provide better matching results as the image bit-rate increases. (3) Although GC and BP algorithms provide good disparity estimates for low bit-rate matching, the performance does not gain too much improvement when additional bits are used. Thus, it is interested to know if it is possible to develop a better variate bit-rate matching scheme by combining the advantages of local and global algorithms.

In the last experiment, we implement a hybrid matching scheme using GC algorithm for low bit-rate initial matching, followed by SAD algorithm for high bit-rate improvement. The disparity estimates computed from 3-bit and 4-bit image pairs are used to derive the binary mask for further iterations with SAD matching. Figure 4 shows the performance comparison of different matching schemes for Tsukuba dataset. The hybrid method with mixed GC and SAD results in the lowest bad matching rate of 11.8% at the cost of increased average image bit-rate (4.71 bpp).

Figure 5 shows the experimental results using Tsukuba, Venus, Cones and Teddy datasets. The first two columns on the top row are the reference view image and groundtruth disparity map. The columns SAD-8, NCC-8, GC-8 and BP-8 are the disparity maps obtained using conventional SAD, NCC, GC and BP algorithms. SAD.VB3, NCC.VB3, GC.VB3 and BP.VB3 indicate the results using variate bit-rate matching starting from 3-bit and 4-bit image pairs.

5   Conclusion

In this paper, we have presented a hierarchical stereo matching approach based on image bit-plane slicing. The variate bit-rate matching technique can be considered as a framework and used to combine with the existing stereo matching algorithms. It is shown that even the image pairs with low intensity quantization are able to produce fairly good disparity results. Based on the disparity estimates derived from low bit-rate image pairs, variate bit-rate stereo matching can be carried out iteratively to improve the disparity quality. Experiments on Middlebury datasets demonstrate that the proposed technique is able to give good results compared to the conventional full bit-rate matching while significantly reduce the average image bit-rate for disparity computation.

Acknowledgement. The support of this work in part by the National Science Council of Taiwan, R.O.C, under Grant NSC-99-2221-E-194-005-MY3 is gratefully acknowledged.

References

1. Bhat, D.N., Nayar, S.K.: Ordinal measures for image correspondence. IEEE Trans. Pattern Anal. Mach. Intell. 20, 415–423 (1998)
2. Min, D., Sohn, K.: Cost aggregation and occlusion handling with wls in stereo matching. IEEE Transactions on Image Processing 17, 1431–1442 (2008)
3. Chen, Y.S., Hung, Y.P., Fuh, C.S.: Fast block matching algorithm based on the winner-update strategy. IEEE Transactions on Image Processing 10, 1212–1222 (2001)
4. Szeliski, R., Zabih, R., Scharstein, D., Veksler, O., Kolmogorov, V., Agarwala, A., Tappen, M., Rother, C.: A comparative study of energy minimization methods for markov random fields with smoothness-based priors. IEEE Trans. Pattern Anal. Mach. Intell. 30, 1068–1080 (2008)
5. Boykov, Y., Veksler, O., Zabih, R.: Fast approximate energy minimization via graph cuts. IEEE Transactions on Pattern Analysis and Machine Intelligence 23, 1222–1239 (2001)
6. Kolmogorov, V., Zabih, R.: Computing visual correspondence with occlusions using graph cuts. In: IEEE International Conference on Computer Vision, vol. 2, p. 508 (2001)
7. Sun, J., Zheng, N.N., Shum, H.Y.: Stereo matching using belief propagation. IEEE Trans. Pattern Anal. Mach. Intell. 25, 787–800 (2003)
8. Ohta, Y., Kanade, T.: Stereo by intra- and inter-scanline search using dynamic programming. IEEE Trans. Pattern Analysis and Machine Intelligence 7, 139–154 (1985)
9. Birchfield, S., Tomasi, C.: Depth discontinuities by pixel-to-pixel stereo. International Journal of Computer Vision 35, 269–293 (1999)
10. Scharstein, D., Szeliski, R.: A taxonomy and evaluation of dense two-frame stereo correspondence algorithms. Int. J. Comput. Vision 47, 7–42 (2002)
11. Hung, Y.P., Chen, C.S., Hung, K.C., Chen, Y.S., Fuh, C.S.: Multipass hierarchical stereo matching for generation of digital terrain models form aerial images. Mach. Vision Appl. 10, 280–291 (1998)
12. Zhang, L.: Fast stereo matching algorithm for intermediate view reconstruction of stereoscopic television images. IEEE Transactions on Circuits and Systems for Video Technology 16, 1259–1270 (2006)
13. Yang, R., Pollefeys, M.: Multi-resolution real-time stereo on commodity graphics hardware. In: IEEE Computer Vision and Pattern Recognition, pp. 1:211–1:217 (2003)
14. Scharstein, D., Szeliski, R.: Middlebury stereo vision page (2002), http://vision.middlebury.edu/stereo

Author Index